Springer Proceedings in Earth and Environmental Sciences

The series Springer Proceedings in Earth and Environmental Sciences publishes proceedings from scholarly meetings and workshops on all topics related to Environmental and Earth Sciences and related sciences. This series constitutes a comprehensive up-to-date source of reference on a field or subfield of relevance in Earth and Environmental Sciences. In addition to an overall evaluation of the interest, scientific quality, and timeliness of each proposal at the hands of the publisher, individual contributions are all refereed to the high quality standards of leading journals in the field. Thus, this series provides the research community with well-edited, authoritative reports on developments in the most exciting areas of environmental sciences, earth sciences and related fields.

V. I. Karev
Editor

Physical and Mathematical Modeling of Earth and Environment Processes—2022

Proceedings of 8th International Scientific Conference-School for Young Scientists

Springer

Editor
V. I. Karev
Antalya, Turkey

ISSN 2524-342X ISSN 2524-3438 (electronic)
Springer Proceedings in Earth and Environmental Sciences
ISBN 978-3-031-25961-6 ISBN 978-3-031-25962-3 (eBook)
https://doi.org/10.1007/978-3-031-25962-3

This Springer imprint is published by the registered company Springer Nature Switzerland AG
The registered company address is: Gewerbestrasse 11, 6330 Cham, Switzerland

Contents

Preliminary Study of the Opportunity to Predict Changes in Rock Samples Inner Structure Caused by Triaxial Loading Based on Stress–Strain Relationship

N. V. Dubinya, D. E. Beloborodov, M. A. Krasnova, A. M. Leonova, and S. A. Tikhotsky

Abstract The paper presents a mathematical model for interpreting the stress–strain curve obtained from triaxial loading tests performed on cylindrical rock samples. The suggested model provides an opportunity to predict tendencies in sample's inner structure changes caused by loading: relative fractions of shear fractures of various spatial orientations are evaluated for the inelastic strain accumulation process. The model is based on non-associated plastic flow law with friction hardening. The model fracture development predictions are proved by measurements of sample's acoustic properties before, during, and after loading. The paper presents the experimental setup used for application and verification of the suggested mathematical model. Limestone samples of the Domodedovo quarry (moscovian stage) were studied following the proposed experimental scheme. The paper presents the interpretation of the obtained experimental results, highlights the main tendencies in fracture development at different loading stages, demonstrates the observed relationship between changes in elastic waves velocities in the studied rocks and evolution of their inner structure. The recommendations on applying the proposed model for solving the geomechanics problems are formulated alongside with the experimental scheme necessary for obtaining the reliable results on fracture prediction in rock masses using the proposed model.

Keywords Geomechanics · Fractures · Plastic flow · Hardening · Triaxial loading

---

N. V. Dubinya (✉) · D. E. Beloborodov · M. A. Krasnova · A. M. Leonova · S. A. Tikhotsky
Schmidt Institute of Physics of the Earth of the Russian Academy of Sciences, Moscow, Russia
e-mail: dubinya.nv@gmail.com

N. V. Dubinya · S. A. Tikhotsky
Moscow Institute of Physics and Technology (National Research University), Moscow, Russia

V. I. Karev (ed.), *Physical and Mathematical Modeling of Earth and Environment Processes—2022*, Springer Proceedings in Earth and Environmental Sciences,
https://doi.org/10.1007/978-3-031-25962-3_1

1 Introduction

Mechanical properties of rock samples subjected to triaxial loading are studied in the current paper. Inelastic deformation is the main focus of the research alongside with induced changes in rock inner structure associated with plastic strain localization effect [1]. Localization of irreversible strain in rock masses can be related to different processes including initiation of new fractures, development of pre-existing natural fractures and loss of stability at weak bedding planes [2, 3]. It is a complicated task to mathematically describe these processes due to their spatial and temporal scaling: microscopic mechanical processes lead to macroscopic effects, so the mathematical models for plastic deformation are to be scalable [4]. At the same time, changes in microscopic inner structure of rocks cause variations of their mechanical, rheological, and filtration properties—this effect is both proved to take place in laboratory conditions [4, 5] and under real-life conditions at hydrocarbon reservoirs [6] and geothermal fields [7]. Prediction of natural fractures in rock masses is closely related to this problem, as natural fractures can be considered as result of irreversible strain localization during the mechanical processes that have taken place for the considered rock. This problem is especially important for hydrocarbon and geothermal fields exploration and development, as fracture presence affects the filtration paths in naturally fractured rocks [8], while knowledge of their stress–strain state allows to predict, which fractures are hydraulically conductive [9].

Current research is aimed at creating and verifying a mathematical model capable of finding the main tendencies in rock inner structure evolution caused by applied loads. The changes of rock samples inner structure are related to macroscopic parameters of the samples, starting, from stress vs. strain curves that can be obtained from experiments. Stress–strain states of the rock samples are limited to triaxial state within this paper. While existing laboratory equipment makes it possible to analyse true triaxial stress states [4], we have studied the simpler, triaxial loading to obtain the various mechanical properties with necessary precision [10] to verify model predictions.

Preliminary results of interpreting experimental data on triaxial loading of rock samples to predict relative development of variously oriented shear fractures using non-associated plastic flow law [11, 12] are presented in the paper. The experiments have been carried out using laboratory equipment belonging to shared research facilities "Petrophysics, geomechanics and paleomagnetism" of the Schmidt Institute of Physics of the Earth RAS [10] using the samples of limestone of the Moscovian stage. Mathematical description of theoretical model, geological description of the studied samples, experimental setup, obtained results, and their discussion are presented below. The opportunities of applying the suggested model to predict natural fracture systems development for reservoir geomechanics problem solutions are discussed in detail alongside with the experimental scheme necessary for model to be established.

2 Mathematical Model

The mathematical model used for analysis of triaxial loading results is based on non-associated plastic flow law [11, 12]. The model itself is discussed in detail in the paper [13]; only its key features important for the current study are described below.

Mohr–Coulomb or Drucker-Prager conditions are often used to describe plastic deformation of rock masses

$$T + \alpha\sigma = c, \tag{1}$$

where $T = (s_{ij}s_{ij})^{1/2}$ is shear stress intensity ($s_{ij} = \sigma_{ij} - \delta_{ij}\sigma$—are components of stress tensor σ_{ij} deviator, δ_{ij} is Kronecker delta; not saturated rocks are considered from now on, so effective stresses are equal to total stresses; Einstein summation notation is utilized for summation indices $i,j = 1, 2, 3$); $\alpha = \sin\varphi$ is friction coefficient (φ is friction angle); $\sigma = (\sigma_1 + \sigma_2 + \sigma_3)/3$ is mean stress (σ_1, σ_2, and σ_3 are maximum, intermediate, and minimum principal stresses respectively, compressive stresses are positive); c is cohesion.

Tensor components of stress increments $d\sigma_{ij}$ and strain increments $d\varepsilon_{kl}$ in inelastic deformation zone, i.e. when condition (1) is fulfilled, can be written as

$$d\sigma_{ij} = E_{ijkl}d\varepsilon_{kl}, \tag{2}$$

where E_{ijkl} are current elastic moduli that can be obtained as [14, 15]:

$$E_{ijkl} = G\left\{\left[\left(\delta_{ik}\delta_{jl} + \delta_{il}\delta_{jk}\right) + \left(\frac{K}{G} - \frac{2}{3}\right)\delta_{ij}\delta_{kl}\right] - \frac{G}{(H+G) + \alpha\Lambda K}\left(N_{ij} + \frac{K}{G}\Lambda\delta_{ij}\right)\left(N_{kl} + \frac{K}{G}\Lambda\delta_{kl}\right)\right\}, \tag{3}$$

where G is shear modulus, K is bulk modulus, H is modulus of plastic hardening, Λ is coefficient of dilatancy, $N_{ij} = s_{ij}/T$.

Coefficient of dilatancy in Eqs. (2) and (3) Λ characterizes inelastic change of volume caused by shear deformation and is defined as the proportional coefficient in $d\varepsilon^{pl} = \Lambda d\Gamma^{pl}$, where $d\varepsilon^{pl}$ is the change in the first invariant of plastic strain tensor (which is equal to the difference between total and elastic strains); $d\Gamma^{pl}$ is plastic shear strain increments intensity: $d\Gamma^{pl} = 2\left[\left(d\varepsilon_{ij}^{pl} - \frac{1}{3}\delta_{ij}d\varepsilon^{pl}\right)\left(d\varepsilon_{ij}^{pl} - \frac{1}{3}\delta_{ij}d\varepsilon^{pl}\right)\right]^{1/2}$.

Plastic hardening modulus H connects increments of mean stress $d\sigma$ and shear stress intensity dT with plastic shear strain increments intensity $d\Gamma^{pl}$ as:

$$d\Gamma^{pl} = \frac{1}{H}(dT + \alpha d\sigma). \tag{4}$$

It is worth mentioning [11, 13], that plastic hardening modulus is not constant if non-associated plastic flow law is used, so it changes with irreversible strain accumulation [11]. Note that the value of hardening modulus depends both on the current stress–strain state and the trajectory of its evolution from the initial state in stress space to the current one.

Generally speaking, both friction coefficient α and cohesion c can be functions of accumulated irreversible strain [11], while hardening modulus depends on α [13]. Friction angle and cohesion hardening cases can be considered separately [11]. In the current study we assume that cohesion does not depend on inelastic strain $c = const$, while friction angle depnds on both mean stress and accumulated shear strain $\alpha = \alpha(\Gamma^{pl}, \sigma)$. in this case integrated form of Eq. (4) is the same as Drucker-Pruger condition (1) with hardening and dilatancy:

$$T + \alpha\sigma = \int\limits_{\Gamma^{pl}} H d\Gamma^{pl} + \int\limits_{\alpha} \sigma d\alpha + c. \tag{5}$$

Given the mentioned assumptions on hardening parameters evolution, Eq. (5) can be rewritten in a form allowing for a construction of an explicit function for hardening modulus calculation from mean stress, friction coefficient, and shear strain intensity increments [13]:

$$H = -\sigma \frac{d\alpha}{d\Gamma^{pl}}. \tag{6}$$

This equation makes it possible to use experimental data from triaxial loading to calculate hardening modulus. The increments of strains and stresses obtained from the loading experiment are connected through Eqs. (2) and (3); elastic moduli are calculated from the linear elastic zone of stress–strain curve; dilatancy coefficient is obtained from the relationship between mean strain and shear strain intensity increments. Only hardening coefficient H and friction coefficient α, depending on each other through Eq. (6), are the single parameters remaining unknown in Eq. (3). Nevertheless, condition (5) fulfilment at every point of loading curve in its inelastic zone makes it possible to define friction coefficient as a function of the current stress state.

This calculation had been completed in paper [13] for triaxial loading of cylindrical samples ($\sigma_2 = \sigma_3 = const$). Coefficient N_α, depending on friction coefficient as $N_\alpha = (1 + \alpha)/(1 - \alpha)$, can be obtained as:

$$N_\alpha = \frac{\sigma_1\sigma_3 + 2c_1^2(1 - \sqrt{c_1^2 + \sigma_1\sigma_3})}{\sigma_3^2} \tag{7}$$

where c_1 is cohesion, found at the first moment of nonlinear behaviour of the stress curve:

$$c_1 = \frac{\sigma_1^{el} - \sigma_3}{2} \tag{8}$$

where ${\sigma_1}^{el}$ is the maximum axial stress in the sample allowing for linear behaviour of axial stress vs. axial strain deformation curve. Solution (8) is used in Eq. (7) under the assumption that inelastic deformation starts when condition (1) is fulfilled for zero friction coefficient α. So, within the proposed model, friction angle φ remains zero until irreversible strain starts accumulating in the rock sample. Then it starts gradually increasing following Eqs. (7) and (8), while Eq. (6) provides corresponding changes in plastic hardening modulus H. As a result, all variables in Eq. (3) can be obtained providing an opportunity to use Eq. (2) to set the direct relationship between stresses and strains in the rock sample.

Paper [13] suggested the match between the real stress vs. strain curves and their reproductions using the described equations as a criterion for model verification. The proposed model was proved to be reliable for plastic deformation process of rock samples subjected to triaxial loading.

Calculation of friction angle φ is important for problems associated with fracture system predictions. Evolution of this angle makes it possible to deal with fractures in the following way. It is clear [11] that friction angle φ stands for the spatial orientations of the planes, where slip occurs during brittle irreversible deformation process. E.g., for a cylindrical sample subjected to triaxial loading (axial stress is maximum principal stress), if linear Mohr–Coulomb angle is applicable, the angle ψ between sample axis and poles to the planes where slip occurs resulting into shear fracture development, is:

$$\psi = \pi/4 + \varphi/2. \tag{9}$$

If friction angle φ is known as a function of accumulated irreversible strain, Eq. (9) can be used to predict the preferable angle of shear fracture development as function of accumulated strain as well. In other words, the proposed model makes it possible to predict the preferable spatial orientations of developing fractures from stress vs. strain curve at any point of loading.

Fracture development intensity can be analysed alongside with it preferable spatial orientation. There are developed energy-based methods of dealing with fracture propagation. According to these approaches, fracture development is associated with change in the curvature of the stress vs. strain curves caused extra energy needed to create extra free surface in the medium [16]. Change in curvature results into increase in inelastic energy which is equal to the difference between total deformation energy $E_{total} = \int \sigma_{ij} d\varepsilon_{ij}$ and elastic energy $E_{elastic} = \int E^0_{ijkm} \varepsilon_{km} d\varepsilon_{ij}$, where ${E^0}_{ijkm}$ are elastic moduli tensor components that can be obtained at linear zone of deformation curve. As a result, loading curve can be interpreted in a following way: inelastic energy can be calculated for each value of friction angle. This energy can be associated with fracture development intensity, as it is closely related to the free surface emerging during shear fracture propagation. The model suggests that every point of loading is characterized by its own friction angle (7) and corresponding preferable fracture development angle (9), while fracture system development intensity is governed by the current fraction of irreversible energy in total energy. So, loading curve interpretation makes it possible to compare differently oriented systems of fractures and deal with the statistical tendencies in fractures distributions for various spatial orientations.

This model application has several limitations. First of all, the whole irreversible energy is associated exclusively with shear fracture development, yet there in general exist other mechanisms of irreversible deformation of rock masses, as paper [2] points out. This limitation can be overcome in future due to calculations of the rock brittleness index [17] allowing for estimation of portion of irreversible energy of free surface increase. Secondly, only triaxial loading case with all loads increments being functions of a single parameter (time) was proved to be a suitable scenario for this model to provide reliable results. We expect to generalize this model for true triaxial loading conditions in future with corresponding changes in Eqs. (7)–(9). Next, the proposed model only gives insights on fractures preferable orientations, but not their sizes and shapes. This limitation can be partially overcome by using theory of effective medium approaches. Finally, only not saturated media are considered in the current study, saturating fluid effect on the rock properties and inner structure is within the scope of the current study.

A set of experiments was carried out for the samples of limestone of the Moscovian stage in order to demonstrate the application of the model to predict fracture system development and prove the reliability of the obtained results.

3 Material Under Study

The samples of limestone of the moscovian stage (C_2m) were studied. A lithological macro description was carried out for saw cuts. According to lithological characteristics, 4 varieties of carbonate rocks were distinguished (Fig. 1).

Group 1 includes dolomitic and dolomitic yellowish-white organogenic-detrital limestones, medium-fine-grained, massive, indistinctly bedded, with psammitic, pelitic and polydetrital structure. Group 2 is represented by calcareous, cavernous, massive, indistinctly bedded dolomite with large leaching voids. Dolomite powder is present in the voids. Group 3 is characterized by white, coarse-grained sorted layered limestone with psammitic and polydetrital structure, organogenic-detrital, predominantly foraminiferal limestone, represented by Fusulinella bocki (Moeler, 1878), Fusulina cylindrica (Moeler, 1878) and some others. There are black siliceous fragments of crinoids, poorly preserved segments, and fragments of brachiopods Choristites sp. Group 4 includes dolomitic yellowish-white organogenic-detrital limestones with siliceous nodules and concretions. The structure is psammitic and pelitic. The rocks are characterized as indistinctly layered and massive.

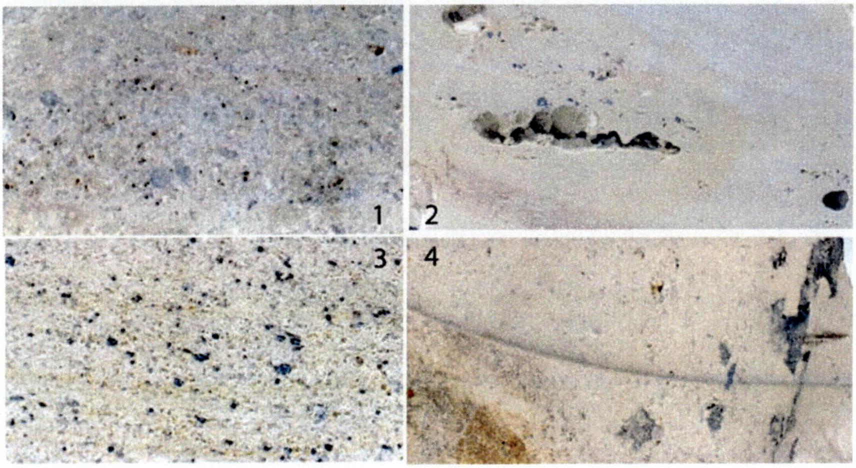

Fig. 1 Differences of carbonate rocks. 1. Dolomitic limestone 2. Cavernous dolomite 3. Foraminiferal limestone 4. Limestone with siliceous nodules

The setting of the formation of the studied carbonate rocks is characteristic of the shallow part of the epicontinental basin of the southern flank of the Moscow syneclise.

4 Experimental Setup

A complex experimental program has been carried out in order to achieve the stated goals of the study. Despite the fact that the mathematical model described above is primarily focused at triaxial loading tests performed with cylindrical samples, verification of its application results required studying dynamic elastic properties as well.

Cylindrical samples (length of 60 mm, diameter of 30 mm) have been made from the described limestones. First of all, their filtration properties were measured: volumetric method was used to measure open porosity; absolute gas permeability was measured using unsteady flow method.

Figure 2 presents the results of filtration properties measurement with distinction between defined groups.

The next step of the experimental procedure following filtration properties measurement consisted in samples analysis with regard to dynamic elastic properties. P- and S-wave velocities were measured along three trajectories: along axis of the sample and two perpendicular diameters [18] providing an opportunity to evaluate anisotropy of each sample before loading.

After these preliminary measurements had been completed, each sample was subjected to uniaxial or triaxial stress state using GCTS RTR4500 testing system

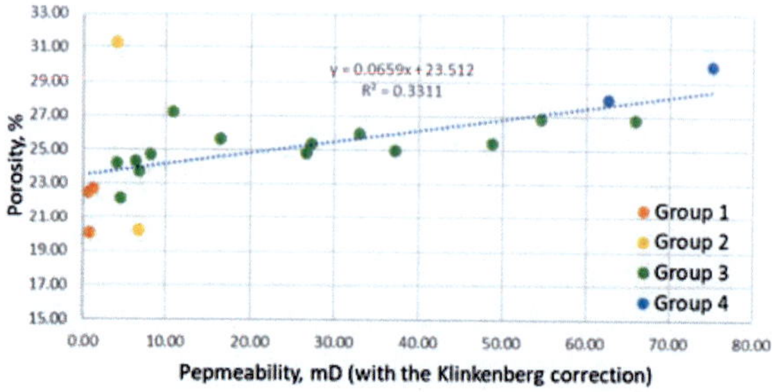

Fig. 2 Interrelationship between porosity and permeability for the studied samples

belonging to shared research facilities "Petrophysics, geomechanics and paleomagnetism" of the Schmidt Institute of Physics of the Earth RAS [10]. This system allows for loading tests of rock samples under normal and in-situ conditions. The overall experimental setup is given in the left part Fig. 3, details are highlighted in its right part.

Each sample has been subjected to triaxial or uniaxial loading with axial stress of σ_a applied to its base; normal radial stress σ_r acts along the side surface. The sample is not saturated, principal stresses acting in the sample are equal to $\sigma_1 = \sigma_a, \sigma_2 = \sigma_3 = \sigma_r$ (the condition of $\sigma_a \geq \sigma_r$ is maintained during the loading procedure). The applied loads were varied during the experiment. Stress and strain tensor components were measured at any loading point. Moreover, *P*- and *S*-wave velocities sensors were installed at top and bottom bases of the sample providing an opportunity to monitor velocities along axis during loading.

Three different schemes of axial stress evolution were considered with respect to typical experiments that are carried out for rock mechanical properties analysis [4, 8]:

1. Unconfined compressive strength test (UCS). Radial stress remains zero for this case, while axial stress is being gradually increased until sample failure.

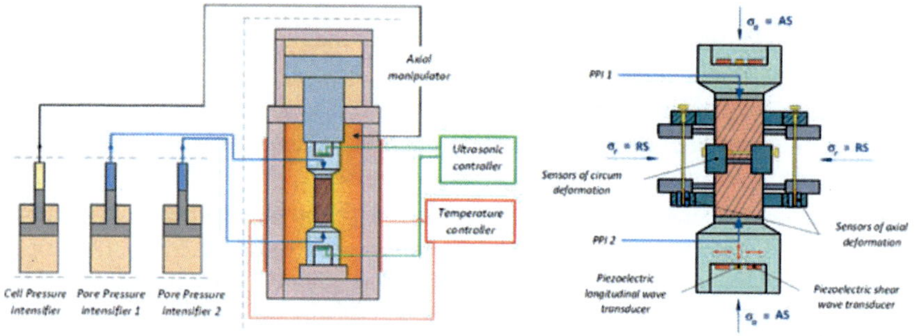

Fig. 3 General scheme of servo-hydraulic test system GCTS RTR-4500 (at the left part of figure) and layout of sensors on a rock sample during geomechanical testing. Abbreviations: PPI—pore pressure intensifier, AS—axial stress, RS—radial stress

2. Single-stage test (SST). Radial and axial stress are being increased simultaneously at the initial step of the loading procedure until reaching a certain value. Afterwards, axial stress is being increased until sample failure.
3. Multi-stage test (MST). After the simultaneous increase in axial and radial stress, axial stress is being gradually increased up to the elasticity limit. Then it is decreased down to the level of radial stress, which remains constant during the stage. The next stage starts with another simultaneous increase in radial and axial stresses of a relatively small value (0.5 MPa), following by another axial stress increase and decrease cycle. The stages are repeated several times; axial stress is increased until sample failure at the final stage.

The samples were unloaded after failure and elastic waves velocities were measured in three directions under atmospheric conditions once more. This procedure provided an opportunity to find the change in sample anisotropy caused by loading and verify the mathematical model described above.

Figures 4, 5 and 6 represent the experimental setups described above. Principal stresses are plotted as functions of time at the left parts of the corresponding figures, and as functions of axial (blue), radial (red), and volumetric (purple) strains at the right parts of the figures.

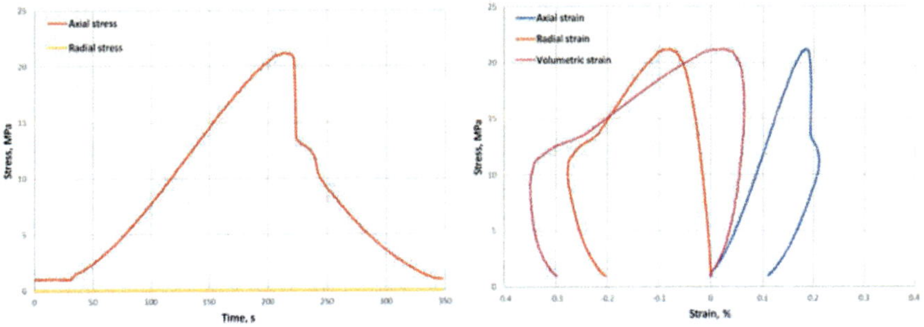

Fig. 4 Typical interrelations of principal stresses on time and strains for UCS test

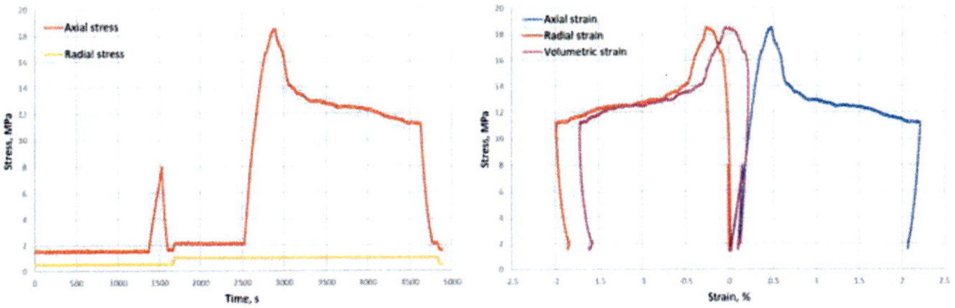

Fig. 5 Typical interrelations of principal stresses on time and strains for SST test

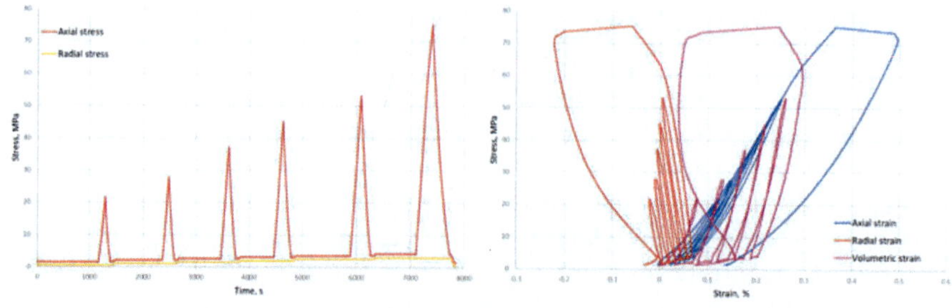

Fig. 6 Typical interrelations of principal stresses on time and strains for MST test

5 Results

The described experimental scheme was successfully applied for the study of the samples identified on the basis of a lithological description to four groups of lithological varieties: 1—dolomitic organogenic-detrital limestones; 2—cavernous dolomites; 3—foraminiferal limestones; 4—dolomitic limestones with siliceous nodules. Based on the geological study, for each sample produced, the spatial orientation relative to the plane of probable bedding was determined: further, the Z symbol marks the samples whose axis is perpendicular to the bedding plane, the X and Y symbols indicate the samples whose axes are in the bedding plane and are perpendicular to each other.

Figure 7 represents the measured compressional V_p and shear V_s elastic waves velocities. Horizontal axis gives the number of the sample and the experimental scheme carried out for the corresponding sample following the proposed notation. Blue colour shows the measured *P*-wave velocities; red colour is used for *S*-wave velocities. Same vertical axis is used for both velocities. Dots represent the measurements performed before loading, arrows give an understanding of velocities changes: they point at the measured velocities in the same directions after loading. Most samples are characterized by total of six measurements: three measurements in three perpendicular directions before loading and three after loading. It was impossible to carry out measurement due to presence of inner inhomogeneities preventing elastic waves from propagation.

The obtained results show that most samples were characterized by low degree of anisotropy before loading, some of them could be considered as isotropic. Triaxial loading of the sample and its failure results in a considerable jump in elastic waves velocities and growth of anisotropy—samples 1-X and 1-Y represent this effect to the highest degree. Their comparison with sample 1-Z belonging to the same group is discussed below.

P-wave velocity absolute value is way more sensitive to changes in inner structure due to loading compared to *S*-wave velocity. This result is in agreement with existing theoretical ideas.

Figure 7 also proves that loading tests tend to decrease the velocities of elastic waves in all directions. It is worth mentioning that velocities measured along radial

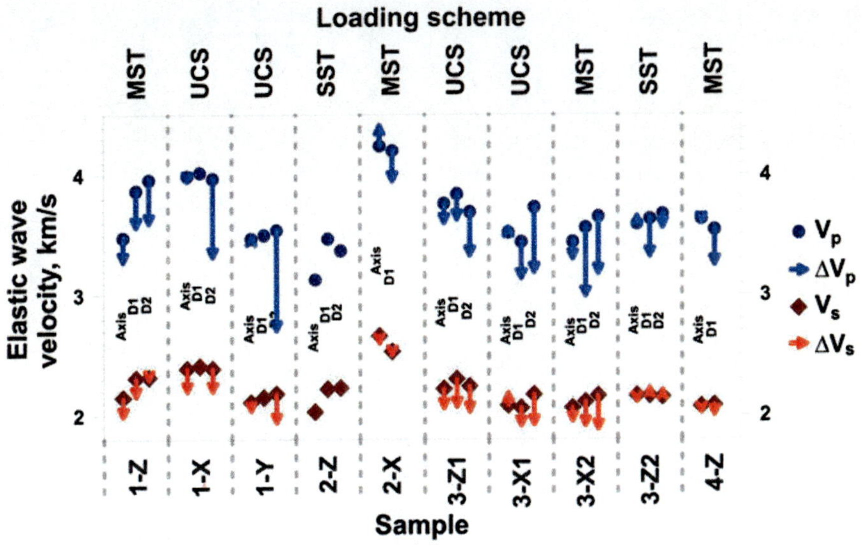

Fig. 7 Dynamic elastic properties measurements results

directions tend to change more than velocities in axial direction. We assume that this effect is associated with development of fractures with planes close to sample axis: the closer fractures are to the axis of the sample, the greater is the difference between changes in velocities in radial and axial directions. At the same time, considerable change in axial velocity suggests a possibility of loss of stability at weak bedding planes [2].

Figure 8 presents the results of experimental results interpretation for typical experiment (sample 3-X2—see Fig. 7 for notation—the last loading cycle is considered) using the proposed mathematical model.

Initial data obtained directly from the experiments are shown in Figs. 8a–c. Figure 8a consists of the obtained axial stress vs. axial (blue curve) and radial (red) strain. Dotted lines give the linearized curves: as it can be seen from Eq. (7) and condition (8), it is important to properly find the elasticity limit for the sample. Slope of theses linearized curves provide necessary information to find elasticity moduli used in Eq. (3). The elastic *P*- and *S*-waves velocities are given as green and red lines in Figs. 8b, c respectively as functions of sample axial strain.

The following procedure was used to apply the proposed mathematical model for inelastic deformation of the sample. Equation (7) was used to find the current friction angle at each time point after elasticity limit—this angle is shown as function of axial strain in Figs. 8b, c. Equation (9) was consequently used to define the preferable angle of fracture propagation ψ—an angle between sample axis and pole to the fractures planes. It should be mentioned that only one angle can be obtained for triaxial loading conditions, as the second angle can be estimated if true triaxial loading conditions are considered. Energetic concept can be used to estimate the relative intensity of

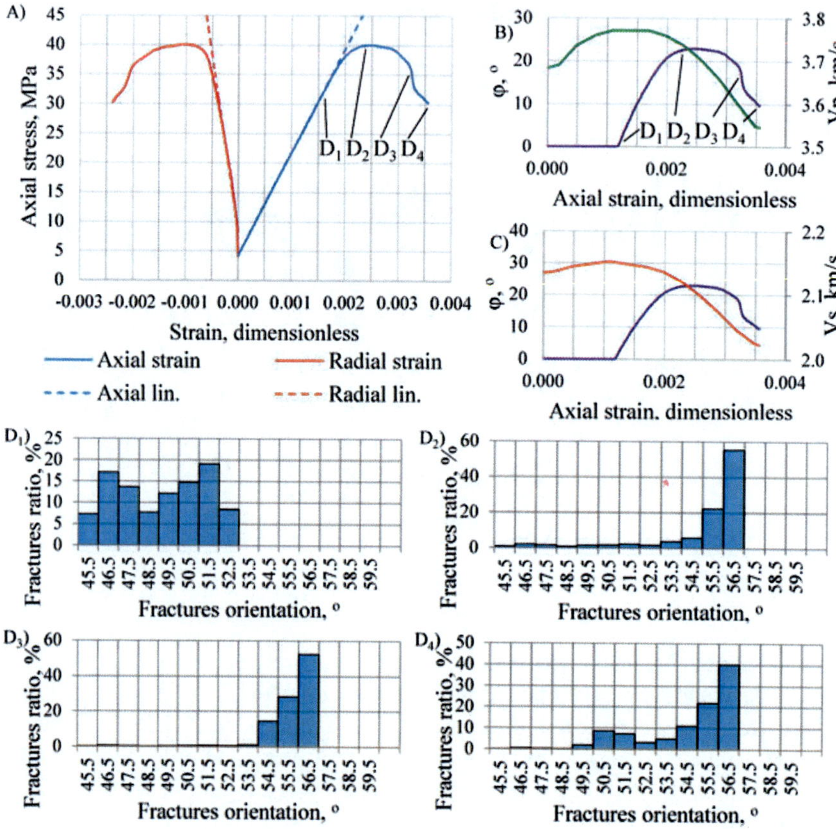

Fig. 8 Triaxial loading test result interpretation

differently oriented fracture systems developing at each point of loading. The possible range of angles ψ was split into discrete sub-ranges ψ_k with equal step $\Delta\psi$ ($\psi_{k+1} = \psi_k + \Delta\psi$). Accumulated inelastic energy was calculated for each sub-range as:

$$E_{inel}|_{\psi\in[\psi_k;\psi_{k+1}]} = \int_{\boldsymbol{\varepsilon}(\psi_k)}^{\boldsymbol{\varepsilon}(\psi_{k+1})} \boldsymbol{\sigma} : d\boldsymbol{\varepsilon} - \frac{1}{2}\big(\boldsymbol{\sigma}(\psi_{k+1}) - \boldsymbol{\sigma}(\psi_k)\big) : \mathrm{E}^{-1} : \big(\boldsymbol{\sigma}(\psi_{k+1}) - \boldsymbol{\sigma}(\psi_k)\big). \quad (10)$$

Component-free treatment of stress $\boldsymbol{\sigma}$ and strain $\boldsymbol{\varepsilon}$ tensor operations is used here for simplification; symbol: stands for tensor double contraction. Tensor $\mathbf{E}^{-1}$used in this equation is the inverse of elastic moduli tensor calculated at the linear zone of stress vs. strain curve. The interval of integration is defined by strain state $\boldsymbol{\varepsilon}(\psi_k)$ corresponding to fracture orientation-governing angle ψ_k (with stress state being defined as $\boldsymbol{\sigma}(\psi_k)$) and strain state $\boldsymbol{\varepsilon}(\psi_{k+1})$ corresponding to fracture orientation-governing angle ψ_{k+1} (with stress state being defined as $\boldsymbol{\sigma}(\psi_k)$). All this energy is assumed to be associated with initiation and propagation of fractures with orientation being between ψ_k and ψ_{k+1}. This allows analysing the statistical aspects of development of fracture systems of different spatial orientations.

Four dots—D_1, D_2, D_3, and D_4—are chosen in Fig. 8a in order to illustrate the described analysis. Evolution of friction angle can be defined for each of these points: Eq. (10) is used to define the whole inelastic energy accumulated in the sample until the considered point of time alongside with its portions accumulated during angle changes within the defined sub-ranges ψ_k. Figure 8D_1–D_4 show the histograms characterizing the statistical properties of differently oriented fractures: horizontal axis gives the angles ψ_k; vertical axis stands for a specific parameter which is the relative fraction of fractures developed during the angle ψ increase from ψ_k to ψ_{k+1}; this fraction is assumed to be equal to the ratio between energy obtained from Eq. (10) and total inelastic energy accumulated in the sample.

The tendency shown in Fig. 8D_1–D_4 remains true for the majority of the studied samples. During the first stages of inelastic loading (point D_1) no prevailing fracture orientation can be distinguished. During inelastic energy accumulation and approaching the failure point (D_2) the angle between the sample axis and fracture poles increases; a local maximum appears for the distribution. This maximum is assumed to be associated with the direction of the major fault causing failure. It is worth mentioning that the proposed model is proved to be reliable for post-failure zone as well [13]: friction angle starts decreasing after faiure. Fractures formed in the zone close to failure start reactivating at this stage of loading (Fig. 8D_3). The final deformation stages (Fig. 8D_4) are characterized by reactivation of fractures formed at the earlier stages of loading: local maxima of fracture fraction in the range between 49 and 52° coincide. The similar tendencies remained true for the other samples without regard to the loading conditions and experimental scheme.

Measurements of the current elastic waves velocities carried out during loading made it possible to evaluate the contribution of changes in sample inner structure into the dynamic elastic moduli. Velocities of both *P*- and *S*-waves velocities can be seen to decrease with inelastic strain accumulation, including the fact that the maxima of waves velocities coincide with the end of linear elastic part of the loading curve—this result is in total agreement with the proposed mathematical model. The rates of velocities decrease depend on the rate of friction angle change proving the reliability of the model.

6 Discussion

Different samples have been interpreted with the help of the proposed mathematical model as well. Some observed tendencies are discussed below.

Figure 9 presents the comparison between the results obtained for similar samples with close dynamic elastic properties—samples 3-X1 and 3-X2 (foraminiferal limestone, Fig. 7) with different loading schemes.

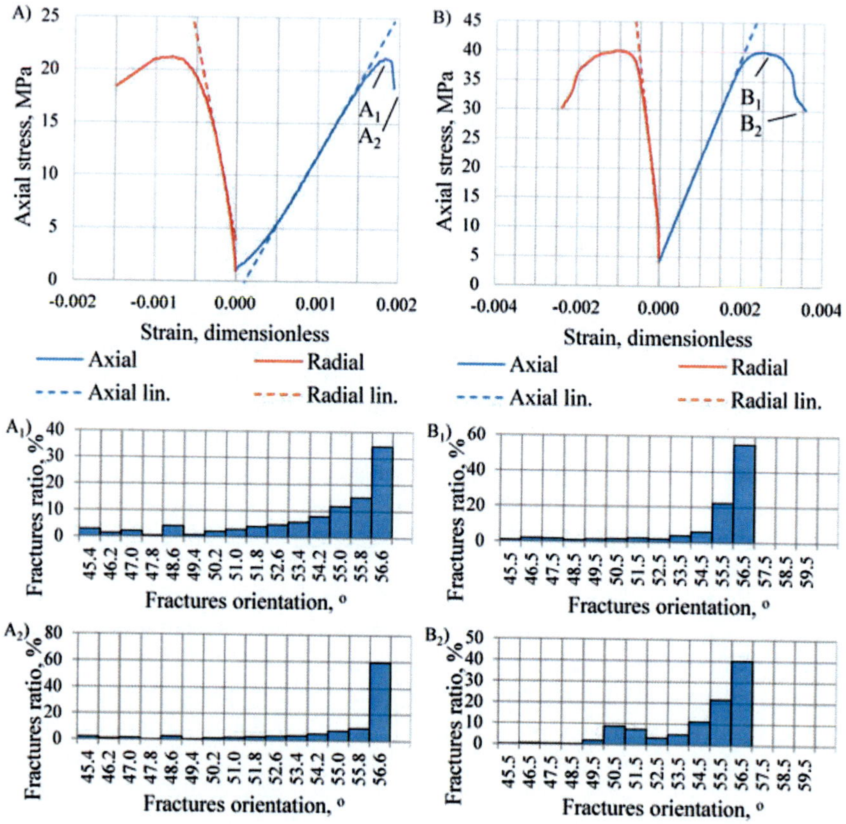

Fig. 9 Effect of preliminary loading on fracture development

The sample 3-X1 (Fig. 9A, A_1 and A_2) was loaded until failure without preliminary loading; sample 3-X2 (Fig. 9B, B_1 and B_2) was loaded following the multistage loading scheme MST. Two points (A_1 and B_1 for samples 3-X1 and 3-X2 respectively) were chosen in the vicinity of failure point in a way similar to Fig. 8; points A_2 and B_2 were chosen in post-peak zone. The corresponding histograms lead to the conclusion that the mean stress level and preliminary loading cycles provide fracture systems with less distinct preferable direction; some of the fractures are reactivated after failure. At the same time, fracture system development without radial stress and preliminary loading provide has a more distinct preferable direction, but failure takes place for one of them. Slip continues along this plane even in post-peak zone—no reactivation of previously formed fractures emerges.

Comparison of three samples 1-Z, 1-Y, and 1-X (highly anisotropic dolomitic limestone, Fig. 7) is shown in Fig. 10. Sample 1-Y was characterized by the lowest elastic waves velocities, sample 1-X had the highest velocities. These two samples were subjected to unconfined compressive stress test (UCS), while the sample 1-Z with the average velocities was subjected to multi-stage triaxial test (MST). The figure

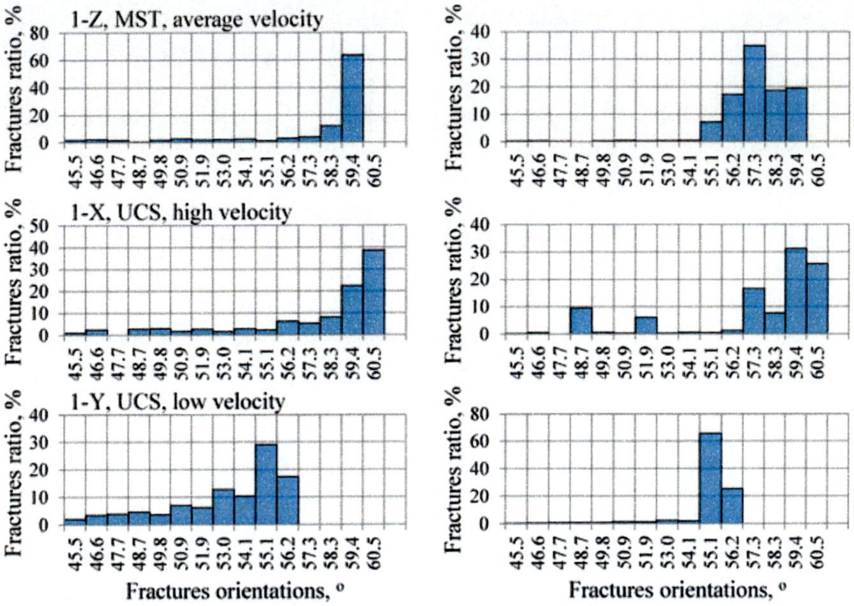

Fig. 10 Relationship between elastic wave velocities and fracture systems development

gives an overview of statistical estimations of the fracture systems development at failure point (left part) and in post-peak zone (right part).

The tendency revealed in Fig. 9 appears to remain true for these samples as well, yet sample 1-X is characterized by local maxima in the left part of the histogram after failure. Moreover, elastic wave velocities seem to have an effect on preferable fracture system development angle: the higher velocities are, the higher is the angle between fracture plane pole and axis of the sample, proving the possibility to use dynamic elastic properties for fracture system development prediction.

7 Conclusions

The paper presents a method to apply non-associated plastic flow law with friction angle hardening in order to predict fracture system development in rock mass according to its stress state evolution.

The limitations of the model are related to the simple loading of non-saturated media case and assumption of all inelastic energy association with fracture development. These limitations can be overcome after modifying the suggested experimental setup. Dynamic elastic moduli analysis step can be expanded in future: measurements along three directions can be replaced by multi-level ultrasonic scanning of the sample along a variety of directions. These measurements application before and after loading provide an opportunity to use effective medium theory to establish the relationship between changes in the rock inner structure and observed variations in dynamic elastic moduli.

Extra information on fractures and verification of the relationship between inelastic energy and relative fraction of fractures can be obtained from triaxial loading tests complemented with acoustic emission registration, so that inelastic energy losses can be quantitatively related to slips along weak planes of known spatial orientations.

The proposed model can be successfully used for the case of true triaxial loading test as well. In general, the developed algorithm provides an opportunity to estimate fracture system development in rock mass for the case of known stress state evolution. As a result, various geomechanical problems suggesting numerical modelling of mechanical behaviour of the rock mass, can be supplemented by oriented fracture systems development.

Acknowledgements The research was performed as part of the State assignment of Schmidt Institute of Physics of the Earth RAS.

References

1. Zuev L.B., Khon Yu.A. Plastic Flow as Spatiotemporal Structure Formation. Part I. Qualitative and Quantitative Patterns // Physical Mesomechanics. 2021. V. 25. P. 103–110.
2. Karev V.I., Klimov D.M., Kovalenko Yu.F., Ustinov K.B. Fracture Model of Anisotropic Rocks under Complex Loading// Physical Mesomechanics. 2018. V. 21. P. 216–222.
3. Zuev L.B., Khon Yu.A. Plastic Flow as Spatiotemporal Structure Formation. Part II. Two-Level Description // Physical Mesomechanics. 2021. V. 25. P. 111–118.
4. Karev V.I., Khimulia V.V., Shevtsov N.I. Experimental Studies of the Deformation, Destruction and Filtration in Rocks: A Review// Mechanics of Solids. 2021. No. 5. P. 613–630.
5. Karev V.I., Kovalenko Yu.F., Sidorin Yu.V. Influence of leaching reservoir rocks of oil and gas deposits on their deformation, strength and filtration properties on the example of chayandinskoye oil gas condensate field [in Russian] // Processes in Geomedia. 2018. V. 18. No. 4. P. 1143–1149.
6. Hickman S.H., Barton C.A., Zoback M.D., Morin R., Sass J., Benoit R. In situ stress and fracture permeability along the Stillwater fault zone, Dixie Valley, Nevada. International Journal of Rock Mechanics and Mining Sciences and Geomechanics Abstracts. 1997. V. 34. No. 3–4. 414.
7. Massiot C., Nicol A., McNamara D.D., Townend J. Evidence for tectonic, lithologic, and thermal controls on fracture system geometries in andesitic high-temperature geothermal field. Journal of Geophysical Research. Solid Earth. 2017. V. 122. No. 8. P. 6853–6874.
8. Zoback M.D. Reservoir Geomechanics, 1st ed. Cambridge: Cambridge University Press, 2007. 490 p.
9. Dubinya N.V. Spatial orientations of hydraulically conductive shear natural fractures for an arbitrary stress state: An analytical study of governing geomechanical factors // Journal of Petroleum Science and Engineering. 2022. V. 212. 110288.
10. Veselovskiy R.V., Dubinya N.V., Ponomarev A.V., Fokin I.V., Patonin A.V., Pasenko A.M., Fetisova A.M., Matveev M.A., Afinogenova N.A., Rud'ko D.V., Chistyakova A.V. Shared research facilities "Petrophysics, geomechanics and paleomagnetism" of the Schmidt Institute of Physics of the Earth RAS // Geodynamics and tectonophysics. 2022. V. 13. No. 2. P. 1–12.
11. Vermeer P.A., de Borst, R. Non-associated Plasticity for Soils, Concrete and Rock. Heron. 1984. V. 29. No. 3. p. 1–64.

12. Garagash I.A., Nikolaevskii V.N. Non-associated flow laws and plastic deformation localization// Advances in mechanics. 1989. V. 12. No. 1. P. 131–183.
13. Garagash I.A., Dubinya N.V., Rusina O.A., Tikhotsky S.A., Fokin I.V. Estimation of rock strength properties from triaxial test data// Geophysical research. 2018. V. 19. No. 3. P. 57–72.
14. Nikolaevskii V.N. Governing equations of plastic deformation of a granular medium// PMM. 1971. V. 35. No. 6. P. 1070–1082.
15. Rudnicki J.W., Rice J.R. Conditions for localization of deformation in pressure-sensitive dilatant materials // Journal of the Mechanics and Physics of Solids. 1975. V. 23. No. 6. P. 371–390.
16. Bieniawski Z.T. Mechanism of brittle fracture of rock: part I—theory of the fracture process. International Journal of Rock Mechanics and Mining Science and Geomechanics Abstracts. 1967. V. 4. No. 4. P. 395–406.
17. Nygard R., Gutierrez M., Bratli R.K., Høeg, K. Brittle–ductile transition, shear fail-ure and leakage in shales and mudrocks. Marine and Petroleum Geology. 2006. V. 23. P. 201–212.
18. Petrov V.A., Nasimov R.M. RF patent no. 2 515 332. 2012.

Estimates of the Periodicity of Atmospheric Blockings Over Kazakhstan in the Spring–Summer Time According to Era 5 Reanalysis Data

A. V. Kholoptsev and Zh. K. Naurozbayeva

Abstract The article estimated the frequency of atmospheric blockings over different regions of Kazakhstan for the spring and summer months. Its regions are identified, where during periods of these phenomena lasting 10 days or more, the greatest increase in average daily air temperatures occurs. This fact may contribute to the occurrence of drought.

Keywords Atmospheric blockings · Atmospheric pressure · Isobaric surfaces · Reanalysis · Air temperatures · Drought

1 Introduction

Atmospheric blockings (AB) of long duration are formed in the spring–summer months over the territories of the temperate climate zone. It is capable of causing drought that threatens the population, agriculture, forestry and water sectors of the economy, and can lead to landscape fires. Therefore, the assessment of the frequency of ABs of different duration for such territories is an urgent problem of meteorology and agroclimatology.

The greatest interest in solving this problem arises in regions where the continental climate and agriculture is the basis of their economy. These include many regions of Kazakhstan. The basis of the economy of Kazakhstan is agriculture (grain farming), and the damage from droughts and fires, as a rule, exceeds the damage caused by other natural hazards [1–5].

Many domestic and foreign authors have devoted their research work to revealing the role of AB in the occurrence of droughts [6–10]. Scientists have found that the risks associated with droughts and landscape fires increase as the duration of AB

A. V. Kholoptsev (✉)
Sevastopol Branch of the State Oceanographic Institute named N.N. Zubov, Sevastopol, Russia
e-mail: kholoptsev@mail.ru

Zh. K. Naurozbayeva
RSE "Kazhydromet" Department of Hydrometeorological Research of the Caspian Sea, Almaty, Kazakhstan

V. I. Karev (ed.), *Physical and Mathematical Modeling of Earth and Environment Processes—2022*, Springer Proceedings in Earth and Environmental Sciences,
https://doi.org/10.1007/978-3-031-25962-3_2

increases. The reason is that during the AB there is no precipitation, in the surface layer of the atmosphere the air temperature rises significantly.

When assessing the risk of drought, scientists consider the characteristics of meteorological conditions—the Selyaninov hydrothermal coefficient (SHC), the Rochevoi index, the Palmer Drought Severity Index (PDSI), the standardized precipitation index (Standardized Precipitation Index, SPI), the Pedia index and others [11–13]. When assessing the fire hazard, the Nesterov method is used [1, 11].

All of these characteristics are very informative, reliable and successfully used in agroclimatology and pyrology. However, they are all local. This fact complicates too much the risk assessment of the considered natural phenomena over large areas. Because it is necessary to take into account the relationship between hydrometeorological processes not only at their monitoring points. When we take into account synoptic information as well as changes in the frequency of macrocirculation processes, which include AB, we can achieve significant success in the development of such forecasts [14, 15].

Modern ideas about the causes of AB and their consequences are presented in the scientific works of many domestic and foreign authors [16–20]. It follows from them that processes of two types can serve as the causes of AB.

The territories of the Urals and Kazakhstan are characterized by blocking of the meridional type. It may be due to the peculiarities of the flow around the Earth's orography by air currents. The reason for the appearance of blocking can also be the large-scale instability of the zonal flow, which leads to the formation of a powerful system of interacting anticyclones and cyclones (split-type blocking) or a powerful ridge and two cyclones (omega-blocking).

Blocking of the meridional type is formed as a result of intrusions of arctic air, which has a low temperature and absolute humidity. The Arctic air mass has a high density and quickly moves to low latitudes, directly above the underlying surface and, at the same time, flows around the unevenness of its relief. The resulting air flow is turbulent. On its upper part, a turbulent boundary layer is formed in which the arctic air is exchanged with the surrounding air. This air exchange leads to the gradual penetration of arctic air into the layers of the atmosphere located above it, which causes an increase in atmospheric pressure in them. The air from these layers is drawn into the Arctic air mass and leads to an increase in temperature and humidity of the air that moves south.

The vertical dimensions of the turbulent boundary layer increase as the arctic air moves to low latitudes. At this time, the Arctic air is additionally heated as a result of heat exchange with the earth's surface. This process leads to a significant decrease in the relative humidity of the air.

Thus, the air mass in a few days reaches the periphery of the subtropical anticyclone. This leads to the formation of a baric ridge that connects this anticyclone with the Arctic anticyclone and stops the Western transfer. The upper boundary of such a baric ridge reaches the tropopause and the lower stratosphere. Along the eastern periphery, Arctic air continues to flow into low latitudes, which leads to the formation of a high and inactive blocking anticyclone here.

Along the western periphery of the formed baric ridge, moist air approaching it from the west moves northward. This causes its thermal transformation and increased precipitation. In a blocking anticyclone, the air warms up quickly, however, its density remains high, because it remains practically dry. This, with a long period of its existence, can lead to drought.

Atmospheric blockings of the meridional type were first studied in the first half of the twentieth century [22–24]. B. L. Dzerdzeevsky considered the location of such blocks in the Northern Hemisphere as one of the main features underlying the typification of the macrocirculation processes occurring here. He established that blockings are capable of existing and maintaining their location for 5–10 days. They affect meteorological conditions more if they exist longer.

Synoptic maps and the results of global atmospheric reanalyses can be used to identify atmospheric blockings [25–28]. However, earlier the values of the frequency of AB of different duration, which occurred in the spring–summer months over the territories of Kazakhstan, were not estimated. The regions where air temperatures increase most significantly during the period of AB existence have not been identified either.

The reason for the insufficient knowledge of AB over the territories of Kazakhstan is the imperfection of the existing methods for their detection [5, 6, 15, 17], which manifests itself quite noticeably where other processes can also lead to an increase in air temperatures in the summer months [3, 29].

Climate characteristics (including AB frequency) are recommended to be assessed over a 30-year time interval. At the same time, under the conditions of accelerating ongoing climate change, the estimates of AB recurrence obtained in this way are not very relevant, since they actually describe not the present of the processes under study, but their past. Considering this, as well as the change in circulation epochs [30–35], it seems more adequate to estimate the frequency of AB, which correspond to the period 2010–2020.

The purpose of the research work is to estimate the frequency for the specified period of AB of one or another frequency occurring over Kazakhstan for the spring–summer months. And also the identification of its regions, where the influence of AB on the thermal regime is strong.

To achieve this goal, the following tasks were solved:

- assessment for 2010–2020 of the frequency of AB of different duration for the spring–summer months over Kazakhstan;
- identification of the regions of Kazakhstan, where the impact of AB on changes in their thermal regime is the strongest.

2 Materials and Methods

To identify AB, the technique proposed by Dzerdzeevsky [22, 23]. In accordance with this technique, detection of AB is carried out based on the results of the analysis of a surface synoptic map, which reflects the distribution of atmospheric pressure reduced to sea level in the Northern Hemisphere.

The decision to detect an AB over the study area is made if there is a baric ridge on the map in the corresponding range of longitudes, extending between the Arctic and the Subtropical anticyclone (approximately between 70 and 40° north latitude). In this case, baric ridges are taken into account, in which the reduced atmospheric pressure exceeds 1015 hPa by some given value L > 0.

The advantage of this technique is its visibility and the possibility of implementation without using a computer. One of its disadvantages is that the analysis of synoptic maps has to be carried out manually, where a person can make mistakes. The second disadvantage of this technique is the possibility of taking other baric formations for AB.

This disadvantage cannot be eliminated even if the relevant reanalysis results are used as factual material. One such reanalysis is ERA 5 (maintained by Copernicus) [25]. Information is presented for all nodes of the coordinate grid with a step of 0.25°, as well as for each hour relating to the period 00.01.01.1959 and up to the present.

Another common method for detecting AB is based on the application of the criterion proposed by Lejenas and Okland [16]. It is based on the fact that during AB, in the troposphere of temperate latitudes of the Northern Hemisphere, the Western transfer of air masses prevails, and the meridional gradient of the geopotential is directed towards the equator.

The value of the modulus of this gradient is maximum in the latitudinal zone 40–60° north latitude. The decision to detect AB is made if, within at least 5 days:

$$\Phi(500) = [I(l - 10) + I(l) + I(l + 10)] < 0 \quad (1)$$

where I(l) = H(l, 40 °C.ш.)—H(l, 60 °C.ш.)—index by Lejenas H. and Okland H., H—isobaric surface height 500 mb, l- longitude.

This method can be done without human intervention and eliminates mistakes. At the same time, the method also does not allow us to exclude the possibility of taking other baric formations for a high anticyclone.

In this research work, for the detection of AB, we applied a technique that combines the above methods and is partially free from the indicated disadvantages. In accordance with it, the decision to detect an AB over the area under consideration is made if, for at least 5 days at all nodes of the reanalysis coordinate grid located between the parallels of 40° latitude and 47° latitude on meridian l:

- reduced to sea level atmospheric pressure exceeds 1015 hPa by a given value L;
- geopotential values of isobaric surfaces of 300, 500 and 850 hPa, where on the considered dates the average monthly values of these indicators are more than 200 m^2/s^2;
- the criterion by Lejenas H. and Okland H. is fulfilled for isobaric surfaces not only 500 hPa, but also 300 and 850 hPa.

The obvious disadvantage of this method is its high complexity. At the same time, the mentioned properties of the meridional-type AB are taken into account in it more fully. Therefore, the probability of making false decisions about the presence of AB may be less.

Taking into account the features of the proposed method for detecting AB, when solving the first task, the ERA 5 reanalysis information on changes in the atmospheric pressure reduced to sea level in the surface layer, as well as the geopotential values of the isobaric surfaces of 300, 500, and 850 hPa, was used as factual material. It was assumed that $L = 5$ hPa.

As an indicator of the frequency of AB in a month, for each meridional stripes 0.25° wide, the total number of days of this month averaged over 2010–2020, during which AB with a long duration were found in it, was estimated. If an AB started in one month and ended in another, when calculating the frequency, the duration of this AB was taken into account for the month to which most of it belonged. Dependences of the frequency of AB (l) were studied for each month, as well as for cases where the minimum duration of AB was 5, 6, 8 and 10 days.

When solving the second task for the points of all regions of Kazakhstan, where the units of Kazhydromet in 2010–2020 carried out systematic monitoring of air temperatures, for the months of April-August of each year, the values of the following indicators were calculated:

T1—the average value of the average daily air temperature, which is determined taking into account all the dates of a particular month related to the period of existence of the AB with a duration of at least 9 days;

T2—the average value of the average daily air temperature, which is determined taking into account all other dates of the same month;

At the same time, the daily data of 15 Kazhydromet stations on air temperature for all months from April to August related to the specified period, the location of which is shown in Fig. 1, was used as factual material.

As can be seen from Fig. 1, the considered stations are located in all regions of Kazakhstan. The Dixon and Smirnov-Grubbs tests were used to assess the quality and homogeneity of the data [36]. According to the Fisher (F) and Student (St) criteria, an estimate of the uniformity of average air temperatures was calculated. According to the Dixon and Smirnov-Grubbs criteria, an assessment was made of the homogeneity of the empirical distribution of air temperatures at meteorological stations. When checking the extremum, it was found that the significance level is more than 10%. Therefore, the considered data can be recognized as homogeneous with a high degree of reliability.

3 Research Results and Analysis

When solving the first task, using the described methodology, all AB that were in the spring–summer months over the territory of Kazakhstan were identified. For each month, the dependencies of the frequency of blockings, the minimum duration of which exceeded one or another value, were studied. Figure 2, as an example,

Fig. 1 Location of hydrometeorological stations

shows the frequency dependences of AB (l) over the territory of Kazakhstan, for April-August, when the minimum duration of blocking was 5, 6, 8 and 10 days.

Figure 2a–e shows that in any month the frequency of AB is the least with a long duration by day. The frequency of AB with a minimum duration of 10 days or more is on average 2–3 cases per month over the entire study area. The frequency of AB with such a duration is maximum in July (up to 5–7 cases).

Figure 2a shows that in April the highest frequency of AB with a minimum duration of 8–10 days is observed in the territories of the Petropavlovsk, Akmola and Central Kazakhstan regions.

From Fig. 2b, it is clear that the highest values of the frequency of AB with a minimum duration of 8–10 days are observed in Western Kazakhstan.

Figures 2c and d show that in June and July the frequency of the considered AB is increased for the western and eastern regions of Kazakhstan.

Figure 2e indicates that in August the frequency of AB lasting 8–10 days is maximum over the western regions of Kazakhstan and decreases towards the east.

From the comparison of Figs. 2a–e, it is clear that the highest values of the frequency of AB exceeding 10 days per month, and hence the likelihood of droughts, are in July. In April, the frequency of such AB increased for the regions of Petropavlovsk, Akmola and Central Kazakhstan. In May, it was increased for the West Kazakhstan region. For June, the frequency of the considered AB is the highest in the territories of the regions of West Kazakhstan, Pavlodar and East Kazakhstan.

In July, a similar feature is typical for the regions of West Kazakhstan, Pavlodar and East Kazakhstan. In August, the highest values of the frequency of the studied AB are observed in the West Kazakhstan region. When solving the second task for each year, the values of T_1 and T_2 are determined for each month and all the points

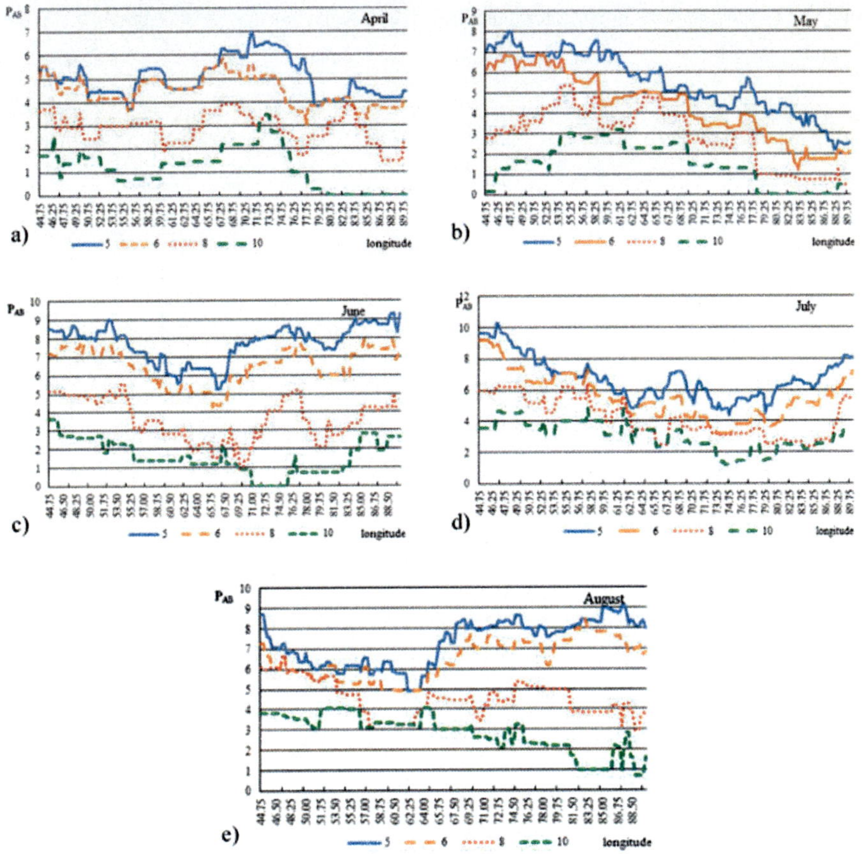

Fig. 2 Dependences of the frequency of atmospheric blockings on longitude in the territory of Kazakhstan. **a** April, **b** May, **c** June, **d** July, **e** August

of the Kazakhstan. For the entire period 2010–2020, April-August were the warmest in 2012. Therefore, for the specified year, Table 1 compares the values of the average air temperature for the period with AB lasting 10 days or more (T_1) and for the period without AB (T_2).

Table 1 shows that in the research years, the average air temperatures, which were estimated for all months for periods with AB with a minimum duration of 10 days, everywhere significantly exceeded the values of this indicator for periods without AB. In April, the T_1-T_2 difference is maximum in 2012 for Uralsk station, 15.7 °C.

In May, the highest values for Petropavlovsk station are 14.8 °C. In June, there were no long periods without AB.

The highest T_1-T_2 values for July were in Zhezkazgan, 6.3 °C.

For August, the maximum value of T_1-T_2 was observed in Aktobe, 8.4 °C.

Table 1 shows that, on average, for all monitoring points, the highest values of T_1-T_2 were in July and August, and the lowest in April.

Table 1 T_1 and T_2 values in 2012, for monitoring points on the territory of Kazakhstan

Region	Station	April		May		June		July		August	
		T_1	T_2	T_1	T_2	T_1	T_2	T_1	T_2	T_1	T_2
West	Aktau	–	8.5	24.0	20.5	23.5	–	27.5	–	30.1	24.3
	Uralsk	18.6	2.9	19.0	–	24.2	–	22.6	–	28.4	20.5
	Atyrau	–	8.4	24.4	–	–	–	27.5	–	32.6	24.6
	Aktobe	18.5	3.7	–	–	27.6	–	–	–	29.7	21.3
North	Kostanay	15.3	2.1	–	–	21.7	–	–	–	–	18.7
	Petropavlovsk	13.3	3.0	22.3	7.5	–	–	24.5	20.1	–	15.9
	Kokshetau	13.4	5.9	21.6	8.6	–	–	25.1	20.0	–	17.8
	Nur-Sultan	–	7.6	22.5	11.9	–	–	27.8	21.8	–	20.3
	Karaganda	–	7.0	19.9	13.3	–	–	24.4	18.4	–	19.6
	Pavlodar	–	7.9	21.4	12.1	–	–	26.2	21.7	–	19.0
South	Shymkent	–	15.9	23.6	16.6	–	–	28.0	24.8	–	26.9
	Taraz	–	13.7	23.1	16.8	–	–	25.9	23.8	–	25.2
	Kyzylorda	–	15.2	25.3	20.3	–	–	31.7	–	29.1	27.4
	Zhezkazgan	–	10.1	22.8	15.1	–	–	29.2	22.9	26.9	23.6
	Turkestan	–	13.8	26.2	19.5	–	–	30.6	25.4	–	28.4
	Almaty	–	–	21.4	16.6			26.2	24.7	–	26.0
East	Ust-Kamenogorsk	–	10.2	–	15.0	–	–	22.3	–	–	–

In other years, this regularity also manifested itself, but the average values of T_1-T_2 were lower. From the comparison of Fig. 2 and Table 1, it follows that the highest values of the T_1-T_2 difference correspond to monitoring points located in the territories of the regions of Kazakhstan, over which the frequency of AB lasting 10 days or more was increased, which contributed to an increase in evapotranspiration and could lead to droughts. It should be noted that in all the years under consideration, almost everywhere during the periods of existence of the research AB, precipitation ceased, which given the long duration of these periods, also contributed to an increase in the risk of droughts.

Thus, when solving the tasks set, it was found that between the obtained estimates of the frequency of AB over different regions of Kazakhstan and the occurrence in their territories of meteorological conditions that contribute to the occurrence of droughts, there is a place to be. This confirms the expediency of taking into account the identified patterns when predicting the risks of hazardous natural phenomena in a particular region. The regions of Kazakhstan were also identified, where the frequency of occurrence of AB lasting 10 days or more, and hence the risks of droughts, in certain months in 2010–2020 were increased.

Consequently, the tasks set are solved, and its goal is achieved.

4 Discussion

The results obtained correspond to the existing ideas about the effect of AB on weather conditions in the territories of the climatic zone [17, 19, 21], as well as about the causes of spring–summer droughts in them.

The novelty of the results obtained is that for the first time estimates of the frequency of AB of various durations for all regions of Kazakhstan. This makes it possible to compare the risks of droughts, extremely high air temperatures and other dangerous meteorological phenomena.

5 Conclusion

As the results of the research showed, the applied method for detecting atmospheric blockings makes it possible to identify periods of time during which air temperatures significantly increase on the territory of Kazakhstan from April to August, for a period of 5 days or more. The features of these periods correspond to existing ideas about the possible consequences of the formation of atmospheric blockings over Eurasia sectors. Therefore, both the ERA-5 reanalysis and the proposed method for detecting atmospheric blockings are quite suitable for their detection over the territory of Kazakhstan.

The actual problem of protecting the population, agriculture and forestry of the regions of Kazakhstan is the development of a methodology for predicting the frequency of spring–summer droughts.

References

1. *Arkhipov E.V., Zalesov S.V.* 2017. Dynamics of forest fires in the Republic of Kazakhstan and their environmental consequences. Agrarian Bulletin of the Urals. Ural State Agrarian University. 4(158). 10–15
2. Yearbooks on natural hydrometeorological phenomena observed on the territory of Kazakhstan // Editions from 1990 to 2015. RSE "Kazhydromet", Almaty
3. *Kozhakhmetov P.Zh., Kozhakhmetova E.P.* 2016. Extreme meteorological events in Kazakhstan in the context of global warming. // Scientific and technical journal "Hydrometeorology and ecology". No. 2 (81). 7–19
4. *Kusainova A.A., Pernebekova M.N.* 2020. Environmental risks of arid climatic conditions in northern Kazakhstan. Materials of the conference "High technologies and innovations in science" SSRI "National Development", St. Petersburg. 83–86
5. Statistical agriculture data of the Republic of Kazakhstan "Agriculture, forestry and fisheries in Kazakhstan 2004–2008", / Statistical collection. 2009. Astana, Statistical Agency of the Republic of Kazakhstan

6. *Baisholanov S.S.* 2010. On the frequency of droughts in the grain-growing regions of Kazakhstan. // Scientific and technical journal “Hydrometeorology and ecology”. No. 3. 27–37
7. *Zolotokrylin A.N., Titkova T.B., Cherenkova E.A., Vinogradova V.V.* 2013. Comparative studies of droughts in 2010 and 2012 on the European territory of Russia based on meteorological and MODIS data. Modern problems of remote sensing of the Earth from space. T. 10. No. 1. 246–253
8. *Zolotokrylin A.N., Titkova T.B.* 1998. Dependence of climate anomalies during the growing season of the forest-steppe of the Russian Plain on large-scale atmospheric circulation. Proceedings of the Russian Academy of Sciences. Ser. geogr. No 5. 121–128
9. *Rysalieva L.S., Salnikov V.G.* 2021. A Study on Atmospheric Drought in Central Asia. Geographic Bulletin, Perm State National Research University. 2(57). 110–120
10. *Suleimenova G.T., Akhmetova S.T.* 2015. Droughts in the territory of Northern and Western Kazakhstan // Scientific and technical journal “Hydrometeorology and ecology”. No. 2(77). 7–15
11. *Zhurina L.L.* 2021. Agrometeorology: textbook / L.L. Zhurina. - 3rd ed., revised. and additional Moscow: INFRA-M. 350 (Secondary vocational education). - ISBN 978-5-16-013877–0. https://znanium.com/catalog/product/1219364 (date of access: 05/19/2022)
12. *Kozhakhmetov P.Zh., Iskakova E.A., Baibazarov D.K.* 2016. Using the Standardized Precipitation Index (SPI) to identify droughts in Kazakhstan. Hydrometeorology and ecology. No 1. 22–31
13. *Ped D.A.* 1975. On the indicator of drought and excessive moisture // Hydrometeorological Center of the USSR. V. 156. 19–38
14. *Podnebesnykh N.V.* 2020. Long-term changes in atmospheric circulation over Siberia // Optics of the atmosphere and ocean. T. 33. No 02. 142–145. DOI: https://doi.org/10.15372/AOO20200209
15. *Cherenkova E.A., Kononova N.K.* 2009. Connection of dangerous atmospheric droughts in European Russia of the XX century with macrocirculation processes // Proceedings of the Russian Academy of Sciences. Ser. geogr. No 1. 73–82
16. *Mokhov I.I.* 2016. Atmospheric blockings and related climatic anomalies. Nonlinear waves. [Electronic resource]. https://docplayer.com/35005034-Atmosfernye-blokingi-i-svyazannye-s-nimi-klimaticheskie-anomalii.html
17. *Shakina N.P., Ivanova A.R.* 2010. Blocking anticyclones: current state of research and forecasting // Meteorology and Hydrology. No 11. 5–18
18. *Drouard, M., Woollings T.* 2018. Contrasting Mechanisms of Summer Blocking Over Western Eurasia, Geophysical Research Letters, 45, 12,040–12,048.https://doi.org/10.1029/2018GL079894
19. *Kononova N. K., Lupo A. R.* 2020. Changes in the Dynamics of the Northern Hemisphere Atmospheric Circulation and the Relationship to Surface Temperature in the 20th and 21st Centuries// Atmosphere. 11(3). 255. https://doi.org/10.3390/atmos11030255
20. *Masato, G., Woollings, T., Hoskins, B. J.* 2014. Structure and impact of atmospheric blocking over the Euro-Atlantic region in present-day and future simulations, Geophysical Research Letters, 41. 1051–1058.https://doi.org/10.1002/2013GL058570
21. *Mokhov I., Timazhev A.* 2019. Atmospheric Blocking and Changes in its Frequency in the 21st Century Simulated with the Ensemble of Climate Models, Russ. Meteorol. Hydrol. 369–377. https://doi.org/10.3103/S1068373919060013
22. *Dzerdzeevsky B.L.* 1968. Circulation mechanisms in the atmosphere of the northern hemisphere in the XX century // Materials of meteorological research, M.: Publishing House of the Academy of Sciences of the USSR and Mezhved. geophys. Committee under the Presidium of the Academy of Sciences of the USSR. 240
23. *Dzerdzeevsky B.L., Kurganskaya V.M., Vitvitskaya Z.M.* 1946. Typification of circulation mechanisms in the northern hemisphere and characteristics of synoptic seasons. // Proceedings of research institutions of the Main Directorate of the Hydrometeorological Service under the Council of Ministers of the USSR Ser. 2. Synoptic meteorology. Issue. 21. M., L., Gidrometizdat. 80

24. *Kononova N.K.* 2009. Classification of circulation mechanisms of the Northern Hemisphere according to B.L. Dzerdzeevsky / rev. ed. A.B. Shmakin; Russian acad. Sciences, Institute of Geography. Moscow: Voentekhinizdat. 372
25. IMERG: Integrated Multi-satellite Retrievals for GPM | NASA [Электронный ресурс]. Режим доступа: https://gpm.nasa.gov/data/imerg
26. *Hersbach H., Dee D.* 2016. ERA5 reanalysis is in production // ECMWF Newsletter. Vol. 147. 7
27. *Hólm E., Janisková M., Keeley S., Laloyaux P., Lopez P., Lupu C., Radnoti G., de Rosnay P., Rozum I., Vamborg F., Villaume S., Thépaut J.-N.* 2020. The ERA5 global reanalysis, Quarterly Journal of the Royal Meteorological Society, 146, 1999–2049.https://doi.org/10.1002/qj.3803
28. *Hoffmann L., Günther G., Li D., Stein O. et al.* 2019. From ERA-Interim to ERA5: the considerable impact of ECMWF's next-generation reanalysis on Lagrangian transport simulations // Atm. Chem. Phys. Vol. 19. 3097–3124
29. Assessment report on climate change in Kazakhstan. 2015. Astana: RSE "Kazhydromet"
30. *Mokhov I.I.* 2021. Extreme atmospheric and hydrological phenomena in the Russian regions: connection with the Pacific decadal oscillation// Reports of the Russian Academy of Sciences. Earth Sciences. T. 500. No 2. 183–188
31. *Mokhov I. I., Akperov M. G., Prokofieva M. A., Timazhev A. V., Lupo A. R., Le Tret E.* 2013. Blockings in the Northern Hemisphere and the Euro-Atlantic Region: Estimates of Changes from Data reanalysis and model calculations // Reports of the Academy of Sciences. T. 449. No 5. 1–5
32. *Bacer S., Jomaa F., Beaumet J., Gallée H., Le Bouëdec E., Ménégoz M., Staquet C.* Impact of climate change on wintertime European atmospheric blocking // https://doi.org/10.5194/wcd-2021-47
33. *Barriopedro D., García-Herrera R., Lupo A. R., Hernández E.* 2006. A Climatology of Northern Hemisphere Blocking, Journal of Climate. 19. 1042–1063https://doi.org/10.1175/JCLI3678.1
34. Central Asia and Caucasus Disaster Risk Management Initiative (CACDRMI): Risk Assessment for Central Asia and Caucasus Desk Study Review (CAC DRMI): Risk Assessment for Central Asia and Caucasus Desk Study Review/ The World Bank, UNISDR, CAREC, GFDRR. 2009. 172
35. *Davini D., D'Andrea F.* 2020. From CMIP3 to CMIP6: Northern Hemisphere Atmospheric Blocking Simulation in Present and Future Climate, Journal of Climate. 33. 10 021–10 038. https://doi.org/10.1175/JCLI-D-19-0862.1
36. *Zaks L.* 1976. Statistical estimation. M.: Statistics. 598

Relations of Interannual Variability of Topography of Water Surfaces of Oceanic Regions with Solar Activity on the Example of Black and Azov Seas

A. V. Kholoptsev and S. V. Palaev

Abstract We studied the relations of "smoothed" in sliding windows of a particular duration, changes in the average annual water surface levels of various sections of the waters of the Azov and Black Seas, as well as solar activity. The example of these seas tested the validity of the hypothesis that the reason for the presence of these connections is the influence on solar activity and the dynamics of the topography of the water surface of some regions of the World Ocean, translational movement in its orbits of the giant planets of the solar system.

Keywords Black and Azov seas · Water surface level · Solar activity · Total momentum · Connections

The effectiveness of engineering protection of the coasts of any regions of the World Ocean from their flooding, which may be caused by an increase in the levels of the water surface (WSL) of the adjacent waters, is largely determined by the accuracy of estimates of the locations of their areas that are most at risk. Therefore, the improvement of methods for predicting the dynamics of the topography of their water surface is an urgent problem not only of hydrography and oceanography, but also of environmental and man-made safety.

The complexity of its solution is largely due to the fact that hydrometeorological factors have the most significant impact on the variations in the WSL of many oceanic regions. Changes in the future characteristics of these factors largely depend on the scenario of further changes in the regional climate, which will actually be realized. Since such a scenario of these changes is not predetermined [1, 2], it is customary

A. V. Kholoptsev (✉)
Sevastopol Branch of the State Oceanographic Institute named after N.N. Zubov, Sevastopol, Russian Federation
e-mail: kholoptsev@mail.ru

S. V. Palaev
Sevastopol State University, Sevastopol, Russian Federation
e-mail: spalaiev@yahoo.com

V. I. Karev (ed.), *Physical and Mathematical Modeling of Earth and Environment Processes—2022*, Springer Proceedings in Earth and Environmental Sciences,
https://doi.org/10.1007/978-3-031-25962-3_3

to develop ultra-long-term forecasts of the dynamics of the topography of the water surface of a particular oceanic region for various alternative scenarios. Such forecasts, as a rule, differ significantly, and it is impossible to give preference to any of them. Therefore, the use of such forecasts in the selection of areas of coasts subject to priority protection seems problematic.

According to existing ideas about the factors of changes in the topography of the water surface of oceanic regions [2–4], some of them do not depend on the scenario of further climate changes, and their dynamics can be predicted with sufficient advance. Given such factors, it is possible to predict changes in the topography of the water surface of the studied region and determine those coastal areas for which the WSL could further increase. Such forecasts will correspond to a scenario in which the characteristics of changes in the regional climate in the future will remain the same as in the period taken into account when verifying the relevant prognostic models.

The justifiability of such forecasts, as well as the feasibility of taking them into account when making any economic decisions, depends on how significant such factors are and how accurate their forecasts, which have the required advance, can be. Therefore, the study of the associations of WSL variations with such factors is relevant.

One of the natural processes independent of the scenario of further climate changes and capable of influencing some natural processes in many regions of our planet is variations in solar activity (VSA) [5].

The reasons that give rise to VSA have not been established, however, the hypothesis that the reason for their cyclicity is movement in their orbits of the planets of the solar system was first expressed by Wolf [6].

Wolf mistakenly explained the connection of the VSA with the movement of planets by the occurrence of tides on the Sun, which are due to the influence of their gravitational fields.

It has now been established [7–9] that due to the movement of planets in their orbits, the resulting forces of attraction of the Sun, as well as their total moment of inertia (TMI), change accordingly. The most significant influence of giant planets.

Under the influence of the forces of attraction to them, the center of mass of the Sun moves relative to the barycenter of the solar system along a complex trajectory. According to estimates [10, 11], in the twentieth century, in this process, the distance between them sometimes exceeded the equatorial radius of the photosphere by 1.5 times (0.6955 million km [12]), which is 1.0433 million km or about 1% of the average distance between the Earth and the Sun. Due to the unevenness of these movements of the Sun, inertia forces arise in its bowels, which affect many physical processes that occur in it. Perhaps this was how VSA could be formed. The latter suggests that **there is a significant link between variations TMI and VSA.**

Due to the movements of the center of mass of the Sun, changes in the distance between it and the Earth occur according to laws that are somewhat different from Kepler's laws. As a result of this influence of TMI variations on the unevenness of the translational movement of our planet, in all geographies, including in the waters of the World Ocean, there are components of inertia forces that affect the location and flow of ocean currents. The latter, as well as the presence of significant statistical links of the VSA with some components of the water balance of a number of oceanic regions [13, 14], suggests that **there are significant statistical links between variations of the TMI and VSA, as well as changes in the WSL of some areas of their water area.**

The validity of the hypotheses put forward was not previously assessed, but it was established that the Azov-Black Sea region also belongs to such oceanic regions.

TMI values can be calculated with high accuracy for decades to come using the methodology [15]. Therefore, if the validity of the hypotheses put forward is confirmed, the connections between the changes in the TMI and the VSA, as well as the TMI and the WSL, could be used in the development of ultra-long-term forecasts of both the VSA and the WSL. Therefore, testing the validity of these hypotheses is of both theoretical and practical interest.

The purpose of this work is to check the fairness of the hypotheses put forward for the Azov-Black Sea region, as well as to identify the locations of sections of the Black and Azov Seas, for which the reliability of the conclusion about the significance of statistical links between variations of the WSL and the VSA, as well as the WSL and the exceeds certain values.

To achieve this goal, the following tasks have been solved:

1. Study of the peculiarities of the links between the states of the VSA and the TMI for the period 1740–2021.
2. Assessment of the significance of statistical links between variations of WSL and VSA, as well as VSA and TMI in the present period (1993–2021).

When solving these problems as an actual material, information on changes in the WSL of various sections of the water area of the Azov and Black Seas, as well as time series of the average annual values of Wolf numbers [16] (characterizing the state of the VSA) and the average annual values of TMI obtained from [15] are used.

Information on WSL changes in the region under study is obtained from the results of global oceanic re-analysis of GLORYS12.v.1 [17], which is supported by Copernicus.

The study method assumed formation of time series of the WSL from the specified actual material.

When solving the first problem, the years for which the extremes of the time series of the VSA and the TMI fell were compared. Correlation analysis of synchronous statistical connections between different fragments of 28-year length is also carried out.

When solving the second problem, the time series of WSL, VSA and TMI were "smoothed" in "sliding windows" lasting from 2 to 8 years. The latter made it possible to weaken the components of the variability spectrum of WSL with periods shorter than the duration of the considered "window," as well as to reduce the values of the ratio of their capacities to the capacities of components with large periods present in the spectra of VSA and TMI [18].

For all smoothed time series obtained by their autocorrelation functions, the number of degrees of freedom and threshold values of their correlation coefficient

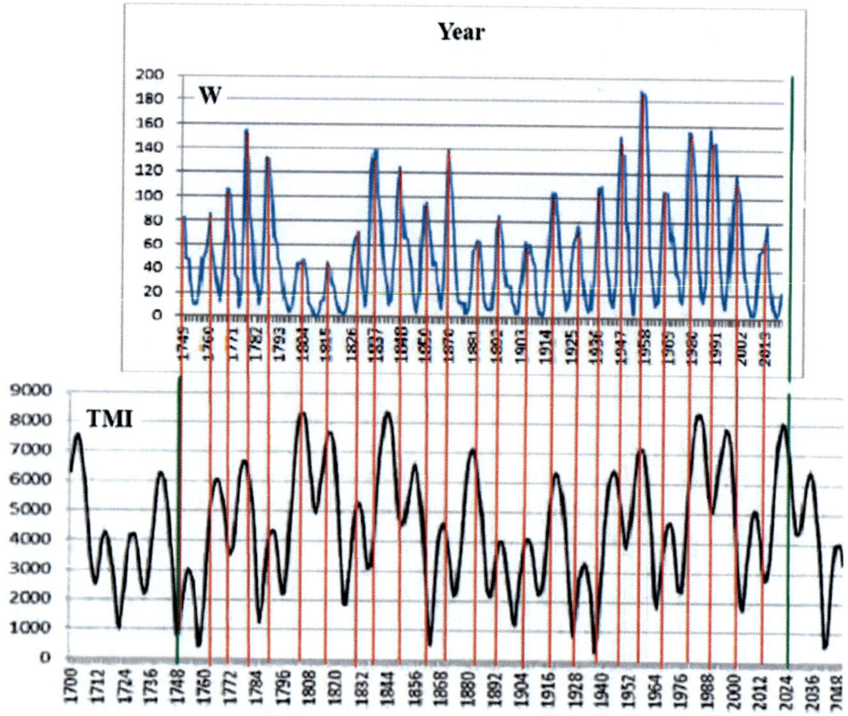

Fig. 1 Long-term changes in the average annual values of Wolf numbers (W), as well as TMI (taking into account information on TMI variations obtained from [15])

are determined, which correspond to the reliability of the conclusion about their significance of 0.85; 0.9 and 0.95 [19].

As a result of solving the first problem, it was established that all extremes of dependence on the time of the TMI and the VSA coincide in time. This is not difficult to see from Fig. 1, in which the mentioned time dependencies are compared with each other.

From Fig. 1, it can be seen that the dependence on the time of the average annual values of Wolf numbers (W), as well as the TMI, are complex fluctuations, in which there is a component with a period close to 11.8 years (the period of revolution of the planet Jupiter). The extremes of both fluctuations practically coincide, which indirectly confirms the validity of the conclusions [8], but the nature of such coincidences depends on the period of time in which they occur.

The highs of the VSA, which corresponded to 1761, 1778, 1788, 1804, 1816 and 1830. corresponded to the highs of the TMI. At the same time, the maxima of the VSA, which fell on 1748, 1769, 1837, 1848 and 1860, meet the minima of the TMI. The highs of the VSA, which took place in 1870, 1883, 1893, 1905 and 1917 again correspond to the highs of the TMI. At the same time, the highs of the VSA observed in 1928 and 1938 coincided with the lows of the TMI. The highs of the

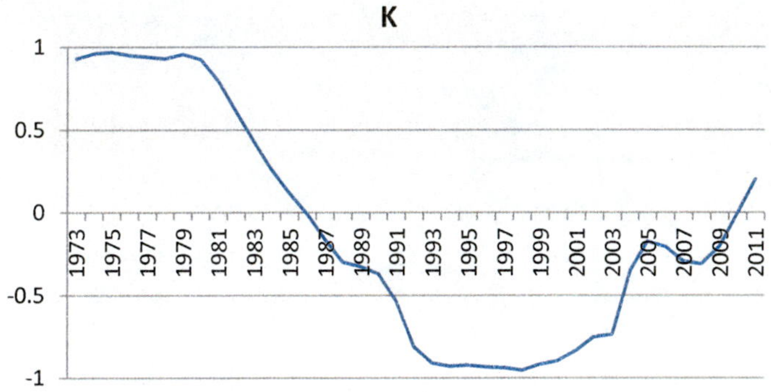

Fig. 2 Dependence of correlation coefficient (K) between time-matched segments, time series of VSA and TMI containing 11 members, on the year of the beginning of these segments

VSA, recorded in 1947, 1957 and 1968, fell on the highs of the TMI. In 1979, the maximum of the BSA corresponded to the intermediate value of the TMI (close to the maximum, but not coinciding with it). In the years 1989, 2000 and 2014 the highs of the VSA corresponded to the lows of the TMI. Nevertheless, in 2022, the maximum of the VSA again coincides with the maximum of the TMI. Consequently, from time to time, communications between the VSA and the TMI are "inverted."

Figure 2 shows the correlation coefficient (K) relation between the time-matched time-series segments of VCA and TMI containing 11 members from the start year of these segments for the period 1973–2021.

As can be seen from Fig. 2, the correlation coefficient values between the time-matched segments, the time series BCA and the 11-member TMI are substantially dependent on the start year of the segments. For the segments of the series in question corresponding to the time periods when the TMI highs are at BSA highs, their correlation is positive and close to 1. Their segments related to periods when the TMI maxima correspond to BCA minima correspond to correlation coefficient values, which are close to K-1.

During both "inversion" periods present in this figure, the correlation coefficient of the corresponding VSA and TMI series segments is not significant, and its sign changes to the opposite. As can be seen from Figs. 1 and 2, "inversion" occurs in the current VSA cycle. Judging by the background of the "inversions" that took place in 1730–2021. It can be assumed that, at least in the next cycle of solar activity, its maxima will coincide in time with the maxima of the TMI, which will clarify the forecasts of the VSA developed using methods [20], according to information about the TMI.

Since time series modeling is effective only taking into account predictors associated with them quite strongly [21], VSA prediction, according to information on the background of TMI variations, is possible only for its cycles in which "inversion" does not occur. Therefore, the development of a technique for predicting "inversions" would be of theoretical and practical interest.

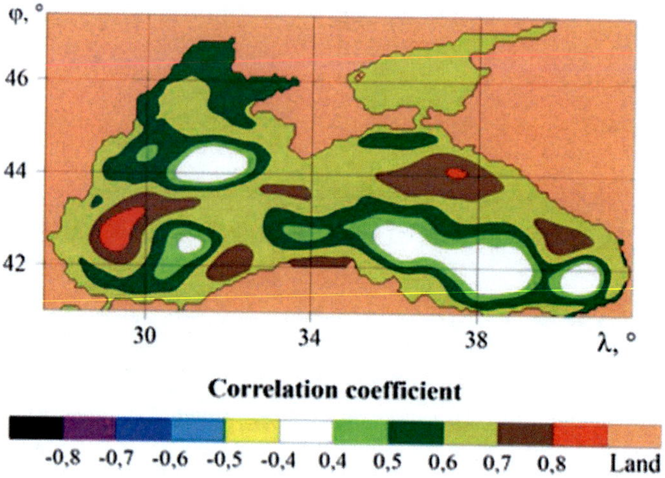

Fig. 3 The regions of the black and Azov Seas, within which the modules of the correlation coefficient of the time series of WSL and VSA, "smoothed" (with a window length of 5 years), exceed certain values

When solving the second problem for smoothed rows of WSL, as well as VSA and TMI TMI corresponding to different lengths of the "sliding window," the locations of sections of the water area of the Azov-Black Sea region, where statistical relations between them have one or another force, are determined. It was established that areas of the water area of the considered seas, where the validity of conclusions about the significance of the connection between the "smoothed" in the "window" with a duration of less than 5 years by the specified rows, exceeds 0.9, were not revealed. With the lengths of this "window" 5 years or more, the desired areas exist.

Figure 3 shows the locations of areas of the Azov and Black Seas, within which the correlation coefficient of the time series of WSL and VSA "smoothed" (with a window of 5 years) exceeds certain values.

As can be seen from Fig. 3, in the entire water area of the studied region, the correlation of time series of WSL and VSA smoothed in a window with a length of 5 years is positive (the degrees of freedom of such series are 7). Areas of the Black Sea water area, for which the validity of the conclusion on the significance of the relationship between the "smoothed" time series of the WSL and VSA exceeds 0.9, (the threshold value of the correlation coefficient is 0.8) exist. They are located in the deep water areas of the Western and Eastern parts of the Black Sea.

Areas of the Black Sea where the reliability of the conclusion on the significance of the correlation of the studied series exceeds 0.85 (the threshold value of the correlation coefficient is 0.7) are more numerous. They are located around the indicated areas. Similar sections of the Sea of Azov are located in Sivash Bay and off its northern coast in the area near the port of Berdyansk.

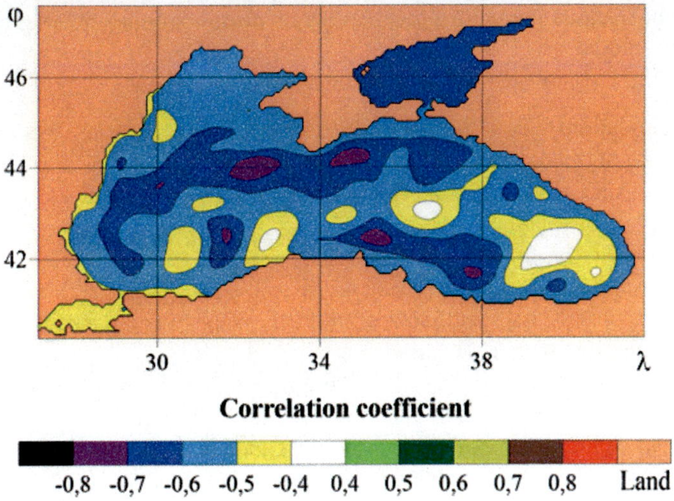

Fig. 4 Areas of the black and Azov Seas, within which the modules of the correlation coefficient of time series of WSL and TMI, "smoothed" with a window length of 5 years, exceed certain values

Figure 4 shows the locations of areas of the Azov and Black Seas, within which the correlation coefficient of the time series of the WSL and TMI "smoothed" (with a window length of 5 years) exceeds certain values.

Figure 4 shows that correlation of "smoothed" time series of WSL and TMI is negative in all areas of the Black and Azov Seas. Areas of the sea within which the validity of the conclusion on the significance of connections between these time series exceeds 0.85 are located in the area of the main Black Sea current. It should be noted that among these areas there are those for which the reliability of the specified conclusion exceeds 0.9.

In the Sea of Azov, areas where the reliability of the conclusion about the significance of ties between the same time series exceeds 0.85 are located off its northern coast, as well as the Taganrog and East Sivash bays. At the same time, no areas where the reliability of such a conclusion exceeds 0.9 have been identified.

From the comparison of Figs. 3 and 4, it follows that the sections of the water area of the Azov-Black Sea region, where the reliability of statistical conclusions about the significance of the connections between the smoothed time series of the WSL, as well as the VSA and the TMI exceed 0.9, exist in the modern period.

Similar patterns are manifested for the time series of the processes studied, which are smoothed out in "sliding windows" 6–8 years long.

From the results presented, it follows that both hypotheses put forward are valid, and the goal of the work is fully achieved.

The results obtained correspond to the ideas about the links between the VSA and the TMI [8]. They have not only theoretical, but also practical significance, determining the relevance of further research aimed at testing the possibility of using

VSA and the TMI as predictor processes, in the ultra-long-term prediction of the WSL of the studied seas.

The new facts established in this work are:

- statistical relationship of VSA, as well as changes in WSL of some sections of the Black Sea is significant (and their correlation is positive) only at time intervals in which VSA maxima correspond to TMI maxima; such coincidences are likely in the next cycle of solar activity;
- the possible reason for the significant statistical relationship of changes in the WSL in such areas of the water area of the studied region, as well as the VSA, which takes place at the specified time intervals, is the causal relationship of both of these processes with TMI variations;
- in addition to such time intervals, there are also significant periods when the maximum VSA corresponds to the minimum TMI, or there is an "inversion" of the studied connections; during such periods, the correlation between the WSL and VSA rows becomes significantly less strong and even changes its sign;
- at such time intervals, the correlation between the variations of WSL and VSA is positive, and with changes in the TMI is negative;
- long-period components, significantly statistically related to variations of the WSL and TMI, are present in long-term changes in the WSL of some areas of the Black and Azov Seas; these components can be extracted by "smoothing" the corresponding time series in a "sliding window" of 5–8 years, and possibly more;
- in 1993–2020 the mentioned areas prevailed in the region of the main Black Sea current;
- the location of the identified sites indicates that the direct reason for the connection of the changes in the Black and Azov Sea with the VSA is the water exchange between the shallow areas of these seas, as well as the deep-sea part of the Black Sea, which is controlled by the TMI variations.

Thus, it was established that variations in solar activity relate to significant factors of time variability in the water surface level of some sections of the Azov-Black Sea region only during periods when the maxima of its cycles coincide in time with the maxima of dependence on the time of the total momentum of the giant planets of the solar system. Statistical relations between these processes are due to the causal relations of each of them with variations in the total momentum of the giant planets of the solar system.

At other periods of time, the maxima of variations in the total momentum of the giant planets of the solar system correspond to the minima of solar activity, or "inversion" occurs, as a result of which the statistical connections between the studied processes are not significant, and their use to predict changes in the level of the world ocean is impossible.

The location of the areas of the Azov and Black Seas in which such areas are localized indicates that the reason for the connection between the studied processes is the influence of the TMI on the inertia forces acting on both their waters and solar activity. The findings suggest that similar patterns may be evident in some other regions of the world's oceans.

The presence of identified areas of its water area in the Azov-Black Sea region indicates the relevance of research aimed at developing a methodology for predicting "inversions" of connections between TMI and VSA.

References

1. *Kotlyakov* V.M., 2012. On the causes and consequences of modern climate change. *Solar-terrestrial physics.* 2012. No. 21. P. 110–114
2. *Malinin V.N.* Ocean level: present and future. St. Petersburg: Russian State Humanitarian University, 2012. – 260 p.
3. *Kaplin P.A., Pavlidis J.A., Selivanov A.O.* Forecast of the Development of the Coastal Zone of the Seas of Russia [including the Arctic] in Conditions of Increasing Their Level and Warming Climate // *Humanity and the Coastal Zone of the World Ocean in the 21st Century*: XX International Conference dedicated to the 90th anniversary of V.P. Zenkovich (4–5 February 2000), Moscow, 2001, p. 16–28.
4. Mörner N.-A. Sea level is not rising. Science and Public Police Institute Reprint Series, Dec.6, 2012. – 26 p.
5. *A.V. Kholoptsev*, Solar activity and forecasts of physical and geographical processes / A.V. Kholoptsev, M.P. Nikiforova / LAP LAMBERT Academic Publishing. Saarbruken (Deutschland) – 2013. – 333 p.
6. Wolf R. // Astr. Mittheilungen. – 1859. – P. 183.
7. *Okhlopkov V.P.* The main periodicity of the movement of the Sun relative to the center of mass of the solar system and solar activity // Bulletin of Moscow State University. *Ser. 3. Physics. Astronomy.* – 2011. – № 6. – C. 138–142.
8. *Ponomareva O.V.* The role of planets and planetary groups in the activity of the Sun//Geophysical monitoring and seismic safety problems of the Russian Far East: in 2 volumes. ***Proceedings of the regional scientific and technical conference.*** Volume 2. Petropavlovsk-Kamchatsky. November 11–17, 2007 Petropavlovsk-Kamchatsky: GS RAS, 2008. – P. 212–216.
9. *Prikhodovsky M.A.* **Analysis of possible causes of Jupiter's influence on the formation of the** solar cycle // *International Journal of Applied and Fundamental Research.* – 2016. – No. 7 (part 4) – P. 547–552.
10. *Dolgachev V.P., Domozhilova L.M., Khlystov A.I.* Some properties of the barycentric movement of large planets and the Sun // *Proceedings of GAISH*, 1991, vol. 62, p. 111–118.
11. *Khlystov A.I., Dolgachev V.P., Domozhilova L.M.* Baricentric movement of the Sun and its consequences for the solar system//Modern global changes in the natural environment. T3. **Factors of Global Change, M.: Scientific World, 2012, p. 62–78.**
12. *Stix M.* The Sun. An Introduction. 2nd Edition. Springer. – 2002. – 490 p.
13. *Babkin V.I.* **Many years of fluctuations in the flow of the Volga, Oka, Don, Dnipro and methods** of its forecast // Izvestia RAS. *The series is geographical*, 2008. № 3. P. 92–98.
14. *Bubin M.N., Rasskazova N.S.* **Rhythmicity of long-term river flow fluctuations as an inte-**gral indicator of climate variability (on the example of the Urals) / Yurginsky Technological Institute. – Tomsk: Publishing House of Tomsk Polytechnic University, 2013. – 279 p.
15. *Sherstyukov B.G.* Oscillatory climate system, resonances, long-range communications, forecasts. Obninsk: FGBU "VNIIGMI-MCD." – 2021. – 222 p.
16. Database The average annual values of Wolf numbers is presented by the Geophysical Center of the Russian Academy of Sciences. World Data Center for Solar-Terrestrial Physics. [Electronic resource]. Access mode. http://www.wdcb.ru/stp/solar/sunspots.ru.html.
17. GLORYS12V1 Reanalysis Database. – URL: https://resources.marine.copernicus.eu/product-detail/GLOBAL_REANALYSIS_PHY_001_030/INFORMATION.

18. *Box J., Jenkins G.* Time Series Analysis. Forecast and management. Issue 2. M.: Mir. – 1974. – 197 p.
19. *Ayvazyan S.A., Mkhitaryan V.S.* Applied statistics and fundamentals of econometrics. M.: Unity, 1998. – 1022 p.
20. *Anishin M.M, Plakhtiy A.D, Mamadou B.A.* Methods for predicting solar activity // Radio communication technique. – 2020. – №4 (47). – P. 57–67.
21. *Draper N., Smith H.* Applied Regression Analysis, 3 Edition.: Wiley Series in Progression and Statistics. – 1998. – 912 p.

Filtration of Salt Solutions Taking into Account the Osmotic Effect

M. M. Ramazanov, N. S. Bulgakova, and S. R. Gadzhimagomedova

Abstract Based on the previously proposed mathematical model of solution filtration with regard the osmotic effect in porous media with semipermeable inclusions, several problems have been solved. Characteristic distributions of salt concentration and pressure are given, estimates of filtration rates caused by osmosis are found. The solutions obtained make it possible to better understand the mechanism of osmotic filtration and may be of great practical importance.

Keywords Osmosis · Filtration of solution · Porous media · Generalized mathematical model

1 Introduction

The phenomenon of osmosis has been known since the eighteenth century. Usually, osmosis is understood as a phenomenon consisting in the fact that if there are aqueous solutions of salts with different concentrations on different sides of a semi-permeable membrane capable of passing only water molecules, water molecules will move through the membrane from a weakly concentrated solution to a more concentrated one. There are forward osmosis and reverse osmosis. Semipermeability is defined as the ability of a material to prevent the passage of a solute without affecting the passage of a solvent. The osmotic effect can manifest itself when a thin membrane is missing, namely in low-permeable porous media. For example, clays can act as semi-permeable membranes and therefore can induce osmotic transport of water [1]. In this case, along with the pressure gradient, the driving force for the solvent can be the salt concentration gradient in the solution (chemical osmosis) [2], the temperature gradient (thermal osmosis) [3], and the electric field gradient (electroosmosis) [4]. In

M. M. Ramazanov · N. S. Bulgakova (✉) · S. R. Gadzhimagomedova
Institute for Geothermal Research and Renewable Energy, Institute for High Temperatures of the Russian Academy of Sciences, Makhachkala, Russian Federation
e-mail: ipgnatali@mail.ru

N. S. Bulgakova · S. R. Gadzhimagomedova
Dagestan State University of National Economy, Makhachkala, Russian Federation

V. I. Karev (ed.), *Physical and Mathematical Modeling of Earth and Environment Processes—2022*, Springer Proceedings in Earth and Environmental Sciences,
https://doi.org/10.1007/978-3-031-25962-3_4

[5], based on the results of laboratory tests, it was shown that the osmosis coefficient of some natural rocks and grouting materials can exceed the filtration coefficient by tens or even hundreds of times. The paper [6] indicates the order of magnitude of the thermal osmosis coefficient within the limits of 10^{-9} to10^{-5} $m^2/(K \cdot day)$.

The study of chemical and thermal osmosis is carried out both theoretically based on the principles of nonequilibrium thermodynamics [7–10] and experimentally in laboratory conditions [11, 12]. The phenomenon of osmosis is used in various scientific and technical fields. For example, in Norway, near the city of Tofte, Statkraft has launched the world's first power plant is a prototype, which uses the osmosis phenomenon to generate electricity, which occurs due to the difference in salt concentration in salty sea water and in fresh water of the fjord. In marine geology, osmosis explains the "buoyancy" effect of ferromanganese nodules [13], which do not sink in loose sedimentary layers on the ocean floor on geological time scales [14, 15].

It is possible that osmosis is directly related to the process of melt migration in the Earth's mantle, meaning the "dissolution-reprecipitation" mechanism [16, 17], where the driving force of melt migration is assumed to be a chemical potential gradient depending on the concentration of the dissolved component [18].

There are a number of scientific areas where the osmotic effect should also be taken into account. For example, when modeling the processes of degradation of permafrost rocks, it is necessary to take into account the peculiarities of the interaction of salt solutions with frozen soils [19]. There are experimental studies of this interaction [20], which indicate a significant influence of the osmotic mechanism of solution transfer together with the processes of adsorption and desorption of salts.

It should be noted that understanding the properties of solution filtration with allowance for the osmotic effect is an urgent task. In this regard, in this work, using a mathematical model of filtration of solutions in a medium with semipermeable inclusions [10], we solved several stationary problems of filtration of salt solutions taking into account the osmotic effect, which make it possible to better understand the mechanism of osmotic filtration and may be of great practical importance. The tasks are considered in order of increasing complexity. The resulting solutions can also be useful in the educational and methodological plan.

2 Stationary Problems

Problem 1 Figure 1 shows a scheme of the problem. There is a porous layer saturated with solution of salt. The specified concentrations and pressures are maintained at the horizontal permeable boundaries of the layer. It is assumed that the concentration of salt at the upper boundary is greater than at the lower, so there is an upward flow of the solution caused by osmosis (concentration gradient). It is required to find a solution to the problem, i.e. the filtration rate of the solutions, as well as the distribution of salt concentration and pressure fields depending on the permeability and other parameters of the problem.

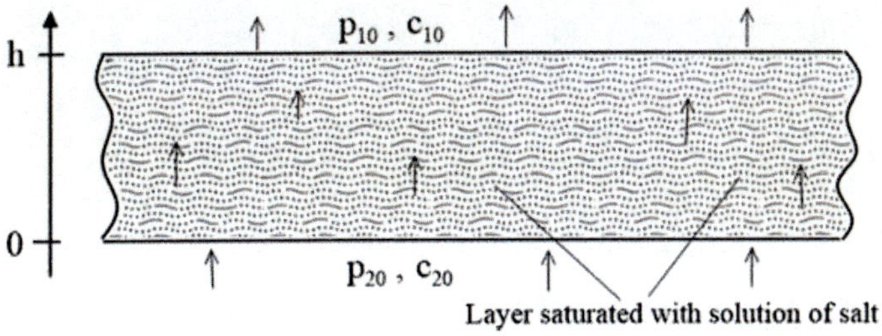

Fig. 1 Staging drawing for the first problem

Solution. Counting the pressure from the hydrostatic one, we have the following equations [10]

$$\begin{aligned} \mathrm{v} &= -\frac{k}{\eta}\left(\rho\frac{\partial\mu}{\partial c}c' + p'\right) \\ \mathrm{v}_1 &= const \\ \mathrm{v}c' &= Dc'' \end{aligned} \tag{1}$$

Here, the prime means the derivative with respect to the z coordinate directed upwards Fig. 1.

Let us introduce the scales of quantities

$$[z] = h, \quad [\mathrm{v}] = -\frac{k}{\eta h}\rho\frac{\partial\mu}{\partial c}c_0, \quad [p] = -\rho\frac{\partial\mu}{\partial c}c_0 \tag{2}$$

Then problem (1) can be written in the dimensionless form as follows

$$\begin{aligned} &\mu_1'' = 0, \quad \mathrm{v} = \mu_1', \quad \mu_1 = c - p \\ &\mathrm{v} = \mathrm{v}_0 = const \\ &Pe\mathrm{v}_0 c' = c'' \end{aligned} \tag{3}$$

$$Pe = \frac{[\mathrm{v}]h}{D} = -\frac{\rho k c_0}{\eta D}\frac{\partial\mu}{\partial c}, \qquad \frac{\partial\mu}{\partial c} < 0$$

Here μ_1 is dimensionless velocity potential.

Boundary conditions

$$\begin{aligned} z = 1: \; c = c_{10}, \; p = p_{10} \\ z = 0: \; c = c_{20}, \; p = p_{20} \end{aligned} \tag{4}$$

By direct substitution, it is easy to verify that problem (3)-(4) has the following solution

$$
\begin{aligned}
&\mathrm{v}_0 = c_{10} - c_{20} + p_{20} - p_{10} \\
&c = c_{20} + (c_{10} - c_{20})\frac{e^{Pe\mathrm{v}_0 z} - 1}{e^{Pe\mathrm{v}_0} - 1} \\
&p = c - \mu_1 = (c_{10} - c_{20})\left(\frac{e^{Pe\mathrm{v}_0 z} - 1}{e^{Pe\mathrm{v}_0} - 1} - z\right) + p_{20}(1 - z) + p_{10}z \\
&\mu_1 = c_{20} - p_{20} + \mathrm{v}_0 z
\end{aligned} \tag{5}
$$

The solution in dimensional form can be written as follows

$$
\begin{aligned}
&\mathrm{v}_0 = \frac{k}{\eta h}\left[-\rho\frac{\partial\mu}{\partial c}(c_{10} - c_{20}) + p_{20} - p_{10}\right] \\
&c = c_{20} + (c_{10} - c_{20})\frac{e^{\mathrm{v}_0 z/D} - 1}{e^{\mathrm{v}_0 h/D} - 1} \\
&p = p_{20}\left(1 - \frac{z}{h}\right) + p_{10}\frac{z}{h} - \rho\frac{\partial\mu}{\partial c}(c_{10} - c_{20})\left(\frac{e^{\mathrm{v}_0 z/D} - 1}{e^{\mathrm{v}_0 h/D} - 1} - \frac{1}{h}z\right)
\end{aligned} \tag{6}
$$

Problem 2 There are two layers, lower thickness and upper thickness, with different permeabilities and thermophysical coefficients (Fig. 2). All layer boundaries are permeable. The pressure and salt concentration are given at the upper and lower boundaries of the two-layer formation. It is required to find a stationary solution to the problem.

In area 1

$$
\begin{aligned}
&\mathrm{v}_1 = -\frac{k_1}{\eta}\left(\rho\frac{\partial\mu}{\partial c}c_1' + p_1'\right) \\
&\mathrm{v}_1 = const \\
&\mathrm{v}_1 c_1 = D_1 c_1''
\end{aligned} \tag{7}
$$

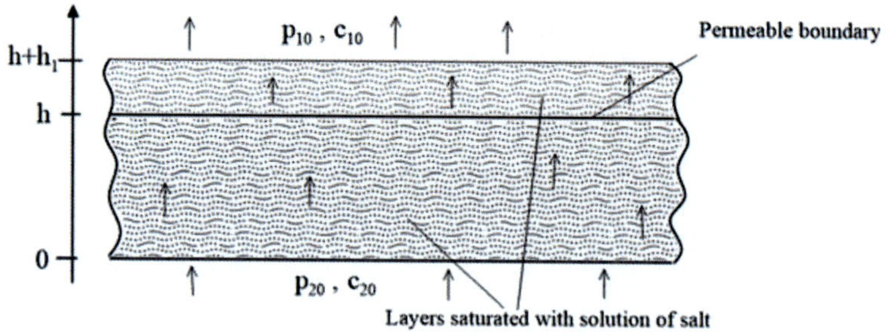

Fig. 2 Staging drawing for the second problem

In area 2

$$\mathrm{v}_2 = -\frac{k_2}{\eta}\left(\rho\frac{\partial \mu}{\partial c}c_2' + p_2'\right)$$

$$\mathrm{v}_2 = const$$

$$\mathrm{v}_2 c_2' \frac{\partial c_2}{\partial z} = D_2 c_2'' \tag{8}$$

We count the pressure from the hydrostatic pressure, the concentration from the concentration at the lower boundary, and introduce the scales

$$[z] = h_2, \quad [\mathrm{v}] = -\frac{k_1}{\eta h}\rho\frac{\partial \mu}{\partial c}c_0, \quad [p] = -\rho\frac{\partial \mu_1}{\partial c}c_0$$

Then

In area 1

$$\mu_1'' = 0, \quad \mathrm{v}_1 = \mu_1', \quad \mu_1 = c_1 - p_1$$

$$\mathrm{v}_1 = const$$

$$Pe\mathrm{v}_1 c_1 = c_1''$$

$$Pe = \frac{[\mathrm{v}]h}{D} = -\frac{\rho k c_0}{\eta D}\frac{\partial \mu}{\partial c}, \quad \frac{\partial \mu}{\partial c} < 0 \tag{9}$$

In area 2

$$\mu_2'' = 0, \quad \mathrm{v}_2 = k\mu_2', \quad \mu_2 = c_2 - p_2, \quad k = \frac{k_2}{k_1}$$

$$\mathrm{v}_2 = const$$

$$Pe\mathrm{v}_2 c_2' = dc_2'', \quad d = \frac{D_2}{D_1} \tag{10}$$

Boundary conditions

$$z = 1 + h_1 : \; c_1 = c_{10}, \quad p_1 = p_{10}$$

$$z = 1 : \quad \mathrm{v}_1 = \mathrm{v}_0, \quad c_1 = c_2, \quad p_1 = p_2,$$

$$Pe\mathrm{v}_0 c_1 - c_1' = Pe\mathrm{v}_0 c_2 - d' c_2$$

$$z = 0 : c_2 = 0, \, p_2 = p_{20}$$

Here, the continuity conditions for the pressure, concentration, and mass flux of the solutions are specified on the common boundary.

Solution in area 1

$$\mathrm{v}_1 = \mathrm{v}_0, \quad \mu_1 = \mathrm{v}_0(z - 1 - h_1) + c_{10} - p_{10}$$

$$c_1 = c_{10} + (c_{11} - c_{10})\frac{e^{Pe\mathrm{v}_0(z-1-h_1)} - 1}{e^{-Pe\mathrm{v}_0 h_1} - 1}, \quad p_1 = c_1 - \mu_1$$

$$p_1 = c_{10} + (c_{11} - c_{10})\frac{e^{Pe\mathrm{v}_0(z-1-h_1)} - 1}{e^{-Pe\mathrm{v}_0 h_1} - 1} - [\mathrm{v}_0(z - 1 - h_1) + c_{10} - p_{10}]$$

Solution in area 2

$$\mathrm{v}_2 = \mathrm{v}_0, \quad \mu_2 = \mathrm{v}_0\frac{z}{k} + c_{20} - p_{20}$$

$$c_2 = c_{20} + (c_{21} - c_{20})\frac{e^{Pe\mathrm{v}_0 z/d} - 1}{e^{Pe\mathrm{v}_0/d} - 1}, \quad p_2 = c_2 - \mu_2$$

$$p_2 = c_{20} + (c_{21} - c_{20})\frac{e^{Pe\mathrm{v}_0 z/d} - 1}{e^{Pe\mathrm{v}_0/d} - 1} - \left(\mathrm{v}_0\frac{z}{k} + c_{20} - p_{20}\right)$$

From the boundary conditions

$$\mathrm{v}_0 = \frac{p_{20} - p_{10} + c_{10} - c_{20}}{\frac{1}{k} + h_1}$$

$$c_{11} = c_{21} = c_\xi = \left(c_{20}\frac{e^{Pe\mathrm{v}_0/d}}{e^{Pe\mathrm{v}_0/d} - 1} + c_{10}\frac{1}{e^{Pe\mathrm{v}_0 h_1} - 1}\right)\left(\frac{e^{Pe\mathrm{v}_0/d}}{e^{Pe\mathrm{v}_0/d} - 1} + \frac{1}{e^{Pe\mathrm{v}_0 h_1} - 1}\right)^{-1}$$

$$p_{11} = p_1(1) = c_{11} - [-\mathrm{v}_0 h_1 + c_{10} - p_{10}]$$

$$p_{21} = p_2(1) = c_{21} - \left(\mathrm{v}_0\frac{1}{k} + c_{20} - p_{20}\right)$$

$$p_{21} - p_{11} = c_{21} - c_{11} - c_{10} - c_{20} + p_{20} - p_{10} - \mathrm{v}_0\left(\frac{1}{k} + h_1\right)$$

The filtration rate in dimensional form is written as

$$\mathrm{v}_0 = \frac{k_1 k_2}{\eta(k_1 h + k_2 h_1)}\left[-\rho\frac{\partial\mu}{\partial c}(c_{10} - c_{20}) + p_{20} - p_{10}\right]$$

Problem 3 There are two layers, lower thickness and upper thickness, with different permeabilities and thermophysical coefficients Fig. 3. The upper boundary of the upper layer and the lower boundary of the lower layer are permeable. The common boundary of the layers is semipermeable, i.e. permeable to water and impermeable to salt. At the upper and lower boundaries of the two-layer formation, the pressure and salt concentration are given. It is required to find a stationary solution to the problem.

We look for a one-dimensional solution of the problem that depends on the z coordinate.

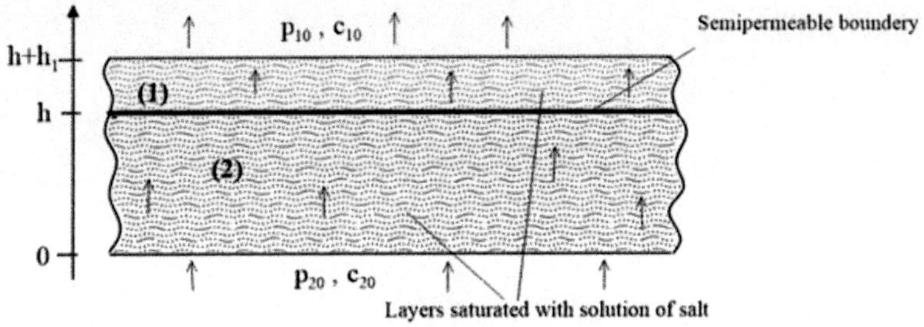

Fig. 3 Staging drawing for the third problem

In area 1

$$\mu_1'' = 0, \quad v_1 = \mu_1', \quad \mu_1 = c_1 - p_1$$
$$v_1 = const$$
$$Pev_1c_1 = c_1''$$
$$Pe = \frac{[v]h}{D} = -\frac{\rho k c_0}{\eta D}\frac{\partial \mu}{\partial c}, \quad \frac{\partial \mu}{\partial c} < 0 \tag{11}$$

In area 2

$$\mu_2'' = 0, \quad v_2 = k\mu_2', \quad \mu_2 = c_2 - p_2, \quad k = \frac{k_2}{k_1}$$
$$v_2 = const$$
$$Pev_2c_2' = dc_2'', \quad d = \frac{D_2}{D_1} \tag{12}$$

Boundary conditions

$$z = 1 + h_1 : c_1 = c_{10}, p_1 = p_{10}$$
$$z = 1 : v_1 = v_0, \quad Pev_0c_1 - c_1' = 0, \quad Pev_0c_2 - dc_2' = 0$$
$$v_0 = \overline{\beta}(c_{11} - p_{11} - c_{21} + p_{21})$$
$$z = 0 : c_2 = 0, \quad p_2 = p_{20}$$

Solution in area 1

$$v_1 = v_0, \quad \mu_1 = v_0(z - 1 - h_1) + c_{10} - p_{10}$$
$$c_1 = c_{10}e^{Pev_0(z-1-h_1)}, \quad p_1 = c_1 - \mu_1$$
$$p_1 = c_{10}e^{Pev_0(z-1-h_1)} - [v_0(z - 1 - h_1) + c_{10} - p_{10}] \tag{13}$$

Solution in area 2

$$\begin{aligned} & \mathrm{v}_2 = \mathrm{v}_0, \quad \mu_2 = \mathrm{v}_0 \frac{z}{k} + c_{20} - p_{20} \\ & c_2 = c_{20} e^{Pe\mathrm{v}_0 z/d}, \quad p_2 = c_2 - \mu_2 \\ & p_2 = c_{20} e^{Pe\mathrm{v}_0 z/d} - \left(\mathrm{v}_0 \frac{z}{k} + c_{20} - p_{20}\right) \end{aligned} \tag{14}$$

From the boundary conditions

$$\mathrm{v}_0 = \frac{p_{20} - p_{10} + c_{10} - c_{20}}{\frac{1}{k} + h_1 + \frac{1}{\overline{\beta}}} \tag{15}$$

$$\begin{aligned} & c_{11} = c_{10} e^{-Pe\mathrm{v}_0 h_1}, \quad c_{21} = c_{20} e^{Pe\mathrm{v}_0/d} \\ & p_{11} = c_{10} e^{-Pe\mathrm{v}_0 h_1} - [-\mathrm{v}_0 h_1 + c_{10} - p_{10}] \\ & p_{21} = c_{20} e^{Pe\mathrm{v}_0/d} - \left(\mathrm{v}_0 \frac{1}{k} + c_{20} - p_{20}\right) \end{aligned}$$

The filtration rate in dimensional form is written as

$$\mathrm{v}_0 = \frac{k_1 k_2}{\eta(k_1 h + k_2 h_1 + k_2 h/\overline{\beta})}\left[-\rho \frac{\partial \mu}{\partial c}(c_{10} - c_{20}) + p_{20} - p_{10}\right]$$

3 Discussion

The calculations in Figs. 4, 5 and 6 are given in dimensionless form for the values

$$Pe = \frac{\mathrm{v}_i h}{D} = 1; 5, h_1 = 1, \quad c_{01} = 1, c_{02} = 0.3, \quad p_{01} = 0, p_{02} = 0, k_2/k_1 = 3$$

Here it is assumed that hydrostatic pressures are maintained at the boundaries of the layer, and the salt concentration at the upper boundary is taken as one.

Figure 4 shows the vertical distributions of the reduced salt concentration in the solution for the three problems solved above, respectively. The dimensionless layer thickness is taken equal to two in all cases. It can be seen that the curves for the first two problems are close despite the fact that in the second case the permeability of the upper layer is half as much. In the third problem, there is a jump in the salt concentration at the boundary of the layers associated with the semipermeability of the boundary. Comparing Fig. 4a, b, we see that the jump in concentration when crossing the semipermeable boundary from bottom to top occurs in the opposite way; in the case of small Peclet numbers, Pe towards an increase, and in the case of large Pe, towards a decrease.

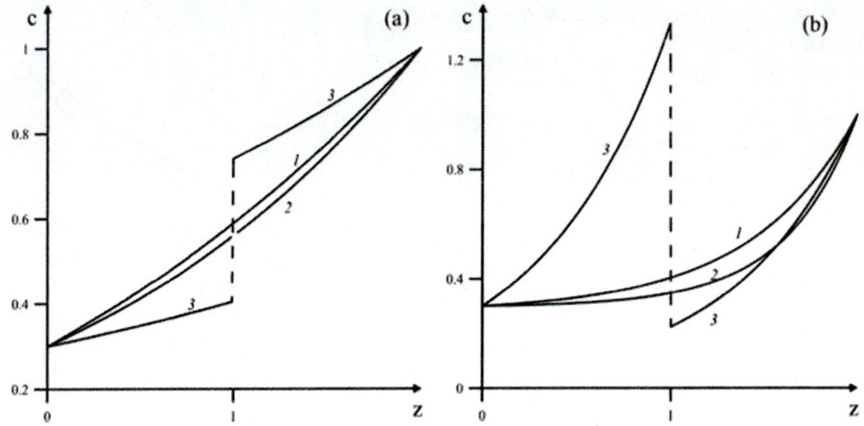

Fig. 4 Vertical distribution of the reduced salt concentration in the solution for three problems, respectively: **a** Pe = 1; **b** Pe = 5

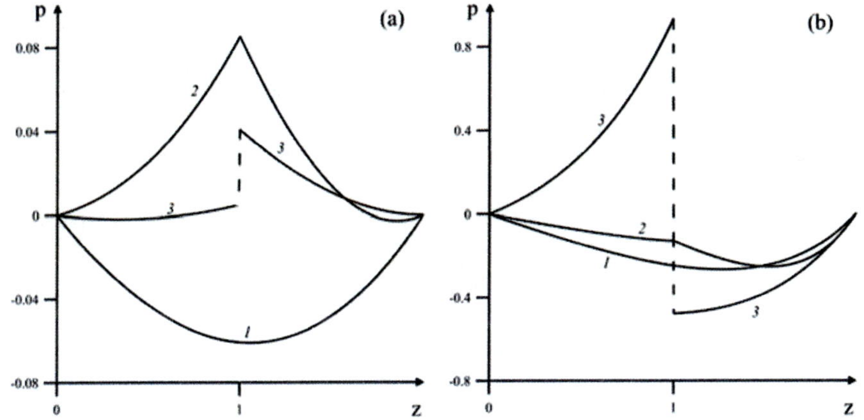

Fig. 5 Vertical distribution of overhydrostatic pressure for three problems, respectively: **a** Pe = 1; **b** Pe = 5

The dimensionless filtration rates in the three problems are respectively equal:

$$v_1 = 0.7; \quad v_2 = 0.53; \quad v_3 = 0.3$$

Thus, in the case of a semi-permeable boundary, the velocity is minimal, other things being equal.

Figure 5 shows the vertical distributions of the overhydrostatic pressure for three problems, respectively, for two different Peclet numbers. As in the case of the salt concentration, in the third problem the pressure suffers a discontinuity at the semipermeable boundary. Moreover, the pressure jump for small Peclet numbers and large

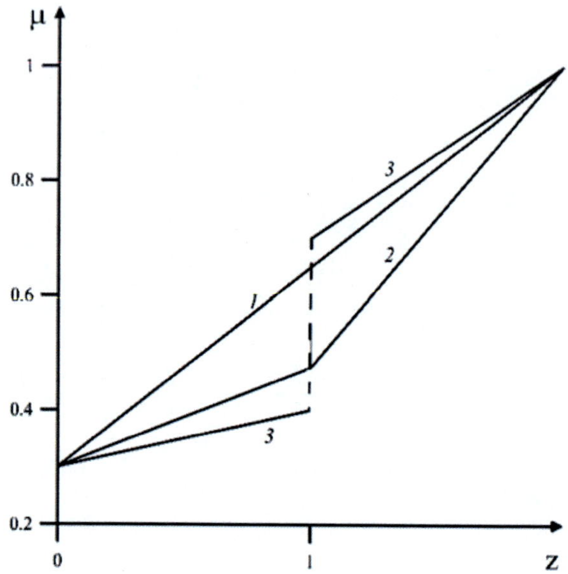

Fig. 6 Vertical velocity potential distribution for three problems, respectively

ones occurs in different directions, just like the concentration. In the second problem, the pressure at the boundary is continuous, but the pressure gradient undergoes a discontinuity, which is associated with different permeability of the upper and lower layers.

Finally, Fig. 6 shows the velocity potential distributions for the three considered problems. The velocity potential also gets a jump at the semipermeable boundary. Note that in the dimensionless form, the velocity potential does not depend on the Peclet number, as can be seen from the solutions obtained above.

We give an estimate of the filtration rate and the Peclet number for the following parameter values:

$$k_i = 10^{-20}\mathrm{m}^2, \quad \eta = 10^{-4}\,\mathrm{Pa}\cdot\mathrm{s}, \quad \rho = 10^3\,\mathrm{kg/m^3},$$
$$\gamma_c = 35\cdot 10^{-3}\,\mathrm{kg/(kg\cdot m)},\ R = 8.3\ \mathrm{J}/mol\cdot\mathrm{K},$$
$$T_0 = 300^{\circ}\ \mathrm{K}, \quad i = 2, \quad m_1, m_2 - \text{molar masses } Na \text{ and } Cl$$

Using the formula,

$$\mathrm{v}_i = \frac{k}{\eta}\rho RT\frac{i}{2}\left(\frac{1}{m_1} + \frac{1}{\mathrm{m}_2}\right)\gamma_c$$

with $D \sim 10^{-9}\,\mathrm{m}^2/s$ for filtration rate, coefficient of chemical osmosis and the Peclet number, we obtain the estimates

$$\mathrm{v}_i \sim 6\cdot 10^{-10}\,\mathrm{m/s} = 2\cdot 10^{-2}\,\mathrm{m/year},$$

$$K_f \sim 5.5 \cdot 10^{-1}\,\text{m}^2/\text{year} = 1.5 \cdot 10^{-3}\,\text{m}^2/\text{day},$$
$$h = 1\,m$$
$$Pe = \frac{v_i h}{D} \sim 1$$

Thus, at these values, the filtration rate is about 2 cm per year, and the Peclet number is about unity. If the layer permeability is one or two orders of magnitude higher, then these values will be, respectively, one or two orders of magnitude higher.

4 Conclusion

Several problems of solution filtration caused by chemical osmosis are considered. In one of the tasks, the influence of not only the concentration gradient (spatial osmosis), but also the presence of a semi-permeable boundary is considered. Characteristic distributions of salt concentration and pressure are given, their jumps at the semipermeable boundary are shown. For the characteristic values of the parameters, estimates of the filtration rates caused by osmosis and the corresponding Peclet diffusion numbers are obtained.

References

1. Kemper W.D. Movement of water as effected by free energy and pressure gradients: II. Experimental analysis of porous systems in which free energy and pressure gradients act in opposite directions // Soil Sci. Soc. Am. Proc. 1961. V. 25. P. 260–265.
2. Reltov B.F., Novitskaya N.A. Osmotic phenomena in cohesive soils with uneven salinization // From. VNIIG. 1954. Vol. 51. pp. 94–122.
3. Dirksen C. Thermo-osmosis through compacted saturated clay membranes // Soil Sci. Soc. Am. J. 1969. 33:821–826.
4. Grundl T. and P. Michalski. Electroosmotically driven water flow in sediments // Water Res. 1996. 301811–818.
5. Voronkevich S.D., Emelyanov S.N., Sergeev V.I. Investigation of filtration-osmotic processes in the creation of dense protective screens // Problems of mechanics of natural processes. Moscow: Publishing House of Moscow State University. 1983. pp.47–63.
6. Soler J.M. The effect of coupled transport phenomena in the Opalinus Clay and implications for radionuclide transport // J. Contaminant Hydrology. 2001. V. 53. P. 63–84.
7. Graham J., Tanaka N., Crilly T., Alfaro M. Modified Cam-Clay modeling of temperature effects in clays // Can. Geotech. J. 2001. V. 38. P. 608–621.
8. Srivastava R.C., Avasthi P.K. Non-equilibrium thermodynamics of thermo-osmosis of water through kaolinite // J. Hydrol. 1975. V. 24. P. 111–120.
9. KeijzerTh.J.S., Loch J.P.G. Chemical osmosis in compacted dredging sludge // Soil. Sci. Soc. Am. J. 2001. V. 65. P.1045–1055.
10. Ramazanov, M. M., Karakin A. V., Lobkovskiy L. I. Mathematical Model for the Motion of Solutions Taking into Account the Osmotic Effect // Doklady Earth Sciences. – 2019. – Vol. 489. – No 1. – P. 1306–1309.

11. Keijzer Th. J. S. Chemical osmosis in natural clayey materials. Geologica Ultraiectina № 196. Ph.D. thesis, Utrecht University. 2000. 170 p.
12. Malusis M.A., Shackelford C.D., Olsen H.W. A laboratory apparatus to measure chemico-osmotic efficiency coefficients for clay soils // Geotechnical Testing Journal. 2001. V. 24. P. 229-242.
13. Ramazanov, M. M., Karakin A. V. Phenomenon of Concretions Occurrence on the Ocean Bottom // Izvestiya, Physics of the Solid Earth. – 2018. – Vol. 54. – No 2. – P. 388–392.
14. Barenblatt G.I., Baturin G.N. On the "unsinkability" of ferromanganese nodules and some features of the ocean bottom layer // DAN. 1989. Vol. 308. No. 1. pp. 183–188.
15. Baturin G.N. Geochemistry of ferromanganese nodules of the ocean. M.: Nauka, 1968. 328 p.
16. Watson E.B. Melt infiltration and magma evolution // Geology. 1982. V. 10. P. 236–240.
17. Hammouda T., Laporte D. Ultrafast mantle impregnation by carbonatite melts // Geology. 2000. V. 28. P. 283–285.
18. Shatsky A.F., Litasov K.D. Conditions for the formation of carbonates and the mechanism of migration of carbonate melts in the Earth's mantle. Novosibirsk: Publishing House of SB RAS, 2015. 247 p.
19. Shakhova N., Semiletov I., Dudarev O., Mazurov A.K., Charkin A., Salyuk A., Kosmach D., Karnaukh V., Chernykh D., Gustafsson O., Sergienko V., Lobkovsky L., Ananiev R., Dmitrevsky N., Meluzov A., Tumskoy V., Koshurnikov A., Gunar A., Grigoriev M. Current rates and mechanisms of subsea permafrost degradation in the east siberian Arctic shelf// Nature Communications. 2017. V. 8. P. 15872.
20. Chuvilin E.M. Migration of ions of chemical elements in freezing and frozen soils // Polar Record. 1999. V. 35. № 192. P. 59–66.

Investigation of Features of Water Circulation in the Northern Part of the Black Sea on the Basis of the Assimilation of Observational Data in the Autumn–Winter Season of 2016

S. G. Demyshev, N. A. Evstigneeva, and O. A. Dymova

Abstract A four-dimensional and energy analysis of hydrophysical fields in the limited area of the Black Sea was carried out on the basis of assimilation of hydrological data in the numerical model (autumn–winter season of 2016). The Rim Current was clearly reconstructed. Multi-scale eddies could be generated between the shore and the Rim Current during some weakening of the wind forcing. Cyclonic and anticyclonic eddies were noticed in the western area, they could merge into larger one during the further calculation. The most energetically significant components in the kinetic and potential energy budget equations were found and analyzed from 15th of November to 5th of December, 2016. Possible mechanisms of circulation features of the Black Sea were determined, analyzing of energy fields (for example, baroclinic and/or barotropic instability). The obtained results were compared with all available data of instrumental measurements.

Keywords Black Sea · Numerical Modeling · Assimilation of Observational Data · Energy Analysis · Mesoscale and Submesoscale Eddies

1 Introduction

Nowadays, Marine Hydrophysical Institute (MHI) of the Russian Academy of Sciences actively continues to conduct oceanographic research in the Black Sea. The obtained experimental data on the distribution of thermohaline characteristics and current velocities expanded the information about the structure and variability of the thermohydrodynamic fields of the Black Sea on the various scales for example, [1–4].

The results, obtained during the 81st cruise of the research vessel (R/V) "Professor Vodyanitsky", showed that the influence of interannual and synoptic variability formed differences in hydrological fields from the climatic norm. The analysis of water circulation during the 87th of the R/V "Professor Vodyanitsky" revealed that

S. G. Demyshev · N. A. Evstigneeva (✉) · O. A. Dymova
Marine Hydrophysical Institute RAS, Sevastopol, Russian Federation
e-mail: naevstigneeva@yandex.ru

V. I. Karev (ed.), *Physical and Mathematical Modeling of Earth and Environment Processes—2022*, Springer Proceedings in Earth and Environmental Sciences,
https://doi.org/10.1007/978-3-031-25962-3_5

western flows prevailed, which were a manifestation of the Rim Current (RC). Several branches of the RC, a clearly defined Sevastopol anticyclone and an anticyclone near the southeastern coast of Crimea were recorded. Investigations of hydrological fields and water dynamics in the northern part of the Black Sea, carried out in 2016 on the R/V "Professor Vodyanitsky", were presented in [3, 4]. The seasonal signal manifested itself in a successive decrease in the average temperature of the upper quasi-homogeneous layer, an increase in the temperature in the core of the cold intermediate layer and an increase in the depths of the lower boundary of the upper quasi-homogeneous layer.

The application of hydrodynamic models with the assimilation of observational data made it possible to use the collected information more efficiently. Continuous dynamics and energy of the marine environment, close to observed, could be reproduced in the survey area.

The problem of the complex use of measurement data of temperature, salinity and current velocity during their assimilation on the basis of the Kalman filter in a hydrodynamic model was considered in [5]. The system of oceanographic data assimilation, developed in the Hydrometeorological Center of Russia, was presented in [6]. Hydrophysical and energy fields were reconstructed in [7] with the help of the MHI hydrodynamic model of the with the assimilation of observational data from surveys of 2007 and 2009 near the coast of the western Crimea and in the region of Sevastopol. The calculated fields of currents were characterized by mesoscale eddy formations and jet currents.

That work continued the cycle of work on the processing of hydrological survey in 2016, the purpose of which was to study the patterns of formation of circulation features depending on the season. Features of Black Sea circulation in the autumn–winter season of 2016 were investigated on the basis of a four-dimensional analysis of hydrophysical fields using temperature and salinity data, received in the expedition on the R/V "Professor Vodyanitsky". Hydrological measurements in November–December of 2016 were made at the stations, which were shown on Fig. 1. To interpret the reconstructed features of the water circulation, the fields of potential and kinetic energy budget components were calculated.

2 Selected Model Parameters and Description of the Data, Measured in 91st Cruise of the R/V "Professor Vodyanitsky"

The detailed specification of the eddy-resolving hydrodynamic model, developed at MHI, was given in [8]. Its original feature was the exact correspondence of discrete energy equations to the difference equations of hydrodynamics. As a procedure for assimilation of data from field observations, the Kalman filter algorithm technique was used. It was based on solving equations for the dispersion of errors in estimates of

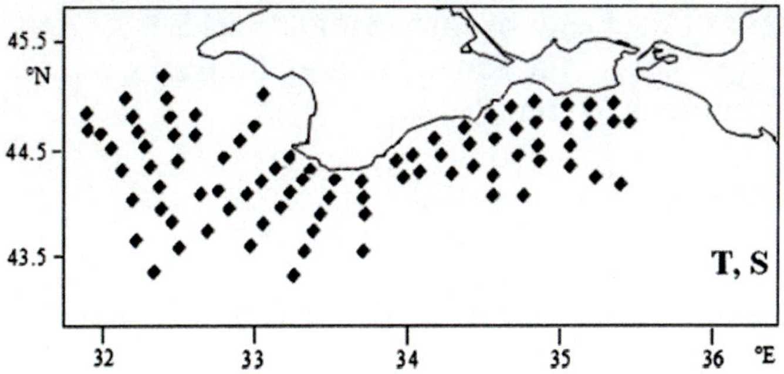

Fig. 1 Station's map in the 91st cruise of the R/V "Professor Vodyanitsky" in the autumn–winter period of 2016

temperature and salinity fields [9, 10]. We also calculated the terms in the equations for the change in kinetic and potential energy [11].

We had set the following model parameters (1.64 km was the horizontal step of the model, 27 vertical horizons, the time step was 96 s, $\nu^H = 10^{16}$ cm^4/s, $k^H = 10^{16}$ cm^4/s were the values of coefficients of turbulent viscosity and horizontal diffusion). The Mellor-Yamada 2.5 parametrization [12] was applied for calculation of coefficients of turbulent momentum exchange and vertical diffusion. The total integration time of the model equations was 20 days (from 15th of November to 5th of December, 2016).

The thermohydrodynamic fields, obtained in our previous calculations for 15th of November, 2016, were used to set the initial conditions. The data of the Greek Center for Atmospheric Predictions SKIRON (1/10°) [13] were used to specify atmospheric effects on the sea surface. Northeast wind acted from 15 to 17th of November with maximum values of velocity 10 m/s, western, northwestern and southwestern winds prevailed from 18th of November to 3rd of December (maximum values could reach 17 m/s).

Hydrological work (105 stations) was carried out using the CTD SBE 911plus complex on the R/V "Professor Vodyanitsky" from 16th of November to 3rd of December (91st cruise), 2016 in the northern part of the Black Sea. Soundings were performed to the depths 30–1000 m.

Our numerical experiment was performed for the entire Black Sea. The area, limited between the meridians 31 and 37° E and parallels 43 and 45.5° N, was chosen for the analysis in the next section.

3 Analysis of the Fields of Currents and Fields of Components of the Kinetic Energy Budget in the Autumn–Winter Season of 2016

In order to identify the most significant energy characteristics, we analyzed the change over time of the integral components of the KE budget equation [11] from 15th of November to 5th of December, 2016 (Fig. 2). $\Pi \leftrightarrow$ E was the work of the buoyancy force, $Diss_{Ver}$(E) and $Diss_{Hor}$(E) was the energy dissipation due to vertical and horizontal internal friction, $\tau \to E = u_0\tau^x + v_0\tau^y$ was the term, determining the contribution to kinetic energy from the wind.

It was found that the principal contribution to the kinetic energy was made by the work of the wind force ($\tau \to E$). The work of the buoyancy force ($\Pi \leftrightarrow$ E) began to exert a predominant influence only with a decrease in wind velocity.

Significant variability of work of the buoyancy force with time was noted from Fig. 2. The values of energy dissipation due to vertical and horizontal internal friction were negative. With an increase in inflow from the wind, the modules $Diss_{Ver}$(E) and $Diss_{Hor}$(E) also increased.

Figure 3 demonstrated the reconstructed fields of currents at a depth of 5 m on 15th, 21st, 25th of November and 5th of December.

Anticyclonic eddies near Sevastopol and along the eastern Crimean coast, as well as small eddies in the western part in the upper horizons, were observed on 15th of November (Fig. 3a).

The RC was reconstructed from 20 to 30th of November (the jet was clearly traced to a depth of 100 m). Its maximum velocity was 52 cm/s at the upper horizon. Due to some weakening of the wind effect from 1st to 5th of December, the RC was less distinguishable (Fig. 3d) and anticyclonic eddies of different scales could be generated and transformed along the eastern Crimean coast.

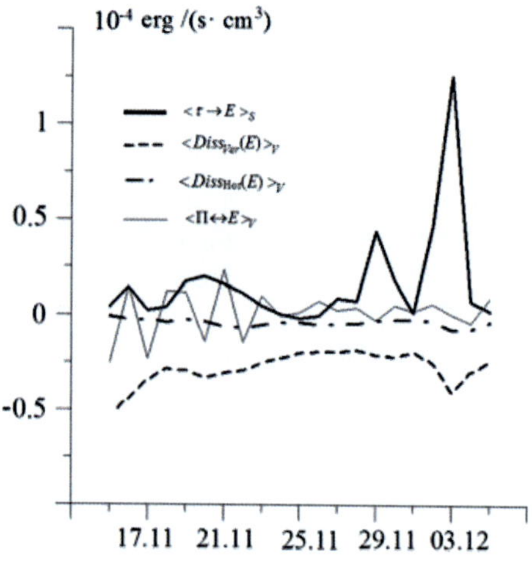

Fig. 2 Change over time of the integral components of the KE budget equation, normalized to the volume of the computational domain, in the autumn–winter season of 2016

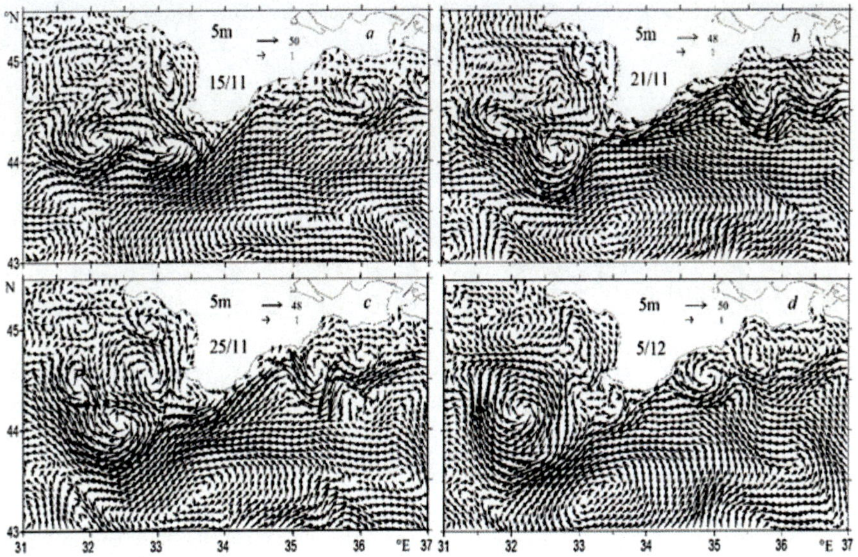

Fig. 3 Calculated fields of currents (cm/s) at a depth of 5 m on 15th of November (**a**), 21st of November (**b**), 25th of November (**c**) and 5th of December (**d**)

A cyclonic eddy with a radius of about 50 km was observed in the center of deep-water region (observed to a depth of 500 m) from 21st of November to 5th of December.

Meandering of the RC could lead to the generation of eddies of different rotation signs in the eastern part (for example, Fig. 3c). We also noted the formation and development of small-scale cyclonic and anticyclonic eddies in the western part (Fig. 3b, c) due to the RC flow around the uneven coastline. Large eddies in the western area (Fig. 3a–c) were observed from 15 to 28th of November, and then they merged into one over the next few days, for example, as in Fig. 3d (5th of December). An anticyclonic eddy near Yalta was noticeable from 30th of November to 5th of December (Fig. 3d).

In order to find out the possible mechanisms of formation of features of circulation, the fields of energy characteristics of the circulation were investigated. Figure 4 demonstrated the reconstructed fields of $\tau \rightarrow E$, $\Pi \leftrightarrow \mathrm{E}$ and $\mathrm{Diss}_{\mathrm{Ver}}(\mathrm{E})$ on 25th of November and 5th of December.

Processes of shear instability evolved in the coastal areas, for example, in the eastern part of the region both on 25th of November and 5th of December (Fig. 4a, b). Negative values of $\tau \rightarrow E$ meant that directions of the wind and currents on the surface were opposite.

Baroclinic instability could be observed along the eastern coast of Crimea and in the western part of polygon (Fig. 4c, d). Zones with negative values of $\Pi \leftrightarrow \mathrm{E}$ expanded (for example, Fig. 4c, d) with some weakening of the wind effect.

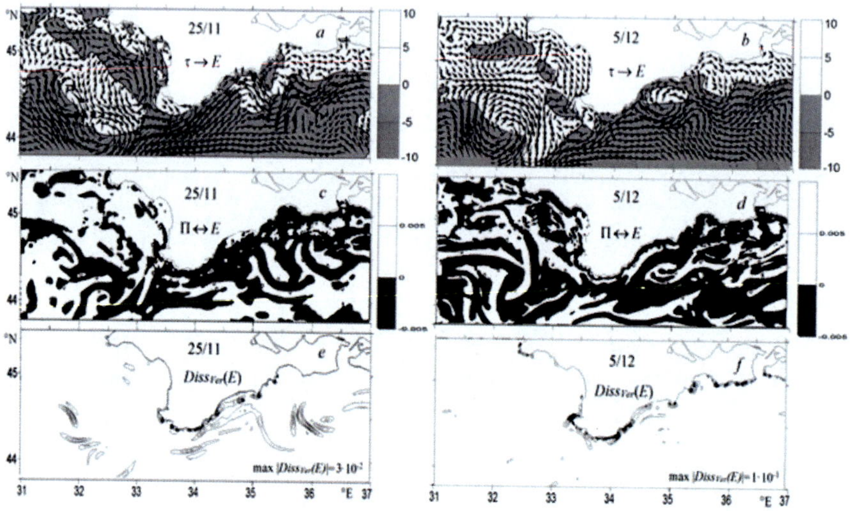

Fig. 4 Calculated fields of currents and energy characteristics on 25th of November and on 5th of December

The eastern coast of Crimea (Fig. 4e and f) and the area, corresponding to the location of the RC jet (Fig. 4e), were characterized by kinetic energy loss due to vertical friction. $\text{Max}|\text{Diss}_{\text{Ver}}(E)|$ was $2 \cdot 10^{-1}$erg /(s·cm^3) on 25th of November, $2 \cdot 10^{-1}$erg /(s·cm^3)—on 5th of December.

The constructed maps of currents supported the results, discussed in [3, 4], according to which the RC intensification to the south-west, the presence of the RC cyclonic meander and the Sevastopol anticyclone on the shelf to the west of Kalamitsky bay were noted.

To summarize the conclusions: the features of water circulation in the limited part of the Black Sea (between the meridians 31 and 37° E and parallels 43 and 45.5° N) were investigated using the MHI model of the and expeditionary data on temperature and salinity (91st cruise of the R/V "Professor Vodyanitsky") in the autumn–winter season of 2016.

It was noted from the analysis of the integral components of the budget equation that the wind force made a significant contribution into kinetic energy, as well as the buoyancy force (with a decrease in wind forcing). Horizontal and vertical internal friction reduced energy.

The RC had a distinct jet character (a maximum of velocity was 52 cm/s, wind velocity 17 m/s). With some weakening of the wind impact, eddies of different scales were reconstructed between the shore and the RC. Anticyclonic eddies with horizontal dimensions about 30 km were observed to the west, near Sevastopol and near Yalta. A cyclonic eddy with a radius of about 50 km was located in the central deep-water part of the region. Small-scale cyclonic and anticyclonic eddy formations were reproduced and developed along the western shore.

Performed energy analysis showed that the processes of shear instability (wind and currents had various directions on the surface) could develop in the coastal zone. Crimean coast could be zone of baroclinic instability (transition from kinetic energy to potential).

Acknowledgements The fields of currents were calculated within the framework of the state assignment on topic No. FNNN-2021-0003 (code "Operational Oceanology"). The energy characteristics of circulation were investigated the framework of the state assignment on topic No. FNNN-2021-0004 (code "Oceanological processes").

References

1. Artamonov JuV, Alexeev DV, Kondratyev SI et al. (2016) Hydrological Conditions in the Western Part of the Black Sea in November, 2015 (based on the Data Obtained in the 81st Cruise of R/V Professor Vodyanitsky). Physical Oceanography, [e-journal] 4: 57–70. https://doi.org/10.22449/1573-160X-2016-4-57-70
2. Artamonov YuV, Skripaleva EA, Fedirko AV et al. (2018) Water Circulation in the Northern Black Sea in Summer, 2016 (Based on the Data Obtained in the 87th Cruise of the R/V "Professor Vodyanitsky"). Physical Oceanography, [e-journal] 25(1): 52–66. doi:https://doi.org/10.22449/1573-160X-2018-1-52-66
3. Artamonov YuV, Skripaleva EA, Alekseev DV et al. (2018) Hydrological Research in the Northern Part of the Black Sea in 2016 (87th, 89th and 91st Cruises of R/V Professor Vodyanitsky). Physical Oceanography, [e-journal] 25(3): 229–234. https://doi.org/10.22449/1573160X-2018-3-229-234
4. Artamonov YuV, Alekseev DV, Skripaleva EA et al. (2016) Peculiarities of seasonal and synoptic variability of the structure of waters in the zone of the Main Black Sea Current in the autumn-winter period of 2016. Ecological safety of the coastal and shelf zones of the sea. 1: 32–43. (in Russian)
5. Knysh VV, Korotaev GK, Mizyuk AI and Sarkisyan AS (2012) Assimilation of hydrological observation data for calculating currents in seas and oceans. Izv. Atmos. Ocean. Phys. 48: 57–73. https://doi.org/10.1134/S0001433812010057
6. Zelenko AA, Vil'fand RM, Resnyanskii YuD et al. (2016) An ocean data assimilation system and reanalysis of the world ocean hydrophysical fields. Izvestiya. Atmospheric and Oceanic Physics. 52(4): 443–454.https://doi.org/10.1134/S0001433816040149
7. Demyshev S, Evstigneeva N (2022) Calculation of Dynamic and Energy Characteristics in the Region of Western Crimea Based on the Observational Data from Hydrological Surveys in 2007–2009 in the Numerical Model of the Black Sea Dynamics Using the Kalman Filter Algorithm. Processes in GeoMedia–Volume V. Springer Geology. Springer, Cham, 59–70.
8. Demyshev SG (2012) A numerical model of online forecasting Black Sea currents. Izv. Atmos. Ocean. Phys. 48: 120–132.https://doi.org/10.1134/S0001433812010021
9. Knysh VV, Moiseenko VA, Chernov VV (1988) Some results of four-dimensional analysis of hydrophysical fields in the Tropical Atlantic. Izv. Academy of Sciences of the USSR. Physics of the atmosphere and ocean. 24(7): 744–752.
10. Korotaev GK, Knysh VV, Lishaev PN and Demyshev SG (2018) Application of the Adaptive Statistics Method for Reanalysis of the Black Sea Fields Including Assimilation of the Temperature and Salinity Pseudo-measurements in the Model. Physical Oceanography, [ejournal] 25(1): 36–51. https://doi.org/10.22449/1573-160X-2018-1-36-51

11. Demyshev SG (2004) Energy of the Black Sea climatic circulation. Part I. Discrete equations for the rate of change of kinetic and potential energies. Meteorology and Hydrology. 9: 65–80.
12. Mellor G, Yamada T (1982) Development of a turbulence closure model for geophysical fluid problems. Rev.Geophys. Space Phys. 20 (4): 851–875.
13. NonHydrostatic SKIRON/Eta Modelling System. http://forecast.uoa.gr/forecastnew.php

Propagation of Tsunami Waves in a *T*-shaped Bay

A. Yu. Belokon, V. V. Fomin, and D. I. Lazorenko

Abstract Based on the *SWASH* numerical hydrodynamic model, a study was made of the penetration of tsunami waves into a *T*-shaped bay, which has the characteristic dimensions and configuration of the Sevastopol Bay. To model the propagation of tsunami waves in the bay, a non-linear hydrodynamic non-hydrostatic *SWASH* model was used. To determine the boundary conditions at the liquid boundary of the computational domain, using the Black Sea tsunami model, we calculated sea level fluctuations at the entrance to the Sevastopol Bay during the propagation of tsunami waves from three hypothetical earthquake foci with magnitude $M = 7$. As a result of the penetration of a tsunami into a *T*-shaped bay of height waves increase by 2–2.5 times compared to the heights on the seashore of the bay. At the same time, at the top of the bay, the amplitude of sea level fluctuations can reach 0.8 m. The most intense fluctuations are observed in the first three hours of the tsunami action. Tsunamis from different foci cause seiche oscillations in the bay with periods of 49.1 min, 19.3 min, 9.2 min, 6.8 min. At the same time, the amplitude-phase characteristics of the tsunami have a significant effect on the redistribution of wave energy between the modes of seiche oscillations. Numerical experiments have shown that the presence or absence of protective breakwaters at the entrance to the bay does not lead to significant changes in the wave field and sea level heights during tsunami propagation inside the bay.

Keywords Numerical simulation · Tsunami · *SWASH* · Sevastopol Bay · South Bay · *T*-shaped Bay

1 Introduction

It is known that tsunamis present the greatest danger when approaching the coast, where their propagation velocity and length decrease, and their height increases significantly. Especially it concerns the penetration of tsunami waves into narrow

A. Yu. Belokon (✉) · V. V. Fomin · D. I. Lazorenko
Marine Hydrophysical Institute of the Russian Academy of Sciences, Sevastopol, Russian Federation
e-mail: aleksa.44.33@gmail.com

V. I. Karev (ed.), *Physical and Mathematical Modeling of Earth and Environment Processes—2022*, Springer Proceedings in Earth and Environmental Sciences,
https://doi.org/10.1007/978-3-031-25962-3_6

bays, straits, channels, river mouths, where the presence of lateral boundaries can lead to focusing of wave energy and an increase in wave heights.

The phenomenon of a tsunami is characteristic not only of the oceans, but also of inland seas, although not with such catastrophic consequences. In the Black Sea, 50 cases of tsunamis are known that have occurred over the past 3000 years [1]. Most of them were of a seismic nature, some of them were of a landslide or meteorological nature. A number of works, some of them [2–7], are devoted to the study of tsunamis in the Black Sea with the help of numerical simulation. However, coastal areas with complex geometries, especially bays and gulfs, require a more detailed study, since when waves penetrate them, significant increases in sea level fluctuations can occur.

According to [8], during a strong earthquake with magnitude $M \geq 7$, which occurred in 103 in the area of the Sevastopol Bay, the water in the sea receded from 500 m to 3–4 km, while the height of the arriving waves was at least 2 m. In [9], an earthquake of magnitude $M = 6.0 \pm 0.2$ was mentioned on November 11, 1869, which caused waves in the sea near Sevastopol. A destructive underwater earthquake of magnitude $M \geq 6.5$ occurred on September 11, 1927, 30 km southeast of Yalta. Unusual waves during calm were observed in different places and after strong shocks. Subsequent weaker shocks with foci near Sevastopol and Balaklava caused water to move away from the coast and solitary waves to run-up to the coast [10]. On December 26, 1939, a strong earthquake $M = 8$ occurred in the city of Fatsa (the coast of Turkey). According to eyewitnesses, the sea receded by 50 m, and then flooded the coast by 20 m. Tsunami waves crossed the Black Sea and were recorded by tide gauges in Sevastopol, where the wave height was 50 cm [8].

The Sevastopol Bay is an estuary-type water area with limited water exchange with the open sea due to the presence of two protective breakwaters [11]. The length of the Sevastopol Bay is 7 km, the width is about 1 km, the depth varies from 20 m at the entrance to 4 m at the top, the average depth is about 12 m. The South Bay adjacent to it has a length of about 2 km. It was shown in [12] that during wind waves, the protective breakwaters installed at the entrance to the Sevastopol Bay have the maximum protective effect in the case of a westerly wind.

The Sevastopol Bay is a partially enclosed basin. In such reservoirs, seiches can occur—standing fluctuations in the mass of water in the entire reservoir or in a separate part of it [13]. They differ from seiches in completely enclosed water bodies in that they can be excited by long waves from the open sea through a liquid boundary. In this case, the loss of wave energy occurs due to its radiation through the open boundary [14]. In such bays, in contrast to closed reservoirs, the Helmholtz mode is generated. This mode usually dominates over all other modes and determines the general character of movements in the bay [13, 15]. The danger of seiches is that they can generate intense periodic currents that pose a threat to coastal infrastructure and ships [16]. In [17], a seiche with a period of about 50 min was identified based on field observations. This seiche is observed almost constantly in the Sevastopol Bay and can be classified as the Helmholtz mode.

Thus, taking into account the above, there is a need to conduct numerical experiments to study the penetration of tsunami waves into a bay similar to Sevastopol and the possibility of seiche formation in the bay.

This article considers a model *T*-shaped bay, which has the configuration and characteristic dimensions of the Sevastopol Bay. The results of numerical simulation of the penetration of tsunami waves into a *T*-shaped bay are presented. Three variants of hypothetical tsunami foci in the Black Sea caused by underwater earthquakes were used as forcing. The spectra generated by the tsunami and seiche oscillations in the bay are compared for different foci.

2 Mathematical Model

To study the tsunami, we used the nonlinear hydrodynamic model *Simulating WAves till SHore* (*SWASH*) [18, 19], which takes into account nonhydrostatic effects. The computational area was a rectangular *T*-shaped basin of constant depth $H = 12$ m, consisting of a large and a small bay (Fig. 1). The large bay had the characteristic dimensions and configuration of the Sevastopol Bay, the small bay had the characteristic dimensions and configuration of the South Bay. Two cases of the computational domain were considered: with the presence of protective breakwaters at the entrance to a large bay and without.

On the western boundary of the computational domain ($x = 0$), sea level fluctuations were set, obtained using the tsunami propagation model for the entire Black Sea [20]. Three cases of tsunami occurrence in the Black Sea with $M = 7$ were modeled. The parameters of tsunami generation foci were determined by empirical formulas [21]. The foci had an elliptical shape and were oriented along the 1500 m isobath, since all known Black Sea earthquakes that led to tsunamis occurred on the continental slope along isobaths not exceeding this depth value. The position of the considered model earthquake foci is shown in Fig. 2.

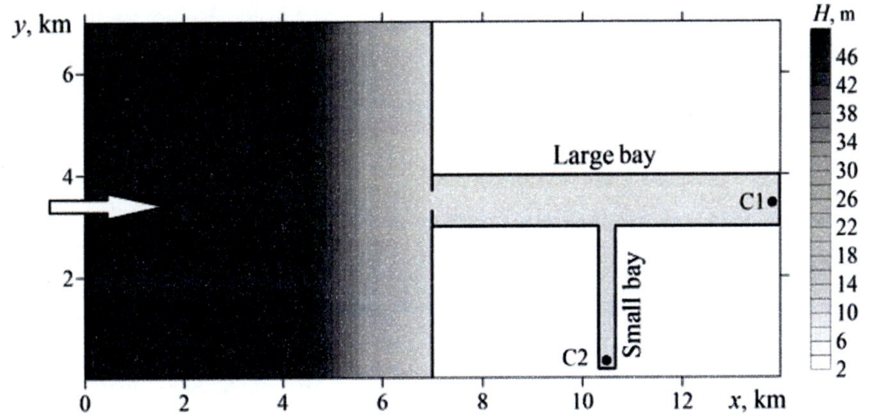

Fig. 1 The relief of the bottom of the computational domain and the position of the sensors C1 and C2

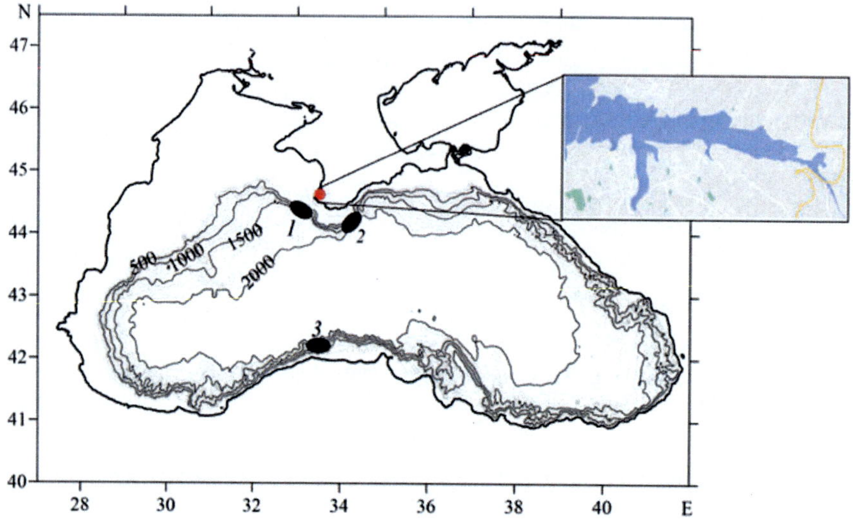

Fig. 2 The position of three hypothetical tsunami foci in the Black Sea: 1—the nearest focus in relation to the Sevastopol Bay; 2—focus similar to the one that caused the Yalta earthquake on September 12, 1927; 3—remote focus. Sevastopol Bay is marked with a red circle

On liquid horizontal boundaries, the radiation condition was used. On solid sections of the boundary, the impermeability condition was set. All calculations using the *SWASH* model were performed for a time period of 9 h with a spatial step of 10 m and a time integration step of 0.5 s. The bottom roughness was taken into account using the Manning parameter $n = 0.019\ m^{-1/3}$ s. The coefficient of horizontal turbulent viscosity was determined by the Smagorinsky formula. Changes in the water level at the tops of the large and small bays were recorded by virtual sensors C1 and C2 (Fig. 1).

3 Results of Numerical Experiments

Sea level fluctuations near the Sevastopol Bay for three tsunami generation foci are shown in Fig. 3 (western boundary of the computational domain, depth 50 m). Their comparison shows that for the nearest tsunami focus (*1* in Fig. 2) the maximum sea level rise on the approach to the bay is 0.3 m, the maximum decrease is 0.15 m; for the tsunami focus located in the Yalta seismically active zone and the distant tsunami foci (2 and 3 in Fig. 2), the sea level deviations were about $\pm$ 0.05– $\pm$ 0.07 m. Such small amplitudes of level fluctuations for foci 2 and 3 are explained by the fact that the Sevastopol Bay is protected by Cape Chersonesus from waves coming from the south and southeast. Thus, the maximum tsunami energy from these foci falls on the southern coast of Crimea. The seashores of the Sevastopol Bay are reached only by the waves that go around Cape Chersonesus.

On Fig. 4 shows the evolution of a tsunami in the bay from focus 1. As can be seen, the wave enters the computational region, where its amplitude increases and its length decreases due to a decrease in depth (Fig. 4a), when approaching the coast, the wave is partially reflected, and partially penetrates into a large bay (Fig. 4b), and then into a small one (Fig. 4c). Further, wave disturbances propagate to the tops of the bays, where they are amplified upon reflection from a vertical wall (Fig. 4d). Reflected waves from both bays move in the opposite direction and form a local rise at the entrance to the small bay (Fig. 4e), which is replaced by a local decrease in sea level (Fig. 4f). In this case, the wave velocity is directed in opposite directions: part of the wave energy leaves the large bay into the open sea through the outlet, and part propagates towards the tops of the large and small bays (Fig. 4g, h). These fluctuations continue for a long time, their amplitude decays over time.

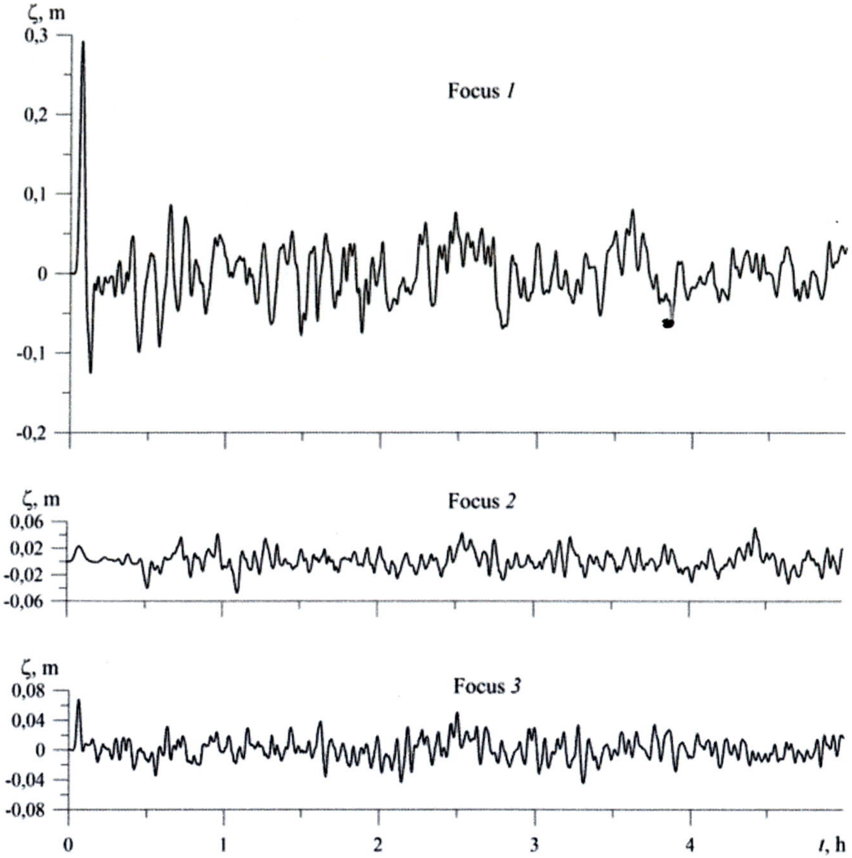

Fig. 3 Sea level fluctuations on the coast of the Sevastopol Bay caused by tsunami foci *1–3*

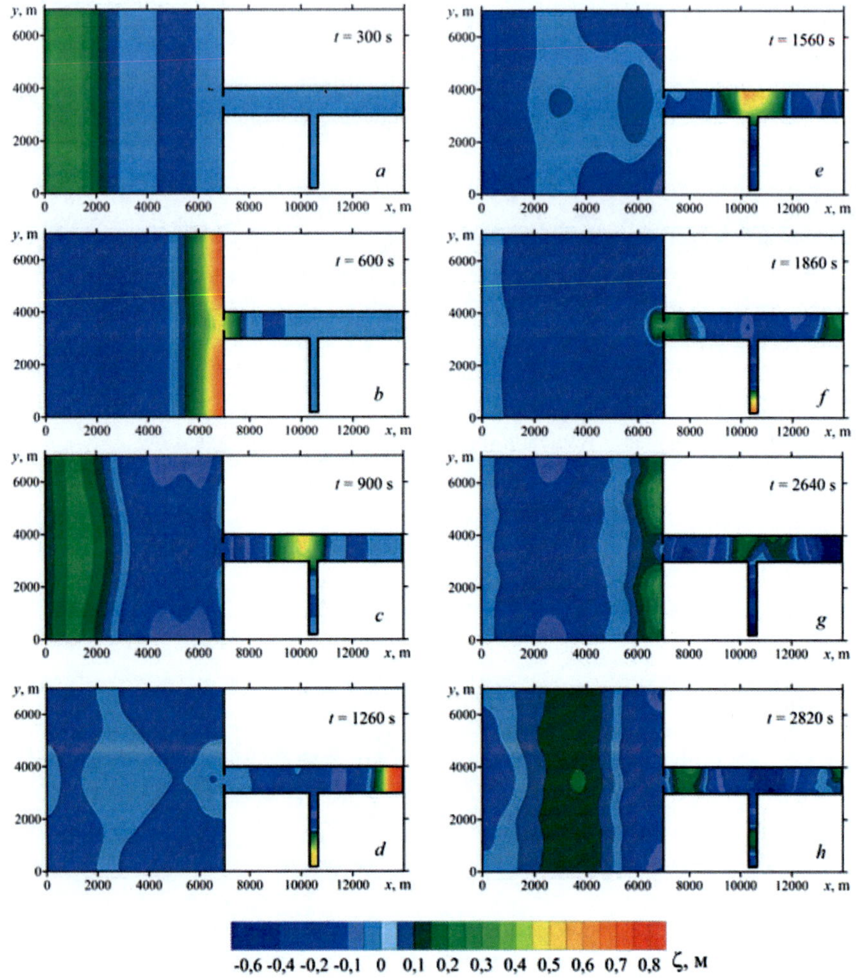

Fig. 4 Tsunami evolution in a *T*-shaped bay for focus *1*

The travel time of tsunami waves from generation focus *1* to the entrance to the bay is 18 min; from focus *2* is 30 min; and from focus *3* is 45 min.

Numerical calculations have shown that the presence or absence of protective breakwaters at the entrance to the bay does not lead to significant changes in the wave field and sea level heights during tsunami propagation inside the bay.

On Fig. 5 shows sea level fluctuations at the tops of the large and small bays (points C1 and C2). As can be seen, for both bays, the wave amplitudes increase by more than 2.5 times compared with the amplitudes at the western boundary of the computational area. The most intense fluctuations are observed during the first three hours of the tsunami, then their amplitude begins to decay.

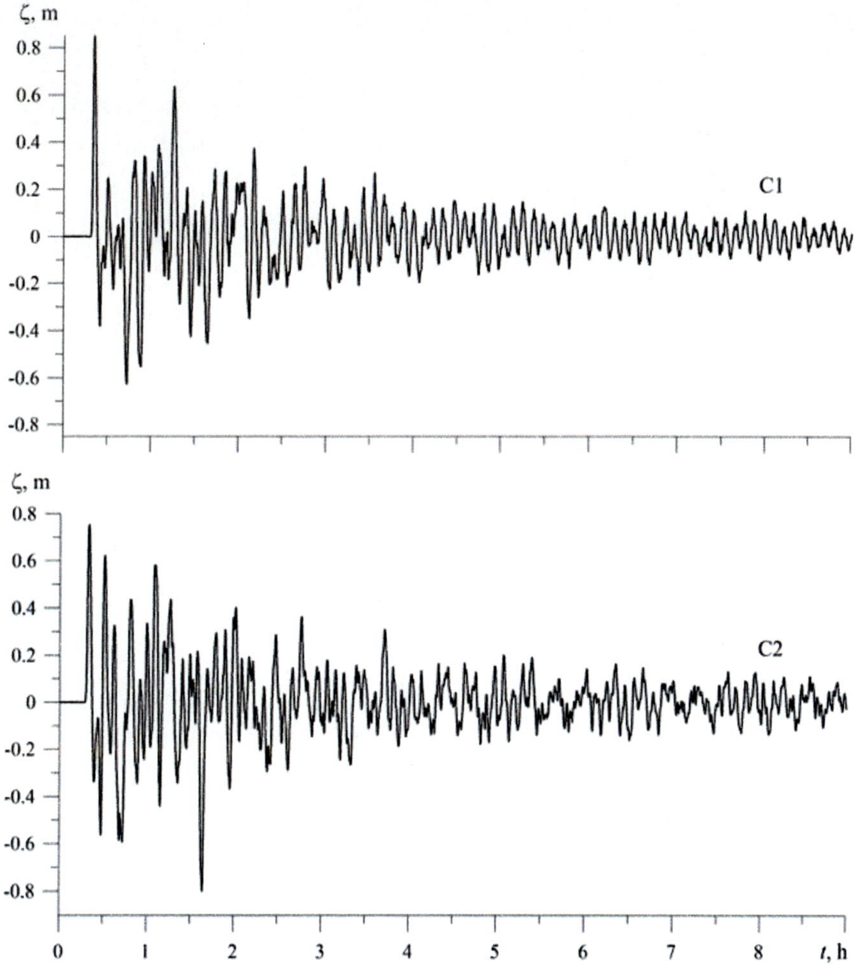

Fig. 5 Sea level fluctuations at points C1 and C2 caused by focus 1

For the remaining two foci, sea level fluctuations at the tops of the large and small bays are also calculated. They are shown in Figs. 6 and 7. The maximum amplitudes of level fluctuations in both cases amounted to 0.1–0.2 m. From here it can be seen that for foci *1* and *3*, the maximum sea level rises are observed during the propagation of the first—the main wave. In the case of focus *2*, the first wave is not maximum, which is a consequence of the anisotropy of the elliptical focus of tsunami generation.

As is known [19], tsunami waves can excite seiche oscillations in bays. To estimate the periods of seiche sea level fluctuations in the basin under consideration, time series at points C1 and C2 were used, 9 h long and with a discreteness of 0.5 s. The spectra were calculated using the scripts given in [22]. On Fig. 8 shows the energy spectra

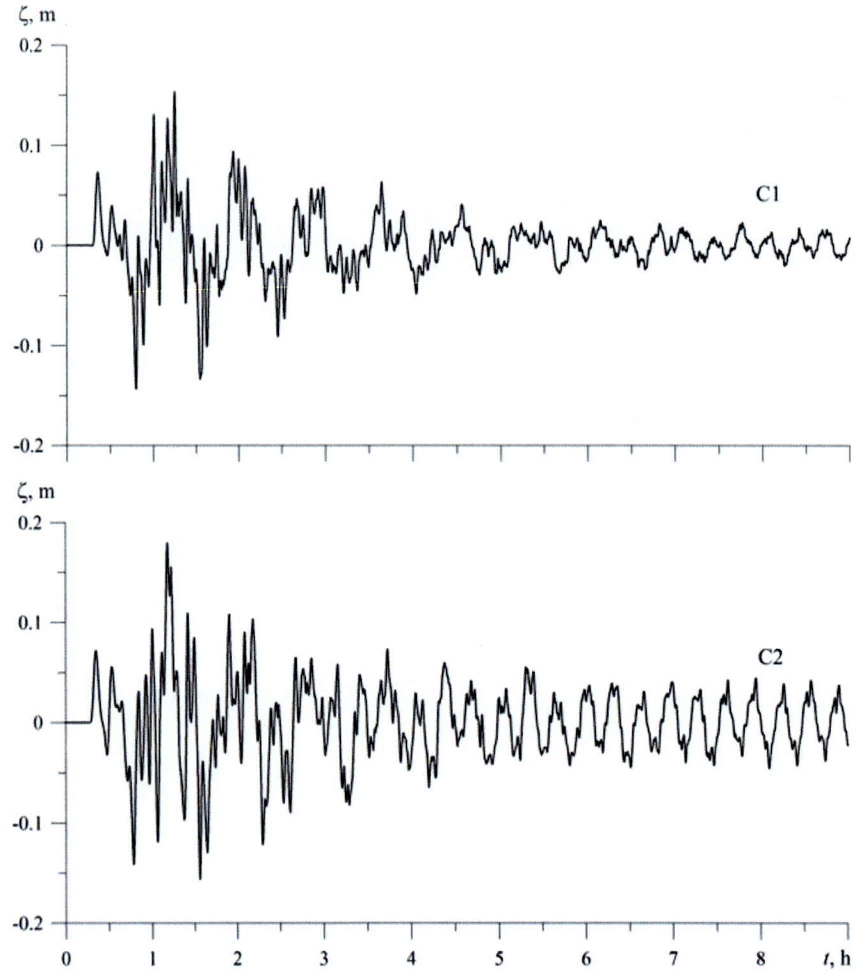

Fig. 6 Sea level fluctuations at points C1 and C2 caused by focus *2*

for points C1 and C2 and three tsunami foci. The spectra show four characteristic peaks with periods: T1 = 49.1 min; T2 = 19.3 min; T3 = 9.2 min; T4 = 6.8 min.

The considered basin is a system of two connected bays. Each bay has its own set of natural oscillation periods. Since the bays are connected, the periods of natural oscillations of both bays should appear in the spectra. Merian's one-dimensional formula [23] and the results of numerical experiments, in which the small bay was not taken into account, were used to determine the belonging of the proper periods to each of the bays. The following was established. Period T1 corresponds to the Helmholtz mode of a large bay, period T2 corresponds to the Helmholtz mode of a small bay. The periods T3 and T4 are identified as the periods of the two-node and three-node large bay modes, respectively. With a nearby tsunami focus, fluctuations

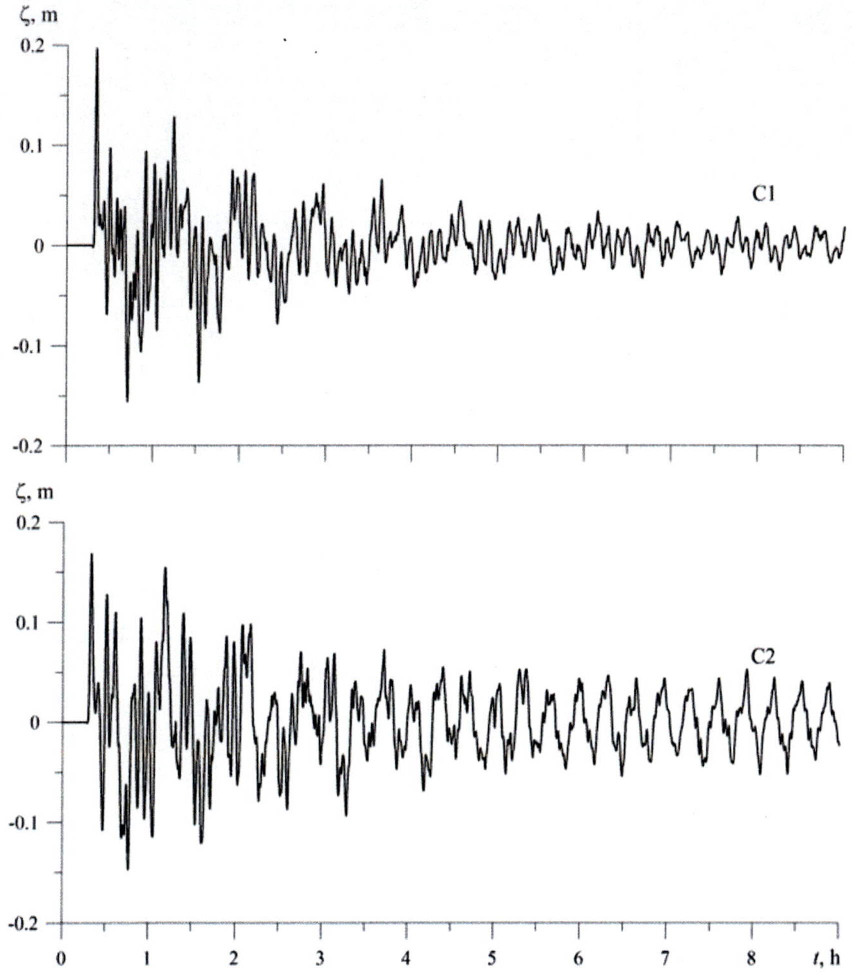

Fig. 7 Sea level fluctuations at points C1 and C2 caused by focus *3*

corresponding to period T4 dominate in a large bay (Fig. 8a). In other cases, the T1 mode dominates (Fig. 8b) or the T1 and T4 modes have the same energy (Fig. 8c). In the small bay, the mode with period T2 prevails in all cases (Fig. 8d–f).

Analysis shows that the spectrum of seiche oscillations has the same periods for different tsunami foci. In this case, the amplitude-phase characteristics of the tsunami have a significant effect on the redistribution of wave energy between the modes of seiche oscillations.

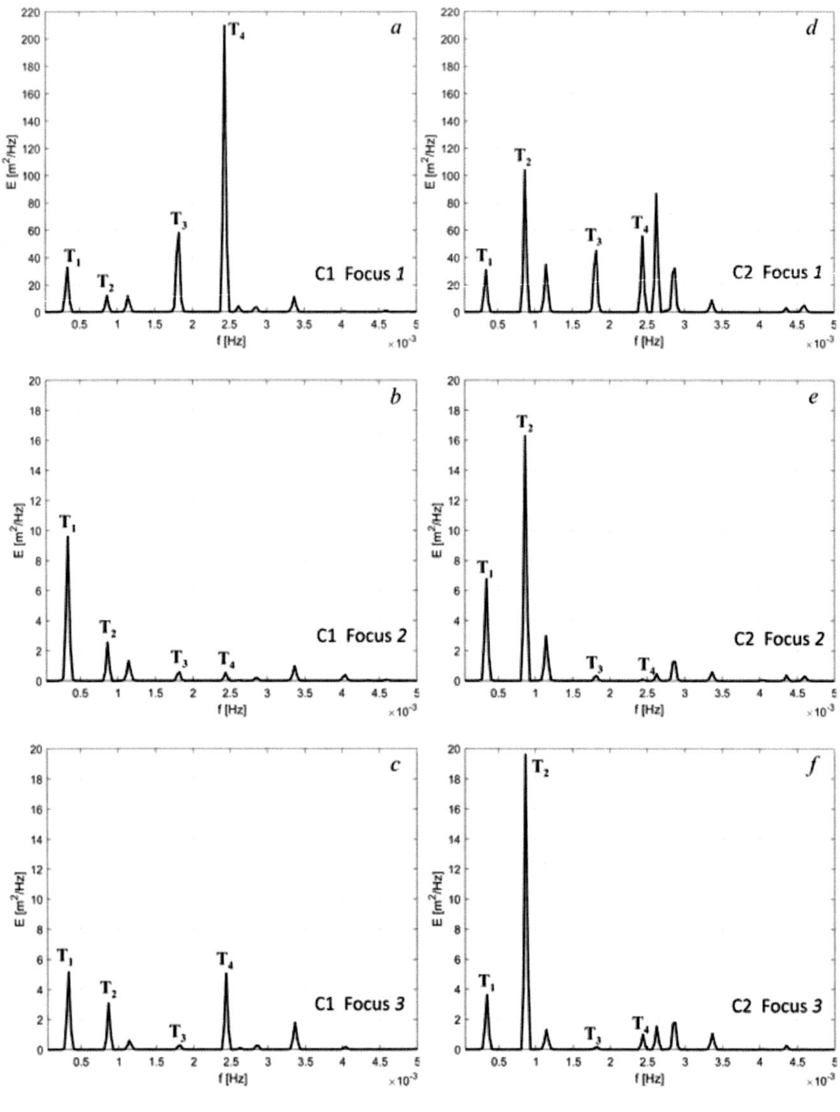

Fig. 8 Spectra of sea level fluctuations in bay tops for three tsunami foci

4 Conclusions

The results of numerical simulation of the penetration of tsunami waves into a *T*-shaped bay, which has the characteristic dimensions and configuration of the Sevastopol Bay, are presented. At the first stage, using the Black Sea tsunami model, the evolution of tsunami waves from three hypothetical foci caused by an earthquake of magnitude $M = 7$ was studied. Time dependences of sea level fluctuations at the entrance to the Sevastopol Bay were obtained. At the second stage, these mareograms were used as boundary conditions on the liquid boundary of the computational domain, for which, using the *SWASH* model, numerical simulation of tsunami wave propagation in the model bay was performed.

It has been established that inside the bay, tsunami heights increase by a factor of 2–2.5 compared to the heights on the coast of the bay. At the top of the bay, the amplitude of sea level fluctuations can reach 0.8 m. The most intense fluctuations are observed in the first three hours of the tsunami.

The seiche oscillations caused by the tsunami have the same periods for different tsunami foci. At the same time, the amplitude-phase characteristics of the tsunami have a significant effect on the redistribution of wave energy between the modes of seiche oscillations.

Acknowledgements The investigation was carried out within the framework of the state assignment on theme No. 0555-2021-0005. Mathematical modeling was performed on the MHI computing cluster (www.hpc-mhi.org).

References

1. Nikonov, A.A., Gusiakov, V.K. and Fleifel, L.D., 2018. Assessment of the Tsunami Hazard on the Russian Coast Based on a New Catalogue of Tsunamis in the Black Sea and the Sea of Azov. *Russian Geology and Geophysics*, 59(2), pp. 193–205. https://doi.org/10.1016/j.rgg.2018.01.016
2. Dotsenko, S.F., 2012. Numerical Modeling of Tsunamis in the Black, Azov and Caspian Seas as a Necessary Element of Regional Tsunami Early Warning Systems. *Ekologicheskaya Bezopasnost' Pribrezhnykh i Shel'fovykh Zon i Kompleksnoe Ispol'zovanie Resursov Shel'fa* [Ecological Safety of Coastal and Shelf Zones and Comprehensive Use of Shelf Resources]. Sevastopol: ECOSI-Gidrofizika, Vol. 2, Iss. 26, pp. 287-300 (in Russian).
3. Dotsenko, S.F. and Ingerov, A.V., 2013. [Characteristics of Tsunami Waves of Seismic Origin in the Black Sea Basin Based on the Results of Numerical Modeling]. ***Morskoy Gidrofizicheskiy*** *Zhurnal*, (3), pp. 25-34 (in Russian).
4. Pelinovsky, E.N. and Zaitsev, A.I., 2011. The Estimation and Mapping of Tsunami Dangerous on the Ukrainian Black Sea Coast. *Trudy Nizhegorodskogo Gosudarstvennogo Tekhnicheskogo Universiteta Im. R.E. Alekseeva*, 3(90), pp. 44–50 (in Russian).
5. Mazova, R.Kh., Kiselman, B.A., Osipenko, N.N. and Kolchina, E.A., 2013. Analysis of Spectral Characteristics of Black Sea Tsunami. *Trudy Nizhegorodskogo Gosudarstvennogo Tekhnicheskogo Universiteta Im. R.E. Alekseeva*, 2(99), pp. 52–66 (in Russian).
6. Zaitsev, A.I. and Pelinovsky, E.N., 2011. Forecasting of Tsunami Wave Heights at the Russian Coast of the Black Sea. *Oceanology*, 51, pp. 907-915. https://doi.org/10.1134/S0001437011050225
7. Yalciner, A., Pelinovsky, E., Talipova, T., Kurkin, A., Kozelkov, A. and Zaitsev, A., 2004. Tsunamis in the Black Sea: Comparison of the Historical, Instrumental and Numerical Data. *Journal of Geophysical Research. Oceans*, 109(C12), C12023. https://doi.org/10.1029/2003JC002113
8. Nikonov, A.A., 1997. Tsunami frequency on the shores of the Black and Azov Seas. *Izvestiya RAN. Fizika Zemli*, Vol. 33, pp. 72-87 (in Russian)
9. Dotsenko, S.F., 1995. Tsunami in the Black Sea. *Izvestiya RAN. Fizika atmosfery i okeana*, Vol. 30, pp. 513-519 (in Russian)

10. Bezushko, D.I., Mironenko, I.N., Murashko, A.V., 2015. Tsunami of the Black Sea coast of Ukraine. *Vestnik Odesskogo natsionalnogo morskogo universiteta,* № 1 (43), pp. 82–90 (in Russian)
11. Stokozov, N.A., 2010. Morphometric characteristics of the Sevastopol and Balaklava bays. *Ekologicheskaya Bezopasnost' Pribrezhnykh i Shel'fovykh Zon I kompleksnoe ispol'zovanie resursov shel'fa,* iss. 23. pp.198-208 (in Russian)
12. Alekseev D.V., Ivanov V.A., Ivancha E.V., Cherkesov L.V., Fomin V.V., 2013. Estimation of the effect of protective piers on wind wave parameters in the Sevastopol bay. *Russian Meteorology and Hydrology*, Vol. 38, iss. 4, pp. 248–255.
13. Rabinovich, A.B., 1993. Long gravitational waves in the ocean: capture, resonance, radiation. *SPb. Gidrometeoizdat*, 325 p. (in Russian)
14. Manilyuk Y.V., Lazorenko D.I., Fomin V.V., 2021. Seiche oscillations in the system of Sevastopol bays. *Water Resources.* Vol, 48, iss. 5, pp. 726–736.
15. Alekseev D.V., Manilyuk, Yu.V., Sannikov, V.F., 2017. Seiche currents in a basin with an open entrance. *Prikladnye zadachi matematiki. Materialy XXV mezhdunarodnoy nauchno-tehnicheskoy konferentcii*, pp. 109–115 (in Russian).
16. Dotsenko, S.F. and Ivanov, V.A., 2010. Natural Catastrophes of the Azov-Black Sea Region. ***Sevastopol: ECOSI-Gidrofizika***, 174 p. (in Russian).
17. Goryachkin, Yu.N., Ivanov, V.A., Repetin, L.N. and Khmara, T.V., 2002. Seiches in Sevastopol Bay. *Trudy Ukrainskogo Nauchno-Issledovatelskogo Gidrometeorologicheskogo Instituta*, (250), pp. 342-353 (in Russian).
18. SWASH User Manual. SWASH version 7.01 / The SWASH team. Delft : Delft University of Technology, 2012. 144 p. URL: https://swash.sourceforge.io/online_doc/swashuse/swashuse.html (date of access: 01.09.2022).
19. Fomin, V.V., Belokon, A.Yu, Kharitonova, L.V. and Alekseev, D.V., 2022. Numerical Simulation of Tsunami Wave Propagation to the Balaklava Bay. *Physical Oceanography*, 29(4), pp. 379-394. https://doi.org/10.22449/1573-160X-2022-4-379-394
20. Bazykina, A.Yu., Mikhailichenko, S.Yu. and Fomin, V.V., 2018. Numerical Simulation of Tsunami in the Black Sea Caused by the Earthquake on September 12, 1927. *Physical Oceanography*, 25(4), pp. 295-304. https://doi.org/10.22449/1573-160X-2018-4-295-304
21. Ulomov, V.I., Polyakova, T.P., Shumilina, L.S., Chernysheva, G.V., Medvedeva, N.S., Savarenskaya, O.E. and Stepanova, M.B., 1993. [Experience in Earthquake Foci Mapping]. *Seismichnost i Seismicheskoe Rayonirovanie Severnoy Evrazii.* Moscow: Institut Fiziki Zemli RAN, (1), pp. 99–108 (in Russian).
22. Winde, H.P., 2012. *Wave Height from Pressure Measurements.* Delft University of Technology, 49 p.
23. Fomin, V.V., Lazorenko, D.I. and Ivancha, E.V., 2017. [Numerical Modeling of Seiches in Balaklava Bay]. *Ekologicheskaya Bezopasnost' Pribrezhnykh i Shel'fovykh Zon Morya,* (3), pp. 32-39 (in Russian).

The Features of Simulation of Radar Altimeter Return Waveform Using Hermite Polynomials

V. M. Burdygov

Abstract Currently, the main approximations of sea surface elevation in remote sensing problems are the Gram–Charlier and Edgeworth distributions. These distributions are expansions of the probability density function into an infinite series over Hermite polynomials, whose coefficients are determined by known cumulants. A limited number of known cumulants leads to distortions at the tails of the distribution. A change in the shape of the reflected pulse of an radio altimeter mounted on a spacecraft is simulated depending on the boundaries of the truncation of the distributions of elevation of the sea surface. The change in the shape of the reflected pulse of the radio altimeter installed on the spacecraft is modeled depending on the boundaries of the truncation of the distribution of the sea surface elevation. The ranges of skewness and kurtosis are chosen on the basis of wave measurement data in different areas of the World Ocean, for skewness the range is − 0.2 to 0.3, for excess kurtosis the range is − 0.4 to 1.1. It is shown that the distribution truncation effect manifests itself in the case when the truncation boundary satisfies the condition $\xi_b < 3$ (here, ξ is normalized to the rms value of the sea surface elevation). The simulation results are compared with sea level calculations in the case when the probability density function of sea surface elevations is described by a Gaussian mixture. A significant dependence of the calculated sea level on the choice of the distribution model is shown. Differences in sea level obtained using the distribution in the form of a Gaussian mixture and the Edgeworth distribution with the same values of the first four cumulants may exceed 20%.

Keywords Space altimeter · Sea surface level · Sea wave · Gaussian mixture · Edgeworth distribution

---

V. M. Burdygov (✉)
Marine Hydrophysical Institute RAS, Sevastopol, Russian Federation
e-mail: burdyugov@mail.ru

V. I. Karev (ed.), *Physical and Mathematical Modeling of Earth and Environment Processes—2022*, Springer Proceedings in Earth and Environmental Sciences,
https://doi.org/10.1007/978-3-031-25962-3_7

1 Introduction

In modern radio oceanography, the problem of increasing the accuracy of determining sea level from spacecraft is one of the most urgent, as evidenced by a large number of publications devoted to its solution (see [1–3] and the bibliography to them). In the present paper, the error caused by the deviation of the distribution of sea surface elevation from the Gauss distribution is considered [4, 5]. The error arises as a result of the fact that the median of the distribution of sea surface elevation does not coincide with the average surface level [6]. The dependence of the shape of the reflected pulse of the radio altimeter on the distribution of elevation of the sea surface, in principle, allows us to solve the inverse problem, to restore the senior statistical moments characterizing the nonlinearity of the wave field [7, 8].

The accuracy of calculating the altimeter return waveform is determined by how correctly the characteristics of the reflecting surface are set, primarily the probability density function of its elevation. Currently, to approximate the probability density function of sea surface elevation, truncated Gram–Charlier or Edgeworth series are used. These series are decompositions by Hermite polynomials, with coefficients calculated from statistical moments or cumulants of surface elevation [9]. The use of Hermite polynomials in truncated series imposes restrictions associated with the appearance of negative values of the probability density function, as well as with the appearance of several local maxima [10–12].

The shape of a short radio pulse reflected from the sea surface can be described by convolution of two functions [13]

$$V(t) = F_1(t) * F_2(t), \tag{1}$$

where t is time, the function $F_1(t)$ is determined by the technical characteristics of the radio altimeter and the spacecraft (the height of the orbit); the function $F_2(t)$ is determined by the distribution of the height of the mirror reflection points; the symbol $*$ means convolution. In turn, the function can also be represented as a convolution of two functions, one of which describes the shape of the probing pulse, the other describes the shape of the pulse reflected from a flat surface [14, 15]. Expression (1) describing the shape of the reflected pulse is integral, so the question arises to which transformation of the leading edge of the reflected pulse of the radio altimeter leads to distortion of the probability density function of surface elevation caused by a low order of truncation of the approximating series.

2 Approximation of the Probability Density Function by Hermite Polynomials

In general, the Gram–Charlier and Edgeworth distributions represent the expansion into an infinite series of the desired function by the derivatives of the function [9]

$$PN(x) = \frac{1}{\sqrt{2\pi}} \exp\left(-\frac{1}{2}x^2\right), \tag{2}$$

In practice, a truncated series is used. If the first four statistical moments are known, after the introduction of normalization for the elevation of the surface, which leads it to a unit variance, the approximations of the Gram–Charlier and Edgeworth distributions have the form, respectively [10]

$$P_{GC}(\xi) = \frac{\exp\left(-\frac{\xi^2}{2}\right)}{\sqrt{2\pi}}\left[1 + \frac{\lambda_3}{6}H_3(\xi) + \frac{\lambda_4}{24}H_4(\xi)\right], \tag{3}$$

$$P_E(\xi) = \frac{\exp\left(-\frac{\xi^2}{2}\right)}{\sqrt{2\pi}}\left[1 + \frac{\lambda_3}{6}H_3(\xi) + \frac{\lambda_4}{24}H_4(\xi) + \frac{\lambda_3^2}{72}H_6(\xi)\right], \tag{4}$$

where λ_3 and λ_4 are cumulants of the third and fourth orders,, $H_3(\xi) = \xi^3 - 3\xi$, $H_4(\xi) = \xi^4 - 6\xi^2 + 3$ and $H_6(\xi) = \xi^6 - 15\xi^4 + 45\xi^2 - 15$ are orthogonal Hermite polynomials of the third, fourth and sixth orders. The normalized random variable is described as

$$\xi = \eta \Big/ \left\langle \eta^2 \right\rangle^{0.5}, \tag{5}$$

where η is the elevation of the sea surface; $\left\langle \eta^2 \right\rangle$ is variance, $\langle\ \rangle$ means averaging.

The expressions in square brackets in (3) and (4) (denote them as R_{GC} and R_E) are respectively truncated Gram–Charlier and Edgeworth series. The multiplier $(2\pi)^{-0.5}\exp\left(-\frac{\xi^2}{2}\right) = P_G(\xi)$ describes the Gaussian distribution. Cumulant $\lambda_3 = \left\langle \xi^3 \right\rangle$ is the skewness and $\lambda_4 = \left\langle \xi^4 \right\rangle$ is excess kurtosis. The distortions of the probability density function caused by the low order of truncation of the approximating series are stronger the greater the deviations of the cumulants from zero values corresponding to the Gaussian distribution.

The conditions of wave formation in different regions may differ significantly, respectively, the parameters characterizing the nonlinearity of the wave field differ. This is especially evident in the coastal zone, where orography has a significant influence on the wind field. Table 1 shows the ranges of changes in cumulants λ_3 and λ_4, determined on the basis of measurements carried out by different researchers in different areas of the World Ocean. Several types of equipment were used for wave measurements: a wave staff, a down-looking radar and a directional wave rider. The variation ranges of λ_3 and λ_4 are determined by the graphic images given in the works indicated in the right table column.

In further analysis, we will consider the ranges

$$-0.2 < \lambda_3 < 0.4,\ -0.4 < \lambda_4 < 1.1. \tag{6}$$

Table 1 Variation ranges of cumulants λ_3 and λ_4

Range λ_3	Range λ_4	Area	Source
– 0.05 to 0.4	– 0.4 to 0.6	North Sea	[18]
– 0.05 to 0.3	– 0.4 to 0.4	Black Sea	[19]
– 0.2 to 0.4		Black Sea	[20]
– 0.05 to 0.25	– 0.3 to 0.5	North Sea	[21]
– 0.1 to 0.4	0.1 to 1.1	North Sea	
– 0.05 to 0.3	– 0.2 to 1.0	Gulf of Mexico	
– 0.2 to 0.3	– 0.3 to 0.6	Sines in Portugal	

The probability density functions $P_{GC}(\xi)$ and $P_E(\xi)$ and, constructed for the limit values of the ranges (6), are shown in Fig. 1.

Since the correlation between cumulants λ_3 and λ_4 is low, four combinations of them are considered here, which are given in Table 2. As can be seen from Fig. 1, with some combinations of cumulants in the areas of function and can take negative values. As can be seen from Fig. 1, for some combinations of cumulants at $|\xi| > 2.5$, the functions $P_{GC}(\xi)$ and $P_E(\xi)$ can take negative values. Previously, it was shown that the appearance of negative values of the function $F_2(t)$ leads to a non-physical effect, which manifests itself in the form of negative values in the calculated form of the reflected radio pulse [5].

The function is obtained by a standard procedure for change of variable probability density function of surface elevation. The transition from $P(\eta)$ to $F_2(t)$ is carried out using a relation linking spatial and temporal coordinates $\eta = \left(c/2\right)t$, where c is the speed of light.

Distributions (3) and (4) are applicable only in the range of variation of a dimensionless random variable $|\xi| < b$, where $b \approx 2.5$. The function constructed on the basis of these distributions is valid for

$$|t| < b(2/c)\langle\eta^2\rangle^{0.5}. \tag{7}$$

3 Numerical Simulation of the Reflected Radio Pulse Shape

Let's analyze how changes in parameter b affect the calculated shape of the short radio pulse reflected from the sea surface. In calculations, we will use the characteristics of the radio altimeter installed on the SEASAT-1 spacecraft. As a parameter characterizing the wave energy, we use a significant wave height H_S, which we assume to be equal to 4 m. The dispersion of sea surface elevations is related to the parameter H_S equation $H_S = 4\langle\eta^2\rangle^{0.5}$. For a given value of the parameter, taking into account that the speed of light is 0.3 m/ns, condition (7) takes the form $|t| < 6.67b$ where t has dimension ns.

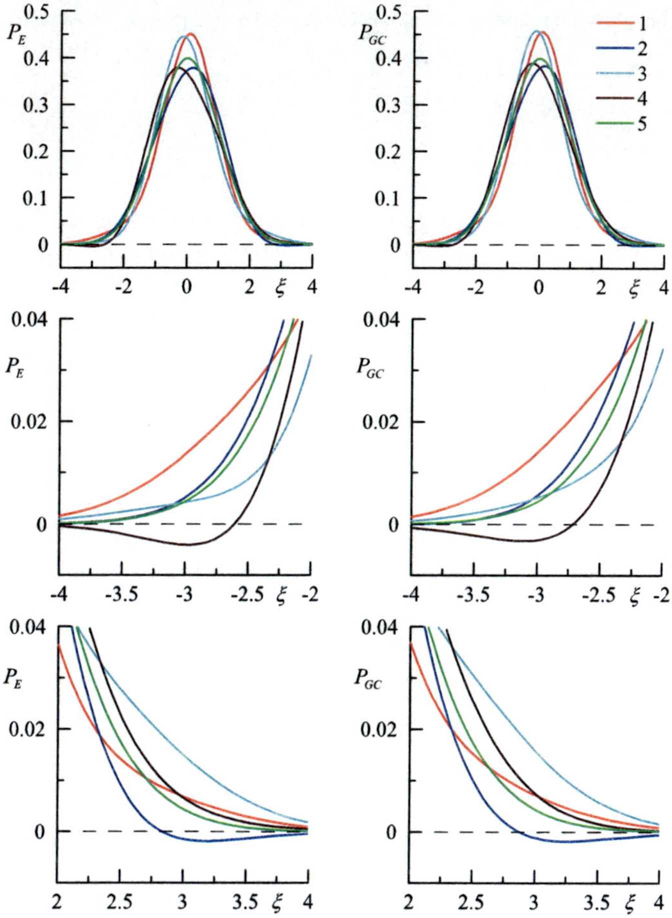

Fig. 1 Probability density functions $P_E(\xi)$ and $P_{GC}(\xi)$. Curves 1–4 correspond to the combinations of cumulants shown in Table 2, curve 5 is the Gaussian distribution

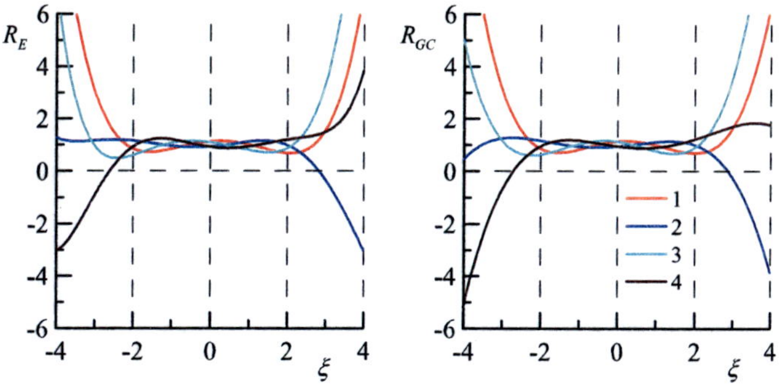

Fig. 2 Truncated rows $R_E(\xi)$ and $R_{GC}(\xi)$. Curves 1–4 correspond to the combinations of cumulants λ_3 and λ_4 shown in Table 2

Table 2 A table of combinations λ_3 and λ_4 for calculating the functions presented in Figs. 1 and 2

Combinations	1	2	3	4
λ_3	– 0.2	– 0.2	0.4	0.4
λ_4	1.1	– 0.4	1.1	– 0.4

The changes in the leading edge of the reflected pulse of the radio altimeter, calculated in the absence of restriction (7), are shown in Fig. 3. Changing the sign of skewness leads to the fact that the error in determining the level of the sea surface also changes the sign.

The error in determining the sea surface level caused by a deviation from the Gaussian distribution is defined as

$$\Delta L = (c/2)\Delta t, \tag{8}$$

where Δt is the displacement of the leading edge of the reflected pulse in its central part relative to the front of the pulse reflected from the Gaussian surface. We introduce truncated distributions

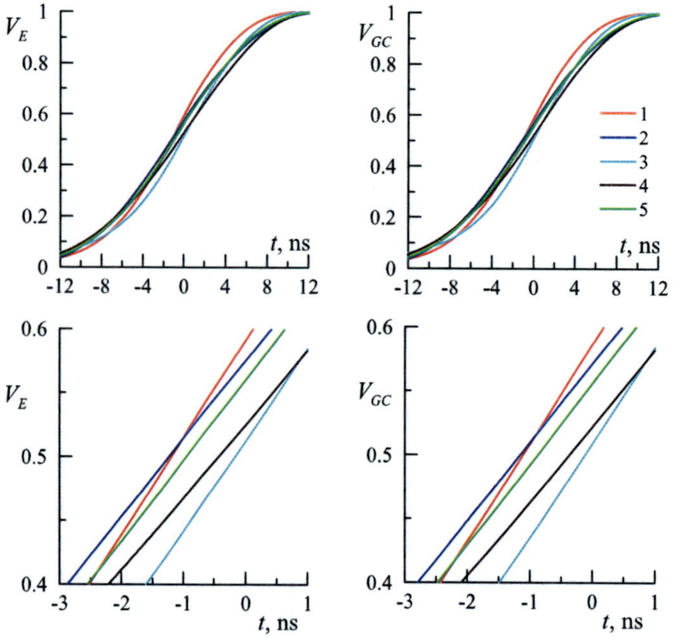

Fig. 3 The leading edge of the reflected radio pulse, constructed according to models (3) and (4). Curves 1–4 correspond to combinations of cumulants λ_3 and λ_4 shown in Table 2, curve 5 corresponds to the Gaussian distribution

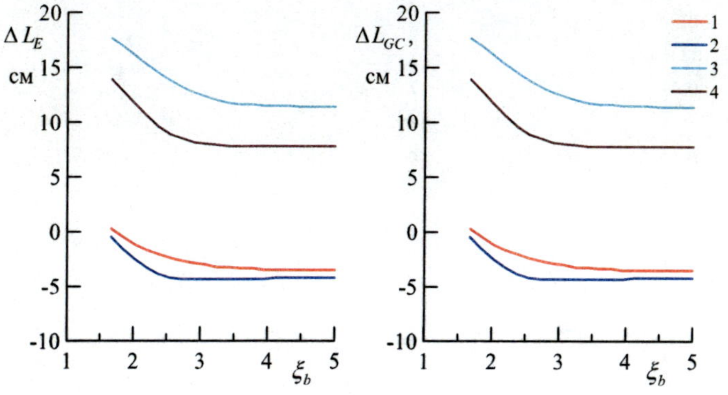

Fig. 4 Dependences of the error ΔL in determining the level of the sea surface on the boundary of the truncation of the distribution of elevations of the sea surface b. Curves 1–4 correspond to the combinations of cumulants λ_3 and λ_4 given in Table 2

$$\tilde{P}_E(\xi, b) = \begin{cases} P_E(\xi) \; if \; \xi \le \xi_b \\ 0 \qquad if \; \xi > \xi_b \end{cases}, \tag{9}$$

$$\tilde{P}_{GC}(\xi, b) = \begin{cases} P_{GC}(\xi) \; if \; \xi \le \xi_b \\ 0 \qquad if \; \xi > \xi_b \end{cases}, \tag{10}$$

and calculate for them the dependence on the truncation boundary ξ_b. The results are shown in Fig. 4.

A comparison of the errors and calculated from two distributions of sea surface elevations shows that at extreme values of asymmetry and kurtosis of surface elevations, the discrepancies in the error values calculated from truncated and non-truncated distributions can be about 30%.

4 Model of Sea Surface Elevation in the Form of a Gaussian Mixture

Taking into account the limitations of the Gram–Charlier and Edgeworth distributions noted above, it is necessary to compare the estimates obtained for them with the results of calculations in the framework of models free from these restrictions. We define the probability density function of surface elevation in the form of a two-component Gaussian mixture $P_S(\eta)$. For a Gaussian mixture, the condition $P_S(\eta) > 0$ is met for all values η. A two-component Gaussian mixture is the sum of two Gaussians with unequal weights and has the form [24]

$$P_S(\xi) = \frac{\alpha_1}{\sqrt{2\pi}\sigma_1} \exp\left(-\frac{(\xi - m_1)^2}{2\sigma_1^2}\right) + \frac{\alpha_2}{\sqrt{2\pi}\sigma_2} \exp\left(-\frac{(\xi - m_2)^2}{2\sigma_2^2}\right), \tag{11}$$

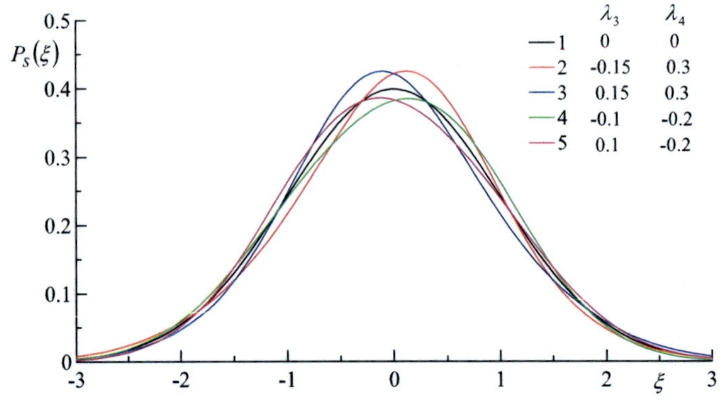

Fig. 5 Probability density functions $P_S(\xi)$ of sea surface elevation in the form of a Gaussian mixture

where α_i is the weight of the i–th component, m_i is the mean, is σ_i^2 the variance. The weight coefficients satisfy the condition $\alpha_1 + \alpha_2 = 1$.

The calculation of the function $P_S(\xi)$ parameters, as in modeling using Gram–Charlier or Edgeworth series, is carried out using known cumulants. The procedure for calculating the parameters of the function $P_S(\xi)$ is described in [25]. A two-component Gaussian mixture poorly approximates the probability density function with significant deviations from the Gaussian distribution [26]. Therefore, here we will limit ourselves to relatively small values λ_3 and λ_4.

Examples of a two-component Gaussian mixture calculated for several values of cumulants and are shown in Fig. 5. The selected values of cumulants lie within the ranges in which they change according to the wave measurements carried out on the Black Sea.

The results of calculations of deviations ΔL of the calculated sea level due to non-zero values and at different levels of truncation of the distributions of sea surface elevation are presented in Fig. 6. When $\xi_b > 2.5$–3 changing the truncation boundary, it practically does not affect the calculated value of sea level. Comparing the dependences obtained using approximations of sea surface elevation $P_E(\xi)$ and $P_S(\xi)$, it should be noted their significant difference in the range.

The condition $\xi_b > 2.5$–3 roughly corresponds to the condition $H > 1.5H_S$. This means that abnormal waves whose height satisfies the condition $H > 2H_S$ [27] do not affect the accuracy of determining the level. Perhaps this is due to the low probability of the appearance of abnormal waves.

It should also be noted a significant discrepancy in estimates ΔL_S and ΔL_E with the same values of statistical moments. This indicates a strong dependence of the calculated level on the choice of a model for the distribution of sea surface elevation.

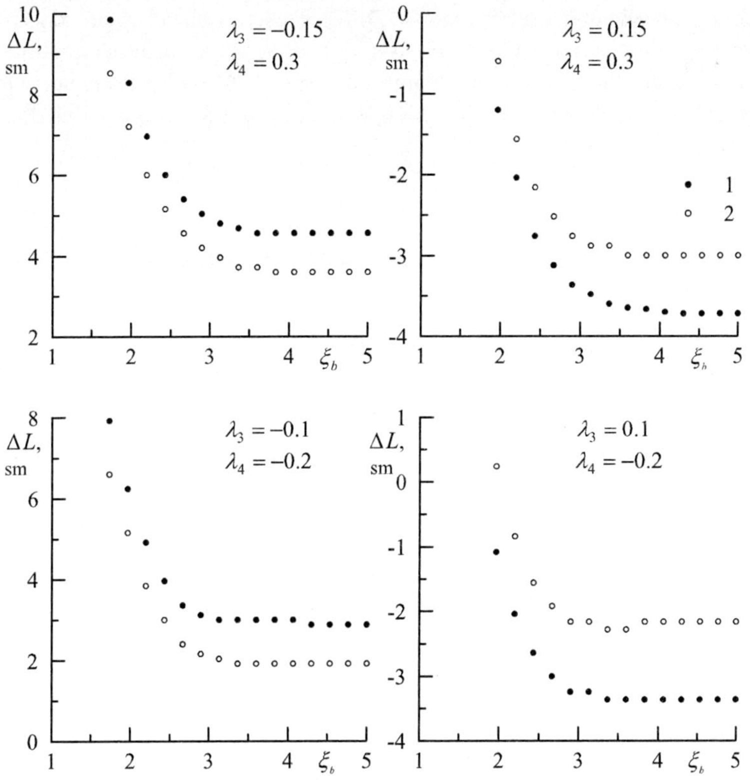

Fig. 6 Dependences of the deviation of the calculated sea level ΔL from the boundaries of the truncation ξ_b of approximations $P_S(\xi)$ and $P_E(\xi)$, 1—Gaussian mixture; 2—Edgeworth distribution

5 Conclusion

The analysis of the error of remote determination of sea level caused by incorrect description of statistical characteristics of the sea surface is carried out. For the analysis, the ranges of changes in skewness and excess kurtosis determined for several water areas of the World Ocean were used.

The currently existing statistical models of sea surface distributions are based on approximations that use Gram–Charlier or Edgeworth series, which are decompositions by Hermite polynomials. The coefficients of the Gram–Charlier and Edgeworth series are calculated from known cumulants. A low order of series truncation leads to distortions at the tails of the distribution.

For Gram–Charlier or Edgeworth series it is shown that the distribution truncation effect manifests itself in the case when the truncation boundary ξ_b satisfies the condition $\xi_b < 3$. The simulation results are compared with sea level calculations in the case when the probability density function of sea surface elevations is described by a Gaussian mixture. A significant dependence of the calculated sea level on the choice of the distribution model is shown. Differences in sea level obtained using the distribution in the form of a Gaussian mixture and the Edgeworth distribution with the same values of the first four cumulants may exceed 20%. The discrepancy in sea level estimates calculated using the Gram–Charlier and Edgeworth series does not exceed 5%.

Acknowledgements This work was carried out as part of a state assignment on the topic No. 0555-2021-0004 "Fundamental studies of oceanological processes which determine the state and evolution of the marine environment influenced by natural and anthropogenic factors, based on observation and modeling methods".

References

1. Longuet-Higgins M.S. (1963) The effect of non-linearities on statistical distribution in the theory of sea waves // J. Fluid Mech. 17(3): 459–480.
2. Jha A.K., Winterstein S.R. (2000) Nonlinear random ocean waves: prediction and comparison with data // Proceedings of the 19th International Offshore Mechanics and Arctic Engineering Symposium. ASME, Paper No. OMAE 00–6125.
3. Guedes Soares C., Cherneva Z., Antão E.M. (2004) Steepness and asymmetry of the largest waves in storm sea states // Ocean Engineering. 31: 1147–1167.
4. Zapevalov A.S., Garmashov A.V. (2022) The appearance of negative values of the skewness of sea-surface waves // Izvestiya, Atmospheric and Oceanic Physics, 58: 263-269. https://doi.org/10.1134/s0001433822030136
5. Glejin J., Sanil Kumar V., Nair T.B., Singh J., Nherakkol A. (2014) Freak waves off Ratnagiri, west coast of India Indian // Journal of Geo-Marine Sciences. 43(7): 1339–1342
6. Didenkulova, I. and Anderson C. (2010) Freak waves of different types in the coastal zone of the Baltic Sea. // Natural Hazards and Earth System Science. 10(9): 2021–2029.
7. Zapevalov A.S., Garmashov A.V. (2021) Skewness and kurtosis of the surface wave in the coastal zone of the Black Sea // Physical Oceanography, 28(4): 414–425.
8. Toloknov Yu.N., Korovushkin A.I. (2010) System of collecting hydrometeorological information // Environmental control systems. (10): 50–53. in Russian.
9. Solov'ev Yu.P., Ivanov V.A. (2007) Preliminary results of measurements of atmospheric turbulence over the sea // Physical Oceanography. Iss. 3: 154–172.
10. Kharif C., Pelinovsky E., Slunyaev A. (2009) Rogue waves in the ocean. Springer-Verlag, Berlin, Heidelberg, 216 p.
11. Fedele, F., Brennan, J., Ponce de León, S., Dudley, J., Dias F. (2016) Real world ocean rogue waves explained without the modulational instability. Sci Rep 6, 27715. https://doi.org/10.1038/srep27715
12. Donelan, M.A., Hamilton, J. and Hui, W.H. 1985. Directional spectra of wind-generated waves. Philos. Trans. Roy. Soc., A315. pp. 509–562.
13. Young, I.R. and Donelan, M.A. (2018) On the determination of global ocean wind and wave climate from satellite observations. Remote Sensing of Environment, 215: 228–241. https://doi.org/10.1016/j.rse.2018.06.006
14. Luxmoore J.F., Ilic S., Mori N. (2019) On kurtosis and extreme waves in crossing directional seas: a laboratory experiment // J. Fluid Mech. 876: 792-817. https://doi.org/10.1017/jfm.2019.575
15. Babanin A.V., Polnikov V.G. (1994) On non-Gaussian wind waves // Physical Oceanography. (3): 79–82. https://doi.org/10.1007/BF02197522

Abnormal Statistics of Sea Waves

A. S. Zapevalov

Abstract The statistics of waves in situations that are relatively rarely observed at sea are investigated. The probability and conditions of the occurrence of situations in which the ratio of the maximum crest and maximum trough during the measurement period (ratio *R*) is less than one are analyzed. The condition $R < 1$ means that the trough is larger than the crest. The data of direct wave measurements obtained in the coastal zone of the Black Sea (depth ~ 30 m) were used. It is shown that the average probability of occurrence of values for an ensemble of situations is 13%. It is shown that as the wave steepness ε increases, the conditional probability $P_\varepsilon = P(R < 1|\varepsilon)$ also increases. The dependence of the conditional probability $P_\zeta = P(R < 1|\zeta)$ on the inverse wave age ζ is observed only for developing waves at $\zeta > 1.5$, at later stages of the wave field development it is absent. There is also a dependence of the probability of occurrence of values on the cumulants of sea surface elevations of the third λ_3 and fourth λ_4 orders (skewness and excess kurtosis). When λ_3 changing from -0.1 to 0.3, the conditional probability $P_{\lambda 3} = P(R < 1|\lambda_3)$ increases by about five times, when λ_4 changing from -0.4 to 0.4, the conditional probability $P_{\lambda 4} = P(R < 1|\lambda_4)$ increases by about one and a half times.

Keywords Remote sensing · Altimetry · Sea surface · Brown model · Distribution of surface elevations

1 Introduction

The problem of statistical description of the sea surface is one of the most important tasks of oceanography. One of the components of this task is the description of anomalous wave statistics, which are relatively rarely observed at sea. In the present work, we restrict ourselves to the analysis of the crest/trough ratio.

The classical nonlinear sea wave has a shape close to a trochoid, a sharp crest and a flat trough. The crest is larger than the trough. This representation of sea

A. S. Zapevalov (✉)
Marine Hydrophysical Institute RAS, Sevastopol, Russian Federation
e-mail: sevzepter@mail.ru

V. I. Karev (ed.), *Physical and Mathematical Modeling of Earth and Environment Processes—2022*, Springer Proceedings in Earth and Environmental Sciences,
https://doi.org/10.1007/978-3-031-25962-3_8

waves suggests that the skewness of sea surface elevations λ_3 always has a positive sign [1]. However, the range of skewness variation following from classical theory is not always consistent with wave measurements. The data obtained in different areas of the World Ocean indicate that there are situations when the skewness of sea surface elevations is negative [2, 3]. According to measurements from a stationary oceanographic platform located on the Black Sea, the probability of occurrence of values is about 15% [4].

Deviations from the classical wave theory were also noted in the analysis of abnormally high waves, i.e. waves whose height is more than twice the significant height. It has been shown that abnormally high waves can have three forms: positive, negative and sign-variable [5, 6]. A wave shape is called positive if the crest is one and a half times larger than the trough; negative is the form in which the crest is one and a half times smaller than the trough. Accordingly, at intermediate ratios, the form is sign-variable. According to wave measurements, 63% of abnormal waves had a positive form, 19.5% negative shape and 17.5% sign-variable shape [6]. The measurements were carried out in the coastal zone of the Baltic Sea, where the depth was 2.7 m. The shallow depth at which these results are obtained leads to the fact that the waveform is influenced by their interaction with the bottom.

In this paper, the relationship between the maximum crest and maximum trough is analyzed for deep water. The data of direct wave measurements performed on the stationary oceanographic platform of the Marine Hydrophysical Institute of the Russian Academy of Sciences are used. The platform is located on the Black Sea near the southern coast of Crimea. The depth at the point where the platform is located is ~ 30 m [7]. For characteristic waves of the Black Sea, the indicated depth corresponds to the condition of deep water. The equipment, measurement conditions, as well as the features of the wind regime in the vicinity of the platform are described in [4, 8, 9].

2 Dependence of Crest/Trough Ratio on Wave Steepness and Inverse Wave Age

Let's analyze how the crest/trough ratio changes in different situations. The crest/trough ratio is denoted as

$$R = C_{\max} \big/ D_{\max}, \tag{1}$$

where $C_{\max}$ is the highest wave crest during the measurement session, $D_{\max}$ is the deepest trough during the measurement session. The procedure for calculating the parameter R is similar to the standard procedure for calculating the abnormality index for extremely high waves, in which the largest values of wave characteristics during a measurement session are determined [10]. In further analysis, we will consider only situations in which a significant wave height exceeds 0.5 m.

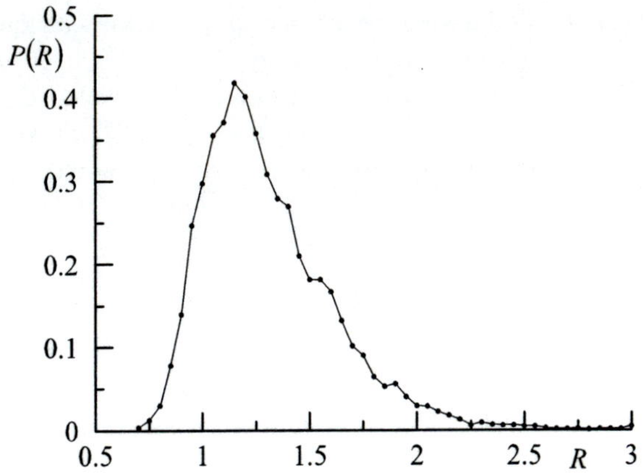

Fig. 1 Probability density function

The values of the parameter R calculated from the wave measurements are in the range from 0.68 to 3.17, with an average value of 1.3. On average, for an ensemble of situations, the probability that a condition $R < 1$ is met for a series of measurements lasting 20 min is 13%. The empirical probability density function $P(R)$ is shown in Fig. 1.

Comparing the results of the study of wave forms obtained in shallow water with the results obtained in deep water, the following can be noted. According to the measurements in shallow water [6], three types of extremely high wave forms were identified. One of the types, called negative shape, corresponds to the condition $R < 1/1.5 \approx 0.667$. When measured in deep water, the values of $R < 0.68$ were not observed, i.e. there are no extreme waves of negative shape in deep water.

A dimensionless measure of the nonlinearity of surface waves is their steepness, which can be defined as [11]

$$\varepsilon = \sigma k_0, \tag{2}$$

where σ is the standard deviation of surface wave elevations, k_0 is the wave number of the peak of the wave spectrum. For gravitational waves in deep water, the wave number k_0 is related to the angular frequency of the spectral peak ω_0 by the dispersion relation $\omega_0^2 = gk_0$, here g is the gravitational acceleration. Hence it follows that Eq. (2) can be rewritten in the form $\varepsilon = \sigma\omega_0^2/g$, which makes it possible to determine the wave steepness from measurements at one point. The wave steepness ε is a small parameter; under field conditions, its values generally do not exceed the level of 0.1 [7].

Another factor determining the nonlinearity of sea waves is their interaction with near surface wind. This interaction is different at different stages of the wave field development. The stage of development of the wave field is usually characterized by the inverse wave age [12, 13]

$$\zeta = U_{10}/C_0, \tag{3}$$

where U_{10} is the wind speed at a height of 10 m, C_0 is the phase velocity of wave with a wave number k_0. An earlier stage of development corresponds to a greater value ζ. It is generally assumed that waves are fully developed if $\zeta_b = 0.83$. We will assume that at $\zeta > \zeta_b$, the wave is windy, and at $\zeta < \zeta_b$, it is swell. Since the swell is almost always observed on the sea surface, this separation can be considered conditional.

We now examine the dependence of the ratio R on two dimensionless parameters of the wave steepness ε and the inverse wave age ζ. The dependences of the ratio R on the ε and ζ are shown in Fig. 2. The dependencies $R = R(\varepsilon)$ and $R = R(\zeta)$ are characterized by a large spread at fixed values of the parameters ε and ζ, respectively. The correlation coefficient between the parameters R and ε is 0.08, between the parameters R and ζ is 0.15. Both the lower and upper limits of the range of variation R practically do not depend on the wave steepness and inverse wave age.

Let's evaluate the conditions on which the probability of occurrence of values $R < 1$ depends. To do this, we will divide the areas in which the values of the wave steepness ε and inverse wave age ζ change into ranges and calculate the conditional probability for each of them. For the wave steepness, the width of the range in

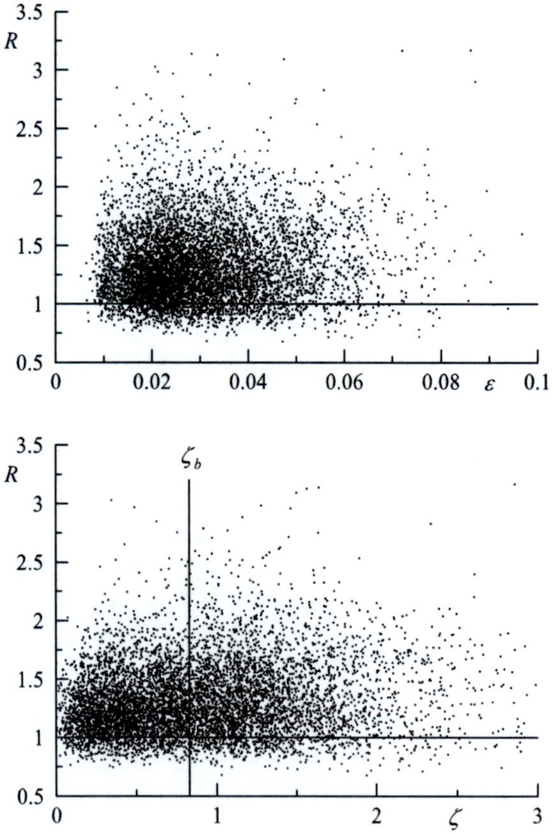

Fig. 2 Dependence of the ratio of the height of the crest and the depth of the trough R on the wave steepness ε and inverse wave age ζ

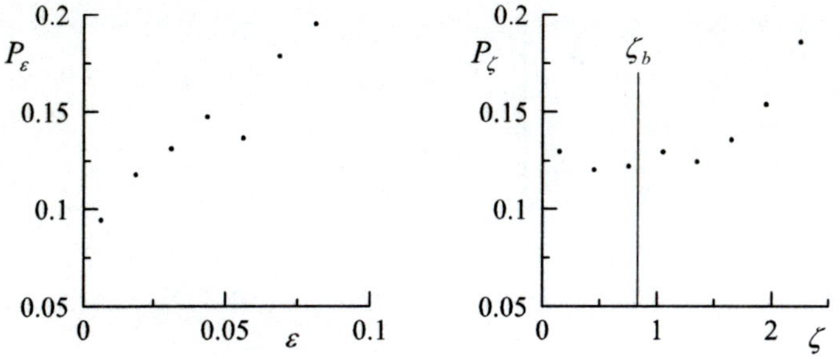

Fig. 3 Conditional probability of occurrence of values of parameter R less than one in different ranges of wave steepness ε and inverse wave age ζ

which the conditional probability $P_\varepsilon = P(R < 1|\varepsilon)$ is considered is 0.0125, for the inverse wave age, the width of the range in which the conditional probability $P_\zeta = P(R < 1|\zeta)$ is considered is 0.3. The results are shown in Fig. 3.

With increasing wave steepness, i.e. with increasing nonlinearity, there is an increase in the conditional probability of situations in which $R < 1$. Depending on the range ε, the conditional probability P_ε varies more than twice. The dependence of the conditional probability P_ζ on the inverse wave age takes place only for developing wind waves, at $\zeta > 1.5$.

Note that the correlation between the parameters ε and ζ depends on the wave regime. For wind waves, the correlation coefficient between and is 0.65; for swell, it is 0.15 [7].

3 Dependence of Crest/Trough Ratio on Skewness and Excess Kurtosis

Along with the wave steepness, the parameters describing the influence of nonlinearity on the geometric characteristics of waves are skewness and excess kurtosis, which can also be used as a measure of nonlinearity [14]. The weak nonlinearity of sea surface waves leads to the fact that the distributions of sea surface elevations deviate from the Gaussian distribution [15]. Skewness λ_3 and excess kurtosis λ_4 are defined as

$$\lambda_3 = \mu_3 \Big/ \mu_2^{3/2}, \tag{4}$$

$$\lambda_4 = \mu_4 / \mu_2^2 - 3, \tag{5}$$

where μ_n is the statistical moment of sea surface elevations of order n.

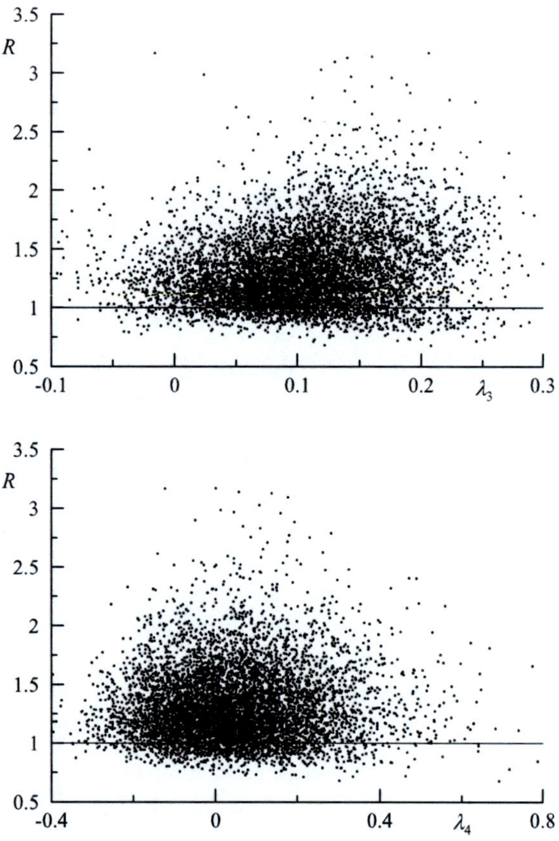

Fig. 4 Dependence of the ratio of the height of the creastand the depth of the trough from the coefficients of skewness λ_3 and excess kurtosis λ_4

Dependencies $R = R(\lambda_3)$ and $R = R(\lambda_4)$ are shown in Fig. 4. As in Fig. 2, there is a large variation in the values of R. The correlation coefficient between the parameters R and λ_3 is 0.18, between the parameters R and λ_4 is 0.4.

Let us consider how the conditional probability $P_{\lambda 3} = P(R < 1|\lambda_3)$ and $P_{\lambda 4} = P(R < 1|\lambda_4)$ of occurrence of values in different ranges of skewness and excess kurtosis changes. The width of the ranges for λ_3 is 0.1, for λ_4 it is 0.15.

As follows from Fig. 5, with large values of the parameters λ_3 and λ_4, the conditional probabilities $P_{\lambda 3}$ and $P_{\lambda 4}$ the occurrence of a situation when $R < 1$ are higher.

4 Conclusion

Earlier studies have shown deviations of the real field of sea waves from its classical representation, in which the wave has a shape close to a trochoid, a pointed crest and

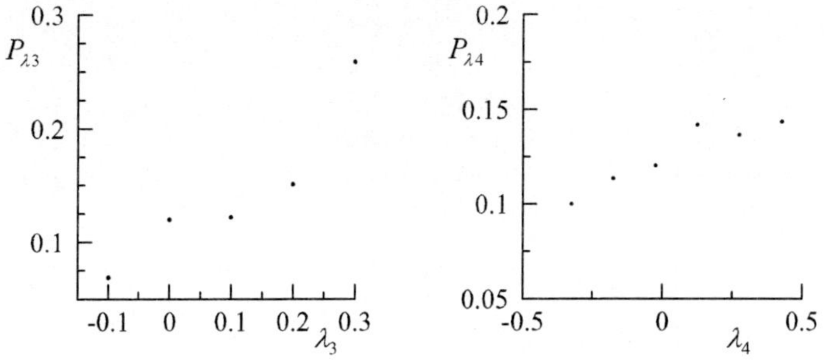

Fig. 5 Conditional probability of occurrence of values of the parameter R less than one in different ranges of variation of the skewness λ_3 and excess kurtosis λ_4

a flat trough. In different regions of the World Ocean, situations were recorded when the skewness of sea surface elevations had a negative sign. There were also situations in which the shape of abnormally high waves was characterized by an excess of the depth of the trough over the height of the crest.

These studies stimulated the analysis presented in this paper of the ratio of the maximum height of the creast/maximum depth of the trough. Data from direct wave measurements were used for the analysis. Wave parameters were determined to record measurements, lasting 20 min. Records were analyzed when the significant wave height exceeded 0.5 m.

According to measurements carried out in the coastal zone of the Black Sea, the probability of occurrence of values $R < 1$ is 13%. As the wave steepness ε increases, the conditional probability $P_\varepsilon = P(R < 1|\varepsilon)$ also increases. The dependence of the conditional probability $P_\zeta = P(R < 1|\zeta)$ on the inverse wave age ζ is observed only for developing waves at $\zeta > 1.5$, at later stages of the wave field development it is absent. There is also a dependence of the probability of occurrence of values on the cumulants of sea surface elevations of the third λ_3 and fourth λ_4 orders (skewness and excess kurtosis). When λ_3 changing from $-$ 0.1 to 0.3, the conditional probability $P_{\lambda 3} = P(R < 1|\lambda_3)$ increases by about five times, when λ_4 changing from $-$ 0.4 to 0.4, the conditional probability $P_{\lambda 4} = P(R < 1|\lambda_4)$ increases by about one and a half times.

Acknowledgements This work was carried out as part of a state assignment on the topic No. 0555-2021-0004 "Fundamental studies of oceanological processes which determine the state and evolution of the marine environment influenced by natural and anthropogenic factors, based on observation and modeling methods".

References

1. Longuet-Higgins M.S. (1963) The effect of non-linearities on statistical distribution in the theory of sea waves // J. Fluid Mech. 17(3): 459–480.
2. Jha A.K., Winterstein S.R. (2000) Nonlinear random ocean waves: prediction and comparison with data // Proceedings of the 19th International Offshore Mechanics and Arctic Engineering Symposium. ASME, Paper No. OMAE 00-6125.
3. Guedes Soares C., Cherneva Z., Antão E.M. (2004) Steepness and asymmetry of the largest waves in storm sea states // Ocean Engineering. 31: 1147–1167.
4. Zapevalov A.S., Garmashov A.V. (2022) The appearance of negative values of the skewness of sea-surface waves // Izvestiya, Atmospheric and Oceanic Physics, 58: 263–269. https://doi.org/10.1134/s0001433822030136
5. Glejin J., Sanil Kumar V., Nair T.B., Singh J., Nherakkol A. (2014) Freak waves off Ratnagiri, west coast of India Indian // Journal of Geo-Marine Sciences. 43(7): 1339–1342
6. Didenkulova, I. and Anderson C. (2010) Freak waves of different types in the coastal zone of the Baltic Sea. // Natural Hazards and Earth System Science. 10(9): 2021–2029.
7. Zapevalov A.S., Garmashov A.V. (2021) Skewness and kurtosis of the surface wave in the coastal zone of the Black Sea // Physical Oceanography, 28(4): 414–425.
8. Toloknov Yu.N., Korovushkin A.I. (2010) System of collecting hydrometeorological information // Environmental control systems. (10): 50–53. in Russian.
9. Solov'ev Yu.P., Ivanov V.A. (2007) Preliminary results of measurements of atmospheric turbulence over the sea // Physical Oceanography. Iss. 3: 154–172.
10. Kharif C., Pelinovsky E., Slunyaev A. (2009) Rogue waves in the ocean. Springer-Verlag, Berlin, Heidelberg, 216 p.
11. Fedele, F., Brennan, J., Ponce de León, S., Dudley, J., Dias F. (2016) Real world ocean rogue waves explained without the modulational instability. Sci Rep 6, 27715. https://doi.org/10.1038/srep27715
12. Donelan, M.A., Hamilton, J. and Hui, W.H. 1985. Directional spectra of wind-generated waves. Philos. Trans. Roy. Soc., A315. pp. 509–562.
13. Young, I.R. and Donelan, M.A. (2018) On the determination of global ocean wind and wave climate from satellite observations. Remote Sensing of Environment, 215: 228–241. https://doi.org/10.1016/j.rse.2018.06.006
14. Luxmoore J.F., Ilic S., Mori N. (2019) On kurtosis and extreme waves in crossing directional seas: a laboratory experiment // J. Fluid Mech. 876: 792–817. https://doi.org/10.1017/jfm.2019.575
15. Babanin A.V., Polnikov V.G. (1994) On non-Gaussian wind waves // Physical Oceanography. (3): 79–82. https://doi.org/10.1007/BF02197522

Emission of Nanoparticles During Loading Rock Salt

S. D. Viktorov and V. P. Malyukov

Abstract Experimental studies of particle emission during loading of rock salt have been carried out. The phenomenon of emission of nanoparticles in the process of loading a sample of rock salt has been discovered. The parameters of a significant emission of nanoparticles under uniaxial compression of a crystalline rock (rock salt) have been determined. The phenomenon of phased emission of nanoparticles during loading of rock salt has been discovered.

Keywords Rock salt · Physical modeling · Sample loading · Nanoparticle emission · Liesegang spherical clay formations · Halitization

Rock salt is a crystalline rock, consisting of crystals with an average size of 3–5 mm (a spar structure, which forms at great depths, has a larger grain size). When developing large-scale models of horizontal mine workings-tanks for storing hydrocarbons in rock salt from an adit in the Khoja-Mumyn salt mountain (Tajikistan) with the hydrodynamic effect of the solution (water supply through the well) [1], after the completion of work and opening the working, spherical clay formations of Liesegang, which, when exposed to a solution, were destroyed into particles, including nanoparticles [2]. Perhaps this is one of the first manifestations of the formation of nanoparticles during the hydrodynamic action of the solution on rock salt with inclusions.

At the Russian Research Center "Kurchatov Institute", the dependence of particle generation upon stretching of a metal rod was experimentally discovered [3]. Later, in the laboratory of IPKON RAS, studies were carried out on the developed installation, the results of which confirmed the formation of particles in the range of 0.3–5.0 μm from the surface of samples of various rocks under uniaxial compression (Fig. 1).

Such a phenomenon is observed in underground workings in the form of detachment of parts of the rock from the walls of the workings and from rocks with high

S. D. Viktorov
Institute for Problems of Comprehensive Development of the Subsoil of the RAS, Moscow, Russia

V. P. Malyukov (✉)
Peoples' Friendship University of Russia, Moscow, Russia
e-mail: v.malyukov@mail.ru

V. I. Karev (ed.), *Physical and Mathematical Modeling of Earth and Environment Processes—2022*, Springer Proceedings in Earth and Environmental Sciences,
https://doi.org/10.1007/978-3-031-25962-3_9

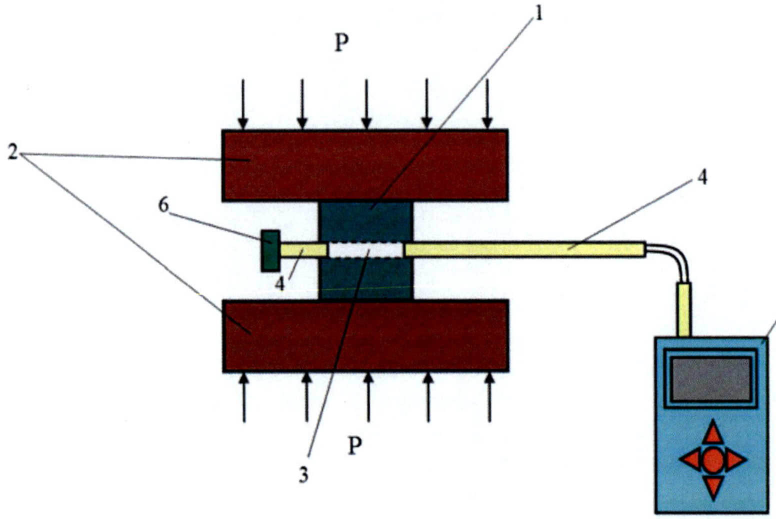

Fig. 1 Schematic diagram of a device for determining the parameters of particle emission during loading of rock samples 1-object of study (rock sample), 2-press plates, 3-through hole created in the object under study, 4-sampling tube, 5-aerosol particle counter, 6-high-performance air filter

natural gas content (including from rock salt [1] with high gas content), and on a microscale this phenomenon is known as fractoemission-shear and detachment of micro- and nanostructural fragments [4–9]. Under the influence of static and dynamic loads, rocks around workings and wells pass into the limit state and can be destroyed under conditions of inhomogeneous volumetric stress states. Using the facility developed at IPKON RAS, the authors carried out research and established new quantitative dependences of nanoemission of particles on the loading of a rock salt core under uniaxial compression by recording the emission of submicron particles [10].

A rock salt sample (core) was subjected to uniaxial stepped compression and the parameters of particle emission were continuously recorded. The results of experimental studies were processed using a computer program developed by S. D. Viktorov. Functional dependence of the number of particles on the relative level of loading is displayed in formula (1).

$$Y = A\left(e^{X^k} - 1\right) \tag{1}$$

where A and k—functional dependence coefficients, X—relative loading level, ($X = \sigma/\sigma^*$); σ—effective compressive stress in the sample; σ*—is the limiting value of the compressive stress.

Figures 2, 3, 4, 5, 6, 7 and 8 show the distribution of the emission of fine particles during uniaxial compression of a rock salt sample at stages 1–7 depending on the magnitude and time of compression when the sample is loaded up to 100 kN (in the

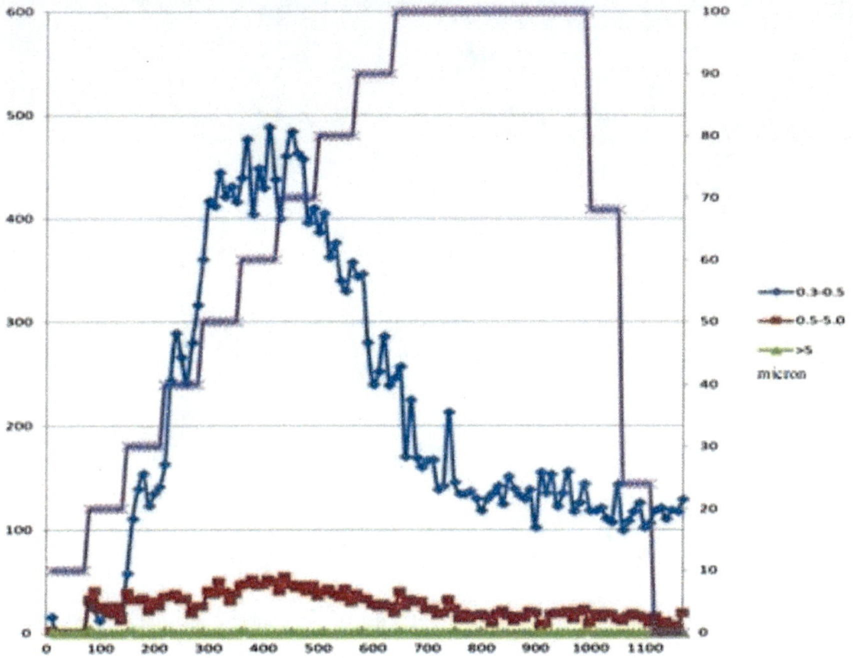

Fig. 2 Graphic representation of the dependence of the emission of the number of particles at the first stage of sample compression

figures on the right), along the abscissa axis—the observation time t, sec, number of particles N, pcs. (left).

At all stages of loading, the maximum release of nanoparticles with a size of 0.3–0.5 μm was noted. This phenomenon differs from previous work performed at IPKON in the study of particle emission during uniaxial compression of various rocks (granite, urtite, dolomite, ferruginous quartzite, dunite, serpentine, limestone, marble, sandstone) with a wide range of particle sizes formed and with a maximum amount of emission particles with a size of 0.5–5.0 μm. The phenomenon of emission of mainly nanoparticles with a size of 0.3–0.5 μm during uniaxial compression of a rock salt sample was discovered—the phenomenon of emission of nanoparticles during uniaxial compression of a rock salt sample.

Figure 9 shows the average number of nanoparticles with a size of 0.3–0.5 μm, formed at stages of loading from 1 to 7: the first stage-206, the second stage-76, the third stage-48, the fourth stage-46, the fifth stage-63, the sixth stage-230, the seventh stage-903. Accordingly, the maximum number of nanoparticles with a size of 0.3–0.5 μm at the stages of loading: the first stage-488, the second stage-123, the third stage-73, the fourth stage-76, the fifth stage-90, the sixth stage-945, seventh stage-966. The minimum number of nanoparticles with a size of 0.3–0.5 μm at the stages of loading: the first stage-12, the second stage-37, the third stage-25, the fourth stage-29, the fifth stage-32, the sixth stage-45, seventh stage-800. The ratio of the

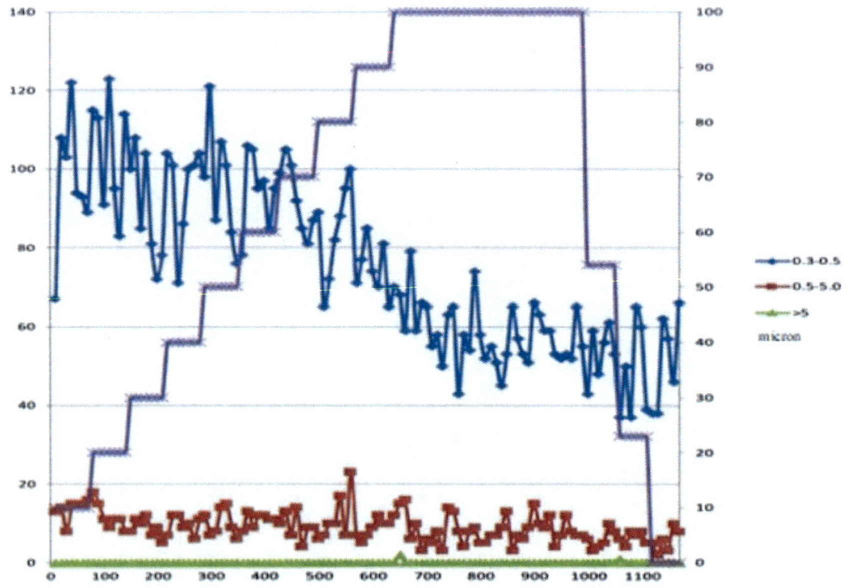

Fig. 3 Graphic representation of the emission dependence of the number of particles at the second stage of sample compression

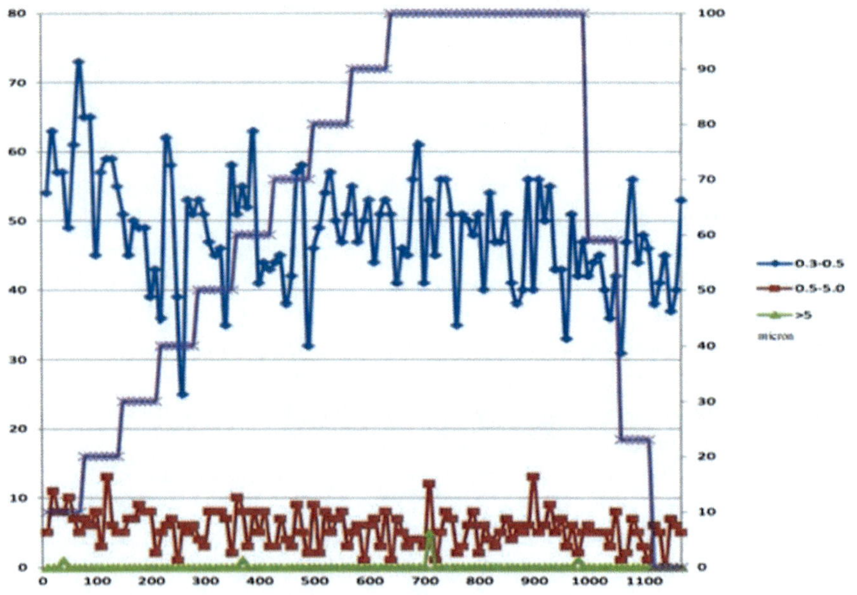

Fig. 4 Graphical representation of the emission dependence of the number of particles at the third stage of sample compression

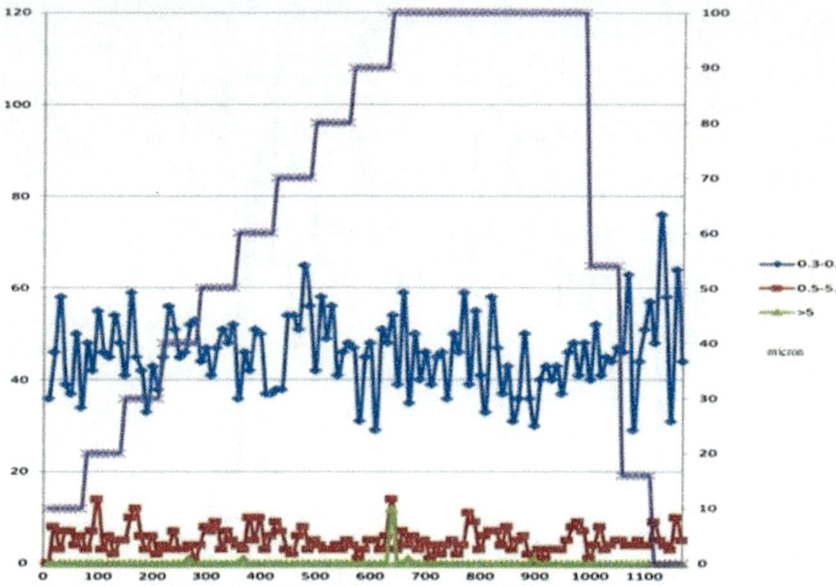

Fig. 5 Graphic representation of the emission dependence of the number of particles at the fourth stage of sample compression

number of nanoparticles with a size of 0.3–0.5 μm to the total number of particles is approximately 93%, plus an additional number of nanoparticles from the range of 0.5 to 5.0 μm.

Under uniaxial compression of a rock salt sample, the predominant formation of nanoparticle emission was noted.

Figure 9 shows the average number of particles with a size of 0.5–5.0 μm, formed at stages of loading from 1 to 7: the first stage-25, the second stage-9, the third stage-5, the fourth stage-5, the fifth stage-5, the sixth stage-20, the seventh stage-61. Accordingly, the maximum number of particles with a size of 0.5–5.0 μm at the stages of loading: the first stage-54, the second stage-23, the third stage-13, the fourth stage-14, the fifth stage-13, the sixth stage-101, the seventh stage-93. The minimum number of particles with a size of 0.5–5.0 μm at the stages of loading: the first stage-2, the second stage-2, the third stage-1, the fourth stage-1, the fifth stage-1, the sixth stage-1, the seventh stage-45.

The average value of the number of formed particles > 5 μm in size is less than unity, therefore this dependence is not graphically presented in Fig. 9.

The maximum number of particles > 5 μm in size formed during loading of the sample was noted at the 4th cycle and amounted to 12.

From the analysis of the graphic dependence presented in fig. It follows from Fig. 9 that under uniaxial compression of a rock salt sample, three stages of emission of fine particles are noted: from the first to the third stage, there is a period of decrease in the emission of particles from the surface of the rock—the stage of mass transfer

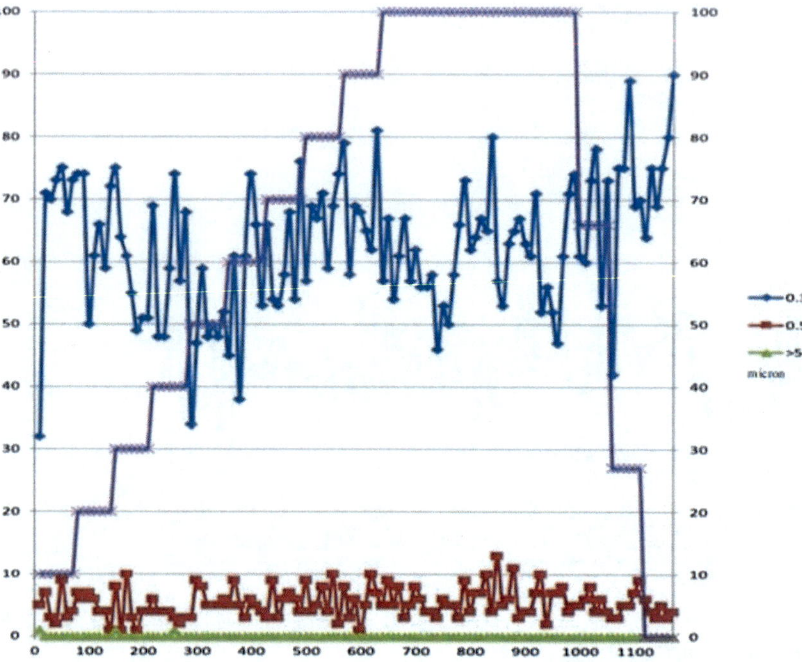

Fig. 6 Graphical representation of the emission dependence of the number of particles at the fifth stage of sample compression

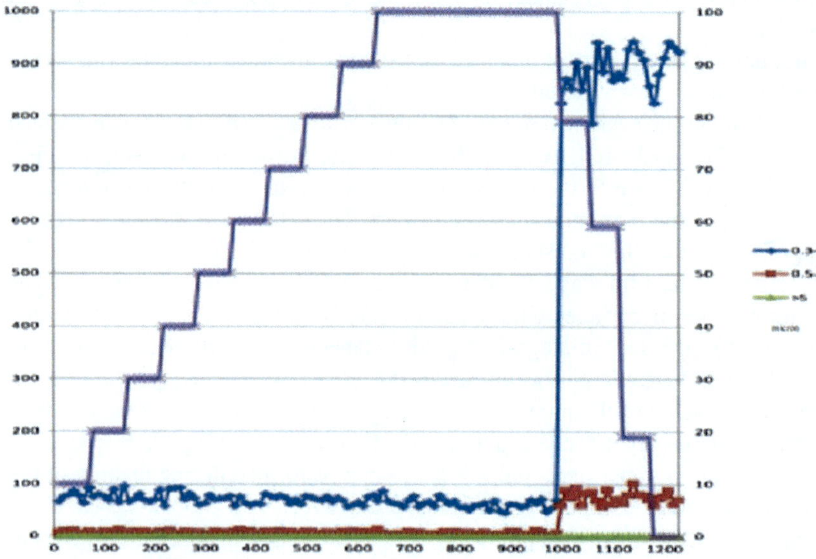

Fig. 7 Graphic representation of the emission dependence of the number of particles at the sixth stage of sample compression

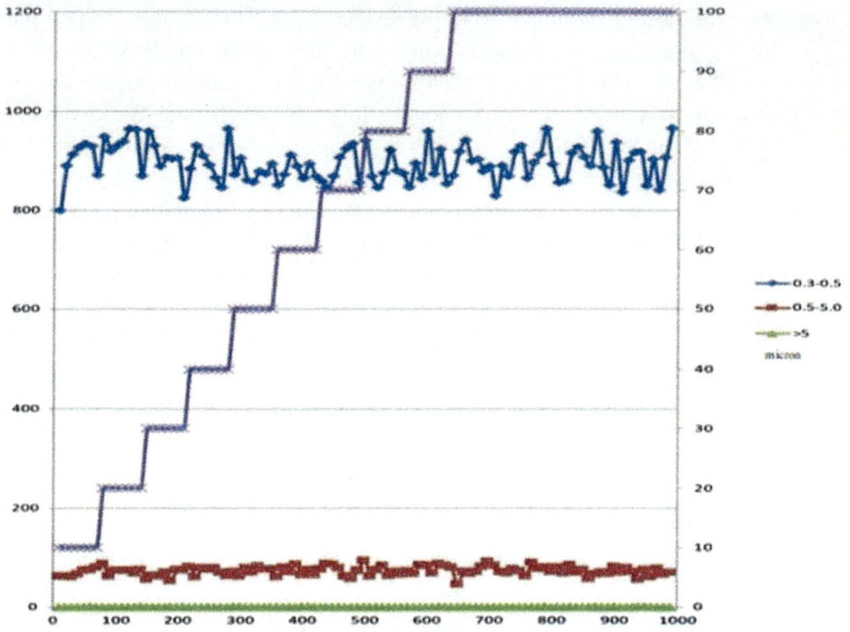

Fig. 8 Graphical representation of the emission dependence of the number of particles at the seventh stage of sample compression

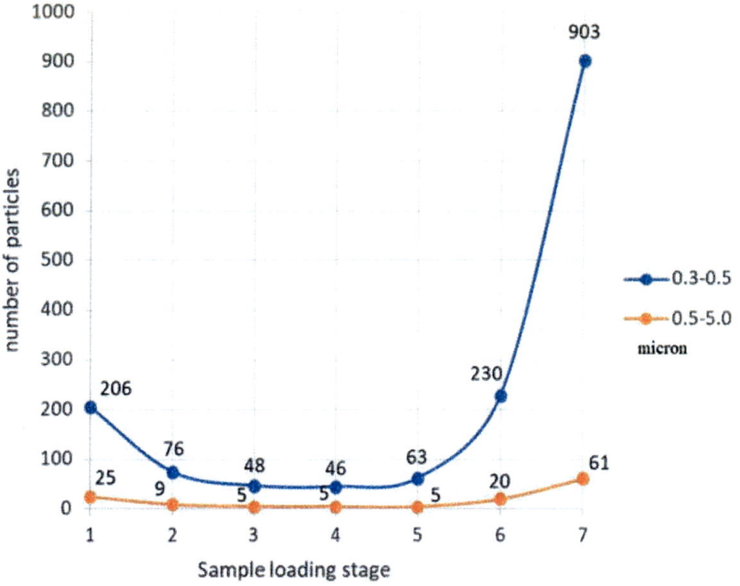

Fig. 9 Distribution of the average emission of fine particles during uniaxial compression of a rock salt sample at stages 1–7 of sample loading up to 100 kN, with the formation of the number of particles N, pcs. (in the figure on the left), along the abscissa axis—the stages of loading the sample

of particles (emission of particles) from the surface of rock salt when loading the sample (initial detachment of particles); from the third to the fourth stage, there is a process of stable emission of particles—the stage of stable mass transfer of particles from the surface of rock salt when loading the sample (stable detachment of particles); after the fourth stage, including 5–7 stages, the process of cracking the surface of the sample under loading (increase in the mass transfer surface or, which is the same—the surface of the emission of particles, the surface of the separation of particles) occurs with a significant increase in the emission of particles—the stage of significant mass transfer of particles from the surface of rock salt during loading sample with cracking (significant detachment of particles during cracking).

The intensity of particle emission during cracking can be characterized by several stages.

When performing hydraulic fracturing (HF) in mining and oil and gas business, it is possible to divide the process into initial hydraulic fracturing (HF-1), hydraulic fracturing (HF-2). A similar process was considered during fluid infiltration into rock salt [9].

By analogy with hydraulic fracturing, when a solution is exposed to rock salt, it is possible to separate the loading process (under uniaxial compression) of a rock sample (rock salt core) into the initial load cracking of the rock (separation of particles during the formation of rock cracking, from the fourth to the sixth stage of loading)-(NRP-1) and load rupture of the rock, from the sixth to the seventh stage (in case of rock failure)-(NRP-2).

Loading of rock salt and other salts in real conditions of mine workings-tanks, brine chambers, as well as salt inclusions in productive hydrocarbon beds [11] and halitization of a number of deposits in Eastern Siberia with the possible formation of nanoparticles is of practical importance. For example, with various combinations of pressure from overlying rocks and back pressure on the surface of the rock salt of the working during construction, operation of the working-tank, with natural loading factors (earthquakes), man-made loading factors (perforation, blasting of various intensity, including nuclear charges).

When salt is loaded with the formation of nanoparticles, nanoparticles can get into the salt solution or the stored product, into hydrocarbons with the corresponding manifestations.

The conducted studies have shown a significant emission of nanoparticles during uniaxial compression of a crystalline rock (rock salt). Phenomenon of phased emission of nanoparticles under rock salt loading discovered.

References

1. Malyukov V.P. Mass transfer and separation of the boundary layer under the influence of an aqueous solution on rock salt // Processes in geomedia. 2020. T. 4 (26). P. 918–930.
2. Malyukov V.P. Formation of Liesegang rings in rock salt with nanoparticles. Mining information and analytical bulletin // Scientific and technical journal. Publishing house "Mining book",

T. 10.-M. 2016. P. 242–248.
3. Aleksandrov P.A., Kalechits V.I., Khozyasheva E.S., Chechueva P.V. Investigation of particle generation during metal rupture // Scientific session of MEPhI, V. 8, 2003.-P. 296–297.
4. Chanturiya V.A., Viktorov S.D., Kochanov A.N., Osokin A.A. Study of the emission of micro- and nano-sized particles during uniaxial compression of rock samples // Physical and mechanical properties of rocks and composites in nano-, micro- and macroscales. Tambov. – V. 4. 2010.-P.5–7.
5. Viktorov S.D., Kochanov A.N., Osokin A.A. Patterns of generation of micro- and nanosized particles during the destruction and deformation of rocks // Mining information and analytical bulletin (scientific and technical journal). 2011. V 1. P. 185–191.
6. Viktorov S.D., Kochanov A.N., Odintsev V.N., Osokin A.A. Emission of submicron particles during deformation of rocks // News of the Russian Academy of Sciences. Physical series. 2012. V. 76. T. 3.P. 388–401.
7. Viktorov S.D., Kochanov A.N. Experimental study of the regularities of formation of submicron particles during the destruction of rocks // Physical and technical problems of mineral development. 2016. V. 5. P. 76–83.
8. Malyukov V.P. Physical processes and technologies for the construction of underground workings-tanks under the hydrodynamic effect of the solution on rock salt: monograph / V.P. Malyukov.-Moscow: RUDN University, 2021.-299 p.
9. Malyukov V.P. Local mass transfers during the creation and operation of underground workings in rock salt // Processes in geomedia. 2022. V. 1 (31). P. 1482–1490.
10. Viktorov S.D., Kochanov A.N. Possibilities of the method of micro- and nanoindentation in the study of physical and mechanical properties of rocks and minerals // Fundamental and applied issues of mining sciences. 2015. V. 2. T. 2. P. 308–313.
11. Kashavtsev V.E. Mishchenko I.T. Salt formation during oil production.-M.: 2004.-432 p.

Retrospective Analysis Methods in the Study of the Existence of Anomalous Processes in Geodynamic Systems

V. K. Kazankov and S. E. Kholodova

Abstract The problem of the methodology for predicting a special nonlinear effect that occurs in the marine environment, called "Rogue waves", is considered. Rogue waves are a phenomenon that cannot be described by means of the apparatus of the linear theory of waves, the existence of which is beyond doubt. The paper presents a formal apparatus that generalizes the concept of a dynamic system, and demonstrates a method for studying a dynamic system based on the analysis of time series.

Keywords Mathematical modeling · Rogue waves · Dynamic processes in geomedia · Dynamic systems · Time series

1 Introduction

Dynamic processes are an integral part of the physical world. Their mathematical models are widely used both in theoretical research and in the introduction of advanced technologies into production. Despite the peculiarities of subject areas, the classical way to describe the dynamics in a system is to use the apparatus of differential equations. Each dynamic system consists of some objects, their states and time flow.

There is a phenomenon that until recently was considered a feature of exclusively hydrodynamic systems, but it turned out that the root cause of its occurrence lies in the structure of the dynamics of the system itself. The phenomenon is called-rogue wave [1, 2], it occurs, as a rule, in systems with nonlinear dynamics. Quite recently, full-scale experiments were carried out, which made it possible for the first time to generate a rogue wave in a pool when a pair of waves intersect at certain angles [3]. The peculiarity of killer waves lies in the suddenness of their appearance and huge size. The unpredictability of their appearance is also due to the fact that the life cycle of a rogue wave is a periodic alternation of an increase and decrease in amplitude. Phenomenologically, the observation of rogue waves of large size can, in turn, be

V. K. Kazankov · S. E. Kholodova (✉)
ITMO University, St. Petersburg, Russia
e-mail: kholodovase@yandex.ru

V. I. Karev (ed.), *Physical and Mathematical Modeling of Earth and Environment Processes—2022*, Springer Proceedings in Earth and Environmental Sciences,
https://doi.org/10.1007/978-3-031-25962-3_10

related to the fact that the nature of such waves is related to the total energy of all waves of a selected continuous medium.

However, the appearance of rogue waves is possible not only in liquid media described by classical hydrodynamics, but also in various quantum mechanical processes that are constant relative to the energy reserve of the system. At the moment, there are several directions in the study of rogue waves: the study of the equations of mathematical physics along with the structure of their solution, as well as predicting the occurrence of rogue waves. The papers [4–8] investigate nonlinear equations corresponding to models of quantum mechanics, the solution of which gives rise to rogue waves. Thus, particular solutions of the Schrödinger equations of a non-linear form provide the necessary, from the point of view of mathematical modeling, structures for constructing a model of rogue waves. Also, the study of solutions to the generalized Hirota equation showed the presence of rogue waves [9, 10].

To structure knowledge about rogue waves, an apparatus was developed that generalizes the concept of a dynamic system, in which it was possible to formulate the necessary conditions for the existence of rogue waves [11]. The paper formulated a hypothesis about the periodic occurrence of rogue waves, which is consistent with the results of [12]. There is a need to form key criteria for predicting the occurrence of rogue waves, taking into account the possible periodicity of their occurrence. Of no less interest is the relationship between the statistical indicators of the amplitudes among the waves and their relationship with each other.

In this paper, it is proposed to confirm the applicability of the theory of rogue waves by considering a dynamical system that does not explicitly obey hydrodynamic laws. It is expected that within the framework of such input conditions it is possible to find dynamical systems in which the structural components necessary for describing rogue waves are easily distinguished. As a research tool, a prognostic model is proposed that allows, using a retrospective analysis, to imagine the possibility of the appearance of rogue waves in the current system. From the point of view of mathematical modeling methods, a class or classes of functions will be proposed that have a number of characteristics sufficient to describe the present model.

2 Mathematical Model

Consider a continuous medium V, consisting of several waves. Let v_i i-th wave. Let us define for each v_i wave a linear functional $J_t : V \to K \subset \mathbb{R}$, for which the parameter $t \in T \subset \mathbb{R}^+$ describes continuous time. Denote by $w_i(t)$ the amplitude of the wave v_i at time t, determined by the formula:

$$J_t(v_i) = w_i(t).$$

Denote by W the set of all estimates $w_i(t)$. Moreover, for any t, the inequality holds

$$\sum_{v_i \in V} J_t(v_i) \le \sup K < \infty.$$

In what follows, we will assume that each wave v_i can transfer energy, and hence increase the height of any wave v_j, where $v_i, v_j \in V$. Such a dynamic structure formally corresponds to a complete graph.

A sequence $u = (v_i, v_j)$ is called connected at time t, if there is an energy exchange between elements v_i and v_j at time. t.

Let in a dynamical system for any element $w_i(t) \in W$ there exists a two-parameter family of closed operators $\mathfrak{D} = \{D_t^u : W \to W\}_{u \in G}$, such that for

$$t = 0 D_0^u = I, \mathrm{t} = 0 \text{ and } \mathfrak{D} \subset C^1(T).$$

The result of the operator D_t^u characterizes the dynamics in the system over time. Let further the behavior of the operator D_t, at time t depends only on time t. Then the operator D_t can be treated as a Markov process. Under the rogue wave we mean the wave v_i, for which the amplitude criterion $\mu(v_i) \ge 2.1$ [13].

Example Let $V = \{A, B, C\}$, then $G = \{AA, BB, CC, AB, BA, CB, BC, AC, CA\}$. The amount of energy in the system $supK = 15$, a $J_\tau(A) = 5$, $J_\tau(B) = 8$, $J_\tau(C) = 2$, $J_t(A) = 7$, $J_t(B) = 3$, $J_t(C) = 5-$ energy estimates of volumes at times τ and t, respectively.

Then $\|D_t^{BA}\|_H = \|D_t^{BC}\|_H = \|D_t^{BB}\|_H = 5$, $\|D_t^{AB}\|_H = \|D_t^{AC}\|_H = \|D_t^{AA}\|_H = 2$, $\|D_t^{CC}\|_H = \|D_t^{CB}\|_H = \|D_t^{CA}\|_H = 3$.

For each x $\hat{D}_t^a, \overline{D}_\tau^b \in \mathfrak{D}$ define the composition of operators as $\hat{D}_t^a \overline{D}_\tau^b = \tilde{D}_{t+\tau}^{a \circ b}$, at $\tau \le t$ and $a \circ b = c \sim u \in G$.

Let $\Delta\tau-$ a period of time taken as a conventional unit, then the following entry describes the dynamic process:

$$(D_{n\Delta\tau}(w))^n = J_t(v_i),$$

where $n \in N$ number of time slots. The value of $\Delta\tau$ can also be interpreted as a sampling parameter of time t, so if $t \to \infty$, then $n \to \infty$. Among the operators $D_t \in \mathfrak{L}$, an ordered hierarchy of classes of operators $[D_t] = \mathfrak{L}_i$ arises with respect to the growth rate of energy estimates.

Let $2w \ne 0-$ a conventional unit of energy measurement that can be registered, then the exact upper bound of the class $\mathfrak{G}$ will be a representative of the class of operators $\mathfrak{L}_1$ such that, if $D_t \in \mathfrak{L}_1$,, then

$$\|D_t\|_H \le \underbrace{2w + \ldots + 2w}_{t} = 2wt = O(t).$$

Since for each conditional time interval $\Delta\tau$ the amount of energy increases by a constant value $2w$, then

$$\sup S = O(t).$$

Consider $sup\mathfrak{L}_1$. Assuming that at time t the amount of energy change is $2w$, for any $D_t \in \mathfrak{L}_1$ the estimate

$$\|D_t\|_H \leq 2w + 2 \cdot 2w + \cdots + t \cdot 2w = (2w + w(t-1))t = w(t+1)t = O(t^2),$$

and for any operator $D_t \in \mathfrak{L}_2$, such that at each moment of time the change in the amount of energy will be $2w$ times greater than at the previous moment of time t, the estimate is correct:

$$\|D_t\|_H \leq 2w + (2w)^2 + \cdots + (2w)^t = \frac{2w\left(1-(2w)^t\right)}{1-2w} \leq Ae^{\alpha t} = O\left(e^{\alpha t}\right),$$

where $A = \frac{2w}{2w-1}$ and $\alpha = \ln(2w)$. An arbitrary operator D_t from the class $\mathfrak{L}_n$„ at ≥ 2 satisfies the estimate

$$\|D_t\|_H \leq \sum_{k=1}^{t} 2w \uparrow^{n-1} k,$$

where «$\uparrow^{n-1}$» – hyperoperator notation in Knuth's annotation [14].

3 Time Series Study

In [15, 16], computational experiments were carried out. The resulting waveform is shown in Fig. 1. We will interpret it as a time series X_t.. We introduce an estimate of the amplitude of each wave according to the following formula:

$$J_t = x_{mx} - \left(x_{mf} + x_{ms}\right)/2.$$

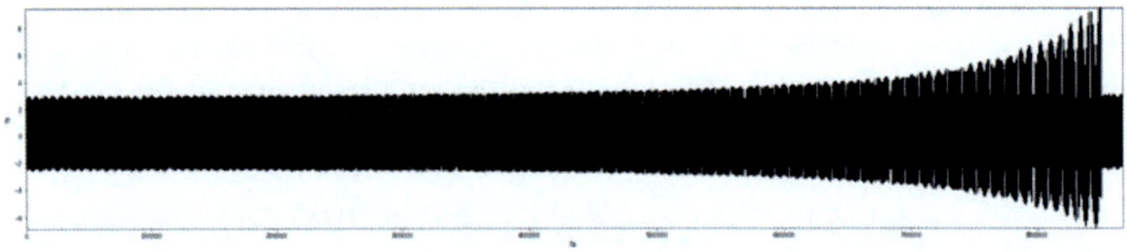

Fig. 1 Visualization of the computational experiment

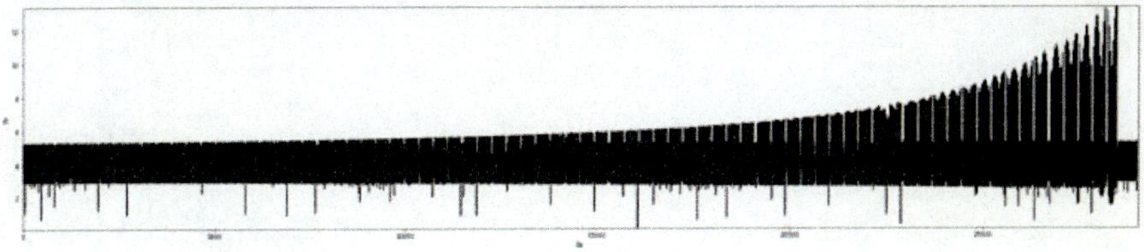

Fig. 2 Wave height

Figure 2 shows the result of converting the time series X_t to Y_t, which displays the change in wave height over time.

According to the amplitude criterion, 25 rogue-waves were recorded in the Y_t time series. For the time series Y_t, we define the estimate of local uniformity $\mu(U_n)$, as the standard deviation calculated for the set $U_n \subset Y_t$, where n—is the number of elements in the set U_n.

The Y_t time series consists of 29,103 elements, and rogue-waves occurred only at time points

$$t = [26925;\ 27164;\ 27198;\ 27232;\ 27403;\ \ldots;\ 28429;\ 28463;\ 28497;\ 28531].$$

Denote by T_r the set of all instants of time when rogue waves occur. Let us introduce a set P consisting of time intervals during which rogue-waves appeared according to the following rule:

$$p_i = t_{i+1} - t_i\, p_i = t_{i+1} - t_i,$$

where $t_i, t_i + 1 \in T_r$.

We assume that the dynamic process contains cycles, which means that the occurrence of a killer wave must be systematic. At the same time, it is much more important that in each cycle there is only one rogue-wave or there are none at all. Such an assumption corresponds to the idea of a possible missed registration of a rogue wave. There are several possible options for choosing the value of the number n. First, let's assume that $n = 67$, since the average of all elements of the set P is equal to 66.917, up to three decimal places, where $n \in N$. Also pay attention to the value 34, which is often found in the set P, so it can be $n = 34$. Note that the average value of the elements of the set P is close to twice the value of the most encountered element, decomposing which into prime numbers, we get 17 and 2, hence $n = 17$.

Figures 3, 4 and 5 below show graphs of $\mu(U_n)$ for $n = 67, 34, 17$. The blue color shows the change in the value of $\mu(U_n)$, denoted as "Origin". "Trend" denotes the result of exponential smoothing "Origin" with parameter $\alpha = 0.02$. The red line marks the moment of registration of the first Killer Wave, while it starts from the value of the indicator $\mu(U_n)$ and ends at the maximum possible value of $\mu(U_n)$.

Figure 5 shows the epoch, which explicitly consists of a cyclic component in the form of a sinusoid and a trend in the form of an exponent. The presented profile

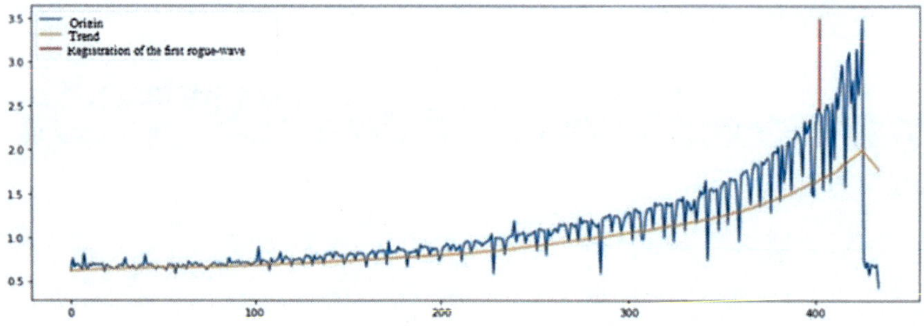

Fig. 3 Change $\mu(U_n)$ over time at $n = 67$

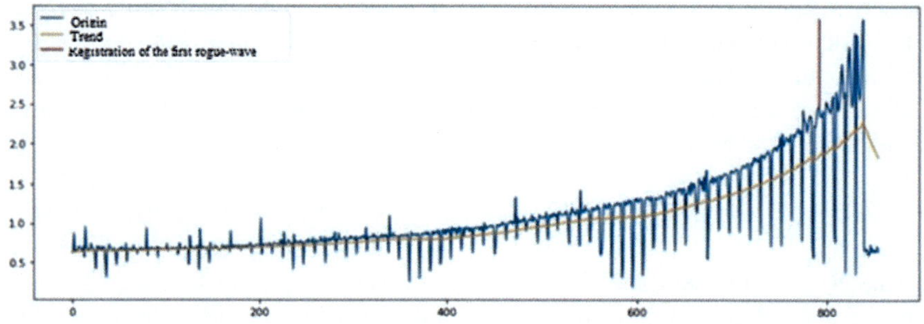

Fig. 4 Change in $\mu(U_n)$ over time at $n = 34$

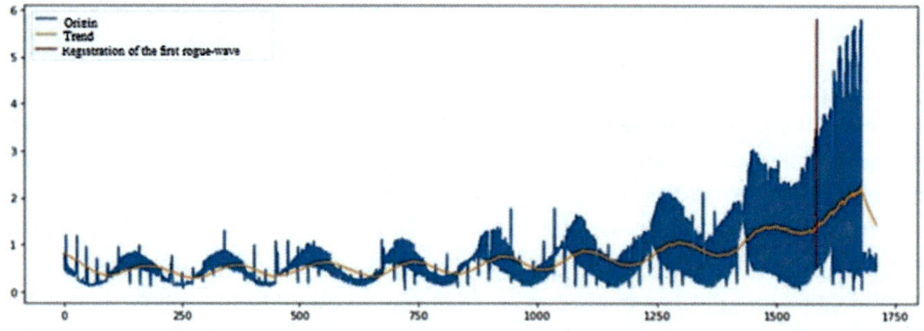

Fig. 5 Change $\mu(U_n))$ over time at $n = 17$

allows us to put forward a hypothesis about the existence of some periodicity in a nonlinear dynamic process.

It is assumed that the occurrence of a rogue-wave should be a systematic phenomenon, that is, for all rogue waves, there is some periodicity. And if this is so, then you can choose n such that the value of $\mu(U_n)$ will be minimal at all times of the occurrence of a rogue wave. The solution of this problem is reduced to the construction of the objective function $\rho(n)$ and its minimization. Let the objective

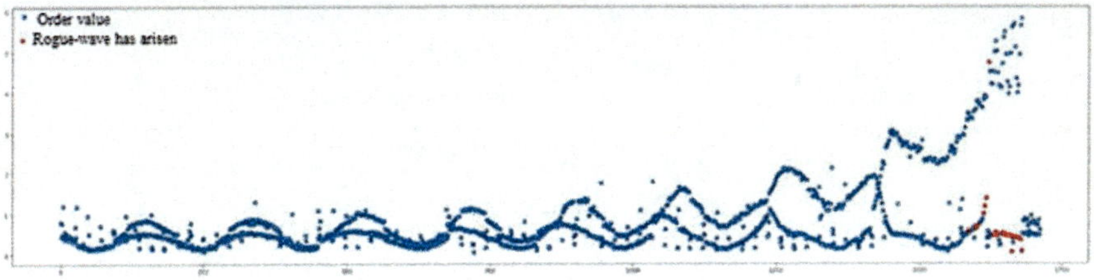

Fig. 6 Dot plot μ(U_17)

function have the form

$$\rho(n) = \frac{1}{25} \sum_{t \in T} (\mu_t(U_n))^2 \to \min .$$

Before starting the optimization, it is necessary to determine the range of acceptable values. Let $n \in [3500]$. By a particle we mean the implementation of the annealing simulation method for the objective function $\rho(n)$. The smallest value of the objective function $\rho(n)$ is achieved at $n = 17$. This value was the most common and had the lowest value of all obtained as a result of launching a swarm of 60 particles.

Figure 6 shows the function $\mu_t(U_n)$ for $n = 17$.

The use of the local uniformity function $\rho(n)$ makes it possible to display the initial time series into another one, which has a more pronounced periodic dynamics. In [17], the Portevin–Le Chatelier effect was studied. During the study of the effect, a methodology was developed for processing time series, using the hypothesis of the periodicity of the occurrence of rogue waves.

Figure 7 shows the wave amplitude distributions obtained from the Y_t time series. The horizontal axis shows the amplitude of the waves in meters, and the vertical axis shows the frequency of occurrence. It can be seen from the figure that the distribution of wave amplitudes resembles a superposition of the normal law with several Rayleigh distributions. Thanks to the use of the central limit theorem for this system, we can conclude that in this dynamic system there is a pronounced dynamics, which is described by the Rayleigh distribution laws. In this case, it is the Rayleigh distribution law that is used to describe the distribution of wave amplitudes in linear wave theory.

It is also worth noting that Fig. 8 shows a graph of the distributions of values, a graph of the function $\mu(U_{17})$, where the frequency of the value is marked on the vertical axis, and the relative value of the function $\mu(U_{17})$, is marked on the horizontal axis. The distribution profile is very reminiscent of the Rayleigh distribution law, from which we can conclude that using the function $\mu(U_n)$ to transform the time series Y_t allows one to go from describing the nonlinear dynamics of the system to some representation of it, consistent with the linear theory of waves.

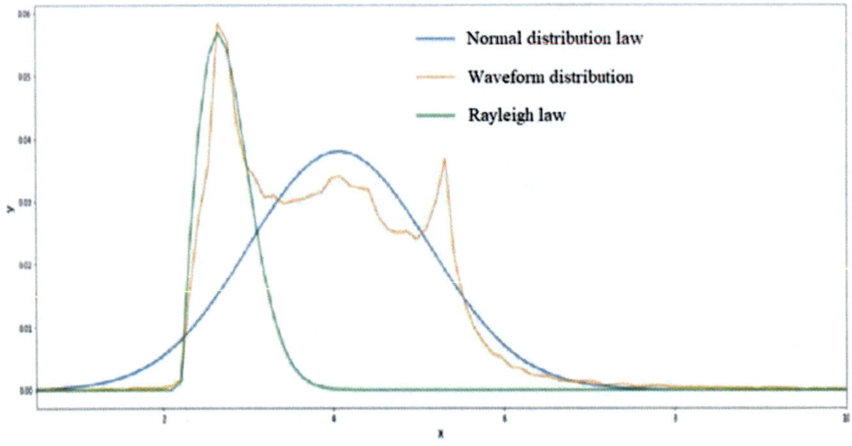

Fig. 7 Distributions of wave amplitudes in the time series Y_t

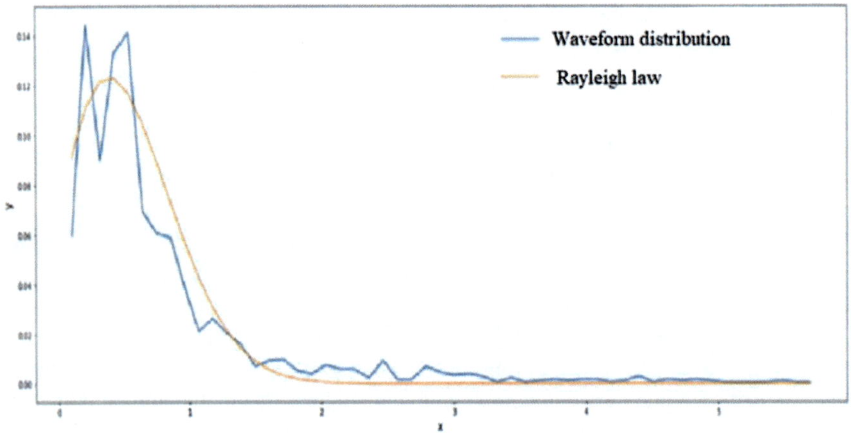

Fig. 8 Distribution of function $\mu(U_{17})$

4 Conclusion

The constructed mathematical apparatus, which allows us to consider an arbitrary dynamic system through the interaction of objects, and their energy exchange, is an opportunity to prove the existence for some differential operators of a special nonlinear effect that occurs in a continuous medium, called rogue waves. The peculiarity of building a model helps to extend the methodology for processing time series to any dynamic system and predict the possibility of occurrence of anomalous phenomena in it.

References

1. Kurkin, A.A. Rogue waves: facts, theory and modeling / A.A. Kurkin, E.N. Pelinovsky. - 2nd ed. - Moscow; Berlin: Direct-Media, 2016. - 178 p. ISBN 978–5–4475–5883–3.
2. Stephane Brule, Stefan Enoch, Sebastien Guenneau.: On the possibility of seismic rogue waves in very soft soils. Geophysics (physics.geo-ph); Pattern Formation and Solitons (nlin.PS); Computational Physics (physics.comp-ph); Fluid Dynamics (physics.flu-dyn)
3. McAllister M. L., Draycott S., Adcock T. A. A., Taylor P. H., van den Bremer T. S. : Laboratory recreation of the Draupner wave and the role of breaking in crossing seas. J. Fluid Mech. (2019), vol. 860, P. 767–786.
4. Akhmediev N. Waves that appear from nowhere: complex rogue wave structures and their elementary particles //Frontiers in Physics. – 2021. – T. 8. – P. 612318
5. Wu, X.Y., Tian, B., Qu, Q.X., Yuan, Y.Q., Du, X.X.: Rogue waves for a(2+1)-dimensional Gross-Pitaevskii equation with time-varying trapping potential in the Bose–Einstein condensate. Comput. Math. Appl. 79 – 2020, P.1023–1030
6. Yang B., Yang J. Rogue wave patterns in the nonlinear Schrödinger equation //Physica D: Nonlinear Phenomena. – 2021. – T. 419. – P. 132850
7. Bang-Qing Li. Hybrid breather and rogue wave solution for a (2 + 1)-dimensional ferromagnetic spin chain system with variable coefficients // International Journal of Computer Mathematics - 2022, Vol. 99, No. 3, P. 506–519
8. Chen Y. Q. et al. Dark wave, rogue wave and perturbation solutions of Ivancevic option pricing model //Nonlinear Dynamics. – 2021. – T. 105. – №. 3. – P. 2539–2548.
9. Liu, J.G., Zhu, W.H.: Multiple rogue wave solutions for (2+1)-dimensional Boussinesq equation. Chin. J. Phys. 67, 492–500 (2020)
10. Liu J. G., Zhu W. H. Multiple rogue wave, breather wave and interaction solutions of a generalized (3+ 1)-dimensional variable-coefficient nonlinear wave equation //Nonlinear Dynamics. – 2021. – T. 103. – №. 2. – P. 1841–1850.
11. Kazankov V., Kholodova S.E., Peregudin S.I. Mathematical Modelling of Anomalous Dynamic Processes in Geomedia // Springer Proceedings in Earth and Environmental Sciences - 2022, P. 63–74
12. Peng W. Q., Pu J. C., Chen Y. PINN deep learning method for the Chen–Lee–Liu equation: Rogue wave on the periodic background //Communications in Nonlinear Science and Numerical Simulation. – 2022. – T. 105. – P. 106067.
13. Kharif C., Pelinovsky E., Slunyaev A.: Rogue waves in the ocean. Berlin: Springer, 2009. 216 p.
14. Knuth, Donald E.: Mathematics and Computer Science: Coping with Finiteness // Science. 194 (4271). (1976), 1235–1242
15. Shamin R.V., Yudin A.V. Modeling the spatio-temporal propagation of rogue waves // Reports of the Academy of Sciences. 2013. V. 448. № 5. P. 592–594
16. Shamin R.V., Gorlenko A.V., Smirnova A.I. Stability issues of rogue waves // Computational technologies. 2013. V. 18. №. 1. P. 96-105
17. Kazankov V.K., Shmeleva A.G., Zaitseva E.V. Unstable plastic flow in structural materials: time series for analysis of experimental data // Физика и механика материалов = Materials Physics and Mechanics - 2022, Vol. 48, No. 2, P. 208–216

The Analysis of the Ice Edge Position Variability in the Arctic Seas Depending on Different Types of Winter Severity

M. S. Teider, N. S. Frolova, N. A. Podrezova, and K. V. Kravtsova

Abstract The article presents the results of research on the ice cover changes in the Arctic basin according to the satellite data. Light cloud cover satellite images of the MODIS spectroradiometer were used to estimate the parameters of the ice cover of the Arctic seas. QueryThe images of the Landsat-8 satellite were used to refine the data obtained from the MODIS spectroradiometer (Terra satellite) because of their better spatial resolution. The classification according to the severity of winter seasons for the White, Kara, Laptev and East Siberian Seas is also described in the article.

Keywords Arctic · Sea ice area · Sea ice concentration · Fast ice width · Sea ice edge · Severity of winter seasons

The introduction of remote sensing from Space has presented new opportunities in the study of the ice cover of the Arctic seas. The use of satellites has undeniable advantages due to their significant spatial coverage tied to a coordinate grid and short time interval of measurements.

It is also worth considering that in the harsh conditions of the Arctic, contact measurements are quite difficult to carry out, and in some areas it is simply impossible. Currently, remote sensing data is an indispensable source of information for conducting production and research activities in the Arctic region. With the help of remote sensing, through the analysis of the ice situation, the issues of navigation support and economic activity and the tasks of determining the total area, concentration and climatology of ice formations are solved. Remote sensing methods are divided into the following types: passive methods which record thermal, visible and natural gamma radiation from the water surface; semi-active methods which analyze the signal received after irradiation of the sea by an electromagnetic radiation source in a wide spectral range; active methods which register reflected radiation, fluorescence, or combination scattering after irradiation by a radiation source at a given spectral composition [1].

M. S. Teider · N. S. Frolova · N. A. Podrezova (✉) · K. V. Kravtsova
Russian State Hydrometeorological University, St. Petersburg, Russia
e-mail: Nadinapod@mail.ru

V. I. Karev (ed.), *Physical and Mathematical Modeling of Earth and Environment Processes—2022*, Springer Proceedings in Earth and Environmental Sciences,
https://doi.org/10.1007/978-3-031-25962-3_11

The visible range is the part of the electromagnetic spectrum that is perceived by the human eye (0.4–0.7 μm). Sensing in this range is carried out using various satellites with different temporal and spatial resolution. Obtained in cloudless and well-lit conditions, the images allow us to identify various characteristics of the ice cover, such as cracks, chips, shape, etc., and thereby contribute to the attribution of ice floes to a certain type. However, there is no possibility of sensing in the conditions of the polar night, and the daytime weather conditions are often not favorable because of thick clouds. This leads to a decrease in possibility of obtaining high-resolution images, and at the same time it becomes difficult to determine the quality and development of ice floes.

The infrared range is the electromagnetic range from 0.78 to 1000 μm. Sea ice chart composition is carried out in the IR range of 10.5–12.5 μm, because spectral differences in absorption, reflection and scattering of radiant energy from ice, snow and water surfaces are maximally pronounced in this range. So, all young types of ice are well recognized.

IR sensors (in the wavelength range over 3 μm) has no restrictions on the time of day. However, thick clouds can become a strong barrier to infrared remote sensing. Temperatures above − 6 °C also have a direct effect on the sensors.

According to the obtained images, it is possible to determine the thickness and some types of ice. When the air temperature drops below zero, it means the following: the thicker the ice, the lower the surface temperature. When the air temperature rises, the ability to determine the ice thickness decreases, because the sea ice surface begins to melt. The water, produced by surface melting, prevents ice from emitting electromagnetic radiation.

In the range 1 mm–1 m the microwave brightness temperature is measured. This parameter depends on the physical properties of water and ice, such as the presence and height of snow cover, its annual evolution, radiation, and the influence of the atmosphere. The water surface has a high reflectivity and strong polarization. The radiation of annual ice is quite strong with weak polarization. Because of their long wavelengths, compared to the visible and infrared, microwaves have special properties that are important for remote sensing. Longer wavelength microwave radiation can penetrate through cloud cover, haze, dust, and rainfall. It is possible to detect microwave energy under almost all weather and environmental conditions so that data can be collected at any time. Microwave sensing includes both active and passive methods. Passive microwave sensing is similar to thermal one. The sensor detects the naturally emitted microwave energy within its field of view. This emitted energy is related to the temperature and moisture properties of the emitting object or surface. Oceanographic applications of passive microwave sensing include mapping sea ice, evaluation of sea ice concentration, and detection of ice hummocks. Active microwave sensors provide their own source of microwave radiation to illuminate the target. These sensors are generally divided into two distinct categories: imaging and non-imaging. The most common form of imaging active microwave sensors is radar. It transmits a microwave (radio) signal towards the target and detects the backscattered portion of the signal. The radar images are used to detect ice (mostly old), determine the thickness of the ice cover and find the presence of certain types of ice formations such as icebergs and ice fields.

Non-imaging microwave sensors include altimeters and scatterometers. Radar altimeters transmit short microwave pulses (more than 1700 pulses per second) and measure the round trip time delay to targets to determine their distance from the sensor. Altimeters measure the thickness of a part of the ice above the water, the thickness and density of snow using the assessment of climatic data [1–5].

The NASA National Snow and Ice Data Center Distributed Active Archive Center (NASA NSIDC DAAC) includes near-global passive microwave data from 1978 to present [6]. These data products are derived from the Scanning Multichannel Microwave Radiometer (SMMR) instrument on the Nimbus-7 satellite and the Special Sensor Microwave/Imager (SSM/I) and Special Sensor Microwave Imager/Sounder (SSMI/S). The data have a number of errors, including underestimation of sea ice concentration of the broken ice during melting, false effects in coastal areas and zones of polar cyclones' movement, as well as the lack of data due to various spots on the images.

Light cloud cover satellite images of the MODIS spectroradiometer were used to estimate the parameters of the ice cover of the Arctic seas. MODIS has a wide field of view which allows us to determine the position of the edge and the residual forms of ice only in summer. Observation of the Arctic in polar winter is difficult because of the long-continuing darkness while the sun is below the horizon. Satellite images in the visible wavebands are not useful during polar nights. The data are selected using EOSDIS Worldview [7]. A combination of 3-6-7 "false" colors according to the RGB model was used. This combination is used to map snow and ice. Snow and ice are very reflective in the visible part of the spectrum (Band 3), and very absorbent in Bands 6 and 7 (short-wave infrared, or SWIR). This band combination is good for distinguishing liquid water from frozen water, for example, clouds over snow, ice cloud versus water cloud. Consequently, the more snow and the thicker ice, the stronger the absorption in the SWIR bands, and the more red the colour. Thick ice and snow appear vivid red (or red–orange), while small ice crystals in high-level clouds will appear reddish-orange or peach. For further work the images from Worldview were decomposed into multiple layers (Fig. 1).

The images of the Landsat-8 satellite were used to refine the data obtained from the MODIS spectroradiometer (Terra satellite) because of their better spatial resolution [8].

Ice concentration is the ratio of an area of ice to the total area of water surface within some large geographic locality. Seasonality has a significant effect on the ice conditions of the Arctic region, therefore the changes in this parameter for every summer month are considered.

The calculations of concentration were performed in the UNESCO Bilko software, where image histograms of pixel intensity corresponding to water, ice and different types of clouds were drawn for each of the Arctic seas on a monthly basis during the summer period. An example is shown in Fig. 2. The month of September was excluded from the investigation because in most seas the water surface was already

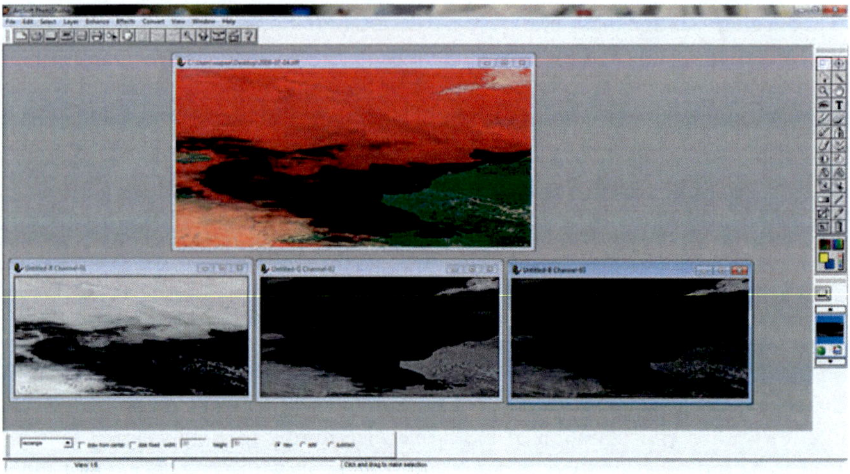

Fig. 1 Terra/MODIS satellite image of the Chukchi Sea on July 4, 2009 (top) and image decomposition in the ArcSoft PhotoStudio program (bottom)

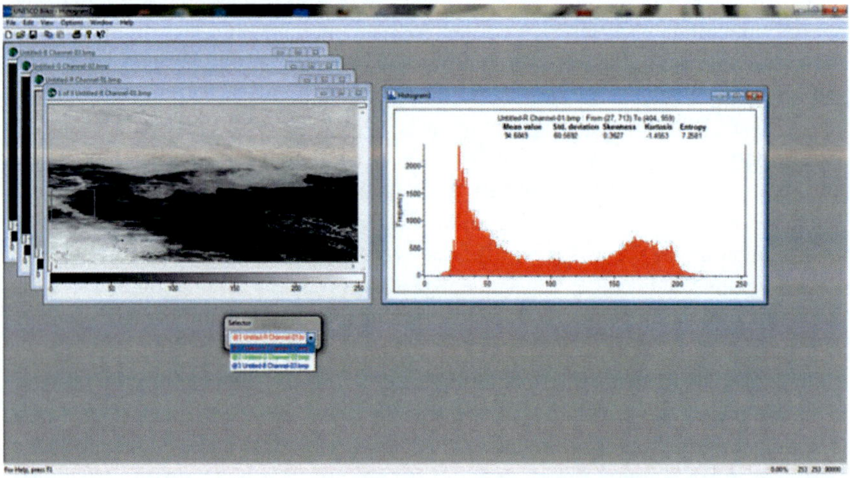

Fig. 2 Process of determining ice concentration in the Chukchi Sea (on the selected fragment) using a Bilko software histogram. Terra/MODIS composite grayscale image with combined bands (July 4, 2009)

ice-free in August. The Lincoln Sea was not included in the calculations, as the trend analysis has shown that the ice cover in this sea has not changed since 1979 [9].

The obtained results of sea ice concentration (in %) in the Arctic region for the period from 2009 to 2019 are recorded in Table 1, where 0 means absolutely clear water free of ice cover, and 100—fast ice.

Table 1 Results of ice concentration (in %) for the Arctic seas during the summer period

Year	Month	Sea											
		East Siberian	Kara	Beaufort	Laptev	Chukchi	Baffin	Barents	White	Greenland	Norwegian	Bering	Labrador
2009	*VI*	95	87	92	89	84	64	50	0	100	0	24	67
	VII	75	57	60	65	70	34	30	0	95	0	0	27
	VIII	0	0	15	30	0	5	0	0	91	0	0	0
2010	*VI*	94	84	88	76	90	62	47	0	100	0	32	57
	VII	80	60	40	59	64	55	33	0	94	0	0	14
	VIII	30	0	5	2	33	19	0	0	89	0	0	0
2011	*VI*	94	85	83	77	95	72	42	0	94	0	13	25
	VII	84	64	61	55	44	34	25	0	90	0	0	6
	VIII	0	0	13	0	0	7	0	0	85	0	0	0
2012	*VI*	88	45	76	72	88	59	34	0	96	0	77	54
	VII	72	0	20	44	67	36	0	0	78	0	0	4
	VIII	45	0	0	0	38	0	0	0	64	0	0	0
2013	*VI*	98	92	87	82	87	65	41	0	94	0	0	63
	VII	84	58	64	54	28	27	5	0	87	0	0	32
	VIII	25	0	0	0	–	0	0	0	81	0	0	11
2014	*VI*	87	91	81	68	55	–	34	0	95	0	48	58
	VII	68	60	33	38	30	48	0	0	85	0	0	20
	VIII	28	0	0	0	10	0	0	0	79	0	0	0
2015	*VI*	84	90	75	78	87	57	48	0	99	0	13	70
	VII	62	55	43	55	41	32	26	0	93	0	0	48

(continued)

Table 1 (continued)

Year	Month	Sea											
		East Siberian	Kara	Beaufort	Laptev	Chukchi	Baffin	Barents	White	Greenland	Norwegian	Bering	Labrador
	VIII	0	0	11	0	0	–	0	0	85	0	0	26
2016	*VI*	88	84	82	75	85	59	20	0	96	0	21	55
	VII	53	42	58	43	52	42	0	0	89	0	0	14
	VIII	15	0	33	0	0	12	0	0	83	0	0	0
2017	*VI*	82	75	84	72	77	52	45	0	90	0	16	67
	VII	61	45	69	44	53	38	21	0	87	0	0	39
	VIII	0	0	0	25	0	14	0	0	83	0	0	18
2018	*VI*	90	89	80	65	80	51	33	0	95	0	0	57
	VII	45	40	62	–	22	35	0	0	88	0	0	34
	VIII	0	0	55	0	0	21	0	0	81	0	0	0
2019	*VI*	80	94	52	67	83	57	47	0	97	0	5	52
	VII	20	0	33	–	35	48	14	0	82	0	0	19
	VIII	0	0	16	0	0	19	6	0	74	0	0	0

June is characterized by the maximum sea ice cover in the Arctic Ocean over the entire summer season. The high-concentration ice extends from the central and eastern part of the Siberian shelf (from the Kara Sea to the East Siberian Sea inclusive) across the entire ocean to Greenland and the Canadian Arctic Archipelago, partly including the Beaufort Sea, where the ice fields break up at this time, and the Greenland Sea shelf area. The Barents Sea ice is characterized by reduced ice concentration (20–45%). The White Sea and Norwegian Sea are completely ice-free. Relatively low concentration is observed in the Baffin and Bering Seas.

In July, total sea ice concentration continues to decrease rapidly: seas influenced by warm Atlantic waters are almost or completely cleared of ice, a significant reduction is also observed on the Siberian shelf. Ice concentration in areas belonging to the USA and Canada changes little, but the trend persists (the index decreases by 10–30%).

August is characterized by an even greater reduction in ice cover: ice has melted in the Barents and Kara Seas, and ice with a concentration of 40% remains only in the north part of the Laptev Sea—almost the entire Russian Arctic sector is free for ship navigation. The highest concentration is noted in the Greenland and Beaufort Seas [4].

Regarding the interannual variation of the parameter, a significant decrease in the Arctic sea ice concentration value is observed from 2009 to 2019. Firstly, during the summer months of this period, the White and Norwegian Seas are absolutely free of ice. The same situation is also expected in the near future in the Bering Sea, where severe ice conditions were observed only in 2012 (77%), and there was almost no ice in any other year during the period under investigation. Secondly, in the Siberian seas, the rate of melting in July has increased significantly, although the figures for June remain about the same. While the 2009 July sea ice concentration is 60–70%, in 2019 it is only 20–30%. The Greenland and Beaufort Seas are also characterized by a decrease in the monthly concentration—it has become about 1.5–2 times less. However, the situation is rather stable in the Baffin and Labrador Seas. It is possible that for these seas, the eleven-year time interval is insufficient for analysis and a longer interval is required, or the area chosen for the study is relatively stable.

A straight line cut through the combined color image was plotted in the UNESCO Bilko software in order to determine the width of the fast ice (Fig. 3). Such lines are called transects. Google Earth Pro was used for checking purposes (Fig. 4). Calculations were made for all seas in the Arctic region for each summer month from 2009 to 2019. The results are summarised in Table 2, where 0 means ice-free water.

June is characterized by maximum fast ice widths. The greatest sea ice melting occurs in July, which is quite noticeable in the seas of the Russian Arctic. In August, the melting rate slows down significantly.

The Greenland Sea is characterized by minimal reductions in the fast ice width. The maximum difference between June and August is 1/3 of the length, while in other areas the ice melts almost completely.

The coordinates of the ice edge position were taken for each of the seas using Google Earth. The results are recorded in Table 3.

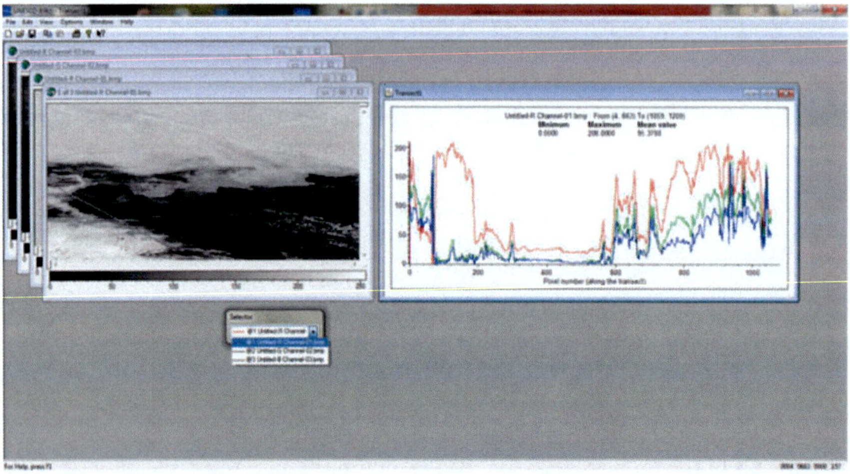

Fig. 3 The process of determining the fast ice width in the Chukchi Sea on a selected transect in the UNESCO Bilko software (July 4, 2009)

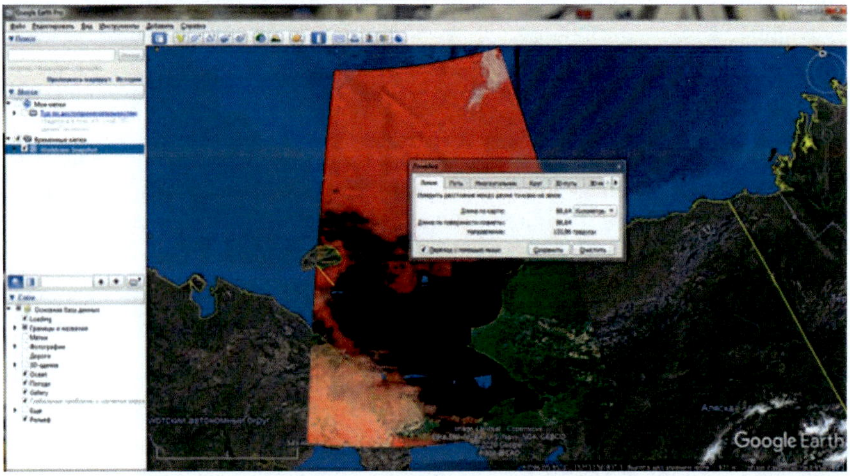

Fig. 4 The process of determining the length of the ice edge in the Chukchi Sea (in the highlighted section) using Google Earth (July 4, 2009)

According to the results obtained, the ice edge is in constant motion –summer melting shifts the southern ice edge closer and closer to the pole. At the same time, melting during the season occurs at different rates, which is determined by the geographical peculiarities of the area. Thus, the most pronounced change in the position of the ice edge during the summer months occurs in the Greenland, East

Table 2 Received results of fast ice width (in kilometers) for the Arctic seas during the summer period

Year	Month	Sea											
		East Siberian	Kara	Beaufort	Laptev	Chukchi	Baffin	Barents	White	Greenland	Norwegian	Bering	Labrador
2009	*VI*	160	247	390	404	211	496	195	0	395	0	47	136
	VII	140	189	129	292	88	156	18	0	383	0	0	11
	VIII	0	0	25	20	0	26	0	0	119	0	0	0
2010	*VI*	177	204	335	400	530	394	74	0	421	0	36	149
	VII	148	84	227	358	200	343	49	0	365	0	0	33
	VIII	73	0	174	5	15	283	0	0	287	0	0	0
2011	*VI*	150	210	357	388	393	499	50	0	391	0	41	87
	VII	115	75	321	282	174	381	47	0	372	0	0	25
	VIII	0	0	171	0	0	35	0	0	281	0	0	0
2012	*VI*	161	43	183	404	323	536	15	0	416	0	64	152
	VII	123	0	12	88	201	84	0	0	374	0	0	55
	VIII	120	0	0	0	130	0	0	0	288	0	0	0
2013	*VI*	223	245	477	282	423	572	158	0	386	0	39	191
	VII	166	166	395	270	119	313	0	0	163	0	0	73
	VIII	85	0	97	0	–	0	0	0	128	0	0	10
2014	*VI*	136	205	389	395	169	–	14	0	392	0	50	126
	VII	11	63	73	140	40	228	0	0	276	0	0	28
	VIII	5	0	0	0	36	0	0	0	259	0	0	0
2015	*VI*	103	158	203	332	243	446	265	0	259	0	5	226
	VII	16	82	187	47	154	365	214	0	196	0	0	193

(continued)

Table 2 (continued)

Year	Month	Sea											
		East Siberian	Kara	Beaufort	Laptev	Chukchi	Baffin	Barents	White	Greenland	Norwegian	Bering	Labrador
	VIII	0	0	149	0	0	–	0	0	133	0	0	90
2016	*VI*	135	38	359	350	335	533	17	0	271	0	37	114
	VII	83	0	321	90	198	246	0	0	178	0	0	104
	VIII	10	0	125	0	0	117	0	0	132	0	0	15
2017	*VI*	86	81	62	401	234	477	230	0	363	0	45	226
	VII	55	35	39	62	90	297	217	0	338	0	0	142
	VIII	0	0	0	20	0	112	0	0	258	0	0	24
2018	*VI*	99	201	345	390	297	676	20	0	261	0	0	216
	VII	67	49	335	–	86	457	0	0	201	0	0	105
	VIII	0	0	263	0	0	0	0	0	167	0	0	0
2019	*VI*	148	138	101	391	239	443	238	0	260	0	98	211
	VII	6	0	52	–	97	143	189	0	257	0	0	15
	VIII	0	0	14	0	0	0	122	0	249	0	0	0

Table 3 Edge position coordinates for the Arctic seas during the summer period

Year	Sea						
	Greenland	Barents	Kara	Laptev	East Siberian	Chukchi	Beaufort
June							
2009	71° 53′ N	73° 12′ N	75° 2′ N	74° 20′ N	69° 44′ N	70° 3′ N	70° 53′ N
2010	72° 29′ N	72° 16′ N	73° 10′ N	72° 17′ N	69° 44′ N	69° 23′ N	72° 18′ N
2011	72° 19′ N	73° 14′ N	74° 26′ N	73° 46′ N	70° 46′ N	71° 26′ N	71° 12′ N
2012	73° 6′ N	74° 33′ N	75° 40′ N	75° 4′ N	72° 1′ N	69° 37′ N	72° 15′ N
2013	74° 39′ N	76° 41′ N	76° 47′ N	74° 49′ N	71° 57′ N	68° 49′ N	70° 28′ N
2014	75° 52′ N	76° 2′ N	76° 11′ N	75° 22′ N	72° 15′ N	71° 51′ N	73° 4′ N
2015	75° 11′ N	81° 11′ N	78° 2′ N	76° 13′ N	73° 0′ N	70° 3′ N	73° 24′ N
2016	76° 24′ N	81° 47′ N	77° 7′ N	75° 1′ N	75° 6′ N	71° 4′ N	72° 58′ N
2017	74° 19′ N	79° 58′ N	79° 51′ N	77° 5′ N	76° 58′ N	71° 43′ N	73° 45′ N
2018	76° 8′ N	82° 8′ N	79° 20′ N	78° 8′ N	77° 16′ N	71° 36′ N	72° 49′ N
2019	77° 9′ N	82° 6′ N	78° 32′ N	79° 13′ N	77° 6′ N	72° 0′ N	73° 5′ N
July							
2009	73° 6′ N	74° 31′ N	75° 45′ N	76° 37′ N	72° 14′ N	71° 37′ N	71° 42′ N
2010	74° 0′ N	76° 33′ N	76° 2′ N	77° 43′ N	73° 24′ N	70° 58′ N	74° 23′ N
2011	74° 29′ N	75° 8′ N	76° 30′ N	77° 15′ N	72° 4′ N	72° 35′ N	71° 48′ N
2012	75° 18′ N	76° 12′ N	77° 1′ N	76° 14′ N	73° 6′ N	71° 36′ N	74° 26′ N
2013	74° 21′ N	77° 11′ N	77° 55′ N	78° 2′ N	75° 11′ N	69° 22′ N	71° 5′ N
2014	76° 43′ N	77° 42′ N	76° 45′ N	77° 13′ N	74° 23′ N	72° 18′ N	74° 33′ N
2015	77° 29′ N	78° 32′ N	78° 54′ N	77° 21′ N	75° 24′ N	71° 45′ N	74° 8′ N
2016	78° 32′ N	81° 52′ N	77° 49′ N	79° 5′ N	76° 3′ N	71° 46′ N	73° 17′ N
2017	80° 9′ N	80° 17′ N	80° 4′ N	78° 43′ N	78° 24′ N	72° 47′ N	73° 53′ N

(continued)

Table 3 (continued)

Year	Sea						
	Greenland	Barents	Kara	Laptev	East Siberian	Chukchi	Beaufort
2018	81° 33′ N	82° 22′ N	81° 7′ N	79° 11′ N	77° 5′ N	73° 35′ N	73° 9′ N
2019	82° 9′ N	82° 11′ N	80° 4′ N	80° 13′ N	78° 13′ N	72° 51′ N	74° 0′ N
August							
2009	74° 11′ N	75° 48′ N	76° 20′ N	77° 10′ N	74° 2′ N	72° 57′ N	72° 6′ N
2010	75° 19′ N	78° 43′ N	76° 58′ N	78° 6′ N	75° 49′ N	71° 43′ N	75° 7′ N
2011	76° 4′ N	77° 33′ N	78° 7′ N	79° 53′ N	74° 54′ N	72° 55′ N	73° 9′ N
2012	77° 13′ N	80° 42′ N	77° 21′ N	77° 51′ N	75° 29′ N	72° 50′ N	75° 5′ N
2013	75° 12′ N	79° 31′ N	77° 5′ N	79° 3′ N	77° 8′ N	70° 3′ N	72° 18′ N
2014	77° 11′ N	80° 1′ N	79° 8′ N	79° 10′ N	76° 45′ N	72° 36′ N	75° 21′ N
2015	78° 9′ N	81° 6′ N	78° 15′ N	80° 17′ N	77° 2′ N	74° 22′ N	74° 59′ N
2016	80° 19′ N	82° 4′ N	78° 37′ N	81° 33′ N	78° 51′ N	73° 5′ N	73° 41′ N
2017	82° 58′ N	81° 31′ N	77° 4′ N	80° 9′ N	79° 2′ N	74° 0′ N	75° 1′ N
2018	83° 57′ N	83° 9′ N	80° 22′ N	81° 12′ N	79° 34′ N	73° 19′ N	74° 1′ N
2019	84° 9′ N	83° 42′ N	82° 20′ N	81° 53′ N	80° 8′ N	74° 3′ N	74° 58′ N
Longitude	16° 55′ W	21° 13′ E	59° 31′ E	134° 23′ E	177° 32′ E	173° 15′ W	127° 26′ W

Siberian and Laptev Seas. The position of sea ice edge shifts on average approximately 458, 449 and 441 km between the sixth and eighth month of the year for the above-mentioned seas.

Regarding the temporal variability on a large scale, namely between 2009 and 2019, it is worth noting that the difference is strongest between August of the first year and the last year. The edge position lines presented in Fig. 5 are relatively similarly located along the entire border in June. At the same time, August is characterized by a stronger shift of the edge towards the north in the eastern Arctic.

We also classify winter seasons of the Russian Arctic seas by their severity. At first, the data, which represent the daily values of atmospheric temperature in the Arctic seas, were used for calculating the cumulative freezing-degree days (CFDD). At the second stage, the results were summarized in dendrograms using Ward's hierarchical clustering method.

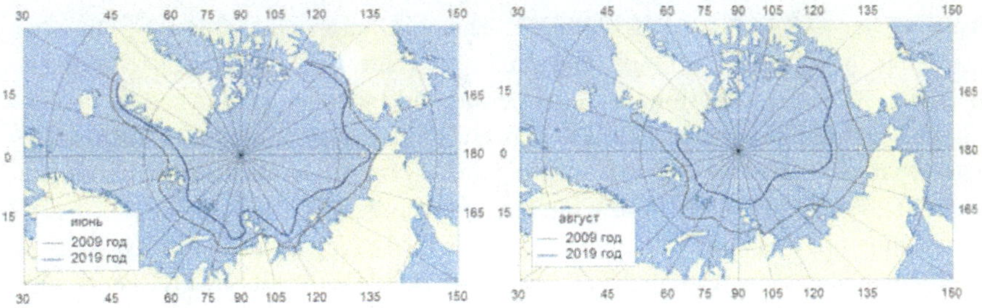

Fig. 5 Arctic residual ice edge position

The White Sea. In this research, we summarized negative air temperatures of winter seasons from Arkhangelsk in the period from 1885 to 2005. As a result, the following classification was obtained according to the severity of winter [10]: mild winter—up to 1052 CFDD, moderate winter—1053–1578 CFDD, severe winter—more than 1579 CFDD. The minimum value recorded in the winter of 1936/37 is 760 CFDD. The maximum value recorded in the winter of 1901/02 is 2288 CFDD. During the study period, moderate winters were observed in 57% of cases, while severe and mild winters—in 33% and 10%, respectively.

The Kara Sea. To classify the severity of winters in the Kara Sea, meteorological data from 1978 to 2017 from the port Dickson were used. As a result, the following classification of winters was obtained [11]: mild winter—up to 3300 CFDD, moderate winter—3301–4500 CFDD, severe winter—more than 4501 CFDD. It should be noted that in the period from 2004 to 2016, there was no severe winter, while a mild winter with a minimum value of 2058 CFDD was recorded in the winter of 2013/14. The maximum value recorded in the winter of 1978/79 was 5426 CFDD. During the study period, mild winters were observed in 46% of cases, severe winters—in 36% and mild winters—in 18%. It should be noted that mild winters have been recorded since 1993–94, while severe winters have not been recorded since 2004.

The Laptev Sea. To classify the severity of winters in the Laptev Sea, weather data from meteorological station Tiksi in the period from 1936 to 2018 were used. As a result, the following classification of winters was obtained: mild winter—up to 4900 CFDD, moderate winter—4901–5650 CFDD, severe winter—more than 5651 CFDD. The minimum value—4454 CFDD was recorded in the winter of 1942/43. The maximum value - 6210 CFDD was recorded in the winter of 1940/41. During the study period, mild winters were observed in 64% of cases, while severe and mild winters—in 20% and 16%, respectively. It should be noted that in the period from 2004 to 2017 there was no severe winter, and in the period from 2013 to 2016 there were only mild winters.

The East Siberian Sea. To classify the severity of winters in the East Siberian Sea, meteorological data from the weather achieve on Wrangel Island were used in the period from 1960 to 2017. As a result, the following classification of winters was obtained: mild winter—up to 3850 CFDD, moderate winter—3851–4500 CFDD, severe winter—more than 4501 CFDD. The minimum value 2649 CFDD was recorded in the winter of 2017/18. The maximum value 4957 CFDD was recorded in the winter of 1975/76. During the study period, mild winters were observed in 57% of cases, while mild winters—in 33% and severe winters—in only 10% of cases. It should be noted that in the period from 1984 to 2017 no severe winters were recorded, and from 2001 to 2017 only mild winters were recorded.

Summing up, we can say that the decrease in the sea ice area and fast ice width is directly related to climate change in the Arctic region. The results are consistent with the presented classification, which shows an increase in mild winters in the Arctic seas during the period under investigation.

References

1. Kostyanoi A.G. Sputnikovyi monitoring klimaticheskih parametrov okeana. Chast' 1. [Satellite monitoring of the ocean climate parameters. Part 1.]. *Fundamental'naya i prikladnaya klimatologiya [Fundamental and Applied Climatology(Russia)]*, 2017, vol. 2, pp. 63–85. (In Russian; abstract in English).
2. Sputnikovye metody opredeleniia kharakteristik ledianogo pokrova morei. Satellite methods for determination of sea ice cover characteristics. Ed by V.G. Smirnov. St. Petersburg, AARI Edition, 2011, 239 p. (In Russian).
3. Repina I.A., Ivanov V.V. Primenenie metodov distancionnogo zondirovaniya v issledovanii dinamiki ledyanogo pokrova i sovremennoi klimaticheskoi izmenchivosti Arktiki. [Application of remote sensing methods in the study of ice cover dynamics and current climate variability in the Arctic]. *Sovremennye problemy distantsionnogo zondirovaniya Zemli iz kosmosa [Modern problems of remote sensing of the Earth from space]*, 2012, vol.9, pp. 89–103. (In Russian).
4. Bukatov A. E., Bukatov A. A., Babii M. V. Prostranstvenno-vremennaya izmenchivost' rasprostraneniya morskogo l'da v Arktike. [Spatial and temporal variability of sea ice distribution in the Arctic]. *Kriosfera Zemli [Earth's cryosphere]*, 2017, vol. 21, pp. 85–92. (In Russian).
5. Smirnov V. G., Bushuev A. V., Bychkova I. A., Zahvatkina N. YU., Loshchilov V. S. Sputnikovyi monitoring morskih l'dov. [Satellite monitoring of sea ice]. *Problemy Arktiki i Antarktiki [Problems of the Arctic and Antarctic]*. Issue 85. St. Petersburg: Gidrometeoizdat, 2010, pp. 62–75. (In Russian).
6. National Snow and Ice Data Center. Available at: https://nsidc.org/data/seaice_index/ (Accessed 02.06.2020).
7. EOSDIS Worldview. Available at: https://worldview.earthdata.nasa.gov (Accessed 10.06.2020).
8. EarthExplorer. USGS. Available at: https://earthexplorer.usgs.gov/ (Accessed 10.06.2020).
9. Teider M.S., Oskotskaia S.A., Frolova N.S., Podrezova N.A. The assessment of Arctic sea ice area changes. Springer proceedings in Earth and Environmental Sciences. 2022, pp. 231–241.
10. Sergeev D.I., Podrezova N.A. Calculation of ice growing in the White Sea in conditions of mild, moderate and severe winter. Meteorologicheskii Vestnik [The Meteorological Bulletin], 2017, vol. 9, no. 2, pp.187-190 (In Russian).
11. Petruchenko A.I., Kruglova E.E., Mechova O.S., Podrezova N.A. Assessment of sea ice changes in the Kara Sea depending on the severity of winter. Trudy IX Mezhdunarodnoi nauchno-prakticheskoi konferentsii «Morskie issledovaniia i obrazovanie» [Proc. of the IX International Scientific and Practical Conference «Marine Research and Education»]. Tver, 2020, pp. 273–274. (In Russian).

Interannual Variability of Ice Coverage in the Area of the Franz Josef Land Archipelago

N. A. Podrezova and M. I. Zemilova

Abstract This paper presents and analyzes the results of the study of ice coverage changes in the area of the Franz Josef Land archipelago according to the data of the Arctic and Antarctic Research Institute archive (AARI World Data Center for Sea Ice, [1]) over the period 1979–2018. Descriptive statistics and trends were used as statistical research methods. The paper also provides a classification by the severity of winter for the study area.

Keywords Arctic · Franz Josef Land archipelago · Area of ice coverage · Winter severity · Classification of winters

For most of the year the straits of the Franz Josef Land archipelago (FJL) are completely covered with fixed first-year ice. The shore begins to freeze in mid-September, and in early October, ice is spreading across the entire water surface. Sea ice has its maximum thickness in the end of the winter period (in April) and can be up to 1.5 m thick in bays and along shores, but in some straits, it can be quite thin (20–40 cm). The melting and destruction of first-year ice are most intense in July and August. The archipelago's straits become ice-free by mid-August [2]. The Atlantic waters have a noticeable warming effect on the areas to the south of the archipelago.

To calculate cumulative freezing-degree days (CFDD), daily data of the average air temperature at the E. Krenkel station on Heiss Island were taken. The data used for obtaining the necessary information is located at the site [3] of the Climate Data Library. Surface water temperature values were also obtained from the E. Krenkel station.

For calculations and plotting, three quasi-homogeneous regions were identified near the FJL archipelago—northern, eastern, southern (Fig. 1). The eastern region, which is significantly affected by the ice carried from the Kara Sea. The southern region, which is affected by Atlantic waters spreading in the northeastern part of the

N. A. Podrezova (✉) · M. I. Zemilova
Russian State Hydrometeorological University, St. Petersburg, Russia
e-mail: Nadinapod@mail.ru

M. I. Zemilova
Saint Petersburg State University, St. Petersburg, Russia

V. I. Karev (ed.), *Physical and Mathematical Modeling of Earth and Environment Processes—2022*, Springer Proceedings in Earth and Environmental Sciences,
https://doi.org/10.1007/978-3-031-25962-3_12

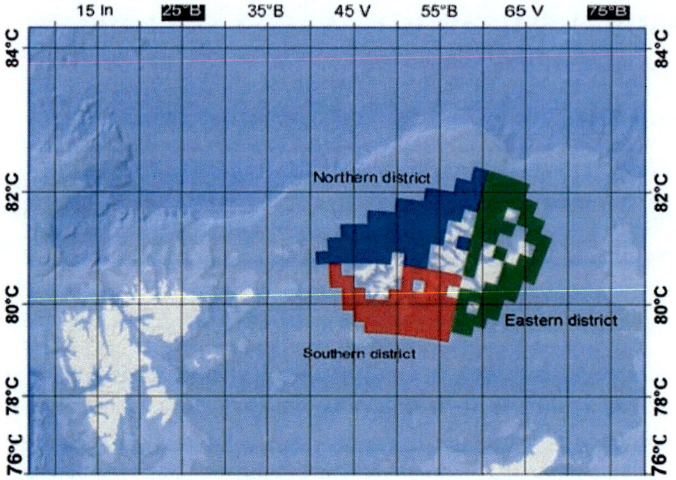

Fig. 1 Quasi-homogeneous areas of the Franz Josef Land archipelago water area

Barents Sea. The northern region, which is influenced by the export of perennial ice from the central part of the Arctic basin [4].

The linear trend of the ice coverage characteristic was calculated for the FJL archipelago water area in three areas—northern, eastern and southern. The trend value in the northern part of the archipelago – 25,000 km^2/year, – 18,000 km^2/year in the eastern part and – 24,000 km^2/year in the southern part.

Figure 2 shows that the ice coverage for all three regions has a negative trend, which also reflects the obtained trend values. The maximum values of ice coverage have been observed in the last century—in 1982 in the northern region and in 1999 in the eastern and southern regions, and the minimum values of ice coverage—in the 2010s—in 2018 in the northern region, in 2012 in the eastern region and in 2016 in the southern region.

The presence of such a trend for all regions of the FJL archipelago is inherent to the Arctic Ocean due to global warming, increasing average air temperature values for the planet Earth [5–8].

Figure 2 shows that over the period 1979–2018 there is a mismatch of maximums and minimums in certain years in different areas. Thus, in 1999 and 2004, ice coverage values in the eastern and southern regions of the archipelago have one of the maximum values in those area, while in the northern region—one of the minimums. In 2001 and 2006, the ice coverage in the northern and southern regions is one of the lowest in that area, while in the eastern region, on the contrary, it's one of the highest.

The average annual monthly values of ice coverage and air temperature are shown in Fig. 3. The ice coverage values during the period from May to September for all regions (northern, eastern, southern) decrease, the minimum ice coverage value is reached in September. On the contrary, the air temperature increases during the indicated period and reaches its maximum values in July. It is noted that there is

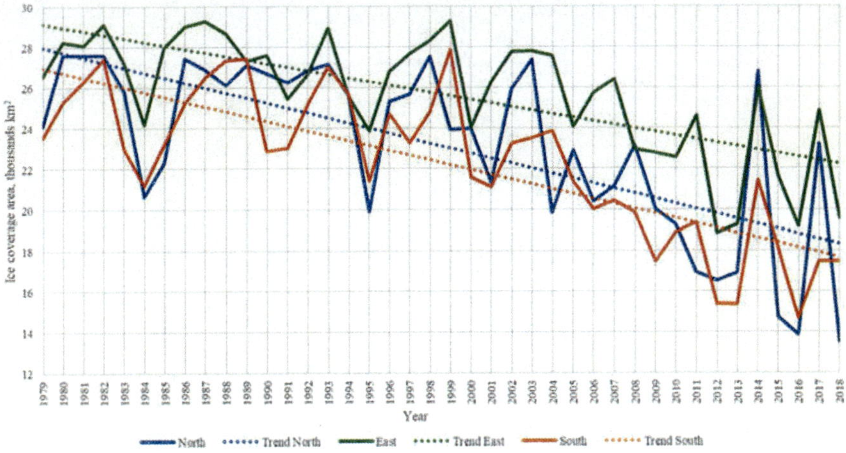

Fig. 2 Annual average ice coverage area (thousands km^2) in three areas of the FJL archipelago and their trends (North, East, South) over the period 1979–2018

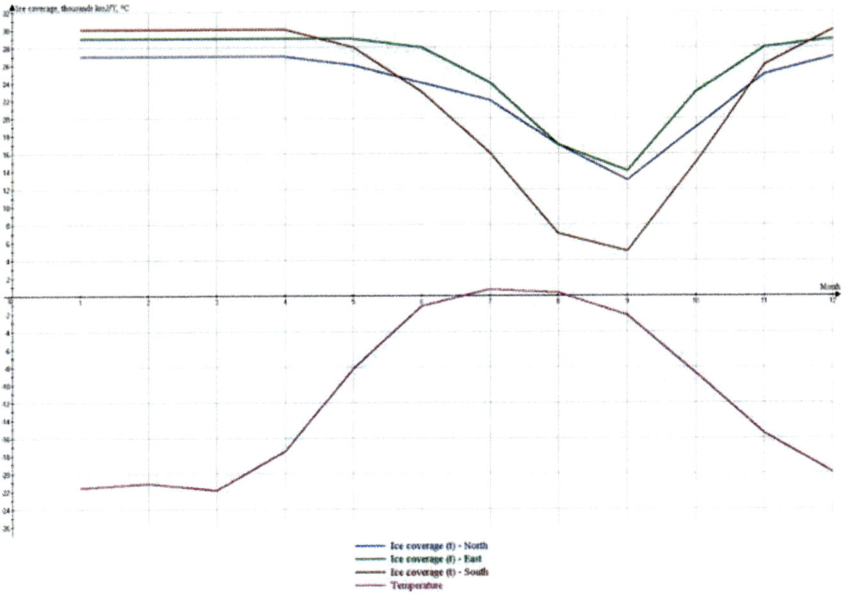

Fig. 3 Graph of the distribution of average monthly values of ice coverage and air temperature in three regions of the FJL archipelago water area for the period 1979–2018

a shift between the maximum for temperature and minimum for ice coverage—the temperature maximum occurs two months earlier than ice coverage minimum.

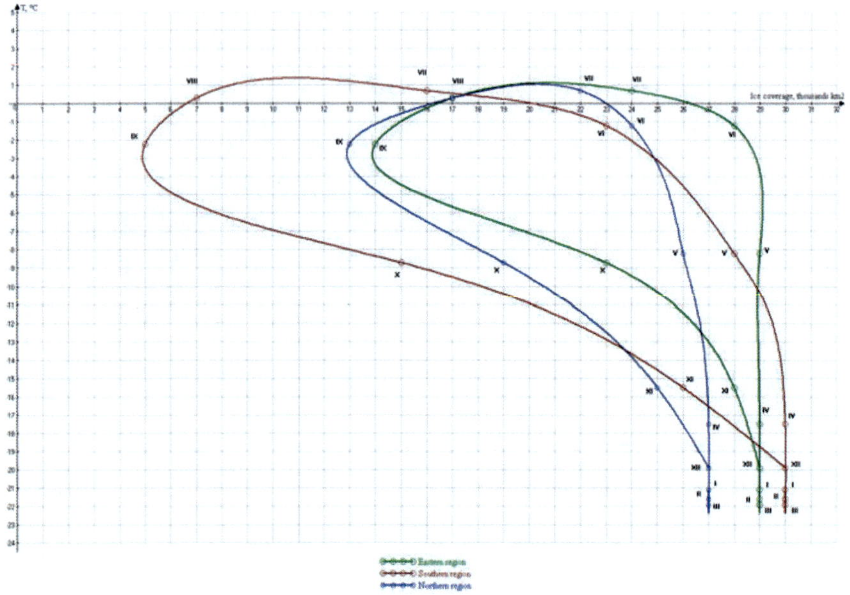

Fig. 4 Hysteresis «loops» average monthly values of ice coverage (thousands km^2) and air temperature for three regions of the FJL archipelago (Roman numerals indicate months)

The detected shift is a characteristic feature for the construction of hysteresis «loops» (Fig. 4). The hysteresis loops for each region are plotted separately and combined in one graph for a more visual representation of the relationship between ice coverage and air temperature. It's noted that the most elongated hysteresis «loop» is presented for the southern region of the FJL archipelago water area. Such a distinctly elongated hysteresis «loop» for the southern region is defined by the presence of minimum ice coverage values in September and maximum ice coverage values in January-March. The presence of such a minimum of ice coverage in September is directly connected with the previously mentioned shift of the temperature maximum and area minimum. This shift between the maximum and minimum indicates the way the relationship between ice coverage and air temperature works: the effect of having a maximum temperature only impacts the area decrease after some time (two months), not immediately (Fig. 3). Therefore, hence the presence of hysteresis—when the ice coverage depends on previous values of air temperature, i.e., it reacts to effect after a certain period of time. The significant, compared to other regions, elongation of the «loop» in the southern region is caused by the presence of a greater number of ice coverage minimums in August and September, and this, as a result, is connected with the presence of maximum temperatures in July. It may be noted that the elongation of the «loop» is least prominent in the northern region, the «loop» for the eastern region is less distinct in comparison with the southern region, but more prominent compared with the northern region.

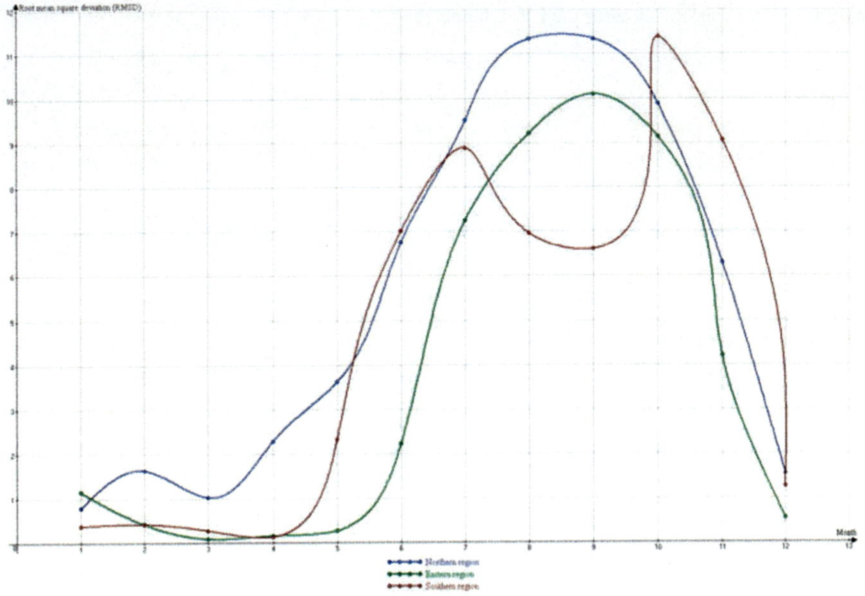

Fig. 5 RMSD of ice coverage for northern, eastern and southern regions of the FJL archipelago water area

The seasonal variability of ice coverage was evaluated by using the calculation of the root mean square deviation (RMSD) for each month for the entire series of observations (1979–2018). Figure 5 indicates that the southern region, compared to the northern and eastern regions, is characterized by a variation in the RMSD values in August and September. The RMSD value for the southern region decreases from July to August, compared to an increase in this value for the other two regions. This is due to the possibility that during the period of ice coverage minimums (August–September) near the FJL archipelago, drifting ice in the southern region can disappear completely. This became particularly frequent from 2000 onwards.

The winter severity for the FJL archipelago was calculated using the cumulative freezing-degree days (CFDD), based on the daily average values of surface air temperature at the Krenkel Observatory station for the period 1979–2018. As a result of calculations for the FJL archipelago the following winter classification was determined: Mild winter—up to 2750 CFDD, Moderate winter—2751–5166 CFDD, Severe winter—more than 5167 CFDD. The winter classification for the years analyzed is given below in Table 1. The minimum value of 2293 CFDD was recorded in the winter of 2011/12. The maximum value of 5502 CFDD was recorded in the winter of 1980/81. Moderate winter are found in 80% of the period analyzed, while severe and mild winters are seen in 10% each. It can be observed that mild winters are observed only in the last decade, while severe winters were recorded only before 1993.

Table 1 Winter classification for the FJL arhipelago

Year	Severity	Year	Severity	Year	Severity
1979–80	Moderate	1992–93	Severe	2005–06	Moderate
1980–81	Severe	1993–94	Moderate	2006–07	Moderate
1981–82	Moderate	1994–95	Moderate	2007–08	Moderate
1982–83	Moderate	1995–96	Moderate	2008–09	Moderate
1983–84	Moderate	1996–97	Moderate	2009–10	Moderate
1984–85	Moderate	1997–98	Moderate	2010–11	Mild
1985–86	Moderate	1998–99	Moderate	2011–12	Mild
1986–87	Moderate	1999–00	Moderate	2012–13	Moderate
1987–88	Severe	2000–01	Moderate	2013–14	Moderate
1988–89	Severe	2001–02	Moderate	2014–15	Moderate
1989–90	Moderate	2002–03	Moderate	2015–16	Mild
1990–91	Moderate	2003–04	Moderate	2016–17	Mild
1991–92	Moderate	2004–05	Moderate	2017–18	Moderate

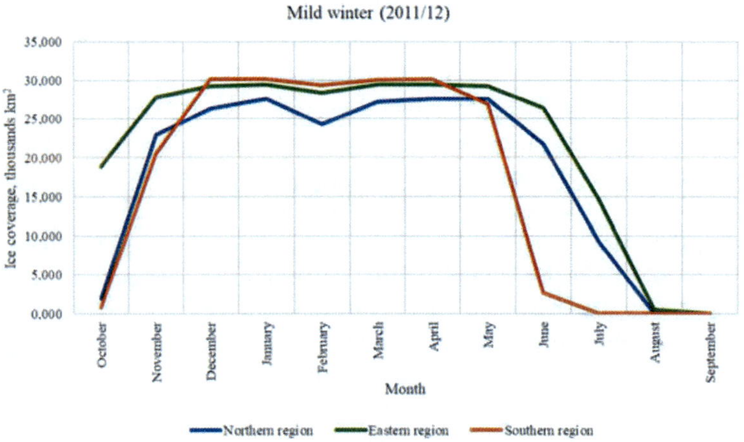

Fig. 6 Distribution of ice coverage (thousand km^2) for three regions of the FJL archipelago for a mild winter (2011/12)

Then, the ice coverage distribution for each region of the FJL archipelago for all types of winter was graphically compiled (Figs. 6, 7 and 8).

Figure 6 shows the mild winter for the period 2011–2012. For this winter type, the CFDD has been calculated since October 2011. The graph represents three water regions of the FJL archipelago—northern, eastern, and southern. It can be observed that the maximum values of ice coverage in the southern region are within the same range as for severe and moderate winters—about 30,000 km^2. Yet, the period over which they are observed has shortened: from December to May for mild winters,

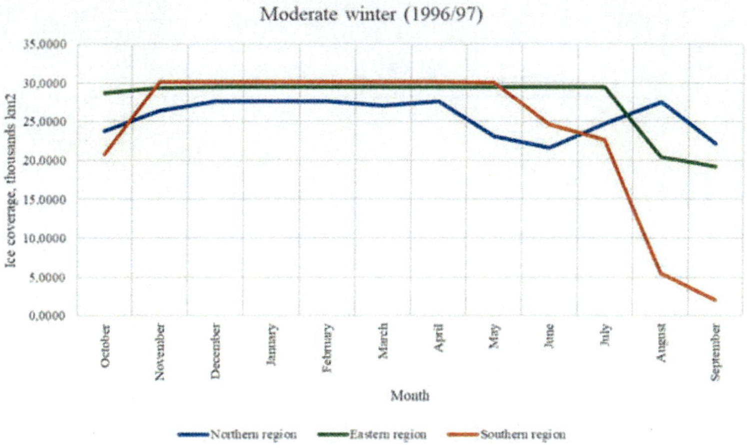

Fig. 7 Distribution of ice coverage (thousand km^2) for three regions of the FJL archipelago for a moderate winter (1996/97)

compared to November–April for severe winters and November–May for moderate winters. The minimum values for the northern region are found in August–September, in comparison with July in a moderate winter, and for eastern region—in August–September, compared to September in a moderate winter. Minimum values in a mild winter in the northern region of ice coverage are equal to 0, in these months ice is not detected (August–September of 2012), while in the same months in the northern region the ice coverage was 22,000–28,000 km^2. In the eastern region in September of 2012 the ice coverage was equal to 0 in a mild winter, while in a moderate one it was equal to 19,000 km^2. Consequently, the decline of ice coverage is more rapid and more pronounced in mild winters, when air temperatures can peak at their local maximums. During a mild winter the southern region, the minimum ice coverage is already found in July of 2012 and is equal to 0, and remains unchanged till September. During a moderate winter in the southern region, the ice coverage only reached its minimum value in September (at the end of the observed period); during a severe winter, the minimum was recorded only at the beginning of the period.

Figure 7 illustrates the distribution of ice coverage in the northern, eastern and southern areas on the FJL archipelago for a moderate winter over the period 1996–1997. The calculation of CFDD for this period began in October 1966. Similar to the severe winter type, the ice coverage in the southern region has the maximum and also the minimum values of all three regions for this period. So, the maximum values of ice coverage are recorded from November 1996 to May 1997, and the minimum value—in September 1997, equal to 2000 km^2. It may be pointed out that the minimum value of ice coverage in the northern region is in June 1997, equal to 22,000 km^2, and then it increases by August and almost reaches its maximum value of 28,000 km^2. In the eastern region, the ice coverage practically remains the same, about 30,000 km^2, from October to July, and by September it reaches its minimum value, equal to 19,000 km^2.

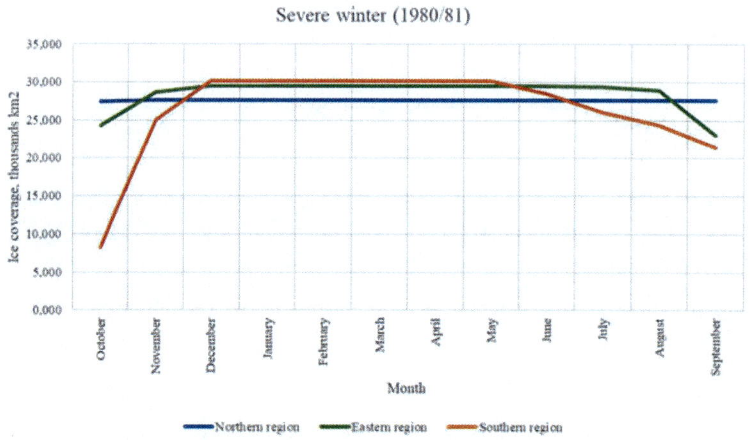

Fig. 8 Distribution of ice coverage (thousand km^2) for three regions of the FJL archipelago for a severe winter (1980/81)

Figure 8 represents the severe winter for the period 1980–81. The changes in ice coverage are spread from September 1980 to August 1981. It can be observed that the ice coverage in the southern region has the maximum and also the minimum values for the indicated period. So, the maximum values of ice coverage are recorded from November 1980 to April 1981 and are equal to 30,000 km^2, and the minimum values of ice coverage—in September 1980 (the start of the CFDD calculation and are equal to 8000 km^2. Ice coverage values in the northern region almost remain the same during the indicated period, in the eastern region they hardly change, and reach their minimum values in September and August of the analyzed winter period.

The comparison of the data on the ice coverage distribution for the regions of the FJL archipelago for severe and moderate winters revealed that the impact if surface air temperature mainly affects the summer-autumn period of the years in question. When temperatures are low (severe winter), the ice coverage values have slight deviations from their average values, while, in moderate winters, a significant drop in the ice coverage by the summer period is observed. For example, in the southern region, ice coverage reduced by 21,000 km^2 from July to September 1997.

To sum up, it can be concluded that the southern region has absolute maximum and minimum ice coverage values, unlike other regions, for all types of winters. As stated before, these values are variable for each type of winter, but the maximum values remain equal to more than 30,000 km^2. From July 2012 there was no ice in the mild winter in the southern region, from August in the northern region and from September in the eastern region.

The occurrence of a negative trend in the changes in ice coverage from 1979 to 2018 in the region in question is also supported by other researchers [9, 10], which corresponds to the currently ongoing global warming trend.

References

1. AARI World Data Center for Sea Ice. URL: http://wdc.aari.ru/datasets/ssmi (Accessed 28.08.2022)
2. Physical-geographical and climatic characteristics of the Franz Josef Land archipelago. URL: https://textarchive.ru/c-1821739- p3.html (Accessed 30.05.2021)
3. Climate data library. URL: aisori.meteo.ru (Accessed 24.05.2021)
4. B.S. Shapkin, A.V. Rubchenya, B.V. Ivanov, A.D. Revina, M.V. Bogryantsev. Multilayer changes in ice coverage in the area of the Svalbard and Franz Josef Land archipelagos. – «Ice and Snow», 2021, vol.61, №1.
5. Teider M.S., Oskotskaia S.A., Frolova N.S., Podrezova N.A. The assessment of Arctic Sea ice area changes. In the collection: Physical and Mathematical Modeling of Earth and Environment Processes. Proceedings of 7th International Conference. Ser. "Springer Proceedings in Earth and Environmental Sciences" 2022. P. 231–241.
6. G. G. Matishov, A. P. Zhichkin. Current trends in the ice coverage in the Franz Josef Land archipelago area. – Papers of the Academy of Science, 2017, vol. 472, № 6, p. 708–711.
7. Teider M.S., Oskotskaya S.A., Frolova N.S., Podrezova N.A. Interannual change in Arctic ice. 2021. № 4 (30). P. 1424–1435.
8. Zhichkin A.P. Ice conditions in the Franz Josef Land region. Trudy Kolskogo nauchnogo centra RAN. Proc. of the Kola Science Center of the Russian Academy of Sciences. 2014, 4: 82–89 [In Russian].
9. Timokhov L.A., Vyazigina N.A., Mironov E.U., Yulin A.V. Climatic changes of seasonal and inter-annual variability of the ice cover of the Greenland and Barents Seas. Problemy Arktiki i Antarktiki. Problems of Arctic and Antarctic. 2019, 65 (2): 148–168. doi: https://doi.org/10.30758/05552648-2019-65-2-148-168. [In Russian]
10. **Alekseev G.V., Kuzmina S.I., Glok N.I., Vyazilova A.E., Ivanov N.E., Smirnov A.V. Influence** of Atlantic on the warming and reduction of sea ice in the Arctic. Led i Sneg. Ice and Snow. 2017, 57 (3): 381–390. [In Russian].

Analysis of Laboratory Hydraulic Fracturing Pressure-time Curves

Elena Novikova and Mariia Trimonova

Abstract Hydraulic fracturing is the process of fracture formation in the reservoir under development by pumping fracturing fluid under pressure through previously prepared well into the formation. Methods of processing the pressure drop curves, that were recorded during preliminary conducted mini hydraulic fracturing process, have always been a subject of interest. Various methods' constant development and application has been going on until now, since such an approach to the data analysis allows to obtain a sufficient number of reservoir and fracture characteristics that significantly simplifies the further development of real deposits. In the course of this study, a series of laboratory experiments on hydraulic fracturing was conducted. The pressure-time data was recorded for each of laboratory experiment. The main aim of the work was to process the obtained dependencies using one of the standard method for analyzing hydraulic fracturing pressure drop curves. A special laboratory setup was used for laboratory modeling of hydraulic fracturing. All the materials for a series of experiments were selected in accordance with similarity criteria. Hence the conditions of laboratory experiments on hydraulic fracturing achieve maximum similarity with the conditions on real field under development. The obtained pressure-time dependencies were analyzed using two standard methods. The G-function method was used as a preliminary stage of the analysis. The focus was on pressure closure determination, as this characteristic played an important role in further analysis. Then the main stage of pressure drop curve processing was implemented. It was aimed at determination of the flow modes that were realized in the model sample with hydraulic fracture after the injection was stopped. Various flow modes can be realized in the model sample due to fluid filtration from the hydraulic fracturing zone into the model sample during the period of time after the injection is stopped. The analysis method considers two different stages: time period before fracture closure and time period after fracture closure. Thus, bilinear and linear flow regimes can be observed in the time period before fracture closure, and pseudo-linear, bilinear and pseudo-radial flow regimes can be observed in the time period after fracture closure. This approach to pressure drop curves analysis allows to determine some characteristics of the model sample and the fracture formed. After special processing of

E. Novikova (✉) · M. Trimonova
Sadovsky Institute of Geosphere Dynamics of Russian Academy of Sciences, Moscow, Russia
e-mail: helenvn97@gmail.com

V. I. Karev (ed.), *Physical and Mathematical Modeling of Earth and Environment Processes—2022*, Springer Proceedings in Earth and Environmental Sciences,
https://doi.org/10.1007/978-3-031-25962-3_13

data related to a time period with a certain flow regime, it is possible to estimate the permeability of the model sample and the fracture length. Since a series of laboratory experiments was conducted, some of characteristics of the used materials are known as preliminary set parameters or as the results of the experiment. Consequently, it is possible to make a comparison of the real and theoretically obtained characteristics of the model sample.

Keywords Hydraulic fracturing · Laboratory experiments · Data analysis

1 Introduction

Hydraulic fracturing is a method of mechanical impact on a productive oil formation by pumping fluid under pressure at a constant rate. It leads to the fracture formation. It is one of the most efficient and widely spread methods of oil production intensification. Hydraulic fracturing is a procedure that requires a lot of money and time expenses. It is caused by the complexity of the preparation process and the duration of the process on the real field. Therefore, before the development of a field, hydrodynamic studies of wells are carried out, as well as mini hydraulic fracturing. These procedures consist of a special set of operations aimed at determining the structure, properties and characteristics of the reservoir being developed. Such analyses allow to obtain time-pressure records at injection and production wells. Numerous methods were developed for analyzing these dependencies. The methods are aimed at assessing some parameters of the formation and the hydraulic fracture. It is possible to determine the size of the developed zone, the size of the fracture, the permeability of the rock or the coefficient of fluid leakage from the fracture into the formation.

Development of various methods for analyzing pressure dependencies has recently been one of the most important issues. These modified methods help to select necessary characteristics of the fracturing fluid and the proppant. It greatly simplifies the process of hydraulic fracturing in real field conditions. Among the existing approaches, the most well-known and widely used methods are the Horner method [1], the Nolte method [2], the Mayerhofer method [3] and the log-log method [4, 5].

All these methods are aimed at analyzing the time-pressure dependencies obtained during hydraulic fracturing. They differ in the approximations used and the algorithms of analysis.

This work was devoted to the analysis of time-pressure dependencies obtained during a series of laboratory experiments on mini hydraulic fracturing. The log-log method was used as the main method for processing experimental data. One of the goals of this study was to identify the regimes of fluid flow during the time period from the moment the injection was stopped until the end of the laboratory experiment. Another important task of this study was to determine the permeability of the model sample in which the hydraulic fracture formed and propagated.

2 The Log-Log Method

The log-log method is one of the most reliable and precise methods for processing pressure-time dependencies obtained during mini hydraulic fracturing. It allows to analyze the processes in the medium under development before fracture closure and after. The main goal of this method contains correct identification of flow regimes that are implemented in a formation with hydraulic fracturing. Thus, a linear flow regime can be observed in the formation before the fracture closure. After the fracture is closed, the following regimes can be observed: pseudo-linear, bilinear and pseudo-radial.

All the regimes described above are determined on the graph of pressure and semi-logarithmic derivative of pressure on time plotted as a function on a logarithmic scale graph. Each of the regimes on such graph is associated with a certain segment of a semi-logarithmic pressure derivative curve that has a characteristic slope.

The linear flow regime is realized before the fracture is closed and considers only the flow inside the fracture. According to the theory under consideration, linear flow regime assumes that the process of fluid reservation in the fracture has stopped, that is, there is an open fracture in the formation and fluid leakage in the surrounding area occurs. At the corresponding period of time the pressure in the fracture is considered proportional to the square root of time [6]. Then, the semi-logarithmic derivative during this period will have a slope of ½. It is possible to estimate the half-length of the fracture, knowing the interval with the realized linear flow regime, plotting the graph of pressure-time data for this period in a rectangular coordinate system. Calculation is done using formula (1) [6].

$$L_f^2 = \frac{qE}{2\pi\left(1 - \nu^2\right)h} \frac{\Delta t}{\Delta p} \tag{1}$$

L_f—half-length of the fracture;
ν—Poisson's coefficient;
E—Young modulus;
q—injection rate;
h—fracture height.

After the supposed fracture closure pseudo-linear, bilinear or pseudo-radial regimes can be observed [4, 7].

In the frame of the theory, it is assumed that a conductive region remains instead of the closed fracture. This region can have different dimensions and characteristics, on which partly depend the type of flow regime realized in the formation. For instance, a pseudo-linear flow regime can be observed in the conditions when the formed fracture was long and did not close, that is, the fracture area still has a sufficiently large conductivity to neglect the flow in the surrounding medium due to leakage. In this case, the pressure will be proportional to the time in degree − 1/2. Then the slope of the semi-logarithmic derivative of pressure on a logarithmic graph during this period will be − 1/2. The bilinear flow regime is realized in a reservoir in the case when the fracture was long and not completely closed. In this case, the simultaneous flow in the fracture and in the formation is taken into account. The pressure in the fracture will be proportional to the time in degree – 3/4. Similar to pseudo-linear regime determination, the necessary time period will be associated with a segment of semi-logarithmic derivative of pressure with the characteristic slope equal to – 3/4. The pseudo-radial flow regime can be observed when the formed fracture was short or the area with residual conductivity is small. Pressure in this case is inversely proportional to the time. Correspondingly, the slope of the semi-logarithmic derivative of pressure on the logarithmic graph will have a slope – 1.

For each flow regime, the corresponding pressure-time dependencies were derived from the equation of unsteady-state fluid flow in a fracture [4, 8, 9]. The derived dependencies are presented in the formulas (2—4), respectively.

$$\Delta p = \frac{V}{16\pi h}\left(\frac{\mu}{\varphi c_t k L_f^2}\right)^{0,5}\left(\frac{1}{\Delta t}\right)^{0,5} \quad (2)$$

$$\Delta p = \frac{V}{4\pi h}(\mu)^{0,75}\left(\frac{1}{\varphi c_t k}\right)^{0,25}\frac{1}{\sqrt{k_f w_f}}\left(\frac{1}{\Delta t}\right)^{0,75} \quad (3)$$

$$\Delta p = \frac{V\mu}{4\pi kh}\frac{1}{\Delta t} \quad (4)$$

V—volume of the fluid injected;
k—permeability of the porous medium;
$k_f w_f$— fracture conductivity;
φ—porosity (of model sample);
μ—fracturing fluid viscosity;
c_t— total compressibility of the model sample.
It is possible to calculate the permeability values of the model medium using proposed dependencies for observed flow regimes.

3 Laboratory Experiment

In the course of the work, the experimentally obtained pressure-time dependencies were analyzed. The purpose of the analysis was to determine the flow regimes realized in a fractured model sample and to estimate values of permeability of the model medium and the half-length of the formed fracture.

To conduct a series of laboratory experiments on mini hydraulic fracturing, a special laboratory setup was used (see Fig. 1a) [10]. It was constructed at the Institute of Geosphere Dynamics. One of the significant characteristic of the setup is the possibility of setting the triaxial loading of the model sample. This allows to achieve

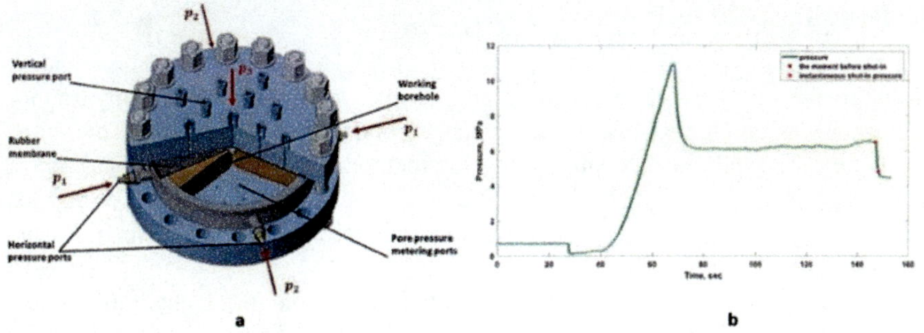

Fig. 1 The scheme of the experimental setup. **a** And the view of pressure-time curve obtained during the laboratory experiment on hydraulic fracturing. **b** The distribution of the applied confining loads to the model sample (p1 and p2—horizontal loads, p3—vertical load) is shown in the left picture **a**

such conditions for conducting laboratory experiments that would be the closest to the conditions of hydraulic fracturing at a real field under development.

The frame of the setup consists of two steel covers in the form of a disk (upper and lower cover-disks), as well as a ring-wall. In total, these details make up a working chamber into which a special mixture is poured. This special material is used for simulation of the medium in which the hydraulic fracture forms and propagates. The dimensions of the working chamber: height-6.6 cm, diameter-43 cm. The dimensions of the solid model sample are identical to the dimensions of the working chamber.

The process of a laboratory experiment on mini hydraulic fracturing is carried out in several stages. First of all, a modeling material is prepared using mixture of gypsum and cement. The characteristics of all materials in the laboratory experiments are selected according to similarity criteria [11]. After that, a model medium is poured into the working chamber of the setup, after complete solidification of the model material, it is saturated with a water solution of gypsum. As a result, a saturated model sample with certain porosity is obtained.

After the preliminary preparation of the model medium, the model sample is covered with rubber membrane, and the upper cover of the setup is closed. Then the triaxial loading on the model sample is performed. The lithostatic pressure is modeled by pumping fluid under pressure into the existing gap between the membrane and upper cover of the installation. Lateral loads are conducted by pumping fluid under pressure into special copper chambers that are mounted in the ring-wall of the setup.

After preliminary preparation, hydraulic fracturing is performed. The fracture is formed as a result of fluid pumping through a prepared borehole in the center of the model sample with a constant rate.

As a result, of series of laboratory experiments, there were obtained horizontal fractures. During each experiment the dependencies of pressure versus time were recorded. An example of obtained pressure-time curve is presented in the Fig. 1b.

4 Results

In the course of the work, a series of laboratory experiments on mini hydraulic fracturing was conducted, the experimentally obtained dependencies of pressure versus time were processed using the log-log method (see Fig. 2). There was not a single experiment in which all existing flow regimes were observed. According to the frequency of detection, the following flow regimes can be distinguished: linear and pseudo-linear.

The log-log method allowed to estimate half-length of the formed hydraulic fractures and calculate permeability values of model samples. The results of the analysis are presented in Table 1.

In the first column of the Table 1 there are presented estimated values of half-length of the fractures. The resulting values exceed the actual size of the fracture. However, with the help of the calculations using data of the linear flow regime, only an approximate estimation can be processed. Such estimates can be considered as rough estimates because during this flow regime the fracture is not fully formed. More precisely, it is assumed that further fracture propagation is possible during this flow regime due to the presence of leakage. Therefore, based on the data obtained, it can be concluded that this method gives an estimate of the order of magnitude of the half-length of the fracture.

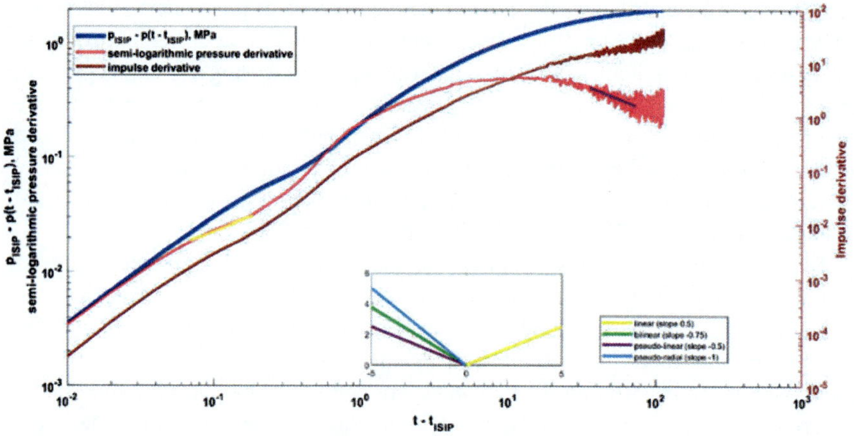

Fig. 2 Application of the log-log method

Table 1 Obtained values of the half-lengths of the fractures and the permeabilities of the model samples

Experiment	Fracture half-length, m	Permeability, md
		Pseudo-linear flow regime
1	0,134	290
2	0,098	3,6

Permeability values calculations did not show any accurate data, because almost in every case the estimated value exceeds the real permeability value of model sample equal to 2 md. At the same time, an interesting pattern is observed. For a horizontal fracture, that did not come to the surface of the model sample, the permeability value turned out to be more corresponding to the real permeability value of the model sample. Whereas for fracture that came to the surface of the model sample, the permeability value exceeded the real permeability value significantly. Obtained values of permeability for pseudo-linear flow regime show that even under almost identical experimental conditions, the results may differ. Overestimated permeability values may also indicate a large impact that an opened fracture in a laboratory experiment can have on the calculated permeability value. This discrepancy can only show that sometimes the presence of the opened fracture in the model sample can dramatically impact the permeable characteristics of the model medium.

5 Conclusions

As a result, two laboratory experiments on mini hydraulic fracturing were conducted. The pressure-time data for each experiment was processed using log-log method.

The calculated values of half-lengths of the fractures formed to be close enough to the actual fracture dimensions. It allows to consider the application of the log-log method as the correct one for evaluating this fracture characteristic.

Significant differences in experimental and calculated permeability values are explained by the presence of an unclosed fracture. The calculations of the study shows that an unclosed fractures in the reservoir significantly affects the flow regime established after fracture is closed, which leads to an overestimated permeability value.

Acknowledgements The research was carried out within the state of Ministry of Science and Higher Education of the Russian Federation (theme № 122032900167-1).

References

1. Horner, D. R.: Pressure Build-up Wells. (1951).
2. Nolte, K. G.: Determination of fracture parameters from fracturing pressure decline. SPE ANNUAL TECHNICAL CONFERENCE AND EXHIBITION. In OnePetro, (1979).
3. Mayerhofer, M. J., Ehlig-Economides, C. A., Economides, M. J.: Pressure-transient analysis of fracture-calibration tests. Journal of Petroleum Technology, 47, 229-234. (1995)
4. Bartko, K. M., Rahim, Z., Ansah, J., Adams, D. M., Soliman, M. Y., and Craig, D. P.: New Method for Determination of Formation Permeability, Reservoir Pressure, and Fracture Properties from a Minifrac Test. (2005).
5. Soliman, MY., Azari, M., Ansah, J., Kabir, C. S.: Design, Interpretation, and Assessment of Short-Term Pressure-Transient Tests. (2004).

6. Hagoort, J.: Waterflood-induced hydraulic fracturing. (1981).
7. Cinco-Ley, H., Samaniego, V. F.: Transient Pressure Analysis for Fractured Wells. Journal of Petroleum Technology. 1981; 33: 1749–1766.
8. Soliman, M. Y.: Analysis of buildup tests with short producing time. SPE Formation Evaluation, 1, 363–371. (1986).
9. Heber Cinco, L.: Transient Pressure Behavior for a Well With a Finite-Conductivity Vertical Fracture. Society of Petroleum Engineers Journal, 18: 253–264, (1978).
10. Trimonova, M., Baryshnikov, N., Zenchenko, E., Zenchenko, P., Turuntaev, S.: The Study of the Unstable Fracure Propagation in the Injection Well: Numerical and Laboratory Modeling. SPE RUSSIAN PETROLEUM TECHNOLOGY CONFERENCE. OnePetro; (2017).
11. de Pater, C. J., Cleary, M. P., Quinn, T. S., Barr, D. T., Johnson, D. E., Weijers, L.: Experimental verification of dimensional analysis for hydraulic fracturing. SPE Production & Facilities. OnePetro, 9(04): 230–238. (1994).

Melting Ice Structure Mechanical Sustainability Numerical Study

D. S. Konov, M. V. Muratov, E. K. Guseva, and I. B. Petrov

Abstract Both thermal and mechanical problems have applications in different practical tasks connected to civil engineering. Dynamical processes in Artic region on land, sea and shelf are an example of such problems. Ice structures including icebergs, ice hummock and artificial ice islands melt, mechanically and thermally interact with each other, environment, man-made structures. This article is devoted to investigation of ways to model those interactions. Physical and numerical models are formulated for mechanical and thermal problems, along with software implementation and test calculations.

Keywords Mathematical modeling · Ice island · Stefan problem · Enthalpy method · Grid-characteristic method

1 Introduction

Arctic is a polar region including number of seas of the Arctic Ocean, archipelagoes, and adjacent continental parts of different northern countries. Due to harsh climate, economical activities are minimal there. In particular, Arctic region of Russia is currently underdeveloped, but it has a potential as a huge source of crude oil, natural gas and other resources. Recent spikes of development cover only a small proportion of the known deposits, which are economically viable to extract using conventional methods and technologies. In order to reduce costs different experimental techniques are implemented and tested [1].

Artificial ice structures are a promising approach to substitute costly capital formations. One of the examples is an artificial island made of ice [2]. Those islands could be used in exploration or maintenance of the deposits. Ice Island in Canadian Beaufort Sea was discussed in [3]. Other tentative uses of ice structures have been proposed, including ships as large as an aircraft carrier out of pykrete [4], which is an ice composite—frozen mixture of pulp and water.

D. S. Konov (✉) · M. V. Muratov · E. K. Guseva · I. B. Petrov
Moscow Institute of Physics and Technology, Moscow, Russia
e-mail: konov.ds@phystech.edu

V. I. Karev (ed.), *Physical and Mathematical Modeling of Earth and Environment Processes—2022*, Springer Proceedings in Earth and Environmental Sciences,
https://doi.org/10.1007/978-3-031-25962-3_14

Obvious advantages of those structures are eco-friendliness and costs. Maintaining their mechanical and thermal integrity is a much more difficult task compared to ordinary constructions. Ice usages are harshly constrained by the fact that ice can melt and change its shape. It's properties also largely vary with temperature, that can cause some issues.

Thermal and mechanical processes in Artic ice islands were studied in [5, 6] and [7] separately. Clearly, thermal processes must be modeled alongside mechanical to develop a comprehensive representation of any ice structure. Ice melts, therefore ordinary heat equation will not suffice, in this work Stefan task is formulated. Ways to numerically solve it are discussed.

Obtaining shape and thermal fields inside those structures enables modelling mechanical problem. Ice is a complex heterogeneous media which can be simulated using different continuum mechanics models. Elastic characteristics and strain-strength properties may vary from computation to computation with temperature change. A method to overcome these difficulties is introduced.

2 Mathematical Formulation

Time scale of thermal interaction is much larger than mechanical one. For ice structures in ordinary Arctic conditions mechanical processes have no significant effect on thermal fields. To make model simpler we may regard thermal processes and melting over a big time period and occasionally check if the structure has remained stable and intact.

To obtain thermal model conservation of energy and Fourier's law should be considered for each phase. Thermal field will be denoted with T, vector field of heat flux is $\boldsymbol{q}$, k and C are thermal conductivity and capacity respectively.

$$\frac{\partial C_i T(r,t)}{\partial t} = div\big(\boldsymbol{q}_i(\mathbf{r},t)\big) \tag{1}$$

$$\mathbf{q_i}(\mathrm{r}, \mathrm{t}) = \mathrm{k_i} \nabla \mathrm{T}(\mathbf{r}, \mathrm{t}) \tag{2}$$

For each phase i those equations have unique constants k and C. Those equations do not describe phase boundary movement and lack fusion enthalpy λ. Special Stefan condition for points $\boldsymbol{r}_\Gamma$ on the boundary, where phase change temperature is achieved, should be attached:

$$\lambda \frac{\partial \boldsymbol{r}_\Gamma}{\partial t} \cdot \mathbf{d}\boldsymbol{S} = \big(\boldsymbol{q}_i - \boldsymbol{q}_j\big) \cdot \mathbf{d}\boldsymbol{S} \tag{3}$$

Another way to formulate the task may be achieved with substitution of the temperature with heat content (or enthalpy).

$$u = \begin{cases} \rho_S C_S T, & T < T_p \\ \rho_L C_L (T - T_p) + \rho_S C_S T_p + \rho_S \lambda, & T > T_p \end{cases} \tag{4}$$

$$T = \begin{cases} u \cdot \rho_S^{-1} C_S^{-1}, & u < \rho_S C_s T_s = u_- \\ T_p\,, & u_- < u < u_+ \\ \frac{u + (\rho_L C_L - \rho_S C_S) T_p - \rho_S \lambda}{\rho_L C_L}, & u > \rho_S C_s T_s + \rho_S \lambda = u_+ \end{cases} \tag{5}$$

Generally, (4) and (5) may be different to account for different substances with arbitrary number of phase transitions and variable thermal capacity. This way two phases of the same substance are connected. Thermal conductivity is left ambiguous in the u range between phases. One way to interpolate it is a linear function:

$$k(u) = \begin{cases} |k_S, & u < u_- \\ |k_S + (k_L - k_S) \cdot \frac{u - u_-}{u_+ - u_-}, & u_- < u < u_+ \\ |k_L, & u > u_+ \end{cases} \tag{6}$$

As a result of those substitutions thermal processes may be described completely with one equation and a pair of initial and border conditions:

$$\frac{\partial u}{\partial t} = \nabla(k(u)\nabla T(u)), \left(\alpha T(u) + \beta \frac{\partial T(u)}{\partial n}\right)|_{\partial\Omega} = \gamma, u\,|_{t=0} = u(T_0(r)) \tag{7}$$

Solving the task provides temperature fields as a function of time. Those fields could be used to acquire elastic parameters. In our model ice elastic characteristics depend on temperature. Berdennicov formula [8] was used to compute Young's modulus:

$$\mathrm{E} = \left(87.6 - 0.21\mathrm{T} - 0.0017\mathrm{T}^2\right) \cdot 10^8 \mathrm{Pa} \tag{8}$$

Constant Poisson coefficient was used $\nu = 0.295$. Changing geometry of the ice structure also should be considered.

Mechanical side of the problem relies on isotropic linear elasticity [9]. Lame coefficients are related to Yung modulus and Poisson coefficient with the following dependence:

$$\begin{aligned} \lambda &= \frac{E\nu}{(1+\nu)(1-2\nu)} \\ \mu &= \frac{E}{2(1+\nu)} \end{aligned} \tag{9}$$

The full system of equations:

$$\begin{aligned} \rho \dot{v} &= \nabla \cdot \sigma + f \\ \dot{\sigma} &= \lambda(\nabla \cdot v) I + \mu\left(\nabla \otimes v + (\nabla \otimes v)^T\right) + F \end{aligned} \tag{10}$$

The velocity vector $\boldsymbol{v}$ and stress tensor $\boldsymbol{\sigma}$ are unknowns. f is external force. For Maxwell viscoelastic model [10] term F is non-zero: $F_{ij} = -\frac{\sigma_{ij}}{\tau_0}$, where τ_0 is relaxation time that equaled to 0.5 s.

Finally, destruction may happen if either static or dynamic criterion is satisfied. The first one is the Mises criterion: the destruction is absent if $0.5 s_{ij} s_{ij} < k^2$, where $s_{ij} = \sigma_{ij} - \frac{\sigma_{ll}}{3}\delta_{ij}$, k is the maximum sheer stress. The second criterion checks if principal stresses exceed the maximum tension in a node [11]. After this a fracture, which affects wave propagation, is placed.

3 Numerical Model and Implementation

In order to solve those equations numerical model should be proposed an implemented as a computer program. To solve Stefan problem Crank–Nicolson methods on a uniform mesh may be used. Two-dimensional and three-dimensional variant for linear parabolic equations are presented (11) and (12). Stencil for two-dimensional case is shown in Fig. 1.

$$
\begin{aligned}
\frac{\tilde{u}_{ml} - u^n_{ml}}{\tau/2} &= \Lambda_{xx} u^{n+1/2}_{ml} + \Lambda_{yy} u^n_{ml} \\
\frac{u^{n+1}_{ml} - \tilde{u}_{ml}}{\tau/2} &= \Lambda_{xx} u^{n+1/2}_{ml} + \Lambda_{yy} u^{n+1}_{ml}
\end{aligned}
\tag{11}
$$

$$
\begin{aligned}
\frac{\tilde{u}_{mlp} - u^n_{mlp}}{\tau} &= \frac{1}{2}\Lambda_{xx}\left(\tilde{u}_{mlp} + u^n_{mlp}\right) + \Lambda_{yy} u^n_{mlp} + \Lambda_{zz} u^n_{mlp} \\
\frac{\tilde{\tilde{u}}_{mlp} - u^n_{mlp}}{\tau} &= \frac{1}{2}\Lambda_{xx}\left(\tilde{u}_{mlp} + u^n_{mlp}\right) + \frac{1}{2}\Lambda_{yy}\left(\tilde{\tilde{u}}_{mlp} + u^n_{mlp}\right) + \Lambda_{zz} u^n_{mlp} \\
\frac{u^{n+1}_{mlp} - u^n_{mlp}}{\tau} &= \frac{1}{2}\Lambda_{xx}\left(\tilde{u}_{mlp} + u^n_{mlp}\right) + \frac{1}{2}\Lambda_{yy}\left(\tilde{\tilde{u}}_{mlp} + u^n_{mlp}\right) + \Lambda_{zz}\left(u^{n+1}_{mlp} + u^n_{mlp}\right)
\end{aligned}
\tag{12}
$$

$$
\Lambda_{xx} u = k_{m+\frac{1}{2}lp} \frac{u_{m+1lp} - u_{mlp}}{h_x^2} + k_{m-\frac{1}{2}lp} \frac{u_{m-1lp} - u_{mlp}}{h_x^2} \tag{13}
$$

Those schemes are stable and using tridiagonal algorithm enables linear computational complexity. But differential operators (13) require thermal conductivity coefficients from the following step. Previous step data is used as a first approach. After that regular time step will be performed. Using just computed u thermal conductivity coefficient can be reevaluated and used to repeat that time step. This way arbitrary accuracy ε can be achieved.

Grid-characteristic method is used to solve hyperbolic system of Eq. (10). First step is reformulation in canonical form for both two-dimensional and three-dimensional cases.

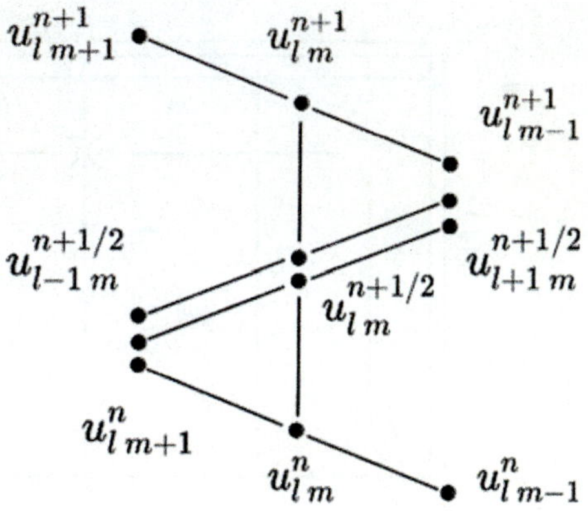

Fig. 1 The Pismen-Rekford scheme stencil

$$\frac{\partial u}{\partial t} + \sum_{i=1}^{N} A_i \frac{\partial u}{\partial x} = 0 \tag{14}$$

After that Riemann invariants are used to get one-dimensional transport equation. Those equations are solved with third-order approximation Rusanov scheme [12] with the Courant number $c_i = \frac{\lambda_i \tau}{h}$ and the current spatial step on structured grids h_i:

$$\begin{aligned}[r_i]^{n+1}_m = [r_i]^n_m + \tfrac{c_i}{2}\left([r_i]^n_{m-1} - [r_i]^n_{m+1}\right) + \tfrac{c_i^2}{2}\left([r_i]^n_{m-1} - 2[r_i]^n_m + [r_i]^n_{m+1}\right) \\ - \tfrac{c_i\left(1-c_i^2\right)}{6}\left([r_i]^n_{m-2} - 3[r_i]^n_{m-1} + 3[r_i]^n_m - [r_i]^n_{m+1}\right)\end{aligned} \tag{15}$$

Monotonicity criterion [13] is applied (which results in decrease of approximation order). If (16) is not satisfied, $[r_i]^{n+1}_m$ is substituted with closest values near characteristics on a previous step.

$$\min\left\{[r_i]^n_1, [r_i]^n_2\right\} \le [r_i]^{n+1}_m \le \max\left\{[r_i]^n_1, [r_i]^n_2\right\} \tag{16}$$

During the step F_{ij} is considered to be zero. Splitting along physical processes is introduced as to account for it to be nonzero. The differential equations are integrated and after each elastic step the computed values are modified: $\sigma_{ij} = \sigma^0_{ij} \exp\left(-\frac{\tau}{\tau_0}\right)$, where τ is the time step. Also coordinates of the nodes are slightly corrected proportionally to their speed.

Program implementation for both tasks allow simple OpenMP parallelization. Distributed memory architectures are availed through MPI package. For explicit Rusanov scheme program structure is straight forward, sending boundary values at boundaries between computational nodes is enough. For implicit Crank–Nicolson methods tridiagonal algorithm should be modified with use of technique similar to [14]. Simplest 2D approach is illustrated on Fig. 2 and requires division of computational domain along the longest edge between computational nodes. N and M are dimensions of the Ω. Every computational node $i = 1, \ldots, P$ contains $N_i \times M_i$

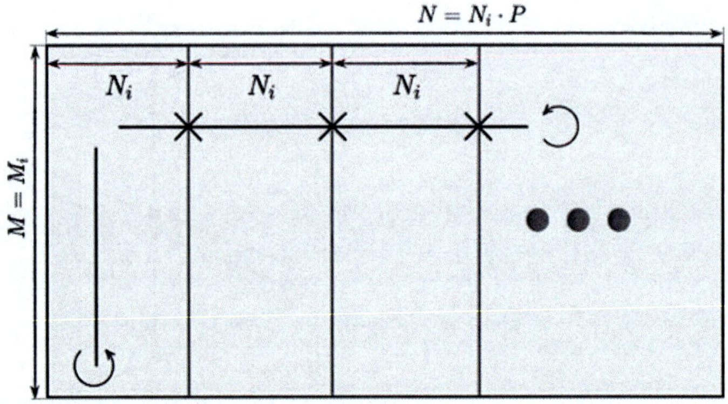

Fig. 2 Tridiagonal algorithm on distributed memory system implementation diagram. Cross indicates communication between nodes

edges. Line indicates sweeps of the tridiagonal algorithm. Crosses indicate pairs of synchronization between nodes.

4 Numerical Experiment Results

Developed model is tested on an ice island—example ice structure. Properties of the substances that are used in all calculations could be seen in Table 1. Barents Sea water and air temperatures (Table 2) are first used to determine thermal processes evolution inside the island, results presented in Fig. 3, computational domain of the task is shown in Fig 4.

Table 1 Thermal and elastic properties of the used materials

#	Substance	$\rho, \frac{kg}{m^3}$	$k, \frac{W}{m \cdot K}$	$C, \frac{J}{kg \cdot K}$	$S_p, \frac{m}{s}$	$S_s, \frac{m}{s}$
1	Ice	917	0.591	2100	3550–3600	1920–1940
2	Seabed	2500	0.8	750	1806	316
3	Water	1000	2.22	4180	1500	–
4	Air	–	–	–	–	–
5	Sedimentary rocks	2500	–	–	2250	1000

Table 2 Average air and water temperature by month

Month	January	February	March	April	May	June	July	August	September
T_a	− 28	− 28	− 27	− 20	− 9	− 1	2	1	− 2
T_w	3	3	3	3	3	5	6	9	7

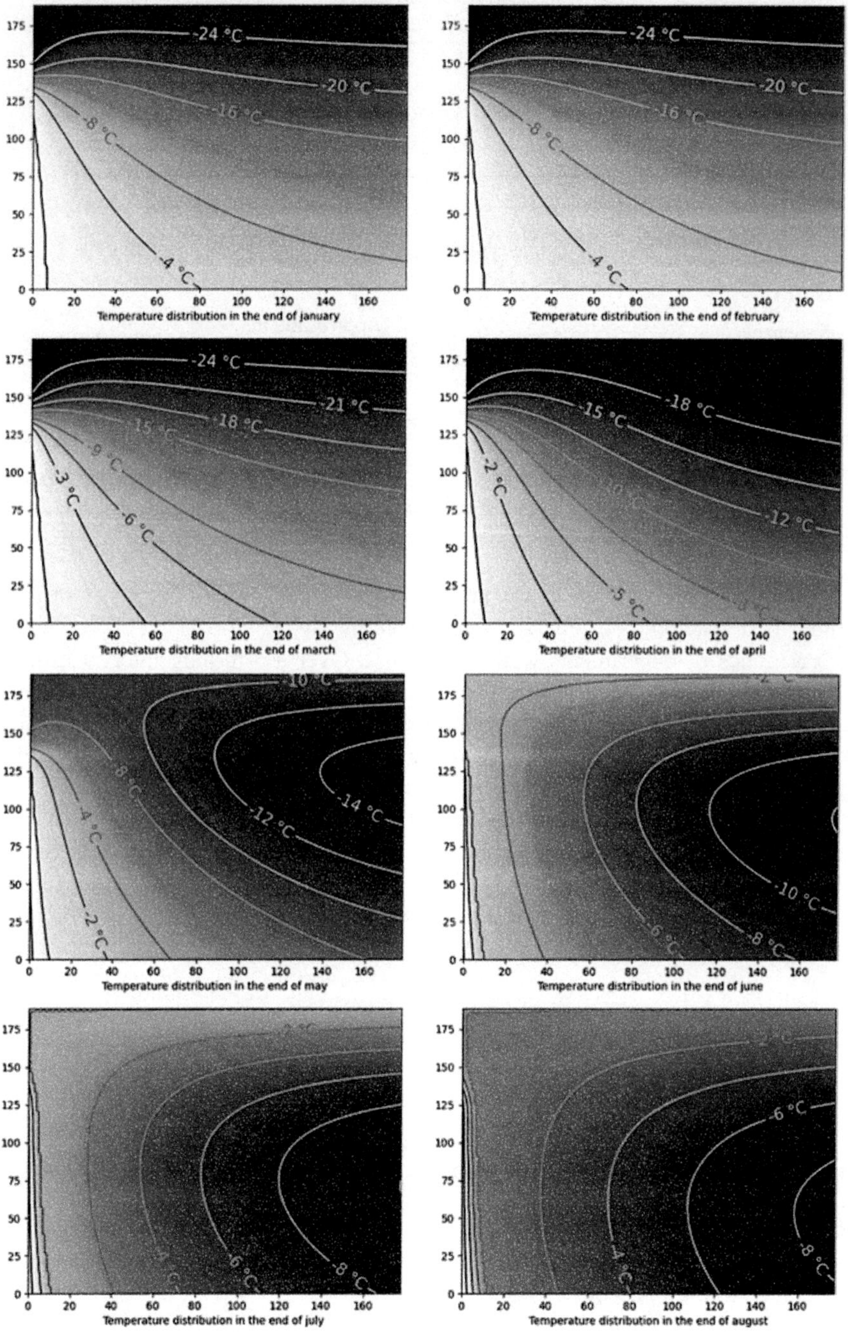

Fig. 3 Temperature distribution on the edge of the ice island

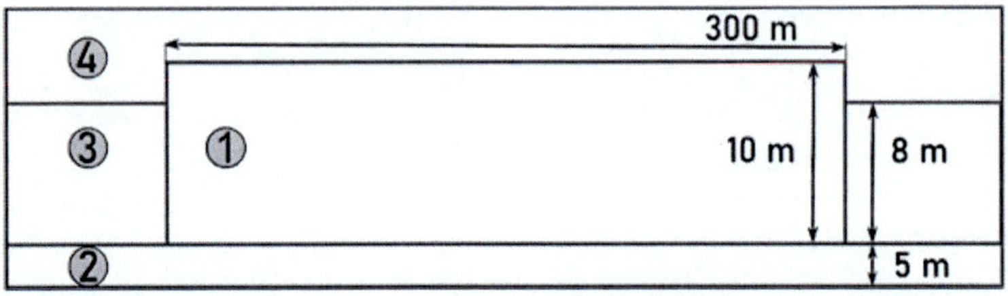

Fig. 4 Scheme of the problem domain in a thermal problem. Indexes are inquired in Table 1

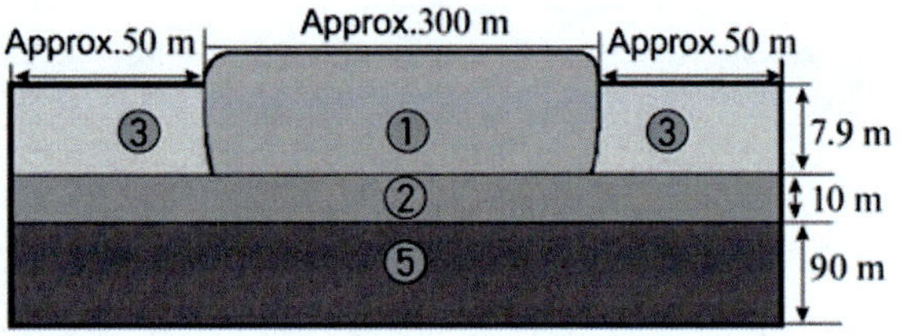

Fig. 5 Scheme of the problem domain in mechanical simulation. Indexes are inquired in Table 1

At the end of each month mechanical stability was examined. Geometry of the island varied, but could be generally represented by Fig. 5.

Two rounds of experiments were carried out. Over a 5-m-long interval in the center of the island two loads were distributed. Firstly, example survey equipment was studied. External force in the scenario was equal to $10^4 N$. Then maximum load each month was calculated. Some visualizations from the experiments are shown on Figs. 6 and 7.

Experiments proved, that island did not collapse under an equipment load. Critical load was shown to be dependent on the month. Split criterion is fulfilled at static loads of $(2.175 \pm 0.025) \times 10^6 N$ for almost all months. In July and August this value turns to be slightly lesser: $(2.125 \pm 0.025) \times 10^6 N$. Ice breaks near the contact with the seabed.

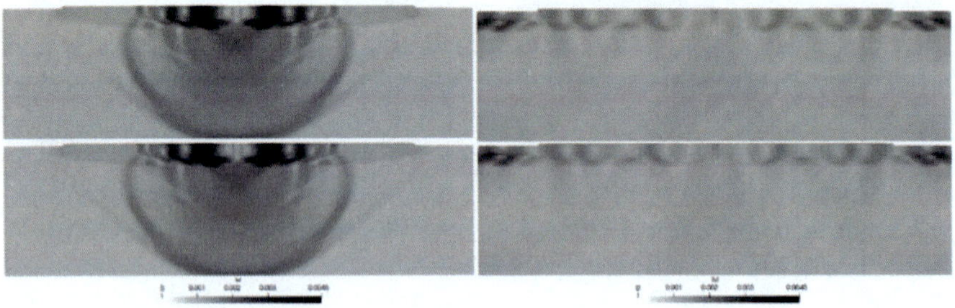

Fig. 6 Velocity module after 0.05 s (left) and 0.5 s (right). Formation at the end of January is represented at the top, at the end of August—at the bottom

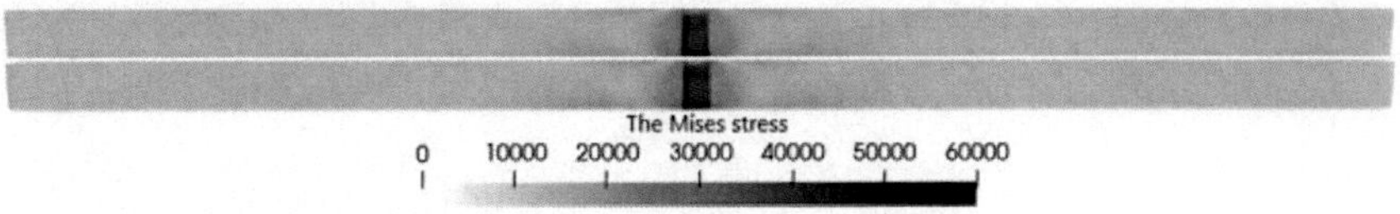

Fig. 7 The Mises stress distribution 0.5 s after the calculation begins. Formation at the end of January at the top, at the end of August—at the bottom

Acknowledgements This work was carried out with the financial support of the Russian Science Foundation, project no. 21-71-10015.

References

1. Gorgutsa E. Yu., Kurilo E. Yu., 2017. Artificial ice islands construction in arctic conditions, *Gidrotekh. XXI Vek*, 4 (32), 54–57
2. Morgan V. et al, 2015. Ice island study. Prepared for minerals management service, US department of the interior. Prepared by C-CORE
3. Riley J. G., 1978. The construction of artificial islands in the Beaufort Sea, *Proceedings of the Annual 7th Offshore Technology Conference*
4. Perutz M. F., 1948. A Description of the iceberg aircraft carrier and the bearing of the mechanical properties of frozen wood pulp upon some problems of glacier flow, *Journal of Glaciology*, 1 (3) 95–104
5. Petrov I. B., Muratov M. V., Sergeev F. I., 2020. Stability analysis of artificial ice islands by methods of mathematical modeling, *Dokl. Math.*, 102 483–486
6. Muratov M. V., Stognii P. V., Konov D. S., Petrov I. B., 2021. Numerical modeling of elastic and thermal dynamical processes in artificial ice island, *Proc. of the 26th int. con. on port and ocean eng. under Arctic cond.*
7. Konov D. S., Muratov M. V., Biryukov V. A., 2022. Numerical solution of the Stefan problem for an artificial ice island, *Russian journal of cybernetics*, 3 (1) 14–19
8. Berdennikov V. P., 1948. Study of the modulus of elasticity of ice, *Tr. GGI*, 7 (61), 13–23
9. Nowacki W., 1970. Theory of Elasticity, *Panstwowe Wydawnictwo Naukowe*, Warszawa
10. Argatov I., 2012. Mathematical modeling of linear viscoelastic impact: Application to drop impact testing of articular cartilage, *Tribol. Int.*, 63 213–225.
11. Favorskaya A. V., Golubev V. I., 2020. Study the elastic waves propagation in multistory buildings, taking into account dynamic destruction, *Smart Innov. Syst. Technol.*, 193 189–199.
12. Rusanov V., 1961. The calculation of the interaction of non-stationary shock waves with barriers, J. *Comput. Math. Phys. USSR*, 1 267–279.
13. Kholodov A. S., Kholodov Ya. A., 2006. Monotonicity criteria for difference schemes designed for hyperbolic equations, *Comput. Math. Math. Phys.*, 46 1560–1588.
14. Lipavskii M., Konshin I, 2022. Parallel implementation of multioperators-based scheme of the 16-th order for three-dimensional calculation of the jet flows, *In Proc. of Int. Conf. RuSCDays*, 1–15.

Wave Asymmetry Evolution in Coastal Zone: Field Data and Xbeach Numerical Modelling Comparison

M. N. Shtremel and D. V. Korzinin

Abstract Wave asymmetry prediction is an important aspect of the coastal zone dynamics studies, as it affects wave-induced sediment transport. In this study comparison of measured during field experiment and predicted with XBeach numerical model wave asymmetry is presented. Reasons for the mismatch between measured and modelled asymmetry are discussed.

Keywords Wave asymmetry · Nonlinear wave transformation · Coastal zone · Sediment transport model · XBeach · Field experiment

While transforming in coastal zone, wave profile and corresponding wave-induced velocities change their shape due to nonlinear wave-wave interaction processes and shoaling. Changes of wave symmetry can be evaluated with the asymmetry parameter, which is defined by the following formula:

$$A_s = \left\langle (\mathrm{H}(u))^3 \right\rangle \Big/ \left\langle u^2 \right\rangle^{3/2}, \tag{1}$$

where u is the free stream velocity, H is the Hilbert transform, angle brackets-averaging.

Changes in the wave velocity asymmetry cause occurrence of wave-induced sediment transport gradients and, consequently, relief deformations and therefore is included in sediment transport models. Impact of wave asymmetry on sediment transport is discussed, for example, in [1]. Changes of wave asymmetry caused by periodical exchange of energy between first and higher wave harmonics may result in sandbar formation and migration [2]. Since the prediction of the relief changes on the sandy shores under the action of different wave conditions is an urgent problem not only of fundamental research, but also of an applied nature, quite a lot of parameterizations have been developed that allow to assess values of wave asymmetry

M. N. Shtremel (✉) · D. V. Korzinin
Shirshov Institute of Oceanology, Russian Academy of Sciences, Moscow, Russia
e-mail: shtremel@ocean.ru

V. I. Karev (ed.), *Physical and Mathematical Modeling of Earth and Environment Processes—2022*, Springer Proceedings in Earth and Environmental Sciences,
https://doi.org/10.1007/978-3-031-25962-3_15

[3–5]. These parameterizations are included in the numerical models that reproduce the deformations of the underwater slope.

In XBeach, which is one of the most popular morphodynamic models, wave velocity asymmetry is calculated there with the parameterization proposed in [6]:

$$A_s = B\cos\psi \tag{2}$$

$$B = p_1 + \frac{p_2 - p_1}{1 + e^{\frac{p_3 - \log Ur}{p_4}}} \tag{3}$$

$$\psi = -90^\circ + 90^\circ \tanh\left(p_5/Ur^{p_6}\right), \tag{4}$$

where Ur—Ursell number, p1 = 0, p2 = 0.857, p3 = − 0.471, p4 = 0.297, p5 = 0.815, p6 = 0.672, B—parameter of total (non-dimensional) nonlinearity, ψ—phase.

Sandbar migration is a complex process that is not always reproduced correctly by XBeach. For example, in [7] it was shown that although the offshore sandbar migration can be adequately reproduced when taking into account variable breaking-wave roller energy model, onshore sandbar migration is still a challenge. This problem may be explained by the shortcomings of the wave asymmetry parameterization. Comparison of the measured and modelled wave asymmetry was done in this research.

1 Materials and Methods

Field data was obtained during "Shkorpilovtsy-2007" field experiment, which took place in the Black Sea coast on the Kamchia-Shkorpilovtsy beach near Varna, Bulgaria (Fig. 1) in September–October 2007. The beach is stable with nearly absent antropogenic impact. The coastline is straight, oriented from NNW to SSE. 200 m long shore-normal research pier is located on the mildly sloping (0.023) sandy beach with medium grain size of the sand (0.3 mm). Stable outer sandbar and dynamic storm inner sandbar were observed during the bottom topography measurements along the pier (Fig. 1).

Outer sandbar was located in 130 m off the shoreline with 3 m depth above its crest. Smaller nearshore sandbar formed at 30–40 m distance from the shore. Free surface elevation was measured along the pier with capacitance type gauges, which allowed to capture wave transformation process in detail. Capacitance wave gauges registered waves with 5 and 20 Hz frequency, time series are 30 min–1 h long. Data in 10 points for two wave regimes was used for the analysis (Table 1): higher and longer regular swell waves and calmer wave regime captured during storm decay phase. During first regime wave breaking of the largest waves occurred above the sandbar (spilling wave breaking type), mass breaking—on 40 m off the shoreline with plunging breaking type. Second wave regime was swell with a long wave fronts

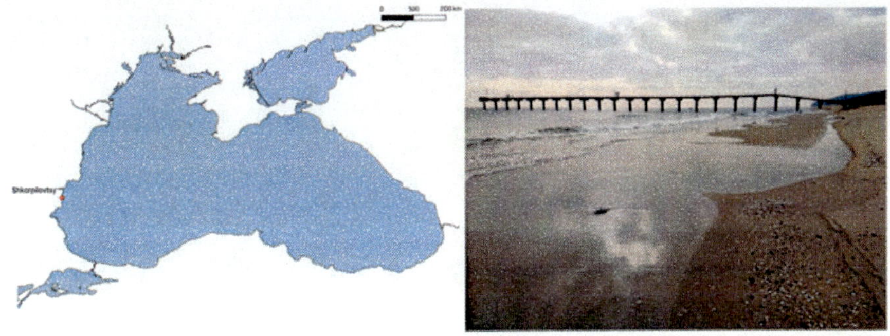

Fig. 1 Left panel—study site location, right—research pier

Table 1 Wave regimes reproduced in the Xbeach model

Initial profile	Wave regime
September 27, 2007	Hs = 1.2 m, Tp = 7 s
September 28, 2007	Hs = 0.7 m, Tp = 6 s

a little angular to pier, about 10 degrees from the south. Waves are breaking in 30 m offshore with plunging type.

Numerical simulations were performed using the XBeach model [8]. The XBeach model is a free-access and open-source model and is designed to reproduce hydrodynamic and morphodynamic processes on sandy shores. In this experiment hydrostatic mode was used, where wave field is calculated as a change in the amplitude of short waves. The model reproduced the profile along the pier where free surface elevation and bottom topography were measured. Wave parameters were recorded on 10 virtual sensors, corresponding to the gauges in the field experiment. Computational grid is 200 m long with mesh size of 1 m was used. Initial depth is 3.9 m. The studied wave regimes (Table 1) were reproduced for 10 h (36,000 s) with 1 Hz output frequency. The values of wave asymmetry on the wave gauges were averaged for the time interval between 500 and 2300 s.

2 Results and Discussion

The most significant differences between calculated and measured asymmetries are observed in the first case (September 27): at a distance of 100–160 m from the shoreline, the asymmetry takes on positive values, when the parameterization incorporated in the model only allows it to take negative values (Fig. 2d). In the second case, the tendencies of the asymmetry evolution along the profile coincide in the measurements and model data, differing only in the absolute value (the model overestimates the values of the asymmetry).

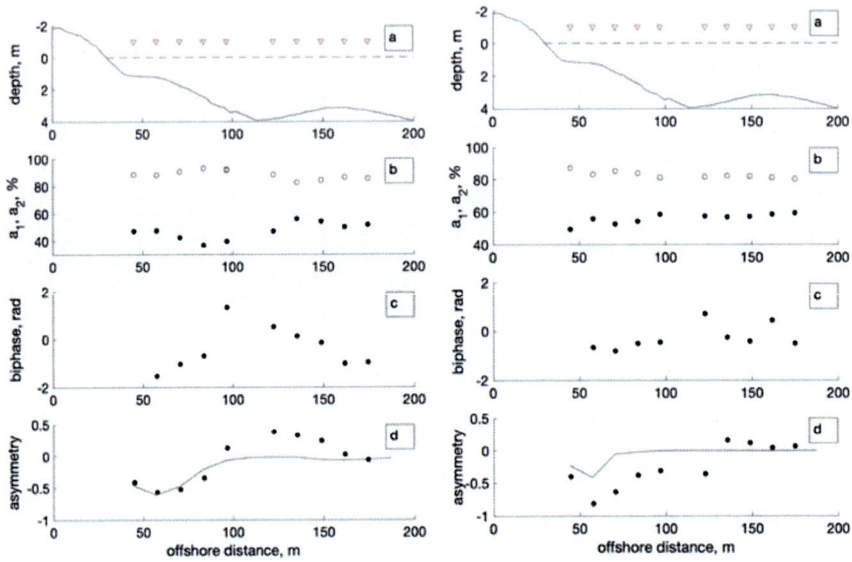

Fig. 2 Left panel—first case (September 27), right—second case (September 28). **a** Mean bottom relief with locations of the wave gauges; **b** first and second harmonic amplitude percentage; **c** biphase; **d** asymmetry: black dots—experimental data, line—model results

Wave asymmetry can be described as parameter that shows if the wave is symmetric with respect to the vertical axis or is closer to the sawtooth shape. In simple model with only two wave harmonics wave shape is determined by relation between first and second harmonic amplitudes and phase shift between them. According to [9] two studied cases correspond to two different scenarios of wave transformation: the first case is characterized by pronounced periodical exchange of energy between first and second harmonics and in the second one second harmonic stays nearly the same, decreasing to the shore (Fig. 2b).

In previous research [10] it was shown that biphase values change in range of $[-\pi/2; \pi/2]$ and strictly follow wave transformation process: starting with $-\pi/2$ in the beginning of second harmonic growth it equals zero together with second harmonic maximum and that turns positive and reaches $\pi/2$ during reverse energy transfer from the second to the main harmonic.

Positive values of biphase correspond to positive asymmetry values (Fig. 2c, d) which are not predicted by the parameterization, implemented in the model.

Occurrence of situations where wave asymmetry takes positive values can be predicted with the criterion proposed in [9]. Such cases are expected in the scenarios with pronounced periodical exchange of energy between first and second harmonic due to triad interactions between wave components. Locations of waves with positive asymmetry values can be determined with the method presented in [11]: waves with such profile are expected in the trough shoreward from the sandbar in the first case.

The method consists in determining the position of the maximum value of the second harmonic amplitude, which is located at a distance of 1.5 of the length of the spatial energy exchange between the first and the second harmonics of the wave motion (Lb). Further, based on the fact that the positive values of the asymmetry are located in the interval between the maximum amplitude of the second harmonic and its minimum, their occurrence can be expected between 1 and 1.5 Lb.

3 Conclusion

Wave asymmetry reproduced by XBeach model differs from the measured one for the first case, where pronounced energy exchange between first and second harmonics of wave motion is observed. In the field data asymmetry takes positive values in the trough before the sandbar, while the in model parameterization asymmetry can only have negative values.

Discrepancies between observed and modelled asymmetry can be explained by the lack of description of backward energy transfer from second harmonic to the first one. Corresponding positive values of biphase and asymmetry are not reproduced as well.

If correct prediction of asymmetry is critical for some reason, location of the positive values of the asymmetry can be predicted using an algorithm that allows to locate the key features of wave transformation: the position of maximum amplitude of the second harmonic, as well as the beat length—length of one full period of energy exchange between harmonics. Positive values of biphase and thereafter of asymmetry will be located between its maximum and shoreward minimum.

Acknowledgements This research was funded by Russian Foundation for Basic Research grant number 20-55-46005.

References

1. *Ruessink, B. vd, Van Den Berg, T. & Van Rijn, L.*, 2009 Modeling sediment transport beneath skewed asymmetric waves above a plane bed. *J. Geophys. Res. Oceans* **114**.
2. *Chapalain, G. & Boczar-Karakiewicz, B.*, 1992. Modeling of hydrodynamics and sedimentary processes related to unbroken progressive shallow water waves. *J. Coast. Res.* 419–441.
3. *Abreu, T., Silva, P. A., Sancho, F. & Temperville, A.*, 2010. Analytical approximate wave form for asymmetric waves. *Coast. Eng.* **57**, 656–667.
4. *Doering, J. C., Elfrink, B., Hanes, D. M. & Ruessink, G.*, 2001. Parameterization of Velocity Skewness under Waves and its Effect on Cross-Shore Sediment Transport. in *Coastal Engineering 2000* 1383–1397 doi:https://doi.org/10.1061/40549(276)107.
5. *Elfrink, B., Hanes, D. M. & Ruessink, B.*, 2006. Parameterization and simulation of near bed orbital velocities under irregular waves in shallow water. *Coast. Eng.* **53**, 915–927.

6. *Ruessink, B., Ramaekers, G. & Van Rijn, L.*, 2021. On the parameterization of the free-stream non-linear wave orbital motion in nearshore morphodynamic models. *Coast. Eng.* **65**, 56–63 (2012).
7. *Rafati, Y. et al.*, 2021. Modeling the hydrodynamics and morphodynamics of sandbar migration events. *Coast. Eng.* **166**, 103885.
8. *Roelvink, D. et al.*, 2009.Modelling storm impacts on beaches, dunes and barrier islands. *Coast. Eng.* **56**, 1133–1152.
9. *Saprykina, Y. V., Kuznetsov, S. Y., Andreeva, N. K. & Shtremel, M. N.*, 2013. Scenarios of nonlinear wave transformation in the coastal zone. *Oceanology* **53**, 422–43.
10. *Saprykina, Y. V., Shtremel, M. N. & Kuznetsov, S. Y.*, 2017. On the possibility of biphase parametrization for wave transformation in the coastal zone. *Oceanology* **57**, 253–264.
11. *Shtremel, M., Saprykina, Y. & Ayat, B.*, 2022. The Method for Evaluating Cross-Shore Migration of Sand Bar under the Influence of Nonlinear Waves Transformation. *Water Switz.* **14**.

Computational Aspects of Solving the Problem of Ekman-Type Wind Currents

V. S. Kochergin, S. V. Kochergin, and S. N. Sklyar

Abstract The aim of the work is to compare the exact analytical solution for a three-dimensional wind flow model with its numerical analogue for testing and analyzing difference discretizations and computational algorithms used in integrating the dynamic model. When integrating the equation for the current function, a multi-parametric family of difference schemes is considered, which, under certain parameters, includes schemes of central difference, directional difference, schemes of A. A. Samarsky and A. M. Ilyin. To find the velocity fields by the found current function, a projection version of the integro-interpolation method (PVIIM) is used. The results of calculations of the current function and integral velocities are compared with an analytical solution. As a result of numerical experiments, it is shown that the best results are obtained when using the Ilyin scheme and when using approximations of derivatives based on the PVIIM method. The implemented testing of difference schemes and algorithms, comparison of the results of calculations with an accurate analytical solution made it possible to implement a reasonable choice of discretizations for integrating the equations of the wind circulation model. The constructed difference discretizations allow, under all other equal conditions, to increase the accuracy of reproduction of the desired numerical solution of the model.

Keywords Dimensionless problem · Wind currents · Current function · Difference schemes · Analytical solution

1 Introduction

Numerical modeling of ocean dynamics in recent years has moved significantly in the direction of describing processes of various nature in models. In part, this is due to an increase in the discretization of the tasks being solved and the use of

V. S. Kochergin (✉) · S. V. Kochergin
Marine Hydrophysical Institute RAS, Kapitanskaya Str. 2, 299011 Sevastopol, Russia
e-mail: vskocher@gmail.com

S. N. Sklyar
American University of Central Asia, A.Tokombaev Str.7/6, 720060 Bishkek, Kyrgyzstan

V. I. Karev (ed.), *Physical and Mathematical Modeling of Earth and Environment Processes—2022*, Springer Proceedings in Earth and Environmental Sciences,
https://doi.org/10.1007/978-3-031-25962-3_16

modern computer capacities. However, the use of new computational schemes and algorithms can give a tangible effect in solving such problems. To test such methods, it is important to have an accurate solution of the problem available, which will allow such a choice. Models of ocean dynamics are quite complex [1]. Analytical solutions exist mainly only for the simplest statements, for example, the Stommel model [2–4]. In [5, 6], such a problem of calculating the velocity field is realized using the method of inversion of a dynamic operator and using computational schemes of a special kind. In [7, 8], the model takes into account the variability of velocity in all three directions, which made it possible to analyze the accuracy of calculating its horizontal and vertical components. The problem in this formulation is also considered in this paper to assess the accuracy of calculating the current function and integral velocities by comparing the result with the obtained analytical solution of the model.

2 The Problem in Dimensionless Form

For a reservoir with a rectangular flat bottom, its surface in the x0y plane is set as follows:

$$\Omega_0 = [0, r] \times [0, q],$$

its depth $H > 0$ is constant. The axes of the Cartesian coordinate system are directed as follows: Ox—to the east, Oy—to the north, Oz—vertically down. In the domain:

$$\Omega = \{(x, y, z)|(x, y) \in \Omega_0, 0 \leq z \leq H\}.$$

Consider the resulting system of stationary equations of motion in dimensionless form:

$$\begin{cases} -lv = -\frac{\partial P^s}{\partial x} + \frac{\partial}{\partial z}\left(k\frac{\partial u}{\partial z}\right) \\ lu = -\frac{\partial P^s}{\partial y} + \frac{\partial}{\partial z}\left(k\frac{\partial u}{\partial z}\right), t > 0, (x, y, z) \in \Omega^0, \\ \frac{\partial U}{\partial x} + \frac{\partial V}{\partial y} + \frac{\partial W}{\partial z} = 0 \end{cases} \tag{1}$$

with boundary conditions:

$$\begin{array}{c} \left\{z = 0, (x, y) \in \Omega^0_{0,}\right\} : k\frac{\partial u}{\partial z} = -\tau_x, k\frac{\partial V}{\partial z} = -\tau_y, w = 0; \\ \left\{z = H, (x, y) \in \Omega^0_{0,}\right\} : k\frac{\partial u}{\partial z} = -\tau^b_x, k\frac{\partial V}{\partial z} = -\tau^b_y, w = 0; \\ \{0 \geq z \geq H, (x, y) \in \partial\Omega_0\} : U \cdot n_x + V \cdot n_y = 0, \end{array} \tag{2}$$

in (2) the dimensionless integral velocities are defined as follows:

$$U(x, y) = \int_0^H u(x, y, z)\mathrm{d}z,\ V(x, y) = \int_0^H v(x, y, z)\mathrm{d}z, \tag{3}$$

and in (1) the following variant of parametrization of bottom friction is used:

$$\tau_x^b = \mu \cdot U,\ \tau_y^b = \mu \cdot V,\ \mu \equiv const > 0. \tag{4}$$

According to the Stommel model, suppose:

$$l = l_0 + \beta \cdot y,\ k = const. \tag{5}$$

3 Analytical Solution

From the first two equations in (1), we exclude pressure gradients using cross-differentiation:

$$\begin{cases} \mu U - lv = -H\frac{\partial P^s}{\partial x} + \tau_x, \\ lU + \mu V = -H\frac{\partial P^s}{\partial y} + \tau_y, \\ \frac{\partial U}{\partial x} + \frac{\partial V}{\partial y} = 0,\ (x, y) \in \Omega^0,\ U \cdot n_x + V \cdot n_y = 0,\ (x, y) \in \partial\Omega_0. \end{cases} \tag{6}$$

We introduce the current function $\Psi(x, y)$ by the formulas:

$$U = \frac{\partial \Psi}{\partial y},\ V = -\frac{\partial \Psi}{\partial x}.$$

As a result, we get the following task for the current function:

$$\begin{cases} \mu\left(\frac{\partial^2 \Psi}{\partial x^2} + \frac{\partial^2 \Psi}{\partial y^2}\right) + \beta\frac{\partial \Psi}{\partial x} = \frac{\partial \tau_x}{\partial y} - \frac{\partial \tau_y}{\partial x},\ (x, y) \in \partial\Omega_{0,}^0 \\ \Psi = 0,\ (x, y) \in \partial\Omega_0 \end{cases} \tag{7}$$

In [9], a solution to the problem was obtained when setting a sufficiently general wind effect:

$$\begin{cases} \tau_x = [F_1 \cdot \cos(r_l x) + F_2 \cdot \sin(r_l x)] \cdot \cos(q_m y) \\ \tau_y = [G_1 \cdot \cos(r_s x) + G_2 \cdot \sin(r_s x)] \cdot \sin(q_p y) \end{cases} \tag{8}$$

where the designations are accepted:

$$\begin{gathered} r_l = \frac{\pi l}{r};\ r_s = \frac{\pi s}{r};\ q_m = \frac{\pi m}{q};\ q_p = \frac{\pi p}{q} \\ l, s = 0, 1, 2, \ldots;\ m, p = 1, 2 \ldots. \end{gathered} \tag{9}$$

Thus, the wind model contains four real: F_1, F_2, G_1, G_2 and four integer: l, s, m, p numerical parameters, the choice of which makes it possible to describe a fairly general wind situation. For example, if $F_1 = \frac{F \cdot q}{\pi}$, $F_2 = G_1 = G_2 = 0; l = 0;$ $m = 1$ we have

$$\tau_x = \frac{F \cdot q}{\pi} \cos\left(\frac{\pi y}{q}\right), \tau_y = 0 \tag{10}$$

and when

$$F_1 = \frac{F \cdot q}{\pi}, F_2 = 0; G_1 = -\frac{F \cdot q}{\pi}; G_2 = 0\ l = 0; m = 1 \tag{11}$$

we have a cyclone over the water area.

4 Basic Formulas for the Stationary Model

Let's put together all the formulas necessary for programming. First we write out the formulas for the stream function. Solving the problem (7) we have:

$$\Psi(x, y) = \Psi_1(x, y) + \Psi_2(x, y);$$

where:

$$\Psi_1(x, y) = \left[C_1 e^{Ax} + C_2 e^{Bx} + D_1 \cdot \cos(r_l x) + D_2 \cdot \sin(r_l x)\right] \cdot \sin(q_m y),$$

$$D_1 = \frac{\mu\left(r_l^2 + q_m^2\right) F_1 + \beta r_l F_2}{\mu^2\left(r_l^2 + q_m^2\right)^2 + \beta^2 r_l^2} \cdot q_m,\ D_2 = \frac{\mu\left(r_l^2 + q_m^2\right) F_2 - \beta r_l F_1}{\mu^2\left(r_l^2 + q_m^2\right)^2 + \beta^2 r_l^2} \cdot q_m,$$

and:

$$C_1 = D_1 \cdot \frac{e^{Br} - (-1)^l}{e^{Ar} - e^{Br}},\ C_2 = D_1 \cdot \frac{(-1)^l - e^{Ar}}{e^{Ar} - e^{Br}};$$

$$A = -\frac{\beta}{2\mu} + \sqrt{\left(\frac{\beta}{2\mu}\right)^2 + (q_m)^2},\ B = -\frac{\beta}{2\mu} - \sqrt{\left(\frac{\beta}{2\mu}\right)^2 + (q_m)^2}.$$

Similar formulas for the second component of the solution:

$$\Psi_2(x, y) = \left[\overline{C}_1 e^{\overline{A}x} + \overline{C}_2 e^{\overline{B}x} + \overline{D}_1 \cdot \cos(r_s x) + \overline{D}_2 \cdot \sin(r_s x)\right] \cdot \sin\left(q_p y\right),$$

where

$$\overline{D}_1 = -\frac{\beta r_s G_1 - \mu\left(r_s^2 + q_p^2\right)G_2}{\mu^2\left(r_s^2 + q_p^2\right)^2 + \beta^2 r_s^2} \cdot r_s,\ \overline{D}_2 = -\frac{\mu\left(r_s^2 + q_p^2\right)G_1 + \beta r_s G_2}{\mu^2\left(r_s^2 + q_p^2\right)^2 + \beta^2 r_s^2} \cdot r_s$$

$$\overline{C}_1 = \overline{D}_1 \cdot \frac{e^{\overline{B}r} - (-1)^s}{e^{\overline{A}r} - e^{\overline{B}r}},\ \overline{C}_2 = \overline{D}_1 \cdot \frac{(-1)^s - e^{\overline{A}r}}{e^{\overline{A}r} - e^{\overline{B}r}};$$

$$\overline{A} = -\frac{\beta}{2\mu} + \sqrt{\left(\frac{\beta}{2\mu}\right)^2 + (q_m)^2},\ \overline{B} = -\frac{\beta}{2\mu} - \sqrt{\left(\frac{\beta}{2\mu}\right)^2 + (q_m)^2}.$$

The integral components of the horizontal velocity vector are obtained by differentiation, in accordance with the formulas:

$$U = \frac{\partial\Psi}{\partial y},\quad V = \frac{\partial\Psi}{\partial x}$$

$$\begin{aligned} U(x,y) =& \frac{\partial}{\partial y}[\Psi_1(x,y) + \Psi_2(x,y)] \\ =& q_m \cdot \left[C_1 e^{Ax} + C_2 e^{Bx} + D_1 \cdot \cos(r_l x) + D_2 \cdot \sin(r_l x)\right] \cdot \cos(q_m y) \\ &+ q_p \cdot \left[\overline{C}_1 e^{\overline{A}x} + \overline{C}_2 e^{\overline{B}x} + \overline{D}_1 \cdot \cos(r_s x) + \overline{D}_2 \cdot \sin(r_s x)\right] \cdot \cos\left(q_p y\right), \end{aligned}$$

$$\begin{aligned} V(x,y) =& -\frac{\partial}{\partial x}[\Psi_1(x,y) + \Psi_2(x,y)] \\ =& -\left[AC_1 e^{Ax} + BC_2 e^{Bx} - r_l D_1 \cdot \sin(r_l x) + r_l D_2 \cdot \cos(r_l x)\right] \cdot \sin(q_m y) \\ &- \left[\overline{AC}_1 e^{\overline{A}x} + \overline{BC}_2 e^{\overline{B}x} - r_s \overline{D}_1 \cdot \sin(r_s x) + r_s \overline{D}_2 \cdot \sin(r_s x)\right] \cdot \sin\left(q_p y\right). \end{aligned}$$

5 Numerical Methods for Solving the Problem

Consider problem (11) for the current function by writing it for an arbitrary right-hand side:

$$\begin{cases} \mu\left(\frac{\partial^2\Psi}{\partial x^2} + \frac{\partial^2\Psi}{\partial y^2}\right) + \beta\frac{\partial\Psi}{\partial x} = \Phi(x,y),\ (x,y) \in \partial\Omega_0^0, \\ \Psi = 0,\ (x,y) \in \partial\Omega_0 \end{cases} \quad (12)$$

Let's construct a numerical method to solve it. For this purpose, in the field $\Omega_0 = [0,r] \times [0,q]$, consider a computational grid Consider in a computational grid:

$$\omega_h \equiv \left\{\left(x_i, y_j\right) | x_i = (i-1)\cdot\Delta x\ ,\ y_j = (j-1)\cdot\Delta y_j\right.$$

$$\left.; i=\overline{1,n};\ j=\overline{1,k};\ \Delta x=\frac{r}{n-1};\ \Delta y=\frac{q}{k-1}\right\}; \tag{13}$$

Let the grid function $\{\Psi_{i,j}\}$ defined on this grid consist of approximate values for the quantities of the $\{\Psi(x_i, y_j)\}$—exact solution. The PVIIM method is applicable to the sequential approximation of Eq. (12), first by the variable "x", and then by the variable "y". Equation (12) is rewritten as:

$$\mu\frac{\partial^2\Psi}{\partial x^2}+\beta\frac{\partial\Psi}{\partial x}=\mathrm{F}(x,y)-\mu\frac{\partial^2\Psi}{\partial y^2}\equiv G(x,y). \tag{14}$$

In accordance with the PVIIM, we consider a test function $\nu(x)$ on an arbitrary grid cell $[x_i, x_{i+1}]$. Equation (14) is multiplied by $\nu(x)$ and the result is integrated across the cell $[x_i, x_{i+1}]$, including in parts (the variable "y", at the same time, is perceived as a parameter), as a result we obtain the following integro-difference identity:

$$\left[\mu\frac{\partial\Psi}{\partial x}\cdot\nu+\Psi\cdot\left(\beta\nu-\mu\frac{\partial\nu}{\partial x}\right)\right]_{x_i}^{x_{i+1}}+\int\limits_{x_i}^{x_{i+1}}\Psi\cdot\left(\mu\frac{\partial^2\nu}{\partial x^2}-\beta\frac{\partial\nu}{\partial x}\right)\mathrm{d}x$$
$$=\int\limits_{x_i}^{x_{i+1}}G(x,y)\nu(x)\mathrm{d}x. \tag{15}$$

When choosing test functions $\nu^{(0)}(x)u\nu^{(1)}(x)$ in the identity (15), we assume that they are solutions of the equation

$$\mu\frac{\partial^2\nu}{\partial x^2}-\beta\frac{\partial\nu}{\partial x}=0, x\in(x_i,x_{i+1})$$

and at the same time satisfy the boundary conditions:

$$\nu^{(0)}(x_i)=1,\ \nu^{(0)}(x_{i+1})=0$$
$$\nu^{(1)}(x_i)=0,\ \nu^{(1)}(x_{i+1})=1$$

This choice makes it possible to zero the integral on the left side of the identity (15). Obviously, for $x\in(x_i,x_{i+1})$

$$\nu^{(0)}(x)=\frac{e^{\beta\Delta x/\mu}-e^{\beta(x-x_i)/\mu}}{e^{\beta\Delta x/\mu}-1},\ \nu^{(1)}(x)=\frac{e^{\beta(x-x_i)/\mu}-1}{e^{\beta\Delta x/\mu}-1}. \tag{16}$$

The integral on the right side of identity (15) will be approximated using the following formulas

$$\int_{x_i}^{x_{i+1}} G(x, y)\nu^{(0)}(x)\mathrm{d}x \approx G(x_i, y) \cdot \int_{x_i}^{x_{i+1}} \nu^{(0)}(x)\mathrm{d}x$$

$$\int_{x_i}^{x_{i+1}} G(x, y)\nu^{(1)}(x)\mathrm{d}x \approx G(x_{i+1}, y) \cdot \int_{x_i}^{x_{i+1}} \nu^{(1)}(x)\mathrm{d}x \tag{17}$$

After substituting $\nu = \nu^{(0)}$ and $\nu = \nu^{(1)}$ into identity (15), one can obtain a multi-parametric family of difference schemes. Note that different variants of the approximation of integrals in (17) may lead to other difference schemes, but we, for simplicity, will focus on the above options. To determine the values $\{\Psi_{i,j}\}$, we have the following family of difference schemes:

$$\begin{cases} \mu\left(\frac{\Psi_{i+1,j}-2\Psi_{i,j}+\Psi_{i-1,j}}{(\Delta x^2)} + \frac{\Psi_{i,j+1}-2\Psi_{i,j}+\Psi_{i,j-1}}{(\Delta y^2)}\right)+ \\ +\beta\left(\frac{1+\theta_{i,j}}{2} \cdot \frac{\Psi_{i+1,j}-\Psi_{i,j}}{\Delta x} + \frac{1-\theta_{i,j}}{2} \cdot \frac{\Psi_{i,j}-\Psi_{i-1,j}}{\Delta x}\right) = \Phi(x_i, y_j), i = \overline{2, n-1};\ j = \overline{2, k-1}; \\ \Psi_{i,j} = 0, i, j \in \partial\omega_h, \theta_{i,j} \in [-1, 1]. \end{cases}$$

Here, $\theta_{i,j}$ is a set of parameters that determines the approximation of the derivatives of $\frac{\partial \Psi}{\partial x}$ at the grid node (x_i, y_j).

By entering the difference $R \equiv \frac{\beta \Delta x}{2\mu}$, the problem can be solved by one of the known iterative methods:

$$\begin{aligned} &\left[\frac{2\mu}{\Delta y^2} + \frac{2\mu}{\Delta x^2}\left(1 + R\theta_{i,j}\right)\right] \cdot \Psi_{i,j}^m \\ &= \frac{\mu}{\Delta x^2}\left[1 + R\left(1 + \theta_{i,j}\right)\right] \cdot \Psi_{i+1,j}^m + \frac{\mu}{\Delta x^2}\left[1 - R\left(1 - \theta_{i,j}\right)\right] \cdot \Psi_{i-1,j}^m \\ &+ \frac{\mu}{\Delta y^2}\left[\Psi_{i,j+1}^{m-1} - \Psi_{i,j-1}^m\right] - \Phi_{i,j}; \\ &i = \overline{2, n-1};\ j = \overline{2, k-1}; \end{aligned} \tag{18}$$

$$\begin{aligned} &\Psi_{i,j}^m = 0, i, j \in \partial\omega_h;\ m = 0, 1, 2 \ldots; \\ &\Psi_{i,j}^0 = 0, \quad i, j \in \partial\omega_h. \end{aligned}$$

We define the value $R \equiv \frac{\beta \Delta x}{2\mu}$, and in terms of this value we write down our formulas. We will solve the last grid problem using one of the iterative method [10]:

$$\begin{aligned} &\left[\frac{2\mu}{\Delta y^2} + \frac{2\mu}{\Delta x^2} \cdot \left(1 + R\theta_{i,j}\right)\right] \cdot \Psi_{i,j}^m \\ &= \frac{\mu}{\Delta x^2} \cdot \left(1 + R\left(1 + \theta_{i,j}\right)\right) \cdot \Psi_{i+1,j}^{m-1} + \frac{\mu}{\Delta x^2} \cdot \left(1 - R\left(1 - \theta_{i,j}\right)\right) \cdot \Psi_{i-1,j}^m \\ &+ \frac{\mu}{\Delta y^2} \cdot \left(\Psi_{i,j+1}^{m-1} + \Psi_{i,j-1}^m\right) - \Phi_{i,j}, i = \overline{2, n-1};\ j = \overline{2, k-1}; \end{aligned}$$

$$\Psi_{i,j}^{m} = 0, i, j \in \partial\omega_h, m = 0, 1, 2, \ldots;$$
$$\Psi_{i,j}^{0} = 0, i, j \in \partial\omega_h.$$

Parameters $\theta_{i,j}$ are usually set as some function of a number R. In our case, the number R is a constant value, so the family of parameters will turn into one parameter $\theta_{i,j} = \theta(R_{i,j}) = \theta(R)$. The function $\theta(R)$ can be selected in various ways, in particular, the following options can be found in literary sources:

1. $\theta(R) = 0$—a scheme with a central difference [10];
2. $\theta(R) = sign(R)$—a scheme with a directional difference [11];
3. $\theta(R) = \frac{|R|}{1+|R|} \cdot sign(R)$—scheme of A. A. Samarsky [12];
4. $\theta(R) = cth(R) - \frac{1}{R}$—scheme of Ilyin A. M. [12, 13].

The choice of a function $\theta(R)$ using options 2–4 guarantees unambiguous solvability of the system of Eq. (1), since in this case the grid operator of this system will be a monotone operator [14].

After the current function is found, integral velocities can be calculated. The approximation of derivatives can be chosen, for example, as follows:

$$\begin{cases} U_{i,j} = \frac{\Psi_{i,j+1} - \Psi_{i,j-1}}{2\Delta y} \\ V_{i,j} = -\frac{\Psi_{i+1,j} - \Psi_{i-1,j}}{2\Delta x}; i = \overline{2, n-1};\ j = \overline{2, k-1}. \end{cases} \quad (19)$$

For convenience, we introduce the notation:

$$\begin{aligned} D_y^+ \Psi_{i,j} &= \frac{\Psi_{i,j+1} - \Psi_{i,j}}{\Delta y},\ D_y^- \Psi_{i,j} = \frac{\Psi_{i,j} - \Psi_{i,j-1}}{\Delta y}, \\ D_x^+ \Psi_{i,j} &= \frac{\Psi_{i+1,j} - \Psi_{i,j}}{\Delta x},\ D_x^- \Psi_{i,j} = \frac{\Psi_{i,j} - \Psi_{i-1,j}}{\Delta x}. \end{aligned} \quad (20)$$

We give further approximations of derivatives consistent with the scheme (18) and obtained on the basis of the projection version of the integro-interpolation method (PVIIM) [5].

For internal grid nodes, including upper and lower horizontal borders:

$$\frac{\partial \Psi_{i,j}}{\partial x} = \frac{1-\theta}{2} \cdot [1 + R(\theta+1)] D_x^+ \cdot \Psi_{i,j} + \frac{1+\theta}{2} \cdot [1 + R(\theta-1)] D_x^1 \cdot \Psi_{i,j} \quad (21)$$

On the left border:

$$\frac{\partial \Psi_{i,j}}{\partial x} = [1 + R(\theta+1)] D_x^+ \cdot \Psi_{i,j} - \Delta x \cdot \frac{1+\theta}{2\mu} \cdot F(x_i, y_j)$$

On the right border:

$$\frac{\partial \Psi_{i,j}}{\partial x} = [1 + R(\theta - 1)]D_x^- \cdot \Psi_{i,j} + \Delta x \cdot \frac{1-\theta}{2\mu} \cdot F\left(x_i, y_j\right)$$

For internal grid nodes, including left and right vertical borders:

$$\frac{\partial \Psi_{i,j}}{\partial y} = \frac{1}{2} \cdot \left(D_y^+ \cdot \Psi_{i,j} + D_y^- \cdot \Psi_{i,j}\right)$$

On the upper horizontal border:

$$\frac{\partial \Psi_{i,j}}{\partial y} = D_y^- \cdot \Psi_{i,j} + \Delta y \cdot \frac{1}{2\mu} \cdot F\left(x_i, y_j\right)$$

On the lower horizontal border:

$$\frac{\partial \Psi_{i,j}}{\partial y} = D_y^+ \cdot \Psi_{i,j} - \Delta y \cdot \frac{1}{2\mu} \cdot F\left(x_i, y_j\right).$$

6 Results of Numerical Experiments

To illustrate, consider as a model object a rectangular pond with a flat bottom with the characteristic dimensions of the Black Sea: $a = 11 \cdot 10^7(sm) = 1100(km)$, $b = 5 \cdot 10^7(sm) = 500(km)$, $D = 2 \cdot 10^5(sm) = 2000(m)$,

where a is the size in longitude, and b—in latitude, D—depth.

Then we have:

$$r = 11,\ q = 5,\ H = 1,\ k = 0.05;$$

$$l_0 = 1,\ \beta = 0.00023;$$

$$\beta = 0(\text{without}\ \beta\ \text{- effect})$$

$$F = \frac{\pi}{2} \cdot 10^{-3},\ \mu = 0.001.$$

The comparison will be made in the following norms:

$$NC\Psi = \frac{\max_{\omega_h}\left|\Psi - \overline{\Psi}\right|}{\max_{\omega_h}\left|\overline{\Psi}\right|} \cdot 100(\%),\ NL\Psi = \frac{\sum_{w_h}\left|\Psi - \overline{\Psi}\right|}{\sum_{w_h}\left|\overline{\Psi}\right|} \cdot 100(\%) \tag{22}$$

where $\overline{\Psi}$ is the exact value obtained by the analytical formula, Ψ is the estimate of the current function calculated using one or another difference scheme. Using similar formulas, we will calculate the norms NCU, NCV, NLU, NLV, characterizing the deviations of the calculated velocities from their exact values. Figure 1 shows the solution of the problem obtained by using an approximation based on Scheme 4. Judging by the values of the norms presented in Table 1, the accuracy of reproduction of the analytical solution is quite high even when using a large grid.

Table 1 shows the values of the norms $NC\Psi$, $NL\Psi$ and with increasing grid nodes. The result was obtained using the Ilyin scheme when calculating the current function, and the velocities are calculated by (19). Numerical experiments have shown that with this approximation, the best result of calculating the current function is obtained compared to other discretizations in the family of schemes (18). The values of the norms NCU, NCV, NLU, NLV, in the case of using the scheme (21) are also presented in the Table 2. An increase in the discretization over space leads to an improvement in the calculation results, but it is not necessary to consider them satisfactory, especially when calculating integral velocities.

Table 2 shows the values of the norms NCU, NCV, NLU, NLV, when using the scheme (21) to calculate the integral velocities. The scheme used is consistent with the approximation (18) and is obtained on the basis of the projection version of the integro-interpolation method (PVIIM). The proposed scheme makes it possible

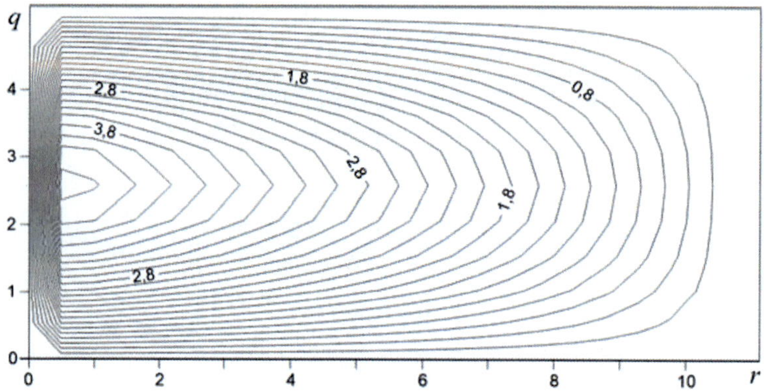

Fig. 1 The current function obtained according to scheme 4 (Ilyin A. M.)

Table 1 Norm values with increasing grid nodes (scheme (19))

Number of grid nodes	$NC\Psi$	$NL\Psi$	NCU	NCV	NLU	NLV
23 × 11	0.30851	0.40652	1.925	2.0367	75.123	67.985
111 × 51	0.01669	0.02123	0.0823	0.08701	30.339	12.717
221 × 101	0.00615	0.00675	0.0226	0.0232	17.027	4.0211
441 × 201	0.00924	0.00711	0.0134	0.0112	9.9514	1.1341

Table 2 Norm values with increasing grid nodes (scheme (21))

Number of grid nodes	NCU	NCV	NLU	NLV
23 × 11	1.925	2.0367	1.0455	0.9950
111 × 51	0.0823	0.0871	0.0489	0.0484
221 × 101	0.0226	0.0232	0.0103	0.0102
441 × 201	0.0134	0.0112	0.0052	0.0053

to obtain a numerical solution with sufficiently good accuracy even on a large grid, which naturally affects the amount of computing resources used.

7 Conclusions

Thus, the results of numerical experiments have shown that a sufficiently accurate solution of the problem for the current function does not guarantee a qualitative calculation of integral velocities. The approach used in this work makes it possible to solve the problem for the current function and calculate the spatial derivatives of this solution to determine the integral components of the horizontal components of the flow velocity. The results of the work can be used in modeling dynamic processes in the waters of the Azov and Black Seas.

Acknowledgements The work was carried out within the framework of the state task on the topic 0555-2021-0005 "Complex interdisciplinary studies of oceanological processes that determine the functioning and evolution of ecosystems of the coastal zones of the Black and Azov Seas" (code "Coastal research").

References

1. Marchuk G.I., Sarkisyan A.S.: Mathematical modeling of ocean circulation, Moscow, Nauka, (1988).
2. Stommel H.: The westward intensification of wind-driven ocean currents, Trans. Amer. Geoph. Un., 29, 202-206(1948).
3. Stommel, H.: Gulf stream., Moscow, IL (1965).
4. Stommel, H.: The GULF STREAM. A Physical and Dynamical Description, University of California Press, (1965).
5. Eremeev V. N.,Kochergin V. P., Kochergin S. V., Sklyar S. N.: Mathematical modeling of hydrodynamics of deep-water basins. ECOSI-Gidrophisika, Sevastopol, (2002).
6. Kochergin V.P., Dunets T.V.: Computational algorithm of the evaluations of inclinations of the level in the problems of the dynamics of basins, Physical oceanography, 3(11), 221 – 232 (2001).
7. Kochergin V.S., Kochergin S.V., Sklyar S.N.:Analytical Test Problem of Wind Currents. In: Olegovna C. (eds) Processes in GeoMedia—Volume I. Springer Geology. Springer, Cham., p.p. 17–25, (2020).

8. Kochergin V.S., Kochergin S.V., Sklyar S.N.: Analytical solution of the test three-dimensional problem of wind flows In: Chaplina, Tatiana (eds) Processes in GeoMedia, Vol. II Springer Geology, Springer,p.p. 65 – 71 (2021).
9. Kochergin V.S., Kochergin S.V., Sklyar S.N.: Analytical solution of the equation for the stream function in the model of Ekman-type flows with variable wind stress in space, Springer Proceedings in Earth and Environmental Sciences. 2022. **C.**147 - 158
10. Rouch P. Computational fluid dynamics.Moscow, Mir, p. 616, (1980).
11. Samarskiy A.A., Nikolaev E.S.: Methods of solving grid equations. Moscow, Nauka(1978).
12. Ilyin A.M.: Difference scheme for a differential equation with a small parameter at the highest derivative . Mathematical notes.6(2),237–248 (1969).
13. Doolan, E. P., Miller, J. J. H. and Schilders, W. H. A.: Uniform numerical methods for solving problems with a boundary layer. Moscow, Mir (1983).
14. Kollatz L.: Functional analysis and computational mathematics.. Moscow, Mir (1969).

Investigation of the Influence of Foam on the Characteristics of Waves in the Framework of Laboratory Simulation

M. I. Vdovin, D. A. Sergeev, Yu. I. Troitskaya, and A. A. Kandaurov

Abstract This paper describes the results of a laboratory investigations of the effect of the foam influence on the characteristics of the wavy surface in a wide range of wind speeds. The experiments were performed on the Termosratified Wind-Wave Flume of IAP RAS. Experiments were carried out for the winds with friction velocities between 0.3 and 1.5 m/s (corresponding to a 10-m wind speed of 12–33 m/s under field conditions). The water-level fluctuations were measured using wave gauges. The surface area covered with the foam was estimated basing on the shadow images of the surface filming by a high-speed camera. It was shown that the presence of foam on a wavy water surface leads to significant modifications in the characteristics of waves: evolution of the frequency and wave-number spectra, and suppression of the short-wave part of the waves (surface smoothing).

Keywords Water–air interface · Laboratory modeling · Airflow · Waves · Surface foam

1 Introduction

In recent years, various field and laboratory studies [1–8] have shown the existence of the effect of saturation of the aerodynamic drag coefficient of the water surface during hurricane winds. One of the possible explanations for the low aerodynamic resistance of the water surface for strong winds can be the effect of surface foam, which modifies the aerodynamic characteristics of the water surface and affects the wind-wave interaction, the processes of momentum, heat and moisture exchange in the turbulent atmospheric boundary layer. Based on the data obtained during aerial imaging [5], it was determined that in a strong wind only a small part of the sea surface is covered with white caps, consisting of dense spots of foam formed as a result of wave breaking. For example, at a wind speed of about 50 m/s, areas with dense foam caps cover less than 5% of the surface. However, at the same wind speed,

M. I. Vdovin (✉) · D. A. Sergeev · Yu. I. Troitskaya · A. A. Kandaurov
Federal Research Center, Institute of Applied Physics, Russian Academy of Sciences, Nizhny Novgorod, Russia
e-mail: maxvdovin@ipfran.ru

V. I. Karev (ed.), *Physical and Mathematical Modeling of Earth and Environment Processes—2022*, Springer Proceedings in Earth and Environmental Sciences,
https://doi.org/10.1007/978-3-031-25962-3_17

the surface is almost completely covered with stripes of less dense foam, which are stretched in the direction of the wind [9]. It was also found that the drag coefficient begins to saturate starting from a speed of 25 m/s, coinciding with the moment when foam streaks begin to form on the surface. Moreover, at a wind speed of about 40 m/s, when the foam strips merge with each other, the drag coefficient begins to decrease.

2 Methods

To study the effect of foam on the surface on the wave characteristics, a laboratory experiment was carried out on Termosratified Wind-Wave Flume which is a part of the Unique Scientific Facility "Complex of Large-Scale Geophysical Facilities" of the IAP RAS. An underwater foam generator was designed to produce foam on the surface (Fig. 1). Its underwater working part consists of two parallel tubes with a diameter of 1 cm and a length of 35 cm, along the entire length of which holes with a diameter of 1 mm and a pitch of 7 mm are drilled in one row. Both tubes are plugged on one side, on the other, compressed air is supplied to one at a pressure of 1.5 atm, and a foam-forming liquid (sodium lauryl sulfate (SLS) solution was used) is supplied to the second, which was supplied under the pressure of a liquid column 1.5 m high. The tubes are wrapped with fine-meshed foam rubber. The working part was located under water (5 mm between the upper boundary of the foam rubber and the undisturbed water level) 1.2 m from the beginning of the flume perpendicular to the side walls. As a result, finely dispersed foam was formed on the surface of the water, which was carried away by the wind induced flow. During the experiment, the level of the SLS solution and the pressure of compressed air were kept constant, which ensured a constant rate of foam generation.

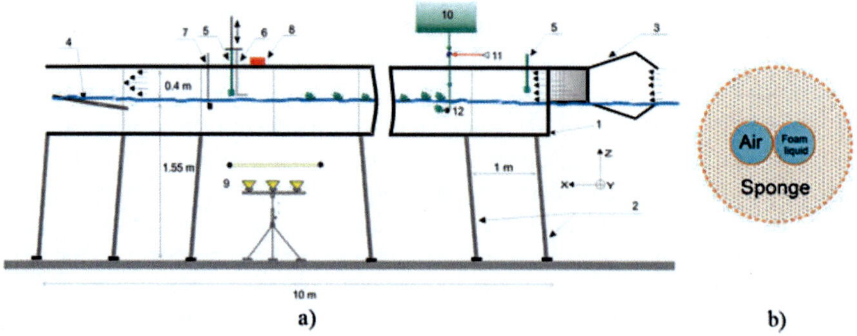

Fig. 1 Schematic diagram of the experimental setup (**a**) and aside cross-section of the working part of the foam generator (**b**). 1—working part of the wind-wave channel, 2—vertical supports of the channel, 3—honeycomb, 4—wave absorber, 5—hot-wire anemometer, 6—Pitot tube, 7—three-channel wire wave gauge, 8—high-speed camera, 9—underwater illumination, 10—reservoir with foaming liquid, 11—line of compressed air, 12—working part of the foam generator

To ensure the effect of foam generator influence on surface waves, special test experiments were carried out, during which only compressed air was supplied to the foam generator without SLS solution. In this mode, bubbles formed only locally in the area of the working part of the foam generator and did not reach the measurement area.

Before and after each series of measurements, the parameters of the water in the tank were monitored to assess the effect of the foaming agent on it. The coefficient of surface tension was measured by the drop method, the viscosity of water was measured using a viscometer. In all cases, the differences from the initial values were within the measurement error. Thus, changes in the physical properties of water in the tank itself and their possible influence on the wind-wave interaction were excluded.

The characteristics of surface waves were measured for six different air flow rates in the air flume and three modes of foam generation: (1) clean water and turned off foam generator; (2) clean water with the supply of compressed air to the foam generation system, but without the supply of a foaming agent; (3) with a fully functioning foam generation system (air + SLS).

To study the characteristics of surface waves in the wind-wave channel, a three-channel string wave gauge was used, which was developed and manufactured at the IAP RAS. The sensor sampling frequency was 100 Hz. The wave gauge consists of three pairs (three channels) of wire resistive sensors, located at equilateral triangle with a side of 2.5 cm. Such an arrangement is necessary to obtain the frequency-angular spectra of waves using the WDM (Wavelet Directional Method) [10] or FDM (Fourier Directional Method) method [4]. In this study, each of the sensors consists of two parallel nickel wires 35 cm long, semi-immersed in water. The distance between the wires is 5 mm. The resistance of a system of two wires and water between them depends on the depth of their immersion, determined by the current elevation of the rough surface. The wind speed in the wind tunnel was measured by profiling scanning using a Pitot tube. The process of measuring and processing the obtained data is described in detail, for example, in [4].

The parameters of the surface foam were studied using the optical shadow method—the rough surface was illuminated from below and filmed with a high-speed camera (Fig. 2). A matte screen was installed under the working section of the flume at a depth of 370 mm, which was illuminated from below by a matrix of 25 (5 $\times$ 5) LED lamps with a power of 10 W each. A "Raptor photonics Cygnet" CMOS camera was installed above the working section (frame size 2048 $\times$ 1088 px, image size 231 $\times$ 123 mm, scale 113 μm/px, average frame rate 12 fps, 3000 frames per recording).

The resulting digital images (Fig. 3) were processed automatically with special algorithms. The proportion of surface area covered with foam was calculated based on the number of pixels that were darker than a certain threshold. To eliminate the influence of noise and small optical inhomogeneities, the original images were divided into regions (the resolution of each of them is 50 times less than the original frame), for each of which the average brightness was calculated. Separately, the background image was calculated as the average brightness value for each pixel based on a recording made on calm water without wind and foam. The background

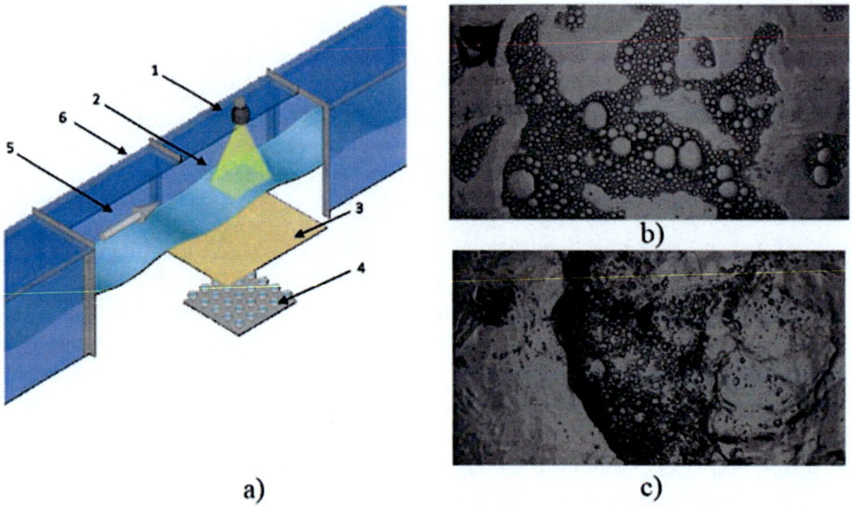

Fig. 2 **a** Schematic diagram of using the shadow method: 1—camera, 2—filming area, 3—matte screen, 4—underwater lamp (matrix of LEDs), 5—wind direction, 6—flume body; **b** example of the obtained image for speed U_{10} = 12 м/с; **c** example of the obtained image for the speed U_{10} = 35 м/с

image was subtracted from each frame, and then a general histogram (brightness distribution of image areas) was constructed for all video frames. For each histogram, you can calculate the proportion of areas with brightness below the threshold. In this case, the threshold was determined by direct calculation of the surface area under the foam using morphological imaging. Thus, the process consisted of several stages: (a) subtraction of the background from the original image; (b) division of the resulting image into smaller areas; (c) binarization of areas by a threshold value; (d) morphological opening was performed by a structural element "disk" with a diameter of 1 pixel with by subsequent morphological closure with a "disk" 21 pixels in diameter; (e) the closed areas of the image were filled using the fill operation. As a result, the ratio of the number of pixels found to the total number of pixels was taken as the fraction of the surface area covered with foam.

3 Results

Based on the results of video recording data processing, the areas of water surface covered with foam were obtained. When the foam generator was operating with a constant flow rate of the SLS solution, the foam coverage area depended on the wind speed in the flume (Fig. 4). At friction velocities less than 0.8–0.9 m/s (U_{10} from 22 to 25 m/s), a decrease in the foam fraction on the surface was observed with an increase in wind speed, while the foam was concentrated near the crests of the waves

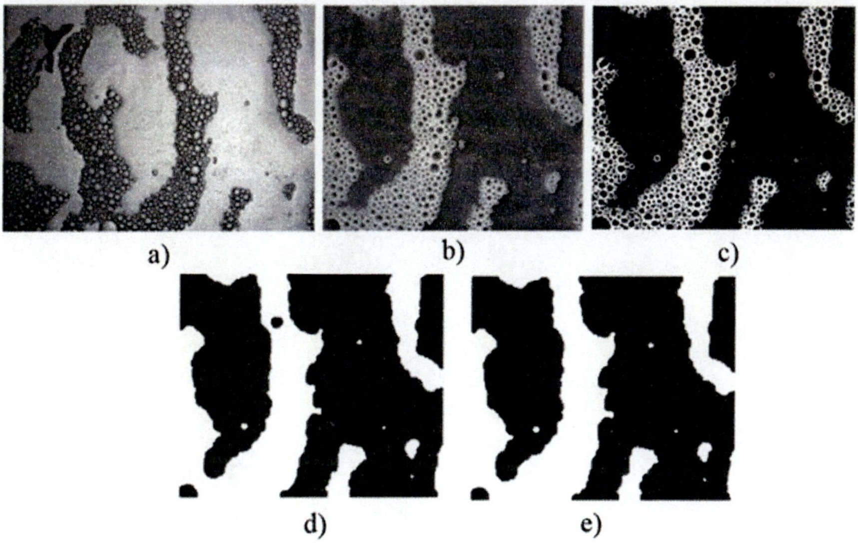

Fig. 3 Stages of morphological analysis to determine the relative area of the foam coverage

(the greater the wind speed, the more foam near the crests, and the less in the troughs of the waves). At friction velocities exceeding 0.8–0.9 m/s, intensive wave breaking was observed, which led to additional natural foam formation, not associated with the activity of the foam generator. As additional tests showed, the amount of foam formed due to wave breaking does not depend on the addition of SLS solution to the flume (at the concentration that was supplied when the foam generator was running) and coincides with the case of pure water.

The isotropic spectra obtained for different modes in terms of frequency and wave number are shown in (Fig. 5a–d). It is shown that the presence of foam on the surface strongly changes the spectra of surface waves. The most striking effect is the suppression of the effect of decreasing peak frequency ω_p and wave number k_p with increasing wind speed. The influence of a running foam generator without the supply of a foaming agent (only compressed air is supplied) on the characteristics of surface waves is not significant (Fig. 5a, d).

The significant wave height was calculated by the following way:

$$H_s = \frac{1}{\frac{1}{3}N} \sum_{m=1}^{\frac{1}{3}N} H_m \quad (1)$$

and mean square slope:

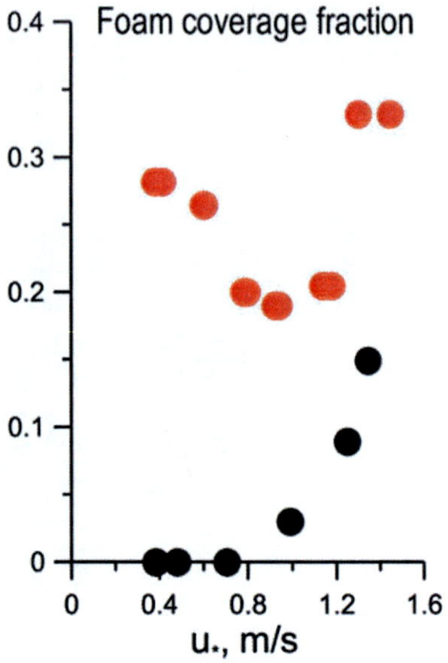

Fig. 4 Foam coverage area (red dots—running foam generator, black dots—clean water)

$$m.s.s. = \int_{k_{min}=0.01\ cm^{-1}}^{k_{max}=1.25\ cm^{-1}} k^2 S(k) dk \qquad (2)$$

The lower limit $k_{min} = 0.01\ cm^{-1}$ was chosen below the smallest wavenumber that was observed in the experiment. And the top one is in accordance with the resolution of the used wave gauge $k_{max} = 1.25\ cm^{-1}$.

It is shown that the presence of foam on the surface leads to a slight decrease in the significant wave height (Fig. 6a). However, in combination with a reduction in peak wave number (Fig. 5b), this leads to a strong decrease in mean square slopes (m.s.s.) (Fig. 6b). Foam on the surface actually leads to the suppression of the short-wave part of the waves. Therefore, one can expect a decrease in the nonlinearity of the field of surface waves and a decrease in the resistance of the waveform in the presence of foam on the surface.

4 Conclusion

Based on the obtained results, it can be argued that the presence of foam on a waved water surface leads to significant changes in the characteristics of waves: a change in the frequency and spatial spectra, and suppression of the short-wave part of the waves (surface smoothing). All this can play a significant role in the problem of surface drag coefficient under strong winds conditions and needs a more detailed study.

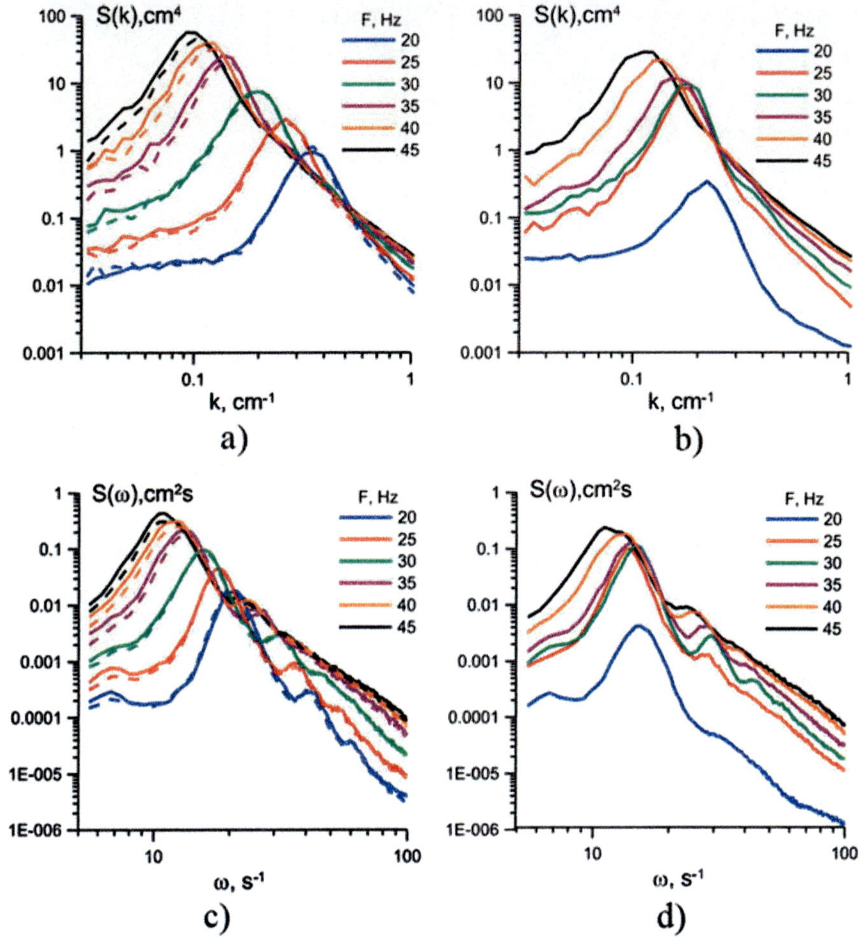

Fig. 5 Isotropic spectra in frequency and wave number obtained for different modes: **a** and **c** clear water (solid line) and foam generation without SLS (dashed line)—to assess the effect of the foam generator operation on waves; **b** and **d**—foam generation with the supply of SLS solution. Lines of different colors correspond to different speed modes of the wind flow in the channel (different frequencies of the blower fan). Here F20–F45 is the frequency of the blower fan, which actually sets the speed mode of the wind flow in the channel (equivalent wind speed at a height of 10 m U_{10} from 12 to 33 m/s)

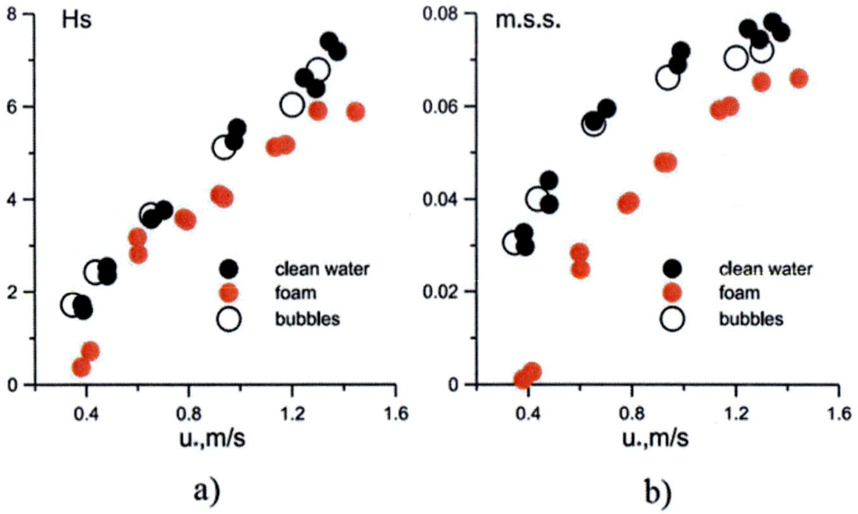

Fig. 6 Dependences of significant wave height (**a**) and m.s.s. (**b**) on friction velocity for different modes: red circles—with artificial foam generation, black circles—clear water, open circles—running foam generator without SLS supply

Acknowledgements This work was supported by the Russian Science Foundation grant 19-17-00209 (experiments) and Russian Foundation Basic Research 21-55-52005 (data processing). The experiments were performed at the Unique Scientific Facility "Complex of Large-Scale Geophysical Facilities" (http://www.ckp-rf.ru/usu/77738/).

References

1. *Powell M. D.* Final Report "Drag Coefficient Distribution and Wind Speed Dependence in Tropical Cyclones" Principal Investigator: Mark D Powell NOAA/AOML, 2005–2007
2. *Jarosz E., Mitchell D. A., Wang D.W., Teague W.J.* Bottom-up determination of air-sea momentum exchange under a major tropical cyclone. Science, 2007, v 315, p. 1707–1709, DOI: https://doi.org/10.1126/science.1136466.
3. *Takagaki, N., S.Komori, N.Suzuki, K.Iwano, T.Kuramoto, S.Shimada, R.Kurose, and K.Takahashi.* Strong correlation between the drag coefficient and the shape of the wind sea spectrum over a broad range of wind speeds. Geophys. Res. Lett., 2012, 39, L23604, https://doi.org/10.1029/2012GL053988
4. *Troitskaya, Yu. I., D.Sergeev, A. A.Kandaurov, G. A.Baidakov, M. A.Vdovin, and V. I.Kazakov.* Laboratory and theoretical modeling of air-sea momentum transfer under severe wind conditions. J. Geophys. Res., 2012, 117, C00J21, https://doi.org/10.1029/2011JC007778
5. *Donelan M.A., Haus B.K, Reul N., Plant W.J., Stiassnie M., Graber H. C., Brown O. B., Saltzman E. S.* On the limiting aerodynamic roughness of the ocean in very strong winds // Geophys. Res. Lett., 2004, v.31, L18306.
6. *Andreas E. L.* Spray stress revised. J. Phys. Oceanogr., 2004, v.34, No 6, p.1429–1440.
7. *Reul N., Branger H., Giovanangeli J.-P.* Air flow structure over shortgravity breaking water waves // Bound.Layer Meteor., 2008. V. 126. P. 477–505.

8. *Bell, M. M., Montgomery, M. T., & Emanuel, K. A.* Air–Sea Enthalpy and Momentum Exchange at Major Hurricane Wind Speeds Observed during CBLAST. Journal of the Atmospheric Sciences, 2012, 69(11), 3197–3222. doi:https://doi.org/10.1175/jas-d-11-0276.1
9. *Holthuijsen, L. H., M. D. Powell, and J. D. Pietrzak.* Wind and waves in extreme hurricanes. J. Geophys. Res., 2012, 117, C09003, https://doi.org/10.1029/2012JC007983
10. *Donelan M.A., Drennan W.M., Magnusson A.K.* Nonstationary Analysis of the Directional Properties of Propagating Waves // Journal of Physical Oceanography, vol. 26, 1996. P. 1901–1914

Determination of Optimal Parameters and Modes of Well Operation in Low-Permeability Reservoirs on a True Triaxial Loading Unit

S. O. Barkov and N. I. Shevtsov

Abstract This article presents the results of a series of experiments on physical modeling of mechanical and filtration processes in productive formations of the Astrakhan gas-condensate field (GCF), which may lead to an increase in permeability of reservoir rock when using the directional unloading method. The experiments were carried out on the Triaxial Independent Load Test System of the Institute for Problems in Mechanics of the Russian Academy of Sciences. The stress–strain state in the vicinity of perforation holes was modeling for both cased and uncased wells. The paper presents the dependence of filtration properties of the studied specimens on the type of the stress–strain state and determines the values of pressure draw-down, leading to an increase in permeability of rocks. The obtained results indicate that the directional unloading method can be successfully applied to uncased wells drilled at the Astrakhan GCF.

Keywords Low-permeability rocks · Filtration properties · Stress–strain state · Enhanced oil recovery · Perforation hole · Directional unloading method

The main problems of geomechanics and applied geology in the oil and gas industry are still the discovery of fields, determination of mineral reserves, engineering of field development systems, projecting the profile of horizontal and directional wells, and modeling various measures aimed at increasing the oil recovery and increasing the completeness of field development. The latter problems today become more and more actual in view of permanent depletion of easy-to-recover reserves and the beginning of active development of hard-to-recover hydrocarbon reserves (HTR). It should also be noted that the development of fields with low-permeability rocks is a rather complicated process not only technically, but also ecologically.

S. O. Barkov (✉) · N. I. Shevtsov
Ishlinsky Institute for Problems in Mechanics RAS, Moscow, Russia
e-mail: sviatoy97@gmail.com

V. I. Karev (ed.), *Physical and Mathematical Modeling of Earth and Environment Processes—2022*, Springer Proceedings in Earth and Environmental Sciences,
https://doi.org/10.1007/978-3-031-25962-3_18

1 Directional Unloading Method

Knowledge of mechanical and filtration characteristics of productive formation plays a key role in creating a geomechanical model of the developed fields, studying the possibility of increasing oil recovery, as well as in ensuring wellbore stability [1]. Flow rate of each well is significantly influenced by filtration properties of rocks that constitute the bottom-hole formation zone (BFZ). In various technological operations during drilling and subsequent operation of a well, possible deterioration of permeability even in a small area near the wellbore can significantly reduce its productivity [2]. It is assumed that permeability reduction is mainly caused by contamination of filtration channels [3], but it should also be noted that the filtration properties of rocks are radically affected by the stress–strain state arising in them [4]. Particular attention is given to studying the properties of rocks comprising the bottom-hole zone, since the characteristics of the formation and the processes that occur in this area determine the productivity of the well.

During well development and operation, the filtration properties of BFZ are negatively influenced by penetration of technological solutions into formation, often resulting in formation of persistent emulsions, as well as siltation and deposition of resins, salts and paraffins. The process of drilling a well significantly changes the conditions of the bedding of rocks, and changes occur in the natural stress field, which lead to both a decrease and an increase in the transverse dimensions of cracks.

The non-uniform stress field formed around the borehole is determined by the rock structure, its depth, the geometry of the well bottom, the operating mode, and also depends on the physical and mechanical properties of the rock mass. During various technological operations, circumferential compressive stresses can act on the well walls, exceeding the values of rock pressure, which can cause destruction of rocks and lead to both decrease and increase their permeability. To date, despite a sufficient amount of experimental studies of core material, the effect of non-uniform stress field in the bottom-hole zone on the permeability of rocks has not been studied sufficiently.

In order to increase the permeability of the productive formation, the Institute for Problems in Mechanics of the Russian Academy of Sciences (IPMech RAS) developed an environmentally safe, effective and cost-effective directional unloading method. The idea of this approach is that by creating the necessary stress state to cause the appearance of a system of micro- and macro-cracks in the vicinity of the borehole, which will thereby increase the permeability of the rock [5]. This stress state can be obtained by lowering the pressure at the bottom-hole and performing a number of technological operations, including the creation of a system of perforation holes on the wall of the well. At the same time, all the values of stresses necessary for specific fields are determined by direct physical modeling on rock specimens with the use of the Triaxial Independent Load Test System (TILTS) developed and designed in the IPMech RAS [6].

2 Testing Facility and Specimen Preparation

A series of experiments was carried out on the TILTS presented in Fig. 1. The development of this system was carried out at the IPMech RAS [7].

TILTS is intended for studying elastic-strength, rheological and filtration properties of rocks. It makes it possible to load cubic rock specimens with faces of 40 or 50 mm independently along each of the three axes. This makes it possible to experimentally simulate any type of stress–strain state occurring in the formation during various technological operations on the well. Also TILTS allows to study the influence of stress–strain state on filtration properties of rocks by continuous permeability measurement during the test [6]. The object of the study are rocks of Astrakhan gas-condensate field (GCF), from which cubic specimens with 40 mm edge were made. Experimental specimens were strongly cemented sandstones with low initial permeability, extracted from the reservoir of Astrakhan GCF from the depth interval of 3780–3790 m. Marking was performed as follows: 1 axis of the specimen coincided with the core axis, orientation of axes 2 and 3 was arbitrary. Four faces of the specimen were covered with impermeable thin polymeric film, two opposite faces were left open for air to pass through the specimen, which is necessary for measuring permeability. A total of 10 specimens were tested. Before the tests, the degree of rock anisotropy was determined by measuring the velocities of elastic longitudinal waves along each of the specimen axes [7]. The results of the sounding of the specimens showed transversal isotropy of the elastic properties of the studied rocks.

This work presents the results of a series of experiments on physical modeling of mechanical and filtration processes in productive formations of the Astrakhan GCF when implementing the directional unloading method, carried out by creating

Fig. 1 Triaxial independent load test system (TILTS)

perforation holes on the wall of the borehole. The specimens were tested using programs corresponding to the case of a perforation hole in a cased borehole [8], as well as in an uncased borehole for the case of a point on the perforation hole surface, distant from the well axis at a distance $r = 1.25\ R$, for a point on the perforation hole surface, distant from the borehole axis at a distance $r = 2\ R$, and for a point at the tip of the perforation hole.

3 Experimental Method

The perforation hole in the cased well is under the effect of an external stress arising in the vicinity of the cased well and coinciding with the natural stress q from the rock pressure deep in the formation. From the solution of the Lamé problem for the effective stresses acting in the soil skeleton, the stress values in the cylindrical coordinate system r', z', φ', associated with the perforation hole, have the form for the cased well [9]:

$$\begin{aligned} s_{r'} &= -(q + p_w)\left(R_h/r'\right)^2 + q + p(r'), s_{z'} = q + p(r'), \\ s_{\varphi'} &= (q + p_w)\left(R_h/r'\right)^2 + q + p(r') \end{aligned} \tag{1}$$

where R_h—radius of the perforation hole, r'—distance from the center of the perforation hole, q—rock pressure, p_w—pressure in the borehole. Compressive stresses are considered negative. Thus, in a cased well the stresses along the surface of the perforation hole are constant and equal to (assuming in (1) $r' = R_h$, $p(r') = p_w$):

$$\begin{aligned} s_{r'} &= 0, s_{z'} = q + p_w \\ s_{\varphi'} &= 2(q + p_w) \end{aligned} \tag{2}$$

The external stress for a perforation hole in an uncased well is no longer the stress from rock pressure deep in the formation q, but the stresses σ_r, σ_φ, σ_z, acting around the uncased well. For an uncased well, the stresses in the vicinity of the perforation hole are determined by the superposition of solutions to the Lamé problem [9] and the problem similar to the Kirsch problem [10]:

$$\begin{aligned} s_{r'}(r) &= 0, s_{z'}(r) = -(q + p_w)(R/r)^2 + q + p_w, \\ s_{\varphi'}(r) &= 3(q + p_w)(R/r)^2 + 2(q + p_w) \end{aligned} \tag{3}$$

The notations are similar to (1). From (3) we can see that the stresses change depending on the distance from the well to the considered point of the perforation hole. Substituting the values we have:

$$To \quad r = R \quad s_{r'} = s_{z'} = 0, s_{\varphi'} = 5(q + p_w) \tag{4}$$

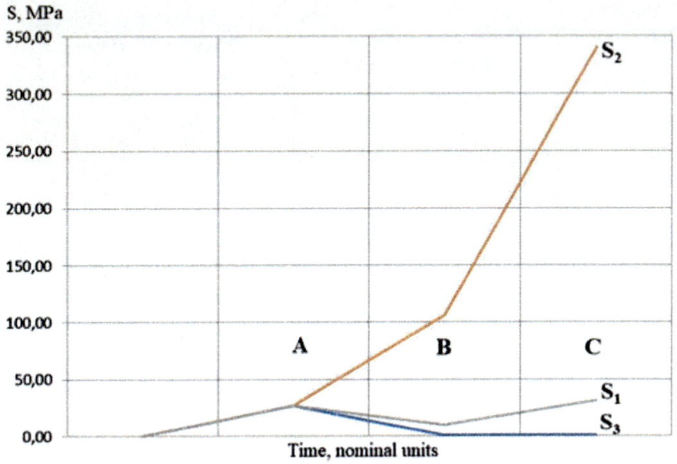

Fig. 2 Load program—point on the surface of the perforation hole, away from the well at a distance of 1.25 well radius, in an uncased well

$$To\quad r = 2R\quad s_{r'} = 0,\, s_{z'} = 3/4(q + p_w),\, s_{\varphi'} = 11/4(q + p_w) \tag{5}$$

From relations (4) and (5) it can be easily seen that the ring compressive stress, which is the maximum in absolute value, the greater the closer the point considered on the perforation wall is to the well. Figure 2 shows the load program view for the case of a point on the perforation hole surface in an uncased well at a distance of 1.25 well radius. Stresses S_1, S_2, S_3—stresses applied to the faces of the specimen along axes 1, 2, 3 of the TILTS loading unit—are plotted along the ordinate axis in MPa. The stress S_1 corresponds to the vertical stress component, S_2 to the annular one, and S_3 to the radial one, i.e. the stresses S_1, S_2, S_3 correspond to the absolute values of the stresses acting in the vicinity of the perforation hole. Point A corresponds to the stresses acting on the soil skeleton before the well is drilled, point B corresponds to the state when the well is drilled and the pressure at its bottom is equal to the formation pressure, point C corresponds to complete drainage of the well.

4 Results

Using relations (2) and (3) we compiled loading programs for specimens for conditions of Astrakhan GCF simulating the change in stresses on the perforation holes surface in cased and uncased wells with decreasing pressure p_w at their bottom-hole. In an experiment with an uncased well, the stress change at a point on the surface of the perforation hole, which is remote from the axis of the well at a distance $r = 1.25\ R$ and $r = 2R$. During the experiments, the deformation of specimens in

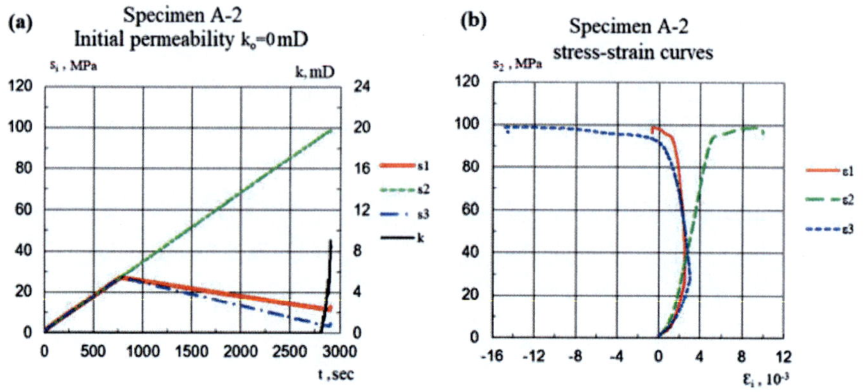

Fig. 3 **a**—Loading program and permeability curve of specimen A-2, **b**—stress–strain curves of specimen A-2

three directions was measured and the change in their permeability in the layering plane was recorded.

Ten specimens were tested at the TILTS facility using physical simulation programs. When tested using a program simulating stress changes in the perforation hole vicinity in a cased well, the specimens deformed elastically throughout the experiment without increasing permeability up to the stresses corresponding to complete well drainage.

A different picture was observed when testing the specimens using the "perforation hole in an uncased well" program, simulating the change in stress at a point on the surface of the perforation hole, remote from the axis of the well at a distance of $r = 1.25\ R$. Figure 3 shows the deformation curves of one of the specimens during the experiment and the change in its permeability. It can be noticed that the specimens fractured at the section of the program AB, not reaching the point B. But at the same time, during the simulation of the point remote from the axis of the well at a distance $r = 2\ R$, the specimens deformed elastically and the failure did not occur, up to the point C of the program, corresponding to complete drainage of the well.

5 Conclusion

In a series of experiments, rock behavior was simulated at the contour of the uncased well, the perforation hole contour: at 1.25 well radius and 2 well radius, and at the tip of the perforation hole. Figures 3 and 4 show plots of stress and permeability dependence on time, as well as strain curves for specimens A-2 and A-4.1, on which points on the perforation hole contour at a distance of 1.25 well radius were modeled. The result of these experiments is the destruction of experimental specimens (Fig. 5

shows the appearance of the specimens after the test—you can see the system of cracks on them) with an avalanche-like growth of permeability, which allows to judge about the possibility of successful application of the directional unloading method on the studied well of Astrakhan gas condensate field. When simulating the tip of a perforation hole in a cased well, the specimens did not fracture until the well was completely dry. These results indicate that lowering the bottom-hole pressure of cased wells with perforation does not increase permeability, whereas perforating uncased wells can result in multiple permeability increases. The experiments also demonstrated the inexpediency of creating perforation holes of long length, as it has been experimentally shown that rock destruction in the perforation hole vicinity occurs at a short distance from the borehole.

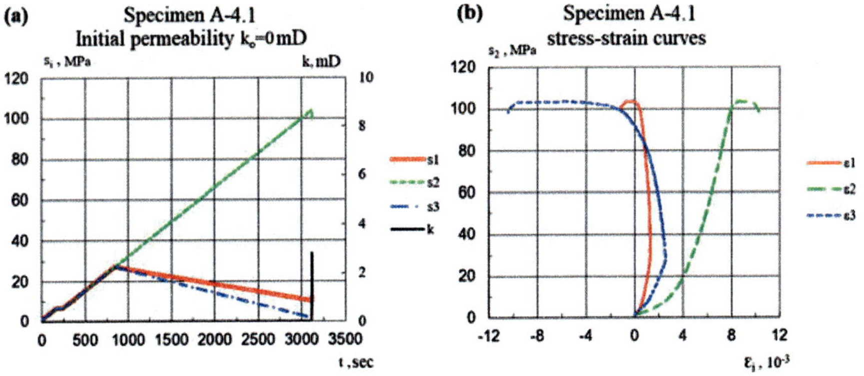

Fig. 4 **a**—Loading program and permeability curve of specimen A-4.1, **b**—stress–strain curves of specimen A-4.1

Fig. 5 Specimens A-4.1 and A-2 after testing

Acknowledgements This work was financially supported by the Russian Federation through the Ministry of Science and Higher Education of the Russian Federation Project No. 13.1902.21.0018 (Agreement 075-15-2020-802).

References

1. Mohammad N., Antonin S., Wan R. Prediction And Optimization Of Fracturing In Tight Gas And Shale Using A Coupled Geomechanical Model Of Combined Tensile And Shear Fracturing // SPE Hydraulic Fracturing Technology Conference, The Woodlands, Texas, USA, February 2012.
2. Boronin S.A., Tolmacheva K.I., Osiptsov A.A. et al. Modelling of Injection Well Capacity with Account for Permeability Damage in the Near-Wellbore Zone for Oil Fields in Western Siberia // SPE Russian Petroleum Technology Conference, Moscow, Russia, October 2017.
3. Reed M.G. Formation Damage Prevention During Drilling and Completion // SPE Centennial Symposium at New Mexico Tech, Socorro, New Mexico, October 1989.
4. Davies J.P., Davies D. K. Stress-dependent permeability: characterization and modeling // SPE J. 2001. 6. P. 224.
5. Kovalenko Yu.F., Karev V.I. Geo-leaching method - A new approach to the problem of increasing well productivity // TEK technologies. 2003. № 1. p. 31.
6. Karev V.I., Kovalenko Yu.F. Triaxial loading system as a tool for solving geotechnical problems of oil and gas production. In M. Kwarsniewski et al. (eds), True Triaxial Testing of Rocks: 301–310. Leiden: CRC Press/Balkema. 2013.
7. Karev V., Kovalenko Y., Ustinov K. Geomechanics of oil and gas wells. Advances in oil and gas exploration and production. Springer International Publishing Cham: Switzerland, 2020. 166 p.
8. Karev V.I., Klimov D. M., Kovalenko Yu. F. Modeling Geomechanical Processes in Oil and Gas Reservoirs at the True Triaxial Loading Apparatus // Phys. and Math. Modeling of Earth and Environment Proc. Springer, 2018. pp. 336–350.
9. Lyav A. Mathematical theory of elasticity. // USSR. 1935. 674 p.
10. Timoshenko S.P. Theory of Elasticity / S.P. Timoshenko, J. Gudier // M.: Nauka. 1979. 560 p.

Method of Liquidation of Hydrocarbon Spills Using Vortex Flow

T. O. Chaplina

Abstract The paper provides an overview of existing methods and sorbents for the elimination of hydrocarbons from the surface of the water, studied their characteristics and the principle of operation of devices that are currently used to eliminate oil spills. An original method of eliminating hydrocarbon spills is proposed, the main element of which is the creation of a controlled vortex funnel in a rotating cylindrical cup equipped. The principle of operation and methods of controlling the modes of operation of the vortex funnel have been confirmed by experimental studies.

Keywords Hydrocarbons · Oil spills · Controlled vortex funnel

1 Introduction

Currently, oil is the most common substance polluting natural waters. Transportation of half of the oil produced on the world shelf is provided by the tanker fleet and is estimated at up to 2 billion tons per year. At the same time, 0.03% of oil and oil products transported by tankers is lost for various reasons. There have been numerous environmental disasters associated with accidents of large-capacity tankers, which were accompanied by large-scale oil leaks and significant consequences for the ecosystems of large areas of the World Ocean [1]. In Russia, the loss of oil and oil products due to emergencies, non-compliance with technological discipline and appropriate safety measures reaches 4.8 million tons annually. At the same time, 30% of oil pollution comes from domestic and industrial waste, 27% from ships, 24% of pollution comes from the ocean floor from natural sources, 12% from tanker and oil platform accidents, 7% from atmospheric precipitation.

To reduce possible negative consequences, special attention should be paid to the study of methods for localization, liquidation of oil spills and the development of an additional set of measures for the collection and disposal of hydrocarbons that have entered the external environment.

T. O. Chaplina (✉)
A.Yu. Ishlinsky Institute for Problems in Mechanics, Prospect Vernadskogo 101-1, Moscow 119526, Russia
e-mail: tanya75.06@mail.ru

V. I. Karev (ed.), *Physical and Mathematical Modeling of Earth and Environment Processes—2022*, Springer Proceedings in Earth and Environmental Sciences,
https://doi.org/10.1007/978-3-031-25962-3_19

The purpose of this work is to study the current state and propose new promising methods for cleaning the sea surface from oil spills.

2 Devices for the Mechanical Collection of Liquid Hydrocarbons from the Water Surface

Currently, there are several methods for eliminating oil pollution from the surface and in the water column. Overview of existing methods for the elimination of hydrocarbons from the surface of the water is given in the works [2–5].

A known device for collecting oil and oil products from the surface of the water, includes a partially flooded drum with floats on the ends of the drum and adjacent to the drum vacuum receiving chamber with a front wall, which is made in the form of a scraper [6]. The disadvantages of this device include the low selectivity of oil collection, with which the predominant amount of water is sucked into the device.

A known method of cleaning the soil and water surface from oil pollution [7], which provides for the use of a device made in the form of cotton-containing sorbent mats, the outer side of which is pre-treated by spraying a thin layer of machine or transformer oil. In this case, a layer of cotton-containing sorbent in the mat is fixed between layers of cotton or synthetic fabric of a rare weave or cotton mesh. The distance between the threads of the fabric or the size of the mesh cell, which fix the sorbent in the mat, does not exceed the particle size of the cotton-containing sorbent. One of the disadvantages of the known device is the low adsorbing capacity, however, the main disadvantage is that cotton absorbs moisture and the mats filled with it sink without exhausting its ability to adsorb oil products. A device is also used to remove oil and oil products from the surface of the water, which includes a rotating drum with a hydrophobic surface and an adjacent system for removing and collecting the product [8]. The disadvantages of this device include a significant effect of ambient temperature on the performance of the device, since during its operation in the cold autumn-spring periods of operation, in winter, at night, due to an increase in the viscosity of the collected oil product, the absorbing capacity of the hydrophobic shell of oil and oil products decreases, which reduces the overall performance of the device.

A device is also known for collecting oil from under the ice cover of a reservoir [9]. A well is drilled in the area where the oil or oil product spot is located in the ice cover, a swirler with a pumping device is immersed through it into the spot area, the swirler is rotated to create a vortex funnel in the water under the ice, which ensures the collection of oil or oil product into it, and the oil or oil product is pumped out from the vortex funnel. Funnels (Fig. 1). The oil or oil product slick is localized by placing an inflatable boom barrier under the ice cover by transporting it by guided torpedoes and/or underwater vehicles or by placing it through wells drilled around the perimeter of the oil or oil product slick. Booms are placed under the ice in an uninflated (deflated) state, and after placement they are inflated. The area formed by the boom and the ice cover is filled with hot air at the final stage of oil or oil product pumping. A known device for collecting oil from the surface of the water includes a floating housing, a swirler with a vertical drive shaft, an exhaust pump and an oil intake pipe connected to it at one end, mounted coaxially with the swirler shaft and covering it, characterized in that, in order to improve the efficiency of the device for by reducing the water content of the collected oil, it is equipped with stuffing box seals and a hollow perforated truncated cone mounted coaxially with the shaft with a large base top and attached to the free end of the oil inlet pipe with the possibility of rotation around the axis of the shaft, while in the places where the perforated cone mates with the oil inlet pipe and the shaft, gland seals [10].

The device is transported over a water surface contaminated with an oil film (Fig. 1). Water from the oil film flows over the edges of the funnel 2 into the device between the floats 1, then flows into the oil-gathering funnel 2 and is discharged through the diffuser 3. Due to the rotation of the shaft 4 with the swirler 5 by the electric motor 6, a funnel is formed. The oil film inside the device is collected by a vortex funnel near the shaft 4 and is pumped out through the oil receiver 7 by the pump 8. Oil enters the oil receiver 7 through the perforations in the cone.

It should be noted the system for collecting oil products from the surface of the water, described in [11] (Fig. 2).

The system for collecting oil products from the surface of the water works as follows. Vessel 5 with jet propulsion 4 comes to the place where there is a localized spot of oil products floating on the surface of the water, and sets the oil intake funnel 1 below the level of floating oil products. Placed at the bottom of the funnel 1, the outlet pipe 2 is connected to the inlet pipe 3 of the conduit of the ship jet propulsion unit 4, and the outlet pipe 6 of the water conduit of the jet propulsion unit is connected to the oil storage tank. It can be, for example, any floating tank, oil tanker, barge, ballast compartment of water-ballast booms, etc. The jet propulsion 4 of the vessel 5 is launched, which, through the inlet pipe 3, begins to pump out, passing through itself, oil products from the surface of the water entering the oil intake funnel 1. Through the outlet pipe 6 of the water conduit of the jet propulsor 4, oil products enter the accumulator. This technical solution is simple in design and has high performance. The use of the same vessel for the installation of oil receiving equipment and pumping

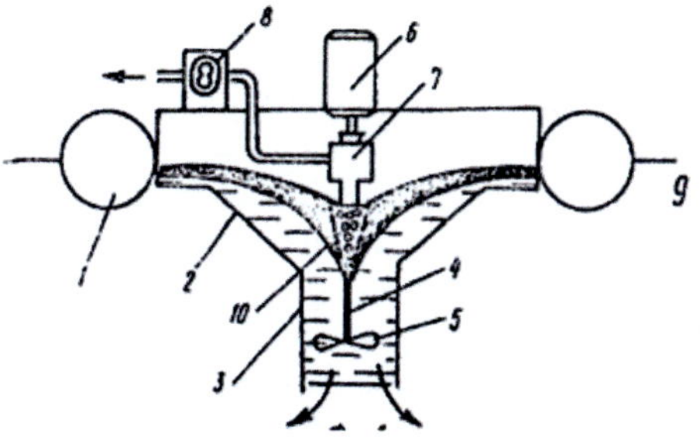

Fig. 1 Device for collecting oil from the water surface [10]

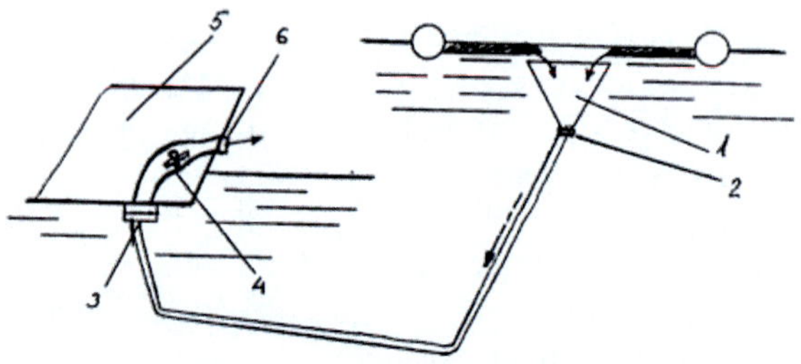

Fig. 2 System for collecting oil products from the water surface: 1—floating funnel with an outlet pipe 2 at the bottom. The upper edge of the funnel 1 is buried under the level of floating oil products. The outlet pipe 2 of the funnel 1 is connected to the inlet pipe 3 of the conduit of the jet propulsion 4 of the ship 5. The outlet pipe 6 of the conduit of the jet propulsion 4 is connected to the storage of oil products (not shown in the drawing). The connection is carried out, for example, using a pipeline to which, if necessary, a hydrocyclone can be connected [11]

of oil products without the use of special pumping equipment significantly reduces the cost of cleaning the water surface from oil pollution, and also allows the use of highly mobile vessels, which is very important to ensure the efficiency of such operations.

Also known is a device for collecting oil from the surface of the water [12], which includes a closed battery of hydrocyclones, evenly spaced around the circumference, interconnected by vertical bridges and combined front and rear walls with inlets expanding upwards, pipelines and oil pipelines connected to coaxially located oil and water collectors in the center of the device, hollow chambers formed by horizontal and vertical bridges, vertical edges formed by the front and rear walls of hydrocyclones, a cover with a breakwater device and a hatch. The device is equipped with external and internal electromagnets, evenly spaced around the circumference of the cover, and cone-shaped electromagnets located on the surface of the hydrocyclones, which are connected by a cable in water- and oil-resistant insulation to the control unit. Invention makes it possible to efficiently collect magnetic oil products of various film thicknesses from a large area by creating a magnetic field, which expands the possibilities of using an oil-collecting device on undulating surfaces of water bodies.

Also used is a method and device for collecting oil products from the water surface [13]. The method includes taking a mixture of oil products with water under vacuum, separating the mixture and then supplying oil products to an oil collector. The intake of a mixture of oil products with water is carried out below the water surface in the cavitation mode and is accompanied by air suction. In the separation process, a directed flow of a mixture of water, dispersed air and oil products is produced.

Despite the intensification of oil production in the Arctic regions, mainly in the offshore areas of the Arctic Ocean, there are still no sufficiently reliable methods and devices for collecting oil spills and oil products from under the ice cover.

It should be noted the method of collecting oil spills from under the ice cover of a reservoir, described in [14]. The invention relates to the field of environmental protection and can be used to collect spills of oil (petroleum products) from under the ice cover, mainly Arctic waters (Fig. 3). The method includes localization of an oil or oil product slick, drilling a well in the ice cover, immersing a swirler with a pumping device through the well into the area of the swirler spot with a pumping device, creating a swirl funnel in the water under the ice by rotating the swirler, which ensures the collection of oil or oil product into it and the subsequent removal of oil or oil product by pumping from vortex funnel into the oil receiver.

Localization of a spot of oil or oil product is carried out by covering the area of localization of the spot, which exceeds the zone of action of the vortex funnel. As the oil or oil product is pumped out, the spot localization area is narrowed to sizes equal to or smaller than the area of action of the vortex funnel. The oil or oil product slick can be localized by placing a boom barrier under the ice cover by transporting it by guided torpedoes and/or underwater vehicles. A boom with a closed contour in the form of a spiral is used, made with the ability to narrow the closed contour by twisting this spiral. The use of the invention will increase the areas of collection and removal from under the ice cover of oil (petroleum products) in flowing and stagnant water bodies.

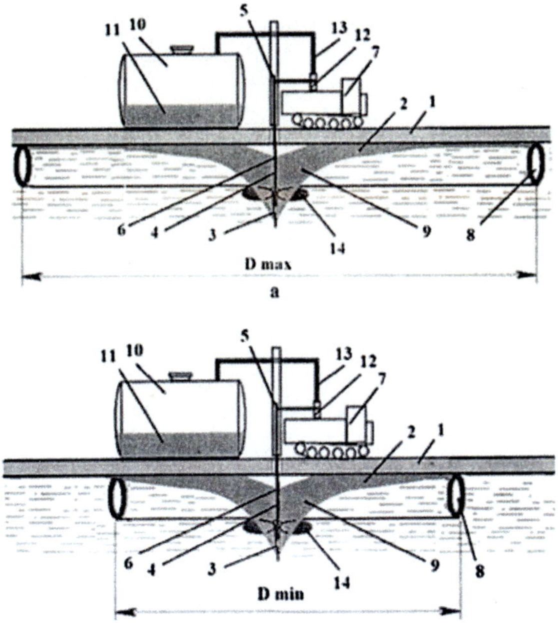

Fig. 3 Method for collecting oil spills from under the ice cover of a reservoir: 1—ice cover, 2—oil or oil products, 3—swirler, 4—hollow shaft of the swirler, 5—pipe for pumping oil, 6—perforations in the hollow shaft, 7—mobile installation for drilling and pumping oil, 8—booms, 9—vortex funnel, 10—tank for collecting oil products, 11—collected oil products, 12—pump, 13—pipeline for pumping oil products into the tank, 14—swirler blades (arrow shows oil pumping direction.) Dmax and Dmin are the maximum and minimum coverage areas of the oil slick [13]

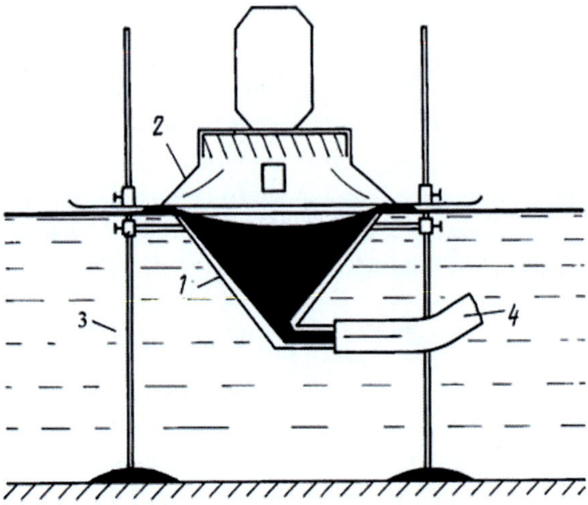

Fig. 4 Oil gathering funnel [15]

We also note a device (Fig. 4) containing an oil-gathering funnel (1) with a cover (2) above it and a system of vertical movement along vertical guides (3) rigidly fixed at the bottom of the reservoir. The removal of the collected oil is carried out through a flexible pipeline (4).

When operating this device, it is necessary to constantly monitor the level of oil on the water and move the funnel in accordance with the position of this level.

3 Cleaning the Sea Surface with a Controlled Vortex Funnel

The author of this paper proposed an original method for cleaning the water surface from liquid pollution [16], when in the volume of a rotating hollow cylinder located below the "water–oil" boundary, water movement is forcibly created with the formation of a vortex funnel in it with continuous pumping of oil products (Fig. 5).

The movement of water is ensured by the rotation of the hollow cylinder, the end of the receiving pipe of the means of pumping out the oil product is placed coaxially inside the hollow cylinder below its upper end by the value $H/2$ (H—is the height of the cylinder). To create an optimal mode of operation of a controlled funnel, express methods measure the thickness of the oil layer, the densities and kinematic viscosities of water and the removed oil products, and the measured values determine the immersion depth h of the upper edge of the hollow cylinder from the «water–oil product» interface, determined by the relationships.

$$h < \frac{H}{\alpha^2 - 1}, \alpha = 1 + \frac{\rho_w}{\rho_o}\sqrt{\frac{v_w}{v_o}} \tag{1}$$

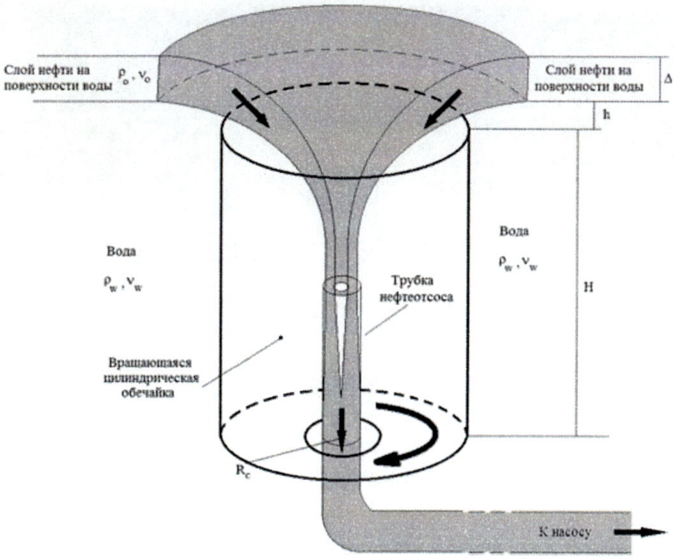

Fig. 5 Scheme of work of the oil collector

where ρ_w, ρ_o—density of water and oil, and v_w, v_0—are their kinematic viscosities.

The optimal angular frequency of rotation of the hollow cylinder ω is determined by the expression

$$\omega = \frac{1}{R_c}\sqrt{g\frac{(H + 2h + \Delta)(\rho_w + \rho_o)}{\left(1 + \alpha^2\right)\rho_w - \rho_o}} \tag{2}$$

where R_c—cylinder radius, Δ—oil layer thickness on the water surface, g—gravitational acceleration.

With known height H and radius R_c of the glass, the measured values of ρ_w, ρ_0, v_w, v_0 according to (1, 2), the required immersion depth and rotational speed are determined. So for $H = 1\ m$, $R_c = 0.04\ m$, $\rho_w = 10^3\ kg/m^3$, $\rho_0 = 860\ kg/m^3$, $v_w = 10^{-6}\ m^2/s$, $v_0 = 6 \cdot 10^{-6}\ m^2/s$ from (1) follows $\alpha = 1.47$ and the immersion depth of the upper edge must be less than $h < 0.85\ m$. We put $h = 0.5\ m$. Then, with the oil product layer thickness $\Delta = 0.1\ m$, the optimal cylinder rotation frequency, according to (2), turns out to be equal to $\omega \approx 10.1$ revolutions per second.

When developing the proposed device, it was experimentally found that the rotation of a hollow cylinder leads to the formation of a vortex funnel inside the volume of the cylinder [17]. Various petroleum products and oils were used in the experiments—sunflower oil, a mixture of sunflower oil and diesel fuel in equal proportions, oil, diesel fuel, etc. It was found that the shape and size of the oil part of the composite vortex depend on a number of parameters—the thickness of the oil layer on the water

Fig. 6 Typical forms of a composite vortex at different speeds of rotation of the glass for various oil products [17]

surface, the density and kinematic viscosity of the water and the removed oil products. As a result of analytical processing of experimental data, it was possible to establish empirical dependences (1, 2) of optimal modes.

Based on the proposed method, a «Device for removing oil products from the surface of water» [18] was patented, in which the means of creating a vortex is made in the form of the mentioned glass, equipped with the necessary mechanisms for moving it along the height. The outputs of the control unit are connected to the movement mechanisms and the cup rotation drive, and the inputs are connected to the density and viscosity meters of water and the collected oil product, the oil product layer thickness gauge and the means for measuring the immersion depth of the top edge of the cup from the «water–oil product» interface.

Empirically, it was found that the optimal location of the end face of the receiving pipe of the means for pumping out oil products in the oil part of the composite vortex with the pump running leads to the selection of only oil from the composite vortex. The resulting oil deficit in the compound vortex is replenished by oil from the surface, which leads to its continuous collection from the water surface. Figure 5 shows the experimentally obtained [16] typical forms of a composite vortex at different speeds of rotation of the glass for various oil products. The rotation frequencies of the experimental glass in the photographs of Fig. 6 were 100, 300, 500 and 700 r/m from left to right, respectively. It can be seen how, with an increase in the rotation frequency, the water surface is cleaned of oil and its localization near the axis of rotation of the glass.

4 Conclusion

The proposed new method and device for collecting oil products from the sea surface, in contrast to the previously developed ones, are distinguished by the use in the control system of the process of collecting the results of preliminary calculations, which ensure the optimal operation of the device. The nature of the movement of oil products on the surface of the vortex funnel is confirmed by experimental studies.

Acknowledgements The work was carried out with the financial support on the topic State assignments 123021700046-4

References

1. Stepanova E.V., Chaplina T.O. Modeling of oil spills on the surface of water and a method for the elimination of surface pollution by hydrocarbons using sheep wool // Teoreticheskaya i prikladnaya ekologiya. 2015. №2. P. 108–115.
2. Karakeyan V.I., Sevryukova E.A. Environmental pollution monitoring: a textbook for secondary vocational education. Moscow: Yurait Publishing House. 2019. 397 p.
3. Bayat A. Oil spill cleanup from sea water by sorbent materials // Chemical Engineering & Technology. 2005. Vol. 28(12). P. 1525–1528.
4. Chaplina T.O., Stepanova E.V. Modern methods and means for spill response oil in the ocean // Springer Geology. Processes in GeoMedia. 2021. Volume III. P. 95–105.
5. Chaplina T.O., Stepanova E.V. Device for collecting liquid hydrocarbons // Utility model patent RU 169140 Publication date: 03/16/2017.
6. Morozov G.I. A device for collecting oil and oil products from the surface of the water. AS USSR N 1625952, MKI E 02 B 15/04. Bull. fig. N5, 1991.
7. Antropova O.N., Leu S.L. Method for cleaning the surface of water and soil from pollution by oil and oil products, Patent RU 2091539
8. Karavaev I.I. A device for removing oil from the surface of water. AS USSR N 1618836, MKI 5 E 02 B 15/04, Bull. fig. No. 1, 1991
9. Mingazhev A.D., Krioni N.K., Davletkulov R.K. Method for collecting oil from under the ice cover of a reservoir, Patent: 2014151348/13, 12/17/2014
10. Gumerov R.S., Karamyshev V.G., Mamleev R.A., Valeev M.D., Akhmadishin R.Z. Device for collecting oil from the surface of the water, 1991, Patent SU 1 654 449 A1
11. Gromov N.G. System for collecting oil products from the water surface, Spasatel Limited Liability Company (RU), 49026 U1, 2005
12. Sokolov E.M., Panarin V.M., Volodin N.I., Pashkov V.P., Lapina O.Yu. Device for collecting oil from the surface of the water, 2005, Patent for invention E02B 15/10
13. Bulgakov B.B., Bulgakov A.B., Gurvich G.A., Petrov A.Yu., Galitsyn V.V. Method and device for collecting oil products from the water surface, 2005, Patent No. 294 417.
14. Bakhtizin R.N., Kuzeev I.R., Mingazhev A.D., Kodak A.A., Gafarova V.A. Method for collecting oil spills from under the ice cover of a reservoir, 2018, Patent No. 2683496
15. Gnatyuk I.V., Zabela K.A. Device for collecting oil products from the surface of the water, 2010, Patent 2049202.
16. Method and apparatus for recovering a substance floating as a sheet on the surface of a liquid mass, 1969, Patent US3635342.
17. Chaplina T.O., Kistovich A.V., Stepanova E.V. Theoretical and experimental studies of the interface between two immiscible liquids in a vortex flow with a free surface // Bulletin of Moscow State University, series 3. Physics. Astronomy, 2016, No. 4. pp. 99–105.
18. Chaplina T.O., Kistovich A.V., Pakhnenko V.P. Utility model patent "Device for removing oil products from the surface of water", RU 2771862, 05/13/2022.

Trends in the Risks of Natural Hazards in the Arctic Zone of Russia, with Further Warming of Its Forest Zone (Case of Eastern Siberia)

A. V. Kholoptsev, R. G. Shubkin, I. U. Sergeev, and A. N. Baturo

Abstract The current trends of changes in average monthly air temperatures in the Arctic and Forest landscape zone of Eastern Siberia, as well as the relationships of these processes with variations in the total area of wormwood, which are formed in the summer months in the ice cover of the seas of Laptev and East Siberia. It is shown that in many territories of the Arctic zone of Eastern Siberia, significant changes in average monthly temperatures of summer months in 2010–2020. As a result, the risks of many natural hazards could not be increased. Therefore, the probable cause of their actual increase is anthropogenic.

Keywords Eastern Siberia · Arctic zone · Forest zone · Average monthly air temperature · Wormwood · Arctic seas · Siberian rivers · Floods · Connections

Abbreviations

TSR	Total radiation of sun—суммарная радиация солнца (ССР)
TWA	Total area of wormwood—суммарная площадь полыней (СПП)
VTWA	Values of total area of wormwood—значение суммарной площади полыней
AMT	Average monthly temperature—среднемесячная температура (СТВ)
Тренд температуры воздуха	Air temperature trend
Долгота	Longitude

A. V. Kholoptsev · R. G. Shubkin (✉) · I. U. Sergeev · A. N. Baturo
State Oceanographic Institute named after N.N. Zubov, 61 Sovetskaya str., Sevastopol 299011, Russia
e-mail: rshubkin@yandex.ru

A. V. Kholoptsev
Siberian Fire and Rescue Academy of the Ministry of Emergency Situation of Russia, 1 Severnaya str., Zheleznogorsk, Krasnoyarsk Territory 662972, Russia

V. I. Karev (ed.), *Physical and Mathematical Modeling of Earth and Environment Processes—2022*, Springer Proceedings in Earth and Environmental Sciences,
https://doi.org/10.1007/978-3-031-25962-3_20

Градусы	Degrees
Широта	Latitude
Коэффициент корреляции	Correlation coefficient

Natural and man-made emergencies in the Arctic zone of Russia in the modern period are increasingly frequent, which causes significant damage to the population, economy and ecosystems not only of the mentioned zone, but also of the entire Russia. Therefore, the development of existing ideas about the causes of their occurrence is an urgent problem of fire and environmental safety.

The solution of this problem is of the greatest interest for the Arctic territories and waters, the natural resources of which are very significant, but in a small part developed by humans. These include many of the northern territories of Eastern Siberia, which are part of the Krasnoyarsk Region and the Republic of Sakha (Yakutia), as well as the adjacent of Laptev and Eastern Siberia seas. The accelerated development of natural resource management systems of the above-mentioned Russian Federation entities provides for «Strategy of development of the Arctic zone of the Russian Federation and provision of national security for the period up to 2035»which was approved by the Decree of the President of the Russian Federation of 26 October 2020 №645.

The development of the natural resources of the above-mentioned Arctic territories and of the seas of Eastern Siberia is inevitably connected with an increase in anthropogenic influence on their landscape complexes, which can lead to an increase in risks associated with the development of a number of dangerous natural processes in them. Regional warming [1–6] can also have an equally powerful impact on such processes.

Therefore, when planning the development of prevention systems, as well as emergency response in the Arctic zone of the Russian Federation, it is necessary to take into consideration the correlation between the effects, which are specific for its various regions, Climate and anthropogenic factors affecting their landscape complexes.

Significant reductions in Arctic sea ice cover, (which is also an important consequence of this warming) [7–11], have been attributed to existing perceptions of the causes of Arctic warming.

In the summer months, the destruction of the ice cover of the Arctic seas begins with the formation of wormwood. Many of them arise at the seams of the rivers that flow into them, after the onset of the floods in their estuary regions. This is facilitated by the increase in the level of the water surface in such areas as the flood, which can break through the stores, as well as by the increase in the speed of the existing drainage currents in them, which carry away the fragments of ice from the shore. In the modern period, on all rivers flowing into the seas in question, the flood begins in the period of the polar day.

The free water surface of the seas absorbs the total radiation of the sun (here and after the TSR) much more actively than their ice cover [12, 13]. Therefore, the formation of wormwood in the ice cover of the seas and the increase in their total area (here and after the TWA) leads to an increase in the flow of heat generated in the waters of these seas while absorbing the TSR.

The resulting heat is consumed by increasing the mean temperature of the operational layer of the sea, melting of the ice sheets, and evaporation of moisture and heating of air. If the amount absorbed in wormwood TSR exceeds their heat loss, TWA increases.

The earlier in the polar day period, in some seas, the longer the cumulative dose of TSR absorbed during the period and the greater the total amount of heat given to them into the atmosphere.

The earlier arrival of a flood in the estuary of any river is a consequence of the earlier beginning of the intense melting of the snow cover in its basin. The latter may be caused by an increase in the average monthly air temperatures (here and after AMT) in the spring months on its territories [14].

The average rate of growth of TSRs is also increased by an increase in the average temperature of the river waters flowing into the sea concerned during the flood season.

The recharge zones of the basins of all rivers flowing into the Arctic Seas of Russia are located in the territory of its Forest Landscape Zone, where in April and May forest fires are a significant factor in increasing AMT.

The above considerations suggest that one of the significant causes of the warming of the Russian Arctic climate for the summer months is the warming of the climate in its Forest Zone for the spring months.

The hypothesis is not obvious because many other processes [2, 15–19] may be responsible for the warming climate for the summer months.

A confirmation of the proposed hypothesis for any territory of the Arctic zone of Russia would make it possible to use information on the terms of the onset of flooding in the recharge zones of the respective rivers in predicting the risks of emergency situations related to climate warming. Therefore, the verification of this hypothesis for the territories of Eastern Siberia belonging to the Arctic and Forest Zone, as well as the waters adjacent to seas, is of considerable theoretical and practical interest. However, it had not previously been tested.

The purpose of the work is to test the proposed hypothesis for the Arctic territories of Eastern Siberia and to assess the current trends in the changes of risks caused by the dangerous natural processes that are possible on them in the summer, further warming of the Forest Zone for the spring months.

It is obvious that the hypothesis put forward can be valid in case of following conditions:

- the influence of inter-annual changes in the total dose of the TRS absorbed by the free water surface of any Arctic seas off the coast of Russia in the summer months, on synchronous variations of the AMT in some territories or areas of the Arctic zone of Russia is significant;
- the year-to-year changes in the TSR dose absorbed in such seas during the summer months are significantly related to the variations of the AMT in the respective territories of the Russian Forest Zone, which occur in the previous months.

The feasibility of the first condition is not obvious, since the flow of heat into the atmosphere over the Arctic zone of Russia from such seas is formed by their entire underlying surface, and not only its areas free of ice. A large amount of heat flows into the atmosphere and from the ice cover of the seas, which also absorbs the TSR (although its albedo is much larger).

Part of the total heat flow from a sea to the atmosphere that forms in its Wormwood, the more the TWA that is dependent on time. The question of whether this part is large enough for a certain summer month to meet this condition for the Laptev and East Siberian seas was not previously considered.

The feasibility of the second condition also needs to be confirmed, because variations in the average monthly values of their TWA (here and after VTWA) the dynamics of AMT in coastal areas for the same months may affect much more strongly.

It should be noted that changes in risks associated with dangerous natural processes in some areas of the Arctic zone of Russia, with further warming of the climate of the respective territories of its Forest for the spring months, It also depended on whether such trends were actually observed in those territories.

Monitoring of climate change in many parts of the Siberian Forest Zone indicates that most of them are experiencing significant warming in the current period [3, 5, 6, 20]. However, the trends in the inter-annual changes of the AMT in the spring months in its territories where they are able to influence the timing of the onset of flooding in the estuarine regions of the Siberian rivers remain open, as the location of such territories has not been identified.

The following objectives have therefore been achieved.

1. Assessment of the significance of the relationship between the inter-annual changes in the monthly doses of the TSR absorbed in the wormwood, which are formed in the summer in the ice cover of the Laptev Sea and the East Siberian Sea, with variations of the AMT over the waters and coasts of Eastern Siberia.
2. Identification of areas of the territory of Eastern Siberia, where inter-annual variations of the AMT for any spring months are significantly related to changes of the VTWA in the ice cover of the Arctic seas off the coast of Eastern Siberia in the following summer months.
3. Assessment of current trends in AMT changes in the identified areas of Eastern Siberia, which correspond to the identified spring months.

In solving these tasks, as a factual material, information was used:

- about the changes in the average daily values of glaciation of different sections of the Arctic seas off the coast of Eastern Siberia, which is obtained from GLORYS12.v.1 reality [21];
- about the variations of the average hourly values of AMT and air temperatures at an altitude of 2 m above various points of the territory of Eastern Siberia and waters of the seas of Laptev and East Siberia, which is presented in the ERA 5 [22].

GLORYS12v.1 is based on the mathematical models of the NEMO [23] family, which are verified by satellite monitoring data. The results under consideration are the daily average of the glacier values of the water bodies of the sites under consideration for each day from 1.01.1993 to 31.12.2020, with a pitch of 5 arc min.

Global mathematical models IMERG [24–26] have been applied for the implementation of ERA-5 recanalization, which are verified from ground-based meteorological and actinometric observations. It contains information on the average hourly values of the insolation intensity of different parts of the Earth's surface and the air temperatures above them in the period from 01.01.01.1979 to 24.31.12.2021. With a coordinate step of 0.25. Both are supported by Copernicus.

Information from the Ocean Information Shared System portal, which is maintained by IDC VNIA (Obninsk) and archival materials, was used in the sample testing of the actual material.

The methodology of the study presupposed the solution of the first problem in three stages. In the first stage, the values of the SPS for each sea and for a certain month were calculated, as the difference between the total area of its water area and the average monthly value of the total area of its ice cover. The areas in question were calculated using GLO YRYR 12 V.1 real time information on changes in average daily values of glaciers of different areas of each sea area, taking into account their latitude. The assumption was that each site represents trapezoid. Parts of the land area were considered to be land.

In the second stage, a monthly dose of TSR was determined for each month under review and absorbed by each wormwood of each sea. The value of this indicator was calculated taking into account the changes in the PPS, as the sum of doses of TSR absorbed in wormwood for each day of the month studied. The latter were calculated taking into account the influence on the albedo of the relevant areas of the sea surface, changes in the height of the Sun above the horizon.

In the third stage, a correlation analysis was carried out between the time series of monthly doses of the TRS and the synchronous series of AMT over the waters of the seas in question and the territories of Eastern Siberia, for the months from May to August. In assessing the significance of the relationships studied, it is assumed that the statistical properties of the processes studied allow the Student criterion to be applied. The links were considered relevant if the validity of such a conclusion exceeded 0.95.

In the second task, the same studies were carried out, assuming that the monthly series of SSSPs lagged behind the average monthly temperature series for the areas in question at 13 months.

In solving the third problem, as quantitative measures of the trend of inter-annual changes of the AMT at an altitude of 2 m over all areas of the Siberian Forest Zone, the values of the angular coefficient of the linear trend (here and after ACLT) of the corresponding time series have been estimated, which correspond to the periods 1979–2020 and 2010–2020.

In assessing the likely changes in the risk of occurrence of natural hazards, which is possible in the summer at different sites of the Arctic zone of Eastern Siberia, it was taken into account that the values of these indicators are proportional to the AMT above [27].

Therefore, in order to obtain the estimates sought, the distribution of AMT series over the June–August months of the ACLT was studied over the whole AMT area and calculated over the given time periods.

It is not difficult to see that the methodology described is based on a number of assumptions that can have a significant impact on the quantitative estimates of the indicators studied. As a consequence, it can be considered to be suitable only for identifying qualitative patterns inherent in the relationships between the processes under study.

Using the above factual material and the described methodology, all tasks have been accomplished. In the first and second stages of solving the first task, for the months from May to August, the dependence on the time of VTWA for both seas off the coast of Eastern Siberia, as well as monthly doses absorbed in their TSR wormwood has been determined.

It has been established that the changes in these indicators for any month are complex variations, with both quasi-two-year and longer-period components in the spectra. These fluctuations for May in the period 1993–2019 occur at almost the same level. For June–August they have increasing trends (values of ACLT of corresponding series are positive, although not significant). At the same time, the average rate of increase of VTWA was highest for the period 2000–2012.

In 2012–2019. Changes of the VTWA for these months, as for May, occurred at almost the same level. It follows that the trends of inter-annual changes of the VTWA in the seas of Laptev and East Siberia, in 2012–2019 changed: a further increase in this indicator and a decrease in the average glaciation of their water area, stopped for the summer months.

In the second stage of solving the first problem, the dependence on the time of doses of the TSR, falling and absorbed in the wormholes of each studied sea, was determined.

Figure 1, as an example, shows the established time dependence of the TSR dose values, falling and absorbed in the wormholes of the entire Laptev Sea, normalized to the maximum values of these values (TSRmax) for 1993–2019 which correspond to the months of May to August.

As can be seen from Fig. 1a, the inter-annual changes in the monthly doses of the TSR entering the Laptev Sea area and being absorbed into the Carpian Sea are also complex variations.

As shown in Fig. 1a, the highest monthly dose of SSR falling on the Laptev Sea area in the months under consideration is 2007–2012. In the current period, the changes in this indicator is a decreasing trend.

Similar features are observed in Fig. 1b. It follows that the change in VTWA trends in 2012 was most noticeable in the inter-annual change of monthly doses of TSR absorbed in the Laptev Sea Polynyas for July and August.

For June, the steady increase in these doses continued after 2012. The latter indicates that the changes in the VTWA in June may have been influenced as the shift of the flood onset dates in the estuary areas of the respective rivers to earlier dates, as well as an increase in the average temperatures of the river waters entering the sea.

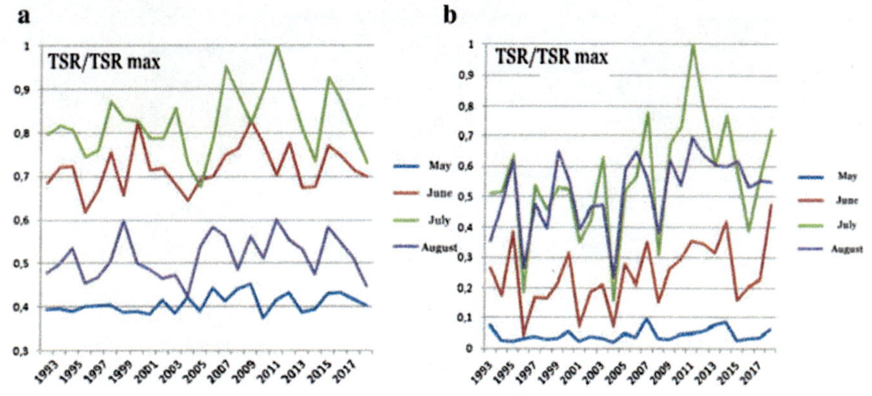

Fig. 1 Time dependencies of TSR doses for the whole Laptev sea area, normalized to their maximum values (TSRmax) for 1993–2019 which correspond to the months of May to August **a** falling **b** absorbed

In the following months, the influence of this factor seems to be weakening, as the bulk of the heat flow from the melting of the ice sheet is formed by absorption of the TSR into the free water surface of the southern areas of the studied seas. The same conclusions are fully valid for the East Siberian Sea.

The results obtained make it possible to associate the revealed inter-annual changes of monthly doses of TSR in the period 2012–2020. not only with appropriate variations of the VTWA, but also with a decrease in the average monthly amount of insolation of the waters of the studied seas, which may be due to increased cloudiness over them.

During the studying of relationships of inter-annual changes in 1993–2019. The total amount of SSR absorbed in the wormwood of the studied seas during the months of May–August, with synchronous variations of AMT at an altitude of 2 m above the surface of the entire Arctic zone of Eastern Siberia, has been identified its areas where these connections were significant. The location of the identified sites is shown in Fig. 2.

It is clear from Fig. 2 that there are many areas of the Earth's surface in the studied region for which the statistical validity of the correlation between the series of AMT above them, as well as the series of monthly doses of TSR absorbed in the Laptev Sea wormwood, exceeded 0.95 the degrees of freedom of such series 28, therefore the corresponding threshold of their correlation coefficient is defined as 0.4.

There are many in the region and at sites for which the validity of such a conclusion was greater than 0.99 (a correlation coefficient threshold of 0.52 was chosen according to this level of confidence). Sites were also identified where the correlation factor of the time series in question was greater than 0.6 and even greater than 0.7 (with a maximum of 0.783).

As you can see from Fig. 2a, for May, the areas of the Earth's surface, where the connections between variations of AMT over them, as well as the changes in monthly doses of TSR absorbed by all wormholes of the Laptev Sea, with a certainty of 0.95

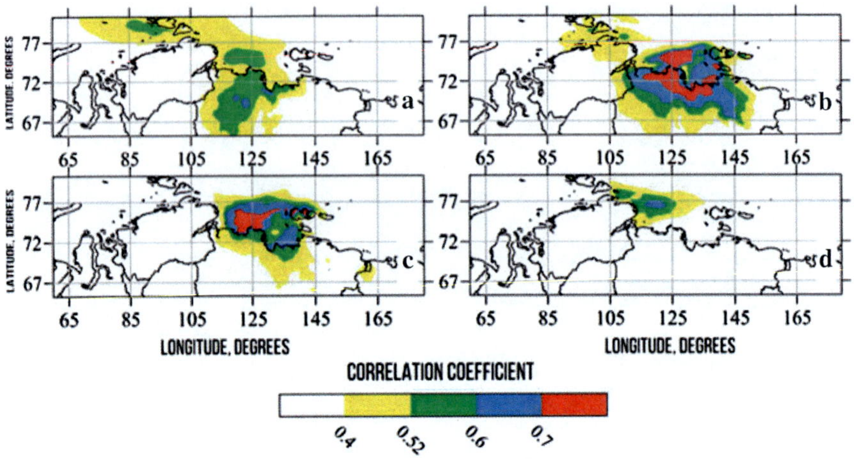

Fig. 2 Land areas of the Arctic zone of Siberia, where the relationships of inter-annual changes in 1993–2020. AMT at a height of 2 m above them, as well as synchronous variations of doses of TSR absorbed by all wormholes of the Laptev Sea, are significant for months: **a** May; **b** June **c** July; **d** August

were considered significant, occupied its entire area, as well as the water area of the north-eastern part of the Kara Sea.

In addition, they were located on the entire continental coast of the Laptev Sea and the entire north-western part of the territory of the Republic of Sakha (Yakutia).

Figure 2b suggests that, as of June, similar areas also included the entire northern part of the Taimyr Peninsula and the Novosibirsk Islands, as well as the western part of the East Siberian Sea. The total area of plots considered for this month is the maximum. The most extensive areas and territories are also the Arctic zone of Eastern Siberia, where the correlation coefficient between the series under consideration exceeds 0.7.

From Fig. 2b it can be concluded that for July the region in question, where the relationships of the inter-annual changes of the AMT are significant, as well as the dose of TSR absorbed by all Laptev Sea wormlands, is somewhat smaller than for June. It includes the territories occupied by the estuary region of the Kolyma River and the southern coast of the East Siberian Sea, which do not include in a similar area for June.

A significant reduction in the size of the area under consideration for August is evidenced by Fig. 2g. It shows that for this month the area includes only the waters of the central and western part of the Laptev Sea, adjacent to the eastern coast of the Taimyr Peninsula.

The results confirm the view [12, 13] that during the Arctic Day period, Absorbed in the Arctic Seas is one of the main sources of heat warming the surface air over many regions of the world, located in its Arctic zone (including over the Arctic

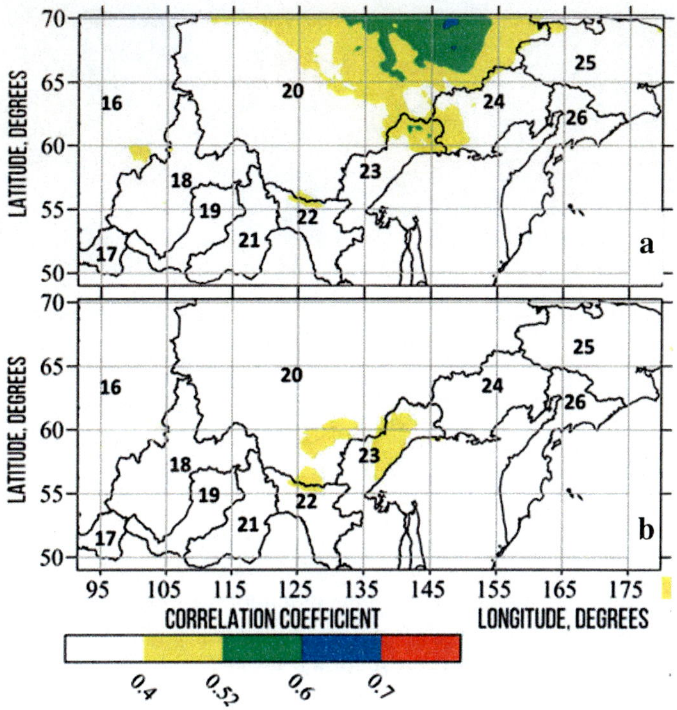

Fig. 3 Areas of the territory of Eastern Siberia, where annual changes in average monthly air temperatures in April and May 1993–2019 were significantly related to the variations of the STWA of the Laptev Sea for the months: **a** May; **b** June

territories of Eastern Siberia). They suggest that the increase in the growth rate of TWA, as well as VTWA, may be a significant cause of warming in the studied region.

The location of sections of the territory of Eastern Siberia, where the inter-annual changes of the AMT in April and May 1993–2020, was revealed in the solution of the second problem, were significantly related to the TWA variations of the Laptev Sea for May, as well as June, shown in Fig. 3.

Figure 3a concludes that a significant factor of the inter-annual changes S for the Laptev Sea in May are the variations of the AMT for April, directly above the territories adjacent to its southern coast. The latter is probably due to the fact that AMT changes over these land areas in May and April are interconnected, and flooding in the estuarine areas of the region does not begin until late May.

Figure 3b shows that the inter-annual changes of the AMT in May over some territories of Eastern Siberia located in the basins of the Lena River and its right tributary Aldan are a significant factor of the inter-annual changes of the TWA of the Laptev Sea for June.

It should be noted that the locations of the identified areas are such that the duration of the periods for which the resulting floods reach the Lena delta is approximately the same and close to 1 month.

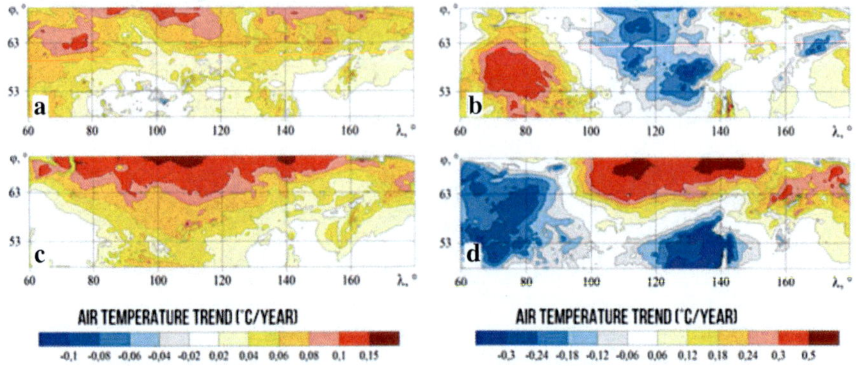

Fig. 4 Distribution over the territory of Siberia of ACLT time series of AMT at a height of 2 m above various points of the Earth's surface, for months: **a** May 1979–2020; **b** May 2010–2020; **c** June 1979–2020; **d** June 2010–2020

Similar links have also been identified between the East Siberian Sea TWA changes for June, as well as the variations of the AMT for May, in the areas of the Kolyma basin, which are as far from the mouth of the most abundant branch of its delta.

Therefore, the hypothesis put forward for the Arctic territories of Eastern Siberia is fair.

The results show that with the further increase of AMT spring months in the identified territories of Eastern Siberia, the values of TWA in the seas of Laptev and Eastern Siberia for June and other summer months could increase. The consequence of the latter could be an increase AMT for the summer months in the Arctic zone of Eastern Siberia, shown in Fig. 4. The latter would increase the risk of occurrence of natural hazards in such territories, which activate when the local climate warms up. At the same time, on the sections of the TWA routes located in the Laptev Sea and East Siberian ice conditions would become more favorable for navigation. However, the fact that such phenomena were possible did not necessarily mean that they actually occurred.

Task 3 has been solved in order to assess the possibility of the occurrence of these phenomena in the Arctic zone of Eastern Siberia.

During the solving, for the months of May and June, the distribution of values of ACLT AMT time series at an altitude of 2 m above the ground, which correspond to the periods 1979–2020, has been constructed throughout Siberia and 2010–2020 and are presented in Fig. 4.

Figure 4a and b show that for both May and June all over Siberia in 1979–2020 AMT have increased (which confirms the validity of [1–8]).

For June, the warming rate was significantly higher than for May and the ACLT values of the AMT series in many parts of Siberia exceeded the selected significance level of 0.1 °C/year. The fastest warming occurred in the north of the 60 s, which confirms the validity of conclusions [1, 2, 4, 8, 10, 11].

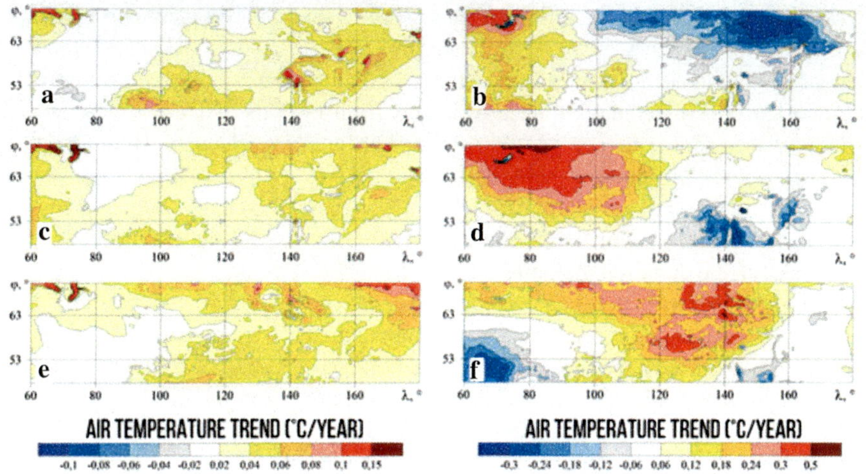

Fig. 5 Distribution over Siberia of ACLT values of AMT series for the months of July–September, which are estimated over time periods: **a** July1979–2020 **b** July 2010–2020 **c** August1979–2020 **d** August 2010–2020; **e** September 1979–2020; **f** September 2010–2020

Figure 4b and d show that in the period 2010–2020 the trends of the studied processes have significantly changed. For May (Fig. 4b) sections of the entire Eastern Siberia for which the ACLT of the AMT series would exceed the corresponding level of significance (0.24 °C/year) have not been identified. At the same time, in the areas of the Stanovoye Plateau and the Steeple Ridge, also in the upper reaches of the Vilyu River, a significant cooling was detected.

At the same time, in the whole territory of Western Siberia, an increase in the AMT was revealed, which in its southern part (Novosibirsk, Tyumen, Omsk, Tomsk and Kurgan regions) was significant.

For June (Fig. 4) significant warming predominated in all areas of Eastern Siberia belonging to the Arctic zone (and in many areas belonging to the Krasnoyarsk Territory and the Republic of Sakha (Yakutia), the values of ACLT AMT exceeded the level of significance). As a result, the risks of natural hazards in such territories have increased. At the same time, for many areas of the Amur basin and areas of Western Siberia there was a significant cooling.

Taking into account the results obtained, the values of the ACLT of the AMT series at an altitude of 2 m above the ground for the months of July–September have been estimated for various parts of the territory of Siberia, which are proportional to the ACLT for the Arctic regions of the risk series of natural hazards caused by climate warming.

Figure 5 shows the distribution of these indicators for the territory of Siberia, the values of which were estimated for 1979–2020 and 2010–2020.

Figure 5a, c and d are clear for all summer months 1979–2020. Risks of occurrence of natural hazards in almost the whole territory of Eastern Siberia have increased (which corresponds to [1, 3, 10]).

Although the ACLT of the AMT series did not reach this level of importance in almost all of this territory, it was significant in some areas.

Figure 5a shows a significant increase the risks of occurrence in the Arctic zone of Siberia of the considered natural hazards in July from 1979 to 2020 are typical only for its north-western part (parts of the territory of the Yamal-Nenets Autonomous Territory), where the values of the ACLT of the AMT series exceed 0.15 C per year (at the selected threshold of significance of 0.1 C per year).

Figure 5b shows that for the period 2010–2020. For July, a significant increase in the risks of the same phenomena was found in the same territories (the ACLT values of the AMT series for them exceed 0.3 C per year (at the selected threshold of significance of 0.1 C per year). At the same time, in many territories of the Republic of Sakha (Yakutia), the risks caused by such phenomena have significantly decreased.

As shown in Fig. 5b, for August the risks associated with natural hazards, for the entire period from 1979 to 2020. Only in the same northwestern districts of the Yamal-Nenets Autonomous Territory increased significantly.

Figure 5 shows that in 2010–2020 a significant increase in the risks associated with the same phenomena occurred in almost all the territories of the Yamal-Nenets and Khanty-Mansiysk Autonomous Okrug and throughout the northern part of the Krasnoyarsk Krai. In the territories of the Arctic zone belonging to the Republic of Sakha (Yakutia), no increase in such risks has been recorded.

As can be seen from Fig. 5d, the risks associated with natural hazards for September 1979–2020 significantly increased in the same territories of the Yamal-Nenets Autonomous Territory, as well as in the northeast of the Republic of Sakha (Yakutia).

Figure 5e shows that for the period 2010–2020 a significant increase in the risk of occurrence of natural hazards related to climate warming was observed both in many territories of the Krasnoyarsk territory and the Republic of Sakha (Yakutia) belonging to the Arctic zone and in the Forest Zone territories (Amur Region and Khabarovsk Territory).

Consequently, in the current period, the risks associated with the occurrence of natural hazards related to climate change in the Arctic zone of Eastern Siberia increase significantly only in June and September (as evidenced by the significant levels of ACLT of AMT ranks over its territory).

In the Forest Zone of Eastern Siberia, they increase most significantly by September. This may explain the decline in the rate of STWA increase in the seas of Laptev and East Siberia, which was identified for the summer months of 2012–2020. As the ice cover of the seas in question increases slightly in July and August, the risks to shipping in the seas are increasing to some extent.

As a result of the revealed features of the changes in 2010–2020 the AMT for July, August and September in many Arctic territories of Eastern Siberia, there in the same years was to be observed stabilization of risks associated with natural hazards.

However, in fact, in these months there has been an increase in both the total number of landscape fires and the area of the areas covered by the fire [28]. Consequently, the results can be used to link the above-mentioned actual elevation of the terrain to the increase in anthropogenic influences on the terrain.

Taking this into account, a set of measures aimed at reducing risks arising in the Arctic zone of the Russian Federation should provide for the improvement of the control and prevention activities carried out here by the EMERCOM of the Russian Federation. This set of measures should provide for the extension of the powers to verify the functional and territorial subsystems of the Unified State System for Prevention and Elimination of Emergency Situations in the Arctic zone of the Russian Federation. Introduction of modern preventive tools takes into account the Arctic climate. Stricter sanctions for offences related to the protection of the population and the territory of the Arctic zone of Russia from emergencies.

The results are fully in line with the existing understanding of the causes and the role of the discharge rivers [1.2]. They show that the effects of the current warming of the climate in the basins of the rivers of Eastern Siberia are an earlier shift of the flowing currents in their estuary regions and a significant increase in their thermal flow. These phenomena lead to an increase in the rate of growth of wormwood in the Arctic seas and the flow of heat from them into the atmosphere, which can cause an increase in AMT over their waters and coasts of the seas.

The following findings are scientifically novel:

1. Identified areas of the Siberian river basins, for which the conclusion about the significance of the influence of inter-annual changes in the thermal regime in May the growth rate of wormwood in the ice cover of the studied seas in June is characterized by a confidence of at least 0.95.
2. Identified areas of the Arctic zone of Eastern Siberia, for which, for the months of May–August, the importance of the relationship between the inter-annual changes of the AMT, with synchronized variations of the monthly dose of the TSR, is equally valid, absorbed by the free water surface of wormwood in the ice cover of the seas of Laptev and East Siberia.
3. It has been proven that an important factor in the dynamics of risks associated with the occurrence of natural hazards in the Arctic zone of Russia for the summer months is the increase of AMT over some territories of its Forest Zone for the spring months.

   Thus, it has been established that:

   1. The warming of the climate now occurring in some parts of Eastern Siberia, located in its Forest Landscape Zone, not only complicates the livelihoods of their population and the functioning of the economy, but also has a significant transboundary impact on risks, associated with dangerous natural processes in its Arctic zone.
   2. This influence manifests itself in an increase in the rate of warming of the climate of the Arctic territories, which contributes to an increase in the frequency of landscape fires and the area covered by fire, and also leads to an increase in the intensity of thermal destruction of the Permafrost, and can also cause man-made and natural emergencies.

4. Problems of emergency prevention in the Arctic and Forest zones of Russia are interconnected, and therefore they need to be addressed in a comprehensive

manner. An essential role in their solution could be played by improving the personnel and logistics support of the units of the Russian Ministry of Emergency situations operating in the territories of the constituent entities of the Russian Federation located on the territory of Eastern Siberia.

References

1. *Head Editor B.A. Morgunov.* - M.: The Scientific World, 2011. http://npa-arctic.ru/rus/sap_ru.html
2. *Frolov I.E., Gudkovich Z.M., Karklin V.P., Smolyanitsky V.M.* Climate Change of the Arctic and Antarctica - the result of natural causes // Problems of the Arctic and Antarctic, 2010, 2 (85). 52–61pp.
3. Roshydromet Second Assessment Report on Climate Change and Its Impact on the Territory of the Russian Federation. Roshydromet, 2014, 1009 p.
4. *Alekseev G.V.* Manifestations and Intensification of Global Warming in the Arctic. Fundamental and applied climatology. 2015. Part. 1. 11–26pp.
5. *Anisimov O.A., Giltsova E.L.*, On estimates of climate change in the regions of Russia in the 20th and early 21st centuries according to observations. Meteorology and Hydrology, 2012. Part 6, 95–107 pp;
6. Report on climate risks in the Russian Federation. St. Petersburg. 2017. 106 p.
7. Shoigu S.K., editor, Atlas of Natural and Technological Hazards and Risks of Emergency Situations, Moscow, Publishing «Feoriya», 2011. 696 p.
8. *Anisimov O.A., Kokorev V.A., Ziltcova E.L.*, Temporal and Spatial Patterns of Modern Climatic Warming: Case Study of Northern Eurasia. Climatic Change, 2013. V. 118, №3, 871–883pp.
9. *Kholoptsev A.V., Zapirin S.A.* Analysis of changes in ice conditions on the Northern Mor way in the late XX- early XXI century/ SUSRF named after S.O. Makarov. 2020. Part 12., 1, 71–84pp.
10. *Mosov I.I.* Features of modern climate change in the Arctic and its consequences. Problems of the Arctic and Antarctic. 2020; 66(4):446–462. https://doi.org/10.30758/0555-2648-2020-66-4-446-462
11. *Overland J.E., Wang M., Walsh J.E.*, Stroeve J.C., Future Arctic climate changes: Adaptation and mitigation times cales, Earth's Future,2013. 2(2) 68–74,
12. Frolov I.E. Oceanography and sea ice. M.Paulsen. 2011. 432p
13. *Agarwal S., Moon W., Wettlaufer J.S.* Decadal to seasonal variability of Arctic sea ice albedo Geophys. Res. Lett. 2011. Part. 38, L20504
14. *Gelfan A., Gustafsson D., Motovilov Y., Arheimer B., Kalugin A., Krylenko I., Lavrenov A.* Climate change impact on water regime of two great Arctic rivers: modeling and uncertainty issues. Climatic change, 2017. Part 141(3). 499–515pp.
15. IPCC (Intergovernmental Panel on Climate Change). IPCC, 2014: Climate Change 2014: Synthesis Report. Contribution of Working Groups I, II and III to the Fifth Assessment Report of the Intergovernmental Panel on Climate Change [Core Writing Team, R.K. Pachauri and L.A. Meyer (eds.). IPCC, Geneva, Switzerland, 151 pp.
16. *Voryshev P.V., Katsov V.M., Meleshko V.P., Alexeyev G.V., Karol I.L., Mirvis V.M.*, Causes of observed climate change. Works of the Main Geophysical Observatory named after A.I. Voeikov, 2014. Part 574, 39–124pp.
17. *Gudkovich Z. M., Karklin V. P., Smolyanitsky V. M., Frolov I. E.* On the Nature and Causes of Earth Climate Change. Problems of the Arctic and Antarctic. 2009. 1 (81). 15–23pp.
18. *V. D. Kovalenko, L. Kizim, A. M. Paprestyuk, V. G. Nikolaev.* Agroclimatic resources of Siberia. Study of the causes of climate variability. Collection of works of AUAAS named after V.I. Lenin, Siberian branch. Novosibirsk, 1987. 103–113pp.

19. *Kotlyakov V. M.* On the causes and consequences of modern climate change. Sol-non-terrestrial physics. 2012. Ex. 21. 110–114pp.
20. *Groisman P.Ya., Blyakharchuk T.A., Chernokulsky A.V., Arzhanov M.M., Marchesini L.B., Bogdanova E.G., Borzenkova I.I., Bulygina O.N., Karpenko A.A., Karpenko L.V., Knight R.W., Khon V.Ch., Korovin G.N., Meshcherskaya A.V., Mokhov I.I., Parfenova E.I., Razuvaev V.N., Speranskaya N.A., Tchebakova N.M., Vygodskaya N.N.* Climate changes in Siberia. Regional Environmental Changes in Siberia and Their Global Consequences. Groisman P.Ya, Gutman G. editors. Dordrecht: Springer Environmental Science and Engineering, 2013. 57–109pp.
21. GLORYS12v.1 [Electronic Resource]. http://marine.copernicus.eu/services-portfolio/access-toproducts/?option=com_csw&view=details&productid=GLOBAL_REANALYSIS_PHY_001_030
22. ERA5 hourly data on pressure levels from 1979 to present. [Electron resource]. Above: https://cds.climate.copernicus.eu/cdsapp!/data analysis mode-era5-pressure-levels? tab=form DOI: https://doi.org/10.24381/cds.bd0915c6
23. *Vichi M., Lovato T., Gutierrez Mlot E., McKeever B.* Ocean Model Connection BMF: NEMO model (core for European ocean modelling). Bologna: BFM Consortium, 2015.
24. IMERG: Integrated Multi-satellite Retrievals for GPM | NASA. https://gpm.nasa.gov/data/imerg
25. *Hersbach H., Dee D.* ERA5 reanalysis is in production // ECMWF Newsletter. 2016. Vol. 147. 7p.
26. *Hoffmann L., Günther G., Li D., Stein O.* et al. From ERA-Interim to ERA5: the considerable impact of ECMWF's next-generation reanalysis on Lagrangian transport simulations Atm. Chem. Phys. 2019. Vol. 19. P. 3097–3124.
27. *Streletskiy D.A., Shiklomanov N.I., Hatleberg E.*, 2012: Infrastructure and a Changing Climate in the Russian Arctic: A Geographic Impact Assessment Proceedings of the 10th International Conference on Permafrost, v. 1, p. 407–412.
28. Federal Forestry Agency remote monitoring information system: official website. Moscow. URL [Electronic text]. Access mode: https://pushkino.aviales.ru/main_pages/index.shtml.

Solution of Problem Questions of Astronomy and Geology Using the Optimized Galactic Model

Azariy Barenbaum

Abstract On the basis of an optimized version of the galactic model, the problematic issues of the Galaxy spiral structure determining, as well as the geology problems associated with presence of various geochronological scales, which differ in age of boundaries the same name stratons are resolved. The galactic model establishes a close causal relationship between the cyclicity of global geological (biotic, tectonic and climatic) processes in Earth history with Sun's movement in Galaxy and bombardments of Solar System by galactic comets. This model has been developed by the author for 30 years, being improved and refined when solving problematic issues of geology and astronomy. There are presented the results of optimization and testing of the model latest version designed to calculate the numerical values of Galaxy spiral structure parameters, which cannot be determined by astronomical methods or accuracy of their measurement is low, as well as for the justification of possibility the constructing a high-precision Phanerozoic and Precambrian geochronological scale, taking into account the intensity of bombardments of Earth by galactic comets. The optimized model is used to refine Galaxy spiral construction and to construct a geochronological scale common for the Phanerozoic and Neoproterozoic on a new physical approach.

Keywords Galactic model · Galactic comets · Galaxy spiral construction · Phanerozoic and Neoproterozoic geochronological scale

1 Introduction

A new approach to solving problematic issues of astronomy and geology based on the Galactocentric paradigm representations is presented [1, 2]. This scientific concept for the first time takes into account an important astrophysical phenomenon—a jet outflow of gas and dust from the nuclear disk of spiral galaxies [1], which was overlooked by astronomers for a long time and therefore it was not taken into account

A. Barenbaum (✉)
Oil and Gas Research Institute RAS, Moscow, Russia
e-mail: azary@mail.ru

V. I. Karev (ed.), *Physical and Mathematical Modeling of Earth and Environment Processes—2022*, Springer Proceedings in Earth and Environmental Sciences,
https://doi.org/10.1007/978-3-031-25962-3_21

by geologists. Taking into account this phenomenon, the galactic model of geological cyclicity [1–13] has been constructed to date, closely linking the key events in the history of Earth and Solar System with Sun motion in Galaxy and the bombardments of our planet by galactic comets during periods of Sun location in the Galaxy spiral arms and jet streams.

The galactic model has been developed, refined and improved by the author for more than 30 years, having undergone a number of modifications. The article presents the results of optimizing the latest version of the galactic model [12, 13], which allows solving two tasks of the paramount importance. One for astronomy, the other for geology:

(1) In astronomy, propose a spiral and physical model of Galaxy that best suits the results of astronomical observations and geology data.
(2) In geology, solve the key problem of stratigraphy, which consists in the simultaneous use of geochronological scales: General Stratigraphic Scale (GSS) adopted in Russia, and International Chronostratigraphic Chart (ICC), which differ in the age of boundaries of the same name stratons. And also to substantiate this decision by proposing a high-precision geochronological scale common to Phanerozoic and Neoproterozoic, based on a new physical principle that takes into account the intensity of bombardments of Earth by galactic comets.

The essence of both problems is explained below.

2 The Problem of the Galaxies Spiral Structure

Spiral galaxies are known [14] to be characterized by a complex configuration and a variety of structures, and their spiral structure can changed at different optical wavelengths. Therefore, studying the structure of spiral galaxies is a very difficult task, which has not yet been solved by astronomers for many spiral star systems. An example of this is our Galaxy. Due to Sun's location in the galactic plane, in which a large amount of light-absorbing dust is concentrated, the question of the spiral structure of our star system is still open in astronomy [15].

Until the early 1990s, it was assumed that our Galaxy has two logarithmic spiral arms [16]. By 2000, astronomers had found out that the Galaxy has not 2, but 4 logarithmic arms [17]. However, later it was shown [18] that the existence of two more Galaxy arms, the positions of which do not coincide with the arms established earlier, is not excluded.

Due to the difficulties in taking into account the absorption of light by dust matter located in the galactic plane, great difficulties also arise in determining Sun's distance $R_{\odot}$ from Galaxy center. Over the past 50 years, the $R_{\odot}$ distance has been reviewed several times and varied within $R_{\odot} = 6.5$–10.5 kpc.

In the 1960s, the distance $R_{\odot}$ was taken to be 7.1 kpc. By 1980, the value $R_{\odot} = 10$ kpc was considered more reliable. In 1985, the IAU General Assembly officially recommends $R_{\odot} = 8.4$ kpc. In the early 2000s, the question of Sun distance from

Galaxy center became so complicated that a number of authors suggested using two different distance scales for different objects: one with $R_{\odot} = 7.5$ kpc, and the other with $R_{\odot} = 8.5$ kpc [19]. At present, this distance, according to the measurements of different authors, is $R_{\odot} = 8.0$–8.5 kpc.

3 Stratigraphy and Its Problems

The first stratigraphic table, as a reflection of the natural stages of the Earth's geological history, was adopted in 1881 at the II session of the International Geological Congress in Bologna, and officially approved in 1900 at the VIII session in Paris. By that time, based on the analysis and generalization of data on the extinction of living organisms in the Phanerozoic (last 570 Myr), geologists established the cyclical nature of global natural processes that took place on Earth in the past. They reflected this fact in a table in the form of a system of nested cycles of different durations (of different ranks) that have a common beginning. The empirical basis for the conclusion about the cyclical nature of the Earth geological development was the identification in the sediments of the rocks of the earth's crust of geological bodies, represented by a complex of rocks with common properties and time of formation, called "stratons". Subsequent studies showed that the main geological events in Earth's history took place at the boundaries of stratons of different ranks.

Since the second half of the last century, isotopic methods for measuring the absolute age of rocks have been widely used in geology. Their application to solve the problems of stratigraphy naturally transformed the stratigraphic table into a geochronological scale, a kind of calendar of all major geological events in the history of our planet. Subsequently, the geochronological scale was constantly refined and detailed [20–22]. This process continues to this day.

In the last two or three decades, as a result of the expansion of the arsenal of stratigraphic methods, the number of unrecorded geological events that require fixation on the geochronological scale has also increased. A number of countries, primarily United States, Canada, China, India and South Africa, studying their own territories, began to make their own adjustments to certain intervals of the geochronological scale. Increasing the detail and accuracy of the scale, led to the fact that some of the previously established boundaries were questioned by the International Commission on Stratigraphy and revised.

The revised boundaries included the boundaries established during the study of geological structure of the vast territory of Russia. However, the Interdepartmental Stratigraphic Committee of Russia did not agree with this decision [21]. As a result, two stratigraphic scales are currently officially recommended for use in our country: the General Stratigraphic Scale (GSS) of Russia and the International Chronostratigraphic Chart (ICC) [22].

Despite the fact that both scales use a single reference system for geological data, they differ markedly in the age of the boundaries of stratons of the same name. First of all, these differences relate to the time of some boundaries of the stages (ages) of the Phanerozoic scale. Since the boundaries of stages (GSS) and ages (ICC) in both scales are the boundaries of stratons of a higher order: systems (periods) and eras (erathems), this leads to a mismatch of the scales as a whole. Attempts to agree on the boundaries of the Phanerozoic in the GSS- and ICC-scales were unsuccessful [21, 22].

The differences between the scales are even stronger in their Precambrian (earlier 570 Ma) part, where both scales differ in the principles of construction. Thus, if Precambrian part of the GSS-scale was built on the basis of an analysis of tectonic–magmatic cyclicity, then the ICC-scale adopted a formal division into approximately equal time intervals 400–300 Ma long, which do not have an official status [22]. The problem is that geologists do not have unanimity of views on the principles for distinguishing stratigraphic units of different ranks, as well as clear rules for choosing the boundaries of stratigraphic units when constructing a scale [21, 22].

These problems are solved by the approach to Phanerozoic and Precambrian geochronological scale construction, based on Galactocentric paradigm representations and using an optimized galactic model.

4 New Approach to the Solution of Problems Based on Galactic Model

This approach to solving the problems of astronomy and geology became possible due to the discovery of the phenomenon of a jet outflow of gas and dust from the center of spiral galaxies [1]. Thanks to this discovery, it was found that spiral galaxies, along with the usual galactic arms twisted into logarithmic spirals, also have a system of gas and dust jet streams. Jet streams flow at a constant speed from the rapidly rotating gas and dust nuclear disk of galaxies and, propagating like a fan in the disk plane, turn out to be twisted in an Archimedes' spirals.

Using data from geology, the author developed a galactic model, according to which our spiral Galaxy has four logarithmic arms and two jet streams. Based on this model, Sun's orbit in Galaxy was calculated and the moments of Sun's intersection of jet streams and galactic arms were calculated. It has been established that during such epochs, Earth and the Solar System as a whole are subjected to powerful bombardments by galactic comets. In Earth's geological history, the times of such bombardments are the epochs of global natural events recorded in the modern geochronological scale. This connection opens up the possibility for us to study the spiral structure and physical processes in Galaxy using geological data, as well as to solve actual problems of geology based on the results of astronomical observations [2].

5 Galactic Model

The model implements the representations [2], according to which, after formation in one of Galaxy arms, the Sun moves in the galactic plane around Galaxy center in a slowly rotating elliptical orbit, crossing time to time of Galaxy jet streams and spiral arms. At such moments, the Solar System planets are subjected to intense bombardments by galactic comets, which bring large portions of energy and cosmic matter to the planets. On Earth, these bombardments are the main cause of global geological events, which are fixed in the stratigraphic (geochronological) scale by its boundaries of different ranks, at that the boundaries rank is determined by intensity of cometary bombardments.

Figure 1 shows the Galaxy spiral structure and Sun's orbit, which best fit the astronomy and geology data. In this galactic model, our Galaxy has 4 identical logarithmic arms of an electromagnetic nature, which rotate uniformly around the Galaxy center, and two Archimedean jet streams flowing from opposite points of a rapidly rotating gas and dusty nuclear disk.

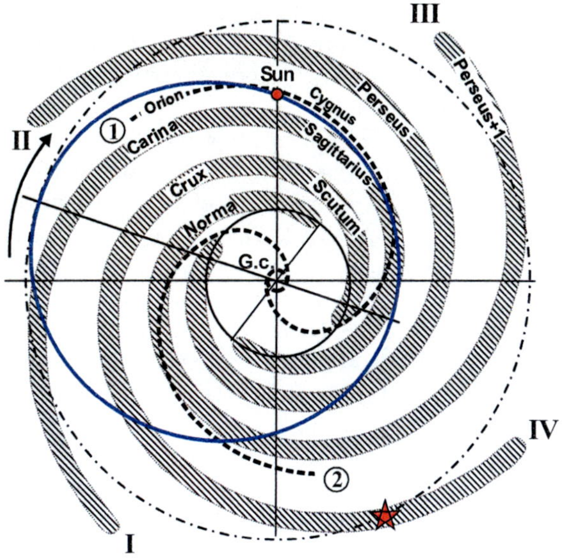

Fig. 1 The current position of the Sun in orbit (ellipse) relative to four galactic arms (Roman numerals) and two jet streams (Arabic numerals in circles) in projection onto the Galaxy plane. The solar orbit is shown for the last revolution of Sun around Galaxy center (G.c.). The small circle in the center, where the jet streams flow out, is the Galaxy nuclear disk. The larger circle where the galactic arms begin determines the size of Galaxy isothermal core. The outer dotted ring corresponds to the Galaxy corotation radius R*. The arrow indicates the direction of Sun moving on orbit, which coincides with rotation of Galaxy and rotation of the apses line (straight line) of solar orbit. The asterisk shows the place in the arm Crux-Scutum (IV) where Solar system was formed earlier

The galactic arms are described by the formula $R(\varphi) = R_{\odot}\{\{\varkappa[\varphi + (2\text{ k}-1)/4]\}$, where: $k = 1$–4 is the arm number; $\varkappa = \text{ctg}(\mu)$ is parameter that characterizes the swirl angle of the arms, uniformly rotated around the Galaxy center with a period T_G. The formula of jet streams has the form $R(\varphi) = V_s(t + \varphi/\omega_d) = V_s t + \rho\varphi$, where V_s is rate of flow out of matter from the nuclear disk, t is the time, $\omega_d = 2\pi/T_d$ is the angular velocity of the disk, T_d is the rotation period; $\rho = V_s/\omega_d$ is the parameter of swirling of jet streams. The disk is inclined to the Galaxy plane at an angle of 22° [23] and processes with the period of rotation of the disk. The value $R^* = \varkappa/\rho$ defines the corotation radius of Galaxy, at which the radiuses of curvature of logarithmic arms and jet streams, swirled into Archimedean spirals are the same.

The gaseous matter flowing out of the nuclear disk spreads like a fan through Galaxy, condensing into dense clouds, comets and stars. These processes are sharply enhanced in intersection zones of jet streams with galactic arms, which partially capture and accumulate gas and dust. At a distance R^* from Galaxy center, where the radii of curvature of jet streams and galactic arms coincide, the gas condensation zone stretches considerably along the arms. Space objects born in the zones of gas-condensation behave differently. Ones of them, formed from the substance of jet streams, move in the radial direction at a speed of $V_s = 300$ km/s and leave our Galaxy forever. Others, which originated from gas and dust of the arms, remain in the arms and subsequently move to independent orbits around Galaxy center. The Sun is one of these objects. The place of its formation in the arm of Crux-Scutum (IV) is indicated by an asterisk on Fig. 1.

The Sun moves in the galactic plane in an elliptical orbit with a large semi-axis (α) and eccentricity (e), characterized by anomalistic (T_R) and sidereal (T_φ) rotation periods around Galaxy center. The line apsid of solar orbit also rotates to the side of Sun movement with some period (T_α). At this the Sun is periodical approaching Galaxy center at a distance of $R_{\min} = \alpha(1-e)$, and it is moving away from it at a distance of $R_{\max} = \alpha(1 + e)$. Simultaneously, the Sun makes vertical fluctuations relative to the galactic plane with a period T_d.

Due to Sun's motion in orbit, as well as the different speeds of nuclear disk and galactic arms rotation, their mutual position changes over time. At that, the Sun quasi-periodically crosses Galaxy spiral arms and jet streams, and during these periods of ~ 2–5 Myr, Solar System is subjected to intense bombardments by galactic comets, at that of varying intensity. In Earth's history such bombardments are epochs of global natural events: tectonomagmatic, biotic and climatic, which are fixed by geologists in the Phanerozoic and Precambrian geochronological scales in form their boundaries, at that of different ranks.

The galactic model allows us to calculate these boundaries age, starting from the rank of systems and above.

6 Galactic Model Optimization Approach

In [10], when analyzing the GSS-1993 and ICC-2020, it was shown that the boundaries rank of Phanerozoic scale is determined by cometary bombardments of different intensity. At that all 15 borders of the systems of Phanerozoic coincide in time with bombardments of Earth by comets of jet streams in the epochs, when Sun was either in one of four galactic arms (12 events) or when jet streams crossed Galaxy plane (3 events). In [11], we optimized the parameters of the galactic model to best fit the calculations to the data of GSS and ICC scales. To this end, we introduced a measure of the "inaccuracy" of the model. For the first 12 events in galactic arms, this measure was the calculated the Sun distance (ΔRi) to the middle of galactic arm closest to it. And for the last three events in Galaxy plane, the Sun distance between its position at a given time and Sun's position at nearest moment time, multiple of period of 25 mln years. Based on all 15 residuals ΔRi, their average root-mean-square value was found.

$$\Delta\overline{\mathrm{R}} = \sqrt{\frac{1}{15}\sum_{1}^{15}\Delta\mathrm{R}_{\mathrm{i}}^{2}} \tag{1}$$

The goal of model optimization was to minimize the value of $\Delta\overline{\mathrm{R}}$, which served as a criterion for the accuracy and adequacy of model, and also an estimate of average width of galactic arms. With a probability of 95%, the width of galactic arms was taken equal to twice the value $\Delta\overline{\mathrm{R}}$.

To limit the number of optimized parameters, all the model parameters included in the formulas for calculating Sun's galactic orbit and the moments of Sun's intersection of jet streams and galactic arms were divided into two groups. The parameters of the first group were considered established in previous versions of the model [1–3, 6, 10], and the parameters of the second group needed to be clarified.

The "established" parameters of the model were considered to be: the galactic model as a whole (Fig. 1), the orbital periods of rotation of Sun and Galaxy. As well as Sun's position in orbit at two points in time: $t_1 = 2 \pm 1$ Ma, when Sun was in the Orion-Cygnus jet stream and $t_2 = 567 \pm 1$ Ma, when Sun crossed the middle line of the Crux-Scutum arm at a distance of the corotation radius from Galaxy center $R^* = \rho/\varkappa$.

The model parameters that needed to be "refined" were: (1) the degree of galactic arms twisting $\varkappa = \mathrm{ctg}\ \mu$, where μ is the angle of twisting of the arms; (2) Sun's distance from Galaxy center $R_{\odot}$; (3) semi-major axis (α) and eccentricity (e) of the Sun's orbit, connected by the condition: $\alpha(1 + e) = \rho/\varkappa = R^*$, 4) initial rotation angles of the line of apsides of solar orbit, as well as of four points of exit of Galaxy arms from the central ring (Fig. 1) and two points of exit jet streams from Galaxy nuclear disk, providing the above conditions at times t_1 and t_2.

Using a special program developed to calculate the times of Sun entry into jet streams and galactic arms, we minimized the $\Delta\overline{R}$ value for each of variable parameters, which varied in wide limits. The calculations were carried out with variations of the parameters: (1) change in $R_\odot$ from 7.8 to 9.0 kpc with a step of 0.05 kpc, (2) change in the angle μ from 70 to 90° with a step of 0.05°, (3) eccentricity of solar orbit changed within $0.34 \le e \le 0.42$ with a step of 0.05. The semi-major axis of the Sun's orbit and the initial angle of rotation of its line of apsides were automatically found.

The program also made it possible to calculate orbital velocity and acceleration of the Sun in Galaxy. In total, more than 100 thousand calculations were performed. The processing of the calculation results showed that there is an optimal set of parameter values that provide the minimum magnitude $\Delta\overline{R} = 0.4$ kpc.

The following values of parameter are optimal: $R_\odot = 8.35 \pm 0.05$ kpc, $\mu = 77.8 \pm 0.05°$, $e = 0.37 \pm 0.005$, $\alpha = 8.373$ kpc, average width of galactic arms at 95% confidence interval $2\ \Delta\overline{R} = 0.8$ kpc.

Let us now analyze the issues of accuracy and adequacy of the constructed galactic model, and also discuss the possibility of using it to solve two problems posed in this article:

Problem 1: Development of dynamic spiral model of Galaxy based on astronomical observations and geological data.
Problem 2: Overcoming the current crisis of stratigraphy is based on a new approach to constructing geochronological scale of Phanerozoic and Neoproterozoic, taking into account cyclic bombardments of Earth by galactic comets of different intensity.

7 Solving Problem 1: Development of the Galaxy Dynamic Spiral Model

The results of several astronomical tests [11] are presented, indicating the high accuracy of constructed dynamical spiral model of our Galaxy.

7.1 The Sun's Distance to Galaxy Center

The dependence of $\Delta\overline{R}$ on the accepted value of $R_\odot$ at optimal values $\mu = 77.8$ and $e = 0.37$ is shown in Fig. 2. The $R_\odot$ values measured by astronomers are also shown (circles with numbers). Our calculation gives the value $R_\odot = 8.35 \pm 0.05$ kpc. This value is in good agreement with the data of astronomers within the error of their measurements. Our calculation prefers a group of points in the region of 8.3 kpc.

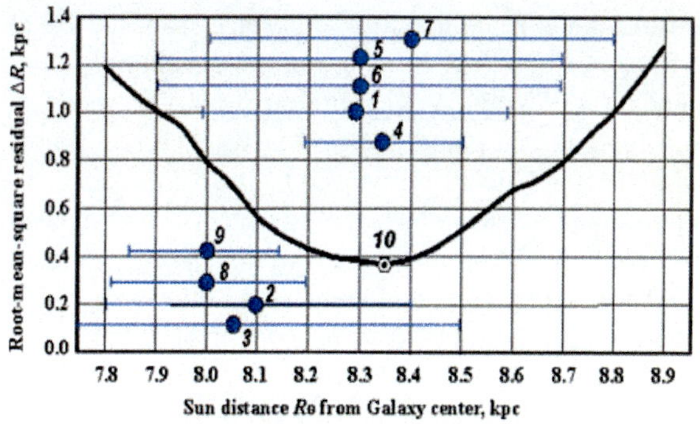

Fig. 2 Dependence of discrepancy $\Delta\overline{R}$ value on Sun's distance to Galaxy center for optimal $\mu = 77.8°$ and $e = 0.37$. Circles with number are measurement results: 1—[24], 2—[25], 3—[26], 4—[27], 5—[28], 6—[29], 7—[30], 8—[31], 9—[32], 10—optimal $R_{\odot} = 8.35 \pm 0.05$ kpc

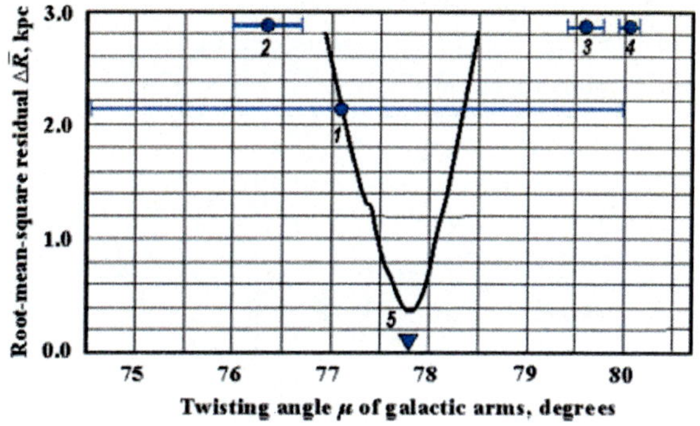

Fig. 3 Dependence of discrepancy $\Delta\overline{R}$ value on twisting angle μ of Galaxy arms for optimal $R_{\odot} = 8.35$ kpc and $e = 0.37$. Circles with number are measure results: 1—[33], 2—[31], 3—[34], 4—[35], 5—optimal twisting angle $\mu_{opt} = 77.8°$ on our calculation

7.2 *The Degree of Twisting of the Galactic Arms*

Figure 3 shows the dependence of discrepancy $\Delta\overline{R}$ on the twist angle of galactic arms μ for optimal distance $R_{\odot} = 8.35$ kpc and Sun's orbit eccentricity $e = 0.37$. In the figure, for comparison, the circles show the angle μ values according to astronomers' measurements. The degree of galactic arms twisting in the astronomers' works was characterized by angle *i*, which is related to angle μ by the relation $i = \mu - 90°$. Astronomers' measurements give angles: $i = -13.6 \pm 0.4°$ [31], $-12.9 \pm 2.9°$ [33], $-9.5 \pm 0.1°$ [35], $-10.6 \pm 0.3°$ [34], which are significantly inferior in accuracy of $\mu_{opt} = 77.8 \pm 0.05°$ measurement in our model.

7.3 *Speed and Acceleration of Sun's Orbital Motion*

Changes in the speed and acceleration of Sun orbital motion cannot be determined by astronomical observations. At the same time, the galactic model makes it possible to make such a calculation. Figure 4 shows the estimated the Sun speed relative to the Galaxy center over the past 600 mln years. The Sun speed changes from $V_{\min}$

= 186.5 km/s at its maximum distance from the Galaxy center $R_{\max} = 11.47$ kpc to $V_{\max} = 332.0$ km/s at its minimum distance $R_{\min} = 5.46$ kpc. The average Sun's speed is 273.1 km/s. At a given time (t = 0), its speed is $V_{\odot} = 240.8$ km/s. Projected onto the galactic plane, the vector $V_{\odot}$ makes an angle $\beta = 19.45°$ with the Sun radius-vector. The current speed of the Sun relative to the Galaxy center was been determined by astronomers. The results of their measurements of $V_{\odot}$ value depend on the accepted Sun distance from the Galaxy center. For comparison, here are the results of several recent experimental studies: $V_{\odot} = 238 \pm 14$ km/s at $R_{\odot} = 8.05 \pm 0.45$ kpc [26], 240 ± 8 km/s at $R_{\odot} = 8.34 \pm 0.16$ kpc [27], 236 ± 6 km/s at $R_{\odot} = 8.3 \pm 0.2$ kpc [36], 240.81 km/s at $R_{\odot} = 8.35 \pm 0.05$ kpc—our calculation.

Figure 5 shows a similar graph of the Sun's acceleration. The graph shows that the Sun's acceleration changes with the same period $T_R = 250$ Myr as the Sun's motion. At the points of the solar orbit $R_{\min}$ and $R_{\max}$, the acceleration turns to 0, changing its sign. At present, the Sun is moving towards the perigalactic point of its orbit with an acceleration of + 1.8 km/s in 1 Myr.

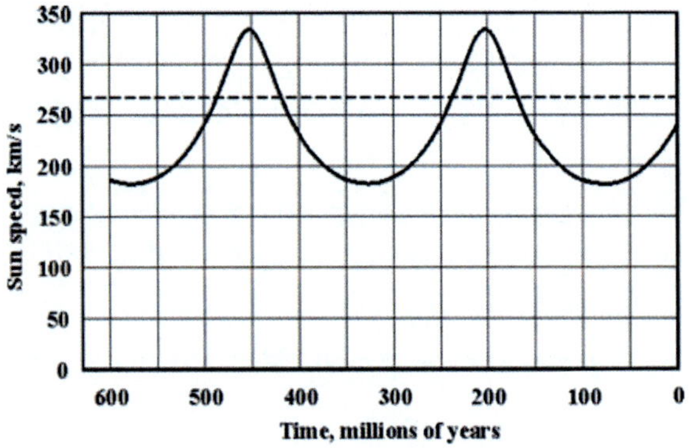

Fig. 4 Change in Sun orbital speed in time. Dotted line is average speed

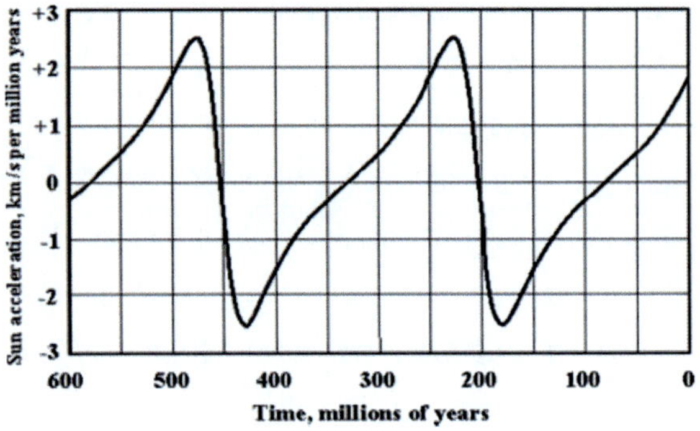

Fig. 5 Change in the acceleration of Sun orbital motion in time

7.4 The Physical Model of the Galaxy. The Problem of Galaxy Evolution

The question of the evolution of our stellar system, in the physical Galaxy model in the form of an isothermal sphere of stars [37], has been raised by the author since the beginning of the 1990s. Focusing on geochronological scale of those years [20], the author believed that in Precambrian the main geological cycle period was shorter than it is now, and increased over time. These ideas are reflected in the works [38–41] and a number of others.

Today, as a result of the development of a more perfect galactic model version, it has become clear that since Solar System formation, neither Sun's orbit, nor Galaxy spiral structure have practically changed. Consequently, the distribution of Galaxy gravitational potential $U(R)$ in region of Sun's orbit localization should have remained unchanged all this time.

According to [2, 7], the distribution function of gravitational potential $U(\mathrm{R})$ in the Galaxy isothermal sphere outside its isothermal core can be represented by an expansion in the asymptotic series

$$\mathrm{U(R)} = \sum_{\mathrm{k=0}}^{\infty} \frac{\mathrm{c_k}}{\mathrm{R^k}}, \tag{2}$$

where c_k are a constant coefficients.

The author has showed [2] that Sun's orbit makes it possible to restrict oneself in formula (2) only the first two terms of expansion in R. In this case, gravitational potential $U(R)$ expression takes the form [42]

$$\mathrm{U(R)} = \mathrm{c_0}\left[1 + \left(\frac{\mathrm{R}}{\mathrm{R_\odot}}\right)^{-1} + \frac{(1-\xi^2)\mathrm{p}}{2\mathrm{R_\odot}}\left(\frac{\mathrm{R}}{\mathrm{R_\odot}}\right)^{-2} + \cdots\right] (\mathrm{cm/s})^2 \tag{3}$$

where: $c_0 = M_0G/R_\odot$ is a constant coefficient, $p = \alpha(1–e^2)$ is an elliptic parameter, $\xi = T_\varphi/T_R$, M_0 is the Galaxy inside the average radius of Sun orbit ($\alpha = R_\odot$), G is the gravitational constant.

In formula (3), first R-dependent term in square brackets defines Sun movement in an elliptical orbit with parameters α and e, and the quadratic term characterizes rotation the line of apsides of solar orbit. In our model, the multiplier in front of this member is equal to 0.091.

The value of $\mathrm{c_0}$ can be calculated using average quadratic velocity of stars in the Galaxy isothermal sphere as $c_0 = {}^2\!/_3\overline{V}^2$, where $\overline{V} = 260$ km/s, and in that case $c_0 = 4.5 \times 10^{14}$ (cm/s)^2 [2]. A can be found by Sun's orbit parameters as $c_0 = L^2/pR_\odot$, where L is the moment of Sun movement amount relative to Galaxy center. Taking $R_\odot = 8.35$ kpc, $e = 0.37$ and $\alpha = 8.373$ kpc, we get $p = 2.58 \times 10^{22}$ cm. Currently, $L = V_\odot R_\odot \mathrm{Cos}(\beta) = 5.094 \times 10^{29}$ cm/s. Substituting the values $R_\odot$, p, and L, we find $c_0 = 4.50 \times 10^{14}$ (cm/s)^2. As you can see, both calculation methods give the same c_0 value.

The Galaxy mass inside the radius of Sun's orbit is found as $M = c_0 R_\odot / G = 0.875 \times 10^{11}\ M_\odot$. This value is in good agreement with astronomers estimates obtained by other methods, which give $M = (1.0\text{–}1.4) \times 10^{11}\ M_\odot$ at $R_\odot = 10$ kpc [16, 23].

8 Solving Problem 2: Construction of the Phanerozoic and Neoproterozoic Geochronological Scale

Below, we discuss the applicability of the galactic model for solving problems of stratigraphy related to construction of scales the Phanerozoic and Proterozoic based on proposed physical approach.

8.1 Phanerozoic Geochronological Scale

An idea of the patterns of cometary bombardments of Earth in Phanerozoic is given in Fig. 6, which shows the trajectory of Sun's motion in Galaxy, and also show the calculated times of Sun entry into jet streams, which are compared with age of boundaries of systems and eras on the GSS-scale [43].

From the analysis of calculation results and data of Phanerozoic scale, it follows that all cometary bombardments in terms of intensity and, accordingly, geological effects for Earth can be subdivided into at least 3 gradations: "weak", "strong", and "very strong".

The "weak" bombardments, corresponding to boundaries of epochs (they are not shown in Fig. 6), are caused by Sun hitting only in jet streams. The "strong" ones, corresponding to boundaries of systems, are caused by Sun presence simultaneously in jet streams and Galaxy arms. We also note that an additional "force" to cometary bombardments is provided by coincidence direction motion of jet streams with Galaxy plane in which Sun moves. Let us especially single out the cases when Sun is in the arms at a distance R^* from Galaxy center, where Earth is subjected to "very strong" cometary bombardments. These events are chosen in the Phanerozoic as the beginnings of Paleozoic (Pz) and Cenozoic (Kz) eras. For Mesozoic era (Mz), this rule is not feasible; therefore, a very powerful bombardment 250 Ma at border of Permian (P) and Triassic (T) was accepted as its beginning.

We see that our calculations are in good agreement with boundaries of systems and eras of the Phanerozoic, which thereby serve as indicators of the times of "strong" and "very strong" cometary bombardments. This conclusion also applies to boundary between Proterozoic and Mesozoic eras. However, since Sun in aphelion of its orbit was not in any galactic arm, geologists timed the Proterozoic-Mesozoic boundary

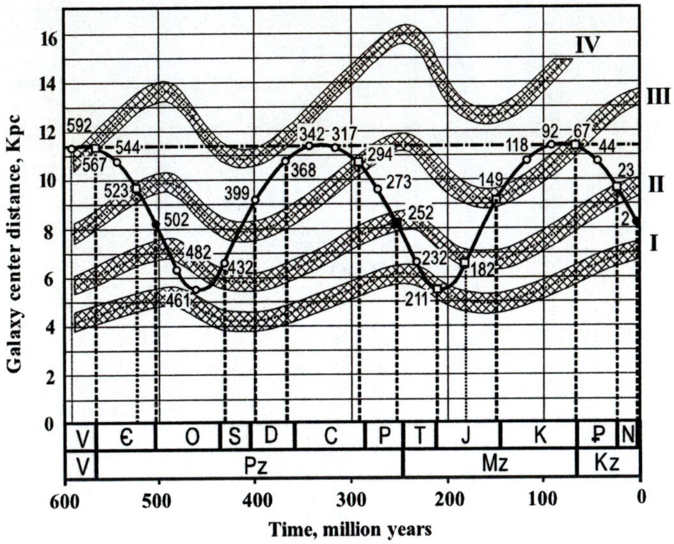

Fig. 6 The sun time position relative to 4 spiral arms (stripes) and its distance to Galaxy center (sinusoidal curve). Roman numerals are Galaxy arms number in Fig. 1. Average arms width is $2\Delta\overline{R} = 0.8$ kpc, where multiplication by 2 corresponds to the 95% confidence interval. Arabic numerals indicate the times (Myr) of Sun entering jet streams (circles) and simultaneously Galaxy arms (squares). The icons corresponding to moments when Sun crosses the galactic plane are filled in black. The horizontal dash-dotted line is corotation radius R^*. The geochronological scale Phanerozoic [43] is shown below, indicating the boundaries of systems (upper row) and eras (lower row)

to a very strong bombardment at the Permian–Triassic boundary in the Perseus arm (II).

The lower boundary of Paleozoic era (Vendian-Cambrian boundary) is also a higher rank boundary between the Proterozoic and Phanerozoic eons. It is known that at this and other boundaries of eons and eras, not only the biota died out, but also serious changes took place in the animal world of our planet. According to the Galactic paradigm, these events were caused by massive fallout of galactic comets, moreover, of other chemical composition [2, 12, 40, 41].

On Fig. 6, first of all, we draw attention to the fact that the model explains well position of most boundaries of systems the Phanerozoic. The exception is the D/C boundary, which in our calculation has the rank of a department boundary, not a system boundary. In this regard, we note that the age of this boundary is considered problematic [44]. The fact is that in period 370–340 Ma (Fransian-Famennian-Tournasian-Visean) a very long mass extinction of living organisms occurred [45]. The reasons for this extinction are still not explained, which is the main reason for the difficulty in choosing the D/C-boundary.

The difficulties of constructing stratigraphic scales GSS [21] and ICC [22] for the Precambrian are especially great. These difficulties, however, are not an obstacle to the galactic model. The resonant nature of the model made it possible to establish

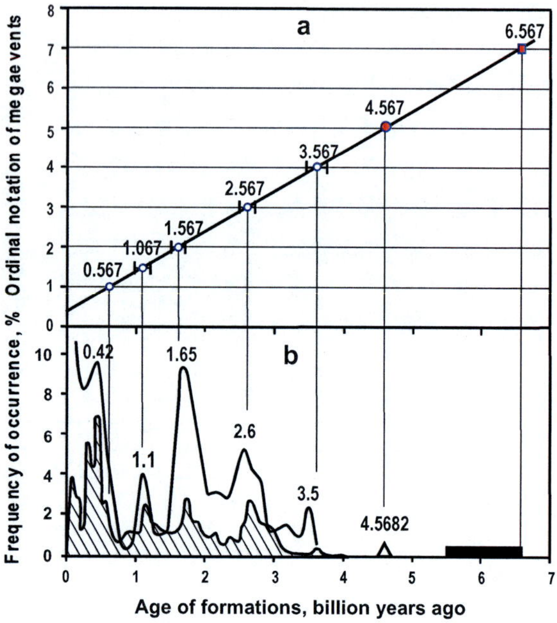

Fig. 7 **a** Megacycles of global mega-events. The numbers are the events time; **b** the actual data used in the construction of the graph (**a**). The curves show the activity of magmatism and ore formation processes (shaded) in Earth history [46]. The triangle is generally accepted age of Solar System [47]. The parallelepiped is the age of iron meteorites Fe II class [48]

[11] that since Solar System formation, Galaxy spiral structure and Sun orbit have hardly changed.

Therefore, the alternation of cometary bombardments of different intensity, which is characteristic of Phanerozoic (Fig. 6), also occurs in Precambrian. However, in Precambrian, in contrast to Phanerozoic, only the strongest cometary bombardments in galactic arms at a distance R^* from Galaxy center are reliably recorded. Most of these bombardments geologists reflected in the Precambrian geochronological scale as global mega-events (Fig. 7).

According to Fig. 7, all mega-events in Earth history occur in galactic arms at a distance R^* from Galaxy center and obeying a general pattern (Table 1).

It follows from (Table 1) that all mega-events occur exclusively in galactic arms and repeat exactly over 2000 Myr. After originating in Crux-Scutum (IV) arm, Solar System repeatedly fell into Galaxy arms, where Earth was subjected to very intense bombardments by galactic comets. However, the power of bombardments in different galactic arms differed significantly. In Carina-Sagittarius (I) and Norma-Perseus + 1 (III) arms, bombardments was weaker than in Perseus (II) arm and even more so in Crux-Scutum (IV).

All events in this arm have highest rank—the borders of eons. We also included here two key events in Earth and Solar System history, information about which was preserved exclusively in meteorite matter [17]. The first event 6.567 Ga was the Sun and planets formation, and the second 4.567 Ga was the repeated cycle of planets formation caused by destruction of the Phaethon planet and appearance an asteroid belt in its place.

Table 1 Relationship between Precambrian mega-events and galactic arms

Galactic arm	Event time, млрд лет	Precambrian boundaries of eons and eras [49]
I. Carina-Sagittarius	1.067 3.067	Neoproterozoic –
II. Perseus	1.567 3.567	Mesoproterozoic Paleoarchean
III. Norma-Perseus + 1	2.067 4.067	– Archean (Eoarchean) (conditional border)
IV. Crux-Scutum	0.567 2.567 4.567 6.567	Phanerozoic (Paleozoic) Proterozoic (Paleoproterozoic) 2nd cycle of planet formation Solar System formation

The facts show [12, 13] that cometary bombardments in Crux-Scutum (IV) arm were not only stronger than in other Galaxy arms, but this arm itself should be twice wide as shown in Figs. 1 and 6. In this case, D/C-boundary (see Fig. 6) takes rank of boundary of systems, and not division boundary of GSS-scale Phanerozoic.

8.2 Neoproterozoic Geochronological Scale

In recent decades, considerable efforts of geologists, geophysicists, climatologists, biologists and paleontologists are aimed at a detailed study of planetary tectonic, climatic and biotic processes in Neoproterozoic, which is perhaps the most problematic period in Earth history, preceding Phanerozoic. The analysis and study of these processes requires at least the development of a reliable geochronological scale Neoproterozoic, which is not available today.

The optimized galactic model makes it possible to build such scale for Neoproterozoic according to the same rules as for Phanerozoic, assuming that Crux-Scutum (IV) arm is more powerful and wider than the other three Galaxy arms. Figure 8 shows the times calculating results of "strong" and "very strong" cometary bombardments for interval time from 1100 to 500 Ma are compared with boundaries of Riphean, Vendian, and Cambrian, where Neoproterozoic includes Riphean and Vendian. In this calculation, in contrast to the previous model version, the width of Crux-Scutum arm is taken equal to 2 kpc, while the remaining three galactic arms are left unchanged.

As we can see, the results of calculations for Vendian and Cambrian agree very well with the scale [43] used for comparison. It should be emphasized that in this scale, as in all stratigraphic (geochronological) scales developed before 1993, the end of Vendian and the lower boundary of Cambrian, i.e., the beginning of Phanerozoic, have an age of 570 ± 10 Myr. Whereas in later scales and modern ICC-2022, Vendian **was replaced by Ediacaran period, the status of which was officially ratified by the** International Union of Geological Sciences (IUGS) in 2004. At that Ediacaran period

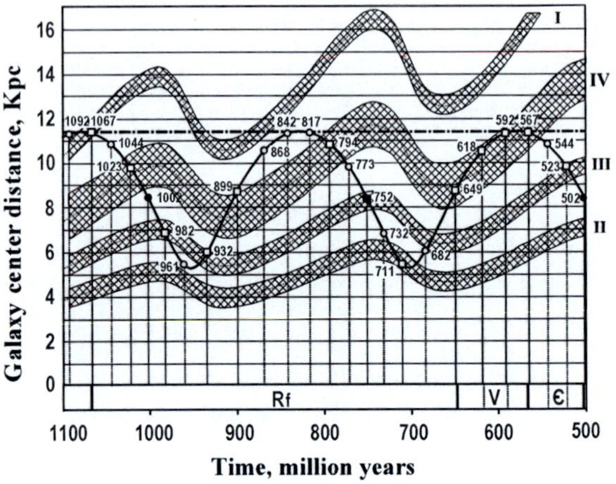

Fig. 8 The sun position relative to 4 galactic arms and its distance to Galaxy center in time interval from 1100 to 500 Ma. The Vendian, Cambrian, and Riphean boundaries are shown in accordance with the GSS-scale [43]. Other designations (see Fig. 6)

beginning is assigned to boundary of 635 Ma, and its end to boundary of 541 ± 1.0 Ma, to which boundary of Proterozoic-Phanerozoic was also transferred.

According to the galactic model, such a transfer of Phanerozoic boundary is completely **unjustified. For the first age value, which in our calculation is 567 Ma,** corresponds to one of the four key events in Solar System and Earth history that occurred in Crux-Scutum (IV) arm (see Table 1). And the event with calculated age of 544 Ma is an ordinary bombardment of Earth by comets of jet stream, of which there were many in Phanerozoic.

Another important remark is that during Vendian, Sun moved within powerful Crux-Scutum arm. Therefore, all comet bombardments in Vendian were "very strong". At the same time, bombardment of 592 Ma was especially strong and almost did not differ from one of 567 Ma, taken as the Phanerozoic beginning.

We also note that galactic comets of two different types fell to Earth in Vendian: comets of jet streams and comets of Crux-Scutum arm. Comets of this arm, as we established [2, 41], had elevated contents of P, Ca, K, Na, S, and other chemical elements of "life" Therefore, we have good reasons to believe that it was these comets that played the key role in Vendian-Cambrian biotic event [41], as well as in next major biosphere rearrangements at Paleozoic, Mesozoic, and Cenozoic boundaries. **It is also difficult to doubt that the biotic events that took place in Vendian, initiated by falls of these comets, served as the basis for Vendian identification as an important** independent period preceding Cambrian.

In the scale GSS-1978 [43], Vendian has rank of a system with beginning of 650 ± 10 Ma and ending of 570 ± 10 Ma, which includes 3 divisions (epochs) with boundaries between them 620 ± 10 Ma and 590 ± 10 Ma. The calculation (Fig. 8) **well confirms all boundaries Vendian, except for their status. Therefore,** in the galactic model, all Vendian boundaries must have a rank not lower than the system. And the rank of 592 Ma boundary should be the same as that of boundary Phanerozoic.

Hence it follows that boundary of Phanerozoic and Precambrian, with which Vendian-Cambrian biotic revolution is commonly associated today, is not an "instant" event [13]. As known, this biotic revolution had two-stage character [50]. At the first stage, from middle to end of Vendian, various soft-bodied invertebrates develop, the first organisms with a silicon skeleton appear, as well as Ediacaran fauna, a significant part of which dies out at Vendian end. At the second stage—Cambrian, a more diverse fauna, close to modern, appears. Its appearance is preceded by a brief Tommotian Stage, during which creatures with simply arranged shells of calcium phosphate spread, variety of which subsequently increases rapidly.

However, stratigraphic position of Tommotian Stage raises many questions [51]. Today, Tommotian Stage is attributed to Cambrian. However, judging by the processes that took place at Vendian-Cambrian boundary, this boundary itself, in the author's opinion, should be considered extended, occupying interval of time 592–567 Ma, which should also include Tommotian Stage [13].

As regards the construction of Neoproterozoic scale as a whole, Vendian in GSS-scale is preceded by Riphean, whose scale can be constructed according to the same rules as Phanerozoic scale. According to Fig. 6, in Riphean cometary bombardments of 711, 752, 794, 899 (?), 932, 1002, 1023, and 1067 Ma took place in galactic arms, i.e. were "strong". Therefore, Riphean can be subdivided into 8 systems, each of which includes from 1 to 4 epochs. Their boundaries are quite well confirmed by the actual data [12, 13].

As the Neoproterozoic boundary in Fig. 8, a "very strong" bombardment in Carina-Sagittarius arm (I) of 1067 Ma is accepted. However, event of 1092 Ma can just as well be considered this boundary.

9 Conclusion

A more advanced version of the galactic model has been developed and tested, which makes it possible to refine the parameters of the Galaxy spiral structure and overcome modern crisis in stratigraphy caused by absence of reliable geochronological scales the Phanerozoic and Neoproterozoic in this science.

The most important new results:

- A dynamical Galaxy model is constructed, which is in the best agreement with results of astronomical observations and geological data. In this model, the Crux-Scutum spiral arm is twice as wide as the other three galactic arms.
- It is concluded that the Sun motion along galactic orbit is in resonance with the rotation of spiral arms and nuclear disk of Galaxy.
- Based on Sun's galactic orbit parameters, distribution of gravitational potential in Galaxy equatorial plane of outside radius of its isothermal core was found. This distribution corresponds to Galaxy physical model in the form of an isothermal sphere of stars, well known in astrophysics.

- An approach to the construction of geochronological scales of the Phanerozoic and Precambrian is substantiated, taking into account bombardments of our planet by galactic comets. Based on this approach, a high-precision geochronological scale, common for the Phanerozoic and Neoproterozoic, was built, which specifies the time of the most important events in Earth's geological history.

References

1. *Barenbaum, A.A.*, 2002, *Galaxy, Solar system, Earth: Subordinate processes and evolution.* PH: GEOS, Moscow [in Russian].
2. B*arenbaum, A.A.*, 2010. *Galactocentric paradigm in geology and astronomy.* PH: LIBROKOM, Moscow [in Russian]
3. *Barenbaum, A.A., Gladenkov, Yu.B., Yasamanov, N.A.*, 2002. Geochronological Scales and the Astronomic Time (State of the Art). *Stratigraphy and Geological Correlation.* V.10, No.2, 3–14.
4. *Barenbaum, A.A. & Yasamanov, N.A.* 2003. Galactic comets as one of leading factors in terrestrial planets tectonic evolution. *Tectonics and Geodynamics of Continental Lithosphere.* Proc. of XXXVI Tectonic Conf. (Moscow, February 04–07, 2003) V.1, PH: GEOS, Moscow. 24–26 [in Russian].
5. *Barenbaum, A.A. & Yasamanov, N.A.*, 2004. Tectonic cycles of Wilson, Bertrand and Stille as result of Earth's bombardments by galactic comets. *Evolution of tectonic processes in Earth's history* (Proc. of XXXVII Tectonic Conf., Novosibirsk, February 10-13, 2004) PH: Siberian Branch RAS, Novosibirsk, V.1, 38–41.
6. *Barenbaum, A.A., Khain, V.E., Yasamanov, N.A.*, 2004. Large-scale tectonic cycles: analysis from the galactic concept positions. *Vestnik MGU. Geol.*, 3, 3–16.
7. *Barenbaum, A.A.*, 2005. "On the galactic version of geochronological scale" by Yu. A. Srlyarov: Reply. *Stratigraphy and Geological Correlation.* V.13 No.5, 135–139.
8. *Barenbaum, A.A.*, 2011. Tectonomagmatic processes in oceans and on continents as indicators of galactic comets fall. *Modern state of Earth Sciences. Proc. Int. conf. commemorating V.E. Khain.* (Moscow, February 1–4, 2011 MGU) MGU Press. 166–177.
9. *Barenbaum, A.A. & Yasamanov, N.A.*, 2004. On possible cause of ice sheets. *Bulieten' Moskovskogo obshchestva ispytateley pripody. Geol.* V.79, Is. 6, 13–21.
10. *Barenbaum, A.A.*, 2018. Geological and meteoric data as a necessary tool for developing an adequate Galaxy spiral model. *Proc. Rus. An. Seminar on Experimental Mineralogy, Petrology and Geochemistry (RASEMPG-2018)*, Vernadsky Inst. Geochemistry and Analytical Chemistry RAS, Moscow. 294–298.
11. *Barenbaum, A.A. & Titorenko, A.S.*, 2020. Galactic model of geological cyclicity: parameter optimization and testing based on geology and astronomy data. *Proc. Rus. An. Seminar on Experimental Mineralogy, Petrology and Geochemistry (RASEMPG-2020)*, Vernadsky Inst. Geochemistry and Analytical Chemistry RAS, Moscow. 210–215.
12. *Barenbaum, A.A.*, 2021. Analysis of the causes of glaciations and $\delta^{13}C$ anomalies of carbonates based on a galactic model. *Physical-chemical and petrophysical studies in Earth sciences*: *Proc. XXII Intern. conf.* Inst. Experimental Mineralogy RAS, Moscow. 22–26.
13. *Barenbaum, A.A.*, 2021. Confirmation of the "Snowball Earth" hypothesis by a galactic model. *Geology of Seas and Oceans: Proc. XXIV Intern. Conf. on Marine Geology*, V.1. Shirshov Inst. Oceanology RAS, Moscow. 21–25.
14. *Vorontsov-Velyaminov, B.A.*, 1978. *Extragalactic astronomy.* Moscow. PH: Nauka, 480 p.
15. *Melnik, A.M.*, 2011. *Kinematics of outer pseudorings and the spiral structure of the Galaxy.* Dis. Dr. Phys.-Math. Sci., Shtenberg Astron. Inst. Moscow, 164 p.

16. *Marochnik, L.S., Suchkov, A.A.*, 1984. *The Galaxy.* Moscow. PH: Nauka, 392 p.
17. *Vallee J.P.*, 2002. *Astrophys. J.*, V.566, №1, 261–266.
18. *Efremov, Yu.N.*, 2005. *Star Islands.* Fryazino: PH: Vek-2, 272 p.
19. *Zabolotskikh M.V. et al.*, 2002. *Astron. Let.* V.28. No.7, 454–464.
20. Harland, W.B, Cox, A.V, Llewellyn, P.G. et. al., 1982. A geologic Time scale. Cambridge Univ. Press.
21. *Zhamoida, A.I.*, 2013. *General Stratigraphic Scale adopted in the USSR–Russia. Its meaning, purpose and improvement.* St. Petersburg: Russian Geological Research Institute. 24 p.
22. *Alekseev, A.S.*, 2015. The International stratigraphic chart: state of the art. *Russian Geology and Geophysics.* V.56, No.4, 524–531.
23. *Space physics. Little encyclopedia* / ed. Sunyaev R.A., 1986. Moscow. PH: Sovetskiy Entsiklopediya. 783 p.
24. *Gillessen S. et al.*, 2009. *Astrophys. J.*, V.707, 2, L114–L117.
25. *Gillessen, S. et al.*, 2013, *Proc. IAU Symp.* 289, 29.
26. *Honma, M. et al.*, 2012, *Publ. Astron. Soc. Japan.* 64, 136.
27. *Reid, M.J. et al.*, 2014. *Astrophys. J.*, 783(2), 130.
28. Zhamoida, A.I., 2015. *Russian Geology and Geophysics.* 56(4), 511–523.
29. *Grijs, R. & Bono, G.*, 2016. *Astrophys. J. Sup. Ser.* 227, 5.
30. *Grijs, R. & Bono, G.*, 2017. *Astrophys. J. Sup. Ser.* 232, 2.
31. *Vallee, J.P.*, 2017. *Astron. Rev.* V.13, Is. 3–4, 113–146.
32. *Camarillo, T. et al.* 2018, *Publ. Astron. Soc. Pacif*. 130, 4101.
33. *Bobylev, V.V.& Baikova, A.T.*, 2014. *Astron. Let.*, V.40, No.12, 773–782.
34. *Rastorguev A.S., et al.*, 2017. Astrophys. Bull. V. 72, No.2, 132–140.
35. *Dambis, A.K,.* et al., 2015. *Astron. Let.*, V.41, No.9, 489–500.
36. *Bobylev, V.V. et al.*, 2016. *Astron. Let.* V.42, No.11, 721–733.
37. *Chandrasekhar S.*, 1948. *Principles of stellar dynamics.* Moscow. PH: In. Lit.
38. *Barenbaum, A.A.*, 1991. Megacyclicity of geological processes and the evolution of the Galaxy. Cycles of natural processes, hazardous phenomena and environmental forecasting. Moscow. PH: Russian Academy of Natural Sci. Is.1, 27–47.
39. *Barenbaum, A.A. & Yasamanov, N.A.*, 1995. Experience in constructing a detailed geochronological scale of Late Riphean based on new ideas about the Galaxy structure // *Doklady Earth Sciences.* V.344, No.5, 650–653.
40. *Barenbaum, A.A.*, 2004. Causes responsible for the "biotic revolution" in the Vendian-Cambrian from standpoint of the comet-galactic hypothesis. *Doklady Earth Sciences.* V.398, No.7, 925–927.
41. *Barenbaum, A.A., Litvinova, T.V., Khain V.E.*, 2009. Geological evidence of the space influence on life evolution. Vendian-Cambrian biotic event. *Uralian Geol. J.*, No.6 (72), 15–39.
42. *Roy, A.E.*, 1981. *Orbital motion.* Moscow PH: Mir. 544 p.
43. *Khain, V.E.* (ed.), 1978. *Geochronological table.* Leningrad: Russian Geological Research Institute.
44. *Nikolaeva, S.V. et al.*, 2016. Devonian-Carboniferrous boundary: new solutions for an old problem. *Bulieten' MOIP. otd. Geology*, V.91. Is.6, 86–74.
45. *McLaren, D.J.*, 1982. *Geol. Soc. Amer. Spec. Paper*, No.190. 477–484.
46. *Pushkarev, Yu.D.*, 1990. *Megacycles in the evolution of the crust-mantle system.* Leningrad, PH: Nauka. 217 p.
47. *Bouvier, A., Wadhwa, M.*, 2010. *Nature Geoscience*. V.3, 637–641.
48. *Sobotovich, E.V.*, 1974. *Isotope Cosmochemistry.* Moscow. PH: Atomizdat, 208 p.
49. *Plumb, K.A.*, 1991. New Precambrian time scale. *Episodes.* V.14, No.2, 139–140.
50. *Sokolov, B.S.*, 1997. *Essays on the formation of the Vendian.* Moscow. PH: KMK LLC. 156 p.
51. *The Tommotian Stage And Cambrian Lower Boundary Problem,* 1969. Trudy GIN RAN. V.206. PH: Nauka, 380 p.

Analysis of Wind Speed Profiles in Hurricane Irma

E. I. Poplavsky, A. M. Kuznetsova, and Yu. I. Troitskaya

Abstract This work is devoted to the analysis of the wind speed profiling method in hurricanes. The profiling method is based on the self-similarity of the wind speed profile in the atmospheric boundary layer. Hurricane Irma is simulated within the WRF LES model using CFSV2 wind reanalysis as initial and boundary conditions. A comparison between the simulated wind speed profiles and the wind speed profiles from the NOAA GPS-dropsondes is performed.

Keywords Numerical simulation · WRF LES · Wind speed profiles · Extreme winds · Hurricane

In the present work, a new method for retrieval the parameters of the atmospheric boundary layer in hurricanes is analyzed. The profiling method based on the self-similarity of the wind speed profiles in the boundary layer is usually used in this case [1]. A similar method was used in earlier laboratory experiments [2] and proved to be applicable to tropical cyclones. The method consists of the approximating of the upper parabolic part of the wind speed profile, one can retrieve the lower logarithmic part. Further on the basis of the logarithmic part, the friction velocity, near-surface wind speed and the aerodynamic drag coefficient are obtained. These boundary layer parameters are important for numerical weather (atmosphere and hydrosphere) models. This method is applicable for most of the average wind speed profiles obtained in a hurricane, however, there are profiles as in Fig. 1, where in the lower logarithmic part of the profile there are two sections with different slopes, which leads to incorrect determination of the 10 m wind speed U_{10} and other parameters. Perhaps this is due to the fact that the GPS-dropsonde does not fall straight down and, deviating during the fall, passes through areas of the hurricane of different intensity. In this work, such wind speed profiles are analyzed in comparison with the wind profiles obtained from the numerical simulation.

Numerical simulation of Hurricane Irma is carried out within the atmosphere model WRF [3] with the large eddy simulation (LES) option. It was important to

E. I. Poplavsky (✉) · A. M. Kuznetsova · Yu. I. Troitskaya
Institute of Applied Physics RAS, Nizhny Novgorod, Russia
e-mail: poplavsky7@gmail.com

V. I. Karev (ed.), *Physical and Mathematical Modeling of Earth and Environment Processes—2022*, Springer Proceedings in Earth and Environmental Sciences,
https://doi.org/10.1007/978-3-031-25962-3_22

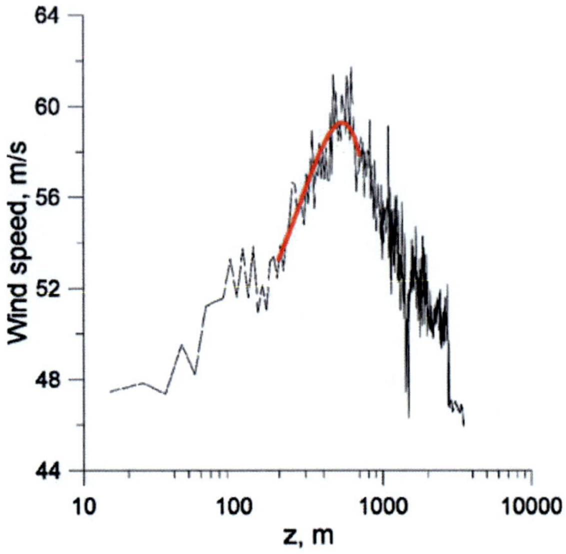

Fig. 1 Average wind speed profile from GPS-dropsondes dropped at a distance of 40–45 km from the center of the hurricane.

achieve a spatial resolution of the vertical wind speed profiles in the simulated hurricane sufficient for a correct comparison with the wind speed profiles from GPS-dropsondes. The structure of the hurricane obtained from the numerical simulation is controlled by those observed in the wind speed reanalysis data (CFSV2 [4] data is used). The simulation is performed on 3 nested domains using CFSV2 wind reanalysis as initial and boundary conditions [5].

For comparison, average wind speed profiles obtained by grouping GPS-dropsondes dropped at approximately equal distances from the center of the hurricane (scatter 5–10 km) are used [6]. From the WRF simulations, the average profiles obtained at the same distance from the center of the hurricane as the groups of GPS-dropsondes from work [6] are used, namely, profiles located in the coordinates of the fall of the dropsondes. The center of the simulated hurricane is determined as the center of the circle obtained for a wind speed of 70 m/s, since the track of the simulated hurricane does not correspond to the track obtained from the NOAA data (www.aoml.noaa.gov). It is a problem that will be checked in the further experiments. The average profile of a group of dropsondes located at a distance of 40–50 km from the center (Fig. 2b) demonstrates high wind speeds, in contrast to the profile obtained from the simulation data, which demonstrates lower wind speeds typical for the center of a hurricane, and not the walls of its "eye". This indicates that the WRF simulations overestimates the size of the eye of the hurricane.

The average profile of the next group of dropsondes (70–80 km from the center of the eye, Fig. 2d) demonstrates a lower wind speed than the simulated profile, since it is located in the region of high winds in the "wall" of the "eye" of the hurricane. Starting from 100 to 110 km, the profiles behave in a similar way (Fig. 2f), and at 150–160 km they almost coincide (Fig. 2h). It should also be noted that the simulated

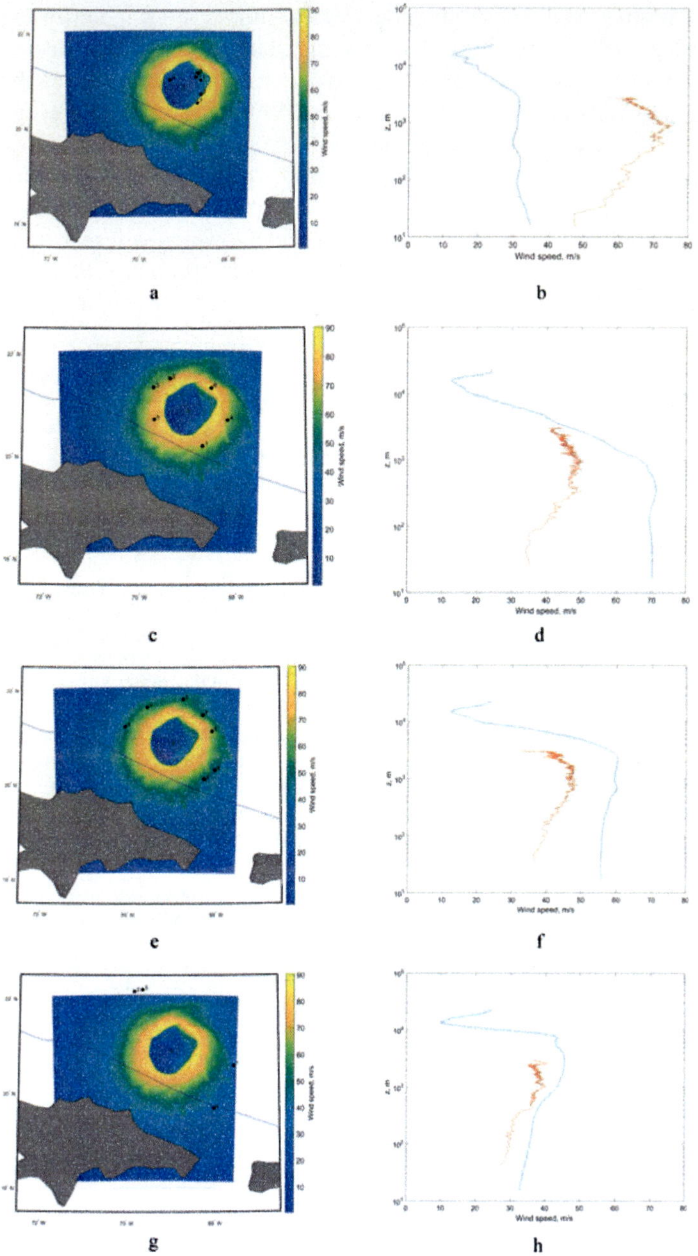

Fig. 2 Surface wind speed distribution (**a**, **c**, **e**, **g**) and comparison of average vertical profiles of a simulated hurricane and from GPS-dropsondes (**b**, **d**, **f**, **h**). The black dots on the graphs on the left show the coordinates of the fall locations of the GPS-dropsondes, the blue curve is the real track of the hurricane. In the graphs on the right, the orange wind speed profile corresponds to the average profile of the group of GPS-dropsondes marked on the map, and the blue profile corresponds to the average profile of the simulated hurricane in the same coordinates. Distance to hurricane center: 40–50 km (**b**), 70–80 km (**d**), 100–110 km (**f**), 150–160 km (**h**)

wind speed profiles do not have a pronounced logarithmic slope in the lower part, in contrast to the profiles from the GPS-dropsondes, but in Fig. 2h it begins to appear.

Thus, significant differences between field data from GPS-dropsondes and simulated data using WRF are observed in the nearest 100–130 km from the center of the hurricane. In the future, it is planned to analyze more simulation result and to improve the difference between the modeled and the observed tracks. It is also planned to analyze the behavior of the average wind speed profiles at different distances from the center of the hurricane and in its different sectors, which will make it possible to determine the advantages of one or another method of grouping GPS-dropsondes. In addition, the self-similarity of the wind speed profiles in the simulated hurricane will be considered.

Acknowledgements The work is supported by RSF grant # 21-77-00076.

References

1. *Hinze, J. O.,* 1959. Turbulence: An Introduction to Its Mechanism and Theory, *1st ed. New York, NY, USA: McGraw-Hill,* 586 pp.
2. *Kandaurov, A.A., Troitskaya, Yu.I., Sergeev, D.A., Vdovin, M.I., Baidakov, G.A.,* 2014. Average velocity field in the air flow over the water surface in laboratory study of the hurricane conditions. *Izvestiya Atmos. Ocean Phys.* V. 50.P. 399-410.
3. *Skamarock W. C., Klemp J. B., Dudhia J., Gill D. O., Barker D. M., Duda M. G., Huang X-Y, Wang W., and Powers J. G.,* 2008: A Description of the Advanced Research WRF Version 3. *NCAR Tech. Note* NCAR/TN-475+STR, 113 pp
4. *Saha, S., et al.* 2011, updated monthly. NCEP Climate Forecast System Version 2 (CFSv2) Selected Hourly Time-Series Products. Research Data Archive at the National Center for Atmospheric Research, Computational and Information Systems Laboratory.
5. *Aleksandra Kuznetsova, Georgy Baydakov, Daniil Sergeev, and Yuliya Troitskaya,* 2019. Estimation of the air-sea exchange processes parameters under modeled hurricane Irma wind conditions. *In OCEANS 2019-Marseille,* pp. 1–4. IEEE, 2019.
6. *Y. Troitskaya, O. Ermakova, N. Rusakov, E. Poplavsky, D. Sergeev and G. Balandina,* 2019. Towards the GMF for Wind Speed and Surface Stress Retrieval in Hurricanes Based on the Collocated Dropsonde Data and Cross-Polarization SAR Images. *IGARSS 2019 - 2019 IEEE International Geoscience and Remote Sensing Symposium,* pp. 4693–4696

Evolution of Approaches to Modelling Geomechanical Processes in Oil and Gas Reservoirs

S. O. Barkov and V. V. Khimulia

Abstract The article describes the main stages of approaches evolution to modeling geomechanical processes in oil and gas science. The paper outlines the key ideas of physical modeling and the leading directions of mathematical modeling used for studying geomechanical processes, production forecasting and evaluation of maximum recoverable oil reserves. Disclosed the main disadvantages and advantages of analytical, statistical and numerical methods, as well as indicating their applicability limits. The analysis of scientific works, devoted to the development of mathematical methods of modeling, dedicated to the prediction of development indicators on the basis of displacement characteristics and decline curves, is carried out.

Keywords Physical and mathematical modeling · Analytical methods · Statistical methods · Numerical methods · Production decline curves · Oil production forecasting

The main tasks of geomechanics and applied geology in the oil and gas industry are still discovering fields, determining mineral reserves, designing field development systems, designing horizontal and directional well profiles, and modeling various measures aimed at improving oil recovery and increasing the completeness of field development. The last of the mentioned tasks becomes more and more actual due to depletion of the easy-to-recover reserves and beginning of the active development of the hard-to-recover reserves. It should be noted that understanding the filtration processes in the reservoir, including the effect of stress on the permeability of the productive formation, is of great importance. It is necessary to understand the nature of fluid movement in the reservoir, as well as the distribution of zones of increased and decreased permeability within the reservoir.

Today, one of the most common methods used to describe processes in oil and gas reservoirs is modeling. Modeling refers to the process of replacing the object under study with its conditional representation, i.e., a model [1]. In our case, the objects are

S. O. Barkov (✉) · V. V. Khimulia
Ishlinsky Institute for Problems in Mechanics RAS, Moscow, Russia
e-mail: sviatoy97@gmail.com

V. I. Karev (ed.), *Physical and Mathematical Modeling of Earth and Environment Processes—2022*, Springer Proceedings in Earth and Environmental Sciences,
https://doi.org/10.1007/978-3-031-25962-3_23

the field, reservoir, well, etc. The model must most accurately reflect the geological structure of the reservoir and fluid movement in it, filtration processes, the dynamics of the well operation, forecast production and oil recovery factor (ORF), of course, within the measurement accuracy. The whole essence of the modeling method comes down to cognition of object properties and processes occurring in it by studying not the object itself, but its model.

A distinction is made between physical and mathematical modeling. Physical modeling is understood as the creation of a laboratory physical model on a quantitatively reduced scale and the subsequent conducting of an experiment on certain setups that reproduce the nature of the phenomena under study. The results obtained in these experiments are then extended to real scales. Physical modeling is used mainly when it is impossible to obtain an exact mathematical description of the phenomenon or process under study, or when a sufficiently large volume of calculations is required, which is difficult to implement.

A similar situation arises, for example, when modeling filtration processes occurring in rocks with anisotropy of elastic and strength properties [2]. A special difficulty of building a mathematical model appears when dealing with layered rocks and directional wells. Note that in order to obtain correct results in physical modeling, it is necessary to observe both geometric and physical similarity of the model and the real object.

Mathematical modeling consists in the construction and study of a mathematical model. The processes and phenomena under study are described by a set of mathematical relations, equations and inequalities reflecting basic laws inherent to these processes and phenomena. Most often, mathematical methods of modeling allow describing only the ideal object or process. The implementation of the connection between the mathematical model and the real object takes place with the help of empirical laws, hypotheses and simplifications. Among all methods of mathematical modeling we can distinguish three main groups: analytical, statistical and numerical.

All these methods individually or in aggregate continue to be used for a comprehensive study of geomechanical processes in oil and gas reservoirs during field development and operation. Let us consider the history of development and application of the above methods in modeling of geomechanical processes.

1 Analytical Modeling Methods

Analytical models are mainly used to describe fairly simple systems, since the application of this method to complex systems causes a number of mathematical difficulties. In order to build an analytical model when describing the properties and parameters of the object under study, we have to resort to significant simplifications, which affects the accuracy of the results obtained. In spite of this, analytical models allow us to find explicit functional dependencies between the quantities sought as well as determine numerical solutions for given initial and boundary conditions. Analytical models are equations or systems of equations represented in the form of algebraic, differential, integral or finite-difference relations with certain initial and boundary conditions [3]. The results of analytical modeling are analytical expressions or equations for the quantities to be sought.

The beginning of studying filtration processes in rocks was laid in the middle of the nineteenth century by French scientists Darcy and Dupuit. The first studies considered filtration of liquids through a fictitious ground, i.e., through a homogeneous medium consisting of balls of equal size, stacked in the same way throughout the entire volume of the porous medium. In their works, empirical dependences of fluid filtration under the action of pressure gradient were proposed. Darcy was able to obtain a formula describing the filtration rate of a fluid that obeys the Navier–Stokes law (Newtonian fluid) in its motion [4].

Dupuit was able to calculate well flow rate in conditions of groundwater inflow [5]. Further development of the general problem of the theory of reservoir fluid filtration to a well was carried out by Zhukovsky, who was able to derive differential filtration equations from the Laplace equation [6]. In subsequent works by Slichter the scientific substantiation of the influence of pore space structure on filtration processes occurring in homogeneous media was added. At the same time Forchheimer noted the great importance of using the potential theory when describing the distribution of filtration flows in a reservoir [7]. At the beginning of the twentieth century, the Reynolds number was discovered, which defined the limits of the Darcy law [8].

In 1934 Leibenzon was able to obtain differential equations of motion for gas and the so-called "fizzy liquid" (carbonated liquid) [9]. Around the same time, the formula linking permeability, porosity and specific surface area in an ideal porous medium was obtained empirically separately by Kozeny and Carman [10, 11]. This formula, characterizing filtration features of reservoir fluid, later became known as Kozeny-Carman formula. At the same time Adzumi publishes his work, in which he talks about the possibility of representing any porous body as a system of parallel cylindrical capillaries with different diameters [12].

All of the above works served as the basis for the formation of a separate science called "underground hydromechanics". Muskat was able to generalize all previous studies in the field of filtration and for the first time wrote down the differential equations of joint motion of oil, water and gas in oil-bearing and gas-bearing strata. We should note the enormous contribution to the development of the filtration theory by Kochina, who proposed calculation methods for determining the flow rate of any directional wells [13].

Since the middle of the twentieth century, the development of filtration theory has continued in two directions. On the one hand, the description of fluid motion in strata is based on the law of conservation of momentum, mass and energy, and on the other hand, the motion of fluids in heterogeneous reservoir strata and bottom-hole zones is considered taking into account the setting of boundary and initial conditions, as well as steady or non-steady flow to the mine site.

2 Statistical Modelling Methods

In the middle of the twentieth century, statistical modeling methods became widespread. Statistical modeling is nothing but multivariate regression analysis, which allows us to establish a statistical relationship of the parameter under study with a set of factors affecting it. Statistical relationship is established on the basis of generalization of available experience, in our case for all time of field development. The main task of statistical modeling is the processing of the accumulated data, for example, on the volume of recoverable oil, for the past period and their further extrapolation [14].

All statistical methods used to predict oil production are divided into three types:

- forecasting using the volumetric method;
- forecasting based on displacement characteristics;
- forecasting based on patterns resulting from analysis of actual oil production data.

Volumetric methods are mainly used to predict watering and current oil recovery of reservoirs. When using this method, the volume of the reservoir covered by water flood is determined in accordance with the research on the position and movement of the water–oil contact front. If we know the oil reserves in the volume of reservoir covered by water flood, we able to calculate the current ORF. If we take into account depletion of reserves at the object under study, we can give prognosis for further growth of both oil recovery and watering of oil deposit reservoir. The essence of this method consists in the assumption that dynamics of water shut-off compositions depends on water flooded formation volume. Brykina and Gattenberger have achieved great success in the application of volumetric method, using the graphical dependence of the current oil recovery on the reserves recovery degree and the water flood zone volume [15]. Surguchev has developed and actively applied the isochronous method of water flooding, in which the construction of graphical dependences of reservoir coverage factor and oil recovery on dimensionless time was carried out [16]. It should be noted, that accuracy of these methods is not high, as it mainly depends on correct determination of water flood volumes and areas. Thus, prediction of development indicators by means of volumetric methods is rather labor-intensive, often giving inadequate results.

Forecasting based on displacement characteristics uses the dependence of some technological development parameters on other parameters. Such parameters usually include accumulated (V_l, V_o, V_w) or current (Q_l, Q_o, Q_w) production of liquid, oil or water, as well as residual or initial recoverable oil reserves. The purpose of this method is to obtain all kinds of approximating dependencies for the subsequent prediction. There are a lot of displacement characteristics, they differ only in technological indicators taken as a basis and dependence between them. All characteristics on the last section of the graph are approximated by a straight line and then extended, assuming that this straightness will be preserved.

There are several groups of methods in prediction based on displacement characteristics. Let us consider in detail two main groups, which determined the further development of these methods. The first group of methods is based on the analysis of dependencies determining the relationship of water–oil ratio (WOR) with the accumulated withdrawals of formation fluids (V_l, V_o, V_w). The basic displacement curves of this group were proposed by Nazarov and Sipachev in 1972 and by Sipachev and Pasevich in 1980 [17].

$$\text{Nazarov} - \text{Sipachev method}: \frac{V_l}{V_o} = aV_w + b.$$

$$\text{Sipachev} - \text{Pasievich method}: \frac{V_l}{V_o} = aV_l + b.$$

Here a and b—parametric coefficients, definitions of which are shown in Fig. 1. It should be taken into account that these dependences approximate to the linear form only on the final section, so to correctly determine the approximation coefficients, the values lying exactly on the final section are chosen. The coefficients are determined by the least-squares method.

These methods describe a direct correlation between the growth of WOR and the growth of water production or recoverable liquid. The longer and more stable the production at a particular site and the greater the accumulated WOR, the more relevant is the use of these methods. For example, the methods of this group show rather good results in case of strong water flooding or active inflow of bottom waters. However, the same approach gives incorrect results when studying exploited objects with low-active waters or when keeping the watercut level at low values. Application of these methods is also impossible when the WOR is reduced. Further, modifications of methods of this group were developed, in which parametric coefficients are determined in different coordinates, which allows to obtain more exact values. Methods of the first group allow predicting maximum recoverable oil reserves at

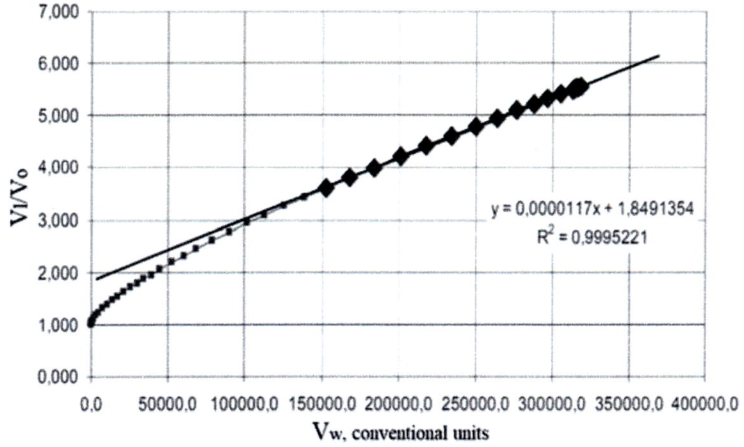

Fig. 1 Determination of coefficients “a” and “b” by the Nazarov-Sipachev method [18]

infinite filtration, as well as recoverable oil reserves for given values of oil content, water content and WOR.

The second group of statistical methods of forecasting of development indicators includes the methods of Maksimov [19] and Sazonov [20]. Their approaches can be applied at earlier stages of production, and the results obtained are practically unaffected by possible corrections in the course of operation. Maksimov studied the process of oil displacement by water from the productive formation. He used a pipe with sand as a model. As a result of experiments the empirical formula for dependence of accumulated water production V_w on accumulated oil production V_o was obtained. The dependence of accumulated fluid production V_l on accumulated oil production V_o was used by Sazonov around the same time.

Maksimov's method:

$$V_w = \beta\alpha^{V_o}, lnV_w = aV_o + b.$$

Sazonov's method:

$$V_l = \beta\alpha^{V_o}, lnV_l = aV_o + b.$$

Here β and α—empirical coefficients, $a = ln\alpha, b = ln\beta$. The drawing is performed in semi-logarithmic coordinates $Y = lnV_l$, $X = V_o$. Further analysis of dependence is carried out in the same way as for the methods of the first group. A comparison of these methods is shown in Fig. 2.

It is noted that the Sazonov method in some cases turns out to be more stable to changes in the process of field development. These methods do not allow to determine the maximum recoverable oil reserves at infinite filtration, but allow to predict the

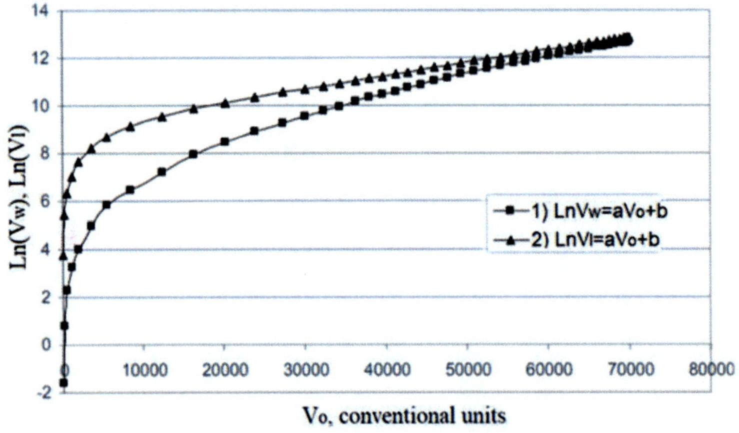

Fig. 2 Comparison of the dynamics of reservoir fluid displacement development using the methods of the second group: Maksimov's method and Sazonov's method [18]

recoverable oil reserves for the given values of oil content or watercut. It is noted that application of methods of the second group gives erroneous results for objects at late stages of operation when implementing methods of oil recovery enhancement.

The main difficulty in application of different methods on the basis of displacement characteristics is substantiation of "base period"—the time by which the coefficients are selected. To summarize, this approach has been applied both in studying the field operation without using the waterflood system [17, 19] and with its application [20].

Now let's consider **forecasting based on patterns resulting from analysis of actual oil production data**. These patterns are used to predict oil recovery and water encroachment processes in fields that have similar geological and physical properties to those under analysis. Among such methods, we can distinguish those that use regression analysis approach. The essence of this approach is to determine the relationship between oil recovery and reservoir properties, as well as some technological parameters of development [21]. Some methods are based on the law of "equal assumptions", which in the first half of the twentieth century was proposed by Bill and Lewis, and a little later by Charnotsky. According to this law for the majority of wells the logarithm of oil flow rate in relation to the initial one changes in proportion to time. Subsequently, the substantiation of this law was presented by Leibenzon. According to his formulation, if two wells have the same flow rate during three years, their flow rates will coincide in the future.

This category of methods also includes methods, which are based on calculation of production decline curves [22, 23]. The essence of the given methods is the supposition, that the average oil production decline of a certain well is rather regular, that is why the curve can be extrapolated for the production forecast. The founder of this approach is Arps, who in 1945 suggested using exponential, hyperbolic and harmonic curves (Fig. 3) to describe the curvature of the rate dependence on time [24].

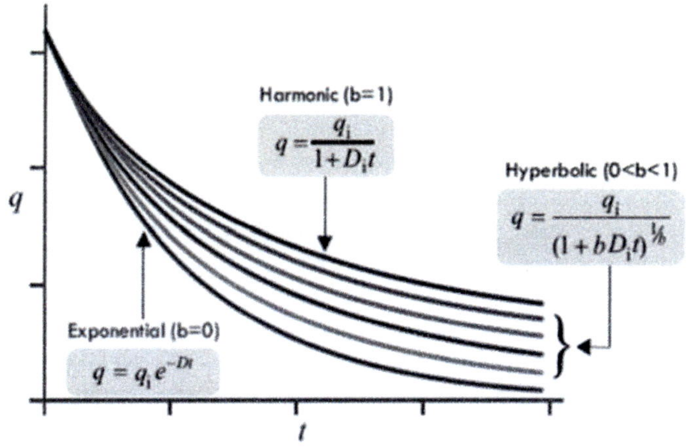

Fig. 3 Three types of Arps production decline: exponential, harmonic and hyperbolic [25]

In general terms, the Arps production decline curves are described by the equation:

$$q_t = \frac{q_i}{(1 + bD_i t)^{1\backslash b}},$$

where q_t—the flow rate at time t (m^3/t), q_i—the initial flow rate (m^3/t), b—the Arps constant for the curve, D_i—the nominal production decline rate (1/t), t—time.

By extending the Arps decline curves, it is possible to estimate the maximum recoverable oil reserves. However, the Arps method is valid only for a pseudo-stable production process, which is observed during long-term stable well operation. Changes in rock filtration properties and oil saturation, for example, can affect the stability of a well. This empirical method is quite easy to use and does not require large amounts of data to be processed for its use and subsequent analysis of the results.

Improvement of Arps method was performed by Fetkovich in 1980 [23]. He showed that the analytical equations for the transient phase of fluid flow in the reservoir and the equations of the pseudo-stable decline curve can be combined and presented on a single logarithmic graph.

Methods based on the construction and analysis of Fetkovich and Arps decline curves are based on the assumption that production is carried out with constant reservoir pressure. In practice, such situations are practically uncommon. Therefore, modern methods take into account changes in reservoir pressure and well flow rate over time.

3 Numerical Modelling Methods

When using numerical methods, the initial data of the problem and the resulting solutions are presented as a set of numbers. Numerical methods are used when it is difficult enough to solve the problem analytically due to the absence of simple explicit dependencies for the system under study. The founders of these methods are Peaceman and Rachford [26].

Application of numerical methods for solving problems of oil and gas industry was started in the middle of the twentieth century. In the first works, the use of these methods made it possible to calculate differential equations of one-dimensional one-phase filtration [27]. This was facilitated by the emergence of a simplified "sand-clay" rock bed model ("reservoir-unreservoir") in the 60 s of the twentieth century. The development of simulation of filtration processes in the reservoir formation was also promoted by the work of Masket [28], in which the formation was represented as a set of current tubes for the first time. Improvement of this representation later served as a basis for the Backley–Leverett model (BL-model), in which capillary effects are not taken into account, as well as the Masket–Leverett model (ML-model), in which the capillary forces are taken into account. The first of these models is applied to near-wellbore zones, and the second—to the main part of the productive formation.

In the mid-1970s, attempts began to describe two-dimensional single- and multiphase filtration. In 1973 Richardson and Stone [29] published a paper in which they presented a detailed analysis of articles devoted to multiphase filtration modelling.

An enormous contribution to two-dimensional modelling of single-phase filtration was made by Peaceman, Vakhitov, Rachford and Carter. The works of Maksimov, Rybitskaya, Katz, and Andriasov [19, 30] are devoted to solving full-scale three-dimensional models. In 1970 Martin obtained the relations in which the vertical capillary-gravity equilibrium in the reservoir was taken into account. The dependences he derived were determined by the initial position of the boundaries between the fluids and the relative permeability's of rocks.

In 1974 methods were finally developed in which vertical distribution is determined mainly by viscous frictional forces rather than gravity-capillary forces. Fractured reservoirs were modelled by Kazemi, Morse and Kleppe. In their works, the fracture reservoir system is represented as a continuous medium, and fluid motion in the fractures was determined by the material balance equation [31, 32].

In the last decade of the twentieth century, stochastic methods of reservoir modelling began to be used predominantly in oil and gas science. If there are no random values among model parameters, it is called deterministic, and if there are, it is called stochastic. Methods of stochastic modelling are mainly used in geological modelling to create the basis of hydrodynamic models, which are deterministic. The main advantage of the hydrodynamic model based on stochastic modelling is the correct representation of the geological structure. The main disadvantage of such models can be called their variability. Most scientists and engineers use stochastic models because they lead to more realistic design metrics.

The modern stage in the development of numerical methods is associated with the development of specialized software that allows calculations to be made in the shortest possible time with minimal resource costs. The latest software also makes it possible to combine information from a variety of data sources and methods into a single model. At the beginning of the twenty-first century, parallel computing on multiprocessor computers was developed, which accelerated the development of numerical methods and allowed to solve large-scale problems.

4 Conclusion

This article describes the main stages of development of analytical, statistical and numerical methods for studying processes in oil and gas reservoirs. The importance of direct physical modeling of geomechanical processes is pointed out. To date, statistical and numerical modeling techniques have been used with the greatest success.

Statistical methods based on the construction and analysis of displacement curves and decline curves are widely used for prediction of production indicators and estimation of reserves. Compared to decline curve analysis, displacement curve analysis gives better results when predicting production in watered oil reservoirs because, in addition to oil production, they also account for fluid and water production. However, both methods are deterministic. They do not allow for uncertainty in the calculations, which can lead to failure to achieve planned production rates. To solve the problem of quantifying uncertainty and assessing risks in production forecasts and final estimates of recoverable reserves, foreign scientists are beginning to apply machine learning methods to traditional methods of forecasting oil and gas production.

The modern stage in the development of numerical methods is associated with the development of specialized software that allows calculations to be made in the shortest possible time with minimal resource costs. The latest software also makes it possible to combine information from a variety of data sources and methods into a single model. Over the last decades, parallel computing on multiprocessor computers has developed, which has accelerated the development of numerical methods and allowed to solve large-scale problems.

Acknowledgements The work was performed within the framework of the state assignment, registration number AAAA-A20-120011690133-1.

References

1. Vasiliev K. K. Mathematical modeling of communication systems: textbook / K. K. Vasiliev, M. N. Sluzhivyi. – 2-edition, revised. and supplementary – Ulyanovsk: ULSTU, 2010. 170 p.
2. Karev V., Kovalenko Y., Ustinov K., 2020. Advances in Oil and Gas Exploration and Production. Switzerland: Springer International Publishing. 2020. 166 p.
3. Gultyaev A.V. Visual modeling in MATLAB environment: training course / A.V. Gultyaev. SPb. 2000. 432 p.
4. Darcy H. Recherches experimentales relatives au mouvement de l'eau dans les tuyaux. Paris. 1857.
5. Dupuit J. Essai et experiences sur le tirage de voitures et sur le frottement de second espece. Paris: Carilian – Goeury. 1837.
6. Zhukovsky N.E. Water infiltration through dams. Experimental-meliorative. Part. Moscow, NKZ, part 3. collection of works, V. 7, M., Gostoptekhizdat, 1956.
7. Barenblatt G.I., Zheltov Yu.P., Kochina I.N. About Basic Representations of Filtration Theory in Cracked Media. – Appl. Math. and Mechanics, V. 24 (5). 1960. pp. 58–73.
8. Pavlovsky N.N. Hydraulic Handbook, 1936. 890 p.
9. Leibenzon L.S. The motion of natural fluids and gases in a porous medium. OGIZ. State Publishing House of Technical and Theoretical Literature, Moscow, 1947. 244p.
10. Carman P.C. Flow of gases in porous media. London, 1956. 182 p.
11. Kozeny, J. Ueber kapillare Leitung des Wassers im Boden. Sitzungsber Akad. Wiss., Wien, 136(2a), 1927. pp. 271–306.
12. Adzumi H. Bull. Chem. Soc. Japan. 12. 304. 1937.
13. Polubarinova-Kochina P.Ya. About inclined and horizontal wells of finite length // Applied mathematics and mechanics. V.20 (1). 1956. p.95–108.
14. Dementev L.F. Statistical methods of processing and analysis of field geological data. – Moscow: Nedra, 1966. 208 p.
15. Gattenberger Yu.P., Dyakonov V.P. Hydrogeological methods of research in exploration and development of oil fields. Moscow: Nedra, 1979. 207 p.
16. Surguchev M.L. Methods of control and regulation of the process of development of oil fields. Moscow: Nedra, 1968. 301 p.

17. Nazarov S.N., Sipachev N.V. Technique for forecasting technological indicators at the late stage of oil reservoir development // Oil and Gas. 10. 1972. p. 41–45.
18. Saveliev V.A., Tokarev M.A., Chinarov A.S. Geological and field methods of oil recovery forecasting: Textbook // V. Saveliev, M. Tokarev, A. Chinarov. Izhevsk: Publishing house "Udmurt University". 2008. 147 p.
19. Maksimov M.M., Rybitskaya L.P. Mathematical modeling of processes of development of oil fields. Moscow: Nedra, 1976. 64 p.
20. Sazonov B.F. Improvement of oil field development technology under water pressure regime. Moscow: Nedra, 1973. 283 p.
21. Shavaliyev A.M. Statistical study of oil recovery and water-oil factor in the fields of the Ural-Volga region // Oil Economy. 12. 1981. p. 30–32.
22. Fetkovich M.J. The Isochronal Testing of Oil Wells // Fall Meeting of the Society of Petroleum Engineers of AIME, Las Vegas, Nevada, 1973. p. 24. SPE 4529. DOI: https://doi.org/10.2118/4529-MS.
23. Fetkovich M.J. Decline Curve Analysis Using Type Curves // J. Pet Technol. V.32. 1980. pp. 1065–1077. DOI: https://doi.org/10.2118/4629-PA.
24. Arps J.J. Analysis of Decline Curves // Trans. AIME, V. 160. 1945. pp. 228–247.
25. Tung L.V. Analysis of the state of oil production in the field "X" Vietnam: master's thesis. National Research Tomsk Polytechnic University, Tomsk, 2018.
26. Peaceman D.W., Rachford, H.H., Numerical calculation of multidimensional miscible displacement. Trans. SPE of AIME, 225. 1962. pp. 327–39 (SPEJ).
27. Smith G. D. Numerical Solution of Partial Differential Equations with Exercises and Worked Solutions, Oxford University Press, New York. 1965.
28. Masket M. The flow of a homogeneous fluid through a porous medium // translated by M.A. Geiman. M.: Gostekhizdat, 1949. 628 p.
29. Richardson J.G., Stone H.L. A quarter century of progress in the application of reservoir engineering, J. Petrol. Technol., 25, 1973. pp. 1371–9.
30. Katz P.M., Andriasov A.R. Mathematical model of three-phase filtration in fractured-porous medium // Collection of scientific papers of VNII. V. 95. 1986. pp. 61–66.
31. Kazemi H. Pressure transient analysis of naturally fractured reservoir with uniform fracture distribution. Soc. of Petroleum Engineers Journal, 1969. pp. 451 – 426.
32. Kleppe J., Morse R.A. Oil production from fractured reservoirs by water displacement, SPE 5084, 49/A Annual Fall Meeting, Houston. 1974.

An Angle Fall of a Drop Onto a Deep Water

A. N. Zotova, Y. I. Troitskaya, and D. A. Sergeev

Abstract A direct numerical simulation of oblique fall of a drop is carried out in Basilisk software package for process parameters corresponding to second regime of drop impact from the work (Reijers et al. in Oblique droplet impact onto a deep liquid pool, 2019) (see (Reijers et al. in Oblique droplet impact onto a deep liquid pool, 2019) Fig. 2b, e). The dynamics of the drop and water surface is obtained, which corresponds to results of (Reijers et al. in Oblique droplet impact onto a deep liquid pool, 2019). Size distributions of secondary droplets are constructed for various maximum mesh refinement levels. To check reliability of the obtained distribution, further calculations with higher mesh resolution are required.

Keywords Drop fall · Direct numerical simulation

1 Introduction

Processes of sea spray formation play important role in interaction of ocean and atmosphere under strong winds. Among other factors, formation of small droplets is caused by falling of previously generated larger drops onto the sea surface. The liquid drop impact onto various types of surfaces has studied experimentally and theoretically [1–8]. In a significant part of the works, the vertical fall of a drop has considered [1–6]. The authors of [7] have studied the oblique fall of a drop onto the surface of a liquid experimentally and have found three regimes of the processes occurring in this case depending on the angle of the drop incidence and the Weber and Reynolds numbers: in the first case, the drop simply merges with the liquid on whose surface it fell, in the second—in the direction of falling of the drop a corona is formed, from which secondary drops are subsequently separated; in the third case, secondary drops are ejected from the corona in all directions. The authors of [8], in agreement with experiment [7], also have described three types of the regime of droplet fall onto the deep water, obtained by numerical simulation, but in this work,

A. N. Zotova (✉) · Y. I. Troitskaya · D. A. Sergeev
Institute of Applied Physics of the RAS, Nizhny Novgorod, Russian Federation
e-mail: anzotova@ipfran.ru

V. I. Karev (ed.), *Physical and Mathematical Modeling of Earth and Environment Processes—2022*, Springer Proceedings in Earth and Environmental Sciences,
https://doi.org/10.1007/978-3-031-25962-3_24

the parameters of secondary droplets resulting from the fall of the primary droplet have not studied. In addition, in the simulation in [8], to save computation time, a symmetry plane has used that divides the falling drop in half, which, according to the authors, can impose restrictions on the development of the Rayleigh–Taylor instability at the edge of the rim of the ejecta sheet. Therefore, these calculations probably cannot provide reliable information about the statistics of secondary drops, since the Rayleigh–Taylor instability can play significant role in their formation. In our work, we carry out direct numerical simulation of the oblique fall of a drop in the Basilisk software package [9], not using a mirror plane, but considering the drop as a whole. The main goal of our work is to obtain the size distribution of secondary drops resulting from the fall of the primary drop, corresponding to the second regime from the work [8] (see [8] Fig. 2b, e).

2 Simulation of Angle Fall of a Drop

The Basilisk software package is used for direct numerical simulation of the oblique fall of drop onto the water surface. Basilisk implements a numerical algorithm that solves the Navier–Stokes equations for incompressible media with variable density and surface tension and combines an adaptive quad/octree spatial discretization and geometrical Volume-Of-Fluid interface representation [10].

We consider the problem of the following configuration (Fig. 1): the cubic domain with size $L_0 = 6$ cm is half filled with water, and the upper half contains air. Drop with a diameter of 2 mm is at a height $h = 2$ cm above the water surface at the initial moment of time. The horizontal projection of the drop velocity is $v_{0x} = 3$ m/s, the vertical projection of the drop velocity $v_{0y} = -1.25$ m/s is obtained from the condition that the vertical component of the drop velocity is equal to zero at a height 10 cm above the water surface. The movement of the drop occur under the action of free fall acceleration. To save computation time, we set periodic boundary conditions along the x coordinate. The coordinates and volumes of each individual water body (each drop) are recorded in time.

For the given initial configuration of the problem, it is obtained that when the drop approached the water surface, its angle of incidence is $\alpha = 65°$, the Weber and Reynolds numbers are, respectively, We = 303, Re = 6600. Taking the splashing number K equal to 104 in accordance with the work [8], we calculate the dimensionless splashing parameter $\mathrm{We}^{1/2}\mathrm{Re}^{1/4}/\mathrm{K} = 1.5$. Thus, our parameters correspond to the second regime of impact behavior of the drop (see Fig. 6 in [8]).

The adaptive mesh refinement method is implemented in used software package—the size of the discretization cell can vary in different areas of the domain depending on the specified parameters. The minimum cell size is set by the maximum level of refinement—MAXLEVEL. The linear cell size corresponding to the maximum level of refinement can be obtained by dividing the domain size L_0 by 2 to a power equal to the value of MAXLEVEL. For example, one of the MAXLEVEL values used in our work is 11, for our domain size $L_0 = 0.06$ m we obtain the minimum linear cell

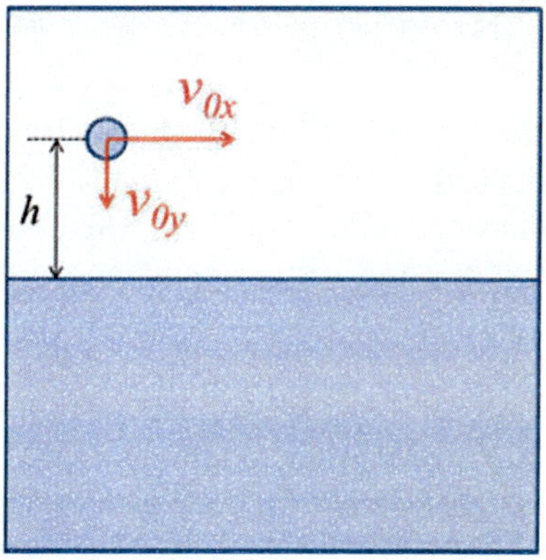

Fig. 1 The configuration of the considered problem: drop is at a height $h = 2$ cm above the water surface, the horizontal projection of the drop velocity is $v_{0x} = 3$ m/s, the vertical projection of the drop velocity is $v_{0y} = -1.25$ m/s

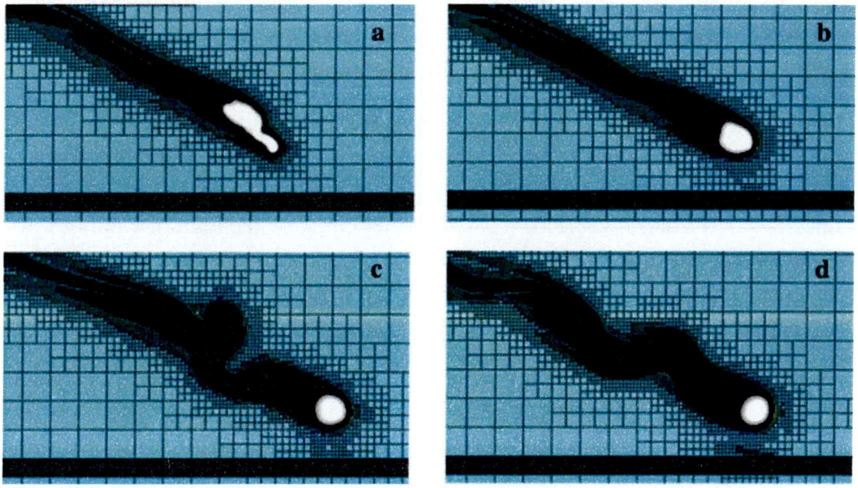

Fig. 2 Simulation of the fall of drop on the water surface: t = 0.013c—drop near the surface (**a**—MAXLEVEL = 10, **b**—MAXLEVEL = 11, **c**—MAXLEVEL = 12, **d**—MAXLEVEL = 13)

size $\Delta l = 0.06/2^{11} = 2.9 \times 10^{-5}$ m. By analogy for MAXLEVEL = 12 minimum linear cell size is $\Delta l = 0.06/2^{12} = 1.5 \times 10^{-5}$ m.

We estimate the minimum resolvable radius $R_{\min}$ as 5 minimum linear dimensions of the cell taking into account the minimum linear size of the cell for each case. Accordingly, the minimum resolvable droplet volume $V_{\min}$ for the case of MAXLEVEL = 11 is 1.4×10^{-11} m^3, for the case of MAXLEVEL = 12 it is 1.8×10^{-12} m^3.

3 Results

The dynamics of the interaction between the fallen drop and the water surface, obtained in our numerical simulation, corresponds to the second type of drop impact behavior described in [7, 8], which is expected based on the values of the parameters corresponding to our formulation of the problem.

In order to determine maximum level of refinement for our problem that we need, we perform calculations with different values of the MAXLEVEL parameter. Figure 2a shows the appearance of flying drop before falling onto the water surface obtained during the simulation with MAXLEVEL = 10. It can be seen that the drop is strongly distorted in flight, which does not correspond to experimental observations. For MAXLEVEL = 11 (see Fig. 2b) the shape of the drop is much closer to that observed in experiments (we compare the appearance of the drop at the same time t = 0.013 s). To understand whether the mesh resolution introduces significant distortions into the simulation result in this case, let's compare this result with the result for an even finer mesh. Figure 2c and d show the simulation results with the maximum levels of refinement MAXLEVEL = 12 and 13. It can be seen that the droplets for the cases MAXLEVEL = 11 and 12 do not differ so much from each other, but differ greatly from the case MAXLEVEL = 10. For the cases (c) and (d) the differences are also small, therefore, in order to understand what value of the MAXLEVEL parameter—11, 12 or 13 we need, we will further compare the size distributions of secondary drops (see Fig. 3) for these cases.

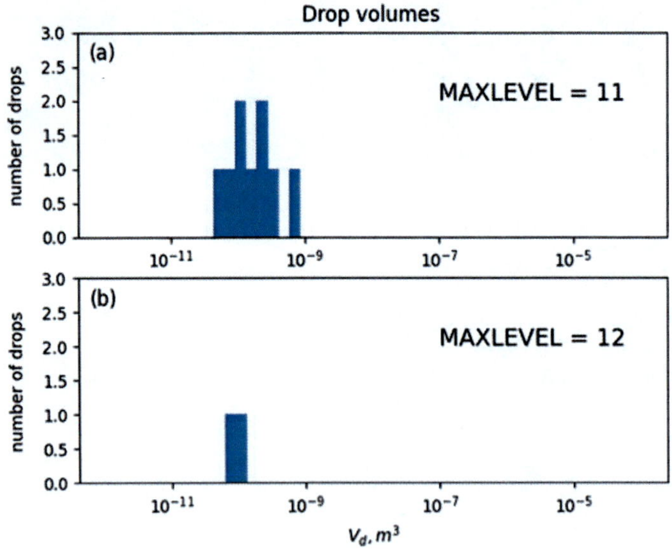

Fig. 3 Volume distribution of secondary drops at the first of the time moments when the number of drops is maximum: **a**—MAXLEVEL = 11, t = 0.022 s, **b**—MAXLEVEL = 12, t = 0.019 s

The size distributions of the secondary droplets formed after the fall of the primary drop are plotted for the cases of MAXLEVEL = 11 and 12 (for the case of MAXLEVEL = 13 simulation is still ongoing) (see Fig. 3). Each distribution is taken at the first of the time moments for which the number of secondary drops is maximum: for MAXLEVEL = 11—t = 0.022 s, for MAXLEVEL = 12—t = 0.019 s. It can be seen that the distributions are very different; in the case of MAXLEVEL = 12 only two drops are formed.

Thus, at the moment, the results of our simulation qualitatively reproduce the primary droplet flying in the air, but to obtain the size distribution of secondary droplets, it is necessary to continue the calculations (currently, the calculation is carried out for MAXLEVEL = 13) and compare the distributions for higher values of the MAXLEVEL parameter.

4 Conclusion

We carried out numerical simulation of the oblique fall of the water drop on the water surface at various maximum discretization levels of the grid—MAXLEVEL. The dynamics of the interaction between the fallen drop and the water surface, obtained in our numerical simulation, corresponded to the second type of drop impact behavior described in the literature, which was expected based on the values of the parameters corresponding to our formulation of the problem. Based on the comparison of the resulting droplet shape during its fall in air for different grid resolutions, the minimum grid discretization parameter MAXLEVEL, necessary for solving our problem, was found. Also, for two values of the MAXLEVEL parameter, size distributions of secondary droplets were constructed. The obtained distributions differ significantly from each other, therefore, in order to obtain a reliable result, we need to continue our calculations, successively increasing the resolution until the size distribution of secondary drops ceases to change.

Acknowledgements Work on direct numerical simulation was supported by the RSF project No. 19-17-00209, statistical processing of simulation results was supported by the RSF project No. 21-19-00755.

References

1. *Yarin, A. L.*, 2006. Drop impact dynamics: splashing, spreading, receding, bouncing. *Annual review of fluid mechanics*, *38*(1), 159–192.
2. Thoraval, M. J., Takehara, K., Etoh, T. G., Popinet, S., Ray, P., Josserand, C., Zaleski, S. & Thoroddsen, S. T., 2012. von Kármán vortex street within an impacting drop. Physical review letters 108(26): 264506.

3. Agbaglah, G., Thoraval, M. J., Thoroddsen, S. T., Zhang, L. V., Fezzaa, K., & Deegan, R. D. 2015. Drop impact into a deep pool: vortex shedding and jet formation. Journal of fluid mechanics 764: R1.
4. *Fedorchenko, A. I., & Wang, A. B.*, 2004. On some common features of drop impact on liquid surfaces. *Physics of Fluids*, *16*(5), 1349–1365.
5. *Singh, A., & Kumar, P.*, 2022. Droplet impact dynamics onto a deep liquid pool of wavy free surface. *Physics of Fluids*, *34*(2), 022107.
6. *Li, Z., Liu, C., & Wan, D.*, 2020. Numerical Simulations of Droplet Impact onto a Pool Surface. In *The 30th International Ocean and Polar Engineering Conference*. OnePetro.
7. Gielen, M. V., Sleutel, P., Benschop, J., Riepen, M., Voronina, V., Visser, C. W., Lohse, D., Snoeijer, J. H., Versluis, M., & Gelderblom, H., 2017. Oblique drop impact onto a deep liquid pool. Physical review fluids, 2(8), 083602.
8. *Reijers, S. A., Liu, B., Lohse, D., & Gelderblom, H.*, 2019. Oblique droplet impact onto a deep liquid pool. *arXiv preprint* arXiv:1903.08978.
9. *Popinet, S.* The Basilisk code: http://basilisk.fr.
10. *Popinet, S.*, 2009. An accurate adaptive solver for surface-tension-driven interfacial flows. *Journal of Computational Physics*, *228*(16), 5838–5866.

Investigation of the Pancake Ice Influence on the Wind–Wave Interaction Within Laboratory Modeling

D. A. Sergeev, A. A. Kandaurov, and Yu. I. Troitskaya

Abstract Presented work concerned on the investigation of the ice cover influence on the wind–wave interaction in the marine atmospheric boundary layer. Unique laboratory experiments on the modeling of influence of pancake type of forming floating ice were carried out on the ring wind-wave facility AEOLOTRON University of Heidelberg. The round rubber pucks were used as artificial ice floes. Experiments were carried out for wide range of wind–wave states and ice concentration. Simultaneous measurements of surface elevation, air flow parameters, and are of artificial floe ice coverage was carried out. For the case of the ice presence, the evolution of the surface had threshold behavior. The obtained threshold of the excitation of long waves (the length is much greater than the average size of the ice elements and the distance between them), depended on the wind speed and ice concentration.

Keywords Pancake ice · Wind · Waves · Laboratory modeling

1 Introduction

In the last decade, the development of the Arctic zone of the world's oceans has become increasingly important in terms of offshore mining (primarily oil and gas), as well as maritime transport (the northern sea route). Just for the polar regions that the greatest difference is noted between the results of forecasting models of the meteorological conditions and wave state with observational data. The most noticeable difference is detected in the regions of the near-marginal ice zone [1]. Therefore, numerical models are being developed that use parametric dependences of momentum, heat, and mass fluxes as boundary conditions on the sea surface, taking into account the influence of ice cover. For example, in the current version of the WAVEWATCH III model it is possible to take into account the influence of ice by connecting parameterizations describing the attenuation of waves under its action (IC) (see [2]), and for modeling atmospheric marine boundary layer in

D. A. Sergeev · A. A. Kandaurov (✉) · Yu. I. Troitskaya
Federal Research Center, Institute of Applied Physics, Russian Academy of Sciences, Nizhny Novgorod, Russia
e-mail: green.pb@gmail.com

V. I. Karev (ed.), *Physical and Mathematical Modeling of Earth and Environment Processes—2022*, Springer Proceedings in Earth and Environmental Sciences,
https://doi.org/10.1007/978-3-031-25962-3_25

polar conditions, a modification of the atmospheric circulation model was developed WRF-ARW: Polar-WRF (see [3]). The parameterizations of atmosphere–sea surface interaction used in these models are based mainly on empirical data, the quality of which ultimately affects the quality of the forecast.

Detailed studies of the effect of the ice influence on the processes of wind–wave interaction in natural conditions are very difficult, especially for the forming and floating types of the ice, including pancake (this type of ice is very common and studied thoroughly, see, for example, [4, 5]). A good alternative here is carrying out laboratory experiments. Previously, only studies of the influence of floating ice on the evolution of the mechanically induced waves (by wavemaker) were performed i.e., experiments were carried out on straight flumes of limited fetches (see [6–9]). In the works [6, 7] grease ice was investigated, and in the work [8] large scale floes 0.5–6 m. Both experiments were carried out on the special ice tanks. In the work [9] artificial floes were used. The most advanced research in this area concerned the modeling of ice breaking processes by waves under various conditions [10].

However, basing on these results there is no possibility taking in to account the floating ice influence on the transfer between atmosphere and ocean. First of all, it concerns a surface rough-ness and momentum fluxes, i.e., in the present models there were no options to modify drag coefficient according to the parameters of the floating ice covering (concentration, typical sizes etc.). Now a-days, only for areas of solid ice (or at least coarse ice) a difference in the turbulent wind stress from the clean water conditions can be considered. For the areas covered by floating ice the drag coefficient is assumed to be close for the clean water surface. That is why it is not taking into account in the boundary conditions in atmospheric models and in the wind induced sources in wave models.

The main problems here concerned with carrying out simultaneous in situ measurements in the system wind-ice-waves and providing laboratory modeling of this system in the laboratory conditions both.

2 Description of Experiment and Methods

This paper presents preliminary results of the laboratory experiments carried out for wind-wave interaction with modeling of pancake ice on the AELOTRON ring wind-wave flume of the University of Heidelberg (see Fig. 1). This facility with inner diameter of 10 m, air flow cross section of 0.6 × 1.4 m, the depth of the water is 1 m. The maximum allowed equivalent wind speeds U_{10} (approximated on the 10-m height according logarithmic law) is 17 m/s. The typical pancake ice floes have a round form with typical width and thickness up to 1 m and 15 cm correspondingly. For the first time, the interaction of the wind with the water surface in the presence of artificial ice elements was simulated, which were rubber round pucks (see Fig. 2) with a diameter of 7 cm and a thickness of 1 cm with a density of about 0.8 kg/m^3. It was impossible to use contact wire wavegauges under the conditions of solid elements of ice permanently presented on the water surface.

Fig. 1 Overall view of the ring wind–wave AEOLOTRON facility

Therefore, measurements of the elevation of the rough surface were performed using a laser wavegauge with a frequency of up to 100 Hz. At the same time, in the time records, the intervals corresponding to the passage of through the laser beam of the wave gauge were filtered and approximated. The part of surface covered with artificial ice were measured simultaneously with water elevation. A special system with shadow imaging of the water surface covered with pucks was used. The image of the top view with under water LED illumination was taken by camera which was processed with special algorithms, to detect area covered by pucks.

The measurements were carried out on the clean water and at three concentrations of artificial ice: maximal, 2/3 of the maximal, 1/3 of the maximal. At the maximum concentration, artificial ice elements covered approximately half of the water surface in the flume. The measurements were carried out for the fan frequency from 16 to 28 Hz, which corresponded to the range of equivalent wind speeds U_{10} from 7 to 16 m/s. The procedure was as follows: at first, at the minimum allowable air flow rate (fan frequency 6 Hz), uniform seeding with artificial ice was performed. Then the wind speed was increased sharply to the desired value.

3 Results and Discussion

For the whole range of allowed wind speeds a monotonic evolution to a stationary state of wavy surface was observed (the higher the wind speed, the more the time of evolution, from 1 to 6 min and longer waves were observed) on the clean water.

Fig. 2 View inside the flume on the water surface covered by artificial pancake ice

In contrast to this for the case of the ice presence, the evolution of the surface had threshold behavior. The obtained threshold of the excitation of long waves (the length is much greater than the average size of the ice elements and the distance between them), depended on the wind speed and ice concentration.

On the Fig. 2 two At all ice concentrations for low wind rates, the development of long waves was not observed for an arbitrarily long wait. When the value of the set wind speed exceeded the threshold value, at first the general wave pattern was similar for light wind conditions, however, after a certain time, long waves spontaneously developed and the situation approached clear water conditions (the wave parameters turned out to be close). On the Fig. 3 the dependence of the significant wave height on the wind speed is demonstrated. The higher the density, the higher the wind speed threshold was. At the same time, the waiting time was sometimes calculated in tens of minutes (maximum one hour), and therefore, the threshold level was determined approximately (for the maximum ice concentration—U_{10} between 11 and 13 m/s, for the intermediate between 10 and 12 m/s, for the minimum between 7 and 9 m/s.

A hypothesis of the that some kind of transition between the mode of drift movement of ice disks on the surface and small ripples, to the mode of generation and development of long waves became possible when at some time place and time areas of clean water, the dimensions of which strongly exceeds the sizes of ice floes could be offered. The higher the density, the higher the threshold for wind speed. Its, could be concluded that, the threshold behavior should be taken in to account in the parameterization of wind source for the wave models and parameters of surface roughness in the atmospheric models.

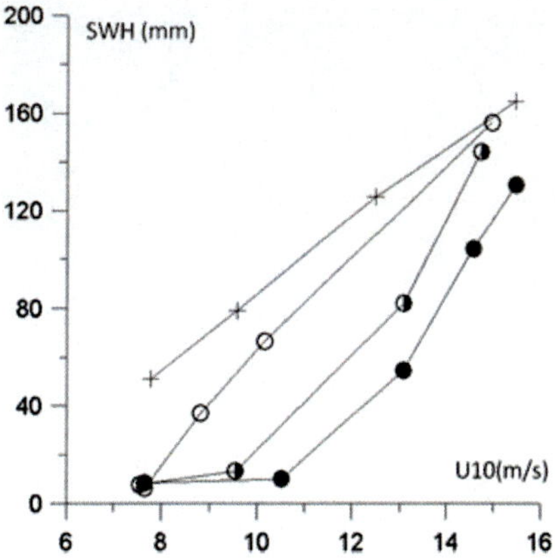

Fig. 3 Significant wave height dependence on the equivalent wind speed U_{10} (on 10-m height). Crosses—clean water, filed circles—maximum ice cover, semi open circles—2/3 ice cover, open circles—1/3 ice cover

Acknowledgements This work was supported by the state agreement № 0030-2021-0007 IAP RAS. The experiments were performed with equipment of the Unique Scientific Facility "Complex of Large-Scale Geophysical Facilities" (http://www.ckp-rf.ru/usu/77738/).

References

1. *Klimov, D.M., Karev, V.I., Kovalenko, Yu. F. & Ustinov, K. B.*, Mechanical-mathematical and experimental modeling of well stability in anisotropic media // *Mech. of Sol.* 2013. 48. 357–363.
2. *Tjernstrom M., Zagar M., Svensson G. et al.* Modelling the arctic boundary layer: An evaluation of six arcmip regional-scale models using data from the Sheba project *Boundary-Layer Meteorology.* 2005.117(2). 337–381
3. *Rogers W. E., Zieger S.* New Wave-Ice Interaction Physics in WAVEWATCH III *Proceedings 22nd IAHR International Symposium on Ice. Singapore. 2014.* http://polarmet.osu.edu/PWRF
4. *Wadhams P.* Ice in the Ocean. CRC Press. 2000. 364 p.
5. *Wilkinson J. P.* Ice dynamics in the central Greenland Sea // *Journal of Geophysical Research.* 2007. 111. C12022.
6. *Martin S., Kauffman P.* A field and laboratory study of wave damping by grease ice // *Journal of Glaciology.* 1981. 27. 283–313.
7. *Newyear K., Martin S.* A comparison of theory and laboratory measurements of wave propagation and attenuation in grease ice // *Journal of Geophysical Research.* 1997 102. 25091–25100.
8. *Herman A., Cheng S., and Shen H. H.* Wave energy attenuation in fields of colliding ice floes – Part 2: A laboratory case study // *The Cryosphere.* 2019. 13. 2901–2914 https://doi.org/10.5194/tc-13-2901-2019
9. *Toffoli A, Pitt J. P. A. Alberello A.and Bennetts L. G.* Modelling attenuation of irregular wave fields by artificial ice floes in the laboratory // *Phil. Trans. R. Soc.* 2022 A.3802021025520210255 https://doi.org/10.1098/rsta.2021.0255
10. *Herman A., Evers K.U., Reimer N.* Floe-size distributions in laboratory ice broken by waves // *The Cryosphere.* 2018. 12. 685–699 https://doi.org/10.5194/tc-12-685-2018

Sequestration of Organic Carbon in Salt Marsh Formations of Lagoons of Sakhalin

Victor V. Afanas'ev and A. B. Faustova

Abstract The oil and gas specialization of the economy, the limited potential of forestry, and the insular position of the Sakhalin Oblast make it very important to assess the role of coastal marine geosystems in the runoff and emission of greenhouse gases (GHGs) to determine their contribution to the overall GHG balance of the territory. The paper deals with the issues of sedimentation in the lagoonal waters of the island. Sakhalin is a coastal wetland with a total area of about 2200 km^2 and a coastline of 2150 km. The change in the area of marches in lagoonal geosystems of various types for the period 1952–2019 is shown. Some mechanisms of the formation of geomorphological forms and sedimentary strata that accumulate a large amount of organic carbon are presented.

Keywords Subarctic coast · Salt marsh · Lagoon sediments · Organic carbon · Peat accumulation

1 Introduction

It is known that bottom sediments and marches of estuarine-lagoonal geosystems are among the most effective components of biosequestration systems on the planet [1–4]. The geochemistry and lithodynamic characteristics of the surface bottom sediments of the Sakhalin lagoons are well studied. The concentrations of organic carbon, depending on the type and location of the lagoon, vary in a very wide range from 0.3 to 5.8% of the dry mass [5, 6]. At the same time, the thickness of watered bottom sediments can reach 4–6 m. Marches are the upper part of tidal silt or silt-sand drylands inhabited by plant communities adapted to periodic salinization [7]. Marching formations on the shores of Sakhalin are represented almost everywhere, in winter, dead vegetation is a grassy mat over the frozen surface of the lagoon

V. V. Afanas'ev (✉)
Institute of Marine Geology and Geophysics, FEB RAS, Yuzhno-Sakhalinsk, Russia
e-mail: vvasand@mail.ru

A. B. Faustova
Sakhalin State University, Yuzhno-Sakhalinsk, Russia

V. I. Karev (ed.), *Physical and Mathematical Modeling of Earth and Environment Processes—2022*, Springer Proceedings in Earth and Environmental Sciences,
https://doi.org/10.1007/978-3-031-25962-3_26

Fig. 1 Lagoon marches of northern Sakhalin

(Fig. 1). However, we still cannot present a complete picture of their cartometric characteristics, morphometric parameters and structure.

And this is despite the fact that the features of sedimentation in march formations of lagoons can be assessed in lagoonal morpholithosystems quite simply and with high accuracy by conventional geological and geophysical methods. Here it seems important to note that marches and internal deltas generally have a two-layer structure. At the base of the lower layer, bottom sediments are presented: rather large in size—if these are internal deltas and finer-grained in marches. Almost everywhere, the upper part of this layer is represented by aleurite-pelitic sediments with a high content of weakly decomposed organic matter of plant origin, mainly the seagrass zostera. The upper layer of marshes and internal deltas is usually represented by weakly decomposed gyttia of sedges, naumburgia, and pondweeds [8].

Considering earlier the cascades of the internal deltas of Sakhalin from the point of view of the sediment budget of the coastal morpholithic system of the lagoon coast, we drew attention to the fact that the tidal internal deltas of the Lunskaya lagoon, which existed before 1952, increased their area over 65 years due to the overgrowth and accumulation of suspended sediments by vegetation [9]. According to updated data, approximately 1.46 times.

Thus, the process of increasing marches in lagoons occurs not only due to the formation of new internal deltas. In this paper, we present an analysis of changes in the areas of marshes in the lagoons of the island Sakhalin.

2 Methods and Results

The methodology of biomorpholithodynamic studies is based on both traditional and modern methods for obtaining and analyzing geospatial and geological-geomorphological information. The work is based on the analysis of arrays of aerial photographs of 1952 and satellite images of 2019, which was performed in the Quantum GIS geoinformation system. The calculations were made on the WGS84 EPSG:7030 ellipse. The resulting attributes were exported to spreadsheets for further processing. The average error in the location of objects on this map is about 5 m, the minimum error is 0.2 m, the maximum error is 18 m, the standard deviation is about 3 m.

At the next stage of the study, in the same QuantumGIS-environment, using standard procedures for analyzing remote sensing materials, we reconstructed the change in march areas for the period 1952–2019.

Figure 2 shows the results of measuring changes in the area of marshes in the delta parts of the rivers flowing into the southern apex of the Chaivo lagoon. The area of marches here increased by 409,841 m^2, or 6117 m^2/year, and the area of marches in the mouth area of the Val and Askasai rivers, located opposite the strait, increased by 651,704 m^2, or 9727 m^2/year. Taking into account the peculiarities of the tidal regime, the data of survey work and drilling, the thickness of the march sediments pushed into the water area of the lagoon is 1.5–1.7 m here. Thus, annually in the march formations of the southern part of the Chaivo lagoon with an area of 24.5 km^2, bounded by the Kleye strait, a sedimentary stratum with a total volume of 23,000–27,000 m^3/year is formed. About 8000 m^3/year of this volume are slightly decomposed plant sediments with a bulk density of 0.19–0.3 g/cm^3. Considering that march gyttia is quite strongly mineralized, the content of organic carbon in it varies within 30–40%.

Thus, when converted to pure carbon, these deposits accumulate approximately 700 tons/year. In the fine-grained facies of marsh deposits, averaging the proportion of organic carbon in the bottom sediments of the Chaivo lagoon to 2%, we obtain an accumulation of pure carbon in the amount of about 500 tons/year.

The total volume of alluvial deposits in the inner part is 5163 m^3/year, and in the mouth area of the Val and Askasai rivers—15,836 m^3/year. But these are mainly suspended sediments or sediments from the open sea, which, to a large extent, must be carried out of the lagoon. Nevertheless, taking into account the peculiarities of lagoonal morpholithogenesis, it can be assumed that about 80% of silty-pelitic sediments remain in the deltaic marshes of lagoonal waters [10]. The share of proper sand-pebble, alluvial traction sediments accounts for only 1384 m^3/year. These deposits have not yet been included in this carbon stock estimate. In general, the decrease in the area of the entire Chaivo lagoon as a result of the formation of marches over the period 1952–2019, amounted to 1.956 km^2. By a simple conversion to pure carbon, it can be established that march gyttia accumulates here approximately 1700 tons/year

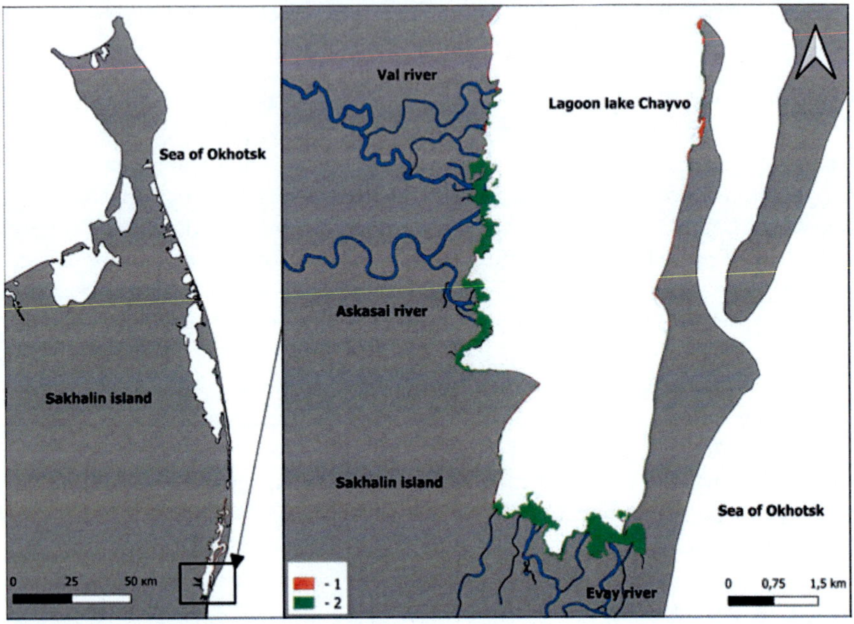

Fig. 2 Change in the water surface area of the Chaivo Lagoon; 1—increase in the area of marches in the delta areas of the rivers flowing into the lagoon

of organic carbon. In the fine-grained facies of marsh deposits, averaging the proportion of organic carbon in the bottom sediments of the Chaivo lagoon to 2%, we obtain an accumulation of pure carbon in the amount of about 2500 tons/year.

Without a doubt, in various lagoons about. Sakhalin runoff properties of marches will vary significantly. So, for example, in the lagoon of the closed type of Lake Ainskoe with an area of 32.2 km^2 of the southern part of the island Sakhalin, the bulk of organic carbon (about 2000 tons/year) is formed and accumulated in bottom sediments. The plant gyttia of the lagoonal marshes accumulates no more than 200 tons/year of pure carbon. Nevertheless, the runoff properties of this lagoon in terms of km^2 are more than one and a half times higher than those of the northern Chaivo lagoon.

3 Conclusion

The proposed model of carbon sink in the march formations of lagoons is based on relatively rough estimates of the carbon parameters of the sedimentation system. It requires development and will be refined. In particular, it is planned to conduct a more detailed study of the bottom sediments of the lagoons, including an analysis of the features of sedimentation in the last phases of the Holocene history. Nevertheless, although the obtained results are very approximate in quantitative terms, they give a clear idea of the features of carbon accumulation by marshes in different types of lagoons. In the future, this will lead to a biohydrogeochemical model of the cycle of matter in estuary-lagoon wetlands of different morphoclimatic zones under different scenarios of the development of the natural environment and technogenic impact. Based on this, it will already be possible to speak about the forecast of changes in the carbon balance in coastal biomorpholithosystems and the creation of artificial zones of accelerated aleurite-pelite and organogenic sedimentation.

References

1. Najjar R.G., Herrmann M., Alexander R., Boyer E.W., Burdige D.J., Butman D. & Zimmerman R.C. (2018). Carbon budget of tidal wetlands, estuaries, and shelf waters of Eastern North America // Global Biogeochemical Cycles, 32(3), P. 389–416.
2. Ouyang X., and Lee S. Updated estimates of carbon accumulation rates in coastal marsh sediments // Biogeosciences 11.18 (2014), P. 5057-5071
3. Spivak A.C., Sanderman J., Bowen J.L., Canuel E.A., Hopkinson C.S. (2019). Global-change controls on soil-carbon accumulation and loss in coastal vegetated ecosystems // Nature Geoscience, 12(9), P. 685–692
4. Villa, Jorge A., and Blanca Bernal. "Carbon sequestration in wetlands, from science to practice: An overview of the biogeochemical process, measurement methods, and policy framework // Ecological Engineering, 114 (2018), P. 115–128
5. Tokarchuk T.N. Geochemistry of Sakhalin lagoons and rational use of their resources: diss. ... cand. geographer. Sciences / Tokarchuk T.N. - Vladivostok: FEGU, 1999. - 135 p.
6. Efanov V.N., Vypryazhkin E.N., Latkovskaya E.M. The current state of bottom sediments of the Busse lagoon (Aniva Bay) // Science and business: ways of development. - 2013. P. 31–36.
7. Zenkovich V.P. Marine geomorphology: Terminological reference book: Coastal zone - processes, concepts, definitions / V. P. Zenkovich. - M.: Thought, 1980. - 280 p.
8. Kafanov V.S., Labai, N.V. Pecheneva Biota and macrobenthic communities of the lagoons of northeastern Sakhalin / A. Yu - Sakh.: SakhNIRO, 2003. - 176 p.
9. Afanasiev V.V., Uba A.V., Levitsky A.I. Migration of Straits and Marine Sedimentation in Lagoons // Geosystems of Transitional Zones. - 2019. - Vol. 3. - No. 3. - P. 310–317.
10. Afanasiev V.V. Morpholithodynamic processes and development of the shores of the contact zone of the subarctic and temperate seas of the North Pacific // Yuzhno-Sakhalinsk: IMGiG FEB RAN. – 2020. 234 p.

Features of Organogenic Sedimentation Within the Coastal Zone of Aniva Bay

V. V. Afanas'ev, A. V. Uba, A. I. Levitsky, and A. B. Faustova

Abstract Many processes are involved in the emission and absorption of greenhouse gases (GHGs), as a result of which the variability of fluxes in various blue carbon systems is very high and cannot always be explained from the standpoint of existing research approaches. Additional large-scale studies are needed to reliably quantify flows and understand the factors causing the variability of GHG flows in coastal marine ecosystems. Including those concerning the issues of determining the geomorphological position of areas with increased intensity of organogenic sedimentation. In this paper, for the first time, the coasts with bench areas on which marches and muddy dries are formed are considered. The change of morphometric parameters and the thickness of sediments of marshes and silty drainages is shown, as they move away from the estuary zone of the river—the source of sediments of the open sea. Some mechanisms of formation of geomorphological forms and sedimentary strata accumulating a large volume of organic carbon are indicated.

Keywords Estuary · Salt marsh · Intertidal mudflat · Bench · Carbon sequestration

1 Introduction

Before the restoration or man-made destruction of coastal ecosystems of blue carbon can be reliably used to mitigate the effects of climate change, it is necessary, along with basic systems of geochemical and biochemical data, to have a clear understanding of the geomorphological situation and conditions of allochthonous inputs of substances that may be involved in the turnover of greenhouse gases.

The research area is located in the northern part of the hall. Aniva, which in geomorphological essence is an estuarine zone of the Tsunai and Susuya rivers. As a rule, the delta systems themselves mainly consist of fresh or oligohaline wetlands

V. V. Afanas'ev (✉) · A. V. Uba · A. I. Levitsky
Institute of Marine Geology and Geophysics, FEB RAS, Yuzhno-Sakhalinsk, Russia
e-mail: vvasand@mail.ru

A. B. Faustova
Sakhalin State University, Yuzhno-Sakhalinsk, Russia

V. I. Karev (ed.), *Physical and Mathematical Modeling of Earth and Environment Processes—2022*, Springer Proceedings in Earth and Environmental Sciences,
https://doi.org/10.1007/978-3-031-25962-3_27

due to the discharge of fresh water through a network of delta channels. River deltas in coastal zones are a potential source of flow, river deltas in coastal zones are a potential source of CH_4 and N_2O flow due to the state of fresh water and high primary production. The fluxes of these gases can potentially compensate either part or all of the benefits from the disposal of carbon (C), nitrogen (N) and phosphorus (P) in these media [1, 2].

The object of the study are marches and silty dries formed on the surface of the bench. The peculiarity of these formations is the almost complete absence of a wave-tidal field at the base of the lower layer of bottom sediments. However, as well as in lagoon marches, the lower part of the sediments are siltopelite sediments with a high content of slightly decomposed organic matter of plant origin, mainly sea grass zostera, and the upper layer of marches and inner deltas is usually represented by a slightly decomposed gittia of sedges, naumburgies and rests [3] (Fig. 1).

These wetland formations are constantly exposed to the tidal effects of marine waters containing large amounts of SO_4, inhibiting methanogenesis [4, 5]. The tides in Aniva Bay are diurnal. The magnitude of the tropical straits is about 1.0 m, the equinoctial ones are 0.5 m. The recorded amplitude span is 2.4 m, the average level (50% security), relative to the zero of the HMS Korsakov post is 1.15 m. The height of the storm surge can reach 1.0 m with a repeatability period of 20–50 years [6].

The aim of the study is to assess the geomorphological position and parameters of the wetlands of the Sakhalin carbon landfill, with a high absorption capacity of climate-modeling components of biomorpholithosystems. The geospatial position of the studied sites is shown in Fig. 2.

Fig. 1 Marshes and silty drainages of the northern part of the Aniva Bay

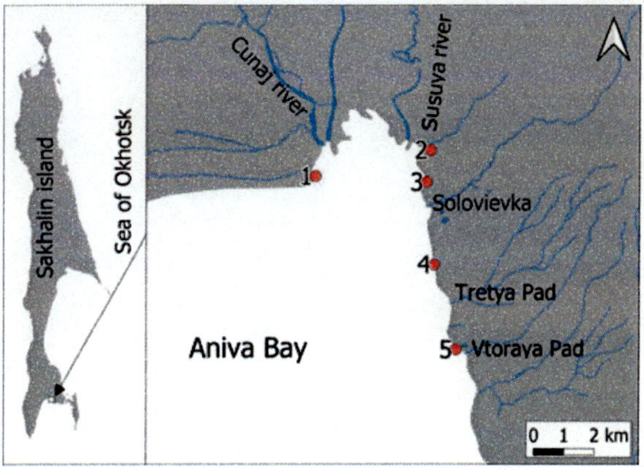

Fig. 2 Location of the studied sites

2 Methods and Results

The methodology of bio-morpholithodynamic studies is based on both traditional and modern methods of obtaining and analyzing geospatial and geological-geomorphological information. All measurements were made by GNSS receivers SOUTH Galaxy G1 (GPS, GLONASS, BEIDOU, GALILEO). Pickets at work sites were filmed in RTK mode (real-time kinematics) from a local base station (BS). The coordinates of the local BS were obtained in static mode from 2 permanent BS of the FT-CORS network located in Korsakov and Yuzhno-Sakhalinsk. The RTK accuracy is within 1 cm in plan and 2 cm in height, the static is no more than 2 cm in plan and 2 cm in height. ITRF2014 (epoch 2010.0) coordinates of BS Korsakov and Yuzhno-Sakhalinsk, height above the ellipsoid WGS84. The obtained heights of the pickets are converted into heights relative to the EGM2008 geoid. An example of orthophotoplanes and a profile along the drilling line is shown in Fig. 3.

Drilling and sampling was carried out using a geoslizer, a set of Edelman soil hand drills (Eijkelkamp) and a Beaker sampler (Eijkelkamp). Samples from columns and sections for the determination of total carbon, organic carbon, and total nitrogen were taken at five-centimeter intervals (Fig. 4). Samples for determining the age of sediments (C14) were taken in the interval of 2 cm.

Figure 5 presents the results of measuring morphometric parameters and the power of wetland formations by profiles in the surveyed areas.

Location No. 1 is located directly in the estuarine zone of the river Tsunai. The thickness of estuarine-lagoon deposits according to drilling data in similar geological and geomorphological conditions in adjacent areas exceeds 100 m [7]. The surface of the muddy drainage is represented in approximately the same altitude interval both for the islets of the branched channel and along the sides of the riverbed. The surface of the march is 0.5–0.6 m above the surface of the muddy drying. The vertical growth rate of the march established by fragments of dated plastic debris is 2.0–2.5 cm/year. The rate of horizontal increase of the march, in some areas exceeds 1.2 m/year.

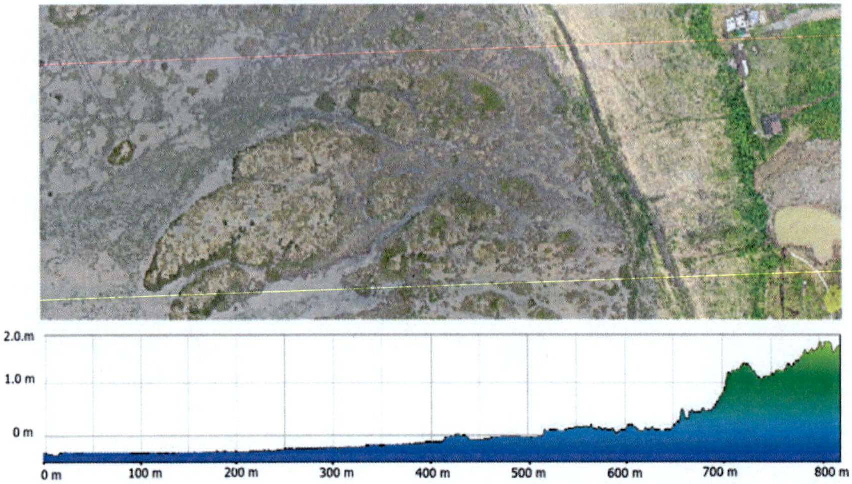

Fig. 3 An example of an orthophotographic diagram and a hypsometric profile along the drilling line at one of the sites

Fig. 4 Sampling from the geosliser cassette

Location No. 2 is located about 800 m from the mouth of the river, Susuya. The thickness of deposits directly in the near–basin part is 1.5–2.0 m. In the context of these sediments, there are 2–3 interlayers of marching slightly decomposed plants. A peat bog extending deep into a 4–5 m sea terrace at 200–250 m, depending on the location of the profile, has been studied at this site. The capacity of the peat

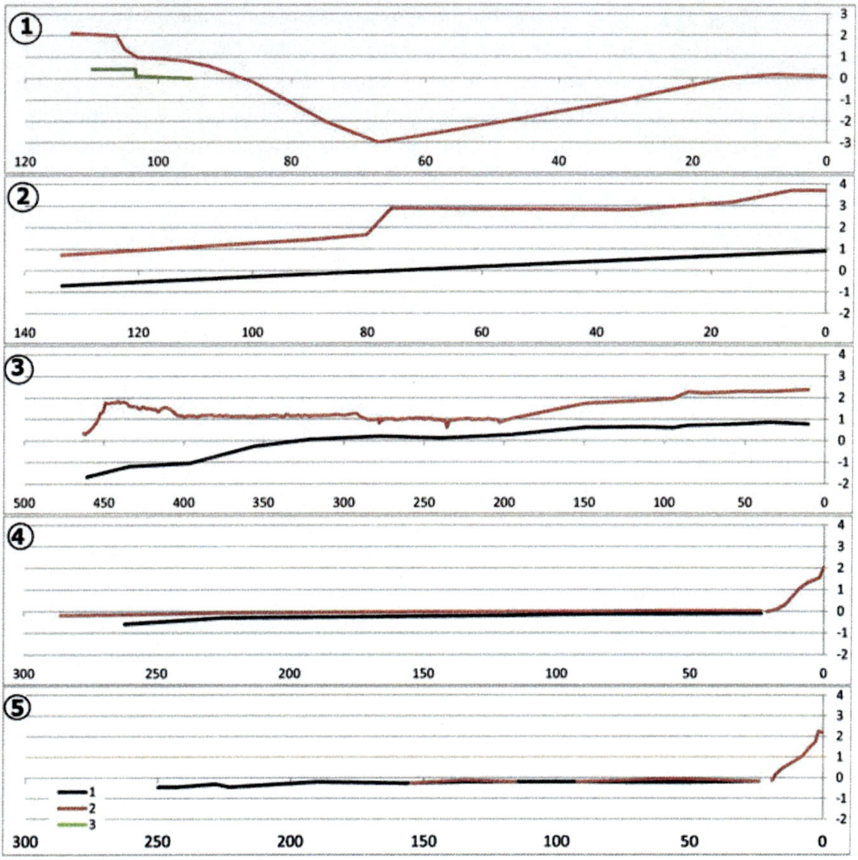

Fig. 5 The change in the thickness of sediments of marches and silty dries along the profiles in the areas marked in Fig. 2: (1) the surface of the bench; (2) the surface of the relief; (3) the surface of the march adjacent to the barrier accumulative form

deposit reaches 1.7 m. The peat bog is underlain by a low-power layer of silt-pelitic sediments.

At location No. 3, the marine part of the bench, with marches at the base of which muddy drying deposits have been uncovered, is blocked by wave-breaking sediments of the coastal accumulative form of the free type [8]. These conditions of coastal-marine sedimentation made it possible to obtain a column in the near-sea zone with the lowest possible elevation of the bench surface. The muddy drainage adjacent to the sandy beach is almost completely covered with sea grass zostera.

The surface of the bench on location No. 4 is blocked throughout by low-power silty sediments, with a maximum of 0.5 m by 250 m more seaward than the beach sediment zone. The surface of the muddy drying when approaching the line of maximum low tide is focally mastered by zostera.

Location No. 5 is characterized by the almost complete absence of silt-pelite deposits with a high content of organic matter on its surface. However, it should be noted that during the dredging and dumping during the period (2005–2007) of the construction of the Prigorodnoye port located more than 15 km along the coastline from the surveyed area, the surface of the bench was covered with a layer of silt-pelite sediment with a capacity of up to 15 cm. However, these sediments were very quickly removed from the surface by wave-tidal processes.

3 Conclusions

Geomorphological data have long been actively used in the analysis of the balance of producers and sinks of GHG in relation to the age dynamics of blue carbon ecosystems [9–11].

Understanding the differences in the accumulation of nitrogen-containing and carbon-containing sediments, not only depending on the removal of sites from the main allochthonous flows of organic matter and its autochthonous regime at the accumulation site, but also on the morpholithodynamic features of the development of the coastal zone, as our data show, is very relevant. Restoration of coastal blue carbon ecosystems that can be reliably used to mitigate the effects of climate change is impossible without understanding the conditions of sedimentation in the coastal zone and morpholithodynamically justified scenarios for their development.

References

1. Wang, D., White, J. R., Delaune, R. D., Yu, Z., & Hu, Y. (2021). Peripheral freshwater deltaic wetlands are hotspots of methane flux in the coastal zone. The Science of the Total Environment, 775, 145784–145784. https://doi.org/10.1016/j.scitotenv.2021.145784
2. C.A. Adams; J.E. Andrews; T. Jickells (2012). Nitrous oxide and methane fluxes vs. carbon, nitrogen and phosphorous burial in new intertidal and saltmarsh sediments. 434(none), 0–. https://doi.org/10.1016/j.scitotenv.2011.11.058
3. Kafanov, V. S., Labai, N. V., Pecheneva Biota and communities of macrobenthos lagoons of northeastern Sakhalin / A. Yu- Sakha.: SakhNIRO, 2003. - 176 p. (in Russian)
4. Nelson J. L., Zavaleta E. S. (2012). Salt marsh as a coastal filter for the oceans: changes in function with experimental increases in nitrogen loading and sea-level rise. https://doi.org/10.1371/journal.pone.0038558
5. Steinmuller, H.E., Hayes, M.P., Hurst, N.R., Sapkota, Y., Cook, R.L., White, J.R., Xue, Chambers L.G., 2020. Does edge erosion alter coastal wetland soil properties? A multi-method biogeochemical study. Catena 187, 104373. https://doi.org/10.1016/j.catena.2019.104373.
6. Hydrometeorology and hydrochemistry of the seas. Vol. 9. The Sea of Okhotsk. Issue 1. Hydrometeorological conditions. St. Petersburg: Hydrometeoizdat, 1998. 370 p. (in Russian)
7. Afanas'ev V.V. Morpholithodynamic processes and development of the shores of the contact zone of the subarctic and temperate seas of the Northern Pacific //Yuzhno-Sakhalinsk: IMGiG FEB RAS. – 2020. 234 p. (in Russian)
8. Zenkovich V. P. Marine geomorphology: Terminological reference: Coastal zone – processes, concepts, definitions / V. P. Zenkovich. – M.: Mysl, 1980. – 280 p. (in Russian)

9. Curado G., Rubio-Casal, A. E., Figueroa, E., Castillo, J. M., 2014. Plant zonation in restored, nonrestored, and preserved Spartina maritima salt marshes //Journal of Coastal Research. T. 30. №. 3. C. 629–634. https://doi.org/10.2112/JCOASTRES-D-12-00089.1
10. Rosentreter, J. A., Al-Haj, A. N., Fulweiler, R. W., & Williamson, P. (2021). Methane and nitrous oxide emissions complicate coastal blue carbon assessments. Global Biogeochemical Cycles, 35, e2020GB006858. https://doi.org/10.1029/2020GB006858
11. Hu, Z., Van Belzen, J., Van Der Wal, D., Balke, T., Wang, Z. B., Stive, M., Bouma, T. J., 2015. Windows of opportunity for salt marsh vegetation establishment on bare tidal flats: the importance of temporal and spatial variability in hydrodynamic forcing

Coastal Marches and Silt Drainage of Sakhalin in the Context of Climate Change

A. B. Faustova and V. V. Afanas'ev

Abstract The role of coastal wetlands as natural sinks of "blue carbon", despite the extremely high rate of its burial, is still greatly underestimated. Moreover, even studies of the modern spatial distribution of marches and silty lands are very fragmentary and far from complete. In light of the growing need for global data on the distribution, size, and sequestration potential of these critical ecosystems, and in order to draw attention to critical geoecological issues, geospatial parameters are considered and an assessment is made of the organogenic sedimentation of coastal marine marshes and mudflats on the island Sakhalin. It is recognized that further research is needed to better quantify greenhouse gas (GHG) fluxes and understand the factors that cause their variability.

Keywords Greenhouse gases · Salt marsh · Intertidal mudflat · Carbon sequestration

1 Introduction

Now, thanks to the 2013 Wetlands Addendum to the 2006 IPCC Guidelines for National Greenhouse Gas Inventories, it is now possible to take into account greenhouse gas (GHG) emissions and removals from wetland uses, such as drainage, reclaiming waterlogging and revegetation, dredging, aquaculture, creation of wetlands for wastewater treatment, etc. [1, 2].

Moreover, coastal marches and silts are already included in the national inventories of the United States and many other countries, they are confident in their cooling effect on the climate [3, 4]. Thus, although scientific evidence is still considered insufficient to provide globally applicable values for GHG emissions and removals from human activities in wetlands.

A. B. Faustova (✉)
Sakhalin State University, Yuzhno-Sakhalinsk, Russia
e-mail: dalmorberegotehnika@mail.ru

V. V. Afanas'ev
Institute of Marine Geology and Geophysics, FEB RAS, Yuzhno-Sakhalinsk, Russia

V. I. Karev (ed.), *Physical and Mathematical Modeling of Earth and Environment Processes—2022*, Springer Proceedings in Earth and Environmental Sciences,
https://doi.org/10.1007/978-3-031-25962-3_28

Fig. 1 Marches of the delta of the Nabil River, which flows into the lagoon of the same name

The 2013 Supplement to the 2006 IPCC Guidelines for National Greenhouse Gas Inventories encourages regional assessments and studies. A total of 5,495,089 hectares of coastal wetlands mapped in 43 countries and territories [5]. The share of Russia, as the authors note, is significantly underestimated in this information array due to the lack of even estimated data for specific regions. The object of the assessment is the geospatial parameters of marches and silty lands formed on the coast of Sakhalin Island. The aim of the study is the geomorphological position, sequestration potential and features of sedimentogenesis, characterized by a high organogenic component.

The paper considers the wetlands of the lagoon waters of about. Sakhalin, the total area of which is about 2200 km^2, the length of the coastline is 2150 km (Fig. 1).

As well as marches and mudflats of estuaries and the Amur Estuary, the total area of which exceeds 163 km^2 (Fig. 2).

This work aims to draw attention to the carbon sequestration potential of wetlands. Estimates of the parameters of organogenic sedimentation of marshes and silty desiccations are constantly updated by us as new information becomes available.

2 Methods and Results

The research methodology is based on modern methods for obtaining and analyzing geospatial information and classical geological and geomorphological approaches, both traditional and traditional. The work is based on the analysis of arrays of aerial and satellite images, which is performed in the geographic information system Quantum GIS. The calculations were made on the WGS84 EPSG:7030 ellipse. The resulting attributes were exported to spreadsheets for further processing. The average

Fig. 2 Muddy drying of the estuarine zone of the Susuya River (Aniva Bay). The inset shows thickets of seagrass zostera in the shore part of the silty dry land

error in the location of objects on this map is about 5 m, the minimum is 0.2 m, the maximum is 18 m, the standard deviation is about 3 m.

At the next stage of the study, in the same QuantumGIS environment, using standard procedures for analyzing remote sensing materials, we calculated the areas of identified coastal biomorpholithosystems. The results of the calculations are presented in Table 1. The age of the deposits was established based on the results of the reconstruction of changes in the areas of marches for the period 1952–2019 and the method of radiocarbon dating of deposits (Fig. 3).

It has been established that march deposits with equal thickness can have completely different ages. Thus, marching soils in the estuary of the Viakhtu River have an age of 430 ± 82 g (FEGU-178), and marches growing into the water area up to 60 cm thick, for example, in Aniva Bay and Nevskaya Lagoon, formed literally in 20–40 years (Fig. 4). Obviously, the geomorphological position of the formed march, expressed mainly in the sediment budget, determines the ratio of the range of vegetation growth and tidal fluctuations. In the case when precipitation cannot form a surface with a flood-free duration, which is necessary to maintain the emerging vegetation, the development of the littoral mudflat does not turn into a march stage even with sufficient nutrient supply [7, 8]. Unfortunately, at present we do not have other data on the vertical growth of marches in our region. If we take the average global rate of carbon accumulation for marshes as 210–218 g/m^2 year, then assuming the area of marshes and silty drains equal to 184 km^2 (lagoons) + 163 km^2 (estuaries and the Amur Estuary), we obtain, according to the most rough estimates, approximately only 70,000 tons/year carbon withdrawn from circulation ocean atmosphere [9]. The estimate is probably greatly underestimated, because only r. The Amur brings 14 million tons of organic matter into the Amur Estuary per year [10, 11].

Table 1 Cartometric parameters of lagoonal reservoirs on Sakhalin

Name	Area (km^2)	Length of the coastline (km)	Name	Area (km^2)	Length of the coastline (km)
Piltyn Bay and Ostox Bay	440.1	269.7	Kolendy Bay (Koldy)	21.6	37.7
Baikal Bay	430.4	147.5	Ekhabi Bay	15.4	16.0
Lake Nevskoe	184.6	179.5	Gulf of Odopty	13.4	26.9
Gulf of Nabilsky	178.8	168.9	Big and Small Chibisan lakes	13.3	22.2
Lake Tynaicha	176.6	93.8	Gulf of Xangyza	9.5	14.0
bay Pomr	153.6	87.7	Lake Changeable	8.6	12.7
Chayvo Bay	112.9	204.5	Gulf of Tront (Тропто)	7.7	16.7
Nyisky Bay	105.0	186.8	Lake Lebyazhye	7.1	20.2
bay.Lynsky	54.2	77.1	Gulf of Ketu	6.8	19.8
lake Big Vavayskoe Small Vavaiskoe	44.9	42.1	Gulf of Viaxty	6.7	21.2
lag. Bycce and lake Vyselkovoe	42.9	33.2	Lake Baklanye	5.2	17.6
оз.Айнское	32.2	47.4	Lake Ptich'ye	3.8	16.2
gulf.Kyegda and gulf.Neypty	30.9	45.2	More than 80 small lagoons	62.3	361.3
Hall Urkt	24.8	27.2	All	2193.2	2213.3

Our preliminary calculations, presented in another article in this issue, suggest that overgrown lagoons can deposit between 80,000 and 120,000 tons/net carbon per year.

We understand that the role of coastal wetlands as natural "blue carbon" sinks can be offset by emissions of methane (CH_4) and nitrous oxide (N_2O). The climatic effect of the flows of these two potent greenhouse gases from marshes and silts in some areas may offset the "cooling effect" of carbon sequestration in coastal marine sediments. However, studies devoted to the study of mechanisms that suppress methanogenesis in the coastal area give hope [12].

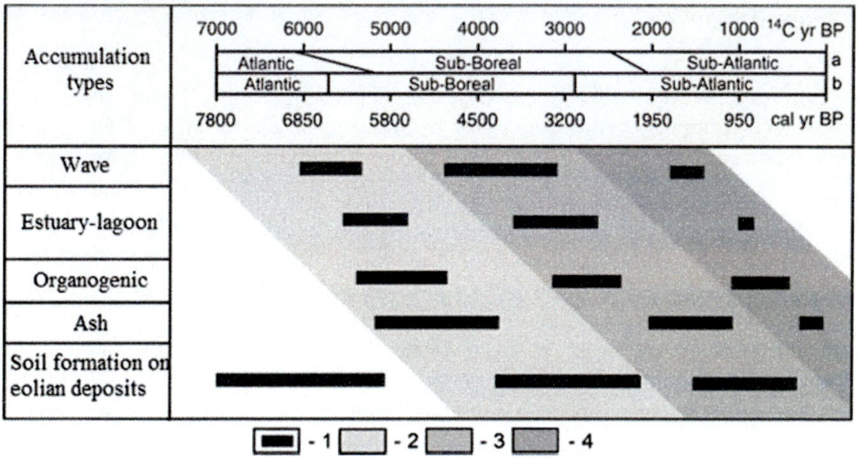

Fig. 3 Sedimentation in the Holocene on the shores of the Far East seas according to C14 data. 1—time intervals; 2, 3, 4—sedimentation cycles; **a** 14C age; **b** calibrated age [6]

Fig. 4 Marsh deposits 60 cm thick formed in Aniva Bay over the past 20 years

3 Conclusion

The conditions of the Sakhalin region make it possible, and in some cases (for wastewater treatment) even necessary to use almost the entire range of technologies for terra reforming the coastal zone and coast to reduce GHG emissions and, accordingly, according to cadastral reporting. The most technologically advanced and suitable for the region first stage is the settlement of existing sludge drylands with algae (mainly zostera) with their subsequent transfer to the surface state and settlement of marshes with plants. No less important is the creation of new areas of sludge dryers as a result of an increase in the rate of sedimentation in estuarine zones and semi-enclosed water bodies. The most acceptable technology is based on the creation of artificial lagoons.

Under anthropogenic impact (in the conditions of the Sakhalin region, this is hydraulic engineering construction, backfilling of new territories in Aniva Bay, etc.), the control and stabilization of methane emissions should be very important when the gas regime of soils of coastal marshes changes. It is also necessary to control and stabilize methane emissions when the gas regime of peat soils changes as a result of drainage, when the groundwater regime changes during the erosion of sea coasts, changes in the water content of the river network during its restructuring and a decrease in precipitation. In addition, it is necessary to stabilize the eroded low-lying seashores, where the thickness of the peat deposit can reach 7–8 m.

Restoration of coastal blue carbon ecosystems that can be reliably used to mitigate the effects of climate change is impossible without understanding the conditions of sedimentation in the coastal zone and morpholithodynamically based scenarios for their development.

These proposals, based on research conducted by SSU, should not only reduce documented GHG emissions, but also identify important opportunities for reducing GHG emissions from coastal wetlands, applicable to the creation of a new GHG inventory sector.

References

1. Solomon S., Qin, D., Manning, M., Averyt, K., & Marquis, M. (Eds.). (2007). Climate change 2007-the physical science basis: Working group I contribution to the fourth assessment report of the IPCC (Vol. 4). Cambridge University Press.
2. Hiraishi T., Krug, T., Tanabe, K., Srivastava, N., Baasansuren, J., Fukuda, M., & Troxler, T. G. (2014). 2013 supplement to the 2006 IPCC guidelines for national greenhouse gas inventories: Wetlands. IPCC, Switzerland.
3. Lovelock, C. E., Adame, M. F., Bradley, J., Dittmann, S., Hagger, V., Hickey, S. M., ... & Sippo, J. Z., 2022. An Australian blue carbon method to estimate climate change mitigation benefits of coastal wetland restoration. Restoration Ecology, e13739. https://doi.org/10.1111/rec.13739
4. Sapkota Y., White J. R. Carbon offset market methodologies applicable for coastal wetland restoration and conservation in the United States: A review //Science of The Total Environment. – 2020. – T. 701. – №. C. 134497. https://doi.org/10.1016/j.scitotenv.2019.134497

5. Mcowen, C. J., Weatherdon, L. V., Van Bochove, J. W., Sullivan, E., Blyth, S., Zockler, C., Fletcher, S. (2017). A global map of saltmarshes. *Biodiversity Data Journal*, (5). https://doi.org/10.3897/BDJ.5.e11764
6. Afanasiev V.V. Morpholithodynamic processes and development of the shores of the contact zone of the subarctic and temperate seas of the North Pacific // Yuzhno-Sakhalinsk: IMGiG FEB RAN. – 2020. 234 p.
7. Cahoon, D. R., Lynch, J. C., Roman, C. T., Schmit, J. P., & Skidds, D. E. (2019). Evaluating the relationship among wetland vertical development, elevation capital, sea-level rise, and tidal marsh sustainability. *Estuaries and Coasts*, *42*(1), 1–15. https://doi.org/10.1007/s12237-018-0448-x
8. Drexler, J. Z., Woo, I., Fuller, C. C., & Nakai, G., 2019. Carbon accumulation and vertical accretion in a restored versus historic salt marsh in southern Puget Sound, Washington, United States. Restoration Ecology, 27(5), 1117–1127. https://doi.org/10.1111/rec.12941
9. Mcleod, E., Chmura, G. L., Bouillon, S., Salm, R., Björk, M., Duarte, C. M., ... & Silliman, B. R. A blueprint for blue carbon: toward an improved understanding of the role of vegetated coastal habitats in sequestering CO_2 //Frontiers in Ecology and the Environment. – 2011. – T. 9. – №. 10. – P. 552–560.
10. Koltunov A. M. et al. Carbonate system of the Amur estuary and adjacent marine areas // Oceanology. - 2009. - T. 49. - No. 5. - P. 694–706.
11. Karetnikova E. A., Garetova L. Spatial and temporal distribution of bacterioplankton and bacteriobenthos in the Amur Estuary and adjacent marine areas // Oceanology. - 2015. - T. 55. - No. 5. - P. 776.
12. Steinmuller, H.E., Hayes, M.P., Hurst, N.R., Sapkota, Y., Cook, R.L., White, J.R., Xue, Chambers L.G., 2020. Does edge erosion alter coastal wetland soil properties? A multi-method biogeochemical study. Catena 187, 104373. https://doi.org/10.1016/j.catena.2019.104373.

Methanotrophic Ability of Cryptogamic Communities of Coastal Ecosystems

V. K. Kadutskiy, S. Yu. Evgrafova, N. N. Lashchinskiy, A. E. Detsura, A. A. Sergeeva, A. V. Zarenkova, and G. K. Zrazhevskaya

Abstract Cryogenic ecosystems are crucial elements of the biosphere, acting as a repository of organic matter on a global scale. However, recent trends in average annual temperatures increasing in these regions are leading to rapid permafrost thaw and promoting the release of organic carbon through microbial degradation and further methane emission into the atmosphere. The methanogenic and methanotrophic microbial communities are the key elements of the methane cycle. We studied associations of methanotrophs with mosses, lichens, and algae in subpolar ecosystems, with the maritime climate since the process of permafrost thawing is most pronounced in coastal areas. In incubation experiments, the methane-oxidizing capacity of methanotrophs associated with cryptogamic communities growing in tundra ecosystems of the Lena River Delta (Yakutia) and King George Island (South Antarctica) was measured. It has been shown, that moss and lichen associates of the studied subpolar ecosystems have a pronounced methanotrophic activity. In addition, increasing moisture changed the methanotrophic ability of the Antarctic cryptogamic communities to the methanogenic one.

Keywords Methanotrophic activity · Methanogenic activity · Cryptogamic communities · Tundra · Arctic · Maritime Antarctic

V. K. Kadutskiy (✉) · S. Yu. Evgrafova · A. E. Detsura · G. K. Zrazhevskaya
Federal Research Center "Krasnoyarsk Science Center SB RAS", Sukachev Institute of Forest SB RAS, Krasnoyarsk, Russian Federation
e-mail: kvkarr@yandex.ru

V. K. Kadutskiy · S. Yu. Evgrafova · G. K. Zrazhevskaya
Sakhalin State University, Yuzhno-Sakhalinsk, Russian Federation

S. Yu. Evgrafova · A. A. Sergeeva · A. V. Zarenkova
Siberian Federal University, Krasnoyarsk, Russian Federation

N. N. Lashchinskiy
Central Siberian Botanic Garden SB RAS, Novosibirsk, Russian Federation

V. I. Karev (ed.), *Physical and Mathematical Modeling of Earth and Environment Processes—2022*, Springer Proceedings in Earth and Environmental Sciences,
https://doi.org/10.1007/978-3-031-25962-3_29

1 Introduction

Permafrost ecosystems are a global repository of organic matter and contain up to 1300 Pg of organic carbon, 800 Pg of which is in the permafrost [1]. At the same time, these eco-systems show that we are highly vulnerable to climate change [1, 2]. The expected degradation of permafrost, as a result of an increase in average annual temperatures, as a result of global warming, will lead to accelerated decomposition of organic material immobilized by low temperatures [1, 3].

The microbial decomposition of organic matter in anaerobic conditions of water-logged soils underlain by permafrost is accompanied by the biological synthesis of methane as a by-product of anaerobic respiration [3].

Methanogenesis is a process carried out by microbial assemblages, including fermenting bacteria and methanogenic archaea. Usually, methanogenesis is carried out in anoxic conditions. Among the ecosystems—sources of biogenic methane are wetland habitats, including permafrost ecosystems [2, 3]. The problem of methane emissions in such ecosystems becomes significant on a planetary scale, since methane is the second most common greenhouse gas after carbon dioxide [4].

After the end of the last ice age, the concentration of methane in the Earth's atmosphere was 600 ppb, but starting from 1800 AD, the content of CH_4 (as well as CO_2 and N_2O) began to increase sharply, and since then has increased by about 0.5–1% per year. Currently, the concentration of methane in the atmosphere is about 1770 ppb [4, 5].

It is noted that in recent years, the increase in the concentration of methane has decreased and stabilized. But, as before, its percentage content in the atmosphere is still high [5]. The potential of methane to cause a greenhouse effect is 23 times higher than that of carbon dioxide. Also, the process of methane emission has a stable positive feedback, i.e. an increase in methane in the atmosphere catalyzes its emission from natural sources [4, 5].

Natural wetlands (23%) and rice fields (21%) are among the main global sources of methane, accounting for almost half of the total methane budget [4–6]. Methane in these biomes is produced exclusively by methanogenic microorganisms [7].

The volumes of methane emissions in natural biogeocenoses are controlled by methanotrophic bacteria, which oxidize part of the formed CH_4. According to various estimates, from 50 to 75% of biogenic methane is oxidized by methanotrophs [8].

Methanotrophic bacteria are representatives of a special group of methylotrophic microorganisms [9]. Methanotrophs are unique due to their ability to oxidize methane and use it as the only source of energy and carbon [9]. In the process of methanotrophy, the key role is played by the enzyme methanemonooxygenase, which exists in two forms of membrane-bound (pMMO) and soluble (sMMO) [10, 11]. The ecology of the habitat of methanotrophic bacteria everywhere, from hydrotherms to ruminant scars [12].

Circumpolar coastal ecosystems attract special attention, since these areas are a serious potential source of biogenic methane due to the thawing of permafrost and erosion of the coastline, as a result of which soil organic matter becomes available for microbial degradation, including for methanogenic communities [13]. In such a situation, methanotrophic bacteria living in circumpolar ecosystems will become **an important component of the equilibrium system, a kind of filter on the way of** emission of increasing volumes of methane [10, 14].

In permafrost ecosystems, methanotrophic bacteria occupy various habitats, such as soil, water, and they are also associated in cryptogamous communities, for example with mosses and lichens [10, 14]. The most studied process of methanotrophy associated with mosses that enter into symbiosis with methanotrophic bacteria. There is information in the literature about the symbiosis of so-called submerged mosses (for example, *Scorpidium scorpioides*) and methanotrophic bacteria, common in ponds of the polygonal tundra of Arctic ecosystems [14]. Bacteria get habitat and protection, and additional carbon dioxide becomes available for moss. According to various estimates, about 32% of moss biomass may account for carbon dioxide formed during the oxidation of methane by methanotrophic associates [14].

The purpose of this work was to study the methanotrophic activity of moss and lichen associates in coastal ecosystems of two circumpolar regions—the Lena River Delta, Yakutia, the Arctic, and the Fildes Peninsula, King George Island, South Shetland Islands, Antarctica.

2 Materials and Methods

The determination of the methanotrophic activity of moss and lichen associates was carried out with samples of cryptogamous communities (mosses, lichens, algae crusts) collected on the Lena River Delta islands (Fig. 1) and King George Island (Fig. 2).

Fig. 1 The Lena River delta on the map of permafrost distribution in Russia

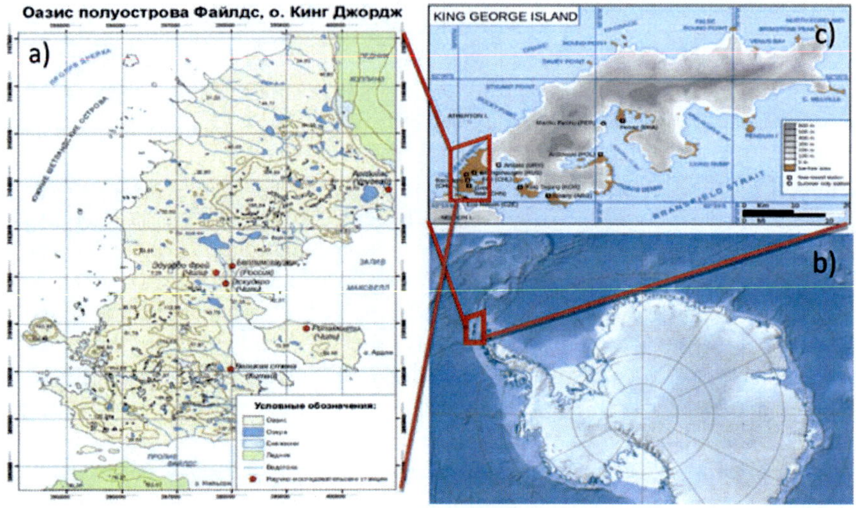

Fig. 2 Fildes Peninsula (**a**), King George Island (**b**), Antarctica (**c**)

Samples were taken in the middle of the growing season (the beginning of August for the Arctic ecosystem and the beginning of February for the Antarctic ecosystem), were placed in sterile bags and stored at a temperature of 4 °C before the experiment.

2.1 Characteristics of the Study sites

2.1.1 Lena River Delta

The ecosystem of the delta is mainly represented by polygonal tundra, with a predominance of sedges and grasses.

Mosses and lichens were sampled on 6 study sites on two delta islands—Samoilovsky Island (5 study sites) and Tit-Ary Island (1 study site).

Study site No. 1 was i laid on the Samoilovsky island (72′22′7.25 N, 126′29′63.66 E), in the zone of dry tundra with a predominance of *Dryas punctata*. Moss-lichen cover—a projective cover of 70%. The average depth of the active layer is 50 cm.

Study site No. 2 was laid on the Samoilovsky island (72°22′11.6 N, 126°30′13.7 E), in a zone of moist tundra with a predominance of *Vaccinium uliginosum*, *Salix polaris*. Moss-lichen cover is a projective cover of less than 20%. The average depth of the active layer is 70 cm.

Study site No. 3 was laid on the Samoilovsky island (72′22′12.92 N, 126′30′3.07 E), in a zone of ice-wedge polygonal tundra with a low center polygons. With the predominance of *Carex rariflora*. Moss-lichen cover is a projective cover of less than 15%. The average depth of the active layer is 40 cm.

Study site No. 4 was laid on the Samoilovsky island (72′22′48.1 N, 126′28′44.3 E), in the zone of the sedge-green swamp. With the predominance of *Carex rariflora*. Moss-lichen cover is a projective cover of less than 10%. The average depth of the active layer is 60 cm.

Study site No. 5 was laid on the Samoilovsky island (72′22′42.1 N, 126′28′28.8 E), in a zone of shrubby thickets, moderately moistened tundra. With a predominance of *Salix polaris*. Moss-lichen cover is a projective cover of less than 10%. The average depth of the active layer is 40 cm.

Study site No. 6 was laid on the Tit-Ary Island, on the shore of temokarst lake, well moistened, with a predominance of *Salix polari*, moss-lichen cover—a projective cover of about 40%. The average depth of the active layer is 60 cm.

The following moss species were collected from the study sites: *Aulacomnium palustre, Hylocomium alaskensis, Rhytidium rugosum, Sphagnum compactum, Tomenthypnum inventions, Dicranum polysetum* and lichens: *Cetraria laevigata, Flavocetraria cucullata.*

2.1.2 Fildes Peninsula, King George Island

The flora of the island is mainly concentrated on the Fildes Peninsula (62°01′27″ S. 58°20′56″ W), an oasis free of glacier, and opening after snow defrosting during summer, and is represented by multiple species of mosses, lichens, and a single species of vascular *Deschampsia antarctica* plants. The places of plant growth are confined to moistened places, floodplains and watercourses, lake shores, places of accumulation of bird guano. Lichens are present in fruticose, foliose, and crustose forms on a partially formed substrate. Algobacterial mats and algae crusts are common in the beds of numerous streams.

The following samples of cryptogamic communities were collected: *Sanionia* sp, *Andreaea* sp, *Warnstorfia* sp., *Campylium* sp., community of mosses *Sanionia* sp. and *Campylium* sp., a community of moss *Sanionia* sp. and lichen *Stereocaulon alpinum*, lichen *Alectoria ochroleuca*, algae crusts.

2.2 Assessment of Methanotrophic Ability by Cryptogamic Communities

Methane consumption in consortia of mosses and lichens and associated microorganisms was studied in laboratory conditions, in incubation experiments, using a Picarro 2201-i gas analyzer (Picarro Inc., USA). The methanotrophic activity was controlled by the shift of the isotopic composition $\delta^{13}C$ in methane.

The samples were placed in gas-tight containers that were connected to a Picarro 2201-i gas analyzer (Picarro Inc., USA). Sampling of air samples from containers was carried out in 3 stages, immediately after placing the sample in the container, after 4 and 24 h.

Since mosses and lichens can grow in both dry and wet habitats in natural ecosystems, the experiment was carried out in two variants: in the first, samples of cryptogamous communities were incubated in an air-dry state, in the second variant, samples were moistened to 60% of the field moisture capacity.

3 Results and Discussions

3.1 *Methanotrophic Activity of Cryptogamic Communities of Lena River Delta Ecosystems*

Studies of the methanotrophic activity of moss and lichen associates growing on the islands of the Lena River Delta have shown that all cryptogamic communities of the ecosystem can consume atmospheric methane, regardless of moisture level, which was confirmed by both the consumption of methane and an increase in the values of $\delta^{13}C$ 4 and 24 h after the start of incubation experiments (Table 1).

The highest methanotrophic ability was demonstrated by the associates of lichen *Cetraria laevigata* and moss *Sphagnum compactum.*

3.2 *Methanotrophic Activity of Cryptogamic Communities of Ecosystems of the Maritime Antarctic, King George Island*

According to the results of incubation experiments with cryptogamic communities of the maritime Antarctic, it was shown that 80% of the studied air-dry samples demonstrated methanotrophic activity at a higher level than recorded in samples taken from the Arctic ecosystem (Table 2). When moistened up to 60% of the water holding capacity, 24 h after the start of the experiment, the methane consumption was replaced by its emission, with simultaneous shift of the composition of stable carbon isotopes in methane, which also indicated a decrease in methanotrophic activity. The exception was the cryptogamic community of mosses *Sanionia* sp. and *Campylium* sp., which had a pronounced methane consumption, both in the air-dry and moistened state. The algal crust also demonstrated a high methanotrophic potential in the air-dry state, but increasing in moisture caused the release of methane after 24-h incubation, which was attended by methanotrophic processes, confirmed by high $\delta^{13}C$ values.

Thus, the potential rates of methane consumption calculated by cryptogamic communities and associates of individual species of mosses and lichens of island circumpolar ecosystems showed that the methanotrophic capacity of the studied samples is comparable in terms of the quantitative level of methane consumption

Table 1 Dynamics of methane release-consumption and the $\delta^{13}C$ isotopic composition shift in methane by cryptogamic communities of the Lena River Delta islands

The test sample	Air-dry samples				Moistened samples (60% of water holding capacity)			
	CH_4, nM m^{-3} g^{-1} h^{-1}		$\delta^{13}C$-CH_4, ‰ VPDB		CH_4, nM m^{-3} g^{-1} h^{-1}		$\delta^{13}C$-CH_4, ‰ VPDB	
	4 h	24 h	4 h	24 h	4 h	24 h	4 h	24 h
Moss *Rhytidium rugosum*	−24.6	−1.3	−56.0	−32.0	−33.6	−7.3	−46.0	−22.0
Moss *Aulacomnium palustre*	−5.7	−0.8	−45.0	−26.0	−15.5	−5.8	−35.0	−16.0
Moss *Hylocomium alaskensis*	−2.4	−1.1	−45.0	−27.0	−18.4	−10.1	−25.0	−17.0
Moss *Sphagnum compactum*	−42.2	−3.8	−56.0	−34.0	−60.5	−20.9	28.9	16.5
Moss *Tomenthypnum inventions*	−13.7	−1.7	−58.0	−53.0	−3.3	−4.4	−52.3	−52.4
Mox *Dicranum polysetum*	−25.1	−7.5	−59.1	−40.0	−93.1	−17.2	−56.1	−41.0
Lichen *Flavocetraria cucullata*	−10.4	−5.5	−52.2	−29.3	−30.8	−8.5	−50.2	−25.3
Lichen *Cetraria laevigata*	−127.6	−26.9	60.0	75.0	−327.6	−56.9	76,0	675,0
Empty container (control)	0	0	−56.2	−55.9	0	0	−56.1	−57.0

but has different trends, depending on the moisture content of habitats. The associates of mosses and lichens of Arctic coastal ecosystems showed methanotrophic activity both in the air-dry and moist state. That is probably because of the low capacity of the active layer in these ecosystems, as a result of which, when the permafrost horizon, which also serves as a water barrier, thaws in the summer, conditions favorable to methanogenesis are created above its surface, and a constant diffusion flow of methane from the soil is created, selectively affecting microbial communities of upper soil and ground cover. In such microbial communities, methanotrophic bacteria are obtained, which retain their activity regardless of the moistening of habitats. Cryptogamic communities of soils of Antarctic maritime ecosystems in our experiments, in the majority, changed the methanotrophic potential to methanogenic, depending on the moisture status. Probably, the limited amount of available substrate in the carbon-poor soils of Antarctica contributes to the rapid change of dominants in microbial communities, which, in turn, contributes to the utilization of all *in situ* available substrates. Moreover, the processes of methane release consumption in

Table 2 Dynamics of methane release-consumption and the $\delta^{13}C$ isotopic composition shift in methane by cryptogamic communities of the Fildes Peninsula

The test sample	Air-dry samples				Moistened samples (60% of field moisture capacity)			
	CH_4, nM m^{-3} g^{-1} h^{-1}		$\delta^{13}C$-CH_4, ‰ VPDB		CH_4, nM m^{-3} g^{-1} h^{-1}		$\delta^{13}C$-CH_4, ‰ VPDB	
	4 h	24 h	4 h	24 h	4 h	24 h	4 h	24 h
Moss *Sanionia* sp. (S.1)	−16.8	−6.2	20.5	15.7	−3.4	7.3	−7.6	−18.1
Moss *Andreaea* sp. (S.4)	−26.5	−11.6	9.7	9.9	−33.1	2.2	1.6	−24.2
Moss *Andreaea* sp. (S.8)	−13.8	−4.6	19.1	12.5	−13.8	14.4	−19.4	−45.2
Moss *Warnstorfia* sp. (S.6)	−10.5	−6.4	13.2	15.3	−28.1	3.5	−13.6	−14.6
Moss *Campylium sp.* (S.9)	−10.1	−0.6	10.1	6.1	23.5	3.4	−18.1	−24.5
Cryptogamous community of mosses *Sanionia* sp. and *Campylium* sp. (S.5)	−77.6	−13.8	122.0	106.2	−67.2	−3.4	16.5	26.7
Cryptogamous community of moss *Sanionia* sp. and lichen *Stereocaulon alpinum* (S.3)	−17.1	−2.3	6.0	2.6	−20.6	8.6	−8.0	−40.4
Lichen *Alectoria ochroleuca* (S.7)	−92.2	−26.3	143.0	142.1	−42.3	2.6	−8.7	−14.2
Algae crust (S.10)	−209.4	−29.9	46.2	8.4	−224.3	177.0	10.3	33.2
Empty container (control)	0	0	−56.3	−57.1	0	0	−57.0	−57.2

this ecosystem can proceed in parallel, as evidenced by the results of incubation experiments with algae crusts.

In the Arctic ecosystem, the highest methanotrophic activity of moss associates was recorded in *Sphagnum compactum*, which is consistent with the data obtained by Danilova and Dedysh [15] for peat acidic sphagnum bogs of the European part of Russia. The authors have shown that an effective methane-oxidizing filter operates in the studied wetlands of the sphagnum-lichen tundra, and its activity is quite high. Other researchers also report on the high methanotrophic activity of bacteria associated with sphagnum mosses in peat soils. Stepnevskaya et al. [16] isolated several strains of endophytic methanotrophic bacteria from the tissues of *Sphagnum magellanicum*. By sequencing the 16S rRNA genes, the strains were identified as representatives of the genus *Methylomonas*. Their methanotrophic activity reached 67.5 μm CH_4 g^{-1} day^{-1}in laboratory conditions. Putkinen et al. [17] also point to the high potential of sphagnum mosses for CH_4 oxidation and, as a consequence, their valuable role as a methane biofilter.

The high methanotrophic activity of the lichen associates of *Cetraria laevigata* in Arctic soils demonstrated by us was also confirmed by studies conducted by Belova and co-authors [18], which showed that in tundra soils with a predominance of lichens in the ground cover, methanotrophic bacteria *Candidatus* Methyloaffinis lahnbergensis are often found. Isolation of *Candidatus* Methyloaffinis-like bacteria remains one of the crucial problems in the cultivation of methanotrophs, and subarctic soils are one of the promising sources for their isolation and research [18].

In our work, the lichens of the Antarctic tundra, as well as the Arctic, have demonstrated significant methanotrophic potential, in particular, the lichen *Alectoria ochroleuca*, which indicates that the soils of both circumpolar ecosystems, with a predominance of lichens in the ground cover, function as sinks of atmospheric methane. Some researchers have shown that methane-oxidizing communities in such soils consist exclusively of high-affinity methanotrophs localized in an upper thin organic surface layer of the soil underlying the lichen cover [15, 18].

Recent studies have shown that methane-consuming bacteria play a crucial role in swamp ecosystems and soil ecosystems. The upper soil horizons have a high potential for methane consumption, and canonical methanotrophs with a high affinity for methane (i.e., capable of oxidizing atmospheric methane) can also use it in higher concentrations [19]. Thus, more and more authors point to a scarce assessment of the contribution of canonical methanotrophs to methane uptake.

4 Conclusions

As a result of the incubation experiments, it was shown that the associates of mosses and lichens of the circumpolar ecosystems (the Lena River Delta and the South Shetland Islands) have pronounced methanotrophic activity. It is generally assumed that Antarctica is not a significant source of methane, however, the presence of active methanotrophs in the cryptogamous communities of King George Island indicates the presence of constant methane flows in the ecosystem, and suggests that Antarctic ecosystems, as well as Arctic ones, can be both a sink and a source of methane. We have noted the change of the methanotrophic ability of the cryptogamic communities of Antarctica to methanogenic with moisture increasing. The high methanotrophic activity of the communities of some lichens of both circumpolar ecosystems and the pronounced methanotrophic activity of the algal crust from King George Island were also demonstrated.

The conducted studies have demonstrated the importance of studying the processes of methane consumption/release by cryptogamous communities of the Arctic and Antarctic in the context of global climate change and the predicted concomitant increase in precipitation in the circumpolar ecosystems.

Acknowledgements The work was carried out with the financial support of the RSF grant No. 21-17-00163. The authors are grateful to the Russian Antarctic Expedition for the comprehensive assistance and support.

References

1. Hugelius, G. et al., 2014. Estimated stocks of circumpolar permafrost carbon with quantified uncertainty ranges and identified data gaps. Biogeosciences. 11: 6573–6593 (2014).
2. Conrad, R., 2007. Microbial ecology of methanogens and methanotrophs. Adv. Agron. 96:1–63.
3. Oremland, R.S., Zehnder, A.J.B., 1988. Biogeochemistry of Methanogenic Bacteria. Wiley: New York, NY, USA, 641–705.
4. Lacis, A., Hansen, J., Lee, P., Mitchell, T., and Lebedeff, S., 1981. Greenhouse effect of trace gases, 1970–1980. Geophys. Res. Lett. 8:1035–1038.
5. Hansen, J., Sato, M., Ruedy, R., Lacis, A., and Oinas, V., 2000. Global warming in the twenty-first century: An alternative scenario. Proc. Natl. Acad. Sci. USA. 97:9875–9880.
6. Knoblauch C. et al., 2018. Methane production as key to the greenhouse gas budget of thawing permafrost. Nature Climate Change. T. 8. – №. 4:309–312.
7. Chen, Y. H., and Prinn, R. G., 2005. Atmospheric modeling of high and low frequency methane observations: Importance of interannually varying transport. Journal of Geophysical Research, 110.
8. Reeburgh, W. S. et al., 2003. Global methane biogeochemistry. In "Treatise on Geochemistry: The Atmosphere". Elsevier-Pergamon, Oxford, UK. 4:65–89.
9. Knief C., Lipski A., Dunfield P.F., 2003. Diversity and activity of methanotrophic bacteria in different upland soils. Appl. Environ. Microbiol. 69: 6703–6714.
10. Hanson R. S., Hanson T. E., 1996. Methanotrophic bacteria. Microbiological reviews. T. 60. №. 2:439–471.
11. Tyurin V.S., Gal'chenko V.F., 1976. Submicroscopic structure of the membrane apparatus of methanotrophic bacteria. Microbiology. V. 45. № 3 P. 503–506. (in Russian)
12. Galchenko V.F., 2011. Methanotrophic bacteria // M.: GEOS. P. 500 p. (in Russian)
13. Trotsenko Yu.A., Khmelenina V.N., 2008. Extremophilic methanotrophs. Pushchino: ONTI PSC RAS. P. 206. (in Russian)
14. Knoblauch, C., O. Spott, S. Evgrafova, L. Kutzbach, and E.-M. Pfeiffer., 2015. Regulation of methane production, oxidation and emission by vascular plants and bryophytes in ponds of the northeast Siberian polygonal tundra. J. Geophys. Res. Biogeosci. 120:2525–2541.
15. Danilova O.V., Dedysh S. N., 2014. Abundance and diversity of methanotrophic Gammaproteobacteria in northern wetlands. Microbiology. Vol. 83. Issue 1–2: 67–76.
16. Stępniewska Z., Goraj W., Kuźniar A., Szafranek-Nakonieczna A., Banach A., Górski A., Pytlak A., Urban D., 2018. Methane oxidation by endophytic bacteria inhabiting Sphagnum sp. and some vascular plants, Wetlands. Vol. 38. Issue 3: 411–422.
17. Putkinen A., Tuittila E.S., Siljanen H.M., Bodrossy L., Fritze H., 2018 Recovery of methane turnover and the associated microbial communities in restored cutover peatlands is strongly linked with increasing Sphagnum abundance. Soil Biology and Biochemistry. Vol. 116:110–119.
18. Belova, S.E.; Danilova, O.V.; Ivanova, A.A.; Merkel, A.Y.; Dedysh, S.N., 2020. Methane-oxidizing communities in lichen-dominated forested tundra are composed exclusively of high-affinity USCα methanotrophs. Microorganisms. 8:2047.
19. Ho A., Lee H.J., Reumer M., Meima-Franke M., Raaijmakers C., Zweers H., de Boer W., Van der Puttend W.H., Bodelier P.L., 2019. Unexpected role of canonical aerobic methanotrophs in upland agricultural soils. Soil Biology and Biochemistry. Vol. 131:1–8.

Estimation of Carbon Stock in Forest Soils of Sakhalin Region

L. V. Mukhortova and D. G. Schepaschenko

Abstract Forests play a significant role in the uptake of atmospheric carbon dioxide and carbon sequestration for a long time. In forests, the long-term carbon depositing mainly takes place in two pools: in tree biomass and in soil organic matter that have different sensitivity to the natural and anthropogenic disturbances. Estimation of these pool sizes and ratio is the first step to the assessment of regional forest carbon budget and prognosis of its feedback to the climate change and disturbances. In this study, we estimated carbon stock in forest soils of Sakhalin region using the information system developed to assess a spatially distributed soil organic carbon with the high resolution (1 km^2). It was found that soil organic matter of forest ecosystems in the region have accumulated about 1230.9 Mt C that is three folds higher than carbon stock in the tree biomass. Forest litter contributes not more than 10% in the northern forest ecosystems and up to 3–4% in the southern forests. Distribution of total carbon stock (live tree biomass + soil organic matter) between above- and belowground pools indicated that 80.0–82.5% of the carbon is allocated in the soil, and forest litter—the component the most vulnerable to disturbances, accounts not more than 5–6% of this stock.

Keywords Forest ecosystems · Soil organic matter · Forest litter · Tree biomass · Carbon stock

1 Introduction

Currently, the forest biome is recognized as one of the important agents of absorption and deposition of atmospheric carbon dioxide [1]. The establishment of carbon stocks

L. V. Mukhortova (✉) · D. G. Schepaschenko
V.N, Sukachev Institute of Forest SB RAS, Krasnoyarsk, Russia
e-mail: l.mukhortova@gmail.com

L. V. Mukhortova
Sakhalin State University, Yuzhno-Sakhalinsk, Russia

D. G. Schepaschenko
International Institute for Applied System Analysis (IIASA), Laxenburg, Austria

V. I. Karev (ed.), *Physical and Mathematical Modeling of Earth and Environment Processes—2022*, Springer Proceedings in Earth and Environmental Sciences,
https://doi.org/10.1007/978-3-031-25962-3_30

in the forests of individual regions is the first step towards assessing the regional carbon budget and is necessary to clarify the contribution of forest ecosystems to the global carbon balance [2].

Long-term carbon deposition in forest ecosystems occurs mainly in two pools: in wood biomass and in soil organic matter. The main long-term reservoir of biomass is stem wood. In this component, carbon is deposited for the entire life of the tree, and after its death, due to the slow rate of decomposition of wood, its release can stretch for tens or even hundreds of years, depending on the conditions in which the trunk falls after dieback [3]. However, this aboveground reservoir of accumulated carbon is very sensitive to various kinds of disturbances of both natural (fires, insects, adverse weather conditions) and anthropogenic (logging) origin. Unlike aboveground woody biomass, its underground part, in the form of coarse tree roots, is more resistant to disturbances, since it is located in the soil body, and the rate of its decomposition and, accordingly, the release of carbon, largely depends on the species of the tree and on the conditions in which this decomposition occurs (soil type, its texture, temperature and water load) [4].

Soil organic matter is the second important reservoir for long-term storage of carbon absorbed from the atmosphere. The turnover time of soil organic matter is estimated in tens and even hundreds of years [5], and its reserves may be comparable or even exceed the carbon reserves in wood phytomass [6]. In addition, this pool of organic carbon is less sensitive to various kinds of aboveground disturbances, compared with wood biomass. During logging, there is mainly a violation of the litter layer. The upper layer of mineral soil, rich in organic matter, is disturbed mainly on the skid trails, which make up a small percentage of the total cutting area [7]. During the passage of fires, even of high intensity, high temperatures do not penetrate deep into the soil and do not significantly affect the pool of organic matter in the mineral layers [8]. Basically, only the upper organogenic horizon is damaged—the forest floor, the turnover time of organic carbon in which is much less, compared with the stable organic matter of the mineral part of the soil. This determines the need to assess the carbon reserves in the forest floor, separately from the mineral layer of the soil, in order to be able to predict changes in the carbon budget in a changing climate and with various disturbances.

Forests are the predominant type of vegetation in the Sakhalin region, according to various authors, they occupy from 65.4 to 85.2% of the total territory [9, 10]. This determines the leading importance of forest ecosystems in the regional carbon budget. The peculiarity of the natural conditions of the Sakhalin region (the Sakhalin Island and the Kuril Islands) is determined not only by the island position of the territory and climatic factors, but also by special geological and geophysical conditions: high seismicity, the presence of active faults of the Earth's crust with increased endogenous geochemical and thermal flows, modern volcanic activity [11]. Due to these geochemical features, soil parent material are characterized by abnormally high contents of many chemical elements [11]. All these factors can affect both the productivity of forest ecosystems in the region and the accumulation of organic carbon in soils.

The purpose of this study was to assess the carbon reserves in the live biomass and soil of forest ecosystems of the Sakhalin region and to identify the features of the formation of these carbon reservoirs in island conditions.

2 The Object and Methods of Research

Sakhalin Island is located near the eastern coast of the Asian continent. It is elongated in the meridional direction by 948 km: its northern tip of the island reaches 54°25′s.w., the southern—45°54′s.w. Most of its territory is medium–high mountains, elongated in the meridional direction [12].

The Kuril Islands, also part of the Sakhalin Region, are the peaks of two parallel underwater ridges: the Greater and Lesser Kuril Ridge. The length of the Greater Kuril Ridge is 1200 km. It consists of more than 30 islands, which are active or extinct volcanoes. The Lesser Kuril Ridge stretches for 105 km and includes 6 islands. There are no young volcanoes on the islands of the Lesser Kuril Ridge. The relief of the islands is flat, sometimes hilly. Land areas rise above the ocean by 20–40 m [12].

The climate of Sakhalin and the Kuril Islands is monsoon. Winter is cold, but wetter and less severe than on the mainland. Summer is cool and rainy [12]. Due to the considerable length from north to south, the complex mountainous terrain and the different temperature regime of the seas, various climatic conditions are created on Sakhalin and the Kuril Islands. The average annual air temperature ranges from − 1.5 °C in the north to + 2.2 °C in the south. The absolute minimum is − 26 °C in the southeast, − 49 °C in the northwest, − 53 °C in the interior of Sakhalin Island. The absolute maximum ranges from + 26 °C in the area of Cape Patience to + 37 °C in the Tymovskaya Valley [13, 14]. The physical and geographical features of the territory determine the uneven distribution of precipitation. Their annual amount increases from 500–600 mm in the north to 800–1200 mm in the south [12].

Winter lasts from November to March. Due to the large amount of snow and deep groundwater level, the soil freezes to a relatively shallow depth: 140–160 cm in the north and in the middle part of the island and up to 40–70 cm in the south. There are areas of permafrost on the North Sakhalin Plain [12].

The structure of the vegetation cover of Sakhalin Island is caused by landscape-zonal differentiation of vegetation. On the territory of the island there are four vegetation subzones, the boundaries between which run in a north-westerly direction and correspond to the isolines of temperatures and humidity [15]: The northern plains of Sakhalin are covered with larch forests (*Larix cajanderi* Mayr) [16], taiga forests with a predominance of Yezo spruce (*Picea ajanensis* Fisch) are common in the central mountainous areas, ex Carrière), the territory south of the isthmus of the Belt is occupied by stands with a predominance of Sakhalin fir (*Abies sachalinensis* (F. Schmidt) Mast.), and the southwestern tip of the island is dark coniferous forests with an admixture of broad-leaved species [10]. In the mountains on the island there is a vertical belt: spruce-fir lichen-green moss forests with shrubs grow in the lower belt, above 500–700 m above sea level. They are replaced in the Eastern Sakhalin

Mountains by Erman's birch forests with shrubs, and in the Western Sakhalin Mountains by Erman's birch forests with bamboo. The peaks of the foothills are covered with Siberian dwarf pine (*Pinus pumila* (Pall.) Regel) [10].

On the northern islands of the Kuril Ridge, forest vegetation is absent or is extremely fragmentary in the form of groups and sparse woodlands of willow (*Salix udensis* Trautv. et Mey.) in river valleys or in the lower part of mountain slopes [17]. In the Southern Kuriles, dark coniferous, Erman's birch, hard deciduous forests are widespread, soft deciduous forests are found. Larch forests are present in the Iturup island. Well-drained teritories are occupied by mixed grass meadows. Marsh vegetation grows on the lowland-shores of lakes, watercourses, sea terraces. Mountain tundra vegetation appears at altitudes of more than 600 m and forms a sparse vegetation stone belt [18].

Heterogeneity of the natural conditions of Sakhalin Island, age diversity and lithological diversity of sediments forming its surface determine the distribution and representation of various types of soils on the island [19]. Under the coniferous forests in the southern part of the island, mountain brown soils are formed. Podzolic soils are widespread under low-density larch forests on the North Sakhalin Plain, which are formed on loose, chemically and mineralogically poor, well-drained rocks. On high sea terraces under dark coniferous forests, sod-humus soils are formed, which frame the southern part of the island, along the sea coast. Meadow soils are formed on high river terraces, and low surfaces are occupied by various types of gley and swampy soils. Peat soils are formed at various geomorphological levels: depressions on watersheds and gentle mountain slopes, on river and sea terraces [19–21]. A characteristic feature of all the soils of the island is their acidic nature, the pH of all soils is on average 4.5–5.0 [14, 21].

Calculations of carbon reserves in the soils of the region were carried out using a specially developed information system that allows to estimate spatially distributed reserves of organic carbon in soils with high resolution (1 km^2). The automated information system for calculating the reserves of organic carbon in soils was created on the basis of: Soil Map of the Russian Federation (scale 1:2.5 million) and data base of typical soil profiles to it [22]; a database of in-situ measurements of organic carbon content in Russian soils (to account for zonal, regional features and register the influence of vegetation type and land use); maps of prevailing vegetation types and land use [23] with a spatial resolution of 1 km^2, maps of natural zones of the Russian Federation [24] (for adjustments of carbon stocks depending on zonal conditions), and the Administrative Map of Russia [24] (for regional adjustments of carbon stocks). A detailed description of the calculation methodology is presented in the work of Schepaschenko et al. [25].

To assess the carbon reserves in the live biomass of woody vegetation in the Sakhalin Region, data on the stocks of stem wood from two accounting systems were used: the State Forest Register (SFR) and the state forest inventory (SFI), combined with remote sensing data [26].

To convert the wood stock (m^3) into carbon reserves in living biomass, the following conversion coefficients were used: the stock in carbon of living biomass of trees, including roots—0.35035; the stock in carbon of aboveground living biomass of trees—0.27923; the stock in aboveground living biomass—0.56131; the biomass of roots to aboveground phytomass—0.288. The carbon content in wood biomass was assumed equal to 50%, for foliage (needles) the carbon content was assumed to be equal to 45% [27].

3 Results and Discussion

Carbon reserves in the soil of forest ecosystems of the research region were calculated separately for the forest litter layer and for the 1 meter layer of mineral soil (Fig. 1).

In the forest ecosystems of the Sakhalin region, the total carbon reserves in the forest litter are estimated at 72.7 Mt. In the mineral part of the soil profile—1158.2 Mt C. In general, about 1230.9 Mt C were deposited in the forest soils of the region. Of these, 878.1 Mt C were deposited under coniferous forests of the region, and 352.8 Mt C were deposited under deciduous forests (Table 1).

An assessment of the spatial distribution of carbon stocks in the litter and soil shows that the largest specific carbon stocks in the litter are characteristic of the tundra zone, which accounts for no more than 0.2% of the territory occupied by forests. The smallest reserves are observed in the North taiga zone, which occupies only 0.1% of the territory. The average carbon reserves in the litter of middle taiga forests are slightly higher than in the southern taiga. The mineral part of the soil is characterized by an almost twofold increase in carbon reserves in the soil profile when moving from north to south (Fig. 2). This distribution of carbon reserves is a reflection of the spread of certain types of soils across the territory: the spread of

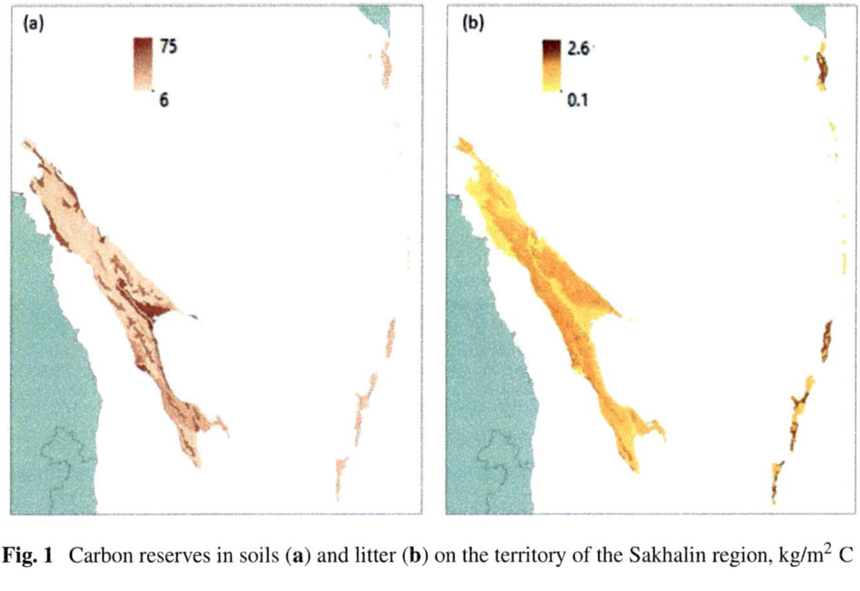

Fig. 1 Carbon reserves in soils (**a**) and litter (**b**) on the territory of the Sakhalin region, kg/m^2 C

Table 1 Carbon reserves in the soil of forest ecosystems of the Sakhalin region

Subzone	Litter, Mt C		Soil 1 m, Mt C		Total, Mt C
	Coniferous	Deciduous	Coniferous	Deciduous	
Tundra	0.19	0.04	1.68	0.44	2.35
Forest tundra and northern taiga	0.03	0.02	0.36	0.20	0.61
Middle taiga	47.47	15.37	599.17	272.06	934.07
Southern taiga	7.12	2.47	222.09	62.16	293.85
					1230.88

organic matter-rich sod-humus soils at the southern tip of the island. Sakhalin, and poor podzolic soils under the northern taiga larch trees in the north.

The contribution of litter to the total carbon reserves in the soil system decreases from 10% in northern ecosystems to 3–4% in southern taiga forests (Fig. 3).

According to our estimates, the specific carbon stock in the soils of the region is 8–18 kg/m^2 in the northern forests, up to 17–24 kg/m^2 C in the middle taiga and 28–32 kg/m^2 C in the southern taiga forests. These values are within the range of variation of carbon reserves in the soils of Sakhalin Island. According to field studies, for peat soils and gleyzems of the northeastern part of the island (middle taiga zone), carbon reserves range from 16.6 to 112.4 kg C/m^2 [28], for podzols of North Taiga larch forests—from 3.9 to 8.6 kg C/m^2 [29], for alluvial soils under floodplain larches—6.5–12.0 kg C/m^2 [30].

In order to assess the contribution of soil to the total carbon reserves in the forest ecosystems of the region, it was necessary to calculate the carbon reserves in the live biomass of woody vegetation. The total area covered by forest vegetation in the Sakhalin region is 6117 thousand hectares. The total stock of stem wood in these

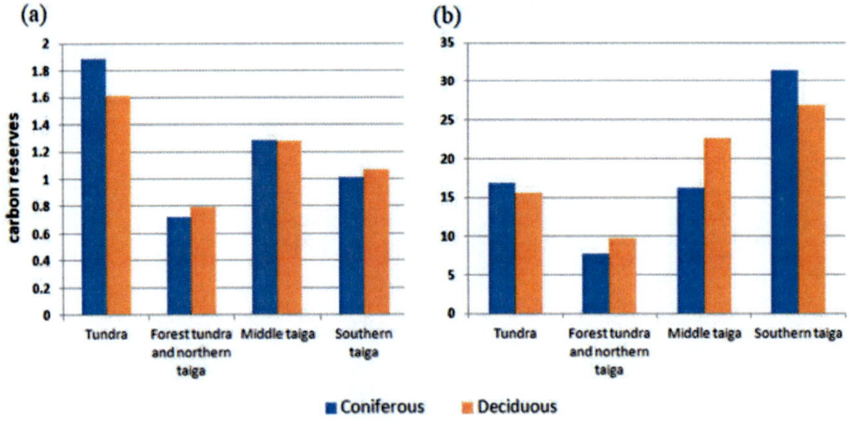

Fig. 2 Specific carbon reserves (kg/m^2) in the litter (**a**) and in the soil (1 m) (**b**) of coniferous and deciduous forests of the Sakhalin region

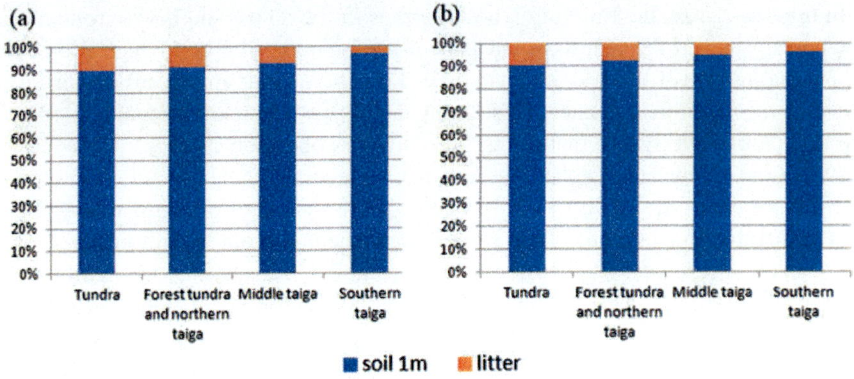

Fig. 3 The contribution of litter to the total carbon reserves in the soils of coniferous (**a**) and deciduous (**b**) forests of the Sakhalin region

forests is estimated at 1168 million m^3 according to SFI and remote data [26]. The area of managed forests of the Sakhalin region is 5608 thousand hectares, and their stock is 1071 million m^3 according to the SFI or 637 million m^3 according to the SFR, which is the main repository of forest information, and is used for national reporting to the IPCC (UNFCCC) [26]. The reason for the almost twofold difference in the stocks of wood given by these two accounting systems may be that the prescription of forest management materials used for SFR for most of the country's forests exceeds 20 years, and there is no acceptable system for updating data [31, 32]. Therefore, in further calculations, we used SFI estimates combined with remote data to estimate wood biomass reserves.

Estimates have shown that in the forests of the Sakhalin region, the total reserves of aboveground wood biomass amount to 665.6 Mt, which is equivalent to 317.7 Mt of carbon. An additional 91.5 Mt C has been accumulated in the underground biomass of woody vegetation. In total, 409.2 Mt C were deposited in the living wood biomass of the region. Coniferous species account for almost 75% of the total biomass reserves in the region, 16% are hardwood deciduous and 3% are softwood deciduous [9]. Accordingly, about 306.6 Mt of carbon is concentrated in the biomass of coniferous forests, while 102.6 Mt of carbon is deposited in deciduous stands.

The total carbon reserves of biomass obtained in our studies are 1.5 times higher than those given in the work of Utkin and co-authors [9], who estimated the carbon pool in the biomass of forests of the Far Eastern Federal District according to SFR records. The estimates of carbon reserves in the soil given by these authors, on the contrary, are slightly higher than those obtained by us (1718.6 Mt C).

The average specific stock of aboveground biomass in the forests of the Sakhalin region is 191 m^3/ha, which is equivalent to 66.9 tC/ha. In total, with the underground biomass of woody vegetation, the specific carbon stock of biomass in the forests of the region is 86.2 tC/ha.

The contribution of soil organic matter to the total carbon reserves ranges from 74% under coniferous forests to 77% under deciduous forests.

In order to assess the ratio of carbon reserves in the above and belowground, it is necessary to add the belowground part of the biomass of woody vegetation to the organic matter of the soil, since it, as well as the organic matter of the soil, is protected from the direct impact of disturbing factors by the thickness of mineral soil. The distribution of organic matter reserves (living wood biomass + soil) between the above and belowground spheres shows that, in general, 80–82.5% of total carbon reserves are concentrated in the soil system, while the share of forest litter—the most vulnerable to violations of the soil profile horizon, accounts for no more than 5–6% of these reserves. For coniferous and deciduous ecosystems, no significant differences were found in the ratio of above and belowground components of total carbon reserves.

In recent decades, the forest ecosystems of Sakhalin Island have been subjected to active forest management and a significant spread of forest fires [10]. Most of the forests of the Sakhalin region are characterized by a moderate and high level of fire danger [33]. Since the most of carbon reserves are allocated belowground, the soil can prevent the rapid loss of carbon by the forest ecosystems of the region as a result of the passage of fires.

A comparison of the data obtained with the forest territories of the continental regions of Siberia and the Far East, located at the same latitudes as the Sakhalin Region, shows that the average specific carbon reserves in the soils of the forest territories of the Sakhalin Region (20.98 kgC/m^2) are slightly higher compared to the inland regions (Zabaikalsky Krai, Republic of Buryatia, Republic of Tyva, Republic of Khakassia, Republic of Altai)—10.6–17.9 kgC/m^2 and Khabarovsk Krai (17.1 kgC/m^2). Comparable specific reserves of carbon in the soil are characteristic only for the Amur Region and the Altai Territory (19.9–21.4 kgC/m^2). At the same time, the contribution of soil to the total carbon reserves of forest ecosystems in all these regions is comparable and amounts to 62.1 to 82.7%, since most of them are characterized by lower carbon reserves in wood phytomass.

4 Conclusions

About 1230.9 Mt C were deposited in the organic matter of forest soils of the Sakhalin region. This is 3 times more than the carbon reserves accumulated in the woody biomass of the forests of the region. The distribution of carbon reserves between living wood biomass and soil does not differ significantly from that characteristic of the continental forests of Siberia and the Far East of the same latitudes. Higher total specific carbon stocks in the forest ecosystems of the Sakhalin region are formed both due to higher carbon stocks in the soil and due to higher productivity of stands, since the average specific stock of wood in these forests is 20–30% higher than in continental areas with a monsoon climate and in some regions with a sharply continental climate.

Acknowledgements Calculations and analysis of carbon reserves in Siberian forests were carried out within the framework of research on the project of State Task No. 0287-2021-0008 "Natural and anthropogenic dynamics of taiga forests of Central Siberia in a changing climate", R&D 121031500339-0.

References

1. Watson R., Noble I., Bolin B., Ravindranath N., Verardo D., Dokken D. Eds. Land Use, Land Use Change, and Forestry: A Special Report of the IPCC. Cambridge: Cambridge University Press, 2000. 375 p.
2. Kurganova I.N., Kudeyarov V.N. Ecosystems of Russia and global carbon budget// Science in Russia. 2012. 5. P. 25–32. [in Russian]
3. Harmon, M. E., Fasth, B. G., Yatskov, M., Kastendick, D., Rock, J., & Woodall, C. W. (2020). Release of coarse woody detritus-related carbon: a synthesis across forest biomes. *Carbon balance and management*, 15(1), 1–21.
4. Mao, R., Zeng, D. H., & Li, L. J. (2011). Fresh root decomposition pattern of two contrasting tree species from temperate agroforestry systems: effects of root diameter and nitrogen enrichment of soil. *Plant and soil*, 347(1), 115–123.
5. Dungait, J. A., Hopkins, D. W., Gregory, A. S., & Whitmore, A. P. (2012). Soil organic matter turnover is governed by accessibility not recalcitrance. *Global Change Biology*, 18(6), 1781–1796.
6. Laclau, P. (2003). Biomass and carbon sequestration of ponderosa pine plantations and native cypress forests in northwest Patagonia. *Forest ecology and management*, 180(1–3), 317–333.
7. James, J., & Harrison, R. (2016). The effect of harvest on forest soil carbon: A meta-analysis. *Forests*, 7(12), 308.
8. Nave, L. E., Vance, E. D., Swanston, C. W., & Curtis, P. S. (2011). Fire effects on temperate forest soil C and N storage. *Ecological Applications*, 21(4), 1189–1201.
9. Utkin A.I., Zamolodchikov D.G., Chestnykh O.V. (2006). Carbon pools and fluxes in forests of Far East federal region.// Coniferous of the boreal zone. 23(3), 21–30. [in Russian]
10. Melky V.A., Verkhoturov A.A., Sabirov R.N., Bratkov V.V. (2019). Analysis of forest land on Sakhalin Island //Bulletin of Ammosov North-Eastern Federal University. Earth Sciences Series, (2), 68–73. [in Russian]
11. Poberezhnaya T.M., Veselov O.V. (2011). Ecological studies at the interface of biological and geological sciences // Bulletin of the Far Eastern Branch of the Russian Academy of Sciences. 6 (160), 43–47. [in Russian]
12. Matyushkov G.V. (2001). On the nature of Sakhalin and the Kuril Islands. Bulletin of the Sakhalin Museum (1), 240–257. [in Russian]
13. Zemtzova A.I. Climate of Sakhalin / A.I. Zemtzova. – L.: Gidrometeoizdat, 1968. – 197p. [in Russian]
14. Romanenko Ya.A. (2014). Physical and geographical characteristics of Sfkhalin Island as a condition for the formation of floras. Amur Scientific Bulletin, (4), 116–128. [in Russian]
15. Sabirov R.N., Melky V.A. Sakhalin Island Forest Productivity Assessment by Remote Sensing Data // Nature Management in the Russian Far East: Proceedings of the Scientific Conference (Khabarovsk, January 19–20, 2006). – Khabarovsk: IVEP FEB RAS, 2006. – 131 c. – P. 90–93. [in Russian]
16. Koropachinsky I.Y., Vstovskaya T.N. (2012). Woody plants of Asiatic Russia. GEO Publishing House. Russian Academy of Sciences, Siberian Branch, Central Siberian Botanical Garden. - 2nd ed. [in Russian]
17. Grishin S.Yu., Barkalov V.Yu. (2009). Vegetation cover of the northern Kurils. Bulletin of the Far Eastern Branch of the Russian Academy of Sciences, (3), 61–69. [in Russian]

18. Ganzey K.S. (2010). Landscapes and physical and geographic zoning of the Kuril Islands. Vladivostok: Dalnauka. 214. [in Russian]
19. Ivlev A.M. (1977). Peculiarities of genesis and biogeochemistry of soils of Sakhalin. Moscow: Nedra. [in Russian]
20. Ivlev A.M. Soils of Sakhalin / A.M.Ivlev. – M.: Nauka, 1965. – 115 p. [in Russian]
21. Poberezhnaya T.M. (2001). Geochemistry of the main types of soils of Sakhalin. Bulletin of the Sakhalin Museum, (1), 301–306. [in Russian]
22. Unified State Register of Soil Resources. Moscow: Ministry of Agriculture of the Russian Federation. Soil Institute named after V.V. Dokuchaev, 2019. http://egrpr.esoil.ru/index.htm [in Russian]
23. Schepaschenko D., McCallum I., Shvidenko A. et al. A new hybrid land cover dataset for Russia: a methodology for integrating statistics, remote sensing and in-situ information // J. of Land Use Sci. 2011. № 6(4). P. 245–259.
24. Stolbovoi V., Mccallum I., 2002. Land Resources of Russia [online]. IIASA & RAS. Laxenburg, Austria. Available from: http://www.iiasa.ac.at/Research/FOR/russia_cd/index.htm
25. Schepaschenko, D.G., Mukhortova L.V., Shvidenko A.Z., Vedrova E.F. (2013). Organic carbon stock in soils of Russia// Russian Journal of Soil Science, (2), 123–123.
26. Schepaschenko, D., Moltchanova, E., Fedorov, S., Karminov, V., Ontikov, P., Santoro, M., ... & Kraxner, F. (2021). Russian forest sequesters substantially more carbon than previously reported. Scientific reports, 11(1), 1–7.
27. Schepaschenko D, Moltchanova E, Shvidenko A, Blyshchyk V, Dmitriev E, Martynenko O, See L, Kraxner F. Improved estimates of biomass expansion factors for Russian forests. Forests. 2018 Jun 1;9(6):312.
28. Lipatov D.N., Scheglov A.I., Manakhov D.V., Brekhov P.T. (2021). Spatial variation of organic carbon stocks in peat soils and gleyzemes in northeastern Sakhalin. Russian Journal of Soil Science, (2), 211–223.
29. Shlyakhov S.A. (2011). Diversity and properties of podzols of Northern Sakhalin./ Vestnik of Krasnoyarsk State Agrarian University, (12), 69–75. [in Russian]
30. Zharikova E.A. (2019). Forest growth properties and ecological significance of soils of floodplain forests of northern Sakhalin. Bulletin of the V.V. Filippov Buryat State Agricultural Academy., (1), 91–97. [in Russian]
31. Shvidenko a.Z., Schepaschenko D.G. (2014). Carbon budget of Russian forests.// Siberian Forest Journal, (1), 69–92. [in Russian]
32. Filipchuk A.N., Malysheva N.V., Zolina T.A., Yugov A.N. (2020). Russia's boreal forests: opportunities for climate change mitigation.// Forestry Information, (1), 92–114. [in Russian]
33. Sheshukov M.A., Brusova E.V., Pozdnyakova V.V. (2008). Modern fire regimes in the forests of the Far East.// Lesovedenie, (4), 3–9. [in Russian]

Refinement of the Model of the Geological Structure of the Southern Urals According to the Peculiarities of the Distribution of the Area of Epicenters of Seismic Events (Methodological Aspect)

M. Yu. Nesterenko, A. M. Tyurin, A. V. Kolomoets, V. S. Belov, and V. P. Petrishchev

Abstract *The object of the study.* Orenburg segment of the Southern Urals. *Materials and methods.* The results of drilling, geological survey, seismic survey, seismic monitoring (epicenters of seismic events, presumably associated with natural factors). The interpretation of the complex of geological and geophysical data is carried out. *Results.* The position of the southern part of the East Kizilsky fault has been mapped. It is shown that the West Kizilsky fault, mapped according to the results of geological survey, does not develop to the south. The geological hypothesis is substantiated: there is a system of diagonal tectonic disturbances of the north-western strike in the region. The zoning of the Magnitogorsk megasynclinorium zone, the most promising for oil and gas, has been carried out. *Methodological conclusions.* According to the peculiarities of the distribution in the region of the epicenters of seismic events associated with natural factors, it is possible to solve regional geological and prospecting tasks, as well as to identify tectonic disturbances that determine the prospects of oil and gas potential of local areas. The reliability of forecasts will increase with the development of the database of observed events and an increase in the accuracy of their localization. The latter can be achieved by installing at least two new seismic stations in the region.

Keywords Geology · Tectonics · Seismic events · Forecast · Southern Urals · Orenburg region

M. Yu. Nesterenko (✉) · V. S. Belov · V. P. Petrishchev
Orenburg Federal Research Center of the Ural Branch of the Russian Academy of Sciences, Orenburg, Russia
e-mail: n_mu@mail.ru

A. M. Tyurin · A. V. Kolomoets
The Center for the Development and Operation of Deposits in the European Part of the Russian Federation, Gazprom VNIIGAZ LLC, Orenburg, Russia

A. M. Tyurin · A. V. Kolomoets · V. P. Petrishchev
Orenburg State University, Orenburg, Russia

V. I. Karev (ed.), *Physical and Mathematical Modeling of Earth and Environment Processes—2022*, Springer Proceedings in Earth and Environmental Sciences,
https://doi.org/10.1007/978-3-031-25962-3_31

1 Introduction

The Orenburg segment of the Southern Urals (Ural Folded System) has been studied by geological survey, gravy-, magnetic and seismic exploration, drilling and mining. A set of the 1:1,000,000 scale state geological map has been compiled (sheet M-40 with valve M-41 [1]). The region includes from west to east the West Ural folding zone, the Central Ural uplift, the Magnitogorsk (Tagilo-Magnitogorsk) megasynclinorium, the East Ural megasynclinorium and the Trans-Ural megasynclinorium (Fig. 1).

The West Ural folding zone in the Orenburg region is represented by the West Ural megasynclinorium. In the practice of geological exploration, this tectonic element is called the advanced folds of the Urals. In the west, they are interfaced through the Surensky uplift with the Pre-Ural deflection. The Central Ural uplift from the west and east is limited by the West Ural and East Uraltaussky faults. Through the latter, it is interfaced with the Magnitogorsk megasynclinorium, bounded from the east by the North Dzhusinsky and Dombarovsky faults. The boundary between the East Ural meganticlinorium and the Trans-Ural megasynclinorium is drawn along the West Saryobinsky and East Saryobinsky faults. The Trans-Ural megasynclinorium in the east, on the territory of the Republic of Kazakhstan, is interfaced with the Tyumen-Kustanai trough.

Seismic events in the Southern Urals recorded by the network of stations of the Kazakhstan National Seismic Center and the Department of Geoecology of the Orenburg Federal Research Center of the Ural Branch of the Russian Academy of Sciences for the period from 2006 to 2020 were taken into account. On average, about 750 seismic events with a magnitude from 0.9 to 3.6 Mb and a hypocenter depth of up to 4.0 km are recorded here per year. For its Orenburg segment, an analysis of seismic events (2230 in total) for 2018–2020 was performed, taking into account data on blasting operations (date, time, geographical coordinates, explosive power and location of mines where explosions were carried out) provided by PJSC Gaisky GOK, JSC ORMET and JSC Kiembaevsky GOK.

The main part of seismic events—2014, coincides in time with the production of explosions [2]. Seismic events that do not coincide with the time of the explosions—216 are highly likely to be natural or man-induced. The latter occurred after the technogenic ones with a time lag sufficient for their separate fixation. However, the great distance of seismic stations from recorded seismic events (up to two hundred kilometers) leads to significant errors in calculating the coordinates of epicenters and parameters, as well as difficulties in determining their nature.

The distribution of epicenters of natural and man-induced events in the region reflects some elements of its geological structure. Based on this, the task of our study was to assess the possibility of refining the model of the geological structure of the Orenburg segment of the Southern Urals according to the peculiarities of the distribution of their epicenters. Taking into account the high uncertainty of the identification of events, explosions that do not coincide with the production of explosions and whose epicenters are located at a great distance from existing quarries and mines

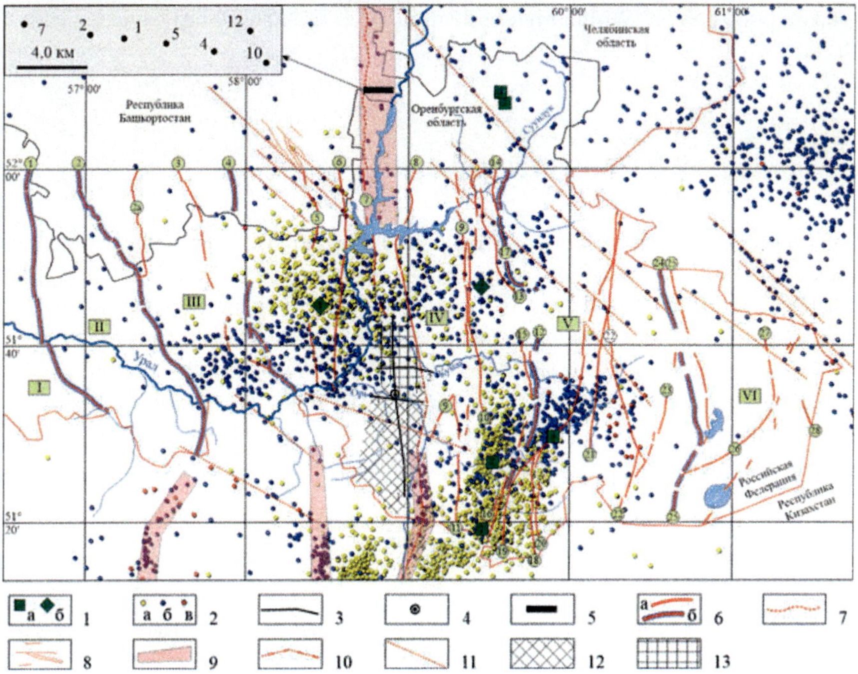

Fig. 1 Orenburg segment of the Southern Urals. The results of the geological interpretation of the features of the distribution over the area of the epicenters of seismic events 2018–2020. 1—quarries (**a**) and mines (**b**) (shown only in the Orenburg region); 2—epicenters of seismic events with a magnitude of Mb below 2.0 (**a**), from 2.0 to 3.0 (**b**), above 3.0 (**c**) units; 3—regional seismic profiles; 4—the position of the well 1 Aschebutakskaya; 5—Ural profile of wells (their position is shown in the inset); 6—regional faults (**a**), faults along which the boundaries of structural and tectonic elements are drawn (**b**) on sheet M-40; 7—Kizilsky regional fault on sheet N-40-XXXV; 8—not basic faults on sheet N-40-XXXV; 9—linear zones of increased seismic activity; 10—forecast position of the East Kizilsky fault; 11—diagonal tectonic disturbances. Zones with minimal (12) and increased seismic activity (13) in the development of Visean-Bashkir age limestones overlain by viscous Jurassic clays. 9–13—results of the interpretation of the authors of the article. Tectonic structures of the Southern Urals: I—Pre-Ural trough; II—West Ural megamonoclinorium (advanced folds of the Urals); III—Central Ural uplift; IV—Magnitogorsk megasynclinorium; V—East Ural meganticlinorium; VI—Trans-Ural megasynclinorium. Regional faults: 1—Surensky; 2—West Uralsky; 2a—Mam-betshinsky; 3—West Uraltaussky; 4—East Uraltaussky; 5—Za-padno-Irendykskу; 6—East Irendyksky; 7—West Kizilsky; 8—East Kizilsky; 9—West-Akzharsky; 10—Vostochno-Akzharsky; 11—Arkhangelsk; 12—Dombarovsky; 13—Vostochno-Terensaysky; 14—Kryklinsky; 15—Jarlinsky; 16—Maldygusaysky; 17—Severo-Dzhusinsky; 18—Zapadno-Anikhovsky (Elenovsky); 19—Sazdinsky; 20—Ushkotinsky; 21—Vostochno-Anikhovsky; 22—Veselovsky; 23—Zheltinsky; 24—Zapadno-Saryobinsky; 25—East-Saryobinsky; 26—Jatiga-rinsky; 27—West-Kungusai; 28—Miylisai. Careers: 1—Spring; 2—Autumn; 3—Kiembaevsky; 4—Belozerskay; 5—Yuzhno Kirovsky. Mines: 1—Gayskaya; 2—Dzhusinskaya

are taken into account. With a high probability, they are natural or technogenically induced.

2 Mapping of Tectonic Disturbances

In the central part of the Magnitogorsk megasynclinorium, two areas with a minimum number of epicenters of seismic events have been identified. One is located north of the Suunduk River, the second is south of the latitudinal riverbed (1). It can be assumed that the events localized in them are mainly related to geological factors.

On sheet N-40-XXXV, the Kizilsky fault is second only to the Main Ural Fault in contrast of manifestation [3]. It is the border between the West Magnitogorsk and East Magnitogorsk structural and formation zones. Along the fault, the Devonian sediments from the west are pushed over the carboniferous. The violation is ancient, but it also has neotectonic activity.

On the territory of Bashkortostan, the Kizilsky fault zone has been studied by the Ural profile, including wells drilled in the 1970s of the last century, in order to assess the prospects of oil and gas potential of the region (Fig. 1). Their depths are from 2407 to 5010 m. The author's version of the interpretation of drilling results [4, 5] is shown in Fig. 2. The thrusts with both western and eastern falling of the displacer planes are highlighted. The Kizilsky fault limits the rise of the same name. It should be noted that the author's version of the interpretation of drilling results according to the Ural profile of Tagirov [6] has fundamental differences from the version of T. T. Kazantseva. According to it, well 5 does not reveal the tectonic block pushed from the east onto the limestones of the Kizil formation.

Within the M-40 sheet, the Kizilsky fault is called the West Kizilsky. According to the Kizilsky fault, the ratio of tectonic blocks correspond to an upsurge, and according to the West-Kizilsky—a steep discharge with a plane of the displacer falling to the east [1]. A complex tectonic zone is confined to the Kizilsky fault from the east (Fig. 2). It corresponds to a linear zone of increased seismic activity with a width of 12 km (Fig. 1). To the east of its boundary, seismic activity is minimal. To the west—the minimum only on the section connected to the northern part of the zone. The southern part of the zone of increased activity overlaps the channel of the Ural River. This is one of the signs of a tectonic disturbance. Thus, the most contrasting tectonic element of the region—the zone adjacent to the Kizilsky fault from the east, was reflected in the field of seismic events.

The zone of development of Mesozoic deposits corresponds to the area with the minimum number of points of seismic events south of the latitudinal channel of the Or river. It was studied in 1991–1992 by the regional seismic survey of MOGT (JSC "Orenburg GE"). Three profiles with a total length of 143 km have been worked out. According to these data, signs of tectonically shielded and lithological hydrocarbon traps have been identified [7, 8].

According to one of the seismic profiles, an anticlinal inflection of the axes of synphase of reflected waves was revealed. In order to assess the prospects of the

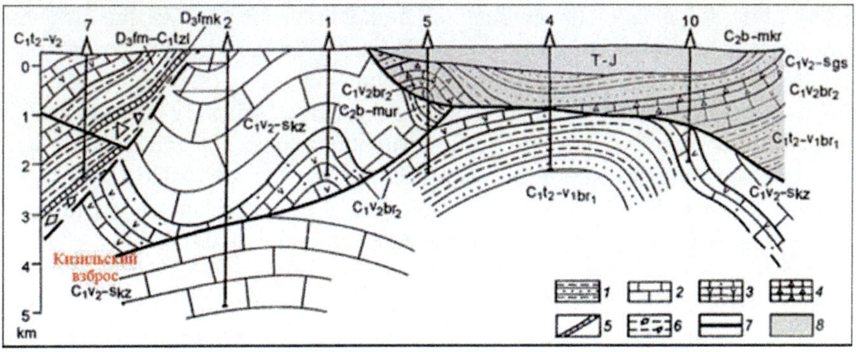

Fig. 2 Schematic geological section through the Kizilsky zone of the Magnitogorsk megasynclinorium according to deep well drilling data (Ural profile) [5]. 1—Sandstones, siltstones, mudstones; 2—limestones; 3—tuff sandstones, tuffs; 4—siliceous limestones; 5—flints; 6—brecciated rocks; 7—thrusts; 8—Gushikhinsky allochthon. Upper Devonian: D_3f mk—Fransky tier, Mukasovsky horizon; D_3fm-C_1t1 zl—Famensky tier—Lower Carboniferous, Lower Tournaisian sublayer, Zilair formation. Lower carboniferous: $C_1 t_2$-v_1 br_1—Upper Tournaisian—Lower Visean sublayers, lower part of the Berezovskaya formation; $C_1 v_2$ br_2—Tula horizon of the Upper Visean sublayer, upper part of the Berezovskaya formation; $C_1 v_2$-s kz—Upper–Viseian sublayer—Serpukhov tier, Ki-Zil formation; $C_1 v_2$-s gs—Upper-Viseian sublayer—Serpukhov tier, Gusikhinsky formation. Middle Carboniferous: C_2b-m ur—Bashkir and Moscow tiers, Urtazim formation; C_2b-m kr—Bashkir and Moscow tiers, Kardailovskaya formation

oil and gas potential of the section in 1993, the parametric well 1 Aschebutakskaya was drilled here. Its depth is 1261.2 m. To a depth of 303.5 m, the section is represented by gray dense viscous clays of Jurassic age with rare layers of sand, below to the bottom—a thickness of massive, thick-layered, organogenic detritus limestones of light gray and gray color. The fauna of the Visean age has been identified in them. There are intervals with intense fracturing in the limestones. In places they are cavernous. The caverns are healed with calcite. According to seismic data, the thickness of the limestone thickness is 1000–1300 m. According to the results of geological survey, their age is Visean-Bashkir [7].

Three seismic profiles are located in a zone with the same type of structure: limestone strata overlain by terrigenous deposits of the Jurassic. The exception is the eastern part of profile 2. In the time section at pickets 190–200 there is a sharp change in the type of wave field. Two tectonic blocks with different seismogeological characteristics are interfaced here, presumably along the East Kizilsky fault. Profile 3 is entirely located in the area with the same type of structure noted above. That is, the fault can be located only to the east of its end. The southern part of the fault is mapped by a linear zone of increased seismic activity. On the territory of the M-40 sheet, it corresponds to the eastern boundary of the development of Jurassic deposits. The development of the West Kizilsky fault to the south according to the data of the seismic survey of the MOGT is excluded.

The Vostochno-Uraltaussky fault is part of the Main Ural Fault system. It is possible that a linear zone of increased seismic activity corresponds to it on the territory of the Republic of Kazakhstan. We do not associate a similar linear zone, which seems to be a southern extension of the West Ural fault, with it. On the territory of the Republic of Kazakhstan, it turns to the south-southeast.

According to the spatial distribution of the epicenters of seismic events, diagonal tectonic disturbances appeared (Fig. 1). Two violations "interrupt" the Ushkotinsky fault. In the southwestern part of sheet N-40-XXXV, they correspond to a system of non-major tectonic disturbances. At the time section of the seismic survey profile 1, at the point of its intersection with the diagonal violation, the specialists of JSC Orenburg GE identified a fault (Fig. 3). In our opinion, a zone of tectonic disturbance appeared here in the seismic field. It is associated with an anticlinal inflection along the roof of the limestone strata and the lower part of the terrigenous deposits of the Jurassic. Perhaps a fault trap of hydrocarbons is localized here.

According to the geodynamic model of Puchkov [9], the Magnitogorsk Island Arc formed in the Devonian in the Paleouralian Ocean opposite the passive margin of the Baltic continent. In the late Devonian, there was an oblique collision of the continent and the island arc. The formation of folds of southeastern vergence took place here [10]. The authors of the last publication worked out this issue when studying the Maksyutovsky metamorphic complex of Uraltau. At one of the stages of the tectonic development of the region, previously formed structures experienced a shift in the

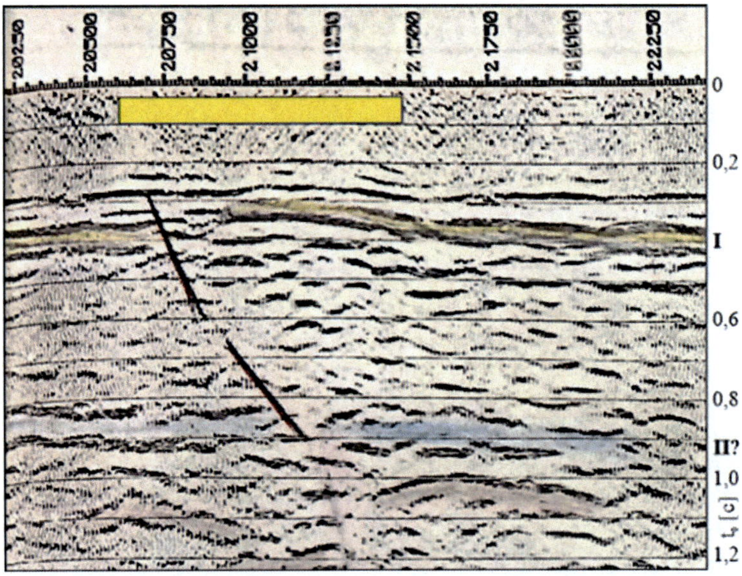

Fig. 3 Magnitogorsk megasynclinorium. A fragment of a time section according to the regional seismic survey profile 1. Reflecting boundaries I and B—the roof and the sole of the limestone strata of the Viseisko-Bashkir age. The black line shows the fault (JSC "Orenburg GE", V. I. Loshmanov, 1993). The yellow rectangle is the area of intersection of the profile with a diagonal tectonic disturbance, taking into account the accuracy of its localization

south-easterly direction. It remains to be assumed that diagonal tectonic disturbances are suture zones between blocks that shifted by different distances.

A model of its tectonic development has been developed based on the results of kinematic analysis of discontinuous faults in the Astafiev crystal-bearing field (Southern Urals) [11]. At a certain stage, shear movements along the uplift (suture zones) of the north-western strike became more active. The author of the publication [12] considered the development models of the Southern Urals at the stage of the Late Paleozoic collision of the Eastern European and Kazakhstan continents. One of the stages of deformations (C2-P) in the Magnitogorsk megazone is shear formation in the mode of left-sided transgression. The axes of the paleo-stress fields in the region have a north-western strike.

Thus, the existence of diagonal tectonic disturbances in the region directly follows from the development models of the Southern Urals. But as a system, they are not highlighted on the M-40 sheet [1].

3 Clarification of Oil and Gas Potential

The geological study of the Southern Urals was carried out mainly in the direction of "Geological mapping and the search for solid minerals". The work on assessing the prospects of the oil and gas potential of the region was carried out in a small volume. In the Orenburg segment of the region, the advanced folds of the Urals were studied by the regional seismic survey of the MOGT [13]. Three regional seismic survey profiles CAN be worked out in the Magnitogorsk megasynclinorium. According to the available geological and geophysical data, scientific substantiation of the prospects of oil and gas potential of the Southern Urals is given in the publications of 2021 [7, 13–16].

Oil and gas occurrences were noted in three wells of the Ural profile. In the core of well 5 (deposits of the Urtazim, Kizil and Gushikh formations), thick oil was present in the pores and cracks, as well as effusions of liquid oil. In the core of well 1 (Berezovskaya formation), oil-stained crack walls are marked. When drilling well 4, the smell of petroleum gas was recorded [4]. When drilling well 2, intensive absorption of the washing liquid was noted in the riphogenic limestones of the Kizil formation at a depth of 1248 m, in the range of 2910–2940 m and at depths of 3085 and 4195 m. This indicates the presence of collectors with good filtration and capacitance properties in its section. Water inflows were obtained from the intervals of 2786–2815 and 3856–3932 m. In well 4 from the deposits of the Kizil formation, its flow rate was 62.4 m3/day. In the section of the well, three reservoir layers were identified in the Kizil and Berezovskaya formations [17]. In well 1 Aschebutakskaya, an inflow of reservoir water was obtained from the interval of 300–320 m, presumably from gravel-pebble deposits lying in the bottom of the Jurassic strata.

The authors of the publication [7] associate the prospects of the oil and gas potential of the Magnitogorsk megasynclinorium primarily with limestone of the Visean-Bashkir age overlain by viscous clays (tire) of the Jurassic. The southern part of the development zone of this type of section is characterized by minimal seismic activity

(Fig. 1), which is a favorable factor for the preservation of oil and gas deposits. In the northern part of the zone, seismic activity is increased. Here it is possible to predict the development of open fracturing in limestone, which is a favorable factor for obtaining high inflows of hydrocarbons. But there are risks that the tightness of the tire is broken. Most likely, the limestone strata in the east is shielded by the East Kizilsky fault. Here, the authors of the article predict tectonically shielded traps. Thus, according to the peculiarities of the distribution of epicenters of seismic events, the megasynclinorium zone, the most promising for oil and gas, is zoned.

4 Conclusions

Based on drilling, geological survey and seismic survey data, the interpretation of the distribution features in the Orenburg segment of the Southern Urals of the epicenters of seismic events, presumably associated with natural factors, was performed. Below are the geological and methodological results.

1. Mapping of the position of the southern part of the East Kizilsky fault has been carried out. It is shown that the West Kizilsky fault, mapped according to the results of geological survey, does not develop to the south.
2. The geological hypothesis is substantiated: there is a system of diagonal tectonic disturbances of the north-western strike in the region.
3. The zoning of the Magnitogorsk megasynclinorium zone, the most promising for oil and gas (limestone of the Visean-Bashkir age, overlain by viscous clays of the Jurassic), was carried out.
4. According to the peculiarities of the distribution in the region of the epicenters of seismic events associated with natural factors, it is possible to solve regional geological and prospecting tasks, as well as to identify tectonic disturbances that determine the prospects of oil and gas potential of local areas.
5. The reliability of forecasts will increase with the development of the database of observed events and increasing the accuracy of their localization. The latter can be achieved by installing at least two new seismic stations in the region.

References

1. Lyadsky P.V., Kvasnyuk L.N., Zhdanov A.V., Chechulina O.V., Shmelkov N.T., Belts G.M., Kurochkina E.S., Olenitsa T.V. State Geological map of the Russian Federation. Scale 1:1000000 (third generation). The Ural series. Sheet M-40 (Orenburg) with valve M-41. // Explanatory note. Saint Petersburg: VSEGEI Cartographic Factory, 2013. 392 p. (in Russian)
2. Belov V.S., Nesterenko M.Yu. Monitoring of geodynamic processes in the eastern part of the Orenburg region // International Research Journal. 2020. No. 98(8). pp. 22–30. https://doi.org/10.23670/IRJ.2020.98.8.037 (in Russian)

3. Montin S.A. State Geological map of the Russian Federation. Scale 1:200000. Second edition. The South Ural series. Sheet N-40-XXXV (Baymak). // Explanatory note. M.: MT VSEGEI. 2015. 188 p. (in Russian)
4. Kazantseva T.T. Geological structure of the Kizilsky zone of the Magnitogorsk megasinform in the light of new data // NTZH Geology, geography and global energy. 2009. No. 3. pp. 27–32. (in Russian)
5. Kazantseva T.T., Valiullin R.A. To the problem of the oil and gas potential of the Magnitogorsk synform // Oil and gas business. 2021. No. 5(19). pp. 6–19. https://doi.org/10.17122/ngdelo-2021-5-6-19 (in Russian)
6. Tagirov I.A. New data on the geological structure of the Magnitogorsk synclinorium // Tectonics and oil-bearing capacity of the Urals and the Urals. Ufa: BFAN USSR. 1978. pp. 7–11. (in Russian)
7. Politykina M.A., Tyurin A.M., Makarov S.E., Petrishchev V.P., Pankratiev P.V., Bagmanova S.V., Kolomoets A.V. Prospects of oil and gas potential of the Orenburg segment of the Magnitogorsk trough // Oil and gas geology. Theory and practice. 2021. № 1(16). (in Russian)
8. Tyurin A.M. The possibilities of seismic exploration of the MOGT in the study of the prospects of oil and gas potential of the Magnitogorsk trough // University complex as a regional center of education, science and culture. Orenburg: Orenburg State University. 2021. 1122–1126. (in Russian)
9. Puchkov V.N. Geology of the Urals and the Urals (topical issues of stratigraphy, tectonics, geodynamics and metallogeny). Ufa: Designpoligrafservice. 2010. 280 p. (in Russian)
10. Golionko B.G., Ryazantsev A.V. Folded structures of the southeastern and western vergence of the Maksyutovsky metamorphic complex of the Southern Urals // Bulletin of St. Petersburg University. Earth Sciences, 2020. No. 65(3). pp. 506–527. https://doi.org/10.21638/spbu07.2020.306 (in Russian)
11. Ogorodnikov V.N., Polenov Yu.A. The role of suture zones of the Urals in the formation of ore and crystal-bearing quartz-vein deposits. // Proceedings of the Institute of Geology and Geochemistry. academician A.N. Zavaritsky. 2018. No. 165. pp. 60–66. (in Russian)
12. Znamensky S.E. Stages of deformations and the mechanism of formation of the structure of the eastern slope of the Southern Urals at the stage of general collision // Geodynamics. Deep structure. The thermal field of the Earth. Interpretation of geophysical fields. Yekaterinburg: Ural Branch of the Russian Academy of Sciences. 2009. pp. 190–194. (in Russian)
13. Politykina M.A., Tyurin A.M., Makarov S.E., Petrishchev V.P., Pankratiev P.V., Bagmanova S.V. Prospects of oil and gas potential of the Orenburg segment of the advanced folds of the Urals // Geology of oil and gas. (2021-b). No. 6. pp. 59–71. https://doi.org/10.31087/0016-7894-2021-6-59-71 (in Russian)
14. Kolomoets A.V., Tyurin A.M., Snachev A.V., Smoleva I.V. Black shales of the Kumak ore field and prospects of oil and gas potential of the Magnitogorsk trough of the Southern Urals // New ideas in the geology of oil and gas. New Reality. Moscow: Moscow University, 2021. pp. 214–217. (in Russian)
15. Tyurin A.M., Politykina M.A., Makarov S.E., Kolomoets A.V. Signs of oil and gas potential of the Central Ural uplift (Orenburg segment) // New ideas in the geology of oil and gas. New Reality 2021. Moscow: Moscow University. 2021-a. p. 605–609. (in Russian)
16. Tyurin A.M., Politykina M.A., Makarov S.E. (2021-b) Oil-producing deposits of the Orenburg segment of the advanced folds of the Urals. Oil and Gas Business, 6(19), 21–33. (in Russian)
17. Popov V.G., Abdrakhmanov R.F. Hydrogeological prerequisites of oil-news of the Kizilsky zone of the Southern Urals // Bulletin of the Institute of Geology of the Komi Scientific Center of the Ural Branch of the Russian Academy of Sciences. 2017. No. 1(265). pp. 3–7. (in Russian)

Development of Mathematical Models for Optimization of the Road Network, Taking into Account Geographical and Natural Factors

Vladimir Valentinovich Nikitin, Vyacheslav Gennadievich Kozlov, Aleksey Vasilyevich Skrypnikov, Galina Anatolyevna Pilyushina, Dmitry Gennadievich Kozlov, Andrey Nikolaevich Bryukhovetsky, Irina Alevtinovna Vysotskaya, and Vladimir Vladimirovich Denisenko

Abstract When developing schemes for the transport development of the forest area, it is of great importance to correctly establish the optimal distance between the main portages. The complexity of choosing the optimality criterion for forest transportation design is determined, on the one hand, by the duration of developing the forest resource base in time and, on the other hand, by the static nature of the mathematical models being developed, in which time is not a variable. The sum of road and transport costs is nothing more than the operating costs for the development of the base, consisting of two components—road and transport. Thus, the minimum amount of road and transport costs for the development of the forest resource base is taken as the optimality criterion in static mathematical models. The presented methodology

V. V. Nikitin (✉)
Mytishchi Branch of Bauman Moscow State Technical University (National Research University), 1, 1st Institutskaya Street, Mytischi 141005, Russia
e-mail: info@aspirans.info

V. G. Kozlov
Faculty of Agricultural Engineering, Department of Operation of Transport and Technological Machines, Voronezh State Agrarian University named after Emperor Peter the Great, 1, Michurina Ave., Voronezh 394087, Russia

A. V. Skrypnikov · A. N. Bryukhovetsky · I. A. Vysotskaya
Dean of the Faculty of Management and Informatics in Technological Systems, Voronezh State University of Engineering Technologies, 19 Revolyutsii Ave., Voronezh 394036, Russia

G. A. Pilyushina
Faculty of Mechanics and Technology, Department of Tribotechnical Materials Science and Materials Technologies, Bryansk State Technical University, 7, 50 Let Oktyabrya Boulevard, Bryansk 241035, Russia

D. G. Kozlov
Faculty of Agricultural Engineering, Department of Electrical Engineering and Automation, Voronezh State Agrarian University named after Emperor Peter the Great, 1, Michurina Ave., Voronezh 394087, Russia

V. V. Denisenko
Department of Higher Mathematics and Information Technologies, Voronezh State University of Engineering Technologies, 19 Revolyutsii Ave., Voronezh 394036, Russia

V. I. Karev (ed.), *Physical and Mathematical Modeling of Earth and Environment Processes—2022*, Springer Proceedings in Earth and Environmental Sciences,
https://doi.org/10.1007/978-3-031-25962-3_32

allows considering the presence of natural, technological, and economic factors. When arranging an optimal road network, it is necessary to take into account the developed complex of intelligent information systems, which allows obtaining an option close to the optimal in terms of economic indicators.

Keywords Unit costs · Logging road · Optimization · Cutting area · Forest transportation

1 Introduction

When studying the issue of optimizing the distances between forest transport routes in the specific conditions of logging encountered in Russia, it is necessary to take into account the following basic provisions:

1. The calculation methodology being developed should pay special attention to the consideration of natural factors and, above all, the requirements of forest management, i.e. take into account the influence of forestry management methods and types of logging for main use on the value of optimal distances between the paths of primary and secondary forest transportation.
2. Due to the concentration of logging operations, it is necessary to take into account the long service life of the tracks. Even skidding tracks are used in these conditions from one to several years, while logging roads for a long time, about several decades.
3. The forest transport road network must ensure both the transportation of wood and the implementation of forestry measures during the entire period of forest exploitation.

2 Methods

When placing a network of paths for primary transportation of wood, the initial data are the width of the cutting area (l_w), the established length, which should be no more than the length of the planning quarter, i.e. within 0.5–0.6 km, as well as the layout of the cutting area of forest reserves provided for cutting [1–3].

Figure 1 shows a scheme for the development of cutting areas with a felling-and-bunching machine designed for the application of this type of machine in narrow cutting areas. According to these schemes, the cutting strips worked up in one trip of the felling-and-bunching machine are located either along the long side of the cutting area or across it.

The latter option requires a sufficiently significant width of the cutting area, as well as additional placement of loading and assembly lines, adjacent to the main portages, located outside the cutting area. With a small distance from the cutting area to the upper warehouse, the bundle assembly is combined with its further transportation to the warehouse. At a considerable distance, it may be economically feasible to

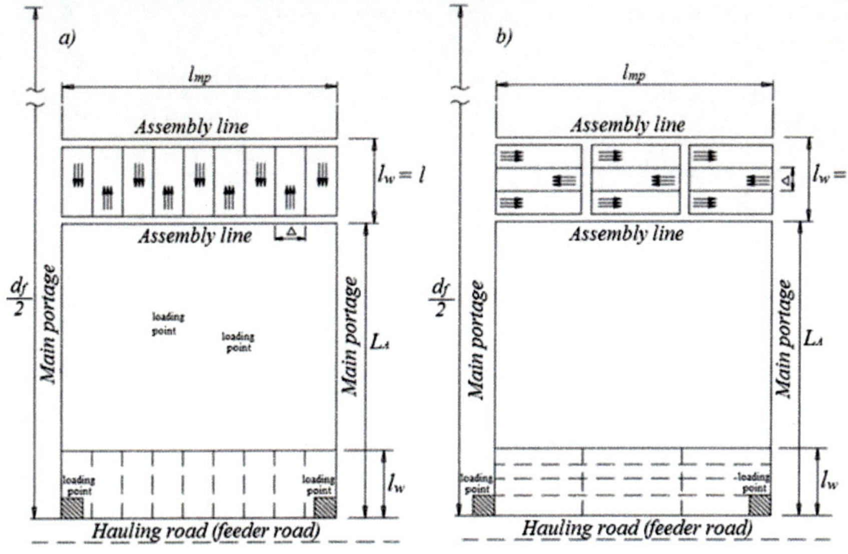

Fig. 1 Schemes for the development of narrow cutting areas

use two types of machines—one machine, working on the rolling cycle, used for assembling bundles, and the other one—for transporting bundles from the cutting area to the warehouse [4].

The length of the cutting strip worked up in one trip of the skidding machine, is determined by the formula:

$$l = \frac{10{,}000 \cdot Q_{wl}}{\Delta \cdot \gamma_l} \tag{1}$$

where Q_{wl} is the working load per trip, m^3; γ_l is the liquid reserve per hectare of cutting area, m^3; Δ is the width of the bundle strip.

In formula (1), the parameter Δ determines the technological factors of the skidding equipment operating in the cutting area.

At a tree stand of 22 m or more in height and at dense soils, it is recommended to take the width of the area worked out by one assembly line within the range from 40 to 50 m. At moist soils, low tree stand (below 22 m), and deep snow cover, this parameter is taken within 30–40 m. At the width of the cutting areas in the range from 40 to 80 m for flat terrain, and 15–30 m for mountainous conditions, it will be enough to place 1–2 assembly portages adjacent to the main one within the cutting area [5–7].

When designing the transport development schemes of the forest area based on the timber-hauling transport network, it is of great importance to correctly establish the optimal distance between the main portages. To do this, we will use the calculation

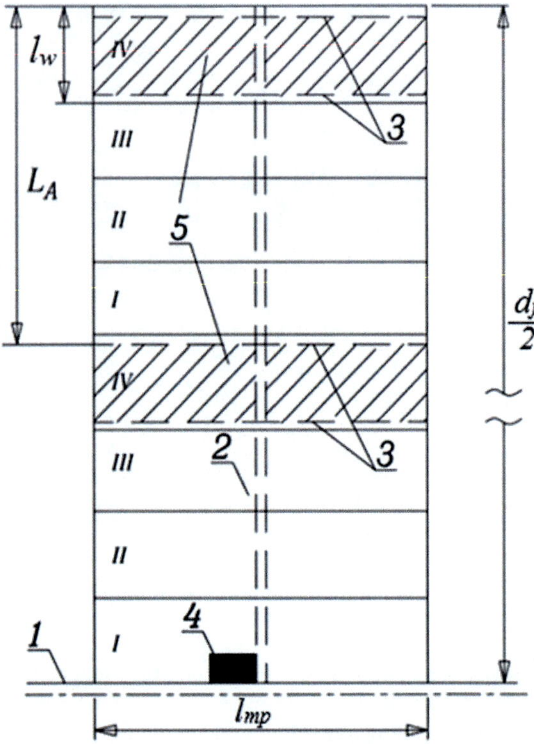

Fig. 2 The design scheme of working out the operational site using forest transport: 1—feeder road; 2—main portage; 3—assembly portages; 4—loading point on logging rolling stock; 5—cutting areas

scheme (Fig. 2), in which the moves of the felling-and-bunching machine are carried out as shown in Fig. 1a and b.

In this diagram, Roman numerals indicate logging areas per single logging year, whose initial areas are located at a L_A distance from each other:

$$L_A = l_w \cdot N \tag{2}$$

where l_w is the width of the cutting area, km; N is the number of logging approaches

$$N = \frac{A}{u}, \tag{3}$$

where A is the period during which it is planned to cut down a mature and overgrown forest, years; u is the period of adjacent cutting areas ($u = 3$–5 years).

According to Fig. 2, on a plot of 0.5 $d_{fr} \times l_{mp}$ km^2 with a reserve $M = 0.5d_{fr} \cdot 100l_{mp} \cdot \gamma_l$, the harvested mass per one cutting approach will be

$$M' = \frac{50d_s l_{mp}\gamma_l}{N} \tag{4}$$

where l_{mp} is the distance between the main portages; d_{fr} is the distance between the feeder roads; γ_l is the liquid stock for 1 ha, m^3.

Transporting this stock will require from skidder tractor to perform the following number of trips

$$n = \frac{50 d_{fr} l_{mp} \gamma_l}{Q_{wl} N} \tag{5}$$

The condition of not exceeding the full operability of the main portage will be fulfilled if the number of trips n, determined by the formula (5), is less than the maximum allowable, i.e.

$$n \leq n_{gon} \tag{6}$$

where n_{gon} is the maximum allowable number of trips of the skidder tractor along the main portage until its destruction (full operability of the main portage).

With known values of n_{gon} and d_s, the greatest distance between the main portages will be determined by the formula:

$$l_{mp} = \frac{Q_{wl} n_{gon}}{100 \gamma_l k_{fr} d_{fr}} \tag{7}$$

The value of n_{gon} is determined by the method described earlier. It should be noted that $|l_{mp}|_{\max}$ should not be greater than the length of the cutting area.

The value of l_{mp} also has restrictions on the ground and soil conditions of the cutting area. The total number of tractor trips for the full development of the cutting area of $l_{mp} \times l_w$ km is determined by the formula:

$$n' = \frac{100 l_{mp} l_w \gamma_l}{Q_{wl}} \tag{8}$$

The operability condition of the loading and assembly portage of the cutting area will be met if $n' \leq n'_{gon}$, where n_{gon} is the maximum allowable number of trips until the destruction of the assembly portage.

If the n', calculated by the formula (8), turns out to be greater than n'_{gon}, then it is necessary to increase the number of assembly portages. In this case, the number of tractor trips along the loading and assembly portage will be:

(a) when skidding on one side of the main portage and increasing the number of loading and assembly portages to two:

$$n' = \frac{50 l_{mp} l_w \gamma_l}{Q_{wl}} \tag{9}$$

(b) when skidding from both sides of the main portage and two loading and assembly portages (Fig. 1):

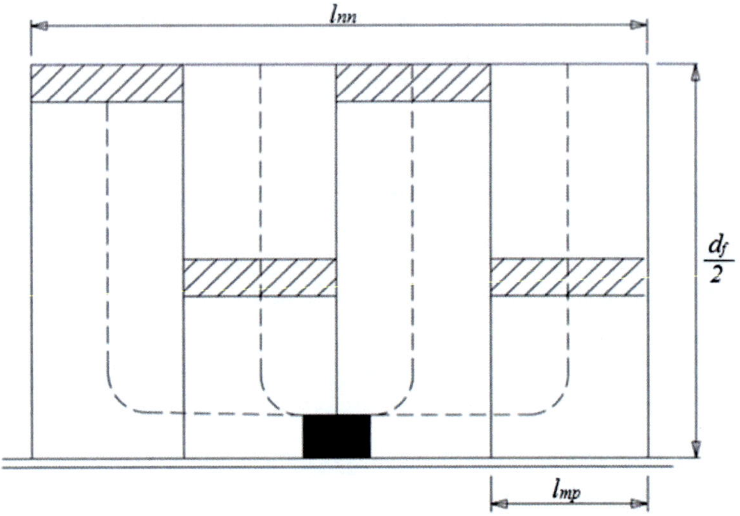

Fig. 3 Calculation scheme for determining the optimal distance between loading points (*l*nn)

$$n' = \frac{25 l_{mp} l_w \gamma_l}{Q_{wl}} \tag{10}$$

In winter, when the frozen soil has a high load-bearing capacity, it is possible to take $l_{mp} = l_{ca}$ (where l_{ca} is the length of the estimated cutting area, equal to 0.5–0.6 km), provided minimal costs for the portages arrangement and maintenance [8].

As can be seen from the diagram in Fig. 2 (the distance between loading points (l_{lp}) was assumed to be equal to the distance between the primary transportation paths (l_{mp}). However, there may also be a solution in which several main portages are adjacent to one loading point, as shown in the diagram (Fig. 3).

Let us determine the distance between the loading points (l_{lp}), at which the sum of the unit costs for the construction and maintenance of the main portages, as well as for primary transportation and the arrangement of the loading point will be minimal.

The length of the main portages, required for working out, can be taken as

$$l_{mp} = \frac{k'_p \left(d_{fr} - 0.5 l_{mp}\right) l_{lp}}{l_{mp}} + k'_p \left(l_{lp} - l_{mp}\right) \tag{11}$$

Considering the above, the unit annual costs for the construction of main portages will be equal:

$$P_{mp} = \frac{k'_p C_{por}}{50 p} \left(\frac{1}{l_{mp}} + \frac{1}{2 d_{fr}} - \frac{l_{mp}}{d_{fr} l_{lp}} \right) \tag{12}$$

where C_{por} is the reduced annual costs for the construction and maintenance of 1 km of portage, rub; l_{mp} is the distance between the main portages, km.

The unit costs for the construction of a loading point are equal to:

$$p_{lp} = \frac{C_{lp}}{50pl_{lp}d_{fr}} \tag{13}$$

Transportation costs for skidding (hauling) to the loading point are:

$$p_m = \frac{120M_{tr}\left(k_1 l_{lp} + k_2 \frac{d_{fr}}{2}\right)k'_p}{(T - t_{pf})k_{fr}Q_{wl}v_{av}} + \frac{M_{tr}\sum t_{it}}{(T - t_{pf})k_{fr}Q_{wl}} \tag{14}$$

where k_1 and k_2 are coefficients depending on the layout of the main portages.

The total unit costs for the types of expenses under consideration are

$$P_\Sigma = P_{mp} + p_{lp} + p_{tr} = \frac{k'_p C_{por}}{50p}\left(\frac{1}{l_{mp}} + \frac{1}{2d_{fr}} - \frac{l_{mp}}{d_{fr}l_{lp}}\right) + \frac{C_{lp}}{50pl_{lp}d_{fr}}$$
$$+ \frac{120M_{tr}\left(k_1 l_{lp} + k_2 \frac{d_{fr}}{2}\right)k'_p}{(T - t_{pf})k_{fr}Q_{wl}v_{av}} + \frac{M_{tr}\sum t_{it}}{(T - t_{pf})k_{fr}Q_{wl}} \tag{15}$$

After the conversion, we get:

$$P_\Sigma = \frac{1}{50p}\left(x + yl_{lp} + \frac{z}{l_{lp}}\right) \tag{16}$$

where

$$x = k'_p C_{por}\left(\frac{1}{l_{mp}} + \frac{1}{2d_{fr}}\right) + E\left(\frac{60k_z k'_p d_{fr}}{v_{av}} - \Sigma t_{it}\right) \tag{17}$$

$$y = E\frac{120k_1 k'_p}{v_{av}} \tag{18}$$

$$z = \frac{C_{lp}}{d_{fr}} - \frac{k'_p C_{por} l_{mp}}{d_{fr}} \tag{19}$$

$$E = \frac{50pM_{tr}}{(T - t_{pf})k_{fr}Q_{wl}} \tag{20}$$

It should be noted that x, y, and z are quantities, independent on l_{lp} at a certain d_s. Solving the problem of finding the extremum of the $P_\Sigma = f(l_{lp})$ function, we get

$$l_{lp} = \sqrt{\frac{z}{y}} = \sqrt{\frac{C_{lp} - k'_p C_{por} l_{mp}}{50 p d_{fr} k_1 k'_p \tau}} \text{ km} \quad (21)$$

If $y \geq \frac{z}{l_{mp}^2}$, than $l_{mp} < l_{lp}$, which means that it is expedient to transport the wood by several portages to one loading point, as shown in Fig. 3.

3 Results

Analyzing the above, it can be noted that the calculation of the average distance between the main portages (l_{mp}) and the average distance between loading points (l_{lp}) must be carried out according to this method with the established preliminary value of d_s, since the arrangement and maintenance costs of the main portages depend on their length and, consequently, on d_s. Besides, in the methodology for determining d_s, the costs of setting loading points are not taken into account. This assumption is quite justified since l_{mp} and l_{nn} largely depend on the equipment used in skidding, which in modern logging equipment changes rapidly, while the road network of permanent feeder roads is calculated for a long period of validity.

Therefore, the optimal distance between the feeder roads (d_{fr}) will be such that the sum of specific, i.e. attributed to 1 m^3 of the exported wood, feeder roads construction and maintenance costs and paths for hauling (skidding) of the wood, for example, main portages, as well as costs for primary transportation and the arrangement of loading points has a minimum value.

Due to the long period of wood reproduction, equal to the full turnover of the felling (on average 100 years), the annual freight turnover of a permanent type feeder road is determined by the average annual productivity of 1 ha of the developed forest area, and the productive forest area, adjacent to feeder road [9].

The unit costs for the construction of paths (feeder roads) of a permanent type, considering the service life of individual elements of the path, are equal to

$$C_{fr} = C_1 \frac{1}{K_1} + C_2 \frac{1}{K_1} + C_3 \frac{1}{K_2} + C_4 \frac{1}{K_3} \quad (22)$$

where C_{fr} is the reduced costs for the construction of 1 km of the feeder road, rubles; C_1 is the costs, associated with the clearing of the road strip, rubles; C_2 is the costs, associated with the construction of the roadbed, rubles; C_3 is the costs, associated with the construction of artificial structures, rubles; C_4 the costs, associated with the construction of road base, rubles; K_1 is the estimated service life of the road, for a permanent road $K_1 = 100$years; K_2 is the estimated service life of artificial structures which can be taken equal to 20–30 years; K_3 is the estimated service life of the road base, which depends on the design of the track and the type of construction materials ($K_3 = 5$–30 years).

Considering formula (22), the unit cost of building one km of the feeder road of a permanent type will be:

$$P_{fr} = \frac{\beta C_{fr} k_p}{100 p d_{fr}} \text{ rub/m}^3 \quad (23)$$

where β is the coefficient considering the construction of feeder roads partly in non-forest areas; k_p is the main road and feeder road development coefficient; p is the annual production from 1 ha of the forest, the average for the full turnover of felling, m^3; d_{fr} is the required distance between the feeder roads, km.

The specific annual costs of maintaining feeder roads in good condition are determined by the formula:

$$P_{mnt} = \frac{\beta F R_{fr} \frac{1}{D} \sin \alpha k_p}{100 p d_{fr}} \quad (24)$$

where FR_{fr} is the cost of the annual maintenance of one km of the feeder road in good condition, considering the costs of medium and capital repairs, rubles; l is the length of the operating section of the feeder road, km; D is the distance from the highway to the border of the raw material base, km; α is the contact angle of the feeder road to the highway, degrees.

The annual costs of maintaining the road in good condition consist of the costs of maintaining the roadbed, artificial structures, and road base, as well as the costs of medium and capital repairs of the road. The most significant annual expenses are those for the maintenance of road base [10–12]. Approximately the service life of road base can be taken as:

- 5 years—for a dirt road reinforced with mineral and organic binders;
- 8 years—for covering made of both gravel and soil-aggregate mixture;
- 10 years—for asphalt-concrete covering;
- 20 years or more—for covering made of cobblestone and cement-concrete mixture.

The cost of average repairs per 1 km during timber cutting cycle of k years is equal to

$$C_{av}\left(\frac{k}{n_{av}} - \frac{k}{n_{cap}}\right) = C_{av} k \left(\frac{1}{n_{av}} - \frac{1}{n_{cap}}\right) \quad (25)$$

where C_{av} is the average cost of repair per 1 km of roads, rubles; n_{av} is the service life of road base from the moment of putting into operation till medium repair, or between two repairs in years; n_{cap} is the operation life of road base until capital repairs, years, k is the period of the full timber cutting cycle in years, k = 100.

Capital repairs are usually carried out when the wear of the road reaches 40% of its construction cost. The cost of capital repairs per one km during the timber cutting cycle is equal to:

$$C_{cap}\frac{k}{n_{cap}} \tag{26}$$

where C_{cap} is the cost of capital repairs.

The sum of costs for medium and capital repairs of one km of the road per timber cutting cycle period is:

$$C_{rep} = C_{av}k\left(\frac{1}{n_{av}} - \frac{1}{n_{cap}}\right) + C_{cap}k\frac{1}{n_{cap}} = k\left[C_{av}\left(\frac{1}{n_{av}} - \frac{1}{n_{cap}}\right) + C_{cap}\frac{1}{n_{cap}}\right] \tag{27}$$

The cost of annual maintenance of one km of a feeder road in good condition, considering the costs of medium and capital repairs, is equal to

$$FR_{fr} = C_{MNT} + C_{av}\left(\frac{1}{n_{av}} - \frac{1}{n_{cap}}\right) + C_{cap}\frac{1}{n_{cap}} \tag{28}$$

where C_{MNT} is the annual cost of road maintenance, rubles.

According to the location scheme of the cutting areas and the scheme of skidding tracks, the development scheme of the forest area employing a network of permanent feeder roads and long-term main portages is shown in Fig. 4.

The unit costs for the construction of main portages are determined as follows. The annual average production from a forest plot with an area of $l_{mp} \times 0.5d_{fr}$ km^2 is:

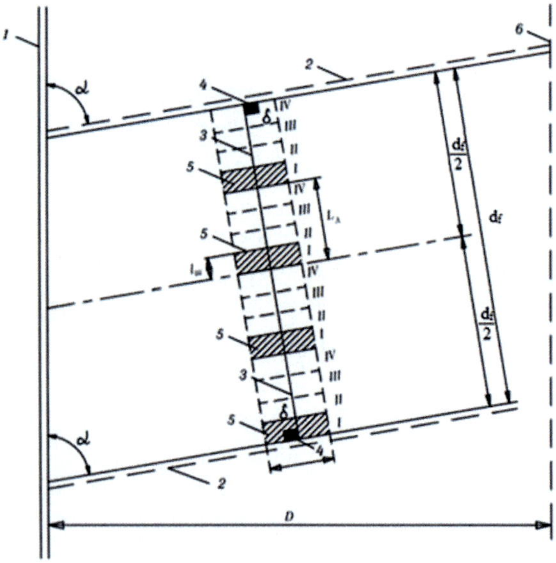

Fig. 4 The scheme of transport development of the forest area employing a network of permanent feeder roads and long-term main portages; 1—main portage; 2—feeder roads; 3—main portages beyond cutting areas; 4—loading points; 5—cutting areas of a single year of logging; 6—the boundaries of the raw material base of the hauling road

$$M_{an} = \frac{0.5 \cdot 100\gamma_l l_{mp} d_{fr}}{A} \tag{29}$$

Accordingly, the unit cost of portage construction is as follows:

$$P_{mp} = \frac{C_{por}\frac{1}{N} \cdot \frac{1}{u} \cdot k_p'}{M_{an}} = \frac{C_{por}k_p'}{100 \cdot 0.5\gamma_l l_{mp} d_{fr} \sin\sigma} \tag{30}$$

The value of l_{mp} is expressed using the formula (30)

$$P_{mp} = \frac{C_{por}\frac{1}{N} \cdot \frac{1}{u} \cdot k_p'}{100\gamma_l \frac{Q_{wl} \cdot n_{add}}{100\gamma_l \cdot k_{fr} d_{fr} \sin\sigma}} = \frac{C_{por}k_p' \cdot k_{fr} \cdot d_{fr}}{Q_{wl} \cdot n_{add} \sin\sigma} \tag{31}$$

where C_{por} is the cost of construction of one km of the portage, rubles; σ is the contact angle of the feeder roads to the hauling road, degrees; k_p' **is the coefficient** of development of the portage line; k_{fr} **is the coefficient considering the method of** timber delivery to the logging road, equal to $k_{fr} = 1$ when skidding on one side, and $k_{fr} = 0.5$—when skidding the timber on both sides of the logging track [15–18].

Production experiments on repairing tractor portages have shown that after the repair of the portage, its performance increases several times compared to the original one. However, the very nature of the repair of the portage by laying tree branches in the track complicates performing repeated repairs. Based on the possibility of performing a portage repair only once, the maximum possible number of tractor passes along the portage, considering its repairs, is:

$$n_{add} = n_{b.r.} + n_{rep} \tag{32}$$

where $n_{b.r.}$ is the possible number of passes of the tractor along the portage before repair; n_{rep} is the same, after repair.

If the cost of one repair of 1 km of portage is equal to (a) rubles, then the total annual repair costs can be expressed by the formula:

$$C_{rep} = \frac{a}{N} \tag{33}$$

The cost of the annual maintenance of one km of the portage during its operation, considering the costs of its repairs during operation, is equal to:

$$FR_{por} = C_{mnt}' + C_{rep} = C_{mnt}' + \frac{a}{N} \tag{34}$$

where C_{mnt}' is the cost of portage maintenance during its operation, rub/km.

It should be noted that during operation, the main portage is used not for its entire length. The average length of the operating section of the portage, with one-sided skidding, is:

$$l_{oper} = d_{fr} - 0.5l_w(N-1) \tag{35}$$

And at two-sided, as shown in Fig. 4, the average length is

$$l_{oper} = 0.5\left[d_{fr} - l_w(N-1)\right] \tag{36}$$

Let us determine the unit costs for the repair and maintenance of the portages at double-sided skidding:

$$p_{cog.b} = \frac{FR_{por} \cdot l_{oper} \cdot k_p^{'}}{\frac{100 \cdot 0.5 l_{mp} d_{fr} \gamma_l}{N}} = \frac{FR_{por} N k_p^{'}\left[d_{fr} - l_w(N-1)\right]}{100 l_{mp} d_{fr} \gamma_l} \tag{37}$$

Instead of l_{mp}, we insert its value according to the formula (24)

$$p_{cog.b} = \frac{FR_{por} N k_p^{'}\left[d_{fr} - l_w(N-1)\right] k_{fr}}{n_{add} Q_{wl} \sin\delta} \tag{38}$$

The unit costs of primary transportation of the timber, considering the skidding from the stump to the main portage, are equal to:

$$p_{tr} = \frac{M_{tr}\left(\frac{120 l_{av}}{V_{av}} + \sum t_{it}\right)}{\left(T - t_{pf}\right) \cdot k_{fr} * Q_{wl}} \tag{39}$$

where M_{tr} is the cost of a tractor shift (including staff salaries), rubles/shift; $\sum t_{it}$ is the total tractor idle time per trip, min; V_{av} is the average speed of the tractor along the main and apiary portage in both directions, km/h; T is the duration of the working shift, min; t_{pf} **is the preparatory and final time, min;** k_{fr} is the labor utilization rate; Q_{wl} is the working load per the tractor trip, m^3; l_{av} is the average distance of transportation, which, considering skidding from the stump to the main portage, can be taken equal to:

$$l_{av} = \frac{k \cdot d_{fr} \cdot k_p^{'}}{\sin\delta} + \frac{l_{mp} k_p^{'}}{4} = \frac{k \cdot d_{fr} \cdot k_p^{'}}{\sin\delta} + \frac{n_{add} Q_{wl} k_p^{'}}{100 \cdot 4 k_{fr} d_{fr} \gamma_l} \tag{40}$$

Considering formula (39), formula (40) has the form

$$p_{tr} = \frac{120 M_{tr} k d_{fr} \cdot k_p^{'}}{\left(T - t_{pf}\right) k_{fr} Q_{wl} V_{av} \sin\delta} + \frac{0.3 M_{tr} n_{add} k_p^{'}}{\left(T - t_{pf}\right) k_{fr} V_{av} k_{fr} d_{fr} \gamma_l} + \frac{M_{tr} \sum t_{it}}{\left(T - t_{pf}\right) k_{fr} Q_{wl}} \tag{41}$$

The total annual unit costs per type of expense under consideration are:

$$p_\Sigma = \frac{\beta k_p C_{FR}}{100pd_{fr}} + \frac{\beta k_p FR_{fr}\frac{l}{D}\sin\alpha}{100pd_{fr}} + \frac{k'_p C_{por} k_{bem} d_{fr}}{n_{add} Q_{wl}\sin\delta} + \frac{k'_p FR_{por} N\left[d_{fr} - l_w(N-1)\right]k_{bem}}{n_{add} Q_{wl}\sin\delta} + \frac{120M_{tr}kd_{fr}k'_p}{(T - t_{pf})k_{fr}Q_{wl}V_{av}\sin\delta} + \frac{0.3M_{tr}n_{add}k'_p}{(T - t_{pf})k_{fr}V_{av}k_{bem}d_{fr}\gamma_l} + \frac{M_{tr}\sum t_{it}}{(T - t_{pf})k_{fr}Q_{wl}} \quad (42)$$

Solving the problem of finding the extremum of the $p_\Sigma = f(d_{fr})$ function and considering that $k_{bem} = 2k$, we get

$$d_{fr} = \sqrt{\frac{\frac{\beta k_p}{100p}\left(C_{FR} + FR_{fr}\frac{l}{D}\sin\alpha\right) + \frac{0.15n_{add}k'_p}{k\cdot\gamma_l}\cdot b}{\frac{k\cdot k'_p}{Q_{wl}\sin\delta}\left[\frac{2}{n_{add}}\left(C_{por} + FR_{por}N\right) + 120b\right]}} \quad (43)$$

where

$$fr = \frac{M_{tr}}{(T - t_{pf})k_{fr}V_{av}} \quad (44)$$

It follows from this formula that an increase in the operability of the primary timber transportation routes, i.e., n_{add} contributes to an increase in the optimal distance between the feeder roads d_{fr}. The value of $\frac{l}{D}\sin\alpha$ determines the location of the raw material base site, for which d_{fr} is determined.

Directly at the highway $\frac{l}{D}\sin\alpha = 0$, and at the border—this value is equal to unity. In this formula, the parameters $k,\ k'_p,\ k,$ and δ characterize the terrain conditions.

Determining the optimal distance between feeder roads in a mountainous area has significant features. First of all, it should be noted that here, in most cases, the forest is skidded along the logging route from only one side, which is taken into account in the formula under consideration using the coefficient k. Besides, in case of a significant steepness of the slope, the skidding lugs should be placed at an angle $\delta < 90°$.

At $\delta < 90°$, the skidding distance increases, and thus the unit costs of primary timber transport increase, which leads to a decrease in the distance between roads [13–15].

To analyze the proposed formulas, we define d_{fr} by taking the following quantities included in the formulas: $k_p = 1.1$; $k'_p = 1.3$; $\beta = 1.1$; $\frac{l}{D}\sin\alpha = 0.5$; $M_{tr} = 36.4\frac{\text{rub}}{\text{shift}}$; $V_{av} = 8\frac{\text{km}}{\text{h}}$; $Q_{tot} = 3\ \text{m}^3$; $T = 480\ \text{min}$; $t_{pf} = 40\ \text{min}$; $k_{fr} = 0.9$; $k = 0.25$; $\delta = 45^0$; $k_{bem} = 0.5$; $\gamma_l = 200\ \text{m}^3$; $n_{add} = 2000$; $C_{por} = 250\frac{\text{rub}}{\text{km}}$; $FR_{por} = \frac{100\ \text{rub}}{\text{km}}$; $N = 4$; $p = \frac{\frac{4\text{m}^3}{\text{ha}}}{\text{year}}$. Based on the data obtained, we have plotted $d_{fr} = f(p)ud_{fr} = f(C_{fr})$ graphs, shown in Fig. 5a and b.

In formula (44), the parameters p and N characterize the method of forest management. With its intensive management, the value p increases. From this, it can be

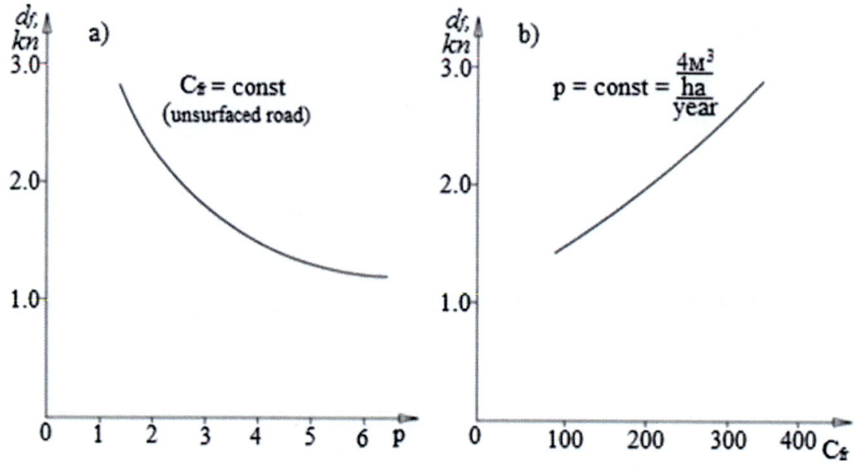

Fig. 5 Dependences of $d_{fr} = f(p)ud_{fr} = f(C_{fr})$

concluded that more intense methods of forest management require a denser road network with the same values of other parameters, as shown in Fig. 5a. This formula also shows the effect of soil conditions on the distance between the forest transport routes.

4 Conclusion

As can be seen from Fig. 5b, an increase in the cost of construction and maintenance of hauling roads leads to an increase in the distance between them. The optimal distance between unsurfaced roads without an improved carriageway in our case is 1.52 km; for unsurfaced roads, reinforced with mineral and organic binders—this distance is 1.77 km, for gravel roads—2.10 km; and for roads with gravel base—2.68 km.

Thus, the calculation according to the proposed methodology takes into account natural, technological, as well as economic factors.

When placing an optimal road network in the forest, using intelligent information systems for this purpose should be considered a promising method, since it allows obtaining an option, close to the optimal in terms of economic indicators. We see further improvement of this method we see in the justification of the number of the initial points of wood concentration that will allow for such road network placement, which ensures the development of forest reserves closest to the optimal skidding distances.

References

1. Babkov V.F. *Dorozhnye usloviya i bezopasnost' dvizheniya* [Road conditions and traffic safety]. (Transport, Moscow, 1993).
2. Birulya A.K. *Ekspluatatsiya avtomobil'nykh dorog* [Motor road maintenance] (Transport, Moscow, 1966).
3. Chernyshova E.V. *Algoritm resheniya zadachi optimal'nogo trassirovaniya lesovoznoi avtomobil'noi dorogi na neodnorodnoi mestnosti* [Solution algorithm of the problem of optimum location of wood transportation road on composite terrain]// Vestnik VGUIT, 2017. Vol. 79, № 2 (72). PP. 113–120.
4. Chernyshova E.V. *Metody formirovaniya tsifrovoi modeli mestnosti pri trassirovanii lesovoznykh avtomobil'nykh dorog* [Formation of digital model of terrain upon location of wood transportation roads] // Sistemy. Metody. Tekhnologii, 2017. № 3(35). PP. 143–148.
5. D. L. Drew, Traffic Flow Theory & Control (McGraw Hill, 1964).
6. Ivanov V.N., Erokhov V.N. *Vliyanie parametrov avtomobil'nykh dorog na raskhod topliva* [The influence of motor road parameters on fuel consumption] // Avtomobil'nye dorogi, 2014. № 8. PP. 10–13.
7. Belyaev A.N., Kozlov V.G., Vysotskaya I.A., Trishina T.V. (2019). Computation of vehicle motion path upon entering a turn. International Journal of Engineering and Advanced Technology, 1, 4527-4531.
8. Kaluzhskii Ya.A., Begma I.V., Kislyakov V.M., Filippov V.V. Primenenie teorii massovogo *obsluzhivaniya v proektirovanii avtomobil'nykh dorog* [Theory of queuing in design of motor roads] (Transport, Moscow, 1969).
9. Khomyak Ya.V. *Proektirovanie setei avtomobil'nykh dorog* [Designing of motor road networks]. (Transport, Moscow, 1983).
10. Kondrashova E.V., Skvortsova T.V. *Sovershenstvovanie organizatsii dorozhnogo dvizheniya v transportnykh sistemakh lesnogo kompleksa* [Improvement of traffic arrangement in transport systems of timber complex] // Sistemy upravleniya i informatsionnye tekhnologii, 2008. № 3.2(33). PP. 272–275.
11. Kozlov V.G. Methods, models and design algorithms of wood transportation roads with consideration for the influence of climate and weather on traffic conditions. Doctoral Thesis. Arkhangel'sk: SAFU, 2017.
12. Kozlov V.G., Skrypnikov A.V., CHernyshova E.V., CHirkov E.V., Postavnichij S.A., Mogutnov R.V. (2018). Teoreticheskie osnovy i metody matematicheskogo modelirovaniya lesovoznyh avtomobil'nyh dorog. Izvestiya vysshih uchebnyh zavedenij. Lesnoj zhurnal, 6, 117–127.
13. Kozlov V.G., Skrypnikov A.V., Mikova E.Yu., Mogutnov R.V., Chirkov E.V. *Formirovanie modeli proektirovaniya sistemy «dorozhnye usloviya – transportnye potoki» i puti ee realizatsii* [Design model of "road conditions – transport flow" system and its implementation] // Lesoinzhenernoe delo, 2018. Vol. 8. № 1 (29). PP. 100–111. https://doi.org/10.12737/article_5ab0dfbe6ece23.91630316
14. Kozlov V.G., Skrypnikov A.V., Mikova E.Yu., Mogutnov R.V., Zelikova Yu.A. *Kompleksnye eksperimental'nye issledovaniya izmeneniya parametrov i kharakteristik dorozhnykh uslovii, transportnykh potokov i rezhimov dvizheniya pod vliyaniem klimata i pogody* [Comprehensive experimental studies of parameters and properties of road conditions, transport flows and traffic modes under the influence of climate and weather] // Lesotekhnicheskii zhurnal, 2018. Vol. 8. № 2 (30). PP. 156–168. https://doi.org/10.12737/article_5b240611858af4.37544962
15. Kozlov V.G., Skrypnikov A.V., Sushkov S.I., Kruchinin I.N., Grigorev I.V., Nikiforov A.A., … Timokhova O.M. (2019). Enhancing quality of road pavements through adhesion improvement. Journal of the Balkan Tribological Association, 3, 678–694.
16. Labudin, B. V., Ivko, V. R., Koltsova, E. I., Skrypnikov, A. V., Kozlov, V. G., Levushkin, D. M., … Zelikov, V. A. (2020). Increasing pit road inclinations at high latitude deposits of solid minerals. ARPN Journal of Engineering and Applied Sciences, 15(19), 2168–2173.

17. Ryabova O.V., Skrypnikov A.V., Kozlov V.G., Tikhomirov P.V. (2021). Studying a geographical environment for road design. Russian Journal of Building Construction and Architecture, 1 (49), 66–78.
18. Sil'yanov V.V., Sitnikov Yu.M. *Raschet skorostei dvizheniya pri proektirovanii avtomobil'nykh dorog* [Prediction of traffic velocities upon designing of motor roads] // MADI, 1974. Issue 72. PP. 47–66.

Mathematical Modeling of Hemodynamic Processes Taking into Account Effects of Border Deformation

S. I. Peregudin, S. E. Kholodova, and K. M. Cherkay

Abstract A mathematical model is investigated that describes the dynamics of blood flow through blood vessels, taking into account the effects of wall deformation. A mathematical model is presented as a model of hydrodynamic processes in a blood vessel. In particular, variants of motion in plane and cylindrical coordinate systems are considered, taking into account the peculiarities of the effects of wall deformation. Functional dependencies are obtained for the amplitudes of the voltage acting on the surface of the blood vessel, which depends on the amplitude, wave number and oscillation frequency. The results obtained allow us to formulate a method for assessing blood dynamics from known values of pulse and pressure.

Keywords Mathematical modeling · Hemodynamics · Vascular movement · Incompressible fluid

1 Introduction

Blood inside a living organism is an integral component in ensuring the normal process of vital activity [1]. Thus, understanding the dynamics of blood in the body is a key problem for the timely prevention of vascular diseases. However, due to the impossibility of visually observing the flow pattern, the question arises of a mathematical description of the movement of blood through the vessels.

The question of describing the dynamics of blood through the vessels, despite the available research, remains relevant. In a large number of presented studies devoted to this issue, the authors often simplify the mathematical model used in their studies, considering the one-dimensional flow of blood through the vessels [2], or offer in their studies numerical modeling of the geometry of the considered vessels and the

S. I. Peregudin (✉)
St. Petersburg State University, St. Petersburg, Russia
e-mail: s.peregudin@spbu.ru

S. E. Kholodova · K. M. Cherkay
ITMO University, St. Petersburg, Russia

V. I. Karev (ed.), *Physical and Mathematical Modeling of Earth and Environment Processes—2022*, Springer Proceedings in Earth and Environmental Sciences,
https://doi.org/10.1007/978-3-031-25962-3_33

movement of blood through them in accordance with complete systems equations of motion for a fluid in space [3–7].

The main problem of compiling a one-dimensional model and its use for practical purposes is insufficient accuracy compared to experimental results, while one of the disadvantages of using numerical methods is the lack of an analytical solution, which complicates the task of determining the qualitative features of blood flow for different values of model parameters. Thus, the question of finding analytical functional dependencies that would describe the movement of blood through the vessels close to the true one remains unresolved, as a result of which a two-dimensional model of blood movement is studied in the present work, in particular, an axisymmetric model of blood movement is considered as in a flat, and in cylindrical geometry, taking into account the presence of wall deformation effects.

2 Problem Statement

Let us consider the motion of a viscous incompressible homogeneous fluid in a cylindrical vessel of variable depth, to the surface of which a surface stress is applied, and analyze the perturbations under which plane-parallel or axisymmetric dynamics are observed.

In the case of plane-parallel motion, we introduce the coordinate system in such a way that the height of the vessel surface relative to the reference level $y = 0$ is given by the function $R + \eta(x, t)$, and we choose the directions of the axes and are parallel to the vessel axis Ox and to the vessel wall, respectively, so that the system Oxy forms a right-handed rectangular coordinate system. In the case of axisymmetric motion, it is assumed that the azimuthal velocity component is equal to zero, all hydrodynamic parameters do not depend on the azimuthal coordinate, and the equation of the vessel surface disturbed by some wave, has the form, where are the cylindrical coordinates.

In the case of axisymmetric motion, it is assumed that the azimuthal velocity component is equal to zero, all hydrodynamic parameters do not depend on the azimuthal coordinate, and the equation for the vessel surface disturbed by some wave $\eta(z, t)$, has the form $r = R + \eta(z, t)$, where r, z—cylindrical coordinates.

The equations of motion of a homogeneous viscous incompressible fluid have the form [8, 9]

$$\frac{\partial \mathbf{v}}{\partial t} + (\mathbf{v},\ \nabla)\,\mathbf{v} = -\frac{1}{\rho}\nabla p + \nu\Delta\mathbf{v}, \tag{1}$$

$$\operatorname{div}\mathbf{v} = 0, \tag{2}$$

where $\mathbf{v}$—speed, p—pressure, ρ—density, ν—kinematic viscosity coefficient.

Continuity Eq. (2) allows us to introduce the stream function ψ by the equalities:

- for plane-parallel motion $v_x = \frac{\partial \psi}{\partial y}$, $v_y = -\frac{\partial \psi}{\partial x}$;
- for axisymmetric movement $v_r = -\frac{1}{r}\frac{\partial \psi}{\partial z}$, $v_z = \frac{1}{r}\frac{\partial \psi}{\partial r}$.

Then the problem of determining the velocity field is reduced to solving the equation for the stream function:

$$\frac{\partial}{\partial t}\Delta\psi + \frac{\partial \psi}{\partial y}\frac{\partial}{\partial x}\Delta\psi - \frac{\partial \psi}{\partial x}\frac{\partial}{\partial y}\Delta\psi = \nu\,\Delta^2\psi;$$

- for axisymmetric movement

$$\frac{\partial}{\partial t}\left(\frac{1}{r}\frac{\partial^2\psi}{\partial r^2} + \frac{1}{r}\frac{\partial^2\psi}{\partial z^2} - \frac{1}{r^2}\frac{\partial \psi}{\partial r}\right) = \nu\,\Delta\left(\frac{1}{r}\frac{\partial^2\psi}{\partial r^2} + \frac{1}{r}\frac{\partial^2\psi}{\partial z^2} - \frac{1}{r^2}\frac{\partial \psi}{\partial r}\right) - \frac{\nu}{r^2}\left(\frac{1}{r}\frac{\partial^2\psi}{\partial r^2} + \frac{1}{r}\frac{\partial^2\psi}{\partial z^2} - \frac{1}{r^2}\frac{\partial \psi}{\partial r}\right).$$

where

$$\psi(x, y, t) = \left(Ae^{kx} + Be^{ly}\right)e^{i(kx+\sigma t)}, \quad l = \sqrt{k^2 + \frac{i\sigma}{\nu}},$$

$$\psi(r, z, t) = \left(C_1 r I_1(\kappa r) + \frac{A\nu r}{\sigma} I_1(\kappa r)\right)e^{i(kz+\sigma t)},$$

where $I_1(\kappa r)$—first-order Infeld function, A, B, C_1—constants for plane-parallel and axisymmetric motions, respectively.

Taking into account the no-slip boundary conditions for the liquid [10]

$$v_x|_{y=R} = 0, \qquad v_y\big|_{y=R} = \tilde{v}_0\, e^{i(kx+\sigma t)}, \qquad v_r|_{y=R} = \tilde{v}_0\, e^{i(kx+\sigma t)}, \qquad v_z|_{y=R} = 0,$$

where $\tilde{v}_0 = -ikv_0$—unknown component of the amplitude of the transverse velocity component on the surface of the vessel, we obtain the following representations of the velocity distribution:

$$v_x = \frac{klv_0}{l-k}\left(e^{k(y-R)} - e^{l(y-R)}\right)e^{i(kx+\sigma t)}, \quad v_y = \frac{ikv_0}{l-k}\left(ke^{k(y-R)} - le^{l(y-R)}\right)e^{i(kx+\sigma t)},$$

and

$$v_r = \frac{ikv_0}{D}(\kappa I_0(\kappa R)I_1(kr) - kI_0(kR)I_1(\kappa r))e^{ikz+\sigma t},$$

$$v_z = \frac{\kappa k v_0}{D}(I_0(kR)I_0(\kappa r) - I_0(\kappa R)I_0(kr))e^{ikz+\sigma t},$$

where $D = kI_0(kR)I_1(\kappa R) - \kappa I_0(\kappa R)I_1(kR)$.

Further, applying the divergence operator to the equation of motion, taking into account the continuity equation and the constructed velocity field, we can conclude that the function describing the pressure is harmonic. Then

$$p(x,z,t) = p_1 e^{ky+i(kx+\sigma t)}, \quad p(r,z,t) = p_1 I_0(kr) e^{ikz+\sigma t}.$$

To determine the dispersion relation, we use dynamic boundary conditions. Let a stress act on the surface of the vessel in the case of plane-parallel motion $\overleftarrow{\tau} = \tau_{yx}\,\widehat{x} + \tau_{yy}\,\widehat{y}$.

Then, on the surface of the vessel $R + \eta(x,t)$ $y = R + \eta(x,t)$, assuming the perturbation $\eta(x,t)$ is small compared to R, the boundary conditions must be satisfied

$$\tau_{yy}\Big|_{y=R} = T_1 e^{i(kx+\sigma t)}, \quad \tau_{yx}\Big|_{y=R} = iT_2 e^{i(kx+\sigma t)},$$

where $\mu = \rho\nu$—coefficient of dynamic viscosity. Let, in the case of axisymmetric motion, a stress acts on the surface of the vessel $\overleftarrow{\tau} = \tau_{rr}\,\widehat{r} + \tau_{rz}\,\widehat{z}$, then the boundary conditions will be satisfied on the surface of the vessel

$$\tau_{rr}|_{y=R} = \tau^2(R) e^{ikz+\sigma t}, \qquad \tau_{rz}|_{y=R} = i\tau^1(R) e^{ikz+\sigma t}.$$

Whence the dispersion relation can be written in the form.

- for plane-parallel motion

$$\sigma(k) = i\nu k^2\left[\frac{P}{2} + \sqrt{\frac{P^2}{4} + \frac{4}{\upsilon}}\right], \qquad P = \frac{4}{\upsilon} - \frac{1}{\upsilon^2}, \qquad \upsilon = \frac{T_2}{T_1};$$

- for axisymmetric movement

$$\sigma(k) = \frac{4\nu}{R^2} + \frac{8\nu\upsilon}{kR^3} - \nu k\left(\frac{3k}{2} + \frac{\upsilon}{R}\right) - \sqrt{\left[\frac{4\nu}{R^2} + \frac{8\nu\upsilon}{kR^3} - \nu k\left(\frac{k}{2} + \frac{\upsilon}{R}\right)\right]^2 + \frac{8\nu^2}{R^2}\left[k^2 - \frac{2I_1(kR)}{I_0(kR)}\frac{kR+\upsilon}{R^2}\right]},$$

$$\upsilon = \frac{\tau^1(R)}{\tau^2(R)}.$$

The dispersion relation makes it possible to plot the dependence of the displacement of the vessel walls on the wave number. Since As $\frac{\partial\eta}{\partial t} \approx \upsilon_y\big|_{y=R}$, then vertical displacement of the vessel:

$$\eta(x,\, t) = \frac{k\nu T_1(k-l)}{l\mu\sigma^2} e^{i(kx+\sigma t)}.$$

Having a complete analytical description of the considered model of blood flow, it is possible to calculate the amplitudes of the stress components T_1 and T_2 on the walls of the vessel, taking into account the possibility of measuring pressure and pulse, namely, the required values will satisfy the system of algebraic equations resulting from the analytical representation of the pressure field and the dispersion relation, the solution of which has the form

$$\frac{T_1}{\tilde{p}} = 1, \qquad \frac{T_2}{\tilde{p}} = \frac{1}{q}\sqrt{4 + \frac{q^2}{1-q}} - 2, \qquad q = \frac{\sigma}{i\nu k^2}, \tag{3}$$

where $\tilde{p}$—pressure amplitude measured on the vessel wall.

It is also possible to estimate the magnitude of the displacement of the walls of the vessel, which, under small perturbations, is associated with the radial component of the velocity by the following formula:

$$\eta(z,\, t) = \int_0^t v_r|_{r=R} dt = -\frac{ik\upsilon_0}{\sigma} e^{ikz+\sigma t}.$$

Thus, a complete analytical description of the considered model of blood flow makes it possible to express the stress amplitudes in terms of pressure, oscillation frequency, and wave number:

$$\begin{aligned} \frac{\tau^1(R)}{\tilde{p}} &= \frac{2x^2}{y^2-x^2}\frac{I_1(x)}{I_0(x)} - \frac{x(x^2-y^2)}{y(y^2-x^2)}\frac{I_1(x)}{I_0(x)}\tau_{rr}\bigg|, \\ \frac{\tau^2(R)}{\tilde{p}} &= \frac{2x^2}{y^2-x^2}\left[\frac{I_1(x)}{yI_0(x)} - \frac{I_1(x)}{xI_0(x)}\right] + 1, \end{aligned} \tag{4}$$

where $x = kR$, $y = \kappa R$, $\tilde{p}$—pressure amplitude measured on the vessel wall.

3 Results

The analysis carried out makes it possible to compile graphical interpretations of the obtained field of hydrodynamic quantities characterizing the dynamic process in a blood vessel, which are shown in Fig. 1—for plane-parallel motion, in Fig. 2—for axisymmetric movement.

An analysis of the graphical dependences of velocities in the case of flat geometry allows us to state that the considered flow forms a velocity profile similar to the parabolic one characteristic of the Poiseuille flow. Thus, a conclusion can be drawn

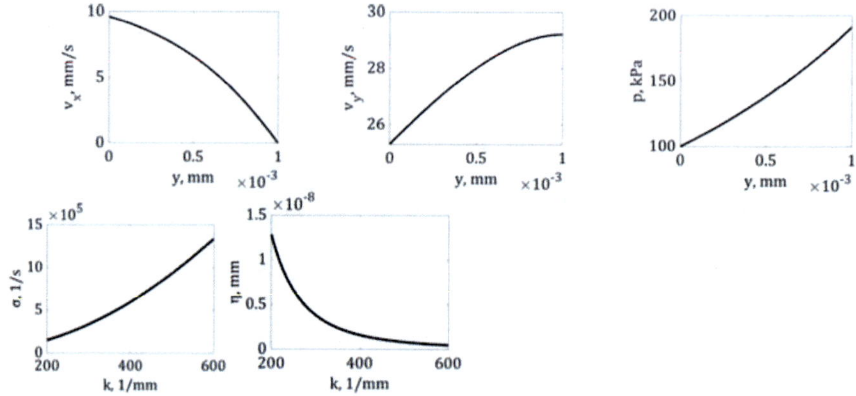

Fig. 1 Graphs of dependences of hydrodynamic quantities in plane geometry

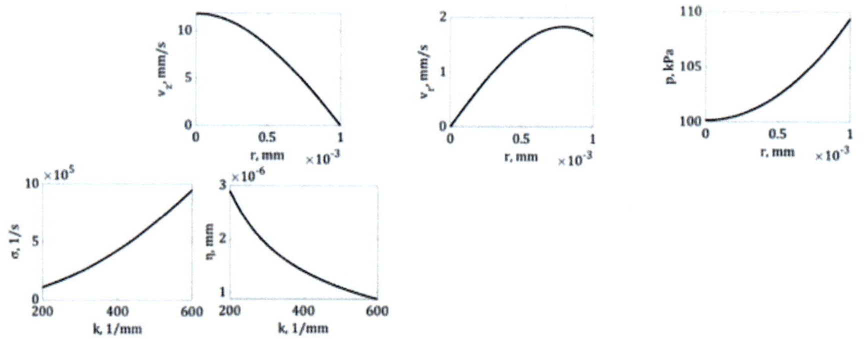

Fig. 2 Graphs of dependences of hydrodynamic quantities in cylindrical geometry

about the laminar flow of blood through the vessels [8, 10]. The graphic dependence of pressure shows that when approaching the walls of the vessel, the pressure of the liquid increases and reaches its maximum value at the wall itself. The plot of the dependence of the vessel wall displacement on the wave number allows us to state that with an increase in the wave number, the value of the vessel wall displacement decreases.

In the case of axisymmetric motion, it can be observed that the resulting velocity distributions form a somewhat different profile compared to plane-parallel motion. Despite this, the obtained profile, as well as for flat geometry, has some similarity with the Poiseuille flow, and, therefore, one can speak of a laminar blood flow through the vessels. From the graph of the pressure dependence, we can conclude: when moving away from the center of the vessel, the pressure of the liquid increases and reaches its maximum at the walls of the vessel. The resulting dispersion curve changes approximately according to a parabolic law. The graphical interpretation of

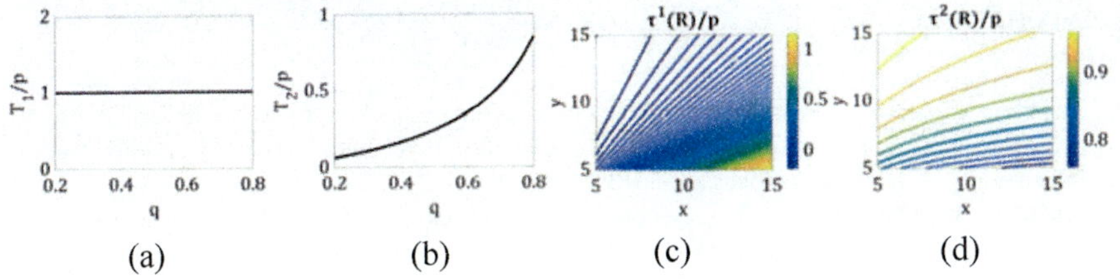

Fig. 3 Graphs of dependences of the ratio of stress amplitudes to pressure amplitude

the displacement of the vessel wall shows that as the wave number increases, the value of the displacement decreases.

The obtained dependences of the stress amplitudes on the hydrodynamic parameters (3), (4) make it possible to construct graphical interpretations, which are shown in Fig. 3a, b—for flat geometry, in Fig. 3c, d—for cylindrical geometry, from the analysis of which it follows that the stress increases in proportion to the pressure inside the vessel.

In the case of flat geometry, an increase in the amplitude of the tangential stress component p with an increase in the value is characteristic, while the amplitude of the normal stress component does not depend on this value at all. In the case of cylindrical geometry, an increase in the amplitude of the tangential stress component with an increase in the ratio of magnitude x to magnitude y is characteristic, while the amplitude of the normal stress component decreases.

4 Conclusions

Modeling of hydrodynamic processes in a blood vessel was carried out. In particular, models of motion in plane and cylindrical geometries are considered, taking into account the presence of wall deformation effects. For each vessel geometry, a boundary value problem for partial differential equations is presented, functional dependences for the hydrodynamic parameters of blood are analytically found, and a dispersion relation is constructed as a function of the oscillation frequency on the wave number. Functional dependencies are constructed for the amplitudes of the voltage acting on the surface of the vessel on the pressure amplitude, wave number, and oscillation frequency, which make it possible to formulate a method for assessing blood dynamics from the measured values of pulse and pressure.

In the future, it is planned to carry out mathematical modeling using elements of the theory of elasticity and dynamics of electrically conductive media [9] in order to take into account the presence of conductive properties of the hemodynamic medium, the thickness and inhomogeneity of the vessel walls, as well as a comparative analysis of the results obtained with experimental data.

References

1. Tregubov V.P., Radichkina A.O. Mathematical modeling of the kinematics of the left ventricle of the human heart during its contraction // Bulletin of St. Petersburg University. Applied math. Informatics. Management processes. 2013. V. 9. No. 4. P. 66–72.
2. Mitsotakis D., Dutykh D., Li Q. Asymptotic nonlinear and dispersive pulsatile flow in elastic vessels with cylindrical symmetry // Computers and Mathematics with Applications. 2018. Vol. 75. No 11. P. 4022–4047.
3. Zakharov Yu.N. Numerical method for predicting hemodynamic effects in vascular prostheses // Siberian Journal of Computational Mathematics. 2019. V. 22. No. 4. P. 399–414.
4. Ovcharenko E.A., Onishchenko P.S., Klyshnikov K.Yu., Ganyukov V.I., Shilov A.A., Vereshchagin I.E., Kokov A.N., Tarasov R.S., Borisov V.G., Zakharov Yu.N., Barbarash L.S. Numerical simulation of hemodynamics during repeated heart valve replacement. Pathology of Blood Circulation and Cardiosurgery. 2019. V. 23. No. 3. P. 30–38.
5. Klyshnikov K.U., Rezvova M.A., Ovcharenko E.A., Glushkova T.V., Kudryavtseva Y.A., Barbarash L.S., Batranin A.V., Nyshtaev D.V., Zakharov Y.N., Borisov V.G. An experimental study of the reinforcing element of a vascular prosthesis fabricated using the internal thoracic artery of large cattle. // Biophysics. 2021. T. 66. № 4. P. 689–699.
6. Selmi M., Belmabrouk M., Bajahzar A. Numerical Study of the Blood Flow in a Deformable Human Aorta // Applied Sciences. 2019. Vol. 9. No 6. P. 1–11.
7. Pontrelli G. Blood flow through an axisymmetric stenosis // Proceedings of the Institution of Mechanical Engineers, Part H. 2000. Vol. 215. No 1. P. 1–10.
8. Wallander S.V. Lectures on hydroaeromechanics. L.: LGU, 1978. 296 p.
9. Kholodova S. E. Dynamics of a rotating layer of an ideal electrically conductive incompressible fluid // Computational Mathematics and Mathematical Physics. 2008. V. 48. No 5. 882–898.
10. Slezkin N.A. Dynamics of a viscous incompressible fluid. M.: State publishing house of technical and theoretical literature, 1955. 521 p.

The Influence of Technical Objects on the Dynamics and Structure of the Environment

A. V. Kistovich

Abstract The problem of the dynamics of the restoration of the stratifying distribution of the sea density after the removal of a disturbance source from a certain spatial region is investigated. Similarities and differences in density distributions and its gradients in cases of salt and temperature stratifications are shown.

Keywords Stratification · Density gap · Kinetic coefficients · Salinity · Temperature

1 Introduction

Any technical object moving on the surface or in the thickness of an aqueous medium perturbs its initial dynamic state. Since the object itself is impervious to all physical fields of the medium, with the exception of temperature and, in part, sound waves, it introduces distortions into the fields of velocity, pressure, impurity concentration and density. Naturally, it also changes the temperature field and the picture of all wave processes in the surrounding space. The presence of different types of boundary layers (viscous, concentration, temperature) on its borders, as well as the disruption of these layers with subsequent entrainment in the satellite trail, introduces additional disturbances, possibly not characteristic of a free medium.

In addition to the actual physical fields, distortions of their production parameters (temperature gradients, pressure, density, vorticity, etc.) occur. As a result, additional flows (convective, advective and diffusive) of the main physical characteristics of the medium arise, which, in turn, leads to a change in its structure. The combination of altered physical fields and disturbances in the structure of the medium leads to a change in the dynamics of seawater. Changing the dynamics of the marine environment requires changing the models underlying the measurement formulas for metrological support of measuring equipment and the measurement process itself.

A. V. Kistovich (✉)
A. Ishlinsky Institute for Problems in Mechanics of the Russian Academy of Sciences, Moscow, Russia
e-mail: kavmendeleevo@mail.ru

V. I. Karev (ed.), *Physical and Mathematical Modeling of Earth and Environment Processes—2022*, Springer Proceedings in Earth and Environmental Sciences,
https://doi.org/10.1007/978-3-031-25962-3_34

Since the technical object does not occupy a stationary position in space, over time it leaves the area of possible measurements and part of the disturbances introduced by it into the medium attenuates, which leads to a gradual return of the medium to its initial state. Since the time between the departure of the object and the moment of measurement can be very significant, the most interesting are long-lived technical concerns.

In general, the attenuation of disturbances occurs due to convective mixing of the elements of the medium and dissipative effects associated with viscous attenuation (for velocity and vorticity fields) and diffusion expansion (for temperature and impurity fields). At the same time, the values of kinetic coefficients substituted into the liquid medium model can be estimated in laminar and turbulent limits.

Since the solution of the problem of dissipation and scattering of disturbances in the medium after the passage of a technical object in it is practically unattainable in an exact formulation, it makes sense to consider approximate models of the dynamics of disturbances.

2 Problem Statement

A model problem on the dynamics of the distribution of a stratifying additive in an incompressible fluid with the sudden occurrence of a local density disturbance is considered. The initial undisturbed liquid is considered stratified according to the linear law

$$\rho_0(z) = \rho_0(1 - z/\Lambda) \tag{1}$$

where Λ is the scale of stratification, z is the vertical axis.

Perturbations in the model are given as follows: at the initial moment of time $t = 0$ of the liquid in an infinite horizontal layer $-H \leq z \leq H$ is suddenly mixed to a homogeneous state (due to the passage of a technical object), characterized by an average density

$$\overline{\rho} = \frac{1}{2H} \int_{-H}^{H} \rho_0(z)\, dz = \rho_0$$

Density discontinuities are formed at the boundaries $z = \pm H$ of the mixed layer, as can be seen from the graph of the new initial stratifying distribution $\rho_{in}(z)$, shown in Fig. 1.

The presence of discontinuities at $z = \pm H$ disrupts the equilibrium in the medium and leads to the appearance of diffusion flows seeking to compensate for the initial disturbance. Since the mixed layer is unlimited in the horizontal direction, the problem is considered in a one-dimensional formulation. In addition, from the

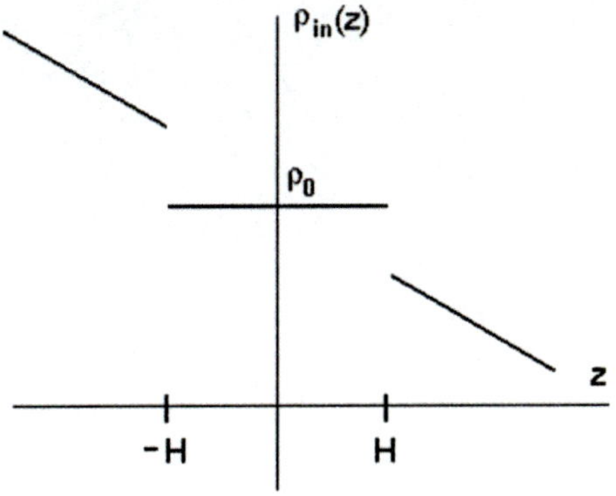

Fig. 1 The form of the initial stratifying distribution $\rho_{in}(z)$

incompressibility condition of the fluid $\nabla \cdot \mathbf{u} = \partial u_z / \partial z = 0$ and the boundary conditions $u_z|_{z=\pm\infty} = 0$, it follows that the velocity field is equal to zero in all space at all times.

In the formulated formulation, the dynamics of the total density of the medium $\rho(z,t)$ is described by the diffusion equation

$$\frac{\partial \rho}{\partial t} - \kappa \frac{\partial^2 \rho}{\partial z^2} = 0, \quad \rho = \rho_{in}(z) + \tilde{\rho}(z,t) \tag{2}$$

where $\tilde{\rho}$ is the perturbation to the new initial $\rho_{in}(z)$ distribution, given, as follows from the statement of the problem, by the relation

$$\rho_{in}(z) = \rho_0 \left[1 - \frac{z}{\Lambda} (\vartheta(-z-H) + \vartheta(z-H)) \right] \tag{3}$$

where $\vartheta(x)$ is the Heaviside function.

Taking into account the limitation of the total diffusion mass transfer in the entire space, the equality of the $\tilde{\rho}$ perturbation to zero at the initial moment of time and the substitution of (3) in (2) make it possible to formulate the problem under consideration in the form of an initial problem for the $\tilde{\rho}$ function

$$\frac{\partial \tilde{\rho}}{\partial t} - \kappa \frac{\partial^2 \tilde{\rho}}{\partial z^2} = -\frac{\kappa \rho_0}{\Lambda} \left[\delta(z-H) - \delta(z+H) + H\delta'(z-H) + H\delta'(z+H) \right]$$

$$\tilde{\rho}(z,0) = 0, \quad \int_{-\infty}^{+\infty} |\tilde{\rho}(z,t)|\, dz < \infty, \quad t \geq 0, \quad -\infty < z < +\infty \tag{4}$$

where δ is Dirac delta-function, the stroke means differentiation by z.

The right part of Eq. (4) is generated by density discontinuities of the new initial stratification distribution $\rho_{in}(z)$, which play the role of sources of density perturbations. The solution for the density perturbation $\tilde{\rho}$ is written as a convolution of the Green function of problem (4) with the source term of the diffusion equation

$$\tilde{\rho}(z,t) = -\frac{\rho_0\sqrt{\kappa}}{2\Lambda\sqrt{\pi}}\int\limits_0^t\int\limits_{-\infty}^{+\infty} K(\zeta)\exp\left(-\frac{(z-\zeta)^2}{4\kappa(t-\tau)}\right)\frac{d\zeta\, d\tau}{\sqrt{t-\tau}}$$

$$K(\zeta) = \delta(\zeta - H) - \delta(\zeta + H) + H\delta'(\zeta - H) + H\delta'(\zeta + H) \quad (5)$$

Integration (5) allows you to present the desired solution in the form of

$$\tilde{\rho}(z,t) = -\frac{\rho_0}{\Lambda}\left[\sqrt{\frac{\kappa t}{\pi}}(A_- - A_+) - \frac{z}{2}(B_- - B_+)\right]$$

$$\mathrm{A}_\pm = \exp\left(-\frac{(z \pm H)^2}{4\kappa t}\right), \quad \mathrm{B}_\pm = \mathrm{sign}(z \pm H)\mathrm{erfc}\left(\frac{|z \pm H|}{2\sqrt{\kappa t}}\right) \quad (6)$$

where $\mathrm{sign}(z) = (-1, z < 0)$ or $(0, z = 0)$ or $(1, z > 0)$.

At the initial moment of time, the perturbation (6) is zero, and at $t \rightarrow \infty$, the total density $\rho = \rho_{in}(z) + \tilde{\rho}(z,t)$ tends to the old initial distribution of stratification $\rho_0(z)$, given by the ratio (1). Thus, diffusion processes gradually restore the undisturbed density distribution. But from the point of view of the issue of registration of moving underwater objects, it is important how quickly it is possible to restore the undisturbed distribution so that the sensitivity of the electrical conductivity sensors does not allow us to draw a conclusion about the passage of an object.

3 Temporal Dynamics of Sea Density Distribution

Below, a series of graphs 1 shows the temporal dynamics of the distribution of sea density after leaving the measurement zone of a technical object with a diameter of $D = 20$ m. Calculations were carried out according to the formula (6) in the range of depressions $z = \pm 20$ m relative to the center of the object. The value of its turbulent limit was chosen as the diffusion coefficient of the salt

$$\kappa \approx 1.35 \times 10^{-6}\,\mathrm{m}^2/\mathrm{s} \quad (7)$$

For the most visual representation of the dynamics of the density distribution in the wake of the object, time counts were taken at moments $t_1 = 10^5$ c $\sim$ 1 day, $t_2 = 5 \times 10^5$ c $\sim$ 6 days and $t_3 = 10^6$ c $\sim$ 12 days. Averaged over the width of the perturbation zone, the relative deviation of the perturbed density distribution from the undisturbed one is ~ 0.5% for the above data. At first glance, such a density deviation cannot stand out steadily against the background of random disturbances of the initial distribution caused by uncontrolled sources, which have values of the same orders of magnitude. The picture, however, changes significantly if we study the behavior of the vertical gradient of the distributions shown in Fig. 2.

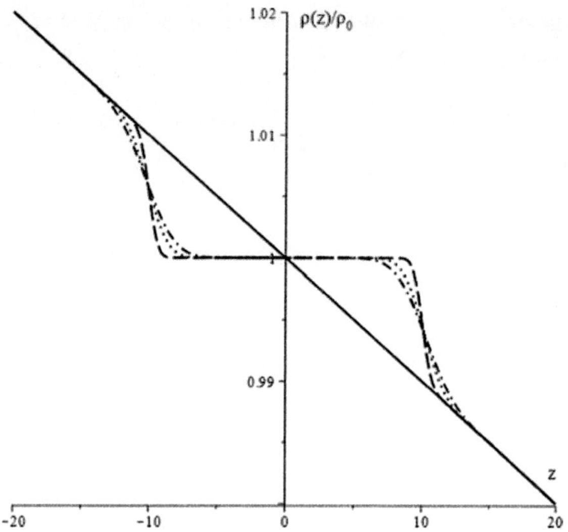

Fig. 2 Density distribution after a day (dash), 6 days (dots), 12 days (dots-dashes). A solid line is an undisturbed distribution

Figure 3 shows the derivatives of the presented distributions along the vertical coordinate z. All calculations were performed for the same parameter values as above.

The analysis of the results presented in Fig. 3 shows that the difference in density gradients in the perturbation zone, even after a long period of time, is so significant (for the presented results—from 3 to 8 times!), which makes it possible to consistently detect the presence of perturbation in the studied area of space when measuring the

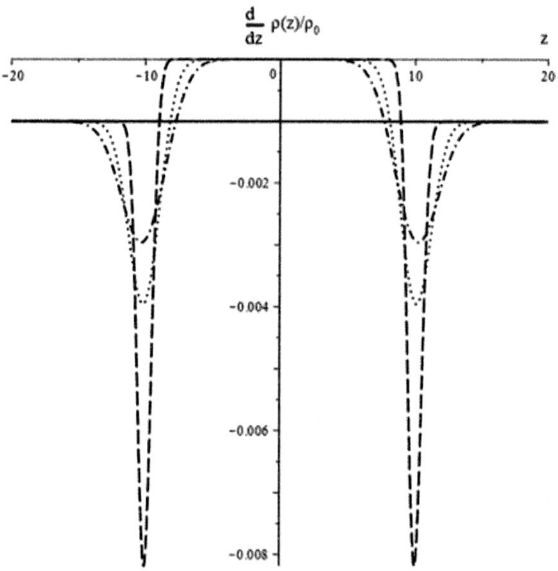

Fig. 3 Density gradient distribution

gradients of physical fields of fine structure. Since random density gradients, after averaging over an ensemble of measurements, will give a value of about several percent of the gradient of the initial undisturbed stratifying distribution, the density gradient resulting from the disturbance of a passing object will stand out steadily against their weak background.

Despite the fact that the presented model of the dynamics of disturbances of the raft is rather rough, it clearly indicates the need to measure the parameters of the fine structure of the marine environment.

4 Temporal Dynamics of Sea Temperature Distribution

In the same formulation as the problem of the dynamics of the density distribution, the problem of the dynamics of the perturbation of the temperature distribution of seawater is considered. The relations describing the solution of this problem are the same as in the previous case, only the coefficient of turbulent temperature transfer is taken as the kinetic coefficient, the value of which is estimated by

$$\chi \approx 1.35 \times 10^{-4}\ \mathrm{m}^2/\mathrm{s} \tag{8}$$

The geometric parameters of the problem do not change, and time counts are taken for two moments $t_1 = 10^5$ c $\sim$ 1 day, $t_2 = 5 \times 10^5$ c $\sim$ 6 days. The reason for such a reduction in the number of compared moments of time is clear from Fig. 4, where temperature distributions are presented.

It can be seen that temperature perturbations are leveled much faster than density perturbations. This is due to the fact that the corresponding kinetic coefficient of

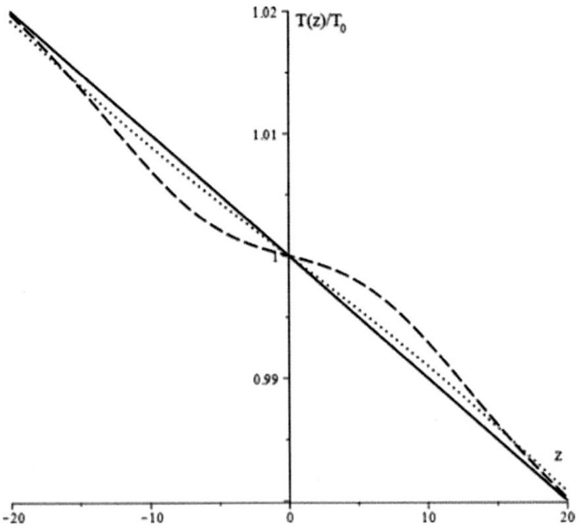

Fig. 4 Density distribution after a day (dash) and 6 days (dots). A solid line is an undisturbed distribution

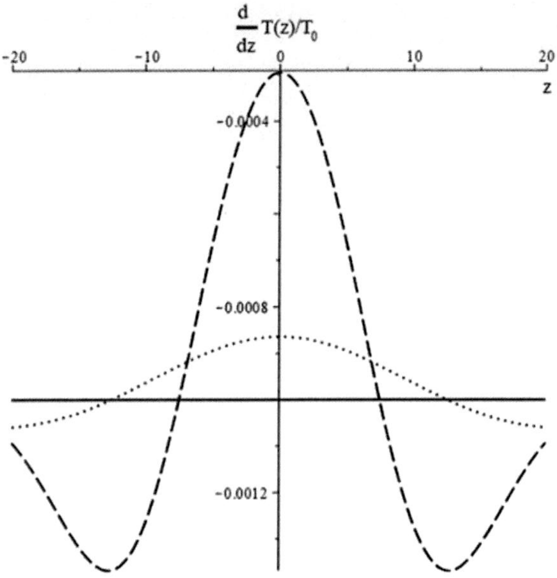

Fig. 5 Temperature gradient distribution

turbulent temperature transfer is two orders of magnitude greater than the salt coefficient. After a day, the relative deviation of the temperature distribution averaged over the disturbance zone to the initial stratifying one is no more than 2%, which cannot be stably recorded against the background of random temperature fields of seawater.

At the same time, just as for the case of density distribution, the field of temperature gradients is more informative, as can be seen from Fig. 5, which shows the dynamics of the value $d\,T(z,t)/d\,z/T_0$, where T_0 is the average temperature of the area mixed by the passage of the object.

It follows from the presented results that after a day the relative deviation of the temperature gradient of the disturbed state from the undisturbed one does not exceed 30%, and after 6 days there is no trace of the introduced temperature distortions.

Nevertheless, the measurement of temperature gradients during the day after the disappearance of the cause of disturbances can reveal the fact of disturbances, of course, with appropriate processing of measuring information, since the noise effect of random temperature gradients in this case is more significant than in the case of salinity gradients.

In the case of daily measurements, the integrated use of temperature and salinity sensors allows, with the correlation processing of data obtained as a result of measurements of fine-structure physical fields, to make more accurate data on the dynamics of the marine environment and to build adequate models of the measuring process.

Acknowledgements The work was carried out with the financial support on the topic State assignments 123021700046-4

A Novel Physical Model to Enhance Precision and Performance of Multidimensional Force Sensors

Maxim Glebovich Ponomarev

Abstract Multidimensional force (load) sensors (transducers) are widely used in modern robotics (including tactile feedback for artificial arm, artificial leg, prosthesis), medical cell tensile experiments, cell mechanical research (Zhou et al. in J Healthc Eng 2017:1587670, 2017 [1]; He et al. in J Healthc Eng 2018:8504273, 2018 [2]), engineering systems to measure three-dimensional forces (feedback strains). For research of multidimensional strain force feedback due to fluid flow pressure developed new physical model. During validation via application of this novel method to the SWMTF (South Western Mooring Tests Facility) 10 times precision increase was obtained for the multidimensional force readings. The method is effective for wide range of measuring equipment based on vector sensors including electronic (strain gauge), piezoelectric, fiber optic load sensors, hydraulic, mechanical (in proving rings) load cells. In particular, the developed method is very effective for accelerated (fast) data processing of loading measurements to achieve real-time sensor-guided robot motions (including those for industrial robot arm, real-time actuators).

Keywords Multidimensional mechanical response to fluid flows · Multi-axis load sensors · Multiaxial fatigue · Force · Sensors · Strain condition monitoring · Transducers · Tri-axis load cell sensors · Vector load sensors · Calibration · Precision · Reliability · Smart sensors · Enhance precision · Overcoming sensor degradation

1 Introduction

This paper deals with yet another example from extensive number of innovative physical models with applications to real world problems outlined in [3–43].

The statistical approach for dealing with measured data has become increasingly popular from the second half of previous century when electronic computers have

M. G. Ponomarev (✉)
Cambridge Research and Consultancy, Cambridge, UK
e-mail: maxim.pv@gmail.com

Adyghe State University, Maykop, Russia

V. I. Karev (ed.), *Physical and Mathematical Modeling of Earth and Environment Processes—2022*, Springer Proceedings in Earth and Environmental Sciences,
https://doi.org/10.1007/978-3-031-25962-3_35

been developed up to the sufficient level to cope with significant data in reasonable time. It has made a contribution to development of concepts for systems of smart sensors [44]. One of the important advantages of the systems that in case of production defects or functional faults happen during or after installation of an instrument there is an option to minimise effect of errors on the data processing stage and avoid costly mechanical reparations.

Despite numerous publications (such as [45–47]) devoted to design and calibration of multi-dimensional load sensors, none was found that addresses the problem of precision when load cell calibration data produces nondeterministic results which could be due to complex manufacturing processes. Furthermore, no publication was found that deals with the problem of outliers for the vector load sensors. Assuming perfect manufacturing any load cell has its specific measurement range, for the multi axis load cells discussed here this was between 2500 and 50,000 N. It was expected that the load cells have a deterministic behaviour within this range whilst below or above this range random behaviour was expected [48–50]. However, outliers have been identified within the measurement range that need to be dealt with in a new manner.

2 Methods to Detect Outliers

There is no universal rule on how to detect and correct outliers (gross errors). Even if someone find a data point at the large distance (more than 4 standard deviations from the expected value) it is not necessary an outlier. The large deviation could be just because of the inappropriate sampling frequency when signal is converted into discrete digital form. In some cases to help understanding it is proposed that there are different processes generating data: one is a core process which gives acceptable data and additional contamination processes generating outliers. Sometimes, it is possible to exclude contamination processes by applying filters, however there is no guarantee that available filtrations would not damage the core signal or all outliers removed completely. New approach to deal with outliers has been described here in general. And this new approach has been tested and validated via application to the calibration data sets for multiaxial load cells.

The data sets contain three-dimensional voltage $\vec{V} = (V_x;\ V_y;\ V_z)$ measurements corresponding to calibration loads $F = (F_x;\ F_y;\ F_z)$, which provides a calibration map over a semi-sphere. In total 5720 data points where taken for the first multi axis load cell (LC1), 5780 for the second one (LC2) and 5760 for the third one (LC3) over the calibration load range and angle settings. At every load point 50 V readings were taken over five seconds at a constant force. To reduce transient effects the first and last five of the 50 readings were ignored, and the remaining 40 readings were used to establish an arithmetic average point for each individual calibration point. The Cartesian coordinate system was introduced to present the arithmetic average points over the calibration load range (1–50 kN) and angle settings discussed above. The resulting voltage readings (V) are presented on the x-, y- and z-axis of Cartesian coordinate system, applying a negative and positive bending range for the x- and y-axis, whilst the z-axis was only calibrated for a positive tension range.

Next stage of the calibration analysis is to determine the resulting vector from the individual x-, y- and z-component and plot relationship between the full absolute voltage $V = \sqrt{V_x^2 + V_y^2 + V_z^2}$ and calibration load $F = \sqrt{F_x^2 + F_y^2 + F_z^2}$. Load instruments are typically designed within the elasticity limit of the material to produce a linear response, therefore it is expected to apply the linear relation:

$$V = kF + V_0, \tag{1}$$

where V is the absolute voltage, V_0 an initial zero offset, k the linear calibration slope and F the calibration load. Figure 1 shows the results of the absolute voltage readings over the absolute calibration loads for LC1, LC2 and LC3. A preliminary screening detected absolute voltage readings outside the linear relation (Eq. 1) for LC1 and LC2, whilst the calibration data points of LC3 are close enough to this relation. Figure 1 shows a physical test for the quality assessment of the three-dimensional sensor system. A spline interpolation line is shown on the scatter plots of full absolute voltages as functions of absolute forces. The full absolute forces were predefined as a discrete set of 20 loads ranging from 1 to 50 kN (that corresponds to horizontal axes). Vertical axes show absolute voltages measured in volts as a reaction to these loads.

No conclusion can be made from Fig. 1 to the extent of data points that are outside the interpolation lines for LC1 and LC2. In order to achieve a reliable calibration curve to obtain force measurements form the voltage readings, these outliers or gross errors need to be quantified and consequently removed or corrected.

The analysis of full force has been proven to be a suitable tool if data points follow the linear relation and if outliers are present, however a more detailed analysis is required to establish location of these outliers in relation to the individual x-, y- and z-axis and load range. As a result an additional statistical analysis of confidence intervals is required for each axis considering the full calibration load range.

To investigate the outliers further, confidence intervals CIv were plotted for the corresponding voltage outputs. The 95% level of confidence, CIv, has been chosen here due to well known accepted requirements for the precision of load cells used in engineering practice, including offshore engineering as described by [51]. Figure 2 shows scatter plots for the 95% confidence intervals ratio in relation to the absolute voltage $V = \sqrt{V_x^2 + V_y^2 + V_z^2}$.

Most meters on the market can process ratiometric measurements. They can measure the input in mV and divide that measurement by the actual voltage being supplied. For instance, we could have an mV measurement of 40.1235 mV and an excitation measurement of 9.9998 V. When displaying in mV/V, one would have 4.01243 mV/V.

The x-axis in Fig. 2 represents the absolute voltage readings V and the y-axis the 95% confidence interval ratio CIv/V. The expression of the 95% confidence interval ratio is providing evidence of good calibration readings if $CIv/V \cong 0.01$. The analysis

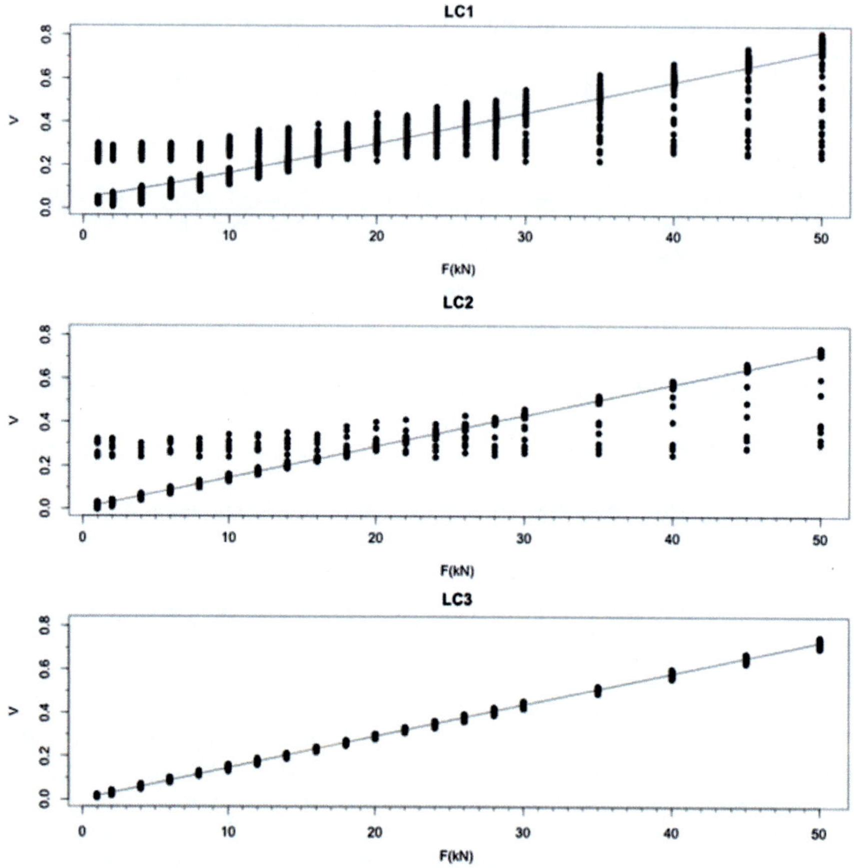

Fig. 1 Physical test for the quality assessment of the multidimensional sensor system. Even with multiple transducers measuring three different force components, there should be a relation between the full force and corresponding voltage measurements which expected to be as close to the shifted linear (Eq. 1) as possible. It shows that while majority of data appeared in the sufficient vicinity of the visually straight lines corresponding to Eq. (1), for the first (LC1) and second (LC2) load cells there are data points with large deviations from the lines and these outliers happen not only for small forces but distributed through the whole load range. Remarkably that the third load cell (LC3) has not shown any large deviated measurements

of LC3 applying the 95% confidence interval methodology resulted in values of $CIv/V \cong 0.01$ over most of the calibration load range. Variations from $CIv/V \cong 0.01$ can only be observed for the lower calibration range, for voltage readings less than 0.05 (V). That could most likely be contributed to the fact that the sensitivity range of the load cell has been exceeded. However, further analysis would be required to state this for certain, addressed in the latter. Additional branches for CIv/V much larger than expected 0.01 can be observed in voltage readings larger than 0.2 V for LC1 and LC2, that could damage validity of these load cells' measurements even

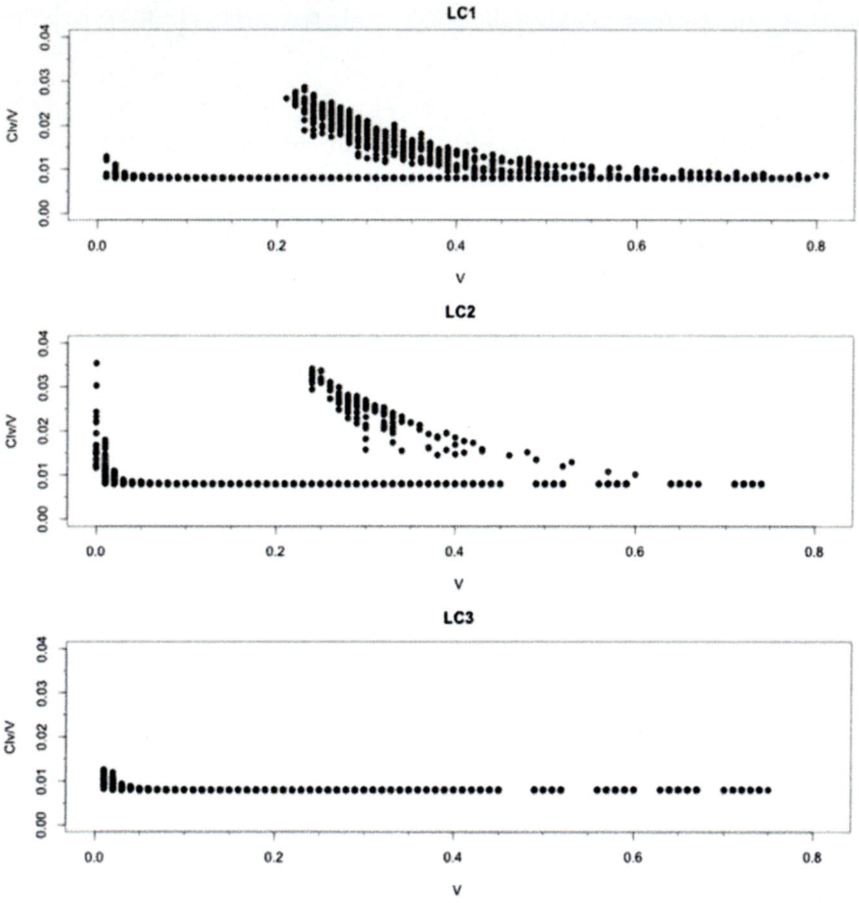

Fig. 2 Test for 95% confidence interval for full absolute voltages data

inside their normal load range. Therefore, these branches of outliers can be observed over a wide calibration range for LC1 and LC2. The outlier branch of LC1 and LC2 have values up to three times of $CIv/V \cong 0.01$, which corresponds to decreasing relative precision (~ $1/(CIv/V)$), visible between full voltage measurements from 0.2 to 0.6 V. Precision means how close multiple measurements are to each other. Precise measurements are both repeatable and reproducible. Relative precision is precision on relative terms, so that it is independent of scale and translation.

The branch of outliers for LC2 is considerably thinner than for LC1, this visual conclusion has to be supported by precise counting of the 95% confidence intervals and a separated analysis of the V_x, V_y and V_z component voltage readings.

In Fig. 3 the absolute 95% confidence interval ratio CIv/V has been separated into its x- y- and z load cell components, namely its component 95% confidence interval ratio's CIv_x/V_x, CIv_y/V_y and CIv_z/V_z. The results of these ratios are plotted vs the correspondent component voltage readings (V_x, V_y, V_z) for each load cell (LC1, LC2, LC3). Again, similar to the absolute 95% confidence interval ratio CIv/V analysis, a relative confidence can be assumed if $CIv_x/V_x = CIv_y/V_y = CIv_z/V_z \cong 0.01$.

Analysing the graphs for the individual CIv_x/V_x, CIv_y/V_y and CIv_z/V_z ratio's it can be observed that all show the larger than 0.01 confidence interval values at component voltage readings below 0.05 V. Again, this can be a result of exceeding the sensitivity range of the load cells, and it can be argued from the graphs that this

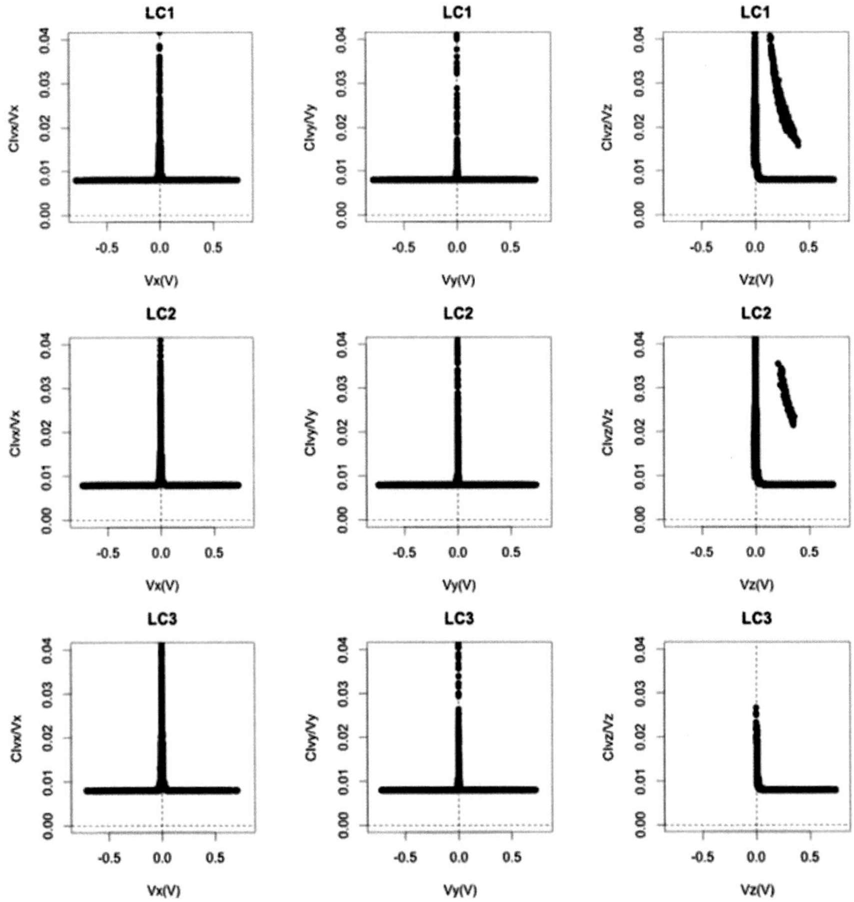

Fig. 3 Component by component (V_X, V_Y, V_Z) 95% confidence intervals plotted for the three load cells LC1, LC2, LC3. Distinctive (with visible boundary) branches in scatter plots for axial V_Z components of data taken from the first (LC1) and the second (LC2) load cells, which correspond to outliers produced by some kind of contamination processes. Namely these contamination processes created about 11.5% of incorrect V_Z data outputs for the first load cell (LC1) and about 2.4% of incorrect V_Z data for the second one (LC2). Remarkably, that cross-sectional (V_X, V_Y) components of these multiaxial load transducers (LC1, LC2) and full calibration data set (V_X, V_Y, V_Z) for the third load transducer (LC3) have not shown any distinctive outliers

is a case for all three axis-components. Furthermore, from the graphs in Fig. 3 it can be identified that a distinctive branch of outliers between voltage readings of 0.2 and 0.6 V, as observed in the graphs for LC1 and LC2 in Fig. 2, appears in the graphs for the V_z component voltage readings only. As a result it can be argued that for all three load cells (LC1, LC2, LC3) the confidence interval is acceptable for the x- and y-axis, whilst outliers would be removed or corrected for LC1 and LC2 in the z-axis.

Physically the appearance of the outliers purely in the z-axis is the result of difference in strain sensitivity, since the measurement in the z-axis uses a relative small axial deformation, whilst for the x- and y-axis bending deformation has been used.

On this stage not only the branches of the outliers are distinguishable in Fig. 3, but also there is a clear boundary that allows counting the outliers. This makes it possible to confirm the quantity of outliers, attempted earlier on visual prediction based on Fig. 4 by precise counting of them.

Table 1 summarises the overall proportions of outliers and acceptable voltage readings, and provides the percentage of outliers for the individual load cells. The percentage of outliers, applying the 95% confidence interval, has been calculated to be 11.5% for LC1, 2.4% for LC2 and 0% for LC3.

It is very important to highlight that the latter described deterministic approach to identify outliers can also be applied to establish the precise location of these. In Fig. 3 the distinctive branches are confined between $0.015 < CIv_z/V_z < 0.04$ (vertical axis) and $0.12\ \text{V} < V_z < 0.41\ \text{V}$ (horizontal axis) for LC1, and between $0.02 < CIv_z/V_z < 0.037$ (vertical axis) and $0.2\ \text{V} < V_z < 0.37\ \text{V}$ (horizontal axis) for LC2. This deterministic approach allows to trace outliers back into a dataset from which a correction model (replacement model) can be derived. This model has been outlined later in this publication.

Figure 4 presents location of traced outliers in relation to the calibration load $\vec{F} = (F_x; F_y; F_z)$ for LC1, LC2 and LC3. These graphs represent the occurrence of outliers related to F_x, F_y and F_z loads when outliers happen in the z-axis. By applying this representation the established 140 outliers for LC2 are scattered around $-0.016\ \text{V} < V_x < 0.022\ \text{V}$ ($F_x \sim 0.00$ kN), $-0.273\ \text{V} < V_y < 0.507\ \text{V}$ ($-19.13\ \text{kN} < F_y < 35.36\ \text{kN}$) and $0.210\ \text{V} < V_z < 0.357\ \text{V}$ ($0.71\ \text{kN} < F_z < 50.00\ \text{kN}$), and for the established 660 outliers for LC1 are scattered around $-0.324\ \text{V} < V_x < 0.265\ \text{V}$ ($-19.13\ \text{kN} < F_x < 19.13\ \text{kN}$), $-0.776\ \text{V} < V_y < 0.715\ \text{V}$ ($-50.00\ \text{kN} < F_y < 50.00\ \text{kN}$) and $0.098\ \text{V} < V_z < 0.402\ \text{V}$ ($0.00\ \text{kN} < F_z < 50.00\ \text{kN}$). The scatter plots show that the clouds of outliers becoming rarefied with increasing loads, which corresponds to decrease of the absolute number of outliers with increasing axial forces in calibration range (up to 50kN). However, this decrease does not mean decreasing probability of outliers for higher loads but it is due to less experimental calibration points obtained for higher axial loads.

Estimated probability of outliers is about 11.5% for LC1, 2.4% for LC2 and close to 0% for LC3, it remains almost the same inside the calibration load range (1–50 kN). A measure of the outliers influence on the precision of the measurements will be evaluated in the following.

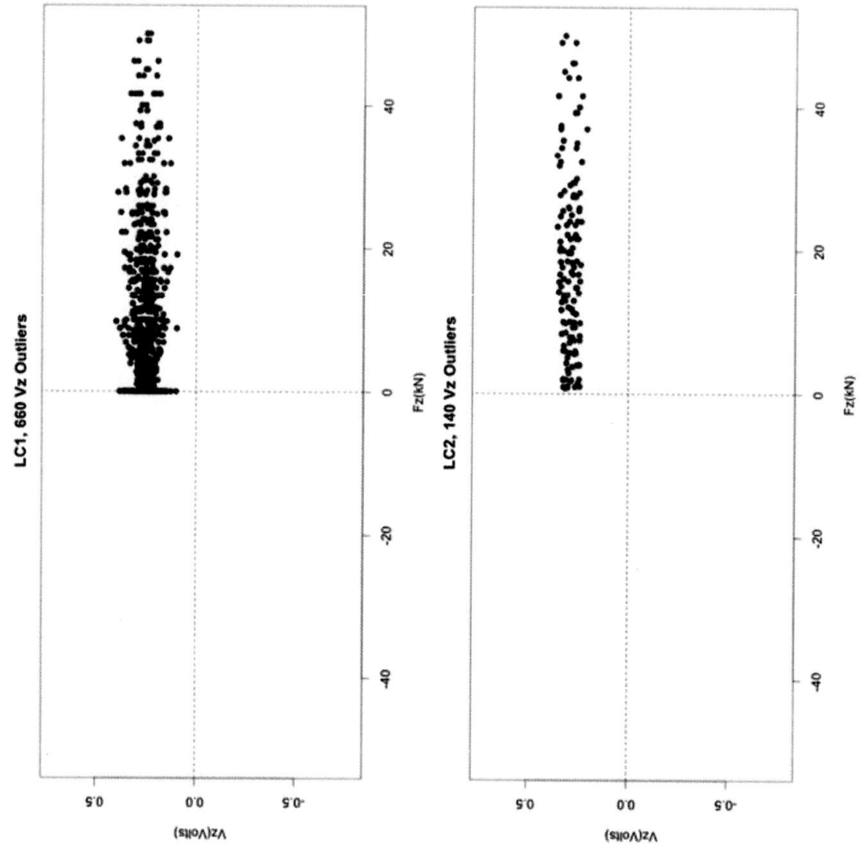

Fig. 4 Outliers happen in the vector load transducers (LC1, LC2) in V_z component not only for small loads but for the loads up to the upper calibration limit (50 kN). The third vector load transducer (LC3) has not shown any distinctive outliers

Table 1 Overall proportion of outliers

	LC1	LC2	LC3
Total calibration data points	5720	5780	5760
Acceptable voltage readings	5060	5640	5760
Outliers	660	140	0
Percentage of outliers (%)	11.5	2.4	0

3 Calibration Equation via Tensor Regression

The tensor regression task is to recover a function g from a collection of input–output data pairs $\{(V_n, F_n)\}_{n=1}^{N}$ generated in experiment

$$V = g(F) + V_0 \tag{2}$$

where the input F and the output V, shift V_0 are tensors of any order.

Present multicomponent load cell designs do not contain an absolute internal reference for measuring forces but they can achieve a certain accuracy only through calibration relative to some known reference. To give reasonable results, a multi-component force sensor should be accompanied by its specific calibration equation which relates the load cell output response to the applied force. A load cell's response is usually expressed in terms of the load force by a polynomial equation:

$$V_i = \sum_{j=x,y,z} s_{ij} F_j + V_{i0} \tag{3}$$

where sensitivity coefficients s_{ij} in Eq. (3) are calculated from the least square fit to the data set including force radial components F_x, F_y and axial one F_z, response of transducer system output voltages V_x, V_y (in corresponding radial directions) and axial V_z. There is a shift $V_{i0} = (V_{x0}, V_{y0}, V_{z0})$ that could be different for each load cell as well. The sensitivity coefficients s_{ij} constitute a sensitivity tensor s_{ij} (i = 1, 2, 3; j = 1, 2, 3).

For load cells containing contamination processes (LC1, LC2), the polynomial Eq. (3) has been fitted to two different sets of data: firstly to corrected data (with removed outliers) and secondly to full set of data. This made it possible to see an influence of outliers on particular components of the calibration Eq. (3).

Below the sensitivity tensors s_{ij} (i = 1, 2, 3; j = 1, 2, 3) are outlined for 3 load cells as s_{ijLC1} (i = 1, 2, 3; j = 1, 2, 3), s_{ijLC2} (i = 1, 2, 3; j = 1, 2, 3), s_{ijLC3} (i = 1, 2, 3; j = 1, 2, 3) the indexes relate the tensors to one of the load cells LC1, LC2, LC3 and an additional FSD (Full Set of Data) index corresponds to the full data set fittings.

The first load cell (LC1) revealed the largest amount of outliers (that is about 10% of the calibration data set), its improved response matrix

$$s_{ijLC1} = \begin{bmatrix} 14.9374 & 0.0154 & 0.1915 \\ -0.4328 & 15.1311 & 0.0657 \\ 0.4160 & 0.1385 & 14.7828 \end{bmatrix} \tag{4}$$

$$\vec{V}_{0_{LC1}} = (-31.0530; -20.2394; -3.0337)\ \text{mV/V}$$

with mean absolute deviation (3.9; 4.1; 3.0) mV/V, standard deviation (5.2; 6.1; 3.8) mV/V.

In contrast its sensitivity matrix for the full set of calibration data (added index 'FSD' means Full Set of Data)

$$s_{ijLC1_{FSD}} = \begin{bmatrix} 14.9416 & 0.0292 & 0.1908 \\ -0.4369 & 15.0632 & 0.0580 \\ 0.5156 & 0.6619 & 13.1017 \end{bmatrix} \tag{5}$$

$$\vec{V}_{0_{LC1_{FSD}}} = (-30.3739; -20.3383; 26.7828)\ \mathrm{mV/V}$$

with mean absolute deviation (4.2; 4.2; 34.7) mV/V, standard deviation (5.4; 6.2; 62.2) mV/V.

After comparison of obtained parameters for improved data set (with excluded outliers) Eq. (4) with corresponding ones for the full data set (FSD) Eq. (5), one could note that indeed the differences for all data related to the axial component (V_z) appears to be relatively large while all corresponding differences for data from the perpendicular components V_x, V_y are within reasonable limits. To be more specific, firstly, z-component of voltage offset V_{0z} changes from about − 3.0 to 26.8 mV/V, it is about 29.8 mV/V of difference while the corresponding differences for both V_{0x} and V_{0y} are less than 1 mV/V. Secondly, the z-main diagonal component is changed from 14.78 to 13.10 mV/V which gives difference about 1.68 mV/V that is much larger than x and y main diagonal components changing by less than 0.1 mV/V. Therefore, it is not surprising that there are comparably high changes in z components of fitting parameters for improved data set relative to corresponding ones for the full data set (FSD), in particular one could note that indeed there are very large differences in the Mean Absolute Deviations and in the Standard Deviations for all data related to the axial component (V_z) with comparably minor changes for the perpendicular components V_x, V_y.

The second load cell (LC2) revealed just about 2.4% of outliers in its calibration data, its improved sensitivity matrix:

$$s_{ijLC2} = \begin{bmatrix} 14.4984 & -0.3992 & 0.0795 \\ 0.3342 & 14.5594 & 0.1476 \\ -0.1332 & -0.3243 & 14.5784 \end{bmatrix} \tag{6}$$

$$\vec{V}_{0_{LC2}} = (0.2445; -6.1245; -6.9407)\ \mathrm{mV/V}$$

with mean absolute deviation (5.4; 5.2; 2.8) mV/V, standard deviation (7.3; 7.3; 3.6) mV/V.

In contrast its sensitivity matrix for the full set of calibration data (Full Set of Data with added index 'FSD')

$$s_{ijLC2_{FSD}} = \begin{bmatrix} 14.4984 & -0.4027 & 0.0815 \\ 0.3342 & 14.5556 & 0.1445 \\ -0.1332 & -0.3098 & 14.1970 \end{bmatrix} \tag{7}$$

$$\vec{V}_{0_{LC2_{FSD}}} = (0.3253; -6.0432; -1.6311)\ \mathrm{mV/V}$$

with mean absolute deviation (5.4; 5.2; 8.4) mV/V, standard deviation (7.3; 7.2; 29.2) mV/V.

Again, it should be noted that there are large differences for all data related to the axial component (V_z) while there is much less of a change in the corresponding data for the perpendicular components V_x, V_y.

Finally, the third load cell (LC3) has not revealed any outliers and its calibration data gives the following sensitivity matrix:

$$s_{ijLC3} = \begin{bmatrix} 13.9315 & -0.1886 & 0.1766 \\ -0.0248 & 14.2542 & -0.0154 \\ -0.0224 & 0.1665 & 14.7517 \end{bmatrix} \quad (8)$$

$$\vec{V}_{0_{LC}3} = (0.2963;\ 1.2889;\ 8.3813)\ \text{mV/V}$$

with quality of fit defined by mean absolute deviation (4.6; 5.1; 3.2) mV/V, standard deviation (6; 7.2; 4.1) mV/V.

The larger numbers on main diagonals of calculated sensitivity matrixes define the larger contribution to any particular component of voltage response (V_x, V_y, V_z) from the corresponding force components (F_x, F_y, F_z), off-main diagonal numbers are due to coupling between voltage readings for different axes.

4 Calculation of Multicomponent Force

Next step is to derive equations for the load cells to calculate unknown forces. In the frame of the multivariate model the force tensor could be calculated as inverse of the corresponding sensitivity tensor. Therefore, it is widely accepted that the unknown force components (F_x, F_y, F_z) could be calculated from the measured voltages (V_x, V_y, V_z) via the following polynomial equation:

$$F_i = \sum_{j=x,y,z} r_{ij} V_j + F_{i0} \quad (9)$$

where i, j are independent indexes taking values from the {x, y, z} set.

The force coefficients r_{ij} constitute a response tensor r_{ij} (i = 1, 2, 3; j = 1, 2, 3), outlined for 3 load cells as r_{ijLC1} (i = 1, 2, 3; j = 1, 2, 3), r_{ijLC2} (i = 1, 2, 3; j = 1, 2, 3), r_{ijLC3} (i = 1, 2, 3; j = 1, 2, 3), which could be calculated by inversing corresponding sensitivity tensors for these 3 load cells as s_{ijLC1} (i = 1, 2, 3; j = 1, 2, 3), s_{ijLC2} (i = 1, 2, 3; j = 1, 2, 3), s_{ijLC3} (i = 1, 2, 3; j = 1, 2, 3):

$$r_{ijLC1} = \begin{bmatrix} 66.9263 & -0.0420 & -0.8591 \\ 1.9330 & 66.0111 & -0.3124 \\ -1.8956 & -0.6110 & 67.6408 \end{bmatrix} \quad (10)$$

$$\vec{F}_{0_{LC}1} = (2.0735;\ 1.3939;\ 0.1399)\ \text{kN}$$

$$r_{ijLC1_{FSD}} = \begin{bmatrix} 66.9078 & -0.0709 & -0.9521 \\ 1.9630 & 66.3210 & -0.2373 \\ -2.3234 & -2.7826 & 65.7032 \end{bmatrix} \tag{11}$$

$$\vec{F}_{0_{LC1_{FSD}}} = (2.0528; 1.4011; -0.1735)\ \text{kN}$$

$$r_{ijLC2} = \begin{bmatrix} 68.8288 & 1.9223 & -0.3951 \\ -1.5469 & 68.5143 & -0.6759 \\ 0.5936 & 1.5517 & 68.5467 \end{bmatrix} \tag{12}$$

$$\vec{F}_{0_{LC2}} = (-0.0077; 0.4135; 0.4903)\ \text{kN}$$

$$r_{ijLC2_{FSD}} = \begin{bmatrix} 68.8259 & 1.9375 & -0.4183 \\ -1.5484 & 68.5337 & -0.6803 \\ 0.5910 & 1.4783 & 68.3361 \end{bmatrix} \tag{13}$$

$$\vec{F}_{0_{LC2_{FSD}}} = (-0.0108; 0.4121; 0.4831)$$

$$r_{ijLC3} = \begin{bmatrix} 71.7003 & 0.9777 & -0.8403 \\ 0.1425 & 70.0384 & 0.0757 \\ 0.1219 & -0.7856 & 67.7489 \end{bmatrix} \tag{14}$$

$$\vec{F}_{0_{LC3}} = (-0.0186; -0.0918; -0.5600)\ \text{kN}$$

Equations (9)–(14) are a complete set of equations to calculate forces using voltage readings for the third load cell (LC3). For the load cells, where the outliers have been detected (LC1 and LC2), there is a need to consider a replacement model for outliers to achieve an acceptable precision.

Estimations of mean deviations for the full calibration data set relative to the full load range (LR = 50 kN) for the model when no outliers considered are summarised in Table 2.

Table 2 Standard deviations (MSD) for the full calibration data set relative to the full load range (LR = 50 kN) clearly indicate substantially lower precision for the axial (F_z) component as compared to the perpendicular components F_x and F_y: for the first load cell (LC1) about 10 times and about 5 times for the second one (LC2)

Straightforward (no outliers considered)	MSD (F_x)	MSD (F_y)	MSD (F_z)
LC1	0.01%LR	0.01%LR	0.1%LR
LC2	0.01%LR	0.01%LR	0.06%LR
LC3	0.01%LR	0.01%LR	0.01%LR

Remarkably that the third load cell (LC3) show same order of magnitude of the MSD for the all three components (compare with [7, 8] Table 1) when MSD calculated after correction of outliers)

The corresponding bias was of the order of 10^{-5}%LR that is much less than the corresponding mean deviations. This table show mean standard deviations of the order 10^{-2}%LR for all three components of the third load cell (LC3) and for the F_x and F_y for LC1 and LC2. The axial component (F_z) reveals 10 times higher mean standard deviation (MSD) of the order of 10^{-1}% of the load range for the first load cell (LC1) and it is about 5 times higher than usual for the second load cell (LC2).

Outlined below is the replacement model which allowed decrease of the standard deviation (MSD) for the F_z by about 10 times (for LC1) and about 5 times (for LC2). This makes the same order of 10^{-2}%LR for the MSD for all force components F_x, F_y, F_z of the all three load cell readings.

Analysis confirms that LC characteristics change with different loads applied.

There is a range for all force components in proximity to zero (starting from about 0.2 kN and rising sharply at 0.01 kN and lower) where relative error become very large, however the corresponding absolute error remains in approximately same range.

5 Physical Model to Improve Readings of Multidimensional Force Response

A general idea is that during calibration a matrix of precision zones would be established for each multidimensional sensor component, so that it could be used to find out which of the component has the largest error around some point in multidimensional precision calibration space. As soon as this precision calibration field is designed with all the vector component errors calculated or estimated for each of its points, it is the turn of the replacement model to identify the vector component with the highest local error and replace its reading with the better one, which has been calculated using the calibration data set. This procedure helps to increase the load response measurement precision both for the particular vector sensor component, and for full multidimensional force reading as well.

As a particular case this physical replacement model could be applied to increase precision of load cells with single-component outliers, which were detected during the confidence interval tests in calibration data for two load cells. These tests show single-component axial voltage (V_z) outliers in calibration data sets from first (LC1) and second (LC2) load cells. Established single-component nature of contamination processes allow to use physical meaning of full vector force together with calibration results to construct the substitution process replacing outliers in experimental data from the multidimensional load cells.

The physical multicomponent replacement model could be introduced for the detected outlier when

$$d\left(F_i^{out}, F_i^{med}\right) = \left|F_i^{out} - F_i^{med}\right| > D_{Fi}^{\max} \tag{15}$$

where F_i^{out} is the outliers i-component of multicomponent force,

$D_{Fi}^{\max}$ is the maximal absolute deviation from F_i^{med}.

If a particular data point detected as an outlier, one has to replace it with

$$F_i^{med} = \left(\frac{1}{k^2}(V - V_0)^2 - F_j^2 - F_k^2\right)^{1/2} \pm D \tag{16}$$

where k and V_0 could be obtained from the prior calibration equation containing voltages and forces in the calibration data set, value and sign of D depends on the standard deviation and direction of absolute deviation. This physical replacement model allows to detect outliers in real-time and replace them with those found with prior calibration.

In previous paper [6] the physical replacement model has been introduced for specific case of one force component $F_{z.}$, namely, for detected outlier when

$$d\left(F_z^{out}, F_z^{med}\right) = |F_z^{out} - F_z^{med}| > D_{Fz}^{\max} \tag{17}$$

where F_z^{out} is the outliers axial (Z) component of the force,

$D_{Fz}^{\max}$ is the maximal absolute deviation from F_z^{med}.

If a particular data point detected as an outlier, one has to replace it with

$$F_z^{med} = \left(\frac{1}{k^2}(V - V_0)^2 - F_x^2 - F_y^2\right)^{1/2} \pm D \tag{18}$$

where k and V_0 could be obtained from the prior calibration equation containing voltages and forces in the calibration data set, value and sign of D depends on the standard deviation and direction of absolute deviation.

6 Validation

This novel physical model tested and validated by application to the calibration data sets for multiaxial load cells installed on the South Western Mooring Test Facility (SWMTF), which is published in separate papers [5–8].

7 Conclusion

New method is described for processing the multidimensional response force data generated by multiaxial load cells, which allow to identify and correct gross errors in sets of measurements. These are then used to identify which vector sensor component has outliers. Within the study a model has been developed for cross-validation of a three-component system [5–7], and extended to more general multidimensional systems in this paper and [8]. In addition, for a case of single component failure a method is introduced that allows obtaining credible multicomponent force measurements even if some components fail and/or have significant outliers.

Results could be used not only for electronic (strain gauge) transducers but also for hydraulic or mechanical (in proving rings, for example) load cells. As a bigger picture it could be considered as a mechanism for self-regulation of sensors systems, when the output improved using the novel physical replacement model and calibration data.

This paper show the cases of outliers in calibration data and suggests methods for minimizing their influence on calibration results, furthermore the novel physical replacement model of detection and excluding or substitution of outliers in future force response data readings is proposed.

This could be considered as a way to get load transducers (load cells) more precise and reliable, and for the cases when mechanical repairs are impossible or too expensive, the new physical replacement model could be adapted to specific transducer configurations to improve measurements. The results show that this new multidimensional physical replacement model significantly decreased the errors of strain control, and that this new model is very effective in controlling multidimensional strain, feedback, and greatly improve the experimental accuracy.

References

1. A Strain Feedback Compensation Method during Cell Tensile Experiments / Zhou R, Yang Y, Zhang W, Zou Y. // J Healthc Eng. 2017. 2017:1587670. https://doi.org/10.1155/2017/1587670. Epub 2017 Jun 18. PMID: 29065572. PMCID: PMC5494114.
2. A Comprehensive Method for Accurate Strain Distribution Measurement of Cell Substrate Subjected to Large Deformation / He H, Zhou R, Zou Y, Huang X, Li J. // J Healthc Eng. 2018 Jan 8. 2018:8504273. https://doi.org/10.1155/2018/8504273. PMID: 29599957. PMCID: PMC5823413.
3. Gun'ko, Yu.F., Ponomarev, M.G. The charged particles emission in a magnetic field // Vestnik Sankt-Peterburgskogo Universiteta. Ser 1. Matematika Mekhanika Astronomiya. 1993. n2. 89–94.
4. Gun'ko, Yu.F., Ponomarev, M.G. The charged particle layer expansion in a magnetic field // Vestnik Sankt-Peterburgskogo Universiteta. Ser 1. Matematika Mekhanika Astronomiya. 1994. n4. 105–108.
5. Ponomarev M., Johanning L., Parish D., Enhancing precision and reliability of tri-axial load cells for mooring load measurements // Proc. of the 3rd International Conference on Ocean Energy. 6 October. 2010 (ICOE2010). ISBN 978-84-693-5467-4.
6. Ponomarev M. A novel physical model to enhance precision and performance of 3-dimensional force sensors // Processes in Geomedia 2022. No2(32). p. 1589–1600.
7. Ponomarev M. Experimental validation of novel physical model for improvement of sensing 3-dimensional fluid flow loads and responses in real sea conditions with South Western Mooring Test Facility (SWMTF) // Processes in Geomedia 2022. No2(32). 1579–1589.
8. Ponomarev M., Johanning L., Parish D., Experimental validation of novel physical model for improvement of sensing multidimensional fluid flow loads and responses in real sea conditions with South Western Mooring Test Facility // Physical and Mathematical Modeling of Earth and

Environment Processes – 2022, Proceedings of 8th International Scientific Conference-School, Springer, ISBN 978-3-031-25961-6

9. Digital radiographic inspection technique for production friendly quality assessment of PM parts / Ponomarev M.G., Kappatos V., Selcuk C., Gan T.-H., Amos M., Halai H., Gierl C., Iovea M. // Powder Metallurgy. 2013. 56. 2. 92–95.
10. PM Dimensional Control: Development of a Digital Radiographic Inspection Technique for Production Friendly Quality Assessment of Powder Metallurgy Parts / M. G. Ponomarev, V. Kappatos, C. Selcuk, T.-H. Gan, M. Amos, H. Halai, C. Gierl, M. Iovea // Proceedings of the International Euro Powder Metallurgy Congress and Exhibition. Euro PM 2012. 2012. Volume 1. Basel. Switzerland. 16 September 2012 through 19 September 2012. Code 105676.
11. Ponomarev M.G., Selcuk C., Gan T.-H. Construction of pattern recognition system optimized for X-ray inspection of plastic electronics and OLED displays // 51st Annual Conference of the British Institute of Non-Destructive Testing 2012. NDT 2012. BINDT 2012. Northamptonshire. United Kingdom. 11 September 2012 through 13 September 2012. Code 107048. pp. 354–360.
12. Defect detection and classification system for automatic analysis of digital radiography images of PM parts / M. G. Ponomarev, C. Selcuk, T.-H. Gan, M. Amos, I. Nicholson, M. Iovea, M. Neagu, B. Stefanescu, G. Mateiasi // Powder Metallurgy. 2014. 57. 1. 17–20.
13. Non Destructive Testing: A Defect Detection and Classification System for Automatic Analysis of Digital Radiography Images of Powder Metallurgy Parts / M. G. Ponomarev, C. Selcuk, T.-H. Gan, M. Amos, I. Nicholson, M. Iovea, M. Neagu, B. Stefanescu, G. Mateiasi // International Powder Metallurgy Congress and Exhibition. Euro PM 2013. Gothenburg. Sweden. 15 September 2013 through 18 September 2013. Code 110976.
14. Preliminary NDT investigation of sintered powder metallurgy parts by high-resolution TDI based X-ray digital radiography / M. Iovea, M. Neagu, B. Stefanescu, G. Mateiasi, H. Halai, M. Amos, M. Ponomarev, V. Kappatos, C. Selcuk, T.-H. Gan, C. Gierl // 2012. Volume 1. European International Powder Metallurgy Congress and Exhibition. Euro PM 2012. Basel. Switzerland. 16 September 2012 through 19 September 2012. Code 105676.
15. PM parts fast in-line X-ray digital radiography / M. Iovea, M. Neagu, B. Stefanescu, G. Mateiasi, A. Clarke, I. Nicholson, M. Ponomarev, V. Kappatos, C. Selcuk, T.-H. Gan // International Powder Metallurgy Congress and Exhibition. Euro PM 2013. European PM Conference Proceedings.1. 2013. European Powder Metallurgy Association (EPMA).
16. A molecular dynamics study of a 5 keV C60 fullerene impact on a two-component organic molecular sample / Ponomarev M.G., Garrison B.J., Vickerman J.C., Webb R.P. // Surface and Interface Analysis. Wiley Online Library. 2011. 43. 1–2. 107–111.
17. Ponomarjov M. G. Imaginary-emission method for modeling disturbances of all magnetoplasma species: Reflecting and absorbing objects in motion through a rarefied plasma at different angles to the ambient magnetic field // Physical Review E. 1996. 54. 5. 5591.
18. Ponomarjov M. G. Disturbances of the ambient magnetoplasma due to its interactions with object surfaces. Imaginary emission method. Far-wake of objects moving through a rarefied plasma at different angles to the ambient magnetic field // Planetary and Space Science. 1995. 43. 10–11. 1419–1427.
19. Ponomarjov M. G. Space flows and disturbances due to bodies in motion through the magnetoplasma // Astrophysics and Space Science. 2000. 274. 1. 423–429.
20. Ponomarjov M. G. Pressure of charged particle flows in ambient magnetic fields // Astronomische Nachrichten. 1997. 318. 3. 187–192.
21. Ponomarjov M. G. 3D time-dependent kinetic simulation of space plasma disturbances due to moving bodies with the ambient magnetic field effect // Advances in Space Research. 2002. 29. 9. 1397–1402.
22. Ponomarjov M. G., Gunko Y. F. Kinetic modeling of charged particle cloud expansion and emission in magnetic and electric fields // Planetary and Space Science. 1995. 43. 10–11. 1409–1418.
23. Ponomarjov M. G. Acceleration and transport of particles in collisionless plasmas: Wakes due to the interaction with moving bodies // Astrophysics and Space Science. 2001. 277. 1. 39–44.

24. Ponomarjov M. G. Outer atmosphere and wake of space objects, kinetic simulation. Disturbances of ambient magnetoplasma due to diffuse reflecting bodies in motion // APS Division of Plasma Physics Meeting Abstracts. 1998. abstract id. U9Q.09.
25. Ponomarjov M. G. Simulation of oscillations in charged particle systems under the ambient magnetic field control // 2nd International Conference. Control of Oscillations and Chaos (COC 2000). St. Petersburg, Russia. 5 July 2000 through 7 July 2000. Code 57512. Proceedings (Cat. No. 00TH8521), 2000, vol. 1. pp. 167–170. https://doi.org/10.1109/COC.2000.873548.
26. Ponomarjov M. G. 3D Collisional Kinetic Simulation of Stratifications and Flute Structures of Plasma Flows and Wakes in External Magnetic Fields // AAS, DPS Meeting. No 32. id.15.10. 2000. Bulletin of the American Astronomical Society. Vol. 32, p. 1022.
27. Ponomarjov M. G., Carati D. Enhanced acceleration of electrons populations by crossing electron cyclotron waves in an ambient magnetic field, Resonant Moments method // Advances in Space Research. 2006. Volume 38. Issue 8. 1576–1581.
28. Ponomarjov M. G. Outer atmosphere and wake of space objects, kinetic simulation. Disturbances of ambient magnetoplasma due to diffuse reflecting bodies in motion // American Physical Society, Division of Plasma Physics Meeting, November 16–20, 1998 New Orleans, LA abstract id. U9Q.09.
29. Ponomarjov M. G. 3D Kinetic Dynamical models of Ionized HII Clouds in External Magnetic Field // Ionized Gaseous Nebulae. 2000. Mexico City. November 21–24. Meeting abstract id. 44.
30. Ponomarjov M. G. Kinetic simulation of stratifications and flute structures of charged particle jets and wakes in the ambient magnetic field // Bulletin of the Astronomical Society. 1999. Vol. 31. No. 4. p. 1157. id.53.12.
31. Ponomarjov M. G. 3D Collisional Kinetic Simulation of Stratifications and Flute Structures of Plasma Flows and Wakes in External Magnetic Fields // 32nd Annual Meeting of the Division for Planetary Sciences. 23–27 October 2000. Pasadena. California. USA. Session 15. Outer Planets IV-Aurorae and Magnetospheres. 2000/10/24. Abstract 15.10. 1022.
32. Ponomarjov M. G. Kinetic Simulation of Magnetic Field Effects on Stratifications and Flute Structures of Space Plasma Flows and Wakes of Bodies // 2001 Joint Assembly American Geophysical Union Spring Meeting. Boston. Massachusetts. USA. Abstracts. 2001. SM52B-04.
33. Ponomarjov M. G., Carati D. Search for optimal 3D wave launching configurations for the acceleration of charged particles in a magnetized plasma: Resonant Moments Method // American Physical Society. 46th Annual Meeting of the Division of Plasma Physics. 15–19 November. 2004. Savannah. GA. MEETING ID: DPP04. Abstract id. RP1.042.
34. Ponomarjov M. G. New Ways of Protecting Astronomical Equipment and Solar Batteries of Spacecrafts // American Astronomical Society, DPS meeting No 29, id.27.02. Bulletin of the American Astronomical Society. 1997. Vol. 29. p. 1023.
35. Ponomarjov M. G., Carati D. Acceleration of charged particles by crossed cyclotron waves, Resonant Moments Method // 35th COSPAR Scientific Assembly. 18–25 July 2004. Paris. France. p. 2851.
36. Ponomarjov M. G., Carati D. Kinetic Simulations of Relativistic Electron Flows in Time-Dependent Electromagnetic Fields // Electron Cyclotron Emission and Electron Cyclotron Heating. Conference Proc. World Scientific Publishing Co Pte Ltd. 2003. 137–142.
37. Ponomarjov M. G. Kinetic simulation of magnetic field effects on wakes of meteoroids imaginary emission method // Meteoroids 2001 Conference. 2001. Proc. 295–300.
38. Ponomarjov M. G., Carati D. Search for optimal 2D and 3D wave launching configurations for the largest acceleration of charged particles in a magnetized plasma, Resonant Moments Method // arXiv 2004. https://arxiv.org/format/physics/0411075. arXiv:physics/0411075 [physics.plasm-ph]
39. Ponomarjov M. G. 3D time-dependent kinetic simulation of turbulent plasma flows under the effect of external magnetic field // American Physical Society. 42nd Annual Meeting of the APS Division of Plasma Physics combined with the 10th International Congress on Plasma Physics October 23–27. 2000 Québec City. Canada Meeting ID: DPP00. abstract id. NO1.007. https://ui.adsabs.harvard.edu/abs/2000APS..DPPNO1007P/abstract

40. Ponomarjov M. G. Kinetic modeling magnetic field effect on ion flows, disturbances and wakes in space plasma // Space plasma simulation: proceedings of the Sixth International School/Symposium. ISSS-6. Garching. Germany. 3–7 September. 2001. Edited by J. Büchner, C.T. Dum, and M. Scholer. Berlin: Schaltungsdienst Lange o.H.G. 2001. p. 328.
41. Ponomarjov M. G., Carati D. Acceleration of electron populations by crossing EC waves in an external magnetic field // 13th Joint Workshop on Electron Cyclotron Emission and Electron Cyclotron Resonance Heating. Nizhny Novgorod. Russia. 2004. Conference paper https://ec13.iapras.ru/papers/Ponomarjov.pdf
42. Ponomarjov M. G., Carati D. Enhanced acceleration of charged particles by crossing electromagnetic waves in a magnetized plasma, resonant moments method // 2004. ffhal-00001926v1f. https://hal.archives-ouvertes.fr/hal-00001926v1
43. Webb R. P., Ponomarev M. Molecular dynamics simulation of low energy cluster impacts on carbon nanotubes // Nuclear Instruments and Methods in Physics Research Section B: Beam Interactions with Materials and Atoms. 2007. 255. 1. 229–232.
44. Brignel J. E. Sensors in distributed instrumentation systems // Sensors and Actuators. 1986. 10. 249–261.
45. Jooa J. W., Naa K.S., Kang D.I. Design and evaluation of a six-component load cell // Measurement. 2002. 32. 125–133.
46. Design and analysis of a column type multi-component force/moment sensor / J. H. Kim, D. I. Kang, H. H. Shin, Y. K. Park // Measurement. 2003. 33. 213–219.
47. Investigation and calibration of a force vector sensor with a calibration artefact / S. Lietz, F. Tegtmeier, D. Röske, R. Kumme, D. Schwind // Proc. of the XIX IMEKO World Congress Fundamental and Applied Metrology. Lisbon. Portugal. 6–11 September 2009. p. 299–302.
48. Carati D., Ponomarjov M. G. Statistical description of currents induced by two electron cyclotron counter-propagating waves // Electron Cyclotron Emission and Electron Cyclotron Heating. Proceedings of the 12th Joint Workshop. Held 13–16 May 2002 in Aix-en-Provence, France. Edited by Gerardo Giruzzi (Association Euratom-CEA sur la Fusion, France). Published by World Scientific Publishing Co. Pte. Ltd., 2003. ISBN #9789812705082, pp. 77–82.
49. Milgram J. H., Triantafyllou M. S., Frimm F.C. Seakeeping and Extreme Tensions in Offshore Towing // Trans. of the Society of Naval Architects and Marine Engineers 1988. 96. p. 35–72.
50. Papazoglou V., Mavrakos S., Triantafyllou M. S. Scaling and Model Testing Cables in Water // J. of Sound and Vibration 1990. 140. p. 103–115.
51. Chakrabarti S.K. Handbook of Offshore Engineering. Elsevier Ocean Engineering Series. 2005. Vol. 2.

Experimental Validation of Novel Physical Model for Improvement of Sensing Multidimensional Fluid Flow Loads and Responses in Real Sea Conditions with South Western Mooring Test Facility

Maxim Glebovich Ponomarev, Lars Johanning, and David Parish

Abstract Multi axis load cells (transducers) are widely used in modern engineering systems to measure multidimensional forces. This paper deals with analysis and calibration methodology for the multi axis load cells, considering them in specific installation South Western Mooring Test Facility as a key instrument to measure the mooring leg tension in its x-, y- and z-component for wave energy device development. Cross-validation model has been applied for the multicomponent system.

Keywords Multidimensional mechanical response to fluid flows · Multi-axis load sensors · Multiaxial fatigue · Force · Sensors · Strain condition monitoring · Transducers · Vector load sensors · Calibration · Precision · Reliability · Smart sensors · Enhance precision · Overcoming sensor degradation

1 Introduction

Present load cell designs do not contain an absolute internal reference for measuring forces but they provide an accuracy of 0.03–0.25% of full scale and are suitable for almost all industrial applications. In its most common form, it consists of a stamped metal sheet attached to a flexible, insulating base [1–4]. The sensor is glued to the object whose deformation is to be studied. As the object deforms, the sheet deforms, causing its electrical resistance to change. Calibration procedure is more difficult for multiaxial load cells due to almost unavoidable cross-couplings (crosstalks) between sensors for different axes (components). If the load cells have been correctly installed

M. G. Ponomarev (✉)
Cambridge Research and Consultancy, Cambridge, UK
e-mail: maxim.pv@gmail.com

Faculty of Engineering and Physical Sciences, Adyghe State University, Maykop, Russia

L. Johanning · D. Parish
University of Exeter, Exeter, UK

V. I. Karev (ed.), *Physical and Mathematical Modeling of Earth and Environment Processes—2022*, Springer Proceedings in Earth and Environmental Sciences,
https://doi.org/10.1007/978-3-031-25962-3_36

and calibrated, accuracy can be within 0.25% of full scale or better, which is acceptable for most weighing processes [5]. A need for analysis of calibration data for the multiaxial load cells from laboratory is outlined here.

It is followed by cross-validation of the calibration, and final validation of the calibrated multiaxial load cells installed on the South Western Mooring Test Facility (SWMTF) in real sea conditions [6] used as mooring load and response test facility. It has been recently installed in Falmouth Bay, Cornwall, UK.

In our study, we proceed from the assumption that the resonant modes of floating wave energy converters (WECs) are close to the excitation modes for obtaining energy from first-order waves. Modes of excitation differ due to the size of the body and the possibility of its application. In the case for moored WEC devices additional complexity is given through the analysis requirement of the coupled system. We can capture many implications that can affect WEC device designs. Therefore, we conclude that a good understanding of the dynamic response of a floating system is required. Reliability, life expectancy, design characteristics, and subsequent cost optimization of WEC devices affect load and response performance results. As a consequence, motion and load characteristics need to be identified during realistic offshore installation to aid the development and design of WECs and to inform fully dynamic numerical simulation models. Realistic in-situ tests can aid to obtain data that can be applied to calibrate and/or validate these simulation models, allowing a higher accuracy of simulations, which consequently could be used to improve WEC designs [7–10].

The SWMTF facilities will allow measuring loads that are experienced in the field through prototype testing, which information can be subsequently used for numerical calibration purpose or to perform accelerated component testing under simulated in-service field conditions. A key instrumentation are three top end multi axis load cells (LC1, LC2 and LC3) [11]. Then it is possible to measure the mooring tension for loads near or under tension. Since no suitable load cell could be identified for this task, multi axis load cells were designed and developed in a collaborative effort between the University of Exeter (UK) and Elite Transducers Ltd. (UK).

The preliminary analysis of the multi axis load cell calibration identified a non deterministic behaviour that required the development of new approach to rectify the calibration results. A new method for enhancing the measurement precision and reliability for tri-axial load cells has been developed with introduction of precision matrix defined with prior calibration by Ponomarev [6,10] for multiaxial load sensors by Ponomarev [12]. Application of this method will be demonstrated for processing data generated by multi-axis load cells which allow to find and correct errors in sets of measurements. Furthermore, the model will be proposed for cross-validation of multicomponent system together with calibration.

2 Validation by Application to the Calibration Data Sets for Multiaxial Load Cells Installed on the South Western Mooring Test Facility

When the model has been built, it is important to check the model's fit to the data, however checks based on internal validation are not equal to prediction testing. To be on the ideal side, the predictive power can only be confirmed by testing on new objects, nevertheless, it is widely accepted, that in majority of cases cross-validation describes sensibly the prediction ability for the first stage testing at least, moreover it could be considered as a necessary condition for the goodness of prediction. In practice, internal validation widely used as the most obvious tool, when the obtained predictor would be tested on the calibration set itself. During cross check of improved calibration matrices with the corresponding improved calibration data sets the averaged bias was of the order of 10^{-5} %LR, that is much lower than the corresponding mean deviations, which are combined in the table below. Table 1 shows enhanced mean precisions for the axial F_z component relative to the full load range (LR = 50 kN), when cross check data with outliers substituted and calculated with improved matrix. Namely, by comparison with the standard deviations (MSD) calculated after calibration with the full data set (see 12 Table 2), there is up to an order of magnitude decrease of the standard deviation for the axial component F_z.

The assessment of the model is presented by the Quality of Prediction figures (Fig. 1), where predicted forces plotted versus corresponding measured forces. Ideally, these scatter plots should be on a diagonal (45°) line, which is plotted over these scatter plots for the easier reference. It presents visually the quality of calibration models for the first (LC1), second (LC2), and third load cell (LC3). It shows that the models seem to be having appropriate scatter ranges over the calibration loads. In addition, it demonstrates on the visual level that the quality of prediction is not the same through the calibration load, but does depend on the value of force measured.

For comparison Fig. 2 shows significantly lower quality for the straightforward calibration made on data with excluded outliers (improved data) but there is no replacement model applied.

The more detailed quantitative measure of the model precision is calculated during cross-checks with plots of mean absolute percentage error (MAPE)

$$MAPE_j = \frac{1}{N}\sum_{i=1}^{N}\frac{\left|\hat{F}_i^j - \tilde{F}_i^j\right|}{\left|\hat{F}_i^j\right|} \times 100\% \quad (1)$$

Table 1 Standard deviations (MSD) over the full load range (LR = 50 kN) for improved calibration

	MSD (F_x)	MSD (F_y)	MSD (F_z)
LC1	0.01 %LR	0.01 %LR	0.01 %LR
LC2	0.01 %LR	0.01 %LR	0.01 %LR
LC3	0.01 %LR	0.01 %LR	0.01 %LR

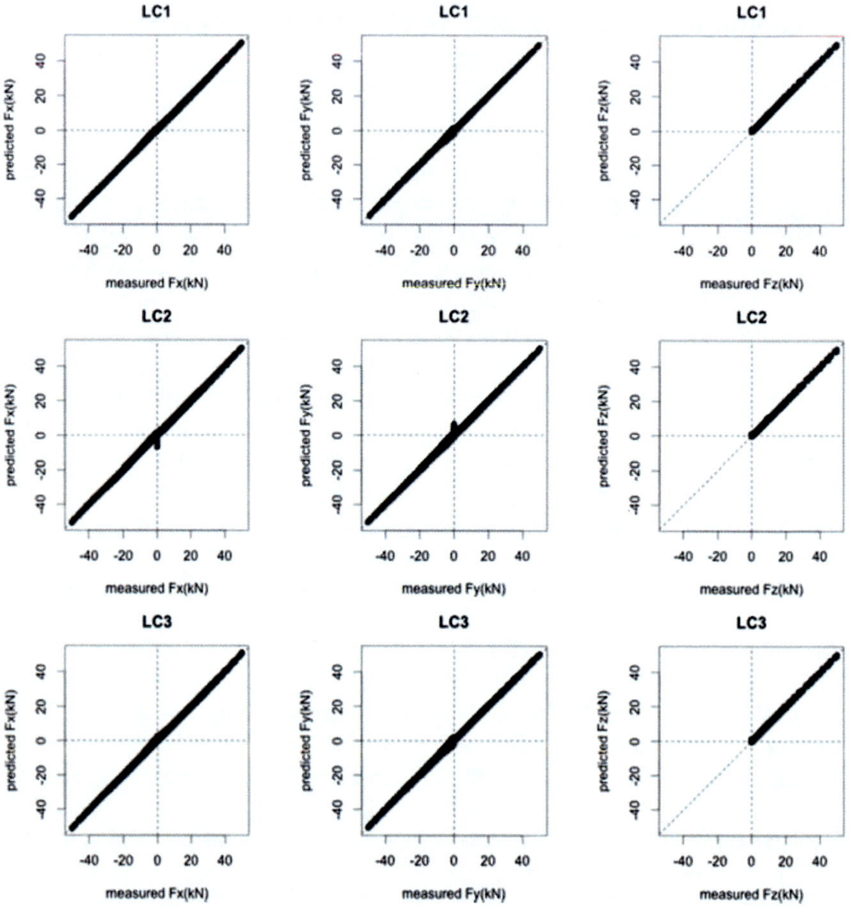

Fig. 1 Quality of the final calibration models for the first (LC1), second (LC2), and third load cell (LC3). It shows that the model seem to be having appropriate scatter ranges over the calibration loads. In addition, it shows that the quality of prediction is not the same through the calibration load range, but does depend on the value of force measured. Namely, there is a limited increase of the scatters range for decreasing loads

where predicted (forecast) force components $\tilde{F}_i^x$; $\tilde{F}_i^y$; $\tilde{F}_i^z$ were checked against corresponding measured (actual) ones $\hat{F}_i^x$; $\hat{F}_i^y$; $\hat{F}_i^z$. Because there is an output for the forces as vector components, it is informative to have the precision plotted for each component separately. Figure 3 shows MAPE, which, as expected, decreasing with increasing corresponding force components. The solid line corresponds to spline interpolation between filled circular dots representing MAPE for the improved calibration. The empty circles represent the MAPE for the case of calibration on the full data set (when no outliers have been taken into account). Not surprisingly, MAPE is not changing for the F_x and F_y components, because of single component nature

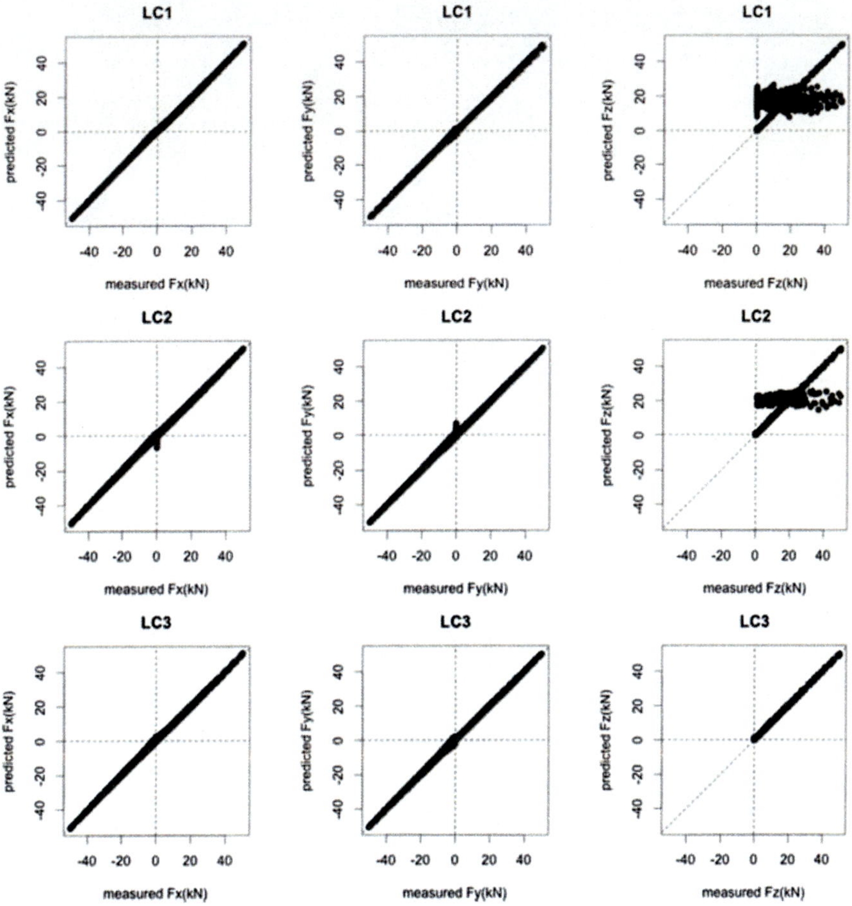

Fig. 2 Quality of the calibration if the replacement model is applied on the first stage only: when outliers were excluded from the calibration data set and calibration made with this improved data set. However, outliers could appear in new measurements, if there is no second stage in replacement model implemented

of the outliers confirmed earlier. It is shown clearly that the replacement model make it possible to decrease the error (MAPE) for the axial force components F_z only. Furthermore it should be noted that after such a decrease the MAPE for F_z becomes very close to the MAPE for the perpendicular components F_x and F_y: estimations show that all the force component measurements reach 10% of MAPE when measured forces' components are 2.5 kN and it becomes 5% at 5 kN.

These estimates are in good agreement with previously made precision estimate for the full force measurements [6]. To get the more obvious visual presentation of improvement factors, the MAPE for F_z are plotted on the different scale on the Fig. 4 shows estimations that the replacement model allows to decrease the MAPE (which could be represented as an estimation of expected relative precision as with mean

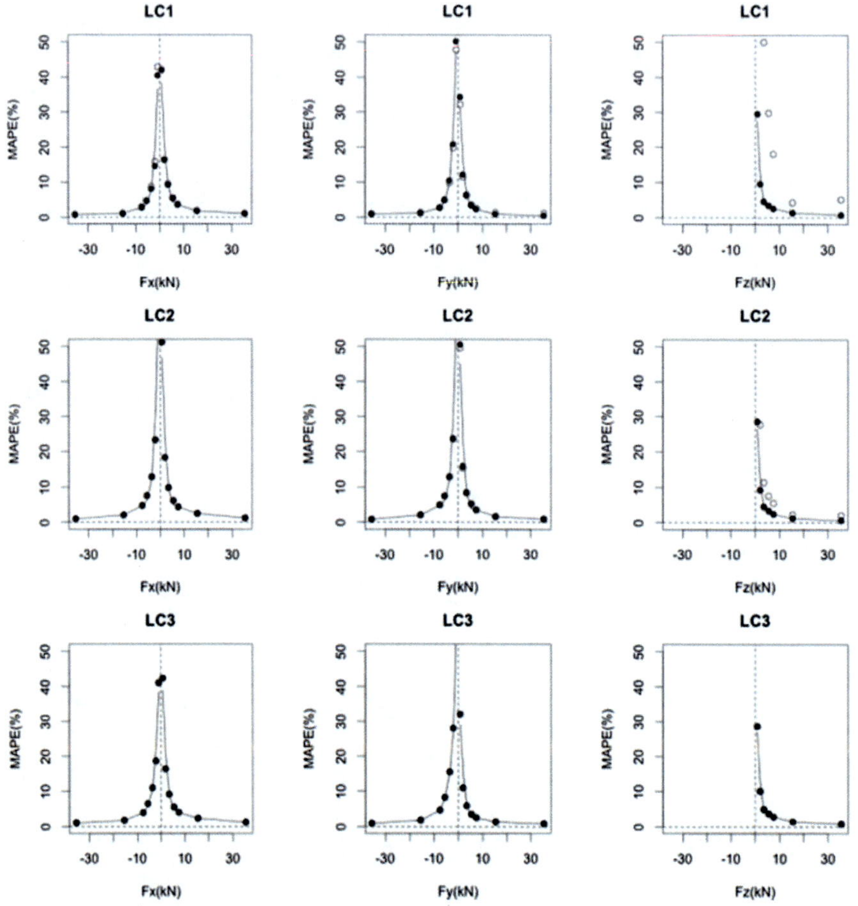

Fig. 3 Mean absolute percentage error (MAPE) as a function of applied forces. For the first and second load cells (LC1, LC2) bold points connected with spline interpolation show the errors for the improved model, empty circles corresponds to the raw data

absolute relative deviation) up to 10 times (for F_z forces < 8 kN) 2 times for (8 kN) (for LC1) and up to 2 times (for LC2) for axial measured component V_z within loads range 1–10 kN.

Figure 5 represents levels of confidence of force readings for the load cell LC1, LC2 (after the correction of outliers) and LC3.

It develops previous results for the confidence levels estimated by Ponomarev [6, 10–12]. It can be seen that for all measured vector force components F_x, F_y, F_z at the low sensitivity range (0–10 kN) the percentage confidence increases, reaching 95% at about 5 kN. At the high sensitivity range (10–35 kN) the confidence percentage approaches 100%. The load cellshave been designed for a measuring range of 5–50 kN to withstand extreme loads. It should be emphasized that this precision in

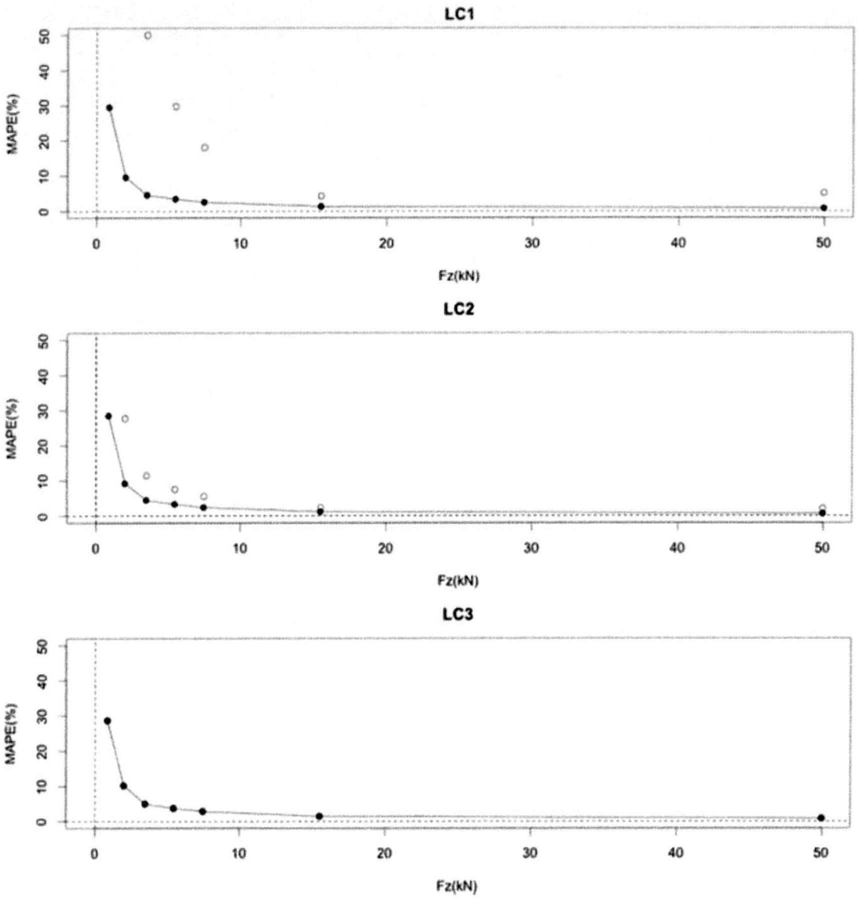

Fig. 4 Model allows to decrease expected relative precision as a mean absolute relative deviation up to 10 times (for F_Z forces < 8 kN) 2 times for (8 kN) (for LC1) and up to 2 times (for LC2) for axial measured component V_Z within loads range 1–10 kN

future measurements can be reached only when replacement model is applied on both first and second stages of corrections.

Up to this point there has been calibration made and cross-checked for the laboratory based equipment. However, it is crucial to be sure that this laboratory calibrated equipment works after installation in real sea conditions as expected. According to project, there is one more tool to verify the work of multi axis load cells as a part of the installed mooring system. Namely, every multi axis load cell attached to the mooring line through the load links, which allow to compare full force readings from the multi axis load cell and connected load link. This projected benefit made it possible to validate the laboratory calibration and data quality from multi axis load cells by comparison with parallel measurements from load links connected.

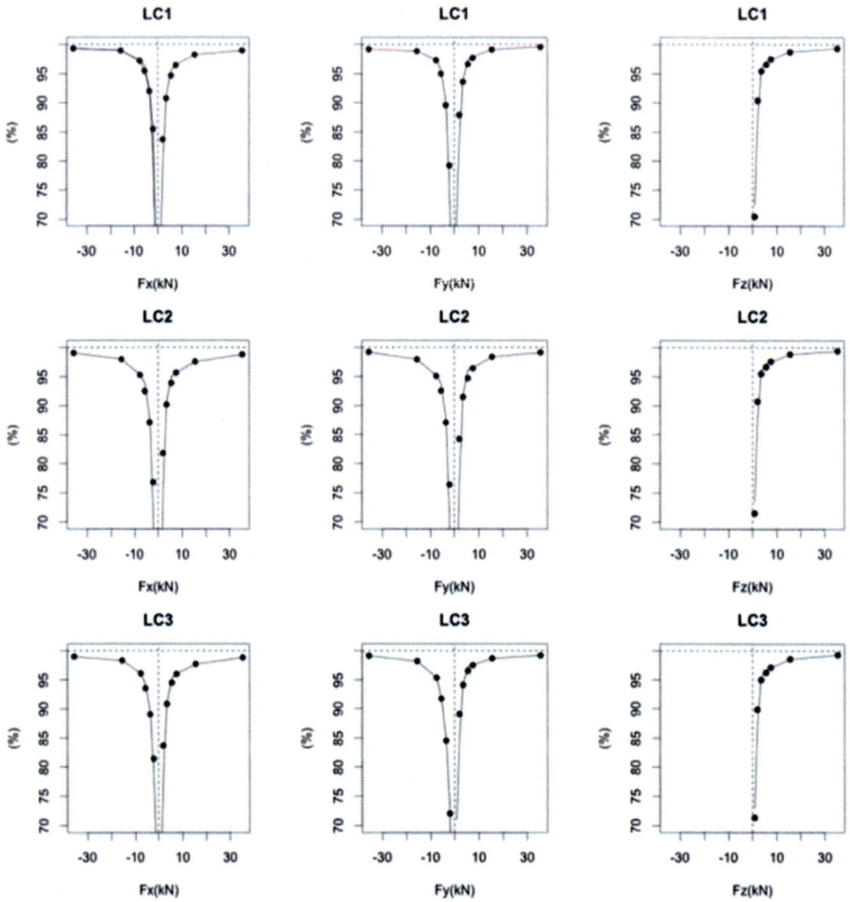

Fig. 5 Confidence levels of force readings for a load cell LC1, LC2 (after the correction of outliers) and LC3. It shows that for all measured vector force components F_x, F_y, F_z at the low sensitivity range (0–10 kN) the percentage confidence increases, reaching 95% with an approximate force of 5 kN. At the high sensitivity range (10–35 kN) the confidence level approaches 99%

Figure 6 confirms proper measurements from calibrated multi axis load cells over the load ranges shown. The solid lines show the full force measured by multi axis load cells, the dotted lines are the measurements from the corresponding load links (for the corresponding linear loads calibrated by the manufacturer). It should be noted that the second load link gives small negative readings two times during this 60 s time slot: around 305 and 326 s. These negative readings could be a result of snapping which can cause softening or hardening of the tension response [10, 11].

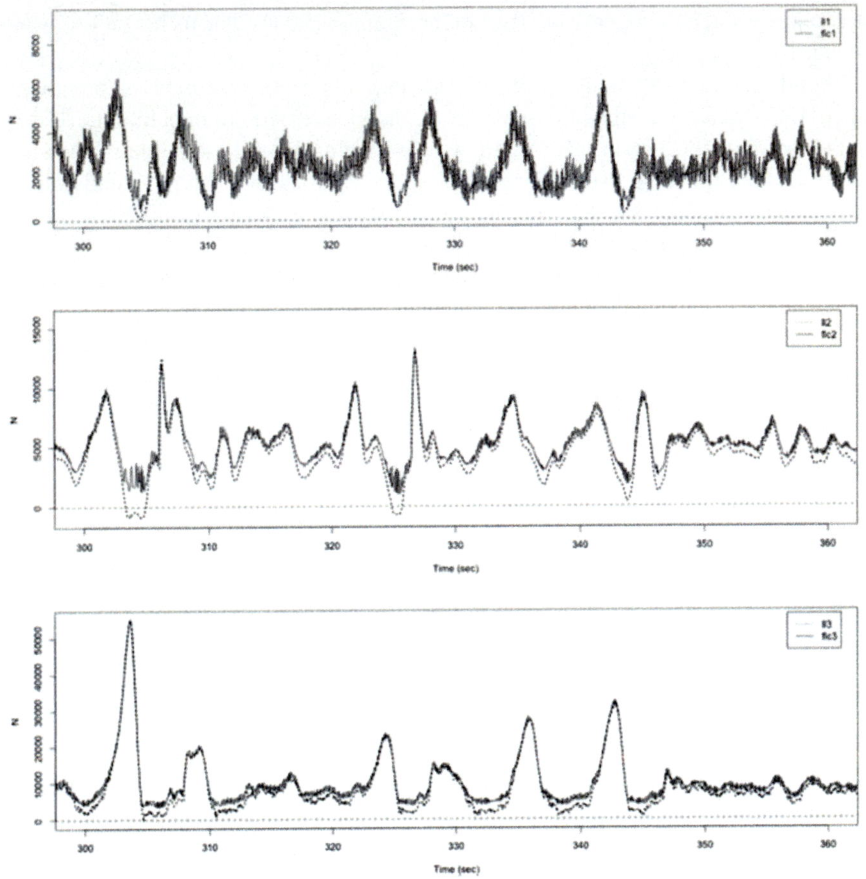

Fig. 6 Validation of measurements from calibrated multi axis load cells over the load ranges shown on the real sea data. The load cells are connected to mooring lines via load links to measure full force in the lines. The solid lines show the full force measured by multi axis load cells, the dotted lines are the measurements from the corresponding load links

3 Conclusion

This paper deals with yet another example from extensive number of innovative physical models with applications to real world problems outlined in [13–31]

Data from calibration experiments of three multi axial Load Cells have been analysed, it has been found that, while data from one load cell (LC3) contains no obvious outliers, two others (LC1 and LC2) have shown records of contamination processes which could alter significantly overall precision of these load cells. The signatures of these contamination processes were detected during the initial stages of data analysis. It was crucial to find that these contamination processes added distorted data for one axial component only (two other orthogonal components haven't shown any outliers).

Confidence interval analysis of voltage measurements (outlined in previous papers [6, 10] has confirmed the presence of distinctive outliers in calibration data for the load cells LC1 and LC2. If there were no exclusion of these outliers it would distort the measurement results. By presenting the calibration result in its individual x-, y- and z-components it was possible to identify that outliers contaminating only the readings in the mechanically less sensitive z-axis.

Therefore it is a case of critically distorted data in one axial component only namely single-component outliers, which could be replaced on the basis of the physical replacement model. This replacement model has been outlined, and detailed precision calculations show 10 time increase in precision of measurement. In addition, this model could detect and correct false snapping measurement readings.

In the interests of reliability analysis it is crucial to note that there are outliers happen even at high loads up to 50 kN as well in both the first, and second load cells. Even with noticing the number of outliers decreasing with increasing axial force component, it has been obtained that the probability of outliers remain almost the same over the load range (1–50 kN), namely it is about 12% for the first and 2–4% for the second load transducer.

For all load cells the sensitivity and calibration matrixes are obtained via least squares and least absolute deviation values according to tensor regression model. The mean absolute percentage errors have been calculated. It is shown that precision is not a constant value for points throughout the range but depends on the load range of the load cell.

The main advantage of proposed calibration method is the ability to detect and remove or substitute outliers, which can cause extremely inaccurate calculation of the forces. The physical replacement model has been suggested that allows tenfold decrease for the mean absolute percentage error (MAPE).

In some cases it could be useful in addition to the filtration of data, because there is no guarantee that any filtration can remove 100% of outliers with no significant loss of precision. The outliers we encountered here are most likely a high frequency, stochastic oscillatory component which could be a result of different noise sources such as electromagnetic pick-up.

The results prove the effectiveness of the novel physical replacement method (outlined in previous papers [6, 10–12]), and show well how the measuring accuracy can be considerably improved [15–25]. From the reliability viewpoint, it contributes with addition of a redundancy which could be used just in case of failing force components [26–31]. The emergency switch could be introduced into the data acquisition system, when the system could adjust itself to produce multidimensional force measurements even with failure of sensors for some of the multidimensional components.

References

1. Barzegar, M., Blanks, S., Sainsbury, B. A., & Timms, W. (2022, May 1). MEMS technology and applications in geotechnical monitoring: A review. Measurement Science and Technology. IOP Publishing Ltd. https://doi.org/10.1088/1361-6501/ac4f00.
2. Xiang, Y. T., Wan, B. F., & Zhang, H. F. (2021). Multiscale and Multiple Physical Quantities Sensor Based on Nonreciprocal Evanescent Wave in the One-Dimensional Photonic Crystals. IEEE Sensors Journal, 21(18), 19984–19992. https://doi.org/10.1109/JSEN.2021.3100403.
3. Mackay, E., de Hauteclocque, G., Vanem, E., & Jonathan, P. (2021). The effect of serial correlation in environmental conditions on estimates of extreme events. Ocean Engineering, 242. https://doi.org/10.1016/j.oceaneng.2021.110092.
4. Joo, J. W., Na, K. S., & Kang, D. I. (2002). Design and evaluation of a six-component load cell. Measurement: Journal of the International Measurement Confederation, 32(2), 125–133. https://doi.org/10.1016/S0263-2241(02)00002-7.
5. Kim, J. H., Kang, D. I., Shin, H. H., & Park, Y. K. (2003). Design and analysis of a column type multi-component force/moment sensor. Measurement: Journal of the International Measurement Confederation, 33(3), 213–219. https://doi.org/10.1016/S0263-2241(02)00044-1.
6. Ponomarev, M., Johanning L., Parish D. (2010). Enhancing precision and reliability of tri-axial load cells for mooring load measurements // Proceedings of the 3rd International Conference on Ocean Energy. 6 October. 2010. (ICOE2010). ISBN 978-84-693-5467-4.
7. Ranganayakulu, S. V., Goud, B. S., Sastry, P. V., & Kumar, B. R. (2015). Calibration of Acoustic Emission System for Materials Characterization. Universal Journal of Materials Science, 3(4), 62–69. https://doi.org/10.13189/ujms.2015.030402.
8. Hsu, W. T., Thiagarajan, K. P., MacNicoll, M., & Akers, R. (2015). Prediction of extreme tensions in mooring lines of a floating offshore wind turbine in a 100-year Storm. In Proceedings of the International Conference on Offshore Mechanics and Arctic Engineering - OMAE (Vol. 9). American Society of Mechanical Engineers (ASME). https://doi.org/10.1115/OMAE2015-42015.
9. Hermawan, Y. A., & Furukawa, Y. (2020). Coupled three-dimensional dynamics model of multi-component mooring line for motion analysis of floating offshore structure. Ocean Engineering, 200. https://doi.org/10.1016/j.oceaneng.2020.106928.
10. Ponomarev M. A novel physical model to enhance precision and performance of 3-dimensional force sensors // Processes in Geomedia 2022. No. 2(32). p. 1589–1600.
11. Ponomarev M. Experimental validation of novel physical model for improvement of sensing 3-dimensional fluid flow loads and responses in real sea conditions with South Western Mooring Test Facility (SWMTF) // Processes in Geomedia 2022. No. 2(32). 1579–1589.
12. Ponomarev M. A Novel Physical Model to Enhance Precision and Performance of Multidimensional Force Sensors // Physical and Mathematical Modeling of Earth and Environment Processes—2022, Proceedings of 8th International Scientific Conference-School, Springer, ISBN 978-3-031-25961-6
13. Ponomarjov, M. G. (2000). Space flows and disturbances due to bodies in motion through the magnetoplasma. Astrophysics and Space Science, 274(1–2), 423–429. https://doi.org/10.1023/A:1026565905268.
14. Ponomarjov, M. G. (2000). Simulation of oscillations in charged particle systems under the ambient magnetic field control. International Conference on Control of Oscillations and Chaos, Proceedings, (1), 167–170. https://doi.org/10.1109/coc.2000.873548.
15. Ponomarjov, M. G. (2002). 3D time-dependent kinetic simulation of space plasma disturbances due to moving bodies with the ambient magnetic field effect. Advances in Space Research, 29(9), 1397–1402. https://doi.org/10.1016/S0273-1177(02)00193-X.
16. Ponomarjov M. G. 3D Collisional Kinetic Simulation of Stratifications and Flute Structures of Plasma Flows and Wakes in External Magnetic Fields // AAS, DPS Meeting. No. 32. id.15.10. 2000. Bulletin of the American Astronomical Society. Vol. 32, p. 1022.

17. Ponomarev, M., & Carati, D. (2006). Enhanced acceleration of electrons populations by crossing electron cyclotron waves in an ambient magnetic field, Resonant Moments method. Advances in Space Research, 38(8), 1576–1581. https://doi.org/10.1016/j.asr.2006.02.015.
18. Ponomarjov M. G. 3D Kinetic Dynamical models of Ionized HII Clouds in External Magnetic Field // Ionized Gaseous Nebulae. 2000. Mexico City. November 21–24. meeting abstract id. 44.
19. Ponomarjov M. G. 3D Collisional Kinetic Simulation of Stratifications and Flute Structures of Plasma Flows and Wakes in External Magnetic Fields // 32nd Annual Meeting of the Division for Planetary Sciences. 23–27 October 2000. Pasadena. California. USA. Session 15. Outer Planets IV-Aurorae and Magnetospheres. 2000/10/24. Abstract 15.10. 1022.
20. Ponomarjov M. G. Kinetic Simulation of Magnetic Field Effects on Stratifications and Flute Structures of Space Plasma Flows and Wakes of Bodies // 2001 Joint Assembly American Geophysical Union Spring Meeting. Boston. Massachusetts. USA. Abstracts. 2001. SM52B-04.
21. Ponomarjov M. G., Carati D. Search for optimal 3D wave launching configurations for the acceleration of charged particles in a magnetized plasma: Resonant Moments Method // American Physical Society. 46th Annual Meeting of the Division of Plasma Physics. 15–19 November. 2004. Savannah. GA. MEETING ID: DPP04. abstract id. RP1.042.
22. Ponomarjov M. G., Carati D. Acceleration of charged particles by crossed cyclotron waves, Resonant Moments Method // 35th COSPAR Scientific Assembly. 18–25 July 2004. Paris. France. p. 2851.
23. Ponomarjov, M., & Carati, D. (2003). Kinetic Simulations of Relativistic Electron Flows in Time-Dependent Electromagnetic Fields (pp. 137–142). World Scientific Pub Co Pte Lt. https://doi.org/10.1142/9789812705082_0022.
24. Ponomarjov, M. (2001). Kinetic simulation of magnetic field effects on wakes of meteoroids Imaginary Emission Method. In European Space Agency, (Special Publication) ESA SP (pp. 295–300).
25. Ponomarjov M. G., Carati D. Search for optimal 2D and 3D wave launching configurations for the largest acceleration of charged particles in a magnetized plasma, Resonant Moments Method // arXiv 2004. https://arxiv.org/format/physics/0411075, arXiv:physics/0411075 [physics.plasm-ph].
26. Ponomarjov M. G. 3D time-dependent kinetic simulation of turbulent plasma flows under the effect of external magnetic field // American Physical Society. 42nd Annual Meeting of the APS Division of Plasma Physics combined with the 10th International Congress on Plasma Physics October 23 – 27. 2000 Québec City. Canada Meeting ID: DPP00. abstract id. NO1.007. https://ui.adsabs.harvard.edu/abs/2000APS..DPPNO1007P/abstract.
27. Ponomarjov M. G. Kinetic modeling magnetic field effect on ion flows, disturbances and wakes in space plasma // Space plasma simulation: proceedings of the Sixth International School/Symposium. ISSS-6. Garching. Germany, 3–7 September 2001. Edited by J. Büchner, C.T. Dum, and M. Scholer. Berlin: Schaltungsdienst Lange o.H.G. 2001. p. 328.
28. Ponomarjov M. G., Carati D. Acceleration of electron populations by crossing EC waves in an external magnetic field // 13th Joint Workshop on Electron Cyclotron Emission and Electron Cyclotron Resonance Heating. Nizhny Novgorod. Russia. 2004. Conference paper https://ec13.iapras.ru/papers/Ponomarjov.pdf.
29. Ponomarjov M. G., Carati D. Enhanced acceleration of charged particles by crossing electromagnetic waves in a magnetized plasma, resonant moments method // 2004. ffhal-00001926v1f. https://hal.archives-ouvertes.fr/hal-00001926v1.
30. Webb, R. P., & Ponomarev, M. (2007). Molecular dynamics simulation of low energy cluster impacts on carbon nanotubes. Nuclear Instruments and Methods in Physics Research, Section B: Beam Interactions with Materials and Atoms, 255(1 SPEC. ISS.), 229–232. https://doi.org/10.1016/j.nimb.2006.11.032.
31. Ponomarev, M. G., Selcuk, C., & Gan, T. H. (2012). Construction of pattern recognition system optimized for X-ray inspection of plastic electronics and OLED displays. In 51st Annual Conference of the British Institute of Non-Destructive Testing 2012, NDT 2012 (pp. 354–360). British Institute of Non-Destructive Testing.

Skimmer Using the Phenomenon of Vortex Cavity Formation

V. P. Pakhnenko

Abstract Various methods can be used to deal with oil spills. A widely used method is when mechanical means are used to remove oil products from the surface of the water. As a rule, booms are used to concentrate spilled oil, after which it is possible to clean the localized slick with the help of skimmers and pump oil into storage tanks. There are many varieties of skimmers, the technical features of which are optimized for specific tasks, different scales of cleaning operations, types of oil products and environmental conditions. This paper describes the requirements for the successful application of skimmers in various situations, as well as a unique new model and its parameters.

Keywords Oil products · Emergency oil spills · Skimmer · Vortex · Model

1 Introduction

The ultimate goal of any oil spill clean-up program is to recover the maximum amount of spilled oil in a cost-effective and environmentally friendly manner. A successful full-cycle oil recovery technique must solve several interrelated tasks from the localization of a significant amount of spilled oil products and their subsequent containment from distribution, to the collection, pumping, accumulation and transportation of collected oil products. In this cycle for the elimination of oil products from the water surface, the functions of collecting and pumping are often performed by skimmers.

All skimmers skim oil instead of water, but their designs vary considerably according to their intended use—for example, offshore or sheltered waters, as well as on land. Skimmers for use on the water are equipped with buoyancy systems or supports, and more complex designs can be self-contained, self-propelled and equipped with several different oil intake systems, can also be equipped with built-in

V. P. Pakhnenko (✉)
A. Yu. Ishlinsky Institute for Problems in Mechanics, Prospect Vernadskogo 101-1, Moscow 119526, Russia
e-mail: pvpmos@gmail.com

V. I. Karev (ed.), *Physical and Mathematical Modeling of Earth and Environment Processes—2022*, Springer Proceedings in Earth and Environmental Sciences,
https://doi.org/10.1007/978-3-031-25962-3_37

oil storage tanks and separators to separate water from oil. When selecting skimmers for a particular application, a number of factors must be considered, the most important of which are the viscosity and adhesive properties of spilled oil products (including any change in the characteristics of oil products due to weathering over time), as well as environmental conditions, such as sea waves and the amount of debris.

In relatively predictable emergencies, such as spills at stationary facilities. For example, in offshore oil depots and oil refineries, the type of oil can be known in advance, so a suitable skimmer design can be selected. Conversely, for different situations and spills of different types of oil products, it is preferable to use a multifunctional skimmer or one that can be adapted to a specific situation. No particular type of skimmer may be suitable for all situations that may occur as a result of an accidental oil spill, so a selection of skimmers may be required for selective use depending on the situation. Accordingly, in order to implement an effective water surface cleaning cycle, it is necessary to determine the intended use and expected conditions for the operation of the skimmer, such as whether the skimmer should be used manually or autonomously, or whether it is necessary that it be built into a ship-mounted oil recovery system with water surface. Once these conditions of use have been established, other criteria can be defined, such as: design features, size, reliability, ease of handling and operation.

2 Oil Product Recycling Cycle and Design Features of Skimmers

Structurally, most skimmers are arranged as follows. The skimmer suction device diverts or collects oil from the water surface, directing it to a pumping system for transfer to a storage tank. Mechanisms for the removal of oil products from the surface most often include oleophilic systems (based on oil sticking to a moving surface), suction systems, gravity weir systems, systems that lift oil products from the surface using mechanical scoops, belts or buckets.

There are many classifications of skimmers. As a rule, they can be divided into two main subgroups: oleophilic and non-oleophilic. The former include skimmers according to the following design features: disk, drum, mop ropes, brush, tape. Non-oleophilic ones include: with vacuum suction, weir, belt, drum. In turn, all of them will have their pros and cons under different weather conditions, with different properties of the water surface (the presence and absence of waves), with various types of oil products that need to be eliminated, with and without various garbage around. It should also be taken into account that different skimmers have different oil collection capacities and may depend on the required auxiliary equipment.

The oleophilic skimmers (Fig. 1) use materials that tend to stick to oil but repel water. The oil adheres to the surface of the material, most commonly in the form of a disc, tape, squeegee, drum, or brush. Which, when rotated, capture oil from the surface of the water. After separation from water, the oil is cleaned or squeezed out of the oleophilic material and enters a special reservoir, from which it is pumped to a storage tank. When using oleophilic devices, the largest ratio of the amount of oil collected and water taken separately or together with oil is usually achieved (oil recovery ratio). They are most effective when picking up medium viscosity oils from 100 to 2000 centistokes. One disadvantage of this method is that low-viscosity oils such as diesel fuel and kerosene do not accumulate on the oleophilic surfaces in a thick enough layer to achieve high recovery efficiency. However, higher viscosity oils such as bunker oil are too sticky for this method and difficult to remove from the surface of the oleophilic skimmer. Also, oil-in-water emulsions can be nearly non-tacky and difficult to pick up with some designs of oleophilic type devices. For example, disc skimmers pass through the emulsion instead of withdrawing it. Oleophilic devices are usually made from polymeric materials, although metal surfaces have also been shown to be effective. It has also been proven that discs and drums with grooved surfaces provide higher oil intake performance than similar devices with a smooth surface [1].

From a theoretical point of view, the simplest design is a suction device in which oil is collected by pumps or vacuum systems directly from the water surface. In particular, mobile vacuum units that combine devices for collecting, accumulating, transporting and separating oil products and water are usually easily accessible near the accident site and, in this form, are convenient for collecting oil onshore or in the coastal zone. The simplest method of collecting oil products is to dip the receiving sleeve with a mesh screen to block the passage of debris directly into floating or stranded oil. The obvious disadvantage of such methods is the following: during such cleaning, along with oil, a large amount of water can be taken. If allowed by

Fig. 1 Oleophilic drum skimmer

Fig. 2 Belt skimmer

regulations and the necessary equipment is available, then excess water should be drained from the tank to optimize the use of the oil storage tank.

Higher selectivity in collecting oil can sometimes be achieved by attaching a weir to the receiving sleeve. When using spillway skimmers, the selective intake of oil products from the water surface is ensured by the action of gravity. By placing the edge of the weir at or just below the floating oil/water interface, the oil flows through the weir and is collected together with a minimum amount of water. Modern types of weir cleaners are equipped with adjustable weir systems, the exact vertical positioning of which is achieved by automatic alignment. None of the spillway skimmers are effective in rough waters, although roughness by itself does not usually interfere with the operation of the skimmer. To eliminate the loss of energy due to friction in the transfer hoses, some spillway skimmers are equipped with an onboard pump so that the collected oil is pumped through the hose, and not just sucked in, however, these methods still have a number of limitations and disadvantages (Fig. 2).

3 New Skimmer Model

Despite the large number of existing skimmers of different designs, intended for different tasks and conditions, they all have both their positive qualities and disadvantages, so some existing designs can be improved and finalized. The proposed further device for collecting oil products from the surface of the water is aimed at improving the efficiency of collecting oil products. The invention relates to a technique for cleaning the water surface from liquid contaminants, mainly from oil products. The device works on the principle of creating a vortex funnel in water, which in turn leads to the separation of water and oil products, which leads to an optimization of the water purification cycle.

This result is achieved by the fact that the skimmer for collecting oil products from the surface of the water includes a cylindrical cup open from above, a means for creating a vortex funnel with a drive for its rotation and a means for pumping out the collected oil products. At the same time, the means for creating a vortex is made in the form of the mentioned glass, which, in turn, has the ability to move in height, and the drive for its rotation is equipped with a control unit. The output of the control unit is connected to the cup rotation drive, and the inputs are connected to various sensors: water and oil product viscosity meters, water and collected oil product density meters, an oil product layer thickness meter and a means for measuring the depth of immersion of the cup upper edge from the «water–oil product» interface. At the same time, the end of the receiving pipe (means for pumping out oil products) is located below the upper end of the cylindrical container by the value $H/2$, where H—cylindrical container height.

This skimmer has a number of distinctive features:

1. implementation of a means for creating a vortex funnel in the form of a cylindrical container, equipped with a drive for its rotation;
2. cylindrical container is equipped with a means of its movement in height;
3. the rotation drive is equipped with a control unit, the output of which is connected to the rotation drive of the cylindrical container, and the inputs are connected to water and oil product viscosity meters, water and collected oil product density meters, an oil product layer thickness meter and a means for measuring the depth of immersion of the upper edge of the cup from the «water–oil product» interface;
4. the end of the receiving branch pipe of the oil product pumping means is located below the upper end of the cylindrical container by the value $H/2$, where H—cylindrical container height.

It was experimentally established that the rotation of the glass in the form of a hollow cylinder leads to the formation of a vortex funnel inside the volume of the cylinder. And it is known from the prior art (see [2–4]) that if oil products are present on the surface of the water, then the separation of water and oil fractions in the vortex occurs. Thus, there is no need to have means, in the form of screws, jet pumps, etc., specially designed to create a vortex funnel, which simplifies the design.

At the same time, equipping the glass with a means of moving it in height allows you to install it submerged to a depth that will provide the most optimal operating conditions for the device, which in turn will have a positive effect on its efficiency.

The essence of the proposed device is illustrated by an implementation example and graphic materials. Figure 3 shows a schematic diagram of the implementation of the device. The basis of the entire oil-collecting structure is the main separate element for cleaning water, which is a cylindrical glass rotating around its longitudinal axis and directed vertically with its open end to the surface of water contaminated with oil products. Inside the cylindrical cup, through a hole in the bottom along the axis of rotation, a tube is introduced with an open end to a predetermined length, designed to drain the oil collected from the surface, the second end of which is connected to the pumping system. The pumping system transports the sucked oil through a flexible hose into a special storage tank.

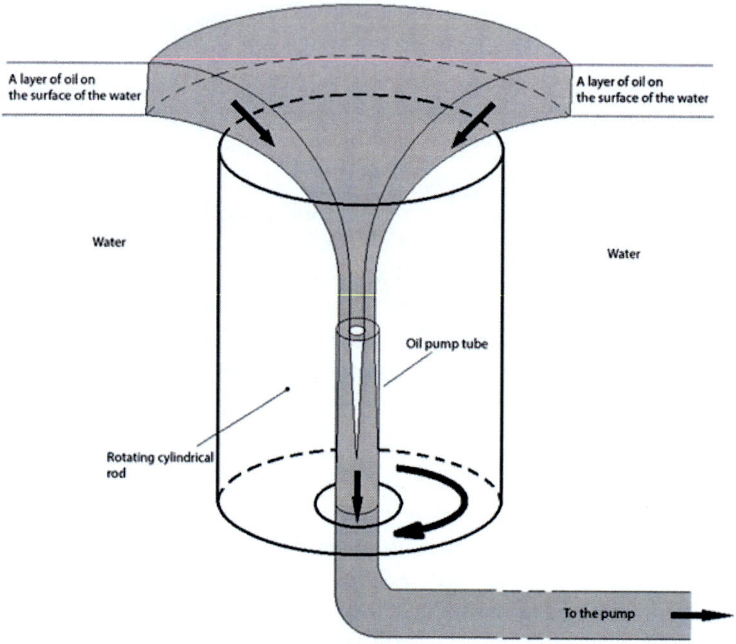

Fig. 3 Model of a skimmer for removing oil products from the surface of the water

The device contains a hollow cylindrical cup, which is equipped with a drive for its rotation with its control unit. The drive can be selected from known and available ones, and a computer equipped with appropriate software can be used as a control unit. Various sensors are connected to the inputs of the control unit. In addition, means are provided for moving the glass in height so that the depth of its immersion can be adjusted. For example, it can be a vertically movable bracket with a device fixed on it, equipped with a special mechanism for its vertical movement, which is fixed on the carrier vessel. All nodes and blocks included in the proposed device are selected from among the known.

The principle of operation of this skimmer is based on the phenomenon of the formation of a vortex cavity, which is formed during the rotation of the bottom or walls (or the simultaneous rotation of the bottom and walls) of a cylindrical shell (Fig. 3). A characteristic feature of such a vortex cavity is the separation of water and oil fractions in a vortex, which was shown in various experimental works [5]. This allows, by selecting the speed of rotation of the shell, to ensure the location of the open end of the skimmer tube in the oil part of the composite vortex, which, when the pump is running, leads to the selection of oil from the composite vortex and its transportation to the collection point through a flexible hose. The resulting oil deficit in the compound vortex is instantly replenished with oil from the surface, which leads to a continuous collection of oil products from the water surface.

To improve the efficiency and speed up the process of cleaning the water surface, it is possible to combine a whole set of such separate skimmers into a chain fixed on a rigid structure (Fig. 4). This design is immersed in water and mounted on a cleaning vessel, which allows the oil recovery process to capture large areas of the water surface.

When performing the oil product collection cycle, in order not to make the cleaning vessel heavier and to ensure its mobility, maneuverability and, if possible, a constant draft for the correct operation of the skimmers, it is supposed to use an additional oil recovery vessel (Fig. 5), to which the collected oil is pumped—water mixture from a contaminated surface.

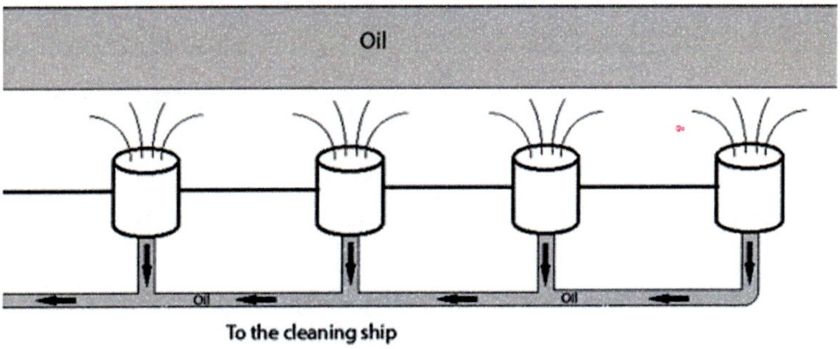

Fig. 4 A chain of skimmers connected in series

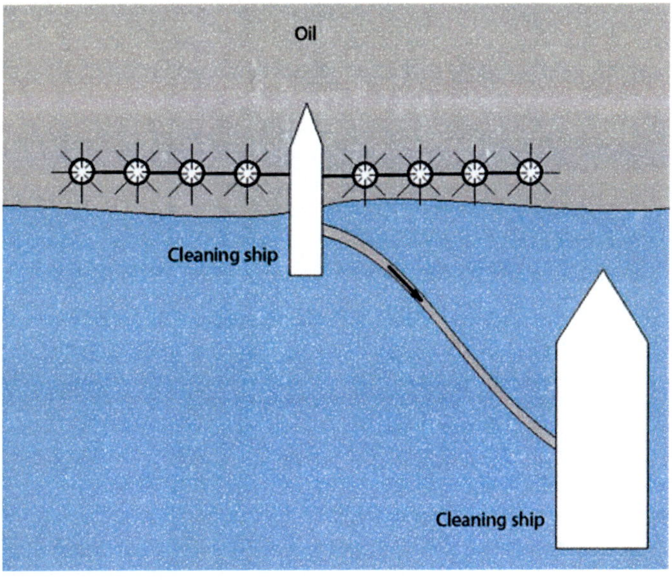

Fig. 5 General principle of operation of the oil recovery system

Accordingly, we obtain the following use of the device. At the site of an oil spill on the surface of the reservoir, the hollow glass of the skimmer is immersed to a certain depth using pre-installed means. Further, using the measurement sensors included in the device, the parameters of the liquid and oil products are determined, which are necessary to calculate the optimal operating conditions. The measurement results are sent to the control unit, which, in accordance with the program available in it, makes the necessary calculations and sets the required depth of the cup placement using the immersion tool. Based on the calculation, a command action is generated on the hollow cup rotation drive, which is set into rotation at the calculated speed.

Since the means for measuring the parameters of liquids and the depth of immersion of the cup, introduced into the device, work online, when they change, the control unit promptly makes changes to the modes of rotation of the cup and its immersion depth. As a result, only the collected oil product without water impurities or with a minimum amount of it enters the oil product pumping device.

It was also experimentally established [6] that the shape and dimensions of the oil part of the composite vortex depend on a number of parameters, namely: the thickness of the oil layer on the water surface, the density and kinematic viscosity of the water and the removed oil products. In the experiments, various petroleum products and oils were used, such as: sunflower oil, a mixture of sunflower oil and diesel fuel in equal proportions, oil, diesel fuel.

As a result of the analytical processing of the experimental data, it was possible to establish empirical dependences of the optimal regimes for the implementation of the water surface cleaning cycle, depending on the properties of the oil product and the state of the aquatic environment of the reservoir on the surface of which this oil product is located. In particular, it was found that the optimal immersion depth h of the upper edge of the hollow glass from the «water–oil product» interface, it is advisable to determine by mathematical dependencies:

$$h < \frac{H}{\alpha^2 - 1}, \quad \alpha = 1 + \frac{\rho_w}{\rho_o}\sqrt{\frac{v_w}{v_o}}$$

where

- h Immersion depth of the upper edge of the cup from the «water–oil product» interface, m;
- H Cylindrical container height, m;
- ρ_w Density of water, kg/m^3;
- ρ_o Oil density, kg/m^3;
- v_w Kinematic viscosity of water, m^2/s;
- v_o Kinematic viscosity of oil, m^2/s,

and the optimal angular frequency of rotation of a cylindrical container ω can be determined from the dependence:

$$\omega = \frac{1}{R_c}\sqrt{g\frac{(H + 2h + \Delta)(\rho_w + \rho_o)}{(1 + \alpha^2)\rho_w - \rho_o}}$$

where

R_c Cylinder radius, m;
Δ Oil layer thickness on the water surface, m;
g Gravitational acceleration, m^2/s.

And in order to ensure optimal conditions for the operation of the skimmer and its high efficiency, it is necessary to obtain data on the parameters included in the above mathematical expressions. Therefore, the device is equipped with density meters for water and the collected oil product, viscosity meters for water and oil products, a thickness gauge for the oil layer and a means for measuring the depth of immersion of the upper edge of the cup from the «water–oil product» interface. The corresponding data is sent to the control unit, processed and the generated control action is fed to the rotation drive of the cylindrical container.

Empirically, it was found that the optimal location of the end face of the inlet pipe of the oil pumping means in the oil part of the composite vortex, (the end of the receiving pipe of the means of pumping out oil products is located below the upper end of the glass by the value $H/2$, where H is the height of the glass) when the pump is running, leads to the selection only oil from the compound vortex and its transportation to the collection point through a flexible hose. The resulting oil deficit in the compound vortex is immediately replenished with oil from the surface, which leads to a continuous collection of oil products from the water surface.

4 Conclusion

The advantages of various skimmers for the collection of oil products from the water surface must be compared with the prevailing conditions, such as sea waves, wind, currents and the location of environmentally sensitive areas. The type of oil to be collected, its viscosity at a certain ambient temperature and any change over time dictate the most efficient type of collection device. The selection of the most suitable skimmer must take into account throughput characteristics, design features, reliability, strength, operation, weight, versatility, power supply, maintenance and costs. A unique model of a skimmer was proposed, the principle of operation of which is based on the phenomenon of the formation of a vortex cavity. This skimmer has unique design features and allows for a more efficient cycle of cleaning the water surface from oil products.

Acknowledgements The work was supported by the Ministry of Education and Science, project 075-15-2020-802.

References

1. Documents & Guides The international tanker owners pollution federation limited, ITOPF, www.itopf.org.
2. Dmitriev V.G., Goncharov V.V. Device for collecting oil products from the surface of the water, Patent RU 2010090C1.
3. Kuznetsov N.P., Tretyakov V.A., Chembrovskaya A.I. Device for collecting oil products from the surface of the water, Patent RU 2010090C1.
4. Slobodyankin I.P. Installation for the removal of oil and oil products from the swampy water surface, Patent SU 458136.
5. Naumov I.V., Chaplina T.O., Stepanova E.V. Investigation of the effect of a vortex flow inductor on the shape of the interface between two immiscible liquids, № 1(23), 2020, pp. 599–610.
6. Kistovich A.V., Chaplina T.O., Analytical and experimental modeling of the hydrocarbon slick form and its spreading on the water surface, Physics of Fluids, vol. 33, 2021.

Studies of the Construction of Horizontal Underground Workings-Tanks Under the Impact of Solution on Rock Salt

V. P. Malyukov

Abstract Studies of the construction of horizontal underground workings-tanks (caverns) were carried out on 4 large-scale models in real conditions with a hydrodynamic effect on rock salt in the Khoja-Mumyn salt mountain (Republic of Tajikistan). Extended horizontal workings with vaulted cross-sections close to equal in size were obtained, as well as analytical dependencies for calculating the technological parameters of the process of constructing horizontal workings in rock salt [1–11].

Keywords Construction of a working-tank · Hydrodynamic impact · Rock salt · Vaulted form · Analytical dependencies · Process parameters · Economic indicators

1 Introduction

To carry out large-scale modeling of the processes of construction of horizontal workings-tanks according to the two-well option, it was planned to create a series of experimental reservoirs-models in the rock salt massif, simulating full-scale production with a volume of 300 thousand m^3 in relation to the mining and geological conditions of the corresponding salt-bearing basins.

Correspondence of the physical processes occurring in the horizontal workings of industrial volumes and workings-models was envisaged to be ensured using traditional similarity criteria, taking into account the values of the mass transfer coefficients of various rock salt surfaces along the contour of the working-model and nature.

The practical construction of experimental workings-models was carried out in the side walls of two adits, previously driven in rock salt on the slope of Mount Khoja-Mumyn at a depth of 150 m.

To create model workings, horizontal wells were drilled, more precisely, gently inclined with a slope towards the bottom of the order of 2°–3° with a diameter of

V. P. Malyukov (✉)
Department of Subsoil Use and Oil and Gas Affairs, Russian Friendship University Peoples, Moscow, Russia
e-mail: v.malyukov@mail.ru

V. I. Karev (ed.), *Physical and Mathematical Modeling of Earth and Environment Processes—2022*, Springer Proceedings in Earth and Environmental Sciences,
https://doi.org/10.1007/978-3-031-25962-3_38

65–70 mm and a length of 11–12 m. The wellhead part of the wells expanded to a depth of 1–1.5 m to a diameter of 110 mm and was cased with pipes with a diameter of 89 mm, equipped with centralizers in two places. The annulus of the wells was cemented to the length of the casing.

To carry out the process of building model workings under the hydrodynamic action of the solution, the drilled wells were equipped with a system of pipes that marked the water supply and brine discharge columns.

In order to ensure that cross sections close in size are obtained by the time the construction of the models is completed, it was envisaged to carry out the process in several successive stages equal in length with a successive increase in the amount of salt extracted at each stage by a given value.

Each stage of the model development was characterized by two main technological quantities: the amount of salt extracted at the stage and the distance from the solvent injection point to the solution withdrawal point, which gradually increased due to the shortening of the pipe stage by the length of the pipe imitating the water supply column.

Model 2D was developed in 5 stages. The amount of salt extracted at the stages, respectively: 9, 13, 18, 25 and 35% of the total.

2 Results

The main technological parameters of full-scale horizontal working-capacity and recalculated parameters for working-model are shown in Table 1.

Rock salt contains insoluble inclusions of clay material (up to 10%). On the surface of rock salt, after the opening of the developed large-scale models, spherical clay formations of the same size (Liesegang rings) located at an equal distance from one another were found.

The linear dimensions of the stages along the length of the horizontal part of the well are assumed to be the same.

To assess the concentration field in the working during the construction process, periodic sampling of the solution was provided directly from the working.

During operation, the water and solution temperature, productivity, solution density, solvent injection point were measured, the pressure in the model was recorded hourly, and samples of the solution were periodically taken from different levels of the model through sampling tubes.

Analysis of the mining parameters of the first working-model showed a certain discrepancy between the calculated parameters and the working shape, a horizontal roof of the working was obtained.

The author of this work supervised the work at the test site for the development of 2, 3, 4 horizontal workings-models. Prior to the work, a set of measures was revised to improve the performance of work, starting with changes to the downhole equipment. An articulated rotary mechanism was used for sampling the solution. Despite the limited depth of casing, there was no leakage of the wells.

Table 1 Main technological parameters of full-scale horizontal working-capacity and workings-models

No.	Parameters	Output capacity	Model 1G	Model 2G	Model 3G	Model 4G
1	Linear scale factor	1	45	40	45	45
2	Working length (m)	300	6.7	7.5	6.7	6.7
3	Production volume (m^3)	300,000	3.3	4.7	3.3	3.3
4	Water supply capacity (m^3/h)	243	0.12	0.15	0.12	0.12
5	Average cross-sectional area (m^2)	1000	0.49	0.63	0.49	0.49
6	Working off time (h/day)	9120/380	192/8.3	228/9.5	192/8.3	192/8.3
7	Number of stages of development	5	6	5	5	5
8	The amount of salt extracted at the stages, in % of the total amount	7; 13; 17; 28; 35	10; 12; 15; 17; 20; 26	9; 13; 18; 25; 35	7; 13; 17; 28; 35	7; 13; 17; 28; 35

When working out 2, 3, 4 models, a vaulted roof of workings was obtained, which indicates the possibility of controlling the technological regime with the design parameters, despite a number of problems. The cross-sections of the working-model obtained during physical modeling in natural conditions have a stable vaulted shape and are close to equal in size. The sectional shape 2 of the horizontal working-model 2G is shown in Fig. 1.

Figure 2 shows the horizontal working-capacity and its cross sections.

Longitudinal-vertical section—section A–A, r—ribbed surface.

After opening the model workings (2G, 3G, 4G), visual inspection revealed that a surface with craters was formed in the upper concave part. In the lower convex part of the workings, the surface is characterized by ribs. Section A–A shows the dynamics of working out and the change in the ratio of different surfaces. Figure 2 shows a wavy convex surface of the lower part of the underground working, "ribbed" with straight ribs perpendicular to the well, located at an angle to the horizon.

Figure 3 shows the three-dimensional form of the "reference" horizontal working-model 2G.

Before working out the "reference" model 2G, a model was worked out, which at the initial stage of development turned out to be leaky and was opened. The obtained form of the model at the initial stage of development, which characterizes

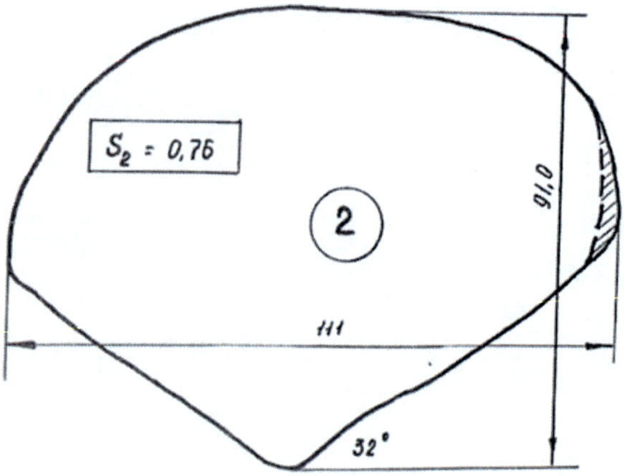

Fig.1 Sectional shape 2 of horizontal workings—model 2G

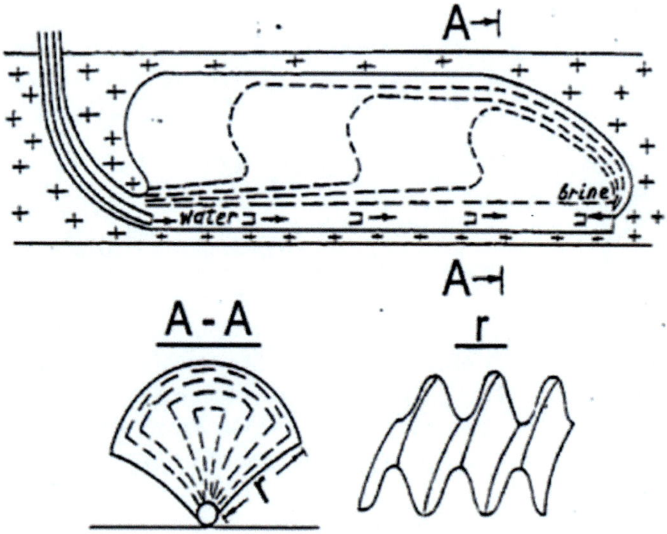

Fig. 2 Underground horizontal working

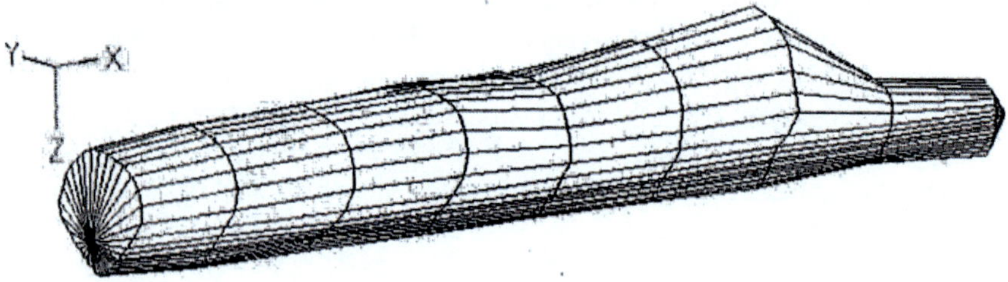

Fig. 3 Volumetric form of horizontal working-model 2G

the entire initial process of mass transfer, is considered. The distinctive shape of the vertical surface of the horizontal model working under the hydrodynamic action of the solution is noted.

The concentration of brine, which is given out of the mine when the position of the solvent input is periodically moved (by successively retreating stages) in the countercurrent solvent supply mode, is determined by the proposed relationship:

$$C = C_H\left[1 - \exp\left(-1.19\frac{1}{Q}\sum_{i=1}^{n} K_i S_i\right)\right] \tag{1}$$

where C_H—saturated solution concentration; K_i—mass transfer coefficient i of rock salt surface; S_i—i surface area; Q—productivity of brine extraction from mine.

The concentration of brine, taking into account the correction for the temperature of the salt dissolution process in relation to the "reference" working-model, is determined by the proposed dependence:

$$C = C_H\left[1 - \exp\left(-1.19\frac{1}{Q}\sum_{i=1}^{n} K_i S_i \frac{T_n - 250.8}{54.4}\right)\right] \tag{2}$$

where: $T_П$—salt dissolution temperature ($T_П > 250.8$ K).

The optimal productivity of the solution supply during the construction of underground workings was determined on the basis of the calculation of capital costs. For this purpose, the main costs for the construction of a horizontal underground working were determined at various capacities, and the optimal one was taken to be the one at which the costs are minimal.

The calculation of the main parameters of the construction process of a horizontal underground working-tank (height, span, cross-sectional area, length, volume, time of creation, the amount and concentration of the issued solution, etc.), which predetermine technical and economic indicators, is made on the basis of» large-scale model 2G with the conversion of these indicators to nature.

In addition to the technological parameters of the construction of a horizontal underground working-tank, the magnitude of pressure losses during the movement of fluid in wells in the pipeline and pressure losses due to local resistances are determined.

The total cost of the processes of building a horizontal working is determined as the sum of cost items: the cost of electricity, wage costs, the cost of removing the solution.

It is determined: the optimal performance of the solution supply based on the minimum cost at the appropriate performance; the geometric volume of the underground gas storage (UGS), taking into account the size of the active stock of the storage, the number of workings for the storage.

The concentration of the dispensed solution and the process time are calculated for the temperature of the full-scale mining process at different capacities.

The cost of the process of building horizontal workings is determined for various solution supply rates for the options: the solution is disposed of using brine-absorbing underground horizons, the solution is disposed of by transferring it to a brine-consuming enterprise. The total geometric volume of the underground storage of natural gas and the number of workings are determined, taking into account the selected value of the active gas reserve.

3 Conclusions

For the first time in the process of working out of a model working in natural conditions, vaulted dome-shaped workings were obtained and samples of the solution were taken from the working to determine the concentration, which increased the scientific significance of the experiment and the reliability of the data obtained. The use of the method of physical modeling in natural conditions makes it possible to calculate the main technological parameters of an underground working of a horizontal type for various natural conditions of the required dimensions.

With the hydrodynamic action of the solution on the surface of the rock salt of the model of a horizontal underground working, separations of the concentration boundary layer occur with greater intensity, since the process goes along the entire length of the working, which leads to an increased concentration of the issued solution compared to the vertical working, worked out by steps limited by the non-solvent. Local mass transfers and, consequently, the total mass transfer during the hydrodynamic action of the solution on rock salt in the process of mining horizontal workings are more intense compared to vertical workings.

References

1. Ivantsov O.M., Golovkin M.Sh. A method for constructing underground tanks for oil, oil products and liquefied gases in rock salt beds. Author's certificate of the USSR No. 140369, 1961.
2. Pozdnyakov A.G., Sidorov I.A., Malyukov V.P., Mazurov V.A. How to create an underground tank. Author's certificate of the USSR No. 972893, 1980.
3. Kunstman A.S., Urbanczyk K.M. Modelling of horizontal cavern leaching main aspects fnd perspectives, SMRI, San Antonio, October 22–25, 1995.
4. Saberian A. A preliminary model horizontal well leaching, SMRI Paper, San Antonio, October 22-25. 1995.
5. Charnauel I., Dump J.G., First Gas de France Horizontal Salt Cavern Experiment, SMRI, New Orleans, Louisiana, USA, Spring Meeting, 1998, pp. 40–52.
6. Kazaryan V.A., Pozdnyakov A.G., Malyukov V.P. The Use of Solvent Recirculation at Target Cavern I, Leaching for Horizontal Cavity Creation and Quality Brine Output. Spring SMRI Conference. 29 April–2 May 2007. Basel, Switzerland. p. 155–164.
7. Liu W., Zhang, Z., Chen J., Fan J., Jiang D., Jjk D., et al., "Physical simulation of construction and control of two butted-well horizontal cavern energy storage using large molded rock salt specimens", Energy, vol. 185, pp. 682–694, Oct. 2019.

8. Zhang Z., Jiang D., Liu W., Chen J., Xie K. Study on the mechanism of roof collapse and leakage of horizontal cavern in thinly bedded salt rocks // Environ. Earth Sci., vol. 78, no. 10, P. 292, 2019.
9. Malyukov V.P. Shepilev A.A. Innovative technologies for the construction of horizontal and two-tier underground reservoirs in rock salt // Processes in Geomedia, 2020. № 4 (26).
10. Malyukov V.P. Mass Transfer and Separation of the Boundary Layer under the Influence of an Aqueous Solution on Rock Salt // Processes in Geomedia, 2020. № 4 (26).
11. Malyukov V.P. Formation of Liesegang rings in rock salt with nanoparticles // Mining information and analytical bulletin. Scientific and technical journal. Publishing house "Mining book", № 10. 2016. P. 242–248.

Influence of Langmuir Circulations on the Intensity of Turbulent Mixing in the Near-Surface Layer of the Sea

M. I. Pavlov and A. M. Chukharev

Abstract The purpose of the article is to describe an experimental study of the effect of Langmuir circulations (CL) on surface turbulence. A technique has been developed for determining the main dynamic characteristics of the CL using instrumental measurements of the flow velocity components with an ADCP DVS-6000 acoustic meter, background horizontal currents using the «Vostok-M» complex, and turbulent fluctuations of hydrophysical quantities recorded by the «Sigma-1» complex. The velocities of descending and ascending flows in the zones of convergence and divergence in coherent Langmuir structures are determined. Based on synchronous measurements of the velocities of currents and turbulent pulsations, the velocity of the transverse displacement of the Langmuir bands in various hydrometeorological conditions is calculated using correlation coefficients. To assess the influence of the studied CL on turbulent exchange, the dissipation rate of turbulent energy ε was calculated. The calculation of ε was carried out from the pulsation components of the current velocity vector using the hypotheses of Kolmogorov and Taylor. The visualization of the Langmuir stripes for photographic and video filming was achieved by spreading paper markers on the sea surface. With a steady wind of 7–17 m/s, markers and algae in the water, sea foam, etc. lined up in clearly visible bands, which made it possible to determine the spatial scales of the CL. To estimate the distribution of the rate of dissipation of turbulent energy over depth, measurements of turbulent characteristics by the «Sigma-1» complex were carried out at several horizons from 1 to 7 m with an exposure time of 15–20 min. All measurements were accompanied by registration of background hydrometeorological conditions (surface wave parameters, wind speed and direction, water and air temperature, etc.). During 9 days of active observations, 74 Langmuir cells were detected. To assess the dynamics of the CL, the following parameters were chosen: the velocity of displacement of structures perpendicular to the direction of the wind, the number of observed cells for the selected observation period; width of convergence and divergence zones and vertical velocities in them. When analyzing the dynamic characteristics of the CL and comparing it with the intensity of turbulence, it was shown that the effect of the

M. I. Pavlov (✉) · A. M. Chukharev
FSBSI FRC Marine Hydrophysical Institute of RAS, 2 Kapitanskaya Str., Sevastopol 299011, Russian Federation
e-mail: mixail.pavlov.1993@mail.ru

V. I. Karev (ed.), *Physical and Mathematical Modeling of Earth and Environment Processes—2022*, Springer Proceedings in Earth and Environmental Sciences,
https://doi.org/10.1007/978-3-031-25962-3_39

CL on turbulence in the near-surface layer depends primarily on the wind speed and direction and on their variability. The conditions under which CLs have a noticeable effect on the turbulent regime are determined. In experiments, the mode was noted when the highest intensity of turbulence from the CL occurs at a wind of about 5 m/s, while the wind direction must be stable, changing by no more than 10°–15° for 10 min. The maximum speed in downdrafts ~ 5–32 mm/s was observed with small changes in wind direction at a speed of 3.7 m/s, in the range 5°–12°. With a sharp change in wind direction by 20°–150°, the number of observed cells and the speed in the convergence zones decrease every 10–40 min. In this case, the dissipation rate becomes much lower, $\varepsilon \sim (1.7 - 8.89) \times 10^{-6}\,\mathrm{m}^2/\mathrm{s}^3$. The developed technique for registering CLs and taking into account their influence on turbulent mixing has shown sufficient efficiency. The results obtained provide new useful information about the role of CL in near-surface turbulence.

Keywords Turbulent exchange · Upper sea layer · Dissipation rate · Stokes drift · Langmuir number · Experimental studies · Langmuir circulations

1 Introduction

Surface turbulence affects many physical processes in the sea: gas exchange, distribution of suspended matter and nutrients, and vertical mixing. Determining the turbulent exchange coefficients, which vary over a wide range depending on hydrometeorological conditions, is an urgent task in oceanology. Insufficient knowledge of physical processes in this layer of the sea leads to the fact that calculations by prognostic models for the upper layer can differ greatly from the actually observed values. Quite a lot of attention is now being paid to Langmuir circulations (CL). Relatively large coherent structures generated by wind and Stokes drift create an instability that leads to turbulent mixing. However, it has not yet been possible to accurately parameterize the CL, since the turbulent mixing in the near-surface layer is immediately affected by a combination of many processes. According to a number of researchers, the inclusion of this mixing mechanism in large-scale models of the interaction between the ocean and the atmosphere will help to significantly improve the objectivity of theoretical calculations.

We consider the theory of Craik and Leibovich [1] as a theoretical basis for the mechanism of CL appearance. In their work, the scientists investigated a system of linearized equations and presented numerical results for weakly non-linear motions in infinitely deep water, assuming «crossed waves» and the invariance of rectified motions along the direction of the wind. The theory was limited to the assumption of time-independent motion, which led to the need for a special definition of the horizontally averaged drift. This limitation was removed by Leibovich [2], who extended the theory to include the temporal evolution of coupled (wind-directed) currents and circulations. The formulation of the problem with the initial value, when the currents and circulations are initially equal to zero and are initiated by the step function of the surface voltage, leads to a well-posed mathematical problem. Assuming that the wave field is invariant in the x (wind) direction and symmetric about the x axis, the problem is independent from x and any resulting circulations must be in the form of spirals. Assuming a constant wind stress for $t \geq 0$, corresponding to the friction velocity u*, the presence of surface waves with characteristic frequency σ, wave number k and characteristic amplitude a, and eddy viscosity ν_T, it was shown that the problem with the initial value depends on one dimensionless parameter

$$La = (v_T^3 k^2 / \sigma a^2 u_*^2)^{1/2} \quad (1)$$

This parameter, which Leibovich called the Langmuir number, expresses the balance between the diffusion rate of longitudinal vorticity and the rate of creation of longitudinal vorticity by stretching the vortex.

As is known, CL consist of two vortices with horizontal axes and oppositely directed rotations, oriented perpendicular to the front of wind waves. These circulations arise in a turbulent fluid when the Stokes drift interacts with vertical shear, and in this case, a vortex force arises [1], which is responsible for the formation of circulations. An external sign of the existence of a CL is clearly visible stripes of foam, debris, or floating objects on the surface. The CL scales can range from a few meters to hundreds of meters, and systems of «nested» CL of different scales are often observed.

Very detailed experimental studies are described in [3], where the CL parametrization is proposed in terms of widely used physical quantities: dynamic velocity and Stokes drift.

In a relatively recent paper [4], both experimental observations of a turbulent boundary layer and modeling of structures in the coastal zone of the ocean were carried out. The authors noted that the existence of a CL increases the role of the vertical transfer of turbulent kinetic energy.

With the development of computer technology, large eddy simulation methods (LES) have become widely used in our time. With the help of LES modeling, the processes of turbulent transport were studied in [5] based on experimental studies carried out using the CBLAST program. The relative contributions of breaking waves and CL in the vertical turbulent transport, calculated using the LES model, corresponded to the observational data.

Important results in theoretical studies of the CL were obtained in a series of works [6–8], where various aspects of the dynamics of the near-surface layer of the sea were also studied on the basis of LES modeling: the effect of swell on the CL, the rate of deepening of the pycnocline, etc.

An analysis of the cited and a number of other works allows us to conclude that CL plays an important role in the vertical exchange in the upper boundary layer of the sea. Therefore, an in-depth study of this phenomenon and its contribution to turbulent mixing is an urgent task.

The question of the effect of CL on turbulent mixing still largely remains open. In [9], in the multiscale model of turbulence for the near-surface layer of the sea, the inclusion of CL was tested in two versions.

(1) According to [10]:

$$P^L = \nu_t\left(\frac{\partial U}{\partial z}\frac{\partial U_s}{\partial z} + \frac{\partial V}{\partial z}\frac{\partial V_s}{\partial z}\right) \quad (2)$$

where, U, V, U_s, V_s—components of the mean current velocity and Stokes drift along the horizontal axes x and y, respectively;

(2) According to [11]:

$$P^L = \frac{\langle w'^2\rangle}{u_+^2} = \begin{cases} 0.398 + 0.48La_{SL}^{-4/3}, & La_{SL} \le 1 \\ 0.64 + 3.50\exp(-2.69La_{SL}), & La_{SL} > 1 \end{cases} \quad (3)$$

where $\langle w'^2\rangle$—rms vertical fluctuations of the flow velocity, u* is the friction velocity in water.

Using the method [11] gives the best result when compared with experimental data. However, it was found that the results of calculations and measurements can differ by several orders of magnitude. This is especially evident when the wind speeds are small: 2–5 m/s. It has also been observed that the variability of wind direction plays an important role. Under similar hydrometeorological conditions, but at different rates of change in wind direction, the experimental values of the dissipation rate can differ significantly. This led to the idea that there may be some regimes that contribute to intense turbulence, or vice versa, the necessary instability does not arise. To clarify these and other issues on a stationary oceanographic platform in the settlement. Katsiveli in 2020–2021 experimental studies of the effect of CL dynamics on the intensity of turbulent mixing were carried out.

2 Equipment and Methods for Studying the Effect of Langmuir Circulations on Turbulent Exchange

Based on the works [12–14], a research technique was developed, which consists in determining the descending and ascending flows in the zones of convergence and divergence in coherent Langmuir structures, using an acoustic current velocity meter DVS-6000 (ADCP) located at a depth of 5 m below the sea surface. In this case, the emitters of the device are directed upwards, measurements are carried out in the depth range of 0.4–4.5 m in 5 layers (bins) 0.8 m thick, measurements are referred to the middle of each layer. The device was placed in a special frame and weighted with a load to minimize natural oscillations and increase the accuracy of measurements (Fig. 1). Also on the horizon of 5 m there is a measuring complex «Vostok-M» (Fig. 2) for recording averaged hydrophysical fields of flow velocity, electrical conductivity and temperature. To estimate the intensity of turbulent processes and the effect on vertical mixing, the rate of dissipation of turbulent energy ε is calculated. The calculation of ε is carried out using the measured fluctuation components of the three components of the flow velocity vector and temperature fluctuations. Measurements of turbulent pulsations of hydrophysical quantities are carried out by the «Sigma-1» complex created at the Marine Hydrophysical Institute [15], the appearance of which is shown in Fig. 3. The mutual arrangement of instruments and a view of the working area of the oceanographic platform, which is located from the sea, can be seen in Fig. 4. The experimental research methodology is as follows: with a steady wind of 7–17 m/s, markers were scattered on the sea surface in the form of cut sheets of thick paper ~ 10 × 15 cm in size, which, in the presence of a CL, lined up in clearly visible stripes. Visual observations of wave breaking and Langmuir bands were accompanied by photography and video recording. The Sigma-1 complex was successively installed at the desired horizon at a depth of 1–7 m with a step of 0.5–1 m to record pulsation characteristics with a duration of 15–20 min. This makes it possible to determine the values of the dissipation rate at different depths during the experiment. During the recording of pulsation values, the descending and ascending flows in the zones of convergence and divergence are recorded by the DVS-6000 m as vertical components of the current velocity.

To study the evolution of Langmuir circulations and the dissipation rate, the «Sigma-1» complex is located at a depth of 2–3 m, and a recording lasting 3–10 h is made. for analysis.

Fig. 1 Flow meter «DVS-6000»

Fig. 2 «Vostok-M» current meter

Fig. 3 Current velocity pulsation meter «Sigma-1»

3 Results of Observations of Coherent Langmuir Structures

During experiments in June 2021 on the oceanographic platform of the Marine Hydrophysical Institute, instrumental methods were used to study the dynamics of Langmuir circulations and their effect on turbulent mixing. To visualize the stripes,

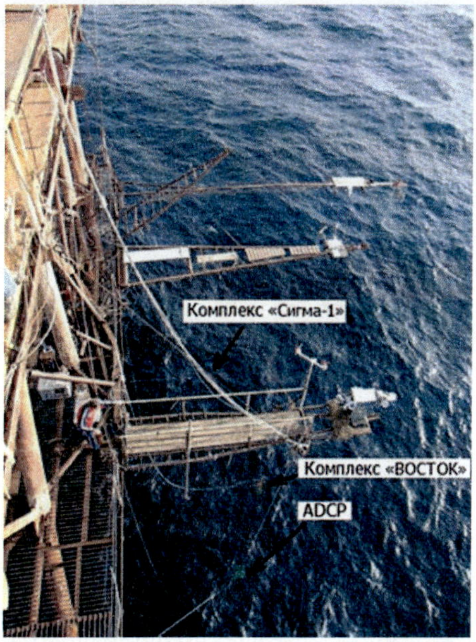

Fig. 4 Installation of meters, top view

cut paper sheets were used as markers. Figure 5 shows the emerging stripes of paper markers.

In the process of conducting experiments on visual observations of Langmuir stripes on the surface, it was established:

- at wind speeds up to 4 m/s, the bands are rather weakly expressed;
- visually noticeable stripes become at a speed of more than 6 m/s;
- with steady wind at speeds of 9–13 m/s, stable, clearly defined stripes appear;

Fig. 5 Visualization of the CL using cut paper markers. Experiment on the oceanographic platform of the Marine Hydrophysical Institute 07/11/2021. Wind speed 17 m/s

- at velocities of 13–20 m/s, the bands are destroyed rather quickly, since with a sharp change in wind direction, the bands do not have time to "rebuild", they are unstable and mix.

With a wind of 17 m/s, however, it was possible to capture clearly pronounced stripes on the surface, which are shown in Fig. 5.

According to the continuous data of the DVS-6000 flow meter for several days, zones of convergence and divergence were identified. Since the CL structures are gradually shifting to the right of the wind direction, in the measurement zone we consistently see the alternation of these zones. Examples of records are shown in Fig. 6 (measurements at a horizon of 0.8 m) and in Fig. 7 (measurements at the horizon 2.5 m). It should be noted that the strict regularity of the alternation of zones in our records is by no means always encountered. The vertical direction of the flow velocity in the zones of convergence and divergence is indicated on the images by red and blue arrows. In our data, as in the measurements of many other authors, the zones of convergence are larger than the zones of divergence in amplitude (velocities are higher), but less in period (in width).

An analysis of the records for all 5 horizons showed that at the 0.8 m horizon, the CL is rather difficult to interpret, since the influence of surface effects and an increased measurement error affect. Structures in convergence zones with the highest descending vertical velocities up to 36 mm/s and periods of 2–6 min are located on

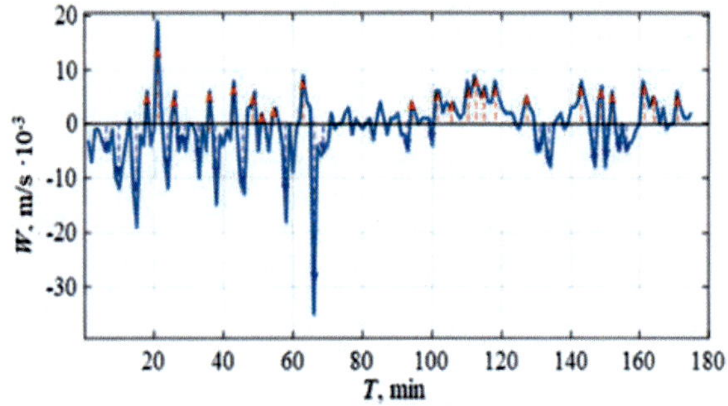

Fig. 6 Registration of zones of convergence and divergence by vertical velocities on the horizon of 0.8 m for 180 min

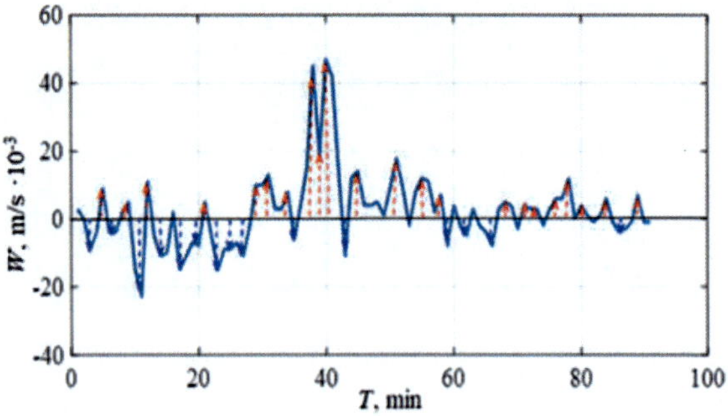

Fig. 7 Zones of convergence and divergence on the record of the vertical velocity component at the horizon of 2.5 m for 100 min

a horizon of 1.7 m. At the horizons of 2.5–3.5 m, the vertical velocities are lower, but the periods are clearly longer: 3–14 min. At the 4.5 m horizon closest to the sensors, the structures are often difficult to interpret due to the small amplitudes of the vertical velocities. It should also be taken into account that the farther the measurement horizon from the sensors, the obviously worse the accuracy, since the distance between the DVS beams increases, which affects the averaging of the vertical current velocity components. Recordings with convergence and divergence zones less than 2 min long and with an amplitude less than 5 mm/s were not studied, since it is not entirely clear whether these segments are part of coherent structures. Therefore, for a more accurate parametrization of the CL, we studied the dynamics of structures at horizons of 1.7–3.3 m, with an amplitude of 5 mm/s and a period of 2 min. In this paper, when describing the measurement results of the DVS-6000, hereinafter we use mm/s as the unit of measurement, in accordance with the units of the DVS data output program.

On the records of horizontal velocities in the convergence zones, an almost twofold increase in velocity was found. In the case of closely spaced cells, the horizontal velocities in the convergence zones are identical. Figure 8 shows records of the vertical current velocity W at a depth of 3 m and horizontal current velocities U, V at a depth of 0.8 m. It can be seen that in the convergence zones on the W plot, the horizontal velocity U has a value of ~ 20 mm/s. Similar results were described in [12]. Using the methods listed above, records for 9 days were analyzed, where 74 CL cells were identified.

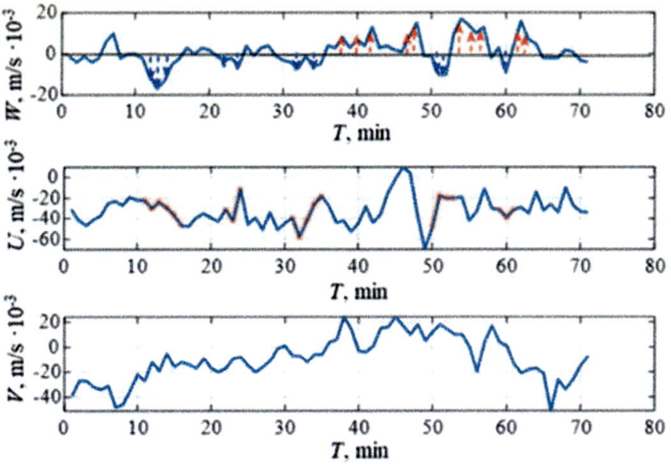

Fig. 8 Registration of zones of convergence and divergence by vertical velocities on the horizon of 3 m for 100 min. All three measured DVS-6000 current velocity components are shown

4 Investigation of Langmuir Circulation Dynamics

To understand how CL affect turbulent mixing in the near-surface layer, the dynamics of coherent structures was investigated and evaluated. The following parameters were chosen for evaluation: the rate of transverse displacement of structures, the number of observed cells during the observation period; width of convergence and divergence zones (for brevity they are sometimes called periods) and multidirectional vertical velocities in them (called amplitudes).

Determination of the displacement velocity: to determine the displacement velocity, we used the relationship between the vertical fluctuations of the current velocity w', measured by the «Sigma-1» complex, and the vertical component of the current W, which was recorded by *DVS*. In the *Matlab* application package, using the cov(x, y, *coef*) function, we calculated the cross-correlation of r.m.s. ripples $\overline{w'^2}$, averaged over 1 min, and W was expected in the recording areas where cells of coherent structures were found. The analysis revealed quite clear peaks of the coefficient of positive and negative cross-correlation, Fig. 9. In general, the level of correlation of the analyzed signals has low values, which can be explained by the remoteness of the instruments from each other; in this case, the cross-correlation functions are usually asymmetric. If we compare the signals W and $\overline{w'^2}$, it can be noted that they are often similar in shape, but, as can be seen in Fig. 10, there are also differences in period and phase shift.

Thus, having determined the time when the signals have a maximum of the correlation function of at least 0.5 and knowing the distance between the meters, the structure displacement velocity was calculated.

Influence of wind: speeds on the surface of the ocean are much less than speeds in the atmosphere. This is mainly due to the difference in densities, since the momentum in water can be transported at a much lower speed than in the atmosphere. The transfer of momentum through the surface can be due to pressure drop when passing through irregularities (waves) or viscous stresses. Comparing the wind dynamics (wind speed at a height of 10 m, change in speed, change in wind direction) with the vertical speed

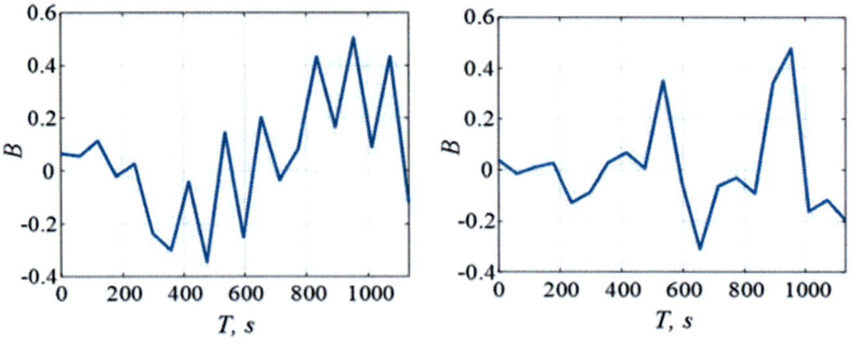

Fig. 9 Cross-correlation coefficients of signals W and $\overline{w'^2}$

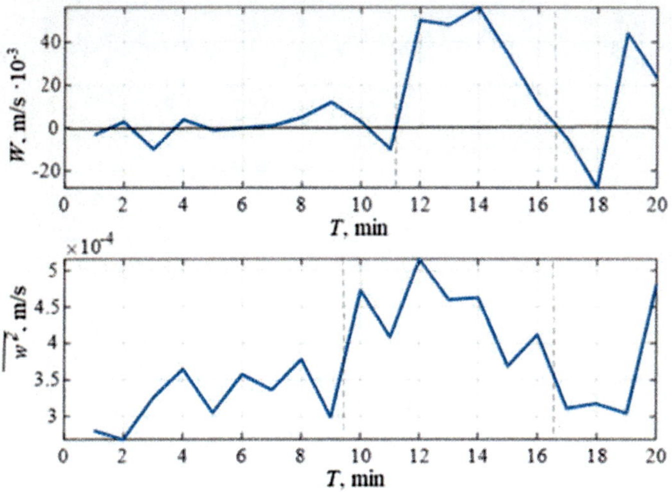

Fig. 10 Comparison of vertical current velocity signals W and averaged vertical current velocity fluctuations $\overline{w'^2}$

of the current in the presence of Langmuir cells, an interesting feature was revealed: the width of the convergence zones is affected by the change in wind speed. Figure 11 shows that the slower the velocity changes by 0.5–1 m/s, the wider the convergence zone. The vertical speed in the convergence zones, according to our observations, is proportional to the wind speed. The dependence is not direct, but in the figure it can be seen that with a sharp increase in wind speed by 1–2 m/s and a change in direction by 20°, the downward speed increases significantly. Convergence zones are marked with blue arrows. Correlated changes in wind and speed characteristics in convergence zones are marked in pink on the graphs.

The analysis of the CL dynamics for 9 days of experimental observations is summarized in Table 1 with the parameters indicated above. Cross-correlation coefficients were also added: (1) vertical current velocity and average wind speed $B_{W\&V10}$; (2) vertical current velocity and change in wind direction angle $B_{W\&\Delta\alpha}$. Highs are in brown, lows are in yellow.

The maximum number of cells, 13, was found at an average wind speed of 7.4 m/s and a change in wind direction by 2°–6° in 10 min. The displacement velocity of the structures was the smallest among all those observed, 0.007 m/s. At the same time, the period of the observed convergence zones was the longest: 3–22 min. The speed in the convergence zones is the highest, 11–32 mm/s. The correlation with the wind and its direction is minimal: $B_{W\&V10} = 0.11$ and $B_{W\&\Delta\alpha} = 1.13$.

The smallest number of cells, 3 over a period of 90 min, was found with the smallest velocity amplitudes in the convergence zones: 3–9 mm/s; in this case, the observation period was also the shortest of the observed 2–3 min. The average wind speed was 3.6 m/s. During this period of CL observation, there was a sharp change in wind direction, every 10 min by 20°–30°, and every 30 min by 120°–250°. The

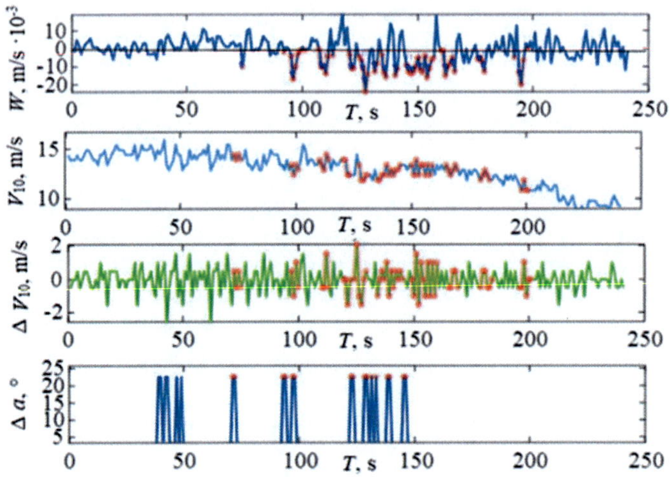

Fig. 11 Comparison of downdrafts W in convergence zones, with wind speed V_{10}, with a change in wind speed ΔV_{10} and with a change in wind direction $\Delta\alpha$

Table 1 Characteristics of the CL in the experiments

Date and time	Number of cells	V disp, м/с	T, min.	$-W$, mm/s	$T(-W)$, min.	$+W$, мм/с	$T(+W)$, min.	$B_{W\&V10}$	$B_{W\&\Delta\alpha}$	V_{10}
11.06.2021 9:08–10:18	5	0.008	70	6–17	4–7	3–17	3–13	0.246	0.166	9.7
11.06.2021 1:08–2:50	4	0.009	100	6–17	4–7	6–66	4–7	0.255	0.38	1.4
11.06.2021 20:47–22:11	5	0.008	80	9–27	2–7	7–16	5–22	0.37	0.28	1.7
13.06.2021 10:28–12:42	8	0.01	200	5–23	3–22	6–51	3–16	0.07	0.51	8.3
11.06.2021 18:18–18:48	3	0.007	90	3–9	2–3	2–27	2–14	0.152	0.27	3.6
20.06.2021 12:33–13:03	11	0.014	240	5–24	2–13	5–19	2–8	0.5	0.32	11.8
17.06.2021 20:30–23:00	11	0.009	150	5–32	2–8	5–51	2–8	0.2	0.07	3.7
18.06.2021 10:03–12:03	13	0.007	120	11–32	2–8	5–27	2–8	0.11	0.127	7.4
18.06.2021 3:23–4:13	9	0.011	53	5–28	2–7	5–20	2–4	0.16	0.5	4.9

vertical component of the current velocity weakly correlated with the wind speed and direction.

The longest period during which Langmuir cells (in the amount of 11 pieces) were continuously observed was 240 min. At the highest wind speed V10 = 11.8 m/s, the structure displacement velocity was also the highest: Vdisp = 0.014 m/s. At the same time, the wind direction practically did not change (no more than 1°–2° for 10 min), and the correlation coefficient $B_{W\&\Delta\alpha} = 0.51$ was also high.

The highest speed in the divergence zone was recorded at night and amounted to 6–66 mm/s with a light wind of 1–2 m/s. The direction of the wind changed on average by 35° every 10 min. Such speeds are most likely associated with nighttime convection processes, when the sea surface cools.

5 Effect of Langmuir Circulations on the Intensity of Turbulent Mixing

Simultaneously with visual observation of the CL on the surface and registration of the current velocity components in the upper five-meter layer, measurements of turbulent fluctuations of three velocity components u', v' and w', temperature fluctuations and electrical conductivity were carried out according to the method described above. To determine the effect of CL on the intensity of turbulent mixing, the rate of dissipation of turbulent energy ε was calculated from the components of velocity fluctuations. Using the method of Stuurt and Grant [16], which was also used in [17], the values of ε were determined at all measurement horizons carried out by the «Sigma-1» complex: from 0.8 to 7 m (Fig. 12).

When analyzing the dependence of ε on hydrometeorological conditions, an important feature was noted: the intensity of turbulence is associated not only with wind speed, but also with the speed of change in wind direction. When the wind direction changed within 1.5 h by $\Delta\alpha = 8°$ on average, the values of ε were significantly greater than at $\Delta\alpha = 2°–4°$, the wind speed in these cases was approximately the same.

The many-hour variability of the intensity of turbulent mixing at a 3 m horizon was also studied. At night, when convective mixing was observed, the dissipation rate was calculated (Fig. 13a), and a comparison was made with the speed and change in wind direction. It can be seen from the figure that the process of convective mixing creates instability, but in our case it slightly affects the intensity of turbulent mixing:

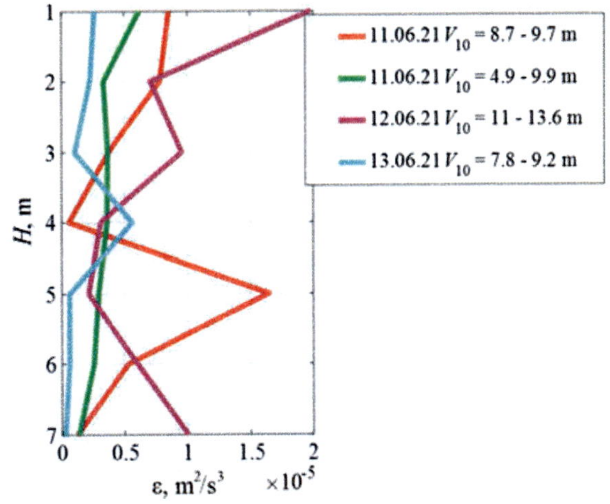

Fig.12 Distribution of the energy dissipation rate ε over depth

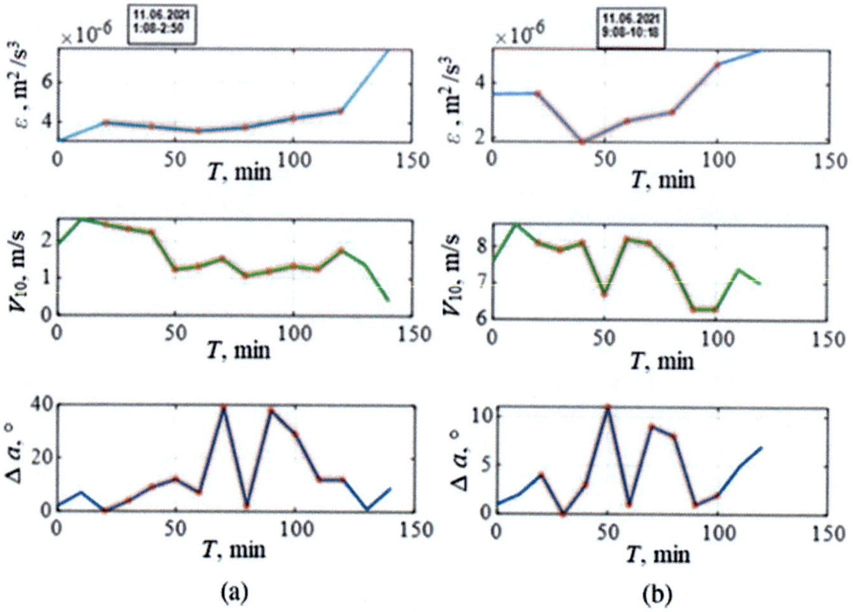

Fig. 13 Distribution of the dissipation rate of turbulent energy ε at a depth of 3 m for 150 min. **a** at wind speed $V_{10} = 1–2$ m/s and wind direction change $\Delta\alpha = 30°$. **b** $V_{10} = 6–8$ m/s and change in wind direction $\Delta\alpha = 5°–10°$

$\varepsilon \sim 4 \times 10^{-6}\,m^2/s^3$. When observing the CL as a whole, ε fluctuates in the range of $1.7 \times 10^{-6} - 8.89 \times 10^{-6}\,m^2/s^3$, with a wind of 3–9 m/s. On (Fig. 13b) shows an example of such measurements. In this case, also when the wind direction changes by 5°–10° at a speed of 2–7 m/s, the intensity of turbulence increases.

The highest intensity of turbulent mixing was found in the evening and at night, which is apparently associated with the development of nighttime convection. On (Fig. 14a, b) show the values of ε, which turned out to be larger than the other measurements by 1–3 orders of magnitude; the wind varied in the range of 2–5 m/s. When the wind direction changed by 25°–45°, the intensity of turbulent mixing began to decrease. Thus, in our experiments, the case was noted when the highest intensity of turbulence from the CL occurs at a wind of about 5 m/s, while the wind direction must be stable, changing by no more than 10°–15° for 10 min. The displacement velocity of the structures was ~ 10 mm/s, the number of cells was 9–11, and the downward flow velocity was ~ 5–32 mm/s. That is, in this mode, the largest number of cells appears and a high downward flow rate is observed at an average displacement velocity of coherent structures.

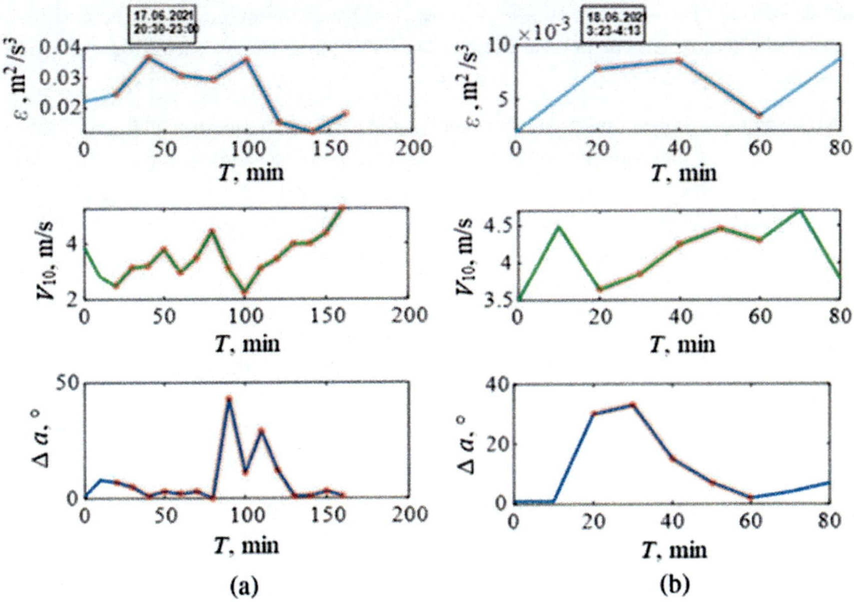

Fig. 14 Change in the rate of dissipation of turbulent energy ε at a depth of 3 m, wind speed V_{10} at a height of 10 m and change in the wind direction angle Δα within 80 min

6 Conclusion

It is well known from observations that the sea surface is extremely variable. This is primarily due to the interaction with the adjacent layer of the atmosphere, which occurs over a wide range of scales. Such a phenomenon as CL can be visually observed in almost any weather, with wind starting from 2 m/s.

In those held in 2020–2021 In experiments at the hydrophysical sub-satellite test site of the MHI RAS, we obtained experimental data on the main dynamic characteristics in the observed CLs. Simultaneously with visual observations of the state of the sea surface and determination of the geometric parameters of the Langmuir strips using paper markers, instrumental measurements of the current velocity components, turbulent characteristics and background hydrometeorological conditions were carried out: wind speed and direction, water and air temperature, wave characteristics, etc.

In the process of research, it was found that the effect of CL on turbulence in the near-surface layer of the sea largely depends on the speed and direction of the wind.

According to visual observations of the Langmuir stripes on the surface:

- at wind speeds up to 4 m/s, the bands are rather weakly expressed;
- visually noticeable stripes become at a speed of more than 6 m/s;
- with steady wind at speeds of 9–13 m/s, stable, clearly defined stripes appear;

- at velocities of 16–20 m/s, the bands are destroyed rather quickly, since with a sharp change in wind direction, the bands do not have time to «rebuild», they are unstable and mix.

A technique has been developed for determining the main dynamic characteristics of the CL, which include: the velocity of displacement of structures perpendicular to the wind, the number of observed cells during the observation period; the width of the zones of convergence and divergence (for brevity they are sometimes called periods) and the vertical velocities in them (amplitudes).

The regularities of CL evolution with changes in the absolute values of the wind speed are revealed:

- with an increase in V_{10}, the average vertical velocity in the convergence zones and the width of the zones increase;
- an increase in wind speed leads to an increase in the speed of displacement of structures to the right of the wind direction. The range of displacement velocities observed in the experiments was from 0.007 to 0.014 m/s;
- with well-defined CL and a stable wind speed of about 10 m/s, a «smooth» change in wind direction (up to 20° for 10–20 min) causes an increase in the displacement velocity, vertical velocity in the convergence zones and the transverse scale of the zones;
- at low wind speeds (up to 2 m/s) and sharp changes in direction (by 30° or more within 10–20 min), the structure displacement velocity is the smallest: about 0.008 m/s.

The developed technique for registering CLs and taking into account their influence on turbulent mixing has shown sufficient efficiency. However, not all important characteristics can be determined in this case, for example, such a parameter as the direction of the CL. It is also possible that the calculation of the rate of transverse displacement of structures also requires refinement and more thorough experimental verification.

At this stage, it was not possible to establish a statistically significant correlation between the intensity of turbulence (the rate of dissipation of turbulent energy ε and the characteristics of the CL in the work at this stage. However, a regime has been identified in which the CL increases turbulent mixing in the near-surface layer—these are sharp changes in the direction of the wind, which, in our opinion, leads to a restructuring of the CL and an increase in the instability of fluid motions. To parameterize this process, and, ultimately, to develop a predictive model, a deeper study of the issue is needed. A detailed calculation of ε in each cell of the structure, more accurate measurements of vertical velocity in downdrafts and updrafts, additional laboratory studies will provide more complete information and statistical data for solving the problem.

Acknowledgements The work was carried out within the framework of the state assignment FNNN-2021-0004 with the financial support of the RSF grant № 22-17-00150. At the expense of the state order FNNN-2021-0004 an expedition to the oceanographic platform was organized in

2021 and a pilot experiment on registration of Langmuir circulations by different measuring devices was carried out.

References

1. Craik, A.D.D., Leibovich, S. A rational model for Langmuir circulation. // J. Fluid Mech. 1976. V. 73. P. 401–426. https://doi.org/10.1017/S0022112076001420.
2. Leibovich, S. On the evolution of the system of wind drift currents and Langmuir circulations in the ocean. Part I. Theory and the averaged current. J. Fluid Mech. 1977. 79:71 5–43. https://doi.org/10.1017/S0022112077001803.
3. Plueddemann J, et al. / Structure and variability of Langmuir circulation during the Surface Waves Processes Program // J. Geophys. Res. Atmospheres · February 1996. V. 101, P. 3525–3543. https://doi.org/10.1029/95JC03282.
4. Kukulka T., Plueddemann A. J. and Sullivan P. P. Nonlocal transport due to Langmuir circulation in a coastal ocean // J. Geophys. Res. 2012. V. 117. https://doi.org/10.1029/2012JC008340.
5. Li S., Li M., Gerbi G.P. and Song J.-B. Roles of breaking waves and Langmuir circulation in the surface boundary layer of a coastal ocean // J. Geophys. Res.: Oceans, 2013. V. 118. P. 5173–5187. https://doi.org/10.1002/jgrc.20387.
6. McWilliams J.C., Huckle E., Liang J.-H., and Sullivan P.P. The wavy Ekman layer: Langmuir circulations, breaking waves and Reynolds stress // J. Phys. Oceanogr., 2012. V. 42. P. 1793–1816. https://doi.org/10.1175/JPO-D-12-07.1.
7. McWilliams J.C., Huckle E., Liang J.-H., and Sullivan P.P. Langmuir Turbulence in Swell // J. Phys. Oceanogr., 2014. V. 44. P. 870–890. https://doi.org/10.1175/JPO-D-13-0122.1.
8. Sullivan P.P., Romero L., McWilliams J.C., and Melville W.K. Transient Evolution of Langmuir Turbulence in Ocean Boundary Layers Driven by Hurricane Winds and Waves // J. Phys. Oceanogr., 2012. V. 42. P. 1959–1980. https://doi.org/10.1175/JPO-D-12-025.1.
9. Chukharev A. M., Pavlov. M. I. Model and Experimental Estimates of Vertical Mixing Intensity in the Sea Upper Homogeneous Layer / Physical Oceanography. – 2021. – Vol. 28. – No 3. – P. 309–325. https://doi.org/10.22449/1573-160X-2021-3-309-325.
10. Wu, L., A. Rutgersson, E. Sahlee. Upper-ocean mixing due to surface gravity waves // J. Geophys. Res. Oceans, 2015. 120. https://doi.org/10.1002/2015JC011329.
11. Harcourt R.R., E.A. D'Asaro Large-eddy simulation of Langmuir turbulence in pure wind seas // J. Phys. Oceanogr. 2008. V. 38(7), P. 1542–1562, https://doi.org/10.1175/2007JPO3842.1.
12. Weller R.A., Price J.F. Langmuir circulation within the oceanic mixed layer. // DeepSea Res. 1988. 35:711–47. https://doi.org/10.1016/0198-0149(88)90027-1.
13. Thorpe S.A., Langmuir circulation. // Annual Review of Fluid Mechanics. 2004.Vol. 36:55–79. https://doi.org/10.1146/annurev.fluid.36.052203.071431.
14. Zedel L., Farmer D. Organised. Structures in subsurface bubble clouds: Langmuir circulation in the open ocean. // J. Geophys. Res. 96:8889–900. https://doi.org/10.1029/91JC00189.
15. Measuring complex "Sigma-1" for the study of small-scale characteristics of hydrophysical fields in the upper layer of the sea / A. S. Samodurov [et al.] // Marine Hydrophysical Journal. 2005. No. 5. P. 60–71.
16. Stewart R.W., Grant H.L. Determination of the rate of dissipation of turbulent energy near the sea surface in the presence of waves // J. Geophys. Res. – 1962. – 67, № 8. – Pp. 3177–3180.
17. Soloviev A., R. Lukas, P. Hacker [et al.]. Near–Surface Microstructure Sensor System Used during TOGA COARE. Part II: Turbulence Measurements // J. of Atmos. and Oceanic Techn. – 1999. – Vol. 16. – Pp. 1598–1618.

On the Solution of Dirichlet Problem for Sobolev-Type Equations for Four-Dimensional Cylindrical Domain

N. V. Kalenova and A. M. Romanenkov

Abstract The paper discusses an initial boundary value problem for four-dimensional, by spatial variables, Sobolev-type equation. The domain, in which this equation is valid, is a classical cylinder in $\mathbb{R}^4$, i.e. one of the coordinate directions, which the cylinder axis is parallel to, is specially singled out; and the cylinder base is located in the subspace set up by the rest of the coordinate directions. It should be pointed out that the spheres from $\mathbb{R}^3$ lie in the cylindrical domain bases. In this case, it is possible to consider solutions with specific radial symmetry by three variables, which makes it possible to obtain precise solutions in the form of Fourier series of Dirichlet boundary value problem for such domain. The theorem of uniqueness is further presented in the paper with the use of integral identities technique.

Keywords Precise solutions of Sobolev equation · Radial solutions · Solution uniqueness

1 Introduction

The following equation is called Sobolev-type equation in literature [1]:

$$\frac{\partial^2}{\partial t^2}(\Delta u) + A^2 \frac{\partial^2}{\partial x_n^2} u = 0, \tag{1}$$

where $A \in \mathbb{R}$, $u = u(X, t)$, $X = (x_1, x_2, \ldots, x_n)$, $\Delta = \frac{\partial^2}{\partial x_1^2} + \cdots + \frac{\partial^2}{\partial x_n^2}$—Laplace operator and $X \in W \subset \mathbb{R}^n$, $n \in \mathbb{N}\backslash\{1\}$.

The issues of asymptomatic behavior of solutions of an initial boundary value problem of Sobolev type equation are studied in the indicated paper [1] from the

N. V. Kalenova (✉) · A. M. Romanenkov
Moscow Aviation Institute (National Research University), Moscow, Russia
e-mail: community.sp@ya.ru

A. M. Romanenkov
Federal Research Center "Informatics and Management" of Russian Academy of Sciences, Moscow, Russia

V. I. Karev (ed.), *Physical and Mathematical Modeling of Earth and Environment Processes—2022*, Springer Proceedings in Earth and Environmental Sciences,
https://doi.org/10.1007/978-3-031-25962-3_40

oscillation point. In paper [2], which is dedicated to non-classical models of mathematical physics, Sobolev-type equation appears when considering the liquid diffusion problem in porous media. This paper contains the overview of results on the properties of equation solutions and theorems of uniqueness for different domains. Paper [3] presents sufficient conditions for the existence of positive solutions of Cauchy and Showalter-Sidorov problems for an abstract linear Sobolev equation. Paper [4] discusses nonlinear Sobolev-type equation from the first by time derivative with time. In this paper, the asymptomatic formula for solving Cauchy problem is obtained by integral estimation method. The methods of integral estimations are very productive for establishing the qualitative behavior of solutions and uniqueness of solutions of corresponding initial boundary value problems. In [5], the conditions of solution uniqueness and correctness of Cauchy problem are given for the class of power growth functions. It should be pointed out that the consideration of models of incompressible viscoelastic liquids results in the necessity of investigating Cauchy problem for Sobolev-type semilinear equations [6].

Let function u be a radial function of its first $n-1$ spatial variables, i.e. $u = u(X,t) = u(r, x_n, t)$, where $r = \sqrt{x_1^2 + x_2^2 + \cdots + x_{n-1}^2}$. $r = \sqrt{x_1^2 + x_2^2 + \cdots + x_{n-1}^2}$. So, we consider the case when the radial symmetry is determined not by spatial variables but only by some specified multitude of these variables. Not limiting the generality, we distinguish variable x_n, the solution radially depends on the remaining variables.

It is known that in the nth space Laplace operator is expressed in radial coordinates by formula [7]:

$$\Delta = \frac{\partial^2}{\partial r^2} + \frac{n-1}{r}\frac{\partial}{\partial r}. \tag{2}$$

2 Solution Formulation in the Form of Fourier Series

After the assumptions made, the initial Eq. (1) can be now rewritten as follows:

$$\frac{\partial^2}{\partial t^2}\left(\frac{\partial^2 u}{\partial r^2} + \frac{n-2}{r}\frac{\partial u}{\partial r}\right) + \frac{\partial^4 u}{\partial t^2 \partial x_n^2} + A^2 \frac{\partial^2 u}{\partial x_n^2} = 0, \tag{3}$$

for which we formulate the first initial boundary value problem for $n = 4$ and W domain of special type:

$$W = \left(X : X = (x_1, x_2, x_3, x_4) \in \mathbb{R}^4, x_4 \in [0; L], \sqrt{x_1^2 + x_2^2 + x_3^2} \leq l; L, l > 0\right),$$

in other words, W is a cylinder in four-dimensional space. So, it is necessary to find the solution of Eq. (1) satisfying Dirichlet's boundary conditions in W domain:

$$u|_{\partial W} = 0$$

and initial conditions:

$$u|_{t=0} = f(X), \quad u_t|_{t=0} = g(X). \tag{4}$$

To solve this problem, let us introduce the supplementary function v for

$$u(r, x_4, t) = \frac{v(r, x_4, t)}{r}. \tag{5}$$

It is possible to obtain the expressions of variables of the initial function u through function v:

$$\frac{\partial^2 u}{\partial x_4^2} = \frac{1}{r}\frac{\partial^2 v}{\partial x_4^2}, \tag{6}$$

$$\frac{\partial u}{\partial r} = \frac{r\frac{\partial v}{\partial r} - v}{r^2}, \tag{7}$$

$$\frac{\partial^2 u}{\partial r^2} = \frac{1}{r}\frac{\partial^2 v}{\partial r^2} - \frac{2}{r}\frac{\partial v}{\partial r} + \frac{2}{r^3}v. \tag{8}$$

Further we substitute the obtained expressions (6)–(8) into Eq. (3):

$$\frac{\partial^2}{\partial t^2}\left(\frac{1}{r}\frac{\partial^2 v}{\partial r^2} - \frac{2}{r^2}\frac{\partial v}{\partial r} + \frac{2}{r^3}v + \frac{n-2}{r}\left(\frac{1}{r}\frac{\partial v}{\partial r} - \frac{v}{r^2}\right)\right) + \frac{1}{r}\frac{\partial^4 v}{\partial t^2 \partial x_4^2} + \frac{A^2}{r}\frac{\partial^2 v}{\partial x_4^2} = 0.$$

After reducing similar summands and multiplying by r, we have:

$$\frac{\partial^4 v}{\partial t^2 \partial r^2} + \frac{\partial^4 v}{\partial t^2 \partial x_4^2} + A^2\frac{\partial^2 v}{\partial x_4^2} = 0. \tag{9}$$

Using the idea from paper [8], we look for the unknown function in the following form:

$$v(r, x_4, t) = T(t)\sin \mu r \sin \lambda x_4, \tag{10}$$

where μ, λ—constants to be defined. Now after substituting the expression for $v(r, x_4, t)$ into Eq. (9) we have:

$$- \sin \mu r \sin \lambda x_4 \left(T''(t) + \frac{A^2\lambda^2}{\mu^2 + \lambda^2} T(t) \right) = 0. \tag{11}$$

Then $T(t) = C_1 \sin\left(\frac{A\lambda}{\sqrt{\mu^2+\lambda^2}} t \right) + C_2 \cos\left(\frac{A\lambda}{\sqrt{\mu^2+\lambda^2}} t \right)$. After that let us point out that the following equalities should be satisfied to fulfill the boundary condition:

$$\sin \mu l = 0, \quad \sin \lambda L = 0, \tag{12}$$

that is $\mu_n = \frac{\pi n}{l}$, $\lambda_m = \frac{\pi m}{L}$, $n, m \in \mathbb{N}$.

For each pair of natural numbers n, m function

$$v_{n,m}(r, x_4, t) = \left(C_{1_{n,m}} \sin\left(\xi_{n,m}(t)\right) + C_{2_{n,m}} \cos\left(\xi_{n,m}(t)\right)\right) \sin\left(\frac{\pi n}{l} r\right) \sin\left(\frac{\pi m}{L} x_4\right), \tag{13}$$

where $\xi_{n,m}(t) = \frac{At}{\sqrt{\frac{L^2}{l^2}\left(\frac{n}{m}\right)^2+1}}$, is the solution of Eq. (9). To find constants $C_{1_{n,m}}$, $C_{2_{n,m}}$ we use the initial conditions:

$$v|_{t=0} = rf(X), \quad v_t|_{t=0} = rg(X), \tag{14}$$

i.e.

$$\sum_{n,m=1}^{\infty} C_{2_{n,m}} \sin\left(\frac{\pi n}{l} r\right) \sin\left(\frac{\pi m}{L} x_4\right) = rf(X), \tag{15}$$

$$\sum_{n,m=1}^{\infty} \left(\frac{A}{\sqrt{\frac{L^2}{l^2}\left(\frac{n}{m}\right)^2 + 1}} \right) C_{1_{n,m}} \sin\left(\frac{\pi n}{l} r\right) \sin\left(\frac{\pi m}{L} x_4\right) = rg(X). \tag{16}$$

After expanding in double Fourier series we have:

$$C_{2_{s,q}} = \frac{4}{Ll} \int_0^L \int_0^l \rho f(\rho, \eta_4) \sin\left(\frac{\pi s}{l} \rho\right) \sin\left(\frac{\pi q}{L} \eta_4\right) d\rho d\eta_4, \tag{17}$$

$$C_{1_{s,q}} = \frac{4}{ALl} \sqrt{\frac{L^2}{l^2}\left(\frac{s}{q}\right)^2 + 1} \int_0^L \int_0^l \rho g(\rho, \eta_4) \sin\left(\frac{\pi s}{l} \rho\right) \sin\left(\frac{\pi q}{L} \eta_4\right) d\rho d\eta_4. \tag{18}$$

Now it is eventually possible to write down the boundary value problem solution:

$$u(r, x_4, t) = \frac{1}{r} \sum_{s,q=1}^{\infty} \left(C_{1_{s,q}} \sin(\xi_{s,q}(t)) + C_{2_{s,q}} \cos(\xi_{s,q}(t))\right) \sin\left(\frac{\pi s}{l} r\right) \sin\left(\frac{\pi q}{L} x_4\right). \tag{19}$$

Theorem "On Uniqueness of Solution" *If the solution of the boundary value problem* (1)–(3) *exists, it will be unique.*

Proof Let us use the standard procedure [7]. We multiply the initial equation by $\frac{\partial u}{\partial t}$ and integrate by W and from 0 to T by the time variable:

$$0 = \iint\limits_{W \times [0;T]} \left(\frac{\partial^2 \Delta u}{\partial t^2} \frac{\partial u}{\partial t} + A^2 \frac{\partial^2 u}{\partial x_n^2} \frac{\partial u}{\partial t}\right) dX dt. \tag{20}$$

At the same time, it should be pointed out that the following identities are available due to Leibniz formula:

$$\frac{\partial^2 u}{\partial x_n^2} \frac{\partial u}{\partial t} = \frac{\partial}{\partial x_n}\left(\frac{\partial u}{\partial x_n} \frac{\partial u}{\partial t}\right) - \frac{1}{2} \frac{\partial}{\partial t}\left(\frac{\partial u}{\partial x_n}\right)^2, \tag{21}$$

$$\frac{\partial^4 u}{\partial t^2 \partial x_n^2} \frac{\partial u}{\partial t} = \frac{\partial}{\partial x_n}\left(\frac{\partial^3 u}{\partial t^2 \partial x_n} \frac{\partial u}{\partial t}\right) - \frac{1}{2} \frac{\partial}{\partial t}\left(\frac{\partial^2 u}{\partial x_n \partial t}\right)^2. \tag{22}$$

Then

$$\frac{\partial^2 \Delta u}{\partial t^2} \frac{\partial u}{\partial t} + A^2 \frac{\partial^2 u}{\partial x_n^2} \frac{\partial u}{\partial t} = \frac{\partial}{\partial x_n}\left(A^2 \frac{\partial u}{\partial x_n} \frac{\partial u}{\partial t}\right) + \sum_{k=1}^{n} \frac{\partial}{\partial x_k}\left(\frac{\partial^3 u}{\partial t^2 \partial x_k} \frac{\partial u}{\partial t}\right)$$
$$- \frac{1}{2} \frac{\partial}{\partial t}\left(\left(A \frac{\partial u}{\partial x_n}\right)^2 + \sum_{k=1}^{n}\left(\frac{\partial^2 u}{\partial x_k \partial t}\right)^2\right). \tag{23}$$

Let us introduce value $E_u(t)$:

$$E_u(t) = \frac{1}{2} \int\limits_W \left(\left(A \frac{\partial u}{\partial x_n}\right)^2 + \sum_{k=1}^{n}\left(\frac{\partial^2 u}{\partial x_k \partial t}\right)^2\right) dX. \tag{24}$$

It should be noted that function $E_u(t)$ can be interpreted as the energy of some system. Then we calculate the integral value (20):

$$0 = \iint\limits_{W \times [0;T]} \left(\frac{\partial}{\partial x_n}\left(A^2 \frac{\partial u}{\partial x_n} \frac{\partial u}{\partial t}\right) + \sum_{k=1}^{n} \frac{\partial}{\partial x_k}\left(\frac{\partial^3 u}{\partial t^2 \partial x_k} \frac{\partial u}{\partial t}\right)\right) dX dt - \int_0^T \left(\frac{\partial}{\partial t} E_u(t)\right) dt$$

$$= \int_0^T \int_W \left(\frac{\partial}{\partial x_n} \left(A^2 \frac{\partial u}{\partial x_n} \frac{\partial u}{\partial t} \right) + \sum_{k=1}^{n} \frac{\partial}{\partial x_k} \left(\frac{\partial^3 u}{\partial t^2 \partial x_k} \frac{\partial u}{\partial t} \right) \right) dX dt$$

$$- \frac{1}{2} \int_W \left(\left(A \frac{\partial u}{\partial x_n} \right)^2 + \sum_{k=1}^{n} \left(\frac{\partial^2 u}{\partial x_k \partial t} \right)^2 \right) \Bigg|_{t=0}^{t=T} dX$$

$$= \int_0^T \int_W \sum_{k=1}^{n-1} \frac{\partial}{\partial x_k} \left(\frac{\partial^3 u}{\partial t^2 \partial x_k} \frac{\partial u}{\partial t} \right) dX dt - E_u(T) + E_u(0)$$

$$= \int_0^T \int_{\partial W} \sum_{k=1}^{n-1} \left(\frac{\partial^3 u}{\partial t^2 \partial x_k} \frac{\partial u}{\partial t} \right) \cos(\nu, x_k) dS dt - E_u(T) + E_u(0) = -E_u(T) + E_u(0).$$

$$E_u(T) = E_u(0). \tag{25}$$

Thus, we found out that if $E_u(0) = 0$, then for all further time moments $E_u(T) = 0$. Now we put forward the standard reasoning for the formal proof of theorem of uniqueness. Let the problem (1), (2), (3) has two solutions: $u_1(X, t)$ and $u_2(X, t)$, besides $u_1(X, t) \neq u_2(X, t)$. Further we introduce the function

$$w(X, t) = u_1(X, t) - u_2(X, t), \tag{26}$$

which satisfies Eq. (1), boundary conditions (2) and trivial initial conditions

$$w|_{t=0} = 0, \quad w_t|_{t=0} = 0. \tag{27}$$

For w we have $E_w(0) = 0$ and by formula (25) $E_w(T) = 0$. Then, using energy estimations [7], we obtain $w(X, T) = 0$, which proves the equality of $u_1(X, T)$ and $u_2(X, T)$.

3 Conclusion

The paper discusses Dirichlet problem for Sobolev equation in cylindrical domain in four-dimensional space. Drawing on specific symmetry of the considered domain, the explicit formulas for solving this problem in the form of double Fourier series by sinuses were obtained for this domain. The theorem of this problem solution uniqueness was proved by the method of energy estimations.

References

1. T. E. Denisova. Asymptomatic behavior of the first initial boundary value problem solution for Sovolev-type equations from oscillation point. Differential equations, 48:2 (2012), 196–206
2. G. A. Sviridyuk, S. A. Zagrebina. Non-classical models of mathematical physics. Bulletin of South Ural State University. Series "Mathematical modeling and programming", 2012, issue 14, 7–18
3. J. Banasiak, N. A. Manakova, G. A. Sviridyuk. Positive solutions of Sobolev-type equations with relatively p-sectorial operator // Bulletin of South Ural State University. Series "Mathematical modeling and programming", 2020. № 2. https://cyberleninka.ru/article/n/pozitivnye-resheniya-uravneniy-sobolevskogo-tipa-s-otnositelno-p-sektorialnym-operatorom (reference date: 02.10.2022)
4. I. A. Shishmarev. On one nonlinear Sobolev-type equation. Differential equations, 41:1 (2005), 138–140; Differ. Equ., 41:1 (2005), 146–149
5. A. L. Pavlov. Cauchy problem for Sobolev-Halpern-type equations in spaces of power growth functions. Mathematical collection, 1993, volume 184, issue 11, 3–20
6. T. G. Sukacheva, A. O. Kondyukov. On class of Sobolev-type equations. Bulletin of South Ural State University. Series "Mathematical modeling and programming", 7:4 (2014), 5–21
7. L. C. Evans. Partial differential equations / Translated from English. – Novosibirsk: Tamara Rozhkovskaya, 2003. – 562 p. ISBN 5-901873-06-8
8. A. M. Romanenkov (2022). On solutions of equation of small lateral oscillations of moving web. Bulleting of Saint-Petersburg University. Mathematics. Mechanics. Astronomy, 9(2), 346–356. https://doi.org/10.21638/spbu01.2022.214

Reasons and Consequences of Sedimentation Layer Formation in Tray Part of Pipes of Waste Water Gravity Flow Networks

O. A. Prodous and D. I. Shlychkov

Abstract Reasons and consequences of sedimentation layer formation in tray part of pipes of waste water gravity flow networks are presented. The formula for calculating the given pipe diameter with internal sedimentation is developed and proposed for use. The comparison of values of hydraulic characteristics of new pipes and pipes with sedimentation in tray part is given on the particular example. The deviation percentage of hydraulic slope values in the rate range from 1.0 up to 3.06 m/s is indicated. The network operation efficiency is estimated based on the proposed technique to predict its utilization duration. It is proposed to work out normative requirements for obligatory hydrodynamic cleaning of waste water gravity flow networks.

Keywords Waste water networks with sedimentation · Hydraulic calculation · Necessity in cleaning and time period norm setting

1 Introduction

The currently applicable normative Code of Practice 32.13330.2012 "Sewage. External networks and facilities" for gravity flows in waste water networks establishes the main requirement—omission of calculated flow rate with self-cleaning rates of waste water flow. This standard fixes the calculated minimal rates of waste water flow depending on the utmost filling degree $\frac{\mathrm{H}}{d_{in}}$, Table 1, Fig. 1.

The analysis of calculated values of waste water flow rates (Table 1) at different filling degrees $\frac{\mathrm{H}}{d_{in}}$ demonstrates that the less $\frac{\mathrm{H}}{d_{in}}$ value in pipes is, the less is the value of minimal gravity flow rate $V_{\min}$ found by the following formula:

O. A. Prodous
INCO-Expert LLC, St. Petersburg, Russia
e-mail: pro@enco.su

D. I. Shlychkov (✉)
Department of Water Supply and Water Disposal, Moscow State University of Civil Engineering (National Research University), Moscow, Russia
e-mail: ShlyichkovDI@mgsu.ru

V. I. Karev (ed.), *Physical and Mathematical Modeling of Earth and Environment Processes—2022*, Springer Proceedings in Earth and Environmental Sciences,
https://doi.org/10.1007/978-3-031-25962-3_41

Table 1 Flow rate depending on filling of pipes $\frac{H}{d_{in}}$

Internal diameter of pipes, d_{in} (mm)	Flow rate V_{min}, m/s at filling $\frac{H}{d_{in}}$			
	0.6	0.7	0.75	0.8
150–250	0.7	–	–	–
300–400	–	0.8	–	–
450–500	–	–	0.9	–
600–800	–	–	1.0	–

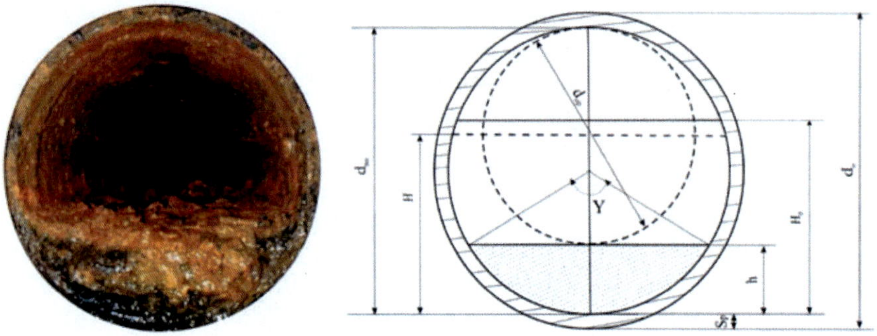

Fig. 1 Fragment of sedimentation in tray part of pipes d_{in}—pipe internal diameter, m; S_p—pipe wall thickness by GOST (State Standard) for particular material type, m; h—sedimentation layer thickness, m (mm)

$$V_{min} = \frac{4 \cdot q}{\pi \cdot \left(d_{in}^{a}\right)^{2}}, \text{ m/s} \tag{1}$$

where

q Waste liquid consumption, m^3/s;
d_{in}^{a} Actual internal diameter of pipes, m.

2 Materials and Methods

d_{in}^{a} value depends on the value of actual rate of gravity flow V_a influencing the possibility of internal sedimentation formation in tray part of pipes as shown in Fig. 1. The sedimentation is possible only on the following condition:

$$V_a \leq V_{min}, \text{ m/s} \tag{2}$$

The value of actual flow rate V_a depends on the sedimentation layer thickness h in tray part of pipes and is characterized by value d_r—reduced internal diameter relying on the sedimentation layer thickness h, found by formula [1]:

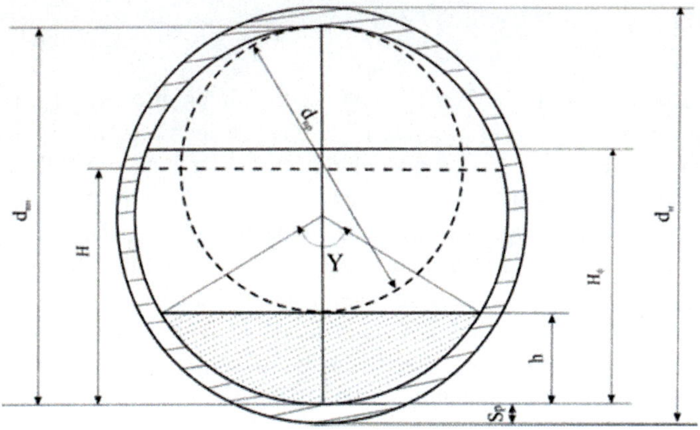

Fig. 2 Filling $\frac{H_{in}^{a}}{d_r}$ in the pipe with layer sedimentation h: d_{in}—internal diameter; d_{ex}—external diameter; S_p—pipe wall thickness; h—thickness of sedimentation layer; H_a—actual filling; H—filling degree in a new pipe; Y—angle between the chords connecting the sedimentation layer boundaries with the pipe center

$$d_r = \sqrt{\left(d_{in} - 2S_p\right)^2 - (d_{ex} - h)^2},\ \mathrm{m} \tag{3}$$

where

d_{ex}	External diameter of pipes by GOST, m;
$(d_{in} - h) = d_r$	Actual reduced internal diameter of a pipe relying on layer thickness h, m.

The reduced internal diameter of the pipe with sedimentation in the tray part—d_r corresponds to the pipe diameter with discharge section area ω (formula (4)) equaled to the free area left to let consumption q pass after the formation of sedimentation layer h in the tray part (Fig. 2).

Let us find the value of $V_a \le V_{min}$ for particular example, at which the process of sedimentation formation in tray part of pipes starts, i.e. condition (2) is violated.

$$\omega_r = \omega_{ex} - \omega_a,\ \mathrm{m}^2 \tag{4}$$

$$\frac{\pi \cdot d_r^2}{4} = \frac{\pi \cdot d_{in}^2}{4} - \frac{\pi \cdot \left(d_{in}^a\right)^2}{4},\ \mathrm{m}^2,\ \text{from which}$$

$$d_r = \sqrt{\left(d_{ex} - 2S_p\right)^2 - (d_{in} - h)^2} = \sqrt{d_{in}^2 - (d_{in} - h)^2}.$$

3 Results

The analysis of formula (3) shows that the reduced internal diameter of pipes d_r decreases during the life cycle "operation" due to the increased layer thickness h (increased filling degree $\frac{H_{in}}{d_r}$). Therefore, the accuracy of hydraulic calculation of pipes with sedimentation in tray part depends on the change in the rate of layer thickness h and waste water physical and chemical composition. The more the value of sedimentation layer thickness h is, the more is the value of filling degree in pipes $\frac{H_{sed}}{d_r}$, since with the unchanged consumption q the dependence $\frac{H_{sed}}{d_r} = f(h)$ takes place, confirming during the hydraulic calculation of pipes the need to take into account the value of layer thickness h influencing the hydraulic slope value i [1].

Let us illustrate it on a particular example.

4 Discussion

4.1 Problem Setting

The consumption $q = 150$ l/s (0.15 m^3/s) moves along the gravity flow pipeline made of concrete pipes with the diameter $d_{in} = 400$ mm (0.4 m). The waste water flow characteristics: contaminants = 350 mg/l, waste water temperature is 12 °C.

Find the reasons and describe the consequences of formation of sedimentation layer with thickness $h = 100$ mm (0.1 m) and calculate the values of characteristics of hydraulic potential of new pipes and pipes with sedimentation h for comparison.

Solution

1. We define flow rate V in a new concrete pipe without sedimentation layer h:

$$V = \frac{4 \cdot q}{\pi \cdot d_{in}^2} = \frac{4 \cdot 0.15}{3,14 \cdot 0.4^2} = 1.19,\ \text{m/s}.$$

2. The currently applicable normative Code of Practice 32.13330.2012 fixes the minimal flow movement rate for pipes 400 mm in diameter as $V_{min} = 0.8$ m/s (Table 1) that is less than the value $V = 1.19$ m/s by 32.77% or in 1.49 times.
3. This means that the norms allow the formation of sedimentation layer thickness h in tray part of pipes changing the actual flow rate mode in the rate range of 0.8 m/s $\le V_a \le$ 1.19 m/s, however, the permissible h value is not indicated.
4. We find the wet perimeter area of a new pipe—ω_n and pipe with the diameter of 400 mm with the sedimentation layer h, corresponding to ω_a:

$$\omega_n = \frac{\pi \cdot d_{in}^2}{4},\ \text{m}^2;\quad \omega_a = \frac{\pi \cdot \left(d_{in}^a\right)^2}{4},\ \text{m}^2.$$

5. Using formula (4) we calculate the value of reduced area of wet perimeter ω_r left to let consumption q pass:

$$\omega_r = \frac{3.14 \cdot 0.4^2}{4} = \frac{3.14 \cdot 0.265^2}{4} = 0.1256 - 0.0708 = 0.0548,\ \text{m}^2.$$

The remaining reduced area of wet perimeter ω_r characterized by value d_r, which is always less than ω_{ex}:

$$\omega_r = 0.0548\,\text{m}^2 < \omega_{ex} = 0.1256\,\text{m}^2,\ \text{by } 56.37\% \text{ or in } 2.29 \text{ times.}$$

Using formula (3), the reduced area of wet perimeter $\omega_r = 0.0548\ \text{m}^2$ is characterized by the value of reduced diameter d_r:

$$d_r = \sqrt{(0.511 - 2 \cdot 0.055)^2 - (0.4 - 0.1)^2} = \sqrt{0.4^2 - 0.3^2}$$
$$= \sqrt{0.16 - 0.09} = \sqrt{0.07} = 0.265$$
$$d_{in}^a = d_{in} - d_r = 0.4 - 0.265 = 0.135\,\text{m}.$$

6. Using the value $d_r^a = 0.135$ we calculate the reduced flow rate value V_{in}^a:

$$V_r^a = \frac{4 \cdot q}{\pi \cdot \left(d_{in}^a\right)^2} = \frac{4 \cdot 0.15}{3.14 \cdot 0.135^2} = \frac{0.6}{0.0572} 10.49,\ \text{m/s}.$$

7. Then we compare the rate values in a new pipe V and pipe with sedimentation layer V_r^a at $h = 0.1$:

$V_r^a = 10.49$ m/s > $V = 1.19$ m/s by 88.66% or in 8.82 times.

This indicates that it is possible to judge about the change in the value of reduced area of wet perimeter ω_r by the change in the actual values of flow rate V_r^a. That is, there is a functional dependence:

$V_r^a = f(\omega_r)$. Then the actual filling in the pipe $\frac{H_{sed}}{d_r^a}$ for the given example is:

$$\frac{H_{sed}}{d_{in}^a} = 0.7 + h = 0.8$$

So, knowing the sedimentation layer thickness h, it is always possible to calculate the value of actual filling degree $\frac{H_{sed}}{d_r^a}$ in a pipe with sedimentation in the tray part. Figure 3 demonstrates the graph of dependence $\frac{H_{sed}}{d_r^a} = f(h)$.

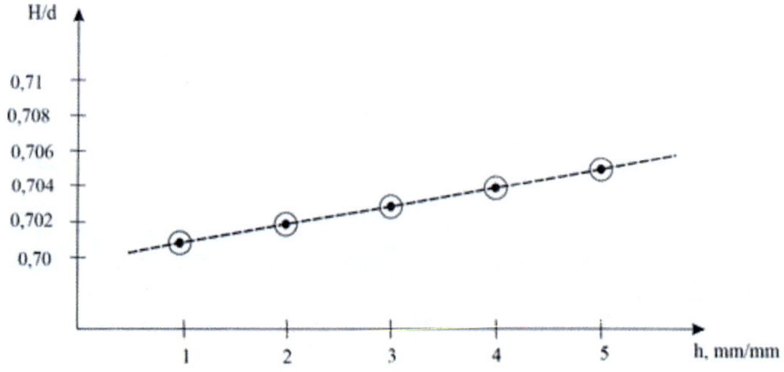

Fig. 3 Graph of dependence $\frac{H_{sed}}{d_{np}} = f(h)$

5 Conclusion

To measure the sedimentation layer thickness h in any gravity flows, the authors of the paper have developed in detail and proposed a special simple device, which allows determining h value with high accuracy [2].

For the conditions of the given example, Table 2 contains the values of hydraulic characteristics of pipes with different h values, based on which the graphs of dependence $i_r = f(h)$ and $i_{in} = f(h)$ are plotted in Fig. 4.

The analysis of values of hydraulic characteristic of pipes for the given example demonstrates that at sedimentation layer thickness $h = 0.15$ m **the network gets clogged** and waste water flow stops. At the same time, the deviation percentage of hydraulic slope values i_r and i_{ex} in the rate range from $V = 1.0 \div 3.06$ m/s is from 82.64% (at $V = 1.0$ m/s) up to 99.86% (at $V = 3.06$ m/s).

That is, the more h value is, the more is i_r value, and it is more possible that the network will be clogged.

Therefore, it is necessary to estimate the operation efficiency of waste water gravity flow network by its operation efficiency coefficient value based on the technique proposed [3]:

$$K_{ef} = \frac{i_p \cdot \left(d_{in}^p\right)^2 \cdot V_p}{i_r^a \cdot \left(d_r^a\right)^2 \cdot V_r^a}, \tag{5}$$

where

i_p, d_{ex}^p, V_p Values of calculated (for new types) from certain material type characteristics of hydraulic potential at the moment of putting the pipeline into operation;

i_r^a, d_r^a, V_r^a Reduced values of actual characteristics of hydraulic potential of pipes with sedimentation layer thickness h in the tray part.

Table 2 Hydraulic parameters of pipes in the example

Hydraulic characteristics of pipes	New pipes		
	Sedimentation layer thickness $h = 0$		
Pipe diameter, d (m)	0.4	0.4	0.4
Flow rate, V (m/s)	1.0	1.5	2.0
Hydraulic slope, i_{in} ⊝ (m/m)	$C = 52.37$	$C = 52.37$	$C = 52.37$
	0.00365	0.00547	0.01459
	Pipes with sedimentation h (m)		
	0.05	0.1	0.15
Actual diameter, d_{a} ⊛	0.35	0.30	0.250
Actual reduced diameter d_r^a ⊕	0.135	0.008	– 0.01 clogging
Actual rate, V_{a} (m/s)	1.56	2.12	3.06
Chezy factor, C ⊜	43.68	27.25	19.23
Reduced hydraulic slope, i_{r} (m/m)	0.03779	3.03	10.13
Percentage of values deviation i_{r} and i_{in}	82.64% or in 10.35 times	99.98% or in 553.93 times	99.86% or in 694.31 times

⊝ $i = \frac{4 \cdot V_{np}^2}{C_r^2 \cdot d_r^a}$, m/m

⊛ $d_r^a = \left(d_{ex} - 2S_p\right) - h$, m

⊕ $d_r = \sqrt{d^2 - (d_a - h)^2}$, m $(d - d_r) = d_r^a$

⊜ $C_{np} = \frac{\left(\frac{d_r^a}{4}\right)^{0,167}}{n}$, $n = 0.013$ is taken for calculations

The operation duration of waste water gravity flow network with sedimentation in the tray part is forecasted based on the network operation efficiency coefficient values given in Table 3 and calculated with the help of dependence (5).

For the considered example

$$K_{ef} = \frac{0.00516 \cdot 0.4^2 \cdot 1.19}{1.17 \cdot 0.135^2 \cdot 10.49} = \frac{0.000982}{0.22368} = 0.004$$

According to the data from Table 3, it is prohibited to operate the waste water pipeline with the diameter of 400 mm with the sedimentation layer thickness $h = 0.1$ m in the tray part, since the value $K_{\text{ef}} = 0.004$ confirms the need in its hydrodynamic cleaning.

Fig. 4 Graph of dependence $i_r = f(h)$ and $i_{ex} = f(h)$

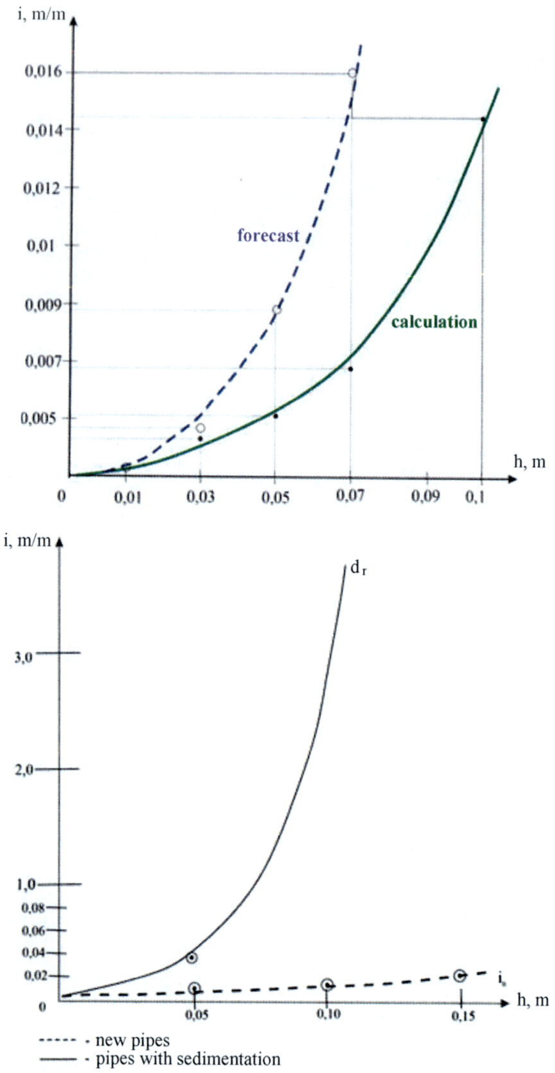

Table 3 Forecast of operation possibility of waste water network

Range of values K_{ef}	Possibility to continue further operation of the network
$0.6 \geq K_{ef} \leq 1$	Possible
$0.5 \geq K_{ef} \leq 0.6$	Hydrodynamic cleaning of the network is required
$K_{ef} \leq 0.5$	It is prohibited to operate the network Hydrodynamic cleaning of the network is required

6 Recommendations

The material presented in the paper allows indicating possible reasons contributing to sedimentation formation in the tray part of pipes and forecasting the consequences of their emergence.

The main reason of sedimentation formation in the tray part of waste water gravity flow networks is the non-uniformity of waste water flow conditioned by:

- flow rate mode when $V_a \le V_{min}$, m/s;
- influence of local resistance hampering the gravity flow (turns, offsets, branch pipes, etc.);
- influence of side connections and drops of network local points;
- line sagging and availability of large mechanical particles in the tray part, which hamper the flow (gravel, crushed stones, tree branches, etc.).

The availability of sedimentation layer h in the pipe tray part (Fig. 1) leads to the following consequences:

1. Increased pipe filling degree $\frac{H}{d_{in}}$ by the sedimentation layer thickness value h.
2. Changed actual wet perimeter area ω_a (formula (4)), and, consequently, decreased free area of the pipe remaining space to let the targeted consumption q pass, which is characterized by the reduced parameter value d_r (formula (3)).
3. Increased values of flow actual rate V_a, and, consequently, increased hydraulic slope values of i_r (loss of head per length unit)—main characteristic of pipe hydraulic potential (V, d, i).
4. Changed values of network operation efficiency coefficient K_{ef} (formula (5)), which is used for forecasting the possibility of further operation and substantiating the necessity in hydrodynamic cleaning of the network.

Thus, the material presented is the foundation for developing normative requirements for obligatory annual hydrodynamic cleaning of waste water gravity flow networks with sedimentation in the tray part of pipes and for continuing research in sedimentation layer morphology and structure depending on the waste water composition [4–16].

References

1. I.I. Vointseva, M.G. Novikov, O.A. Prodous. Prolongation of operation period of pipelines of water supply systems from concrete and cast-iron pipes // Scientific and technical journal “Engineering Systems “AVOK – North-West”, № 1, 2019. - P. 44–47.
2. I.I. Vointseva, T.Yu. Nizhnik, T.V. Strikalenko, A.I. Baranova. Anticorrosion properties of disinfecting reagents based on polyhexamethylene guanidine hydrochloride // Water: chemistry and ecology. 2018. № 10–12 (117). P. 99–108.
3. O.A. Prodous, M.G. Novikov, G.A. Sambursky, A.A. Shipilov, L.D. Terekhov, P.P. Yakubchik, V.A. Chesnokov. Recommendations on reconstruction of used metal water supply pipelines

from steel and grey cast-iron // Publishing house: LLC "Svoe Izdatelstvo", Saint-Petersburg – Moscow. 2021. – 40 p.

4. Ya.I. Blyashko. Use of sewage effluents of water channels of pressure and gravity flow water ducts and discharges of thermal power plants to produce electric energy // Glavny energetik. 2020. № 6. P. 35–44.
5. O.A. Prodous. Dependence of research duration of metal water supply pipelines on thickness of sedimentation layer on internal surface of pipes // Proceedings of XV International scientific and technical conference "Yakovlev's Readings". March 19, 2020. Moscow: Publishing house of MISI-MGSU. 2020. – P. 113–117.
6. O.A. Prodous. Methodology for estimating operation duration of metal pipelines of water supply and waste water systems // Water purification. Water preparation. Water supply. 2021. № 1 (157). P. 4–10.
7. O.A. Prodous, M.G. Novikov, G.A. Sambursky, A.A. Shipilov, L.D. Terekhov, P.P. Yakubchik, V.A. Chesnokov. Recommendations on reconstruction of used metal water supply pipelines from steel and grey cast-iron // Publishing house: LLC "Svoe Izdatelstvo", Saint-Petersburg – Moscow. 2021. – 36 p., illustrated.
8. O.A. Prodous, D.I. Shlychkov. Comparative analysis of calculated dependencies for hydraulic calculation of waste water gravity flow networks // University News. Investments. Construction. Real estate. 2021. V. 11. № 3. – P. 462–469 https://doi.org/10.21285/2227-2917-2021-3-462-469.
9. O.A. Prodous, D.I. Shlychkov. On changes in values of hydraulic characteristics of pressure sewage collectors from steel and cast-iron pipes with internal sedimentation // University News. Construction. 2020. № 12 (744). P. 70–77.
10. V.A. Orlov. Energy saving as a result of reconstruction of water supply networks by trenchless methods // In collection: Urgent problems of construction branch and education. Proceedings of I National Conference. 2020. P. 866–870.
11. R.V. Chupin. Optimization of developing waste water systems. Monograph. Irkutsk: Publishing house of Irkutsk State Technical University. 2015. – 418 p.
12. N.F. Fedorov, L.E. Volkov. Hydraulic calculation of sewage networks // Publishing house "Stroiizdat", 4th edition, revised. Leningrad. 1968. – 252 p.
13. F.A. Shevelev, A.F. Shevelev. Tables for hydraulic calculation of water supply pipes. Reference book // Publishing house "Bastet". Moscow, 2014. - 382 p.
14. D.I. Shlychkov. Problems of technical condition of functioning water supply systems // Innovations and investments. 2020. № 4. P. 207–210.
15. D.I. Shlychkov. Energy-saving as an integral part of technical and economic efficiency // Opcion. 2019. V. 35. № Special Edition 24. P. 1626–1636.
16. Carsten Ulrich Schwermera, Wolfgang Uhlabc. Calculating expected effects of treatment effectivity and river flow rates on the contribution of WWTP effluent to the ARG load of a receiving river // Journal of Environmental Management. 2021 № 288 P. 2-14.

The Influence of Strong Nonequilibrium on Multifractal Scaling of Two-Dimensional Distributions of Seismic Energy

I. R. Stakhovsky

Abstract Spatial distributions of seismic energy (seismic energy fields) in the vicinity of the epicenter of the October 30, 2020 Aegean Sea Earthquake ($M \approx 7$) have been investigated by the multifractal analysis methods. It is shown that the singularity spectrum of the seismic energy field before the main shock of this event had undergone significant widening and asymmetry changing. From the physical point of view, these changes can be explained by seismogenerating system transition into strongly nonequilibrium state before the main shock. Quantitative characteristics of the revealed effects which can be used in the methods of seismogenerating medium current state monitoring are proposed.

Keywords Earthquakes · Seismic energy · Multifractal · $f(a)$-spectrum asymmetry · $f(a)$-spectrum widening

1 Introduction

Dynamic system evolution from the equilibrium state to destruction includes the transitions into weakly nonequilibrium, and then—strongly nonequilibrium states. According to the theory of nonequilibrium systems [1], these states differ by the character of fluctuations: in equilibrium state fluctuations are absent, in weakly nonequilibrium state fluctuations emerge but die out, in strongly nonequilibrium state fluctuations grow, while the emergence of infinite fluctuation is equivalent to destruction. The indicated transitions are accompanied by radical structural reconstructions of the system, therefore in most cases for their detection the structural analysis of the system is effective, including multifractal one [2–4].

In seismic systems the seismic wave focal point, located in the lithosphere massif, is inaccessible for direct investigation, therefore the main means of receiving direct information about processes taking place in the preparing source of large-scale earthquake is the analysis of small-scale seismicity of its focal area. It is known that

I. R. Stakhovsky (✉)
Schmidt Institute of Physics of the Earth of Russian Academy of Sciences, B. Gruzinskaya St., 10, 123242 Moscow, Russia
e-mail: community.sp@ya.ru

V. I. Karev (ed.), *Physical and Mathematical Modeling of Earth and Environment Processes—2022*, Springer Proceedings in Earth and Environmental Sciences,
https://doi.org/10.1007/978-3-031-25962-3_42

spatial distributions of seismic epicenters are self-similar and can be modeled by multifractal measures [5–7]. Thus, the indicators of seismogenerating system transition into strongly nonequilibrium state should be sought in variations of multifractal structure of spatial distributions of seismic epicenters and fluctuations of seismic activity. In [8, 9] such indicator was found and called "the effect of seismic field $f(a)$-spectrum widening" (it is described further). The term "seismic field" is used here for brief designation of multifractal measure, modeling spatial distribution of earthquake epicenters.

The effect of seismic field $f(a)$-spectrum widening precedes the main shock of strong earthquake and can be considered as a method of monitoring the current state of seismogenerating medium, however seismic data for its accurate determination have to be highly representative. In the analysis of seismic field, the lack of data leads to artificial scaling break, so this fact actualizes the search for methods of decreasing the dependence of analysis results on the lack of data.

It is further demonstrated that the period of seismogenerating system transition into strongly nonequilibrium state can be also revealed under conditions of data shortage by replacing the analysis of spatial distribution of seismic epicenters with the analysis of spatial distribution of seismic energy. For brevity the multifractal measures modeling spatial distribution of seismic energy will be designated below as "seismic energy fields". The analysis of seismic energy field provides additional advantages when it is necessary to differentiate the seismogenerating system, caught up by the process of strong earthquake preparation, from the system, in which such preparation has not started yet.

2 Seismic and Seismic Energy Fields

In the work [8] it was found that during the last $\sim$ 2.5 years before main shocks of strong earthquakes in their focal areas the spatial distributions of epicenters of weak seismicity have demonstrated the fluctuations of seismic activity, i.e., local short-term peaks or attenuations of seismic events generation. Due to high intermittency peculiar to seismic process in general, these fluctuations cannot be singled out as an independent research object, however in multifractal measures modeling the distribution of seismic epicenters, the fluctuations of seismic activity appear in cumulative form as the increase of seismic field $f(a)$-spectrum width (i.e., as the increased difference between the maximum and minimum values of seismic field singularity indices) in comparison with the earlier field. It should be noted that according to the theory of nonequilibrium systems [1] the described result can be considered as theoretically expected.

This effect allows to differentiate the steady seismic regime, i.e., weakly nonequilibrium state of seismogenerating system, from the transient regime, i.e., strongly nonequilibrium state preceding the global destruction. In the steady regime, multiple but "weak" (in the scale of the system investigated) events do not destruct the system global integrity, do not change boundary conditions determining the system evolution, and do not lead to changes in its dynamics. In the transient regime, all the considered system is caught up by the process of event preparation, which will further result in the system destruction, i.e., by the process of preparing "strong" (in the scale of the system investigated) earthquake. In the steady regime, the singularity spectra of seismic fields are subject only to weak variations; in the transient regime, before the global destruction the singularity spectra significantly widen.

The investigation of the effect of seismic field $f(a)$-spectra widening requires large data volumes. Theoretical consideration of the issue of constructing $f(a)$-spectrum for the distribution of single-type events leads to the estimations of the required data volumes, which predominantly exceed the capabilities of modern seismic catalogs [10]. At the same time, there is a firmly established power-law dependence between seismic energy and number of events inducing this energy (Gutenberg-Richter law) [11]. Thus, if t measures modeling spatial distributions of seismic epicenters, i.e., seismic fields, are multifracls, then the measures modeling spatial distributions of seismic energy are also subjected to multifractal organization.

After covering spatial distributions of seismic energy by scale grid which is ordinary used in multifractal analysis, the total energy of seismic events within the grid box due to Gutenberg-Richter law is determined not by the number of events but by the energy of several events with maximum magnitude values. The rest of the events, irrespectively of their quantity, add only minor corrections to the total sum. In the process of constructing $f(a)$-spectra of seismic energy field this fact qualitatively decreases the dependence of calculation accuracy on the number of events available for research.

In accordance with the physics of the process, the variations of interconnected seismic and seismic energy fields have to be synchronous. Besides, the overall numbers of events in the grid boxes usually differ on the order of magnitude by not more than one-two orders when overall values of seismic energy of these events can differ by four-five orders and more, what may be considered as sort of multiplier of the variations under investigation. Due to these peculiarities of seismic energy fields, their $f(a)$-spectra can be constructed rather accurately even under conditions of lack of data, which makes impossible to construct full $f(a)$-spectra of seismic fields. Thereby, the seismic process variations, connected with the transition of seismogenerating system into strongly nonequilibrium state preceding the main shock of strong earthquake, can be more likely detected in seismic energy fields than in seismic ones. This is further demonstrated by the analysis of small-scale seismicity preceding the October 30, 2020 Aegean Sea Earthquake, which has become one of the strongest earthquakes in recent years in Europe.

3 Initial Data

The Aegean Sea Earthquake occurred on 30.10.2020 at 11:51:27 (UTC) in the region of a so-called Hellenic Arc, i.e., in the collision zone of African and Anatolian lithospheric plates. Its magnitude was $M \cong 7.0$, hypocenter depth—$H \approx 20$ km. The epicenter proved to be in the sea strait between the Greek island Samos and western coast of Turkey, and had the coordinates 26.78° N, 37.89° E. The earthquake was accompanied by numerous victims and destructions of engineering constructions [12], as well as tsunami [13].

In this research the seismicity preceding the October 30, 2020 Aegean Sea Earthquake was studied within the 100×100 km polygon with sublatitudinal and submeridional orientation of sides and center coinciding with the epicenter of this event. The polygon territory is directly controlled by two seismic networks: Hellenic Unified Seismological Network [14] and Turkish National Seismic Network [15]. Several seismic stations of both networks are located on the polygon territory. Both networks transmit data to International Seismological Centre, ISC in the UK [16] where they (together with data of other seismic networks) are systemized in ISC-catalog [17].

The data from ISC-catalog about the seismic events, preceding the October 30, 2020 Aegean Sea Earthquake, for the period 10.05.2013 ÷ 30.10.2020 within the boundaries of the polygon under investigation were used in the calculations. Within this period the polygon territory was characterized by weak seismicity with the maximum magnitude of events $M = 4.7$ and depths of hypocenters in the range $H < 25$ km (with singular exceptions). The checks run demonstrated the compliance of ISC-catalog data within the polygon borders with the earthquake recurrence law (Gutenberg-Richter law) in the magnitude range $1.5 \leq M \leq 4.7$, the data on weaker events were not used in the calculations as unrepresentative. The total volume of ISC-catalog data with the indicated parameters amounted to 3428 events.

4 Calculation Technique

Magnitudes of seismic events systematized in catalogs are connected with the energy of seismic waves by empirical dependencies, among which the Bath formula is one of the most popular [18]:

$$\lg E = 5.24 + 1.44M, \tag{1}$$

where E—seismic energy (J), M—magnitude by surface waves. A number of alternative formulas of magnitude connection with energy are known, which result in somewhat different energy absolute values, however, only the parameters of energy scaling independent of energy absolute values will be calculated further.

In the course of calculations, the investigated polygon (investigated system) was covered by two-dimensional renormalizing scale grid of non-overlapping square boxes. The grid boxes were enumerated by index $i\,(i = 1, 2, 3\ldots)$. The spatial distribution of seismic energy was modeled by measure P whose content in the i th box p_i was estimated with the help of normalization:

$$p_i = E_i / E_0, \tag{2}$$

where E_i—total value of seismic energy in the i th box, i.e., the sum of seismic energies determined according to expression (1) for all events in the i th box, E_0—total value of seismic energy of all events in the sample, i—box sequence index. Let us construct the partition function:

$$Z_q(r) = \sum_{i=1}^{N} p_i^q(r), \quad q \in \{-\infty, +\infty\}, \tag{3}$$

where q—measure moment order, r—grid box size (scale), N—total number of nonempty boxes of the grid. The following correlation is true for self-similar measures [2]:

$$\sum_{i=1}^{N} p_i^q(r) \propto r^{-\tau(q)}, \tag{4}$$

where $\tau(q)$—cumulant-generating function:

$$\tau(q) = \lim_{r\to 0}\left[\ln \sum_{i\to 0}^{N} p_i^q(r) / \ln(1/r)\right]. \tag{5}$$

The singularity spectrum of the investigated measure can be constructed with the help of Legendre transformation:

$$a = -\frac{\mathrm{d}}{\mathrm{d}q}\tau(q), \tag{6}$$

$$f(a) = aq + \tau(q). \tag{7}$$

where a—singularity index, $f(a)$—singularity spectrum, however, in the conditions of data deficiency the smoothing procedures of Legendre transformation are undesirable. Let us differentiate the function (3):

$$\frac{\mathrm{d}}{\mathrm{d}q}\left(\sum_{i=1}^{N} p_i^q(r)\right) = \sum_{i=1}^{N} p_i^q(r) \ln p_i(r). \tag{8}$$

Combining the expressions (5)–(8) and using (4), after some transformations we can obtain the following equations [19]:

$$a(q) = \lim_{r\to 0} (\ln r)^{-1} \sum_{i=1}^{N} \tilde{p}_i(q, r) \ln p_i(r), \tag{9}$$

$$f(a(q)) = \lim_{r\to 0}(\ln r)^{-1}\sum_{i=1}^{N}\tilde{p}_i(q,r)\ln\tilde{p}_i(q,r), \tag{10}$$

where

$$\tilde{p}_i(q,r) = p_i^q(r)/\sum_j p_j^q(r). \tag{11}$$

In this paper $f(a)$-spectra of seismic energy fields were constructed using Eqs. (9) and (10). Equations (9) and (10) allow to construct $f(a)$-spectrum of multifractal measure P without intermediary procedures of Legendre transformation. Limits in (9) and (10) were estimated as coefficients of linear regressions $\sum_{i=1}^{N}\tilde{p}_i(q,r)\ln p_i(r)$ and $\sum_{i=1}^{N}\tilde{p}_i(q,r)\ln\tilde{p}_i(q,r)$ versus $\ln r$. Scale changes r ($r = r_1, r_2, r_3\ldots$) were realized by grid renormalizing. In numerical calculations q values were sorted out discretely with step $\Delta q = 0.1$ in the range of values $-20 \le q \le 20$.

5 Calculation Results

The ISC-catalog data used in the calculations were split into three samples: the 1st sample covered the time period from 10.05.2013 to 31.07.2015 and contained 1142 events, the 2nd sample covered the time period from 01.08.2015 to 31.12.2017 and contained 1133 events, the 3rd sample covered the time period from 01.01.2018 to 30.10.2020 and contained 1153 events. The third sample characterized the seismic process directly preceding the main shock of the October 30, 2020 Aegean Sea Earthquake (the Aegean Sea Earthquake itself was not included into the sample). From physical considerations connected with empirically estimated sizes and time of forming the preparation areas of earthquakes with magnitude $M = 7$, the volume of the 3rd sample could not be increased by extending the polygon sizes or time period duration covered by the sample, also the magnitude range was limited by the representativeness threshold of catalog. The volumes of the 1st and 2nd samples were selected approximately equal the third sample volume to avoid errors in calculations caused by different data representativity.

These data are not enough to construct full $f(a)$-spectra of seismic fields, besides, the measure content in scale grid boxes is principally limited by the impossibility for fractional amounts of events less than one to be placed in the boxes. Due to the scaling break, even incomplete right branches of $f(a)$-spectra of seismic fields prove to be distorted. In order to track changes in the seismogenerating system state, let us refer to the analysis of seismic energy fields.

Figure 1 demonstrates $f(a)$-spectra of seismic energy fields constructed with the data of the 1st and 2nd samples. In this case, we managed to construct relatively full $f(a)$-spectra. Let us consider numerical characteristics of $f(a)$-spectra:

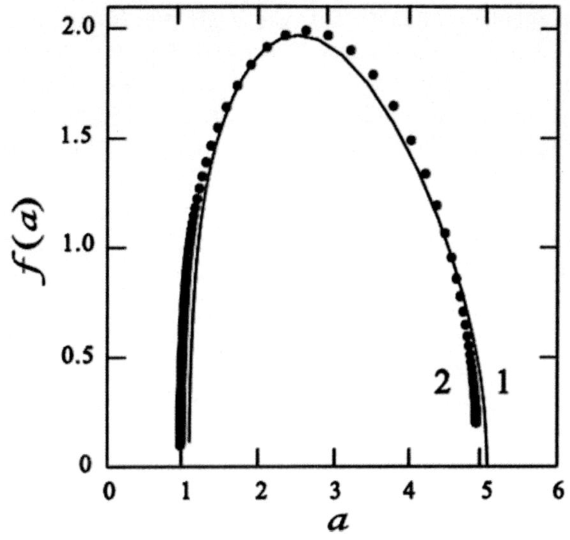

Fig. 1 Singularity spectra of seismic energy fields based on the data of the 1st and 2nd samples. The solid line (designated with 1)—$f(a)$-spectrum constructed by the data of the first sample, dots (number 2)—$f(a)$-spectrum constructed by the data of the second sample

$$\Delta a = a_{\max} - a_{\min}, \tag{12}$$

where Δa—expansion (width) of $f(a)$-spectrum, $a_{\max}$ and $a_{\min}$—maximum and minimum values of singularity indices of the investigated fields, respectively, and

$$\mathbb{A} = \frac{a_{\max} - 2a_{ex} + a_{\min}}{a_{\max} - a_{\min}}, \tag{13}$$

where $\mathbb{A}$—parameter of spectrum asymmetry, a_{ex}—singularity index value corresponding to the position of $f(a)$-spectrum extreme point. The parameter of asymmetry $\mathbb{A}$ determined in such a way can have the values in the range $-1 < \mathbb{A} < 1$. The signature of $\mathbb{A}$ indicates the shifting direction of the spectrum extreme point relatively to the center of range Δa. For a symmetric spectrum $\mathbb{A} = 0$. Subscripts at parameters $\mathbb{A}$, Δa and $f(a)$ will be further used to identify them with the corresponding data samples.

Sets of events in the 1st and 2nd samples have no intersections, however, the values of extreme points of $f(a)$-spectra (monofractal dimensions of seismic energy fields), as shown in Fig. 1, are approximately the same $f(a_{ex})_1 \approx f(a_{ex})_2$. The expansions of $f(a)$-spectra during this period were $\Delta a_1 = 3.97\ldots$, $\Delta a_2 = 3.94\ldots$, i.e., they practically did not change either. The ranges of singularity indices for both fields are quite similar. The seismic process took place rather quietly during this period and $f(a)$-spectra of seismic energy fields do not demonstrate the growth of fluctuations of seismic activity.

However, $f(a)_1$ and $f(a)_2$ spectra are not identical and are characterized by different curvature of branches and different asymmetry degree. Both spectra in Fig. 1 are asymmetric, their extreme points are shifted to the side of lower values of singularity indices. This physically means that during the whole period of time covered by both samples, from 10.05.2013 to 31.12.2017 the seismic flux in the investigated area consisted of events with low magnitude values ($M \leq 3$), events with higher magnitude values ($M > 3$) occurred rarely. For the spectrum constructed by the data of the first sample $\mathbb{A}_1 = 0.27\ldots$, for the spectrum constructed by the data of the second sample $\mathbb{A}_2 = 0.15\ldots$.

Comparison of $f(a)$-spectra of seismic energy fields by their expansion values Δa_1 and Δa_2 gives the ground for conclusion that in the period from 10.05.2013 to 31.12.2017 the seismic process in the investigated polygon proceeded in the steady regime. The comparison of $f(a)$-spectra by their asymmetry parameters $\mathbb{A}_1$ and $\mathbb{A}_2$ demonstrates that the seismic process during this period also comprised internal structural dynamics.

Figure 2 demonstrates $f(a)$-spectra of seismic energy fields constructed by the data of 1st and 3rd samples. As seen from Fig. 2, in the time period from 01.01.2018 to 30.10.2020 the seismic process character changed significantly.

The changes refer to the expansion of $f(a)$-spectrum of seismic energy field directly preceding the October 30, 2020 Aegean Sea Earthquake, as well as to its asymmetry. The values of $f(a)$-spectra extreme points in Fig. 2 still differ insignificantly $f(a_{ex})_1 \approx f(a_{ex})_3$. However, the expansion of $f(a)_3$ spectrum is $\Delta a_3 = 5.05\ldots$, which exceeds Δa_1 value by 27%. Figure 2 demonstrates the effect of multifractal field $f(a)$-spectrum widening but detected already not in the seismic but seismic energy field. At the same time, the position of $f(a_{ex})_3$ spectrum extreme point shifted relatively to the center of Δa_3 range to the side of high values of singularity indices, i.e., the share of events with magnitudes $M > 3$ in the sample increased. Parameters $\mathbb{A}_1$ and $\mathbb{A}_3$ of spectra $f(a)_1$ and $f(a)_3$ have already different signatures, $\mathbb{A}_3 = -0.07$.

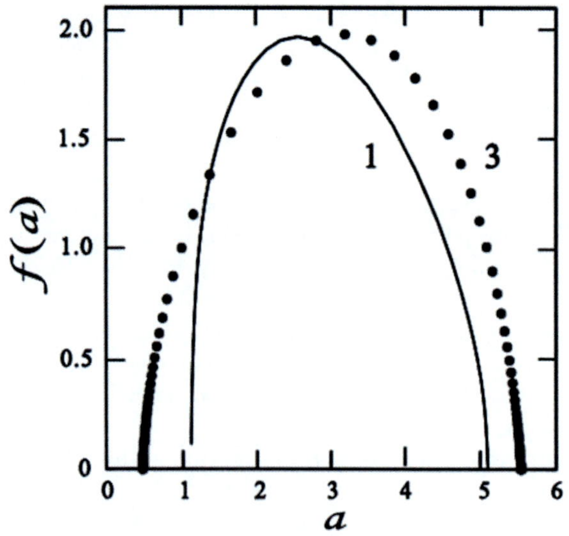

Fig. 2 Singularity spectra of seismic energy fields based on the data of the 1st and 3rd samples. The solid line (designated with 1)—$f(a)$-spectrum constructed by the data of the first sample, dots (number 3)—$f(a)$-spectrum constructed by the data of the third sample

Thus, in the time period from 01.01.2018 to 30.10.2020 the seismic process ceased being quiet, the steady seismic regime was replaced by the transient one. The peaks and attenuations of seismic activity resulted both in the growth of maximum values and decrease in the minimum values of singularity indices of seismic energy field, what is demonstrated by the expansion of its $f(a)$-spectrum. The relative increase of strong energy fluctuations, i.e., events with increased magnitude in seismic flux resulted in the signature inversion of asymmetry parameter $\mathbb{A}$. The seismogenerating system turned into the strongly nonequilibrium state. The seismic process in the transient regime naturally resulted in the strong earthquake.

At the same time, it should be pointed out that all seismicity preceding the October 30, 2020 Aegean Sea Earthquake can be referred as so weak that neither foreshocks nor any other physical phenomena able to contain the prognostic information were revealed before this event. Nevertheless, the prognostic information was contained in the scaling structure variations of seismic energy fields, i.e., the investigation of these variations can be considered as a tool for analyzing a seismic process fine structure, which cannot be currently analyzed by other methods yet.

In [8, 20] it is proposed to quantitatively estimate the effect of $f(a)$-spectra widening with the help of functional δS:

$$\delta S = S^{**} - S^{*} = \int_{a^{**}_{\min}}^{a^{**}_{\max}} f^{**}(a)\mathrm{d}a - \int_{a^{*}_{\min}}^{a^{*}_{\max}} f^{*}(a)\mathrm{d}a, \quad (14)$$

where S-area of coordinate space enveloped by the spectrum (and another characteristic of the spectrum expansion), the superscripts in the form of two asterisks refer to the spectrum of seismic energy field preceding a strong earthquake, the superscripts in the form of one asterisk refer to the spectrum of earlier seismic energy field. Functions $f(a)$ in the integration area are continuous, smooth and limited, i.e., widening parameter δS can be always estimated numerically. In this case, by spectra $f(a)_1$ and $f(a)_3$ it is $\delta S = 1.45\ldots$ (by spectra $f(a)_2$ and $f(a)_3$ it is $\delta S = 1.25\ldots$). The functional δS is convenient for quantitative estimation of the effect, first of all, because it takes into account not only the expansion but also the shape of $f(a)$-spectra.

As we can see, the effect of $f(a)$-spectrum widening can be considered as a quantitative characteristic of the preparation process of a strong earthquake. However, at the same time, the investigation of seismic energy fields also provides a qualitatively new opportunity in comparison with seismic fields—analysis of $f(a)$-spectra asymmetry.

6 Discussion and Conclusions

Detection of the strong nonequilibrium period preceding a strong earthquake, requires the seismic process parameterization and reduction of its complex physical manifestations to quantitative parameters allowing the comparison of preparation processes of different seismic events. Such parameters also include quantitative characteristics of the effect of seismic energy field $f(a)$-spectra widening, many aspects of which (though not all) can be discussed and interpreted today quite substantively.

The results of this investigation demonstrate that monofractal dimensions of seismic energy fields are insensitive to the preparation of critical transition (destruction) of seismogenerating system, i.e., the sets of points of energy dissipation in the lithosphere are mainly stationary. The dissipation points (weakened points of the lithosphere, fault nodes, focal centers) induce seismic events repeatedly not migrating in space. In other words, seismic energy fields (the same way as seismic) are multifractal fields with localized singularities. This aspect has the principal meaning in the mathematical simulation of a seismic process, which is often performed with the help of multiplicative cascade procedures and, consequently, requires their adequate selection.

The effect of seismic energy field $f(a)$-spectrum widening is determined by the behavior of higher moments of the multifractal measure, modeling spatial distribution of seismic energy. Physically, this means that the multifractal structure of seismic energy spatial distribution is formed due to energy activity of localized dissipation points in the seismogenerating system. Changes in the system state, i.e., the transition from the weakly nonequilibrium state into the strongly nonequilibrium one, are reflected in the energy activity of the dissipation points. During the system transition into the strongly nonequilibrium state the energy activity of the dissipation points is revealed in the growth of fluctuations, at the same time the total amount of energy induced by the seismic flux can increase not very significantly. The effect of seismic energy field $f(a)$-spectrum widening in the integral form combines many details of quite complex evolution process of seismogenerating system in strongly nonequilibrium state, however the investigation of these details requires higher quality of seismic data.

Nevertheless, the possibility to describe the effect by scalar characteristics (for example, δS, though alternative numerical characteristics of the effect can be proposed) allow to consider the seismic scaling as the seismogenerating system macroparameter, providing an opportunity, in a certain sense, to measure the process of system forthcoming to the moment of global stability loss. In this context, not only the numerical value of parameter δS can be useful but also the change of signature of parameter $\mathbb{A}$.

As demonstrated above, the change of seismic regimes can be accompanied by the change of asymmetry of seismic energy field $f(a)$-spectrum. The asymmetry of $f(a)$-spectrum of seismic energy field is connected with the lack of strong energy fluctuations in the steady regime and growth of fluctuation range in the transient regime, thus resulting in the inversion of parameter $\mathbb{A}$ signature. The time moment characterized by the value $\mathbb{A} = 0$, i.e., by symmetry of seismic energy field $f(a)$-spectrum, can be considered as some reference moment of earthquake preparation process. Previously, when analyzing seismic fields, it was impossible to reveal the asymmetry of $f(a)$-spectra due to their incompleteness.

Thus, the effect of seismic energy field $f(a)$-spectrum widening even with the current representativity of seismic data can be used both for the detection of the fact of preparation processes of strong earthquakes in seismogenerating medium and also for the quantitative description of these processes. Nevertheless, there is a number of questions connected with details of this effect, which can be answered only after increasing the representativity of seismic data, i.e., this effect requires further investigation.

References

1. Prigogine I. From being to becoming. San Francisco. W.H. Freeman and Co. 1980. 200 P.
2. Mandelbrot B. Multifractal measures, especially for geophysicist // PAGEOPH. 1989. V. 131. № 1–2. P. 5–42. https://link.springer.com/journal/24/volumes-and-issues/131-1.
3. Takens F., Verbitski E. Multifractal analysis of dimensions and entropies// Regular and chaotic dynamics. 2000. 5(4). P. 361–382. https://doi.org/10.1070/RD2000v005n04ABEH000154.
4. Fractal Analysis and Chaos in Geosciences (Ed. by Ouadfeul S.). 2012. London: IntechOpen. 184 p. https://doi.org/10.5772/3309.
5. Geilikman M.B., Golubeva T.V., Pisarenko V.F. Multifractal patterns of seismicity // Earth and Planetary Science Letters. 1990. V. 99. № 1/2. P. 127–132.
6. Hooge C., Lovejoy S., Pecknold S., Malouin F., Schertzer D. Universal multifractals in seismicity // Fractals. 1994. V. 2. № 3. P. 445–449.
7. Stakhovsky I.R., Belousov T.P. Scale Invariants in Seismotectonics // Dokl. Akad. Nauk. 1996. V. 347. № 2. P. 252–255.
8. Stakhovsky I.R. Widening of the $f(a)$-Spectra of Seismic Fields in Preparation Zones of Strong Earthquakes // Izvestiya, Physics of the Solid Earth. 2002. Vol. 38. No. 2. PP. 156–160.
9. Stakhovsky I. R. Scale invariance of shallow seismicity and the prognostic signatures of earthquakes // Physics-Uspekhi. 2017. V. 60. № 5. P. 472–489. https://doi.org/10.3367/UFNe.2016.09.037970.
10. Eckmann J.P., Ruelle D. Fundamental limitations for estimating dimensions and Lyapunov exponents in dynamical systems // Physica D. 1992. V. 56. P. 185–191.
11. Gutenberg, B., Richter, C. F. Magnitude and Energy of Earthquakes // Annali di Geofisica. 1956. V. 9. № 1. P. 15–20.
12. Altunisik A.C., Atmaca B., Kartal M.E., Gunaydin M., Demir S., Ulusan A. Assessment of Structural Damage Following the October 30, 2020 Aegean Sea Earthquake and Tsunami // Journal of Earthquake and Tsunami. 2021. V. 15. N. 06. 2150029. https://doi.org/10.1142/S1793431121500299.
13. Dogan, G.G., Yalciner, A.C., Yuksel, Y., Ulutas E., Polat O., Guler I., Sahin C., Tarih A., Kanoglu U. The 30 October 2020 Aegean Sea Tsunami: Post-Event Field Survey Along Turkish Coast // Pure Appl. Geophys. 2021. 178. 785–812. https://doi.org/10.1007/s00024-021-02693-3.
14. D'Alessandro A., Papanastassiou D., Baskoutas I. Hellenic Unified Seismological Network: an evaluation of its performance through SNES method // Geophys. J. Int. 2011. 185. 1417–1430 https://doi.org/10.1111/j.1365-246X.2011.05018.x.
15. Cambaz M.D., Ozer M., Gunes Y., Ergun T., Ogutcu Z., Altuncu-Poyraz S., Koseoglu A., Turhan F., Yilmazer M., Kekovali K., Necmioglu O., Kalafat D., Cakti E., Ozener H. Evolution

of the Kandilli Observatory and Earthquake Research Institute (KOERI) Seismic Network and the Data Center Facilities as a Primary Node of EIDA // Seismological Research Letters. 2021. 92(3). P. 1571–1580. https://doi.org/10.1785/0220200367.
16. International Seismological Centre (2021), Summary of the Bulletin of the International Seismological Centre, January–June 2019, 56(I), https://doi.org/10.31905/M8L1R7WI.
17. International Seismological Centre (20XX), On-line Bulletin, https://doi.org/10.31905/D808B830.
18. Bath M. Introduction to seismology. Basel and Stuttgart. Birkhauser Verlag. 1973. 350 P.
19. Chhabra, A.B., Meneveau, C., Jensen, R.V., and Sreenivasan K.R., Direct Determination of the $f(a)$-Singularity Spectrum and Its Application to Fully Developed Turbulence // Phys. Rev. A. 1989. V. 40. № 9. pp. 5284–5294.
20. Stakhovsky I.R. Variations of seismic scaling before strong earthquakes. Processes in GeoMedia. V. 1, Springer Geology, Cham, Springer Nature (Switzerland), 2020. C. 271–277. https://doi.org/10.1007/978-3-030-38177-6_29.

Geomechanical Modeling During Solving Problems Related to Development of Hard-to-Recover Hydro-carbonate Deposits

V. I. Karev, Yu. F. Kovalenko, and K. B. Ustinov

Abstract The developed geomechanical model to describe deformation, fracture and filtration processes occurring within bottom-hole zone of production well is presented, and the results of mathematical modeling for a particular gas-condensate field are provided. The optimal constructions of the bottom-hole zones were chosen on the base of the conducted analysis. It was demonstrated, a significant increase in flow rate can be achieved by choosing the optimal bottomhole design.

Keywords Low-permeability rocks · Enhanced oil recovery · Perforation hole · Mathematical modeling · Directional unloading method

1 Introduction

Due to reduction of the part of easy-to-recover hydro-carbonate deposits, the necessity arises to develop deposits with reservoirs having low permeability characteristics and located at huge depths. Development of such deposits cause significant technological difficulties, since applying the traditional methods including hydrofracture often becomes non-effective. This leads to necessity to develop new methods of increasing wellbores productivities and methods of reducing the risks of negative effects on the bottom-hole zones. For developing such methods geomechanical modeling has to be used.

A geomechanical and filtration model of processes within the bottom-hole zone of a reservoir that accounted their mutual influence was developed and adopted to conditions the particular hydro-carbonate deposit. The model includes description of elastic deformation, transition to inelastic state, as well as description of elastoplastic deformation of reservoir rocks after transition to non-elastic state. The model is detailed described in [1–4]. A short mathematical formulation is provided below.

V. I. Karev (✉) · Yu. F. Kovalenko · K. B. Ustinov
Ishlinsky Institute for Problems in Mechanics of the Russian Academy of Sciences, Moscow, Russia
e-mail: wikarev@ipmnet.ru

V. I. Karev (ed.), *Physical and Mathematical Modeling of Earth and Environment Processes—2022*, Springer Proceedings in Earth and Environmental Sciences,
https://doi.org/10.1007/978-3-031-25962-3_43

2 Mathematical Model

The stress–strain state and fluid flow is described by the laws of poro-elasticity [5, 6] and Darcy's law. In case of inelastic (plastic) deformations the system of equations may be written as follows

$$(\kappa_{ik} p_{,i})_{,k} = 0 \tag{1}$$

$$\sigma_{ij,i} = 0 \tag{2}$$

$$s_{ij} = \sigma_{ij} + \alpha_P p \delta_{ij} \tag{3}$$

$$s_{ij} = \Lambda_{ijkl} \varepsilon_{kl}^E \tag{4}$$

$$\varepsilon_{ij}^T = \varepsilon_{ij}^E + \varepsilon_{ij}^P = \frac{1}{2}\left(u_{i,j} + u_{j,i}\right) \tag{5}$$

Here p is the pore pressure; κ_{ik} is permeability tensor generally depending on coordinate and the stress state; σ_{ij}, s_{ij} are components of the total and effective (acting on solid skeleton) stress tensors; ε_{ij}^T, ε_{ij}^E, ε_{ij}^P are components of the total, elastic and plastic strain tensors; u_i are components of displacement vector; Λ_{ijkl} are components of elasticity tensor; $0 \le \alpha_P \le 1$ is Bio's coefficient, characterizing the porosity structure (note, that more precise consideration requires to consider Bio's coefficient as a second rank tensor). Summation is supposed for repeating indexes, indexes after the comma stand for derivative over the corresponding coordinate.

Before fulfilling the criterion of elastic–plastic transition $\varepsilon_{ij}^P = 0$, and Eqs. (1)–(5) accompanied by the boundary conditions form the closed system. Inelastic deformation is described by using a non-associated law of plastic flow with yield function F and plastic potential Q written in the form of a modified Hill's [7, 8] law for anisotropic plasticity in the form of [9], respectively (see also [1])

$$d\varepsilon_{ij}^P = \frac{\partial Q}{\partial s_{ij}} \frac{\partial F}{\partial s_{km}} ds_{km} \left(H \frac{\partial Q}{\partial s_{pq}} s_{pq} \right)^{-1}, \tag{6}$$

$$\begin{aligned} F = {} & \left(G_{(23)}^0 (s_{22} - s_{33})^2 + G_{(13)}^0 (s_{11} - s_{33})^2 + G_{(12)}^0 (s_{11} - s_{22})^2 + 2L_{(23)}^0 s_{23}^2 \right. \\ & \left. + 2L_{(13)}^0 s_{31}^2 + 2L_{(12)}^0 s_{12}^2 \right)^{1/2} + \left(B_{(1)}^0 s_{11} + B_{(2)}^0 s_{22} + B_{(3)}^0 s_{33} \right) - A(k) \end{aligned} \tag{7}$$

$$\begin{aligned} Q = {} & \left(G_{(23)}^0 (s_{22} - s_{33})^2 + G_{(13)}^0 (s_{11} - s_{33})^2 + G_{(12)}^0 (s_{11} - s_{22})^2 + 2L_{(23)}^0 s_{23}^2 \right. \\ & \left. + 2L_{(13)}^0 s_{31}^2 + 2L_{(12)}^0 s_{12}^2 \right)^{1/2} + \left(B_{(1)}^1 s_{11} + B_{(2)}^1 s_{22} + B_{(3)}^1 s_{33} \right) \end{aligned} \tag{8}$$

$$dk = s_{ij} d\varepsilon_{ij}^{p} \tag{9}$$

Here $H \equiv -\partial F/\partial k = \partial A/\partial k$ is a material characteristic, obtained from experiments. At first approximation it may be supposed to be a constant $H = E_p^{-1}$, where E_p is the plasticity modulus; components of effective stress tensor s_{ij} are written in coordinate frame, related to the material axes of isotropy; $G^0_{(ij)}$, $L^0_{(ij)}$, $B^0_{(i)}$, $B^1_{(i)}$ are material constants; k is the parameter of hardening, related to the work of plastic strains (isotropic hardening). The above introduced yield function F and plastic potential Q differ only by parts related to normal stresses governed by constants $B^0_{(i)}$, $B^1_{(i)}$. Therefore, the "non-assotiativity" covers only volumetric part of the law, which may be considered in accordance with the original concept of the origin of volumetric inelastic strains related to the intensity of shear strains (dilatancy) proposed by Reynolds [10]. For isotropic case the model is reduced to Drucker-Prager model.

$$F = \sqrt{\frac{1}{6}\left((s_2 - s_3)^2 + (s_1 - s_3)^2 + (s_1 - s_2)^2\right)} + \alpha(s_1 + s_2 + s_3) - A(k) = 0,$$
$$A(0) = \tau_s \tag{10}$$

$$Q = \sqrt{\frac{1}{6}\left((s_2 - s_3)^2 + (s_1 - s_3)^2 + (s_1 - s_2)^2\right)} + \alpha'(s_1 + s_2 + s_3) \tag{11}$$

Here s_1, s_2, s_3 are the principle effective stresses; α, α', τ_s are constants of Drucker-Prager model. Equations (1)–(9) together with the boundary conditions form a closed system. The above equations were implemented into a FEM code.

3 Instrumentals and Some Experimental Results

All experiments were carried out using the Triaxial Independent Load Test System (TILTS) developed in the IPMech RAS [1, 11]. The tests were made on cubic specimens with edge of 40 mm. The loading programs corresponded to the stress state occurred within the well bottom-hole zone. The loading program and dependence of permeability for a typical sample of studied rock are depicted on Fig. 1a; the deformation curves are depicted on Fig. 1b.

The samples revealed no pronounced anisotropy of elastic, plastic and filtration properties, therefore the isotropic variant of the model were used. The constants obtained in experiments (averaged over the set) and used in modeling are $E = 3.25 \times 10^4$ MPa; $\nu = 0.28$; $\alpha_P = 0.8$; $\tau_s = 15$ MPa; $\alpha = 0.24$; $\alpha' = 0$ (no pronounced dilatancy was observed); $E_p = E = 5 \times 10^2$ MPa. The dependence of permeability on stresses was taken as a piecewise function corresponding to that shown in Fig. 1a.

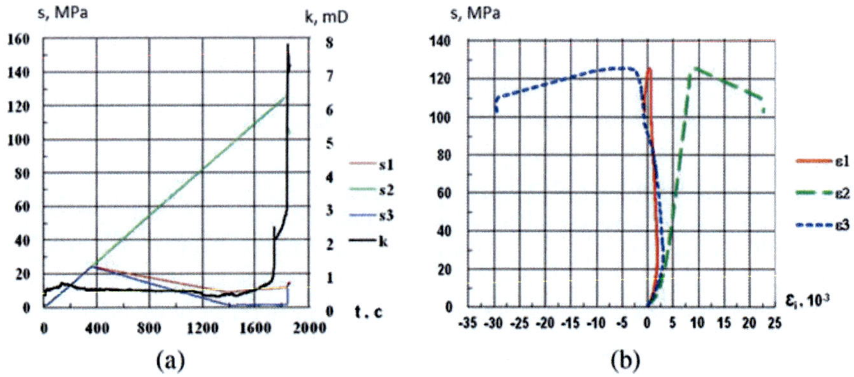

Fig. 1 **a** Loading program and permeability curve; **b** stress–strain curves

4 Results of Modeling

When adapting the model to the conditions of specific low permeability reservoirs, deformation and filtration processes occurring in bottomhole zones were simulated for various types of bottomholes: open, cased with perforations, open with perforations. Perforation holes of various geometry were considered for both cased and uncased wells with perforations(their lengths and diameters were varied). Isolines for stress intensity around a perforation hole in cased and uncased well calculated according to the described above algorithm are shown on Fig. 2a, b. The calculations were carried out for the following boundary conditions: $q = 40.25$ MPa is total normal stress at the external contour with radius equal 100 m (uniform normal pressure, corresponding to rock pressure); $p_0 = 17.5$ MPa is pore pressure at the external contour corresponding to hydrostatic pressure of the fluid; $p_w = 0$ MPa is pressure in the hole (the minimal possible well pressure); $q_w = q_h = p_w$ are the total normal stresses at the contours of the wellbore and the holes for uncased wellbore; $q_w = 0.5q_0$, $q_h = p_w$ are the total normal stresses at the contours of the wellbore and the holes for cased wellbore, respectively.

The figures demonstrate that in case of the uncased wellbore the stress intensity around the boreholes increase are sufficient to initiate inelastic deformation for producing an additional crack system resulting in the increase in permeability, while in case of the cased well the level of stress intensity is not high enough. The calculation demonstrated also, that the zones of the highest stress intensity appeared around the boreholes in the vicinity of the wellbore, therefore increasing of the holes length would not increase the sizes of these zones. Meanwhile, increasing of the holes diameters would result in increasing the sizes of these zones. Thus it is concluded, that for the most pronounced effect it is desirable to make large amount of short and wide boreholes.

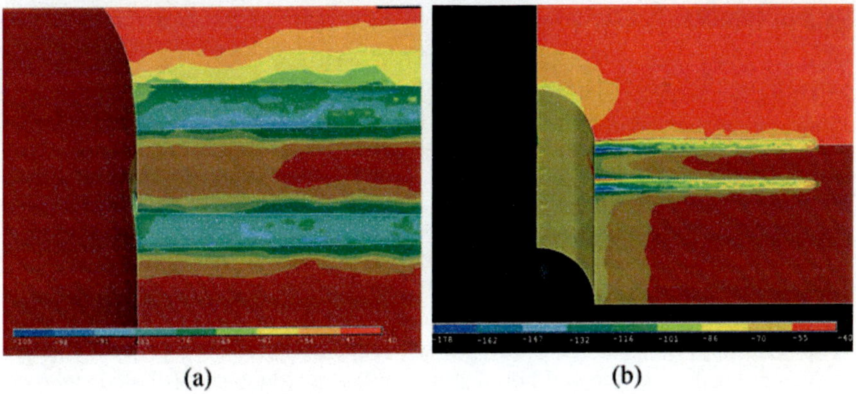

Fig. 2 Isolines of stress intensity. **a** Around a perforation hole in cased wellbore; **b** around a perforation hole in uncased wellbore

The filtration flows and well flow rates were also calculated. It was demonstrated that the increase in the well flow rate into an uncased wellbore with perforation holes comparing to the well flow rate into a cased wellbore with perforation holes increase up to 15%.

It has to be noted, that the change of stress–strain state due to formation of the inelastic zones might lead to further initiation of secondary cracks accompanied with further increase in permeability.

5 Discussion

It has been demonstrated that for uncased wellbore with perforation holes inelastic deformation occur only near the perforation holes near the wellbore, while the stresses near the wells of the wellbore are not high enough to initiate them. Therefore, perforation of uncased wellbore allows initiating the process of crack formation and increase permeability in the bottom-hole zone. Neither for uncased wellbores without perforation holes, nor for cased wellbores with perforation holes the sufficient magnitudes of stresses to cause inelastic deformation accompanied by crack formation and increase in permeability were achieved for the conditions of reservoirs of deposits in question.

On the base of the geomechanical approach in the Laboratory of Geomechanics of the Institute for Problems in Mechanics RAS a new technology of increasing productivity of well and gas boreholes, namely, the directional unloading method, was developed and passed the industrial approbation successfully [1]. The method consists in increasing permeability of reservoir rocks in the bottom hole zone due to creating systems of artificial micro and macro-cracks serving as additional filtration channels by the directional unloading of the bottom hole zone from rock pressure.

6 Conclusion

The developed geomechanical model to describe deformation, fracture and filtration processes occurring within the bottom hole zone of a producing well is presented. The results of mathematical modeling for a particular gas condensate field are provided.

These results demonstrated that design of cased wells with perforation holes as well as uncased wells without perforation holes does not result in increase in permeability, while design of uncased wells with perforation holes results in essential increase in permeability. The analysis demonstrated that the increasing the lengths of the holes does not affect the increase in permeability, since the stress concentrations governing the appearance of inelastic zones with enhanced permeability occur within zones of conjugation of the wellbore and the holes. On the base of the conducted analysis optimal design of the bottom hole zone was chosen: uncased wellbore with wide, short perforation holes. It was also demonstrated, that the potential increase in well flow rate may reach an essential percentage by choosing the optimal design.

Thus, it may conclude, that geomechanical modeling of the processes of deformation, filtration and fracture within productive layers may serve as a basis for developing new effective technologies of increasing productivity and oil and gas recovery of low permeable reservoirs.

Acknowledgements This work was financially supported by the Russian Federation through the Ministry of Science and Higher Education of the Russian Federation Project No. 13.1902.21.0018 (Agreement 075-15-2020-802).

Literature

1. *Karev V., Kovalenko Y., Ustinov K. 2020.* Mechanical and mathematical, and experimental modeling of oil and gas well stability. *Advances in Oil and Gas Exploration and Production. 2020.*
2. *Karev V.I., Kovalenko Yu.F., Zhuravlev A.B., Ustinov K.B.* Model of filtration to a wellbore accounting the dependence of permeability on stress state. *Processes in Geomedia.* 2015. № 4: 34–44. (In Russian).
3. *Klimov D.M., Karev V.I., Kovalenko Yu.F., Ustinov K.B.* 2006. Mathematical and physical modeling of rock failure during drilling inclined wells. *Technology TEK.* № 5: 22–27. (In Russian).
4. *Zhuravlev A.B., Karev V.I., Kovalenko Y.F., Ustinov K.B.* The effect of seepage on the stress-strain state of rock near a borehole. *Journal of Applied Mathematics and Mechanics.* 2014. V. 78. No. 1: 56–64.
5. *Biot M A* Le problème de la consolidation des matières argileuses sous une charge. Ann. Soc. Sc. de Brux Ser. 1935, B **55:** 110–3.
6. *Christianovich S A, Jeltov Yu P* O gidravlicheskom razryve neftenosnogo plasta. Izv. AN SSSR. OTN. 1955. **5:** 3–41.
7. *Hill R* A theory of the yielding and plastic flow of anisotropic metals. Proc. Roy. Soc. London A. 1948. 193: 281–97.
8. *Hill R* The Mathematical Theory of Plasticity New York: Oxford University Press. 1983.

9. *Caddel R M, Raghava E S, Atkins A G* A yield criterion for anisotropic and pressure dependent solids such as oriented polymers. 1973. 8 1641–6.
10. *Reynolds O.* On the dilatancy of media composed of rigid particles in contact, with experimental illustrations // Philosophica Magazine. 1885. Series 5. No. 20 (127): 469–481.
11. *Karev V.I., Kovalenko Yu.F.* Triaxial loading system as a tool for solving geotechnical problems of oil and gas production. In M. Kwasniewski et al. (eds). True Triaxial Testing of Rocks. 2013. Leiden: CRC Press/Balkema. 301–310.

Micro-CT Analysis of Fractures and Permeability Changes in Low-Permeability Rocks After True Triaxial Loading

V. V. Khimulia and V. I. Karev

Abstract This article presents the results of complex studies of changes in the internal structure and evolution of filtration properties of rocks in the Astrakhan gas-condensate field after physical modeling the implementation of the method of directional unloading of the reservoir. The experiments were carried out on the Triaxial Independent Load Test System of the Institute for Problems in Mechanics of the Russian Academy of Sciences. Scan of one of the samples on a high resolution computed tomography scanner CT-MINI was performed. A digital model of the internal structure of the rock was obtained. Different methods of estimating the finite permeability of a sample using numerical simulation of filtration flows in a rock structure obtained from CT scanning are reviewed and compared. Filtration flow velocity fields and sample permeability values were calculated. It is shown that the use of different approaches to permeability modeling based on scan results can lead to significantly different results. Research results allows to conclude that the method of directional unloading of reservoir can be successfully applied to the rocks of Astrakhan gas condensate field, allowing to significantly improve the filtration properties of rocks in the vicinity of a well.

Keywords Low-permeability rocks · Permeability · Filtration properties · Pore space structure · Single phase filtration · Computed tomography (CT) · Stress strain state · Enhanced oil recovery · Directional unloading method

1 Introduction

With the accelerating depletion of easily recoverable hydrocarbon reserves, the primary focus is moving towards the development and exploitation of complex and unique reservoirs. In this connection the major operating companies are attempting to produce as much as possible from the already exploited fields, which does not require such high costs. At present, research related to improving oil and gas recovery of

V. V. Khimulia (✉) · V. I. Karev
Ishlinsky Institute for Problems in Mechanics RAS, Moscow, Russia
e-mail: valery.khim@gmail.com

V. I. Karev (ed.), *Physical and Mathematical Modeling of Earth and Environment Processes—2022*, Springer Proceedings in Earth and Environmental Sciences,
https://doi.org/10.1007/978-3-031-25962-3_44

reservoirs and increasing the percentage of field depletion becomes very important [1]. When modeling various measures aimed at improving reservoir recovery, it is important to understand the changes in the deformation and filtration properties of the productive formation under complex stress conditions [2]. In real geotechnical conditions in rocks and soils, the stresses acting in the three directions can be significantly different. In order to simulate the stress–strain state of geomaterials in real conditions, true triaxial loading units are designed, which are able to independently and simultaneously change the stresses or deformations along each of the three axes. Such installations allow to conduct physical modeling of geomechanical processes, occurring in the rock mass during certain technological operations. Such installations are used to study deformation, strength, filtration and rheological properties of rocks under conditions of unequal three-axis compression and serve as an indispensable tool to determine the parameters of mathematical models created for calculations of deformation and fracture processes of geomaterials taking into account their properties anisotropy [3].

Bottomhole zone filtration properties have a significant impact on well productivity. The wellbore walls during various technological operations can be affected by stresses that can cause rock failure, as well as lead to both lowering and increasing their permeability [4].

One of the main nondestructive methods of studying the internal structure of rocks, significantly supplementing the direct laboratory tests, is X-ray CT scanning. X-ray tomography of heterogeneous media is based on the reconstruction of the spatial distribution of the linear attenuation coefficient of X-rays by means of computer processing of the projections of the substance layer obtained by radiation imaging. The difference in the attenuation coefficient is determined by the difference in density and chemical composition of the elements that make up the studied substance [5]. As a result of such reconstruction the internal three-dimensional microstructure of the sample is visualized. Linear dimensions of elements, their number, sphericity, anisotropy are determined using special software [6]. Modern software also allows the segmentation of the obtained image of a heterogeneous material, i.e. the separation of its components: the simplest example for rocks is the separation of the rock material and the pore space in it. After that, it is possible to perform numerical simulation of various processes on the resulting structure, including the process of filtration through the pore space of the rock [7].

2 Research Methodology

In order to increase permeability of productive formation on the basis of stress–strain state control in the vicinity of the well, IPMech RAS developed an environmentally friendly, cost-effective and efficient way — method of directional unloading reservoir (DUR method). The essence of this method is a controlled pressure decrease in the well, which causes in its vicinity the appearance of stresses that lead to the formation of micro- and macro-cracks, thereby increasing the permeability of the rock [2, 8].

All necessary for specific fields values of stresses are determined by direct physical modeling deformation and filtration processes, occurring in reservoir during well operations, on rock samples using Triaxial Independent Load Test System (TILTS), developed in IPMech RAS [2, 9].

TILTS is designed to study the elastic, strength, rheological and filtration properties of oil, gas, coal, and ore deposits rocks. It allows to load cubic rock samples with edges of 40 or 50 mm independently along each of the three axes. This makes it possible to experimentally recreate the stress–strain state of any type occurring in the formation during various technological operations in a well. The main feature of TILTS is the possibility to study the influence of stress strain state on rock filtration properties by continuous permeability measurement during the test.

This paper presents the results of complex studies of changes in the internal structure and evolution of filtration properties of rocks in the Astrakhan gas condensate field when modeling the implementation of the directional unloading method. Different methods of estimating the finite permeability of a specimen using numerical simulation of filtration flows influence on a rock structure obtained from CT scan are reviewed and compared. Astrakhan gas condensate field is one of the unique and difficult to develop fields [10]. The peculiarity of the field is a complex structure of the geological section, abnormally high reservoir pressure, heterogeneity of distribution of filtration capacitive properties over the area and the profile [11].

Experiments were carried out on the unique TILTS on samples cut from the core material of the field. Four faces of the sample were covered with an impermeable thin polymer film, two remaining opposite faces were left open for air to pass through the sample. During the experiments, a physical modeling of the directional unloading method implementation was performed for the conditions of an uncased well with a perforation hole. After that, the sample was placed in a ProCon X-Ray CT-MINI high-resolution micro-CT scanner of IPMech RAS to study the internal structure of the sample. A digital model of the rock sample was created based on the analysis of the obtained images. The system of cracks that appeared in the sample was analyzed. Mathematical modeling of the filtration process on the obtained digital rock sample model was performed using the GeoDict software.

A loading program was generated to simulate on TILTS stress strain state on the perforation hole surface in an uncased well at the top point at the distance of 1.25 well radius. The stress state in the vicinity of the perforation hole in the uncased well can be represented by the superposition of solutions to the Lamé problem [12] and the problem similar to the Kirsch problem [13]. In this case, the stress state can be described by three the principal stresses: coaxial with the well $S_{z'}$, circular $S_{\theta'}$ and radial $S_{r'}$ (Fig. 1). In the TILTS loading unit these stresses correspond to S_1, S_2, S_3 applied to the faces of the sample. The stress S_2 corresponds to the stress component $S_{z'}$, S_1 corresponds to the component $S_{\theta'}$, and S_3 corresponds to the radial component $S_{r'}$ (Fig. 2).

As already said, X-ray computed tomography methods were used to quantitatively and qualitatively investigate the internal structure of the sample, as well as to evaluate the final permeability. The sample was scanned using a high-resolution X-ray scanner CT-MINI [14]. The total weight of the scanner is 570 kg. The dimensions of the device

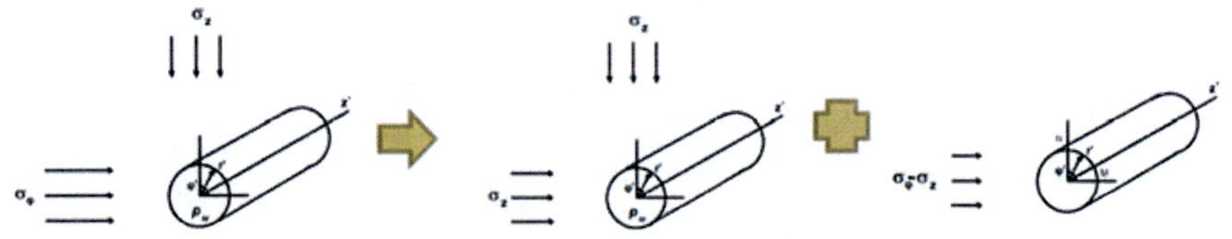

Fig. 1 The stress state in the vicinity of the perforation hole in the uncased well is the superposition of solutions to the Lamé problem and the problem similar to the Kirsch problem

are 1300 × 850 × 600 mm. The device has a rigid base (monolithic granite slab) for practical elimination of the influence of temperature drift and accurate installation of X-ray optical and other system components inside the device, which provides accuracy and stability of measurements in the whole range of working space. A design scheme with vertical placement of the core for research is provided inside: the maximum height of the loaded sample is 200 mm, the maximum diameter of the loaded sample is 200 mm. The maximum weight of the loaded sample is 5 kg.

The optical scheme of the scanner allows to set a distance of 335 mm from the tube to the object. For this purpose, a precision positioning system (manipulator) is built into the design. The detector has the ability to move perpendicularly to the main axis of the system within a range of ± 25 mm. The manipulator allows for 360° rotation of the sample around its axis with an angular position reproducibility accuracy of 1.5 angular seconds.

The high resolution, microfocus, closed-loop X-ray tube has an adjustable output voltage range of 20–90 kV and an adjustable current range of 10–160 μA. The maximum output power of the tube is 8 W. The smallest focal spot size is 5 μm. The

Fig. 2 Actual loading program of specimen A4.1

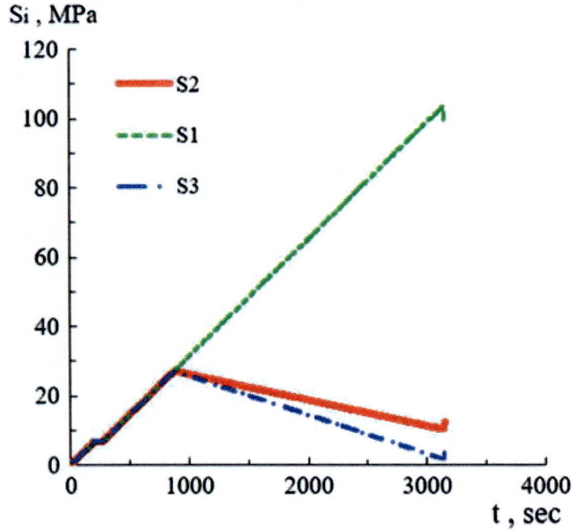

high-sensitivity, low-noise X-ray detector has a pixel count of 2940 × 2304. Pixel size is 49.5 μm. The size of the active (sensitive) area is 146 × 114 mm.

Parameters of the scan performed on the specimen A4.1: source characteristics 90 kV/89 μA/8 W, exposure time 0.5 s, voxel size 44.117 μm. Reconstruction of the scan was performed using VGSTUDIO program. 3D image processing was performed by means of Geodict Math2Market GmbH software. Such image processing methods as brightness correction, NLM filter application, image segmentation were used.

3 Experimental and Simulation Results

The values of initial permeability of the sample were obtained in the experiment, strain curves were plotted. The initial permeability of sample A4.1 was less than 1 mD and almost did not change during the experiment. The sample was elastically deformed up to the value of the maximum principal stress S_1 equal to 80 MPa. The fracture of the sample occurred at a stress S_1 equal to 103 MPa. After the test, a system of macro-cracks was formed in the sample, acting as new filtration channels, but the final permeability value could not be registered during the experiment due to depressurization of the sample during cracking. A photo of the specimen after the tests is shown in Fig. 3a.

Such image processing methods as brightness correction, application of NLM filters, image segmentation were used after scanning. To guarantee avoid scanning artifacts and to remove the material boundary with air space from consideration, the 3D image of the sample was cropped on all faces of the cubic sample. Image segmentation was performed using the Global Tresholding method, since only macrocracks are of research interest. Figure 3b shows the internal structure of the sample obtained

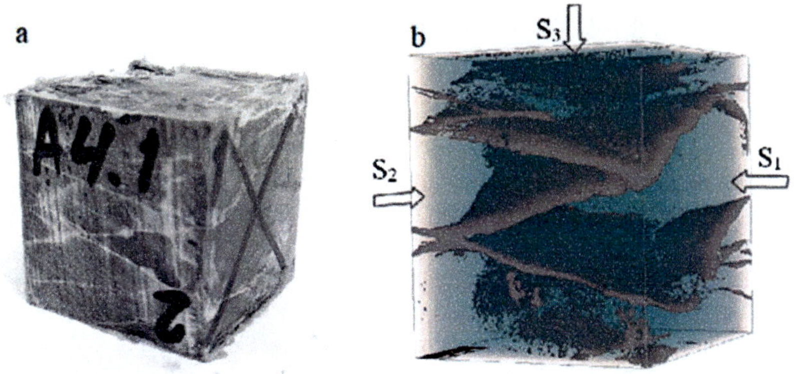

Fig. 3 **a** Sample A4.1 after tests. **b** Part of the obtained digital structure of sample A4.1

after segmentation: cracks and pores distributed over the image volume are shown in gray.

The orientation of the resulting fracture system can be explained successfully, for example, by the theory of conjugate fractures of Gzovsky [15], which is used in the study of reservoir structures [16]. Acute angle between the cracks corresponds to the quadrant (sector) of compression, and the bisector of this angle—the axis of the maximum compressive principal normal stress S_1. The bisectors of adjacent obtuse angles correspond to the axis of the minimum compressive stress S_3. Also this space geometry agrees with the inverse method of the main sections of the stress ellipsoid using the drawing of bisecting planes.

The digital rock model was used for mathematical modeling of filtration flow in order to estimate the final permeability. Filtration fields were calculated using the FlowDict module of the GeoDict package. FlowDict predicts the physical mean flow velocity for a given pressure drop [17] and calculate permeability of a porous structure by applying Darcy's law. This module is capable of calculating incompressible, stationary Newtonian flows based on Navier–Stokes equations with various approximations. In order to recreate the permeability measuring conditions analogous to the experimental ones the boundary conditions along the flow measuring direction were chosen—periodic, and along the other faces—No Slip. The calculation was performed in the direction of the stress S_3, i.e. along the rock layering. Figure 4 shows in gray the fracture system formed in the sample and presents part of the calculated filtration paths along the fractures as a flow velocity distribution for two models: Stokes and Navier–Stokes, respectively. A total of nine filtration flow calculations were performed for different models and their parameters. The results of modeling are presented in Table 1: for each calculation the system of equations to be solved, the pressure drop, the condition of exit from the calculation cycle (calculation error), the obtained permeability in mD are indicated. Despite the fact that the main fractures run along the two directions in the sample, the calculated permeability in the perpendicular direction was 227.4 mD within the Navier–Stokes equation model and about 1.32 D within the Stokes model for a pressure drop of 0.1 bar. In order to compare the calculated value with the experimentally measured one, the sample was repeatedly placed in the TILTS facility for physical evaluation of the final permeability. Due to the technical specifics of the experiment and preparation of specimens, the final conditions differed from the mathematically simulated ones: there is irreversible sealing of part of the filtration channels by an impermeable membrane, as well as the need for all-round compression when measuring the actual permeability, which leaded to a partial fracture closure. Given these differences, the minimum value of measured permeability is estimated at 150 mD.

As the pressure drop decreased, the Navier–Stokes model gave ever higher absolute values of permeability. For the same pressure drop within the same model, the simulation results coincide within the margin of error. For different boundary conditions in the flow direction (periodic/symmetric) with the same pressure drop, the results are the same within the margin of error, but require different amounts of resources and time. Boundary conditions "symmetric" require more resources and time.

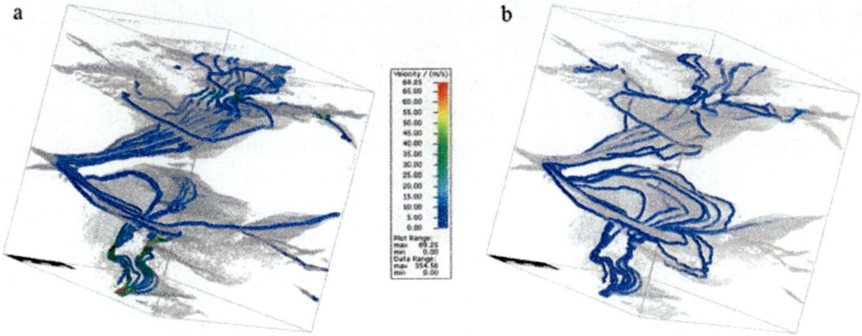

Fig. 4 Representation of selected filtration flows for Stokes (**a**) and Navier–Stokes (**b**) models

Table 1 Results of the simulation of filtration flow through the obtained structure

Filtration model	Pressure drop (atm)	Cycle exit condition, error bound	Permeability (mD)
Stokes	0.05	0.05	1324
Stokes	0.1	0.05	1324
Navier–Stokes	0.05	0.1	383
Stokes	0.05	0.01	1543
Navier–Stokes	0.1	0.1	225
Navier–Stokes	0.1	0.05	227.4
Navier–Stokes/symmetric	0.1	0.05	247.6
Navier–Stokes	0.05	0.05	332.4

The difference in computational results for the two models is predictable and is due to the differences in the mathematical assumptions contained in the models. The computational model using Stokes equations is most applicable for small pressure drops, but has the undeniable advantage of lower computational resource consumption. The Navier–Stokes model provides a more precise estimate at higher pressure drops, but requires careful optimization calculations to successfully complete the calculations: the calculations using these methods do not always complete successfully. To optimize calculations it is necessary to use special procedures that use more memory, processor resources or require more time.

One of the ways to solve the problem of high resource consumption is to choose a smaller simulation domain size, i.e., to choose a smaller representative volume. In the case of multiphase filtering modeling or filtering through multicomponent structure, the selected volume should become even smaller. Otherwise, high-performance cluster and supercomputer resources are required. It is worth noting that differences in the calculated and physically measured permeability values are inevitable even under perfectly matched conditions: the quality of the input image (image resolution) and post-processing (artifacts, filtering) have a significant influence on the calculation result [18]. By reducing the voxel size, recognizable borders of fractures and pores are clarified, and the proper scanning will minimize the presence of image artifacts, which affect the recognizability of the internal space. The selection of image filtering parameters affects the procedural merging of pixels for the subsequent segmentation and lower resource consumption when modeling different kinds of processes on the resulting structure. On the other hand, image smoothing reduces the distinguishability of the pore space boundaries, which will affect the computation. An important disadvantage of CT in general is the ambiguity of segmentation of the obtained images, i.e. assigning to the image pixels the labels of different materials and substances (for example, rock/air distinction). Depending on the chosen method of segmentation and preliminary work with the image, the distribution of pores, fractures and refinement of their boundaries may differ, affecting the final appearance of the resulting structure (and sometimes quite significantly). And of course the models behind the computational methods can produce different results depending on the simplifications and approximations underlying them.

4 Conclusion

The paper describes the results of physical modeling of deformation and filtration processes in the rocks of the Astrakhan gas condensate field when implementing the method of increasing filtration properties of low-permeability rocks—the method of directional unloading of a reservoir. Experimentally obtained curves of changes in the filtration and deformation characteristics of rocks when simulating controlled pressure reduction in uncased wells with a perforation hole at a distance of 1.25 of the well radius. Scanning of one of the samples on a high-resolution micro-CT scanner CT-MINI was performed. A digital model of the internal structure of the rock was obtained. Filtration properties of rocks were digitally analyzed using experimental data of micro CT scanning. Filtration flow velocity fields and sample permeability values along the filtration direction were calculated. The results of calculations obtained within the framework of different filtration models with varying calculation parameters were compared. It is shown that the use of different approaches to permeability modeling based on scanning results can lead to significantly different results. The choice of optimal computational model and its parameters should be carried out depending on the considered volume, available computational resources, filtration flow parameters and the type of obtained structure, which in turn depends on many factors, such as type of rocks under study, parameters of performed scanning, image processing, etc. Research results allows us to conclude that the method of directional unloading of reservoir can be successfully applied to the rocks of Astrakhan gas-condensate field, allowing to significantly improve the filtration properties of rocks in the vicinity of the well.

Acknowledgements This work was financially supported on the topic State assignments 123021700046-4

References

1. Kang W., Zhou B., Issakhov M., Gabdullin M. Advances in enhanced oil recovery technologies for low permeability reservoirs // Petroleum Science. 2022. V. 19. Is. 4. pp. 1622–1640.
2. Karev V., Kovalenko Y., Ustinov K., 2020. Advances in Oil and Gas Exploration and Production. Switzerland: Springer International Publishing. 2020. 166 p.
3. Karev V.I., Khimulia V.V., Shevtsov N.I. Experimental studies of the deformation, destruction and filtration in rocks // Mechanics of Solids. 2021. V. 56. № 5. pp. 613–630.
4. Davies J.P., Davies D.K. Stress-dependent permeability: characterization and modeling // SPE J. 2001. 6. P. 224.
5. Lichun J., Mian C. 3D imaging of fractures in carbonate rocks using X-ray computed tomography technology // Carbonates and Evaporites. 2014. V. 29. I. 2, pp. 147–153.
6. Diaz M., Yeom Kim K., Yeom S., Zhuang L., Min K., Kraub F., Giese R., Alexandrakis C., Buske S. Surface roughness characterization of open and closed rock joints in deep cores using X-ray computed tomography // International Journal of Rock Mechanics and Mining Sciences. 2017. V. 98. pp. 10–19.
7. Gerke K.M., Korost D.V., Karsanina M.V., Korost S.R., Vasiliev R.V., Lavrukhin E.V., Gafurova D.R. Modern approaches to pore space scale digital modeling of core structure and multiphase flow // Georesources. 2021. 23(2). pp. 197–213. https://doi.org/10.18599/grs.2021.2.20.
8. Kovalenko Yu. F., Karev V. I. Method of geo-leaching - A new approach to the problem of increasing the productivity of wells // Technologies of Fuel and Energy Complex. 2003. 1. pp. 31–35. (in Russian)
9. Karev V., Kovalenko Y. Triaxial loading system as a tool for solving geotechnical problems of oil and gas production. True triaxial testing of rocks. 2013. Leiden: CRC Press/Balkema: 301–310.
10. Komarov A. Yu., Pushkareva D. A. Comprehensive study of reservoir properties of middle and upper-jurassic deposit of the Astrakhan gas condensate field. Territoriya Neftegaz. 2021. 7–8: 36–43. (in Russian).
11. Revina A., Revina N. Analysis of results of geological and technical measures and their impact on component recovery. Vestnik of Astrakhan State Technical University. 2022. 1: 43–49. (in Russian).
12. Lyav A. Mathematical theory of elasticity. USSR. 1935. 674 p. (in Russian).
13. Timoshenko S.P. Theory of Elasticity / S.P. Timoshenko, J. Gudier // M.: Nauka. 1979. 560 p. (in Russian).
14. CT-MINI by ProCon X-Ray GmbH available at https://procon-x-ray.de/en/ct-mini.
15. Gzovsky M.V. Fundamentals of Tectonophysics. Nauka. Moscow. 1975. 536 p. (in Russian).
16. Toksarov V.N. Strategy and processes of development of georesources: Collection of scientific papers / edited by A.A. Baryakh / Perm. Mining Institute of the Ural Branch of the RAS. 2018. pp. 119–121.
17. Kling, T., Huo, D., Schwarz, J.-O., Enzmann, F., Benson, S., and Blum, P.: Simulating stress-dependent fluid flow in a fractured core sample using real-time X-ray CT data // Solid Earth. 7: 1109–1124. https://doi.org/10.5194/se-7-1109-2016. 2016.
18. Zakirov T., Galeev A. Absolute permeability calculations in micro-computed tomography models of sandstones by Navier-Stokes and lattice Boltzmann equations. International Journal of Heat and Mass Transfer. 2019. 129. pp. 415–426. https://doi.org/10.1016/j.ijheatmasstransfer.2018.09.119.

Investigation of Wave Breaking by Radar Measurements in the Laboratory Modeling

G. A. Baydakov, N. S. Rusakov, A. A. Kandaurov, D. A. Sergeev, and Yu. I. Troitskaya

Abstract The article describes a series of experiments to study the scattering of microwave radiation by a water surface. Measurements of the NRCS of the water surface were carried out for co- and cross-polarizations in a wide range of wind speeds, including hurricane ones. As part of laboratory modeling, a study was made of the scattering of microwave radiation by wave breaking, for which regular artificial wave breaking was organized at a given point, a specially developed optical method was used to control the wave breaking parameters, with side illumination of the survey area. The dependence of the power of the scattered signal on the area of the wave breaking was obtained.

Keywords Laboratory modelling · Remote sensing · Cross-polarization · NRCS · Wave breaking

1 Introduction

One of the most important tasks of monitoring the state of the World Ocean and the marine boundary layer of the atmosphere is the restoration of the surface wind speed and its direction according to active remote sensing data. Microwave radars located on artificial satellites of the Earth irradiate the water surface and determine its normalized radar cross-section (NRCS) by the power of the backscattered signal. To do this, the so-called geophysical model functions (GMF) are used, they are empirical relationships between the NRCS of the water surface and the parameters of the air flow. Initially, for remote sensing, radiation was used at polarizations that were consistent in emission and reception, and for such devices, a set of GMF was proposed [1–4]. However, in this case, despite the high intensity of the scattered signal, the received power tends to saturate with increasing wind speed, which makes it impossible to restore this parameter at high speeds. To solve this problem, they began to use a cross-polarized signal, the power of which remains sensitive to changes in wind

G. A. Baydakov (✉) · N. S. Rusakov · A. A. Kandaurov · D. A. Sergeev · Yu. I. Troitskaya
Institute of Applied Physics (IAP) RAS, Nizhny Novgorod, Russia
e-mail: baydakov@ipfran.ru

V. I. Karev (ed.), *Physical and Mathematical Modeling of Earth and Environment Processes—2022*, Springer Proceedings in Earth and Environmental Sciences,
https://doi.org/10.1007/978-3-031-25962-3_45

speed in a wide range of winds, including hurricanes [1, 2, 5–10]. However, under conditions of high wind speeds, when the water surface is characterized by complex rheology (collapse of surface waves, foam coating, spray, suspended air bubbles), the physical mechanisms underlying the process of microwave signal scattering are not well understood.

Measuring the parameters of the wind flow and waves in strong winds in field conditions is technically complex and dangerous, and to accumulate a large amount of data in various conditions, many such experiments are required. Under the conditions of laboratory modeling, it is possible to ensure high repeatability and controllability of measurement conditions, and modern techniques make it possible to accurately describe the ongoing processes even with hurricane winds. In particular, it is possible to measure the wind speed profile near the water surface itself, and the use of optical methods makes it possible to accurately determine the proportion of water surface coverage by whitecaps. In the present work, a laboratory simulation of the scattering by a breaking wave crest of a microwave signal was carried out in order to create semi-empirical models that qualitatively and quantitatively describe the properties of the signal scattered by the water surface during hurricane winds.

2 Experimental Setup and Data Processing

Experimental studies of the scattering of microwave radiation by breaking waves were carried out at the Wind Wave Tank (TSWiWaT) of the IAP RAS (Fig. 1). The length of the channel is 12 m, the working depth of the liquid layer is 1.5 m with the deepening of the side walls of 0.5 m, the cross section of the channel varies from 0.7 × 0.7 m at the inlet to 0.7 × 0.9 m at the outlet. The variable height of the channel compensates for the increasing height of the developing boundary layer, allowing to save the effective cross-sectional area of the channel, which allows to eliminate the pressure gradient, that is, the velocity on the channel axis does not change. The maximum air flow speed on the channel axis reaches 33 m/s, which, in terms of natural conditions, corresponds to the wind speed at a standard meteorological height of 10 m, about 50 m/s. At the beginning of the channel there is a wave generator that generates a train of three waves with a frequency of 1.04 Hz and a length of about 1 m every 18 s. Directly in front of the study area, an underwater inclined plate was installed, simulating an exit to shallow water and providing multiple waves breaking in the zone of radar illumination and optical observations. The correct depth of the plate was predetermined for each wind speed and changed during the experiment in order to ensure a sufficiently intense wave breaking and, at the same time, prevent the wing trailing edge from being exposed. To control wave parameters and synchronize measurements, a resistive-type wire wave gauge was installed in front of the plate.

To study the dynamics of a breaking wave crest under the action of a wind flow, a series of measurements was carried out for 5 values of wind speed from 17.6 to 38.4 m/s. The measurement of the air flow parameters was carried out by the gradient method. To do this, using a scanning device in the working section on the

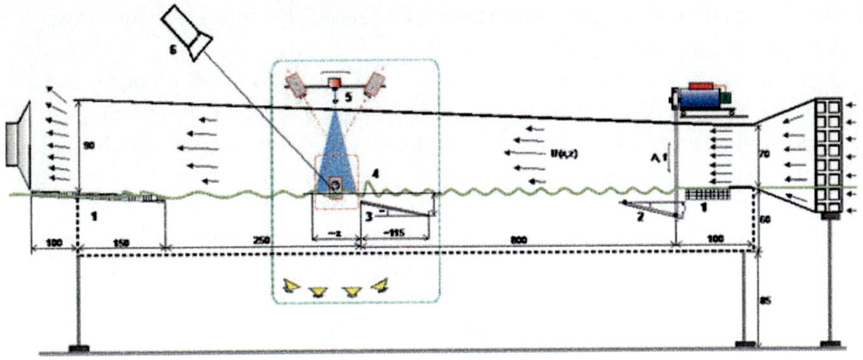

Fig. 1 Facility scheme. 1—beaches-wave damper, 2—wave generator, 3—inclined plate simulating the exit to shallow water, 4—breaking wave, 5—optical system, 6—Doppler X-band scatterometer

channel axis, the air flow velocity profiles were measured using an S-shaped Pitot tube connected to a differential pressure gauge (range of measured velocities up to 40 m/s with an accuracy of 0.15 m/s). The determination of the parameters of the atmospheric boundary layer was carried out according to an algorithm based on the self-similarity of the velocity defect profile in the air channel proposed in [11].

To determine the characteristics of the waves in the channel, measurements were taken by an antenna of three resistive wave recorders located at the vertices of an equilateral one with a side of 2.5 cm, the data sampling frequency was 200 Hz. Such a system makes it possible [11] to reconstruct the spatiotemporal spectra of surface waves: from the wave frequency, the modulus of the wave vector, and the angle relative to the wind direction. The upper limit of the wave number spectrum is determined by the distance between the wave gauges in the antenna and is equal to 1.25 rad/cm.

Microwave measurements were carried out using a coherent X-band Doppler scatterometer with a wavelength of 3.2 cm with simultaneous reception of linear polarizations. A pyramidal horn with a square section of 224 mm × 224 mm and a length of 680 mm was used as a receiving-transmitting antenna of the scatterometer. This choice is determined by the need to obtain a sufficiently wide radiation pattern, the width of which is 9°, and at the same time to localize the side lobes as much as possible to simplify the screening of areas, the reflection from which could introduce significant errors in the measurements. The antenna was equipped with an orthogonal polarization separator (OMT) with a polarization separation of more than 40 dB. The measurements were made at two co- and two cross-polarizations (VV, HH, VH, HV).

The observation window in the cover of the wind-wave channel had dimensions of 40 cm × 40 cm, the angle of incidence varied from 30° to 50° towards the wind, the distance to the middle of the measurement area was 3.15 m, the cover of the working part was made of radio-transparent material (polystyrene) 11.2 mm thick. With such parameters, the illumination spot on the water surface had a width of up to 58–68 cm, i.e., less than the width of the channel, while due to the drop in the directivity pattern to the signal from near-wall regions, the power was more than 6 times lower than on the channel axis.

The absolute value of the radar cross section (RCS) of the rough water surface at co-polarizations was determined by comparing the scattered signal with the signal reflected from a reference calibrator with a known RCS value—metal balls with a diameter of 6 and 4 cm, the RCS of which was determined analytically. To calibrate cross-polarizations, a secondary calibrator was used in the form of a thin wire 19.2 cm long, oriented at different angles in a plane perpendicular to the radiation. In the vertical and horizontal position, the signal maximum was observed at HH and VV polarizations, and at an inclination angle of 45°, the same values were observed for all polarizations (HH, VV, HV, VH) and amounted to 1/4 in power from the maximum of HH and VV.

Preliminary measurements of the NRCS of the water surface were carried out for various wind speeds without the use of an inclined plate and a wave maker, that is, in the mode of natural generation of waves by the wind and their breaking. An analysis of the measurement results (Fig. 2) showed that at small angles of incidence, the dependence of the NRCS on co-polarizations on the wind speed does not change within the measurement error, while for cross-polarizations there is a steady increase in the NRCS with increasing wind speed, the trend to saturation at wind speeds U10 more than 50 m/s is also not observed.

Comparison of the spectral distribution of wind waves by phase velocities with the spectral power density of the received radar signal, reduced to the Doppler velocity of the scatterers (Fig. 3), showed that an important contribution, especially at high wind speeds, to scattering is made by elements whose speed significantly exceeds not only the speed of the dominant waves, but also the short-wave ripples generated on the slopes of the long waves. It was concluded that such elements can only be collapsing ridges and the lambs formed by them. It was noted that for co-polarizations, the peak of the spectrum is already and more shifted towards an increase in the speed of the scatterers, while for cross-polarizations, the maximum captures both fast elements and the speed of the dominant waves.

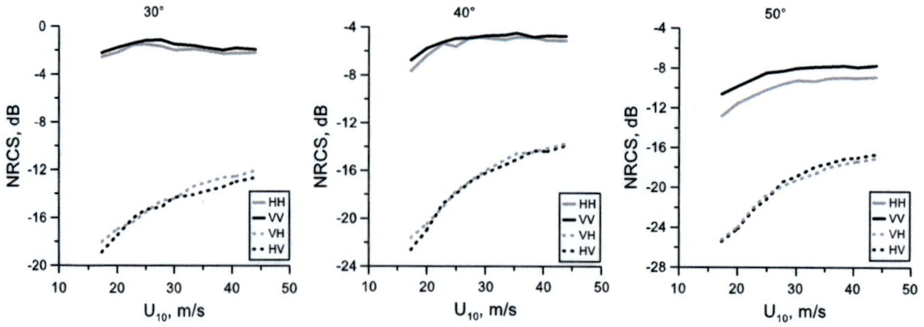

Fig. 2 Co-pol and cross-pol NRCS of water surface versus wind velocity, for different incident angles: 30°, 40°, 50°

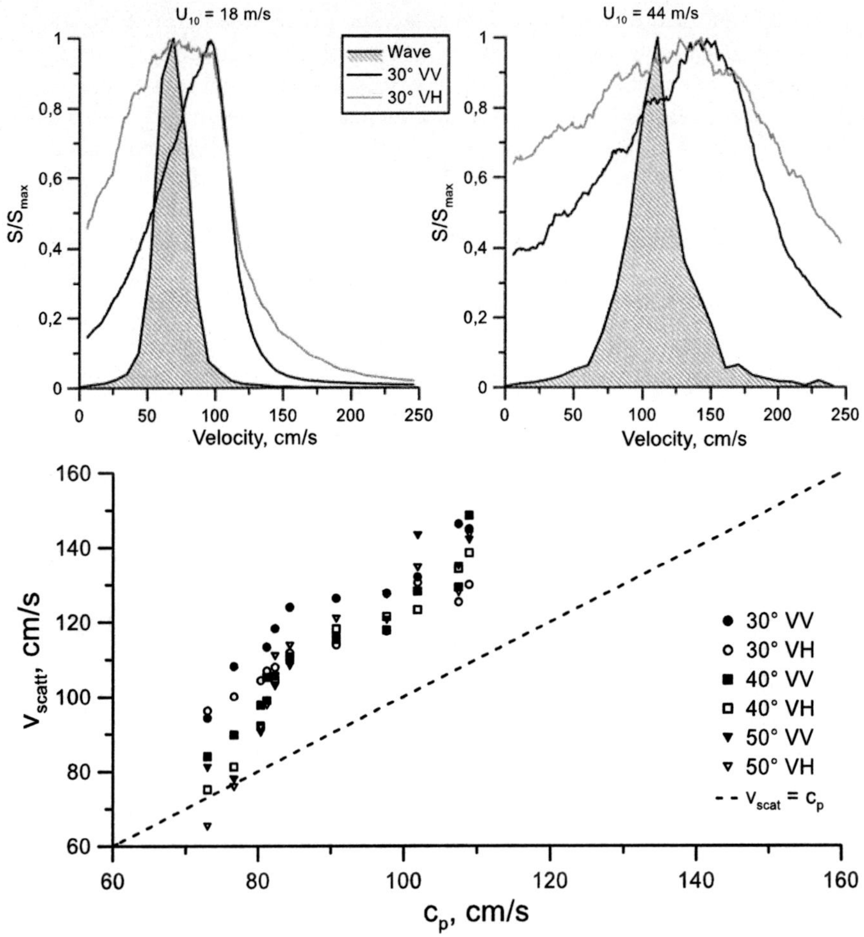

Fig. 3 At the top: filtered Doppler spectra of co-pol (black line) and cross-pol (gray line) X-band microwave return for upwind looking at different incidence angles versus equivalent velocity of the Bragg scatters, and along-wind phase velocity spectra of surface waves (the silhouettes). At the bottom: Doppler spectra weighted mean versus the along-wind phase velocity of dominant waves

It was found that when a breaking wave crest passes on the water surface, structures are generated that have a characteristic size on the order of a few centimeters, and therefore have a significant RCS for the radiation frequency used in the experiment, which manifests itself in the spectrogram as an increase in the intensity of the spectrum components. In strong winds, when artificial generation of waves occurs against the background of regularly breaking wind waves, it has less effect on the scattered signal. In this case, after passing through the studied area of the breaking crest of a long wave, backscattering from the water surface decreases compared to the background value observed during chaotic breaking.

Regardless of radar experiments, optical measurements of the area of the water surface covered with collapses were carried out using a high-speed camera. As a result of processing video frames, the time dependences of the area covered by the foam in the irradiated area were obtained for all combinations of setup parameters. For this, two Cygnet CY2MP-CL-SN cameras with 50 mm lenses were used. Shooting from two synchronized cameras was used to eliminate the false detection of glare in the form of white spots as collapses. The cameras were installed at a height of 273 cm from the surface of the water at 89 cm from each other. The image scale is 277.3 μm/px, the size of the image received from each camera is 1024 × 1920 pixels, which corresponds to a platform on the water of 284 × 532 mm. To obtain the proportion of collapses over the entire area studied in the experiment, a system of two chambers was moved over the channel. To illuminate image areas from the side of the measurement zone, a matte screen was placed on the side wall, which was illuminated by powerful LED lamps to create a uniform light source that occupied the entire side wall of the section.

Synchronization of the time dependences of the power of the signal received by the scatterometer and the data of optical measurements of the area of the water surface covered with collapses was based on the recording of a string wave recorder installed in front of the inclined plate in all experiments, and for a more accurate determination, correlation analysis was used. After the time alignment procedure, the dependences of the received signal power on the breaking area were obtained (Fig. 4), demonstrating the high sensitivity of the cross-polarized signal to the presence of whitecaps on the water surface, which can be approximated by a linear function. Moreover, at shallower angles, this sensitivity turned out to be stronger. At co-polarizations, the dependence was not revealed due to the high level of the signal from non-breaking small ripples compared to the signal from the surface covered with whitecaps.

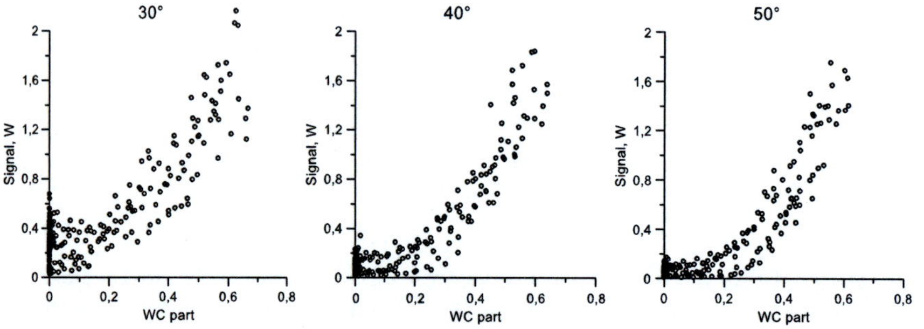

Fig. 4 Dependence of the received power on the whitecap coverage at various wind speeds on cross-pol for three incidence angles: 30°, 40°, 50°

3 Conclusion

In this work, the dependences of NRCS on wind speed were measured for three angles of incidence of microwave radiation. It is shown that at a wind speed of more than 20 m/s, co-pol NRCS does not change its value within the measurement error. At the same time, the cross-pol NRCS on wind speed has a monotonic dependence without a tendency to saturation at maximum winds.

A comparison of the Doppler spectra of the scattered signal and the spectral distribution for surface waves showed that an important contribution to the scattering of the radar signal is made by the collapse of the crests of the dominant waves. At the same time, it was shown that co-pol NRCS increases sharply with an increase in the proportion of active whitecaps coverage and loses sensitivity to its further increase. At the same time, cross-polarization is characterized by a monotonic dependence of the received signal power on the fraction of the water surface covered by wave breaking and associated whitecaps. This indicates that whitecaps contribute to the critical factor that ensures the sensitivity of the cross-polarization NRCS to changes in wind speed under extreme conditions.

When a long breaking wave passes, after reaching the peak power on the maximum area of the white cap coverage, the power value drops below the background observed in the presence of only wind waves, the water surface smoothes out and its NRCS decreases, which is clearly seen from the dependence of the received signal power on time. At the same time, it should be noted that at different stages of the passage of the collapse ridge, both wind wave zones and smoothing zones simultaneously fall into the radar coverage zone of illumination. The obtained results will allow us to determine the contribution of whitecaps to the scattered signal, which can be considered linearly dependent on the proportion of whitecaps coverage.

Acknowledgements The experiments were performed at the Unique Scientific Facility "Complex of Large-Scale Geophysical Facilities" IAP RAS (http://www.ckp-rf.ru/usu/77738/). Investigations were carried out with the financial support of Russian Science Foundation (project 22-77-00076). The technique for processing radar signals was developed as part of the project of Russian Science Foundation (project 21-17-00214).

References

1. P.W. Vachon and J. Wolfe, "C-band cross-polarization wind speed retrieval", IEEE Geosci. Remote Sens. Lett., vol. 8, no. 3, pp. 456–459, May 2011, https://doi.org/10.1109/LGRS.2010.2085417.
2. B. Zhang and W. Perrie, "Cross-polarized synthetic aperture radar: A new potential measurement technique for hurricanes", Bull. Am. Met. Soc., vol. 93, no. 4, pp. 531–541, Apr. 2012, https://doi.org/10.1175/BAMS-D-11-00001.1.

3. W. J. Donnelly, J. R. Carswell, R. E. McIntosh, P. S. Chang, J. Wilkerson, F. Marks and P. G. Black, "Revised Ocean Backscatter Models at C and Ku Band under High-Wind Conditions", J. Geophys. Res., vol. 104. no. C5, pp. 11485–11497, May 1999, https://doi.org/10.1029/1998JC900030.
4. M.A. Donelan, B.K. Haus, N. Reul, W.J. Plant, M.Stiassnie, H. C. Graber, O. B. Brown and E. S. Saltzman, "On the limiting aerodynamic roughness of the ocean in very strong winds", Geophys. Res. Lett., vol. 31, no. 18, L18306, Sept. 2004, https://doi.org/10.1029/2004GL019460.
5. G.-J. van Zadelhoff, A. Stoffelen, P. W. Vachon, J. Wolfe, J. Horstmann and M. B. Rivas, "Scatterometer Hurricane Wind Speed Retrievals Using Cross Polarization", Atm. Measur. Tech. Discuss., vol. 7, no. 2, pp. 7945–7984, Aug. 2013, https://doi.org/10.5194/amtd-6-7945-2013.
6. B. Zhang, W. Perrie, J.A. Zhang, E.W. Uhlhorn, and Y. He, "High-resolution hurricane vector winds from C-band dual-polarization SAR observations", Journal of Atmospheric and Oceanic Technology, 31(2), pp. 272–286. 2014 https://doi.org/10.1175/JTECH-D-13-00006.1.
7. F. Fois, P. Hoogeboom, F. Le Chevalier and A. Stoffelen, "An analytical model for the description of the full-polarimetric sea surface Doppler signature", J. Geophys. Res., vol. 120, pp. 988–1015. 2015 https://doi.org/10.1002/2014JC010589.
8. Vachon, P. W., and Wolfe, J.,"C-band cross-polarization wind speed retrieval," IEEE Geosci. Remote Sens. Lett. 8(3), 456–459 (2011).
9. Hwang, P. A., Stoffelen, Avan Zadelhoff, G.-J., Perrie, W., Zhang, B., Li, H., and Shen, H., "Cross-polarization geophysical model function," J. Geophys. Res. 120, 893–909 (2015).
10. Yu. Troitskaya, V. Abramov, G. Baidakov, O. Ermakova, E. Zuikova, D. Sergeev, A. Ermoshkin,V. Kazakov, A. Kandaurov, N. Rusakov, E. Poplavsky, M. Vdovin, "Cross-Polarization GMF for High Wind Speed and Surface Stress Retrieval", J. Geophys. Res., vol. 123, I. 8, pp. 5842–5855, August 2018.
11. Y.I. Troitskaya, D.A. Sergeev, A.A. Kandaurov, G.A. Baidakov, M.A. Vdovin, and V.I. Kazakov, "Laboratory and theoretical modeling of air-sea momentum transfer under severe wind conditions", J. Geophys. Res., 117, C00J21, June 2012, https://doi.org/10.1029/2011JC007778.

The First Results of the Study of Sequestration Properties of Coastal Marine Biomorpholithosystems (Sakhalin Region)

V. V. Afanas'ev and A. B. Faustova

Abstract The study of the first stage is completed by determining the geospatial position and sequestration capacity of coastal biomorpholithosystems. The general conservative estimate of carbon sequestration by lagoonal biomorpholithosystems alone, in terms of CO_2 equivalent, is at least 500,000 tons/year. Taking into account the sequestration properties of marshes and droughts of northwestern Sakhalin, the figure is likely to increase several times. The work of the next stage of the study of coastal marine biomorpholithosystems, along with geospatial inventory, determination of dynamic parameters and material composition of carbon-fixing sediments, involves the creation of a system for obtaining data on the level of greenhouse gases in the atmosphere, water, and soil gas exchange. Particular attention will be paid to the creation of ground experimental industrial sites within the ground, supralittoral zone of the Sakhalin State University carboniferous range, where technologies will be tested and artificial zones of accelerated silty-pelitic and organogenic sedimentation will be created.

Keywords Estuary · Salt marsh · Intertidal mudflat · Carbon sequestration

1 Introduction

Geomorphological data are actively used in the analysis of the balance of producers and sinks of greenhouse gases (GHGs), including in relation to the dynamics of blue carbon ecosystems [1]. Flows of nitrogen-containing and carbon-containing sediments, due to the high variability of the dynamics of the environment of the sedimentation basin, as a rule, are clearly pronounced intermittent and uneven. The main allochthonous fluxes of organic matter (mainly from land) are essentially modeled

V. V. Afanas'ev (✉)
Institute of Marine Geology and Geophysics, FEB RAS, Yuzhno-Sakhalinsk, Russia
e-mail: vvasand@mail.ru

V. V. Afanas'ev · A. B. Faustova
Sakhalin State University, Yuzhno-Sakhalinsk, Russia

V. I. Karev (ed.), *Physical and Mathematical Modeling of Earth and Environment Processes—2022*, Springer Proceedings in Earth and Environmental Sciences,
https://doi.org/10.1007/978-3-031-25962-3_46

by its autochthonous regime at the place of accumulation. In addition, depending on the morpholithodynamic trends in the development of the coastal zone, the reserves of organic matter formed in earlier epochs can again enter the ocean–atmosphere circulation. For example, the erosion of the low peaty shores of the Sakhalin Amur Estuary causes about 2 thousand tons of pure carbon to enter the ocean–atmosphere circulation.

The aim of the study is to assess the geomorphological position and parameters of organogenic sedimentation within the uppermost part of the supralittoral zone, marshes and mudflats. Mainly thin silty and pelitic material brought to the silty dry land during the tide phase or by river runoff falls on its surface. Due to these sediments and weakly decomposed organic matter of plant origin, mainly the seagrass Zostera, this surface enters an altitudinal range already favorable for terrestrial vegetation. In our case, these are mainly sedges and pondweeds [2]. The accumulation of peat-gley earth layers at the first stage of formation occurs at a very high rate, according to our data, up to 3 cm/year. It is known that march growth rates strongly decrease with increasing age [3].

The paper deals with the issues of sedimentation in the lagoonal waters of the island. Sakhalin—coastal wetlands, the total area of which is about 2200 km^2, and the length of the coastline of lagoons is 2150 km. Also, in relation to the sequestration of organic carbon, an analysis of the main processes of coastal-marine sedimentation in the estuary zone of the Tsunai and Susuya rivers, which is part of the Sakhalin University marine carboniferous range, is presented. It should be noted that we have not yet begun to study more than 100 km^2 of marches and about 200 km^2 of muddy drylands of northwestern Sakhalin [4]. Meanwhile, it is known that The Amur annually brings about 14 million tons of water into the Amur Estuary and adjacent water areas. Corg [5, 6].

2 Methods and Results

The methodology of biomorpholithodynamic studies of this reconnaissance stage is based on both traditional and modern methods for obtaining and analyzing geospatial and geological-geomorphological information. All measurements were made with SOUTH Galaxy G1 GNSS receivers (GPS, GLONASS, BEIDOU, GALILEO). Pickets at work sites were filmed in RTK (real-time kinematics) mode from a local base station (BS). The coordinates of local BSs were received in static mode from 2 permanent BSs of the EFT-CORS network located in Korsakov and Yuzhno-Sakhalinsk. At the same time, the RTK accuracy is within 1 cm in the plan and 2 cm in height, the statics are no more than 2 cm in the plan and 2 cm in height. The ITRF2014 (epoch 2010.0) coordinates of the BS Korsakov and Yuzhno-Sakhalinsk are taken as reference, the height above the WGS84 ellipsoid. The received picket heights are recalculated into heights relative to the EGM2008 geoid.

Sampling from the silty drying deposits was carried out using a Beaker sampler (Eijkelkamp). Drilling and sampling from the march deposits was carried out using a geoslicer, a set of soil hand drills by Edelman (Eijkelkamp). Sediment analyzes were carried out in the Laboratory of Chemical and Biological Research of Sakhalin State University. Samples from cores and sections for the determination of total carbon, organic carbon, and total nitrogen were taken at five-centimeter intervals. The analysis of arrays of aerial photographs of 1952 and satellite images of 2019–2020 was carried out in the Quantum GIS geoinformation system. The calculations were made on the WGS84 EPSG: 7030 ellipse. The resulting attributes were exported to spreadsheets for further processing. The average error in the location of objects on this map is about 5 m, the minimum error is 0.2 m, the maximum error is 18 m, the standard deviation is about 3 m. march areas for the period 1952–2019–2020.

Figure 1 shows the location of the areas where the studies were carried out.

Currently, the characteristics of the horizontal growth of marches have been determined for the following objects; Chaivo Lagoon, Nyiski Lagoon, Nabil Lagoon, Lunsky Lagoon, Nevskoye Lagoon and Ainskoe Lagoon, Salmon Bay (Table 1).

It should be clarified that the surveyed part of Salmon Bay is an estuary-delta formation with an underwater swell separating it from the main water area, a kind of barrier form, and most of the time, by the type of water exchange, it is a kind of crypto lagoon.

Thus, the area of the water area of the surveyed reservoirs is approximately one third of the area of all lagoons on the island. Sakhalin. The increase in the area of marshes in these lagoons was 10.317 km^2 or 197.173 m^2/year. Accordingly, the reserves of organic carbon recorded naturally in marsh soils, when converted to CO_2 equivalent, are approximately 20–25 thousand tons/year [7].

The volumes of carbon annually fixed in the bottom sediments of the closed Ainskoye lagoon and the Nyiskiy lagoon are about an order of magnitude higher than the Corg volumes recorded in marshes, in the Chaivo lagoon and in the Lososy Bay by about one and a half times. Thus, the overall conservative estimate of carbon sequestration by lagoonal biomorpholithic systems alone, in terms of CO_2 equivalent, is at least 500,000 tons/year. Taking into account the sequestration properties of marshes and droughts of northwestern Sakhalin, the figure is likely to increase several times.

3 Conclusion

There is little doubt that bottom sediments and marches of estuarine-lagoonal geosystems are among the most effective components of biosequestration systems on the planet [8–11]

The proposed model of fixation of organic material in the march formations of the lagoons of the region is based on relatively rough estimates of the parameters of the sedimentation system and requires development and refinement. We have not

Fig. 1 Map of the actual material

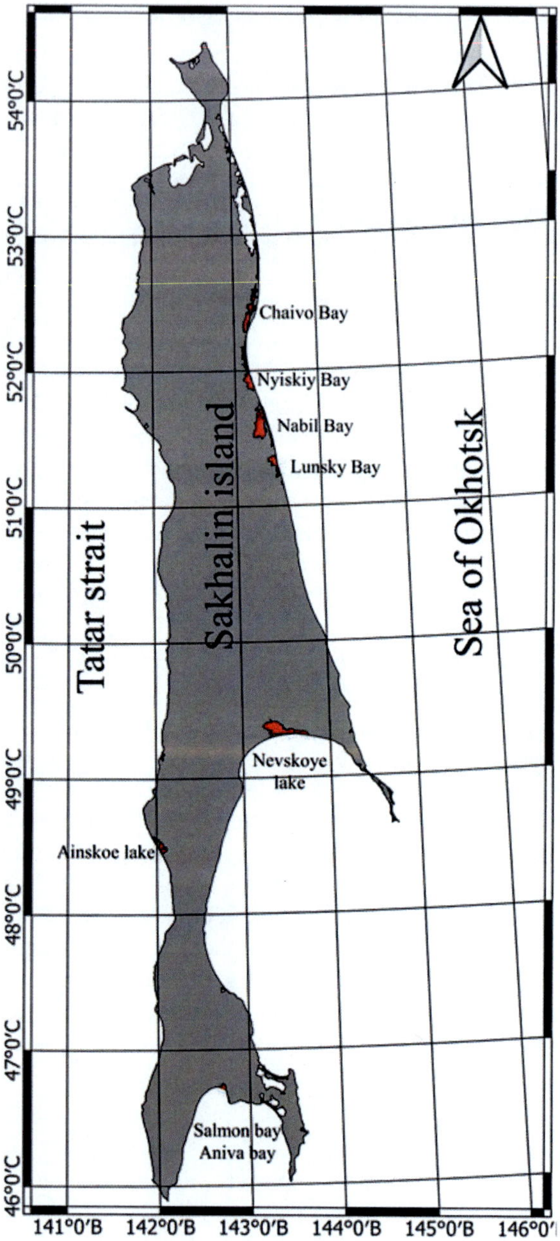

Table 1 Main geospatial parameters of lagoons, solid runoff and dynamics of marches

Lagoon	Water surface area km^2		Change in the area of the water surface km^2			Area change m^2/year	Suspended sediment runoff (thousand tons/year)
	1952	2018–2022	Нарастание маршей	Размыв маршей	Результат		
Chaivo	109.600	107.645	2.379	0.424	1.955	28,765	27.0
Nyiskiy	106.567	104.302	3.059	0.794	2.265	37,750	560.0
Nabil	164.700	162.380	2.388	0.068	2.320	39,320	19.0
Lunsky	53.900	52.739	1.399	0.238	1.161	20,368	7.2
Lagoon Lake Nevskoye	185.127	182.808	3.021	0.701	2.319	34,107	18.2
Lagoon Lake Ainskoe	32.075	31.778	0.856	0.559	0.297	4368	43.7
Salmon Bay	–	–	2.255	0.045	2.210	32,495	100.8
Sum	651.969	641.652	15.357	2.829	10.317	197,173	775.9

yet even approached the solution of the issues of biohydrogeochemical cycle of matter in estuary-lagoon wetlands, which require high-tech methods for obtaining and processing information. We understand that the climatic effect of methane (CH_4) and nitrous oxide (N_2O) fluxes from marshes and mudflats in some areas may offset the "cooling effect" of carbon sequestration in coastal marine sediments, despite the fact that studies devoted to the study of mechanisms that suppress methanogenesis in the coastal area give hope [12].

The work of the next stage of the study of coastal marine biomorpholithosystems, along with the continuation of the process of geospatial inventory, the determination of dynamic parameters and the material composition of carbon-fixing sediments, involves the creation of a system for obtaining data on the level of greenhouse gases in the atmosphere, water and soil gas exchange. Particular attention will also be paid to the creation of ground experimental industrial sites within the ground, supralittoral zone of the Sakhalin State University carboniferous range, where technologies will be tested and artificial zones of accelerated silty-pelitic and organogenic sedimentation will be created. Nature itself tells us in which direction to move (Fig. 2).

Moreover, this practice already exists [13, 14].

Fig. 2 Mud dry surface overgrown with sedges—the first stage of marsh formation

References

1. Hu, Z., Van Belzen, J., Van Der Wal, D., Balke, T., Wang, Z. B., Stive, M., Bouma, T. J., 2015. Windows of opportunity for salt marsh vegetation establishment on bare tidal flats: The importance of temporal and spatial variability in hydrodynamic forcing. Journal of Geophysical Research: 120.7 C. 1450–1469. -doi.org/https://doi.org/10.1002/2014JG002870
2. Kafanov, V.S., Labay, N.V., Pecheneva Biota and macrobenthic communities of the lagoons of northeastern Sakhalin / A. Yu - Sakh.: SakhNIRO, 2003. - 176 p.
3. Temmerman S., Govers G., Wartel S., Metre P. Modelling estuarine variations in tidal marsh sedimentation: response to changing sea level and suspended sediment concentrations // Mar. Geology. 2004. V. 212. № 1-4. P. 1-19.
4. Afanasiev V. V., Ignatov E. I., Chistov S. V. Morphology and dynamics of the banks and bottom of the strait of the Nevelsk area for the design of a permanent railway crossing // Smolensk: Moscow State University, 2008. - 128 p.
5. Koltunov, A. M., Tishchenko, P. Ya., Zvalinsky, V. I., Chichkin, R. V., Lobanov, V. B., Nekrasov, D. A. . Carbonate system of the Amur Estuary and adjacent marine areas // Oceanology. - 2009. - T. 49. - № 5. - P. 694–706
6. Karetnikova E. A., Garetova L. Spatial and temporal distribution of bacterioplankton and bacteriobenthos in the Amur Estuary and adjacent marine areas // Oceanology. - 2015. - T. 55. -№ 5. - P. 776–776.
7. Lipatov, D.N., Shcheglov, A.I., Manakhov, D.V., Brekhov, P.T. Sakhalin // Soil Science. – 2021. – № 2. - P. 211–223.
8. Najjar, R. G., Herrmann, M., Alexander, R., Boyer, E. W., Burdige, D. J., Butman, D. & Zimmerman, R. C. (2018). Carbon budget of tidal wetlands, estuaries, and shelf waters of Eastern North America. *Global Biogeochemical Cycles*, *32*(№ 3), P. 389–416.
9. Ouyang, X., and S. Y. Lee. "Updated estimates of carbon accumulation rates in coastal marsh sediments." *Biogeosciences* 11.18 (2014): P. 5057-5071

10. Spivak, A. C., Sanderman, J., Bowen, J. L., Canuel, E. A., & Hopkinson, C. S. (2019). Global-change controls on soil-carbon accumulation and loss in coastal vegetated ecosystems. *Nature Geoscience*, *12*(№ 9), P. 685–692.
11. Villa, Jorge A., and Blanca Bernal. "Carbon sequestration in wetlands, from science to practice: An overview of the biogeochemical process, measurement methods, and policy framework." *Ecological Engineering* 114 (2018): P. 115-128.
12. Steinmuller, H.E., Hayes, M.P., Hurst, N.R., Sapkota, Y., Cook, R.L., White, J.R., Xue, Chambers L.G., 2020. Does edge erosion alter coastal wetland soil properties? A multi-method biogeochemical study. Catena 187, 104373. doi.org/https://doi.org/10.1016/j.catena.2019.104373.
13. Broome, S.W.; Seneca, E.D., and Woodhouse, Jr., W.W., 1988. Tidal salt marsh restoration. Aquatic Botany, 32, P. 1–22
14. Cadier, C., Bayraktarov, E., Piccolo, R., & Adame, M. F Indicators of coastal wetlands restoration success: a systematic review //Frontiers in Marine Science. – 2020. – T. 7. – P. 600220. https://doi.org/10.3389/fmars.2020.600220

New Research on Characteristics RLC Circuit

Parfentev Nikolay Andreevich and Trukhanov Stepan Vikentievich

Abstract Currently, there is no textbook on physics or electrical engineering that has a complete description of the amplitude frequency responses of a real RCL circuit. It is shown that the frequency characteristics of the RCL circuit have a special point at which the modulus of the complex resistance is not dependent on resistance. We investigated the application domain of the real model of the circuit. Shown a comparison of different schemes of resonant circuit. Also shown a diagram of the mechanical model with similar properties.

Keywords RCL circuit · Mechanic resonance · Frequency · Resistance

1 Introduction

In textbooks and manuals on electrical engineering [7–10] is not carried out a full analysis of the properties RCL circuit. The model the parallel RLC circuit had studied in [1–6] Frequency characteristics of the circuit contain the special point located at a frequency equal to $\frac{f_r}{\sqrt{2}}$, where the impedance does not depend on resistance. In these work investigated the existence region of a module of the impedance of the circuit and phase characteristics. In the study of the imaginary components of the impedance detected another singular point, located at the resonant frequency. Valid component impedance has the shape close to the hyperboloid of rotation. Remains open the question of the physical interpretation of observed phenomena. In this work, the authors continue original research on the oscillatory circuit, which complements the information known from many textbooks and articles on electrical and radio

P. N. Andreevich (✉)
All-Russian Research Institute for Optophysical Measurements (VNIIOFI), Ozernaya street, 46, 199361 Moscow, Russian Federation
e-mail: nic_parfenium@mail.ru

T. S. Vikentievich
Moscow State University of Civil Engineering (MGSU), Yaroslavskoye shosse, 26, 129337 Moscow, Russian Federation
e-mail: svtrou@mail.ru

V. I. Karev (ed.), *Physical and Mathematical Modeling of Earth and Environment Processes—2022*, Springer Proceedings in Earth and Environmental Sciences,
https://doi.org/10.1007/978-3-031-25962-3_47

engineering. As a result of the research, the research has been studied. properties of a complete circuit model that includes additional inductance resistance and capacitor leakage resistance.

2 Analysis of the Contour Model

Consider two models of the contour built according to the schemes (Fig. 1). The model with a parallel connection A is the closest to reality, since a real inductor always has an active resistance, and real high-quality capacitors have a very high leakage resistance, which can be ignored in calculations.

A model with a series connection B is doubly impossible, since it contains an ideal inductance and an unrealistically bad (large leakage) capacitor. Apparently, because of these circumstances, this scheme has not been previously investigated.

However, both models are described by similar equations when the impedance is calculated for model A and complex conductivity for model B.

Для модели A impedance is easy to calculate by the formula:

$$\overline{Z} = \frac{\overline{R} + j\overline{\omega}}{j\overline{\omega}\overline{R} + \left(1 - \overline{\omega}^2\right)}, \tag{1}$$

$\overline{Z}$ and $\overline{R}$—complex and active resistance normalized by the characteristic resistance of the circuit is $\sqrt{\frac{L}{C}}$, and $\overline{\omega}$ is the frequency (circular) normalized by resonance frequency. The relative frequency equal to $\frac{1}{\sqrt{2}}$, the formula takes

$$\overline{Z} = \frac{\overline{R} + \frac{j}{\sqrt{2}}}{j\frac{\overline{R}}{\sqrt{2}} + \frac{1}{2}} \tag{2}$$

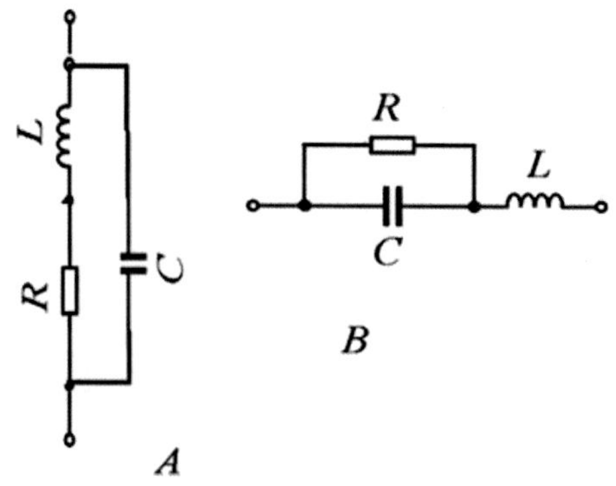

Fig. 1 Two types of the contour RLC

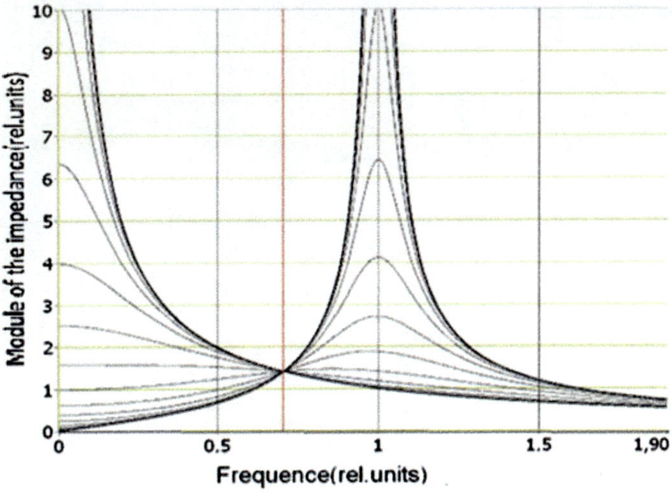

Fig. 2 The dependence of the module of the impedance (rel.units) in the circuit frequency

It is obvious that the module of relative impedance at this frequency does not depend on the resistance and equal to $\sqrt{2}$.

The vector phase of the semicircle at $\overline{R} = 0$ equal $\overline{\frac{j\overline{\omega}}{(1-\overline{\omega}^2)}}$, when $\overline{R} \to \infty$ equal $\frac{1}{j\overline{\omega}}$.

In Fig. 2 presents the dependence of the нормированного импеданса, the frequency at various values of relative resistance (at unity voltage).

In Fig. 2 clear the area of existence of the current is bounded by two asymptotes $\frac{\omega}{|1-\overline{\omega}^2|}$ at $R = 0$ when and $\frac{1}{\overline{\omega}}$ at $\overline{R \to \infty}$.

Note that the graph shows lins of the type of curves—resonant at low values of active resistance and aperiodic at high values of active resistance.

These characteristics coincide with the conductivity in model B.

All the curves of an impedance module (and asymptotes) have the general point with value of the relative module $\sqrt{2}$ at the relative frequency equal $1/\sqrt{2}$. The existence region of the module from above is limited by the second asymptote at low frequencies, and from below the first, and for frequencies is higher critical on the contrary.

Note that the product of frequency stable points on the value of the module is equal to the product of the resonance frequency to the characteristic impedance. It is possible that this fact will be useful to solve the main problem (Fig. 3).

At the critical point corresponding to the impedance stability, the graph is a circle with a radius equal $\sqrt{2}$ (Figs. 4 and 5).

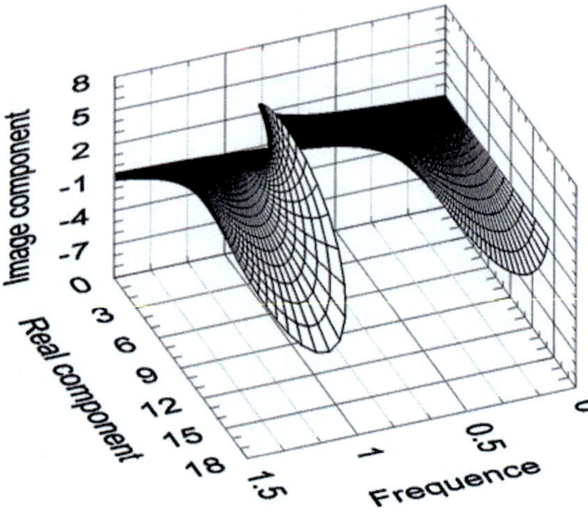

Fig. 3 3D image of real and imaginary impedance contour

3 The Imaginary Part of the Impedance

The components of the impedance calculated depending on the relative frequency and the relative resistance in the course of studying this issue.

$$Z = \frac{\overline{\omega}\left(1 - \overline{r}^2 - \overline{\omega}^2\right)}{\left(1 - \overline{\omega}^2\right)^2 + \overline{r}^2\overline{\omega}^2}, \tag{3}$$

where $\overline{\omega}$ is normalized frequency, $\overline{r}$—is the normalized resistance. The computation results are shown in Fig. 6. Features include a special point located at the resonant frequency. The imaginary component does not depend on the resistance and equal to -1 at this point.

The three asymptotes form the region of existence of imaginary components

1. asymptote $-\frac{1}{\overline{\omega}}$,when $\overline{r} \to \infty$,
2. asymptote $\frac{\overline{\omega}}{1-\overline{\omega}^2}$,when $\overline{r} \to 0$,
3. cut the y-axis from 0 до $-\infty$.

Two asymptotes are marked by fat lines in Fig. 4. Special points investigated in these characteristics: the extreme of the imaginary components of the frequency. The results are shown in Fig. 7. The maximum of the imaginary part of impedance is present only in the range of relative resistances from 0 to 1. The minimum of imaginary components is present throughout the range of relative resistances. The characteristic point associated with the factor $\sqrt{2}$, which arises in the analysis of the dependence of impedance on frequency. The minimum is the resonance frequency at this value of relative resistance, as well as at zero resistance value. When the resistance is $1/\sqrt{2}$, the minimum is at the maximum possible frequency in excess

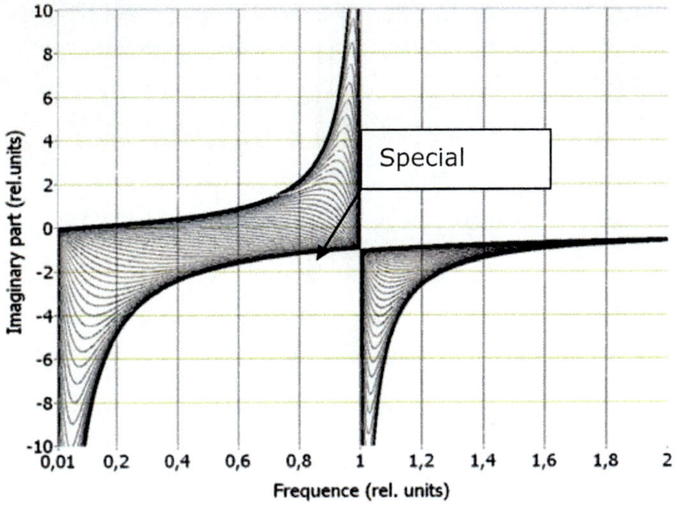

Fig. 4 Dependence of the imaginary components of impedance against frequency and normalized resistance

of approximately 1.1719 times the resonant frequency. The last digit occurs in the numerical solution of the cubic equation.

4 The Real Part of the Impedance

The real component of the impedance of such a circuit—Re, normalized by the characteristic loop resistance, is calculated according to the formula

$$Re = \frac{\overline{r}}{\left(1 - \overline{\omega}^2\right)^2 + \overline{r}^2\overline{\omega}^2} \tag{4}$$

where $\overline{\omega}$ and $\overline{r}$ are the normalized parameters introduced earlier. The result of calculation (Fig. 8)

The two asymptotes form the region of existence of real component:

1. asymptote—$\frac{1}{2\overline{\omega}\left|1-\overline{\omega}^2\right|}$
2. the x-axis.

The first asymptote has a local minimum at the relative frequency $\frac{1}{\sqrt{3}}$. The real part of the impedance is equal $\frac{3\sqrt{3}}{4}$ at this frequency (Fig. 5).

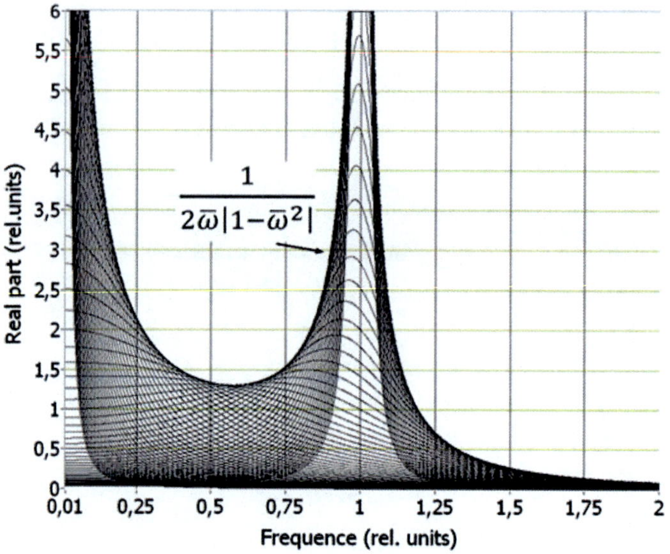

Fig. 5 Real part of impedance RCL circuit

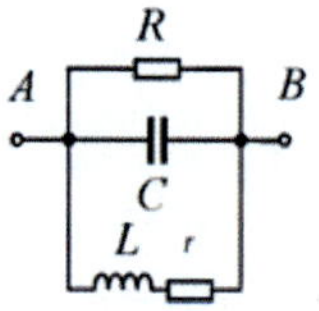

Fig. 6 General scheme for the parallel circuit

5 Model with Capasitor Liekage Resistor

Turning to a model of the circuit, sequentially connect the inductor and an additional resistance $\overline{r}$ (normalized by the characteristic resistance).

The complex conductivity for a parallel circuit may be represented as:

$$\overline{Y} = \frac{\left(1 - \overline{\omega}^2\right)\overline{Y_r} + j\overline{\omega}}{j\overline{\omega}\overline{Y_r} + 1} + \overline{Y_R} \tag{5}$$

It is easy to notice that the formula (5) coincides with (1) when connected to inductance of the additional resistance—$\overline{r}$.

In Fig. 7 the calculations of the impedance circuit when the magnitude of $\overline{r}$ r = 0.5

The region of existence of the impedance contour in the General case, limited to three asymtote:

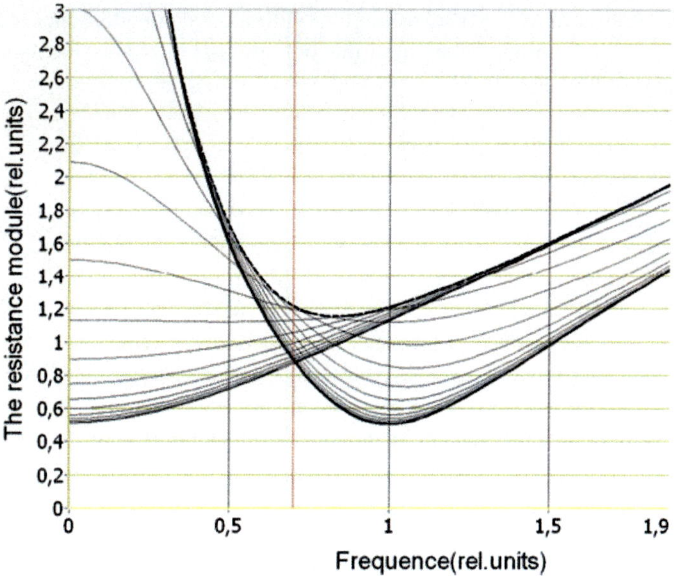

Fig. 7 The impedance of the circuit when $\overline{r} = 0.5$

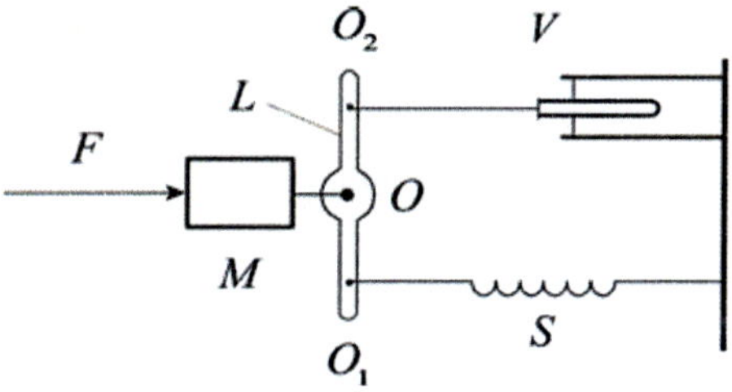

Fig. 8 Mechanical model of a parallel electrical circuit. F—the variable power, S—spring, O_1—hinge spring, L—weightless lever, O—the main hinge, O_2—hinge of the second link, M—solid body V—element viscous friction

1. Asymptote $\overline{Z}_{\min} = \sqrt{\overline{r}^2 + \overline{\omega}^2}$ for range of frequencies from 0 to $\frac{\overline{\omega}}{\sqrt{2}}$.
2. Asymptote $\overline{Z}_{\min} = \sqrt{\overline{r}^2 + \left(\frac{1-\overline{\omega}^2}{\overline{\omega}}\right)^2}$ for frequency $\frac{\overline{\omega}}{\sqrt{2}}$ to ∞.
3. Asymptote $\overline{Z}_{\max} = \sqrt{\overline{r}^2 + \left(\overline{\omega} - \frac{1}{2\overline{\omega}}\right)^2} + \frac{1}{2\overline{\omega}}$

It is obvious that in this case there is a particular point on the frequency $\frac{\overline{\omega}}{\sqrt{2}}$. It is interesting to note that when $\overline{r} \to 0$ the third asymptote has two values of the derivative at the point $\frac{\overline{\omega}}{\sqrt{2}}$.

The local minimum of the third asymptote with increasing additional resistance increases from $\frac{\overline{\omega}}{\sqrt{2}}$ to infinity. When the value of $\overline{r}$ is equal to $\sqrt{2}$ is a local minimum at the resonant frequency.

6 Mechanical Analogues

A mechanical analog of the considered type of the circuit shown in Fig. 8.

In a horizontally located model, there are no frictional forces between the elements of the model and the underlying surface. It is also assumed that the connection between the deformation and the applied force occurs according to the line law, as a consequence of the assumed small displacement. The mechanical model consists of an upper subsystem, a massive body M and the bottom containing spring and viscous element.

Weightless spring S (low mechanical losses), is characterized by the coefficient of flexibility is the inverse of stiffness.

If the forces applied to the spring and the element with viscous friction are equal, the system is like an electrical oscillatory circuit to the branches of which the same voltage is applied. The addition of the displacements of both branches of the model corresponds to the addition of currents in the path. A lever with an extreme low mass, rotating relative to the point O, ensures the fulfillment of both conditions formulated above (Fig. 8).

The equality of forces and moments of forces acting on both arms of the system corresponds to the first and second equilibrium conditions of a solid.. In case of equal arms of the lever forces applied to the links will be equal to half the force F applied to the system. The speed of movement of the axis тcalculated as the average speed of movement of a spring and an element with viscous friction:

$$U_L = \frac{Us + Uv}{2} \tag{6}$$

Us—speed link with a spring, and the *Uv*—speed link with viscous element. In the result, the resonance frequency of the mechanical system is equal to

$$\omega_r = \frac{2}{\sqrt{mC}} \tag{7}$$

The total mechanical resistance of the system in complex form coincides with f.3. In this case the calculated resistance in the formula should be 4 times higher viscous resistance level of the lower subsystem.

Properties of an ideal mechanical system consisting of a spring and a load on a horizontal smooth surface are paradoxical. In the case when the force applied to the system will be determined by the formula $F_t = \pm F_0 \cos(\omega t)$, where ω is the resonant frequency, after instantaneous compression (stretching) of the spring, the point of application of the force will be fixed.

7 Conclusions

1. It is shown that the singular point exists in the frequency characteristics of the circuit.
2. Determined the region of existence of the impedance of the series resonant circuit.
3. It is revealed that both types of contour are described by a single mathematical form of resistance in the serial circuit are changed to the conductivity in parallel.
4. The model of the mechanical system, analogous to the serial circuit.

References

1. Parfentev N.A. Paradoxes the Model of an RLC Circuit. Engineering Physics Volume 2, Issue 3, June 2017, Pages: 49–53
2. Parfentyev N. A., Parfentyeva E.N., Sevastyanov S. I. Pecularity of frequency characteristics of a parallel oscillatory contour. News of higher education institutions. Electronics No. 62014, page 110.
3. Parfentyev N. A. Invariant one model of a mechanical system with viscous friction, Electronic journal technical acoustics. 2014.
4. Parfentyev N. A., Parfentyeva N.A., Truchanov S. V.. On the singular point of the frequency characteristics of oscillatory systems. Natural and technical Sciences, №9–10, 2014, c. 31.
5. Parfentyev N.A.,Parfentyeva E.N.,Krivolutsrfya N.V. Amazing properties of a parallel electrical circuit. XIX International Scientific and Practical Conference. Russia St. Petersburg, 19–20.02.2016 page.11–15
6. Parfentev N.A. New paraljxes the model of an RLC circuit. Science of Europe. N12. Vol.1, page 19–23
7. MEASUREMENT AND CALIBRATION HANDBOOK Precision Measurement Equipment Laboratory... L. SERIES RCL EL, XL EA, ZT (Inductively). Circuit Analysis: Theory and Practice 5th Edition.
8. Horowitz and Hill, the Art of Electronics, Second Edition, Cambridge University Press, 1989. ISBN 0521370957.
9. US Navy, Basic Electrity, Dover, 1970. ISBN 0486209733.
10. http://www.wiley.com/college/comer/ ISBN 0471410160.

Inelastic Deformation of Rocks with Deformation and Strength Anisotropy

K. B. Ustinov

Abstract Problems, related to various types of anisotropy of rocks for elastic and inelastic deformation are considered. Two types of strength anisotropy of rocks are described. Generalizations of the model of plastic deformation of rocks due Drucker and Prager, as well as applicability of associated and non-associated plastic flow lows, are discussed.

Keywords Strength anisotropy · Non-tensor functions · Non-associated flow rule · Dilatancy

1 Introduction

Generalizing the huge amount of accumulated experimental data on rock plasticity, one may conclude that most essential peculiarities of inelastic deformation of rocks (excluding hard brittle rocks, for which using theories of plasticity is irrelevant), different from plasticity of metals, are: (i) dependence of the start of inelastic deformation not only on shear, but also on volumetric stresses; (ii) loosing of co-axiality between strain tensor increment and stress tensor, which makes using of traditional theories of plastic flows inadequate for description of those processes; (iii) existence of deformation and strength anisotropy.

Rocks are usually heterogeneous media, often having layered texture. That is almost compulsory for sedimentary and metamorphic rocks. Effective properties of such media are described in the approximation of transversal isotropic continua with the plane of isotropy being parallel to layering. For properties described by 2-nd rank tensors (thermo-conductivity, electro-conductivity, thermal expansion, etc.) components of the corresponding tensors are expressed through two independent values, characterizing the corresponding properties in the plane of isotropy and along the normal to this plane). For properties described by 4-nd rank tensors, such as elasticity, the components of the corresponding tensors are expressed in terms of five

K. B. Ustinov (✉)
Ishlinsky Institute for Problems in Mechanics of the Russian Academy of Sciences, Moscow, Russia
e-mail: ustinoff127@mail.ru; ustinov@ipmnet.ru

V. I. Karev (ed.), *Physical and Mathematical Modeling of Earth and Environment Processes—2022*, Springer Proceedings in Earth and Environmental Sciences,
https://doi.org/10.1007/978-3-031-25962-3_48

independent constants. The role of various parameters that describe the degree of elastic anisotropy was discussed in [1]. However, some important properties, first of all, properties characterizing strength, do not possess tensor nature.

2 Typical and Atypical Strength Anisotropy of Rocks

Usually, strength anisotropy of rocks is attributed to existence of weekend planes related to layering or to systems of existing cracks [2–5]. In such a case, dependence of strength under compression on the angle between the applied stresses and the plane of layering (plane of isotropy) possesses a minimum, corresponding approximately 30^0 and two maximums, corresponding to compressive stresses acting along and normally to this plane, the second of which being usually more pronounced [5, 6]. However a number of later experimental data, obtained in laboratory of Geomechanics of IPMech PAS, reviled that such dependencies are observed not for all types of rocks [7, 8]. Three independent sets of experiments carried out on porous weak Senoman sandstones [8] reveal atypical anisotropy: monotonous dependence of compressive strength on the angle between the maximal applied compressive stress and the plane of bedding with the maximal strength corresponding to loading normally to bedding and the minimal strength corresponding to loading along the bedding plane with no extremums in between. The hodographs of strength for the typical and atypical strength anisotropy is schematically shown on Fig. 1.

Existence of anisotropy of rocks of such a type suggests an alternative view on some geomechanical problems, such as stability of wellbores and interpretation of log measurement of wellbore breakouts while determining the in situ stress state [8]. Existence of anisotropy of such a type had an indirect evidence followed by an analysis of shapes of breakouts of holes in coal [9].

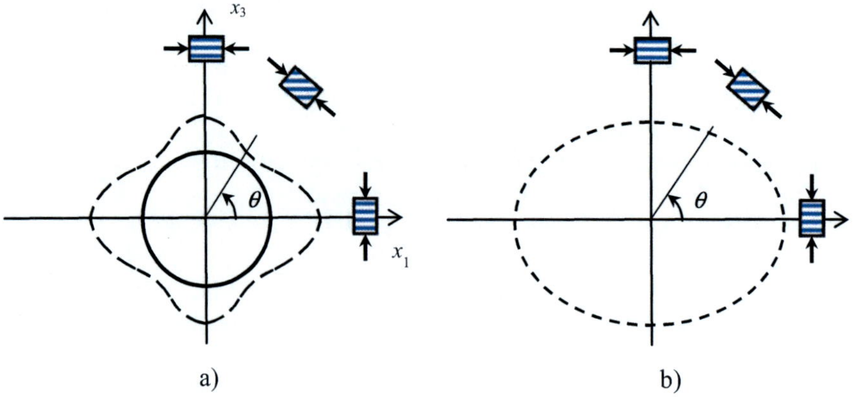

Fig. 1 Hodographs of the critical stresses for rocks with typical **a** and atypical **b** anisotropy

Accounting for strength anisotropy is necessary for control of stability of wellbores [10–14], and for analysis of data of wellbores breakouts used for estimations of magnitudes and directions of in situ stresses [4]. The described atypical anisotropy is expected inherent to weak porous rocks, therefore it has to be accounted while processing hydro-carbonate deposits and underground gas storages (UGS).

3 A Variant of System of Equations

Reaching the critical state, like the ones described above, does not implies immediate fracture, the rocks, especially soft sedimentary rocks forming reservoirs of hydrocarbonate deposits, continue to deform. A promising method for description of this inelastic deformation is using a properly modified theory of plastic flow.

Consider, as an example, a system of equation describing elasto-plastic deformation of anisotropic rocks with non-associated law and isotropic hardening with yield function F and plastic potential Q written in the form of a modified Hill's [15, 16] law in the form of [17], (see also [5]).

$$\sigma_{ij,i} = 0 \tag{1}$$

$$\sigma_{ij} = \Lambda_{ijkl}\varepsilon_{kl}^{E} \tag{2}$$

$$\varepsilon_{ij}^{T} = \varepsilon_{ij}^{E} + \varepsilon_{ij}^{P} = \frac{1}{2}\left(u_{i,j} + u_{j,i}\right) \tag{3}$$

$$d\varepsilon_{ij}^{P} = \frac{\partial Q}{\partial \sigma_{ij}}\mathrm{d}\lambda, \tag{4}$$

$$d\lambda = \frac{\partial F}{\partial \sigma_{km}} ds_{km}\left(H\frac{\partial Q}{\partial \sigma_{pq}}\sigma_{pq}\right)^{-1}, \tag{5}$$

$$\begin{aligned} F =&\Big(G^{0}_{(23)}(\sigma_{22} - \sigma_{33})^{2} + G^{0}_{(13)}(\sigma_{11} - \sigma_{33})^{2} + G^{0}_{(12)}(\sigma_{11} - \sigma_{22})^{2} \\ &+2L^{0}_{(23)}\sigma_{23}^{2} + 2L^{0}_{(13)}\sigma_{31}^{2} + 2L^{0}_{(12)}\sigma_{12}^{2}\Big)^{1/2} \\ &+ \left(B^{0}_{(1)}\sigma_{11} + B^{0}_{(2)}\sigma_{22} + B^{0}_{(3)}\sigma_{33}\right) - A(k) \end{aligned} \tag{6}$$

$$\begin{aligned} Q =&\Big(G^{0}_{(23)}(\sigma_{22} - s_{33})^{2} + G^{0}_{(13)}(\sigma_{11} - \sigma_{33})^{2} + G^{0}_{(12)}(\sigma_{11} - \sigma_{22})^{2} \\ &+2L^{0}_{(23)}\sigma_{23}^{2} + 2L^{0}_{(13)}\sigma_{31}^{2} + 2L^{0}_{(12)}\sigma_{12}^{2}\Big)^{1/2} \\ &+ \left(B^{1}_{(1)}\sigma_{11} + B^{1}_{(2)}\sigma_{22} + B^{1}_{(3)}\sigma_{33}\right) \end{aligned} \tag{7}$$

$$dk = \sigma_{ij}d\varepsilon_{ij}^{P} \tag{8}$$

Here σ_{ij} are components of stress tensors; ε_{ij}^T, ε_{ij}^E, ε_{ij}^P are components tensors of the total, elastic and plastic strain; u_i are components of displacement vector; Λ_{ijkl} are components of elasticity tensor. Isotropic hardening, governed by parameter k, related to the work of plastic strains (8). Function $H \equiv \partial A/\partial k$ is a material characteristic to be obtained from experiments. In many cases it may be set constant $H = E_p^{-1}$, where E_p is the plasticity modulus; $G_{(ij)}^0$, $L_{(ij)}^0$, $B_{(i)}^0$, $B_{(i)}^1$ are material constants. The introduced yield function F and plastic potential Q differ only by parts related to normal stresses governed by constants $B_{(i)}^0$, $B_{(i)}^1$. Therefore, the "non-association" relates to volumetric part only. For isotropic case the model is reduced to Drucker-Prager model. Using the associated law would result to unrealistic volumetric stains, which was, probably, the main reason for introducing the non-associated flow rule [18, 19].

Also isotropic work hardening law is used in (8), kinematics hardening, or a combination of both may also be used [5]. Kinematic (translation) hardening law may be also considered as a way to introduce initial or induced anisotropy [5].

Note, that contrary to elasticity, where increments of elastic strains depend on increments of stresses, according to the theory of plastic flow, which is supported by enormous amount of experimental data, increments of plastic strains are proportional to the derivatives of plastic potential over corresponding components of stress tensor. Hence, ratio of increments of elastic strains depends on the increments of stresses, while ratio of increments of plastic strains depends on the components of stresses in the actual state. In other words, contrary to elasticity, the direction of the increment of plastic strain in the strain space is prescribed by the current stress state and does not depend on the direction of the increment of stresses in stress space.

Note also, that plastic potential Q in (4) is written in terms of stresses, i.e. it is written in stress space, while the left part of Eq. (4) relates to the strain space. According to the traditional approach the left-hand and right-hand parts (both 2-nd rank tensors) are related through a scalar, $d\lambda$. However, the most general relation between two 2-nd rank tensors is through a 4-th rank tensor. Thus, introducing a tensor of plastic moduli M_{ijkl} we may write an analogous of (4) a kind of associated law, but using a tensor relation between the left-hand and right-hand parts:

$$d\varepsilon_{ij}^P = \mathrm{M}_{ijkl} \frac{\partial F}{\partial \sigma_{ij}} d\lambda, \tag{9}$$

Therefore, instead of using plastic potential, at least written in the form (7), and formally different from the yield function (6), one may use a "pseudo-associated law" (9) by choosing a properly defined tensor of plasticity M_{ijkl}.

4 Deformation Under High Compressive Stresses

Under relatively low stresses the yield function in a projection of stress space to $\sigma\tau$ plane form a monotonous function, schematically depicted as curve ABCD on Fig. 2,

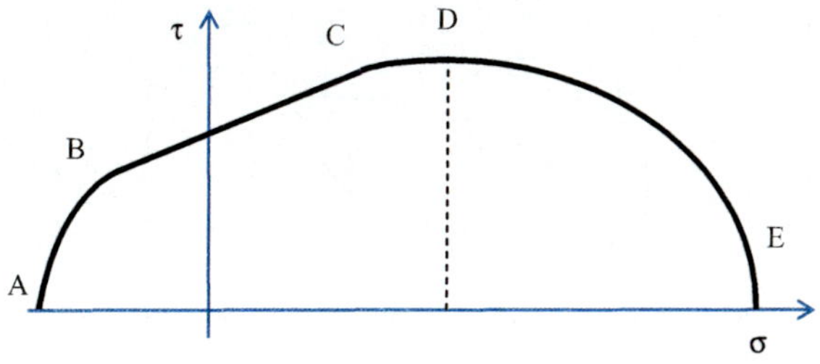

Fig. 2 Yield function for rocks describing behavior under high compressive stress

almost linear part BC being described by Coulomb-Mohr or Drucker-Prager criteria. However, fracture, or more generally, transition to inelastic deformation may take place under higher compressive stress and relatively small shear stress mainly due to closure of pore space. This process, especially pronounced in high-porous rocks is described by part DE of the yield curve on Fig. 2. Various variants of models are suggested to describe this descending part of the curve, like "cup" models, in which the yield line is approximated by piece-wise function [20–22], or Cam-clay models approximating the whole curve as an ellipse [23–25].

5 Inelastic Volumetric Deformation

Contrary to plasticity of metals, for which both magnitude of plastic volumetric strains and influence of volumetric stress on plastic deformation, and are usually negligible, it is not the case for rocks. Influence of volumetric stresses on elastic–plastic transition is accounted for in Coulomb-Mohr and Drucker-Prager criteria; existence of volumetric strains during inelastic deformation is a well-known fact. To describe the inelastic volumetric strain Reynolds [26] proposed the concept of dilatancy, according to which the increment of the volumetric inelastic strain is proportional to the increment of intensity of shear inelastic strains. It implies that inelastic volumetric strains occur due to rearrangement of the structure, caused by shear strains. This concept is in agreement with the spirit of the theory of plastic flaw: the ratio of increments of volumetric and shear strains depends on the acting stress, rather than their increments, which is accounted for by formulae (4)–(8).

Inelastic volumetric strains may be also negative, corresponding to compaction. This phenomenon in usually manifests itself in porous rocks [22]. When using the associated flow rule, the AD part or the yield function (Fig. 2) would associate with inelastic expansion (dilatancy), DE part would associate with inelastic compaction, point D would correspond to pure shear deformation. However according to a non-associated flow rule, the transition between militancy and compaction should not necessarily coincide with the point D.

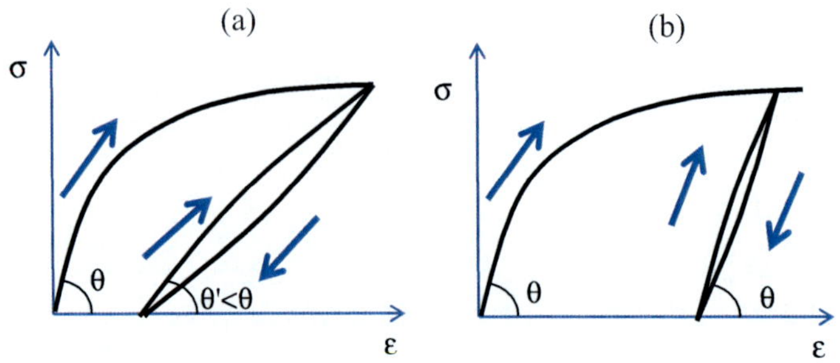

Fig. 3 Typical deformation curves **a** for rocks; **b** for metals

6 Problems Related to Using Plastic Flow Laws for Rocks

In spite of evident efficiency of using plastic flaw theories for description deformation of soft rocks there are some fundamental questions.

1. Any law of plastic flow implies hardening, which means that during unloading and following cycles of loading the material deform elastically with the initial elastic constants up to the maximal level of stresses achieved in the previous cycles. However, deformation of rocks often reveals another type of deformation. During the following cycles elastic moduli are often less than the initial ones, which implies some changes in rock structure, like a kind of damage (Fig. 3). Therefore, deformation of rocks is often described as a softening rather than hardening process [27].

2. Another problem may arise when specifying the yield function for anisotropic rocks. Due to a strong dependence of the critical stresses on the orientation, the yield function (as well as plastic potential) written in space of stresses may loose convexity. Itself the problem is not so drastic, because such a behavior reflects the real mechanical processes. However, the corresponding mathematical model and algorithms may become unstable. Note, that no such problem occurs when dealing with rocks possessing the atypical strength anisotropy mentioned above.

7 Conclusion

Two types of strength anisotropy of rocks, namely the traditionally considered weakening along the planes of bedding, and an atypical anisotropy, manifesting itself as monotonous dependence of compressive strength versus the angle between the applied compression and the plane of bedding, with the minimal strength corresponding to loading along the bedding plane, are discussed. It was pointed out that accounting for strength anisotropy of both types is necessary for control of stability of wellbores while processing hydro-carbonate deposits and UGSs, for analysis of data of wellbores breakouts used for estimations of in situ stresses.

Theory of plastic flow may be used for description of inelastic deformation of rocks, although with some caution. The theory allows the possibility to account for deformation and strength anisotropy (Hill's anisotropic model, a model accounting two mechanisms of fracture). It was shown that application of non-associated laws appears more preferable to compare with the associated law, because the results agree better with the experimental data.

Acknowledgements This work was financially supported by the Russian Science Foundation № 22-11-00273

References

1. *A. B. Zhuravlev and K. B. Ustinov,* On Values Characterizing the Degree of Elastic Anisotropy of Transversely Isotropic Rocks; Role of Shear Modulus Mechanics of Solids, 2019, Vol. 54, No. 6, pp. 958–967
2. *Goodman R.E.* 1980. Introduction to rocks mechanics. New York, John Wiley and Sons, 478 p
3. *Germanovich, L.N., Galybin, A.N., Dyskin, A.V., Mokhel, A.N., Dunayevsky V.* 1996.Borehole stability in laminated rock. In G. Barla (Ed.). Prediction and Performance in Rock Mechanics and Rock Engineering, Torino, Italy ed., V. 2, P. 767–776). CRC Press/Balkema.
4. *Zoback, M.D.* Reservoir Geomechanics. 2007. Cambridge University Press, Cambridge, ISBN-978–0–521–77069–9. 443p. https://doi.org/10.1017/CBO9780511586477
5. *Karev, V; Kovalenko, Y; Ustinov, K.* 2020. Geomechanics of Oil and Gas Wells Springer 2020 184p DOI: https://doi.org/10.1007/978-3-030-26608-0
6. *Singh M., Samadhiya N. K., Kumar A., Kumar V., Singh B.* 2015. A nonlinear criterion for triaxial strength of inherently anisotropic rocks. *Rock Mech. and Rock Eng.* 48(4). 1387-1405.
7. *Klimov D.M., Karev V.I., Kovalenko Yu.F., Ustinov K.B.* 2022. On atypical strength anisotropy of weak sandstones. Doklady RAS. Mathematics, Informatics, Control, 506: 108–112. In Russian
8. *Kovalenko Y.F., Ustinov K.B., Karev V.I.,* 2022. Geomechanical analysis of breackouts in wellbores. *Izvestia RAS MTT,* №6: 157–172. In Russian. Accepted for publication.
9. *Kaiser P.K. Guenot A, Morgenstern N.R.* Deformation of small tunnels. IV. Behaviour during failure // Int. J. Rock Mech. Min. Sci. Geomech. Abstr. 1985. V. 22. P. 141–152.
10. *Karev, V., Kovalenko, Y., Ustinov, K. 2020.* Mechanical and mathematical, and experimental modeling of oil and gas well stability. *Advances in Oil and Gas Exploration and Production. 2020.*
11. *Karev V.I., Kovalenko Yu.F., Zhuravlev A.B., Ustinov K.B.* Model of filtration to a wellbore accounting the dependence of permeability on stress state. *Processes in Geomedia.* 2015. № 4: 34-44. (In Russian).
12. *Karev V.I., Kovalenko Y.F., Ustinov K.B.* Modeling deformation and failure of anisotropic rocks nearby a horizontal well *Journal of Mining Science. 2017. V. 53. № 3. P. 425-433..*
13. *Klimov D.M., Karev V.I., Kovalenko Yu.F., Ustinov K.B.* 2006. Mathematical and physical modeling of rock failure during drilling inclined wells. *Technology TEK.* №5. 22 – 27. (In Russian).
14. *Karev, V.I., Klimov, D.M., Kovalenko, Y.F. et al. Fracture Model of Anisotropic Rocks under Complex Loading. Phys Mesomech 21, 216–222 (2018). https://doi.org/*https://doi.org/10.1134/S1029959918030050
15. *Hill R* 1948 A theory of the yielding and plastic flow of anisotropic metals Proc. Roy. Soc. London A 193 281–97

16. *Hill R* 1983 The Mathematical Theory of Plasticity New York: Oxford University Press
17. *Caddel R M, Raghava E S, Atkins A G* 1973 A yield criterion for anisotropic and pressure dependent solids such as oriented polymers J of Materials science 8 1641–6.
18. *Nikolaevskii VN.* Governing equations of plastic deformation of a granular medium. Journal of Applied Mathematics and Mechanics. 1971;35:1017-1029
19. *Nikolaevskii VN.* Mechanical properties of soils and plasticity theory. In: Solid Mechanics. 1972;6:5–85.
20. *Issen KA, Rudnicki JW.* Theory of compaction bands in porous rock. Physics and Chemistry of the Earth, Part A: Solid Earth and Geodesy. 2001;26:95-100.
21. *Grueschow E, Rudnicki JW.* Elliptic yield cap constitutive modeling for high porosity sandstone. International Journal of Solids and Structures. 2005;42:4574-4587.
22. *Garavand A., Myasnikov A.V., Stefanov Y.P., Bakeev R.A., Rebetsky Y.L.* 2020. Numerical modeling of plastic deformation and failure around a wellbore in compaction and dilation modes. International journal for numerical and analytical methods in geomechanics. V 44, 6. P. 823–850.
23. *Roscoe K., Burland J.B.* On the generalized stress-strain behaviour of wet clay // Proceedings of Engineering Plasticity. 1968. P. 535 – 609.
24. *Kuznetsov S.V.* Cam-clay models in mechanics of granular materials // Mechanics and Mechanical Engineering. 2017. Vol. 21. № 4. P. 813 – 821.
25. *Kuznetsov S.V.* 2020. Critical State Models in Non-Cohesive Media Mechanics (Review). Mechanics of Solids, 55, № 8, c. 1423–1431.
26. *Reynolds O.* On the dilatancy of media composed of rigid particles in contact, with experimental illustrations // Philosophica Magazine 1885. Series 5. № 20 (127). pp. 469–481.
27. *H.E.Read, G.A.Hegemier.* Strain softening of rock, soil and concrete — a review article. Mechanics of Materials. V. 3, Is. 4, December 1984, P. 271–294.

Carbon Sequestration in the Coastal Marine Biomorpholithosystems of the Salmon Bay of Aniva Bay

Victor V. Afanas'ev, E. M. Latkovskaya, A. V. Uba, and A. I. Levitsky

Abstract Determining the carbon sequestration capacity of coastal marine ecosystems in Russia is a process that has just begun. The category of priority tasks includes issues related to determining the geomorphological position and geospatial parameters of areas with an increased intensity of organogenic sedimentation. In this paper, for the first time, numerical values of the dynamics of silt dredges and marshes in the estuary zone of the sea bay are given. The potential of the main processes of coastal-marine sedimentation in relation to the sequestration of organic carbon is considered. It is assumed that the carbon sink capacity of the estuarine zone of the Tsunai and Susuya rivers, which is about 9 thousand tons in CO_2 equivalent, will be adjusted upwards.

Keywords Estuary · Salt marsh · Intertidal mudflat · Carbon sequestration

1 Introduction

The study area is located in Salmon Bay, the northern part of the hall. Aniva, which, according to its geomorphological essence, is the estuarine zone of the Tsunai and Susuya rivers. The marshes formed in this deltaic system are mainly composed of oligohaline wetlands due to fresh water discharges through the delta's canal network. Like other elements of the morpholithic system of deltas, wetland formations and mudflats are subject to significant rearrangements. The volume of suspended sediment runoff at a compacted silt density of 1.2 t/m^3 is about 80 thousand m^3/year. The tides in Aniva Bay are diurnal. The value of tropical straits is about 1.0 m, and that of equal straits is 0.5 m. The recorded amplitude range is 2.4 m; repeatability 20–50 years [1].

V. V. Afanas'ev (✉) · A. V. Uba · A. I. Levitsky
Institute of Marine Geology and Geophysics, FEB RAS, Yuzhno-Sakhalinsk, Russia
e-mail: vvasand@mail.ru

V. V. Afanas'ev · E. M. Latkovskaya · A. V. Uba · A. I. Levitsky
Sakhalin State University, Yuzhno-Sakhalinsk, Russia

V. I. Karev (ed.), *Physical and Mathematical Modeling of Earth and Environment Processes—2022*, Springer Proceedings in Earth and Environmental Sciences,
https://doi.org/10.1007/978-3-031-25962-3_49

The aim of the study is to determine the geomorphological position and dynamics of marshes and silt drains of the Sakhalin Carboniferous Polygon, as well as to estimate the content of organic carbon in these climate-modeling components of biomorpholithosystems.

2 Methods and Results

The methodology of biomorpholithodynamic studies is based on both traditional and modern methods for obtaining and analyzing geospatial and geological-geomorphological information. All measurements were made with SOUTH Galaxy G1 GNSS receivers (GPS, GLONASS, BEIDOU, GALILEO). Pickets at work sites were filmed in RTK (real-time kinematics) mode from a local base station (BS). The coordinates of local BSs were received in static mode from 2 permanent BSs of the EFT-CORS network located in Korsakov and Yuzhno-Sakhalinsk. At the same time, the RTK accuracy is within 1 cm in the plan and 2 cm in height, the statics are no more than 2 cm in the plan and 2 cm in height. The ITRF2014 (epoch 2010.0) coordinates of the BS Korsakov and Yuzhno-Sakhalinsk are taken as reference, the height above the WGS84 ellipsoid. The received picket heights are recalculated into heights relative to the EGM2008 geoid.

Analyzes of bottom sediments were carried out in the Laboratory of Chemical and Biological Research of the Institute of Economics and Safety of the Federal State Budgetary Educational Institution of Higher Education "Sakhalin State University". The moisture content of the samples was determined on an MOC-120 Shimadzu analyzer according to GOST R ISO 11465 [2]. The content of organic carbon was analyzed by the spectrophotometric method according to Tyurin after the oxidation of organic carbon with a chromium mixture [3, 4] on a Shimadzu UV-1800 UV spectrophotometer at a wavelength of 590 nm. Before analysis, air-dry sediment samples were ground and sifted through a sieve with a mesh size of 0.25 mm [3]. The interfering effect of chlorides was eliminated by washing. The Corg concentration in g/kg was determined from the calibration dependence. The accuracy index (the limits of the relative error at a confidence level $P = 0.95$) is 20%.

The reliability of the results was assessed by analyzing a standard sample of continental sedimentary deposits (GSO 5358-90/5367-90) with a certified total carbon content $Ctot = 3.6\%$. To recalculate the obtained values of organic carbon in dry sediments per wet mass and per area, the recommendations from [5] were used. Sampling from the silty drying deposits was carried out using a Beaker sampler (Eijkelkamp). Samples from cores and sections for the determination of total carbon, organic carbon, and total nitrogen were taken at five-centimeter intervals. Analysis of arrays of aerial photographs of 1952 and satellite images of 2020 was carried out in the Quantum GIS geoinformation system. The calculations were made on the WGS84 EPSG: 7030 ellipse. The resulting attributes were exported to spreadsheets for further processing. The average error in the location of objects on this map is

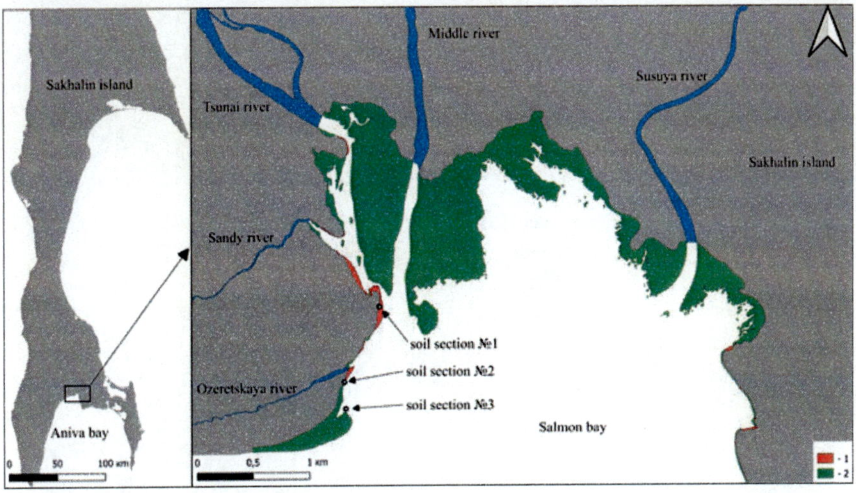

Fig. 1 Growth and erosion of marching terraces in Lososey Bay and location of soil sections (Aniva Bay, Sakhalin). 1—blur, 2—increase

about 5 m, the minimum is 0.2 m, the maximum is 18 m, the standard deviation is about 3 m.

At the next stage of the study, in the same QuantumGIS-environment, using standard procedures for analyzing remote sensing materials, we reconstructed the change in march areas for the period 1952–2020. The geospatial position of the studied areas and areal changes in the march surfaces are shown in Fig. 1.

It has been established that during the period from 1952 to 2020, 45,113 square meters of low marching terraces were destroyed in the studied area, mainly due to channel reconstructions of the river. Tsunai. During the same period, 2,254,795 square meters of silt drains passed into the category of marches. As is known, the reserves of organic carbon recorded in peat-gleyzems are 20.9–42.7 kg/m^2 (Lipatov et al., 2021) respectively. By simple calculations, it is possible to establish the amount of organic carbon recorded in the sediments of the Salmon Bay marches, which is 46182—94,353 tons or about 1000 tons/year. Figure 2 shows soil sections of marches. Marching peat-gley earth deposits in sections No. 3 and No. 2, 6o and 90 cm thick, respectively, were formed after 1952. Moreover, the plastic bag at the base of the peat deposits of section No. 3 dates back to the beginning of the twenty-first century. Section No. 1 characterizes a march of an earlier age, overlain by floodplain deposits.

For the content of organic carbon, the sediment cores of silt desiccations were tested (Fig. 3).

A sample of just two columns does not yet make it possible to draw any conclusions. And the data obtained should be classified as estimated (Table 1).

However, it should be noted that, in general, the results on the content of organic carbon in the sediments of silty drying are somewhat lower than the data obtained in the study of bottom sediments of lagoons [6–8].

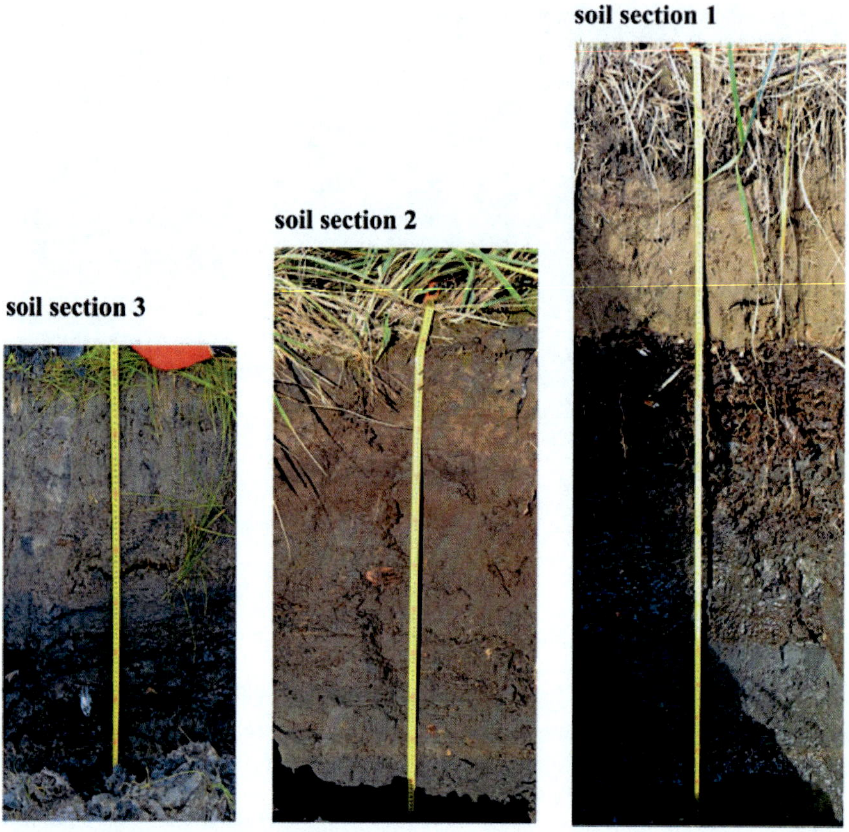

Fig. 2 Soil sections of march formations (location in Fig. 1)

It can be assumed that, in contrast to the bottom sediments of the lagoons, there is practically no autochthonous organic carbon in the deltaic deposits of the silty dry land located directly in the estuary zone of the river. Thus, if we determine the volume of suspended sediments that are deposited in the estuary zone, then we can estimate the amount of organic carbon that remains in the estuary sediments. The maximum possible average long-term volume in this case will be about 2 thousand tons of Corg.

3 Conclusion

We understand that the restoration and reshaping of coastal blue carbon ecosystems, which can be reliably used to mitigate the effects of climate change, is impossible without understanding the conditions of sedimentation in the coastal zone

Fig. 3 Muddy drying (**a**), sediment column of muddy drying (**b**)

and morpholithodynamically substantiated scenarios for their development [9, 10]. However, it should be taken into account that the data obtained mainly on the basis of geomorphological analysis of sedimentation processes are probably greatly underestimated. Because the autochthonous regime of organic matter flows is not taken into account.

Therefore, the carbon sink capacity of the estuarine zone of the Tsunai and Susuya rivers, which is about 9 thousand tons in CO_2 equivalent, will probably be adjusted upwards.

Table 1 Results of determination of Corg in sediments of silty drying

Sample name and sampling interval	Corg, g/kg dry.gr	Corg, % dry.gr	Humidity, W%	Corg, g/kg wet ground	Copr, % wet ground	Copr, mg/cm^3, wet ground	Copr, g/m^2, wet ground
A1 (0–4)	12.80	1.28	36.96	8.07	0.81	12.98	0.13
A1(4–8)	13.08	1.31	34.27	8.60	0.86	14.20	0.14
A1(8–12)	8.29	0.83	33.54	5.51	0.55	9.17	0.09
A1(12–16)	15.30	1.53	38.91	9.35	0.93	14.76	0.15
A1(20–24)	13.17	1.32	34.32	8.65	0.87	14.28	0.14
A1(24–28)	9.14	0.91	35.61	5.89	0.59	9.59	0.10
A1(28–32)	16.19	1.62	41.43	9.48	0.95	14.62	0.15
A1(32–36)	9.96	1.00	32.02	6.77	0.68	11.44	0.11
A1(36–40)	7.24	0.72	32.34	4.90	0.49	8.25	0.08
A1(40–44)	11.20	1.12	34.05	7.39	0.74	12.22	0.12
A1(44–48)	6.44	0.64	33.26	4.30	0.43	7.17	0.07
A1(48–52)	4.05	0.40	28.60	2.89	0.29	5.05	0.05
Column №2							
A2 (0–4)	14.17	1.42	31.93	9.65	0.96	16.30	0.16
A2(4–8)	7.58	0.76	28.38	5.43	0.54	9.52	0.10
A2(8–12)	22.67	2.27	43.04	12.91	1.29	19.62	0.20
A2(12–16)	13.18	1.32	36.01	8.43	0.84	13.69	0.14
A2(16–20)	7.91	0.79	32.22	5.36	0.54	9.04	0.09
A2(20–24)	13.21	1.32	23.85	10.06	1.01	18.53	0.19
A2(24–28)	15.12	1.51	32.01	10.28	1.03	17.36	0.17
A2(28–32)	12.81	1.28	33.33	8.54	0.85	14.24	0.14
A2(32–36)	12.09	1.21	32.77	8.13	0.81	13.63	0.14
A2(36–40)	17.41	1.74	34.35	11.43	1.14	18.86	0.19
A2(40–44)	22.97	2.30	39.62	13.87	1.39	21.75	0.22
A2(44–48)	14.17	1.42	38.22	8.76	0.88	13.91	0.14
A2(48–52)	24.24	2.42	33.41	16.14	1.61	26.88	0.27
A2(52–57)	3.54	0.35	24.25	2.68	0.27	4.91	0.05

References

1. Hydrometeorology and hydrochemistry of the seas. T. 9. Sea of Okhotsk. Issue. 1. Hydrometeorological conditions. St. Petersburg: Gidrometeoizdat, 1998. 370 p.
2. GOST R ISO 11465–2011 Soil quality. Determination of the mass fraction of dry matter and the mass ratio of moisture by the gravimetric method.
3. GOST 26213–2021 SOILS. Methods for determining organic matter.
4. FAO. 2021. Standard Operating Procedure for Soil Organic Carbon Tyurin Spectrophotometric Method. Rome //https://www.fao.org/3/cb4757ru/cb4757ru.pdf

5. Methodological guide to the analysis of organic matter of bottom sediments / ed. E. M. Zaslavsky. M.: Publishing House of VNIRO, 1980. 64 p.
6. Tokarchuk T. N. Geochemistry of Sakhalin lagoons and rational use of their resources: diss. ... cand. geographer. Sciences / T.N. Tokarchuk. - Vladivostok: FEGU, 1999. - 135 p.
7. Efanov V. N., Vypryazhkin E. N., Latkovskaya E. M. The current state of bottom sediments of the Busse lagoon (Aniva Bay) // Science and business: ways of development. - 2013. P. 31–36.
8. Kafanov, V.S., Labay, N.V., Pecheneva Biota and macrobenthic communities of the lagoons of northeastern Sakhalin / A. Yu - Sakh.: SakhNIRO, 2003. - 176 p.
9. Curado G., Rubio-Casal, A. E., Figueroa, E., Castillo, J. M., 2014. Plant zonation in restored, nonrestored, and preserved Spartina maritima salt marshes //Journal of Coastal Research. T. 30. №. 3. C. 629–634.- doi.org/https://doi.org/10.2112/JCOASTRES-D-12-00089.1
10. Rosentreter, J. A., Al-Haj, A. N., Fulweiler, R. W., & Williamson, P. (2021). Methane and nitrous oxide emissions complicate coastal blue carbon assessments. Global Biogeochemical Cycles, 35, e2020GB006858. - doi. org/https://doi.org/10.1029/2020GB006858

Organogenic Sedimentation in the Nyivo Lagoon

Victor V. Afanas'ev, A. V. Uba, A. I. Levitsky, and A. B. Faustova

Abstract The assessment of the role of coastal-marine geosystems in the runoff and emission of greenhouse gases (GHG) to determine their contribution to the overall balance of GHG in the territory of the Sakhalin Region is very relevant due to the limited potential of forestry and the geographical position of the region. The process of studying the emission and absorption of greenhouse gases (GHGs) in coastal marine ecosystems began with determining the geomorphological position of areas with an increased intensity of organogenic sedimentation, i.e. lagoons, studying the dynamics of marshes and silty lands, determining the main parameters of sedimentation in these elements of lagoonal biomorpholithosystems. The paper deals with the issues of sedimentation in the Nyivo lagoon, which receives the maximum amount of solid alluvial runoff for the region and for which the highest rates of horizontal growth of marshes are noted.

Keywords Sakhalin Island · Salt marsh · Intertidal mudflat · Organic carbon · Peat accumulation

1 Introduction

The object of this study is the marshes and mudflats of Nyisky Bay (essentially, lagoons), the formation of which is associated with the subboreal phase of the Middle Holocene transgression [1]. A notable feature is that the bay area of the Nyivo lagoon is a zone of convergence of the resulting sediment flows of the southern and northern directions of the lithodynamic system of the lagoon coast of northeastern Sakhalin [2].

During the period 1927–2014, the distal end of the spit shifted in the N–NW direction by more than 2200 m. At the same time, for the period up to 1977, there

V. V. Afanas'ev (✉) · A. V. Uba · A. I. Levitsky
Institute of Marine Geology and Geophysics, FEB RAS, Yuzhno-Sakhalinsk, Russia
e-mail: vvasand@mail.ru

A. B. Faustova
Sakhalin State University, Yuzhno-Sakhalinsk, Russia

V. I. Karev (ed.), *Physical and Mathematical Modeling of Earth and Environment Processes—2022*, Springer Proceedings in Earth and Environmental Sciences,
https://doi.org/10.1007/978-3-031-25962-3_50

were about 1360 m of growth of the spit in the indicated direction, i.e. 27 m/year, for the period 1977–1985 about 490 m or 61 m/year and 1985–2014 about 360 m i.e. 12 m/year. According to the calculated data characterizing the wave climate of the last 20 years, the resulting flow here is currently acting in a southerly direction and the previously eroded northern spit is increasing its area [3].

The rapid movement of the strait makes it difficult to forecast the evolution of the morphodynamic system of the Nyisky Bay for the next decades, taking into account possible changes in the relative sea level. The question of the relationship between the rate of coastal subsidence and the volume of marine and alluvial sediments filling the lagoon basin remains open. The diffusion transport of sediments between the elements of the system cannot yet be described in terms of real concentration. Thus, the main object of our interest, the dynamics of marshes and silty lands, serves as a good indicator of the preservation or degradation of the entire lagoonal biomorpholithic system, including changes in its sequestration properties.

A characteristic feature of the Nyivo lagoon is the maximum river runoff for all Sakhalin lagoons and, accordingly, the volume of solid runoff, which is more than 500 thousand tons/year, which is about 20 times more than the runoff of river sediments into the comparable Chaivo lagoon [4]. As a result, the lagoon is a shallow basin almost completely filled with sediments with a high content of weakly decomposed organic matter of plant origin, mainly the sea grass Zostera [5]. Figure 1 shows an area with a typical change from the shallow part of the lagoon inhabited by zostera to a silty dry land with rare clumps of sedges, which, in turn, passes into the marsh surface. The level of the march surface is 30–60 cm higher than the surface of the silty drying.

The modern area of the water surface of the Nyivo lagoon is 104.3 km^2, which is about 2.3 km^2 less than in 1952. Almost the entire decrease in the area is associated with the transformation of silty drylands into the surfaces of marches.

A typical soil section of the march is shown in Fig. 2. The thickness of the peat-gley soil horizon is about 60 cm, the exposed thickness of gley soils is up to 1.0 m.

2 Methods and Results

The methodology of biomorpholithodynamic studies is based on both traditional and modern methods for obtaining and analyzing geospatial and geological-geomorphological information. The work is based on the analysis of arrays of aerial photographs of 1952 and satellite images of 2020, which was performed in the Quantum GIS geoinformation system. The calculations were made on the WGS84 EPSG: 7030 ellipse. The resulting attributes were exported to spreadsheets for further processing. The average error in the location of objects on this map is about 5 m, the minimum is 0.2 m, the maximum is 18 m, the standard deviation is about 3 m. changes in the areas of marches for the period 1952–2020 When calculating the carbon stocks in the sediments of marshes, silty desiccations and bottom sediments, the published data and the authors' own results were used.

Fig. 1 Lagoon shallow water turning into dry mud

Figure 3 and Table 1 present the results of measuring changes in the area of marches in the delta parts of the rivers flowing into the Nyivo lagoon. The area of the marches of the lagoon for 1952–2020 increased by 2,265,623 m^2, or by 33,318 m^2/year.

Stocks of organic carbon recorded in the northeast of the island. Sakhalin, in peat-gleyzems it is 20.9–42.7 kg/m^2, in gleyzems it is 16.6 kg/m^2 [6]. Accordingly, 700–1400 tons/year of organic carbon will be accumulated in the soils of marshes, which in terms of CO_2 equivalent is 2100–4200 tons/year. The reserves of organic carbon in the gleyzems of the base of the marches are about 550 tons/year, respectively, or about 2000 tons/year in terms of CO_2 equivalent.

The solid runoff of rivers into the Nyisky lagoon is about 500 thousand tons/year of suspended sediments. Taking the density of compacted silts as 1.2 t/m^3, we get that the volume of sediment supplied by rivers to the lagoon is about 420 thousand m^3/year. And if these sediments are simply evenly distributed over the entire area of the lagoon, a layer about 4 cm thick will be obtained. It is in this layer that the content of organic carbon is 1.6–2.2% per natural sediment [7]. What part of organic carbon is brought by rivers, and what part is autochthonous, given the widespread distribution of zoster thickets, it is not yet possible. However, under favorable conditions, i.e. the minimum removal of suspended sediments to the open sea. Bottom sediments can contain up to 8–10 thousand tons/year of organic carbon or up to 30 thousand tons/year in terms of CO_2 equivalent.

Fig. 2 Typical soil section of a march

3 Conclusion

Geomorphological approaches have already shown their effectiveness in analyzing the balance of producers and sinks of greenhouse gases, especially in relation to the dynamics of marches and silt lands [8–10].

We acknowledge that the lack of a detailed facies analysis of sedimentation in the intertidal zone of the lagoon raises many questions. However, the main goal of this stage is a kind of geomorphological reconnaissance, in the form of analyzing geospatial information using modern methods and assessing the main biomorpholithic complexes of the lagoons. Work on a detailed study of carbon turnover in lagoonal biomorpholithosystems has already begun. And if we were able to estimate already at this stage the sequestration capacity of the Nabil lagoon, which can reach 36 thousand tons/year in terms of CO_2 equivalent, this article deserves attention.

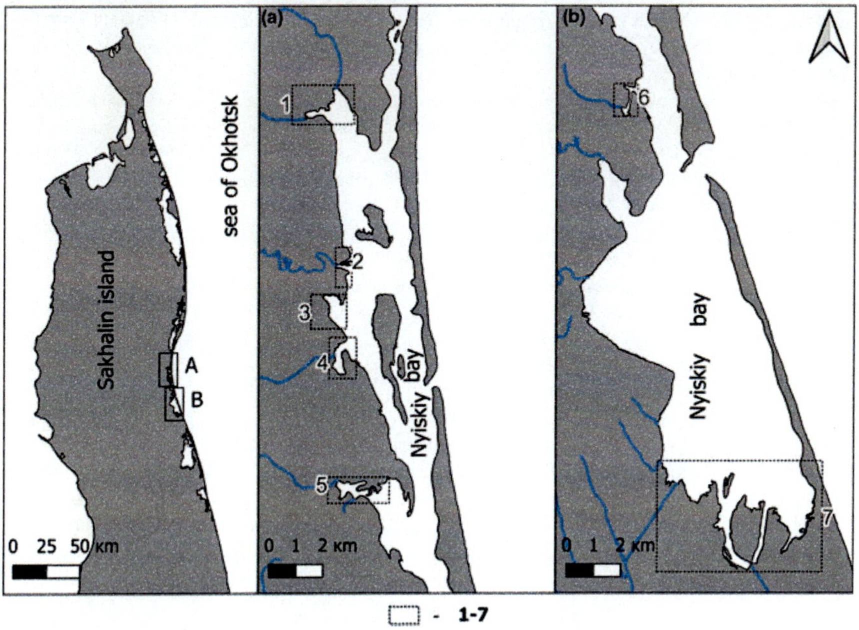

Fig. 3 Change in the area of the water surface of the Nyisky bay **a**—northern part of the bay, **b**—southern part of the bay, 1–7—calculation area and site number

Table 1 Results of measuring changes in the area of marches for the polygons shown in Fig. 3

Section	Accumulation km^2	Abrasion km^2	River mouths
1	0.273,205	0.000,691	Small Tapauna, Patience, Tapauna
2	0.113,086	0.006,769	Dougie
3	0.223,060	0.000,764	–
4	0.189,795	0.000,802	Tomy
5	0.139,151	0.026,199	Nelbuta
6	0.061,497	0.002,041	Small Bauri
7	1.246,068	0.159,888	Tym, Jimdan
Total	3.059,729	0.794,106	

References

1. Afanasiev V.V. Evolution of the coasts of the seas of the Far East in the Holocene // Evolution of the coasts under conditions of ocean level rise. M.: Publishing House of Moscow State University, 1992. P. 160–174.
2. Afanasiev VV Morpholithodynamic processes and development of the shores of the contact zone of the subarctic and temperate seas of the North Pacific // Yuzhno-Sakhalinsk: IMGiG FEB RAN. – 2020. 234 p.

3. Leontiev I. O., Afanasiev V. V. Dynamics of the lagoon coast of northeastern Sakhalin on the example of the Nyisky Bay system and the Plastun spit // Oceanology, 2016, volume 56, no. 4, P. 618–626
4. Bobrik K. P., Brovko P. F. River runoff and sedimentation in the lagoons of Northern Sakhalin // Materials on hydrology of rivers in the BAM zone and the Far East. T. XX. issue 3. L.: Gidrometeoizdat, 1986. P. 439–444.
5. Kafanov, V. S. Labai, N. V. Pecheneva Biota and macrobenthic communities of the lagoons of northeastern Sakhalin / A. Yu - Sakh.: SakhNIRO, 2003. - 176 p.
6. Lipatov, D.N., Shcheglov, A.I., Manakhov, D.V., Brekhov, P.T. Sakhalin // Soil Science. – 2021. – no. 2. - P. 211–223.
7. Tokarchuk T. N. Geochemistry of Sakhalin lagoons and rational use of their resources: diss. ... cand. geographer. Sciences / T.N. Tokarchuk. - Vladivostok: FEGU, 1999. - 135 p.
8. Curado G., Rubio-Casal, A. E., Figueroa, E., Castillo, J. M., 2014. Plant zonation in restored, nonrestored, and preserved Spartina maritima salt marshes //Journal of Coastal Research. T. 30. №. 3. C. 629–634.- doi.org/https://doi.org/10.2112/JCOASTRES-D-12-00089.1
9. Rosentreter, J. A., Al-Haj, A. N., Fulweiler, R. W., & Williamson, P. (2021). Methane and nitrous oxide emissions complicate coastal blue carbon assessments. Global Biogeochemical Cycles, 35, e2020GB006858. - doi. org/https://doi.org/10.1029/2020GB006858
10. Hu, Z., Van Belzen, J., Van Der Wal, D., Balke, T., Wang, Z. B., Stive, M., Bouma, T. J., 2015. Windows of opportunity for salt marsh vegetation establishment on bare tidal flats: The importance of temporal and spatial variability in hydrodynamic forcing. Journal of Geophysical Research: 120.7 P. 1450–1469. -doi.org/https://doi.org/10.1002/2014JG002870

Experimental Determination of Permissible Drawdown for Gas Field Wells on a True Triaxial Loading Unit

N. I. Shevtsov and S. O. Barkov

Abstract One of the typical complications arising during the operation of wells in formations formed by weakly cemented rocks is sand production, when the sand from the reservoir enters the well together with the fluid. To study this issue, it is necessary to conduct experimental modeling of conditions in specific wells. This work is devoted to this method of studying the sand problem in gas fields. A unique research installation triaxial independent loading test system developed in IPMech RAS—TILTS—was used for experimental modeling of gas field bottom-hole zone. Nine specimens were made from core material of real gas field. Samples were used for physical modeling of the bottomhole formation zone. As a result of the studies, a significant anisotropy of strength properties of the studied rocks has been revealed. Due to anisotropy of strength properties, rock destruction along the contour of the well will be uneven—it will begin at the points corresponding to the upper and lower points of the contour of the horizontal well in the formation, and then will continue in this direction, forming a characteristic borehole collapse. Also, from the above data, we can conclude that the maximum pressure draw-down that can be safely given in the wells in the considered field should not be more than 0.5 MPa. Also, during all tests there was observed a decrease in permeability of samples, nevertheless final permeability remained big enough for effective operation of wells.

Keywords Physical and mathematical modeling · True triaxial loading · Pressure draw-down · Sand carryover · Geomechanical modeling

---

N. I. Shevtsov · S. O. Barkov (✉)
Ishlinsky Institute for Problems in Mechanics RAS, Moscow, Russia
e-mail: sviatoy97@gmail.com

N. I. Shevtsov
e-mail: red3991@ya.ru

V. I. Karev (ed.), *Physical and Mathematical Modeling of Earth and Environment Processes—2022*, Springer Proceedings in Earth and Environmental Sciences,
https://doi.org/10.1007/978-3-031-25962-3_51

1 Introduction

One of the typical complications arising during the operation of wells in formations formed by weakly cemented rocks is sand production, when the sand from the reservoir enters the well together with the fluid. This leads to formation of sand plugs and sand carry-over along the further technological chain, which leads not only to flow rate reduction, but also to wear-out of downhole and surface equipment and, consequently, to increase of economic losses due to reduced productivity, expenses for current and overhaul repair of wells, cleaning fluid from mechanical impurities.

In the literature, the main cause of sand outflow into the well is the destruction of the bottomhole formation zone, which is a consequence of the occurrence of stresses in the rock during filtration of formation fluids (gas, water and their mixtures). Bottomhole zone destruction occurs when these stresses exceed the rock strength limit [1–6]. In this regard, the main directions of preventing the destruction of the borehole zone are considered to be the reduction of underbalance on the formation by reducing the flow rate of the well and creation of screens at the filtration surface by installing sand-holding filters.

Despite the fact that the problem is actively being studied, at the moment there is no sufficiently substantiated theoretical model of reservoir failure in the borehole zone. To study this issue [7, 8], it is necessary to conduct experimental modeling of conditions in specific wells. This work is devoted to this method of studying the sand problem in gas fields.

2 Experimental Method

A unique research installation triaxial independent loading test system developed in IPMech RAS—TILTS—was used for experimental modeling of gas field bottomhole zone [9]. The installation TILTS allow loading rock samples in the form of a cube with 40 mm edge along any loading trajectory independently in each of the three directions. The possibility to load the sample independently in each of the three axes is provided by the original kinematic scheme used in the construction of the loading unit, which allows the pressure plates to move closer to each other in three directions without obstructing each other. At the same time, it was necessary to ensure that there was no friction between the plates, which is a technical challenge.

The maximum pressure of the working fluid in the hydraulic cylinder of the unit is 20 MPa, and the hydraulic cylinder develops a force of 500 KN. This makes it possible to create stresses up to 280 MPa on samples with an edge of 40 mm.

TILTS use a four-channel control system with electrohydraulic converters (servo valves). Three channels of the control system are used to control power hydraulic cylinders, one channel is redundant. Dynamometers installed in the rods of the power hydraulic cylinders or displacement sensors are used as feedback sensors in the control system. Possibility of controlling the loading process either by force or displacement in each of the three channels allows realization of practically any loading (deformation) trajectory of a specimen, including the fracture process. This makes it possible to determine elastic moduli and strength characteristics of anisotropic rocks by loading rock samples according to special loading programs.

An important peculiarity of the TILTS installation is a possibility to study the dependence of rock permeability on the value and type of the acting stresses. It is known that rock permeability can both decrease and increase (irreversibly) depending on the occurring stresses in them. The type and level of these stresses are determined by the bottom hole design (casing presence or absence, perforation type, etc.) and pressure draw-down created at the bottom hole. TILTS installation allows to simulate these conditions on rock samples and continuously record the change in their permeability. For this purpose, one pair of pressure plates, which form the loading assembly, has a perforation for air supply and exhaust. The four faces of the sample, parallel to the filtration axis, have a latex-based shell, which ensures their tightness at relatively large deformations of the sample up to the formation of magistral cracks.

The capabilities of TILTS unit are not limited to determination of deformation, strength and filtration characteristics of rocks in conditions of complex stress state. Its main advantage lies in the possibility of direct physical simulation of deformation and fracture processes as well as filtration processes in oil reservoirs under the influence of real stresses occurring during drilling, development and operation of wells. Specimens for testing at the TILTS unit are produced on a specially designed machining complex that includes two machines—stone-cutting and grinding. This complex allows to produce cubic specimens with high accuracy and any orientation relative to the borehole axis.

The object of the research is core material, selected from a real gas field, from which 20 samples were made. Part of them was used to determine the elastic and strength properties of the rock, part of them was used for physical modeling of the bottomhole formation zone.

Figure 1 schematically shows a vertical section of a horizontal uncased well and the radial and annular stresses acting in its vicinity at two points M and N. Compressive stresses are assumed to be negative, as is common in deformable solid mechanics. To determine the type and degree of anisotropy of the studied rocks, the velocities of longitudinal ultrasonic waves in three mutually orthogonal directions were measured in the samples before testing on the TILTS facility. The velocities were measured in three directions: along axis 1, i.e., along the core axis, and along axes 2 and 3 in two mutually perpendicular directions in the horizontal plane. Measurements of elastic wave velocities showed that the presented samples are isotropic. For conditions of isotropic medium with elastic properties, the values of stresses acting on the borehole wall are calculated from the solution of Lame's problem [10]. Using this solution programs of experimental samples loading have been compiled. Since it is planned to produce in the studied field by means of inclined wells, which are close to horizontal wells in productive strata, for physical modeling of deformation and destruction processes of bottomhole reservoir rocks, the programs of core sample loading corresponding to the stresses actually occurring in the vicinity of horizontal wells during their operation were developed.

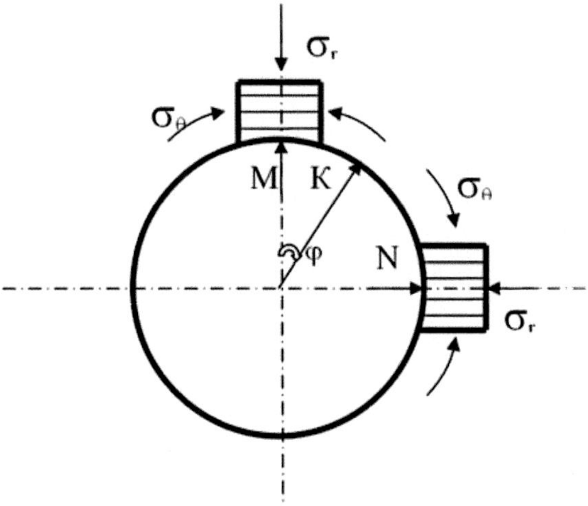

Fig. 1 Circular and radial stresses σ_θ, σ_r, acting in the vicinity of the horizontal well at the upper and lateral points *M* and *N*

3 Experiment

One of the main factors determining rock failure in the vicinity of a well is the anisotropy of strength properties of the reservoir rocks in which the well is conducted. Anisotropy of strength properties of reservoir rocks is mainly related to the presence of bedding planes in the formations, since bedding planes are usually the planes of weakening. This explains the fact that rock failure in the vicinity of the well largely depends not only on the magnitude of stresses arising in this area, but also on the orientation of the stress ellipsoid relative to the bedding planes. In particular, the simulation of the pressure reduction process at the horizontal well bottom, performed on reservoir rock samples from various fields, revealed that the value of stresses, at which the samples were destroyed, significantly depends on the location of the sample on the contour of the horizontal well, Fig. 1. Destruction of samples located at the upper point M on the contour of the well occurred at significantly lower values (and, accordingly, lower values of pressure draw-down) than samples located at the lateral point N.

At the point M, the pressure drop in the borehole unloads the rock in the vertical direction, i.e., in the direction in which it has the lowest strength. In addition, it is subjected to even greater compression along the layers by annular stress, which also contributes to rock delamination in the vertical direction. This is the type of rock fracture observed in most experiments. At point N, a different picture is observed. Here, the rock is unloaded along the bedding, i.e., in the direction where rock strength is highest, when simulating a pressure decrease in the borehole. Moreover, the horizontal layers are compressed by large annular stresses, which also increases the rock's resistance to fracture. As a result, in the vicinity of point N, rock failure does not occur at significantly higher depressions.

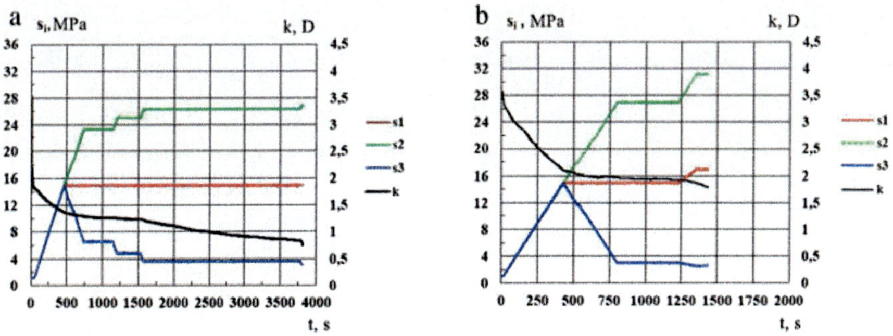

Fig. 2 Stress-time dependences and permeability curves for samples M1.5.1—upper point *M* (**a**) and M1.5.2—lateral point *N* (**b**)

Therefore, in the course of experiments on a part of samples was simulated location of samples on the contour of the horizontal well at point M (top), and on a part—at point N (side). Permeability of the sample at points N and M was also measured along different axes of the sample: at point M—along the core axis, and at point N—perpendicular to it, that is, in both cases in the direction of the borehole radius.

As an example, consider specimens M1.5.1 and M1.5.2, cut from the same piece of core, which were tested according to programs corresponding to the lateral and upper point of the horizontal well. Figure 2 shows the loading programs and permeability curves for these specimens. It is easy to see that the failure of M1.5.1 occurred at a lower value of maximum compressive stress than M1.5.2, which corresponds to a lower pressure draw-down in the well. For M1.5.1, the simulated borehole draw-down was 0.5 MPa, while M1.5.2 was 1.8 MPa. Qualitatively similar results were obtained on the entire sample array, which may indicate that the upper point of the borehole contour is more dangerous in terms of wall failure. The data for the rest of the samples of the series of experiments are given below.

Sample number	Sampling depth, m	Experiment	Result	Permeability
M1.4	972.0	Well, top point	Collapsed at S2 = 29.0 MPa Pressure draw-down $\Delta Pc = 1.35$ MPa	Initial 8.8 D Final 5.9 D
M2.4.2	973.0	Well, lateral point	Collapsed at S2 = 27.6 MPa Pressure draw-down $\Delta Pc = 1.48$ MPa	Initial 10.4 D Final 9.2 D
M2.4.3	973.0	Well, top point	Collapsed at S_2 = 23.0 MPa Pressure draw-down $\Delta P_c = 0.5$ MPa	Initial 8.12 D Final 5.3 D

(continued)

(continued)

Sample number	Sampling depth, m	Experiment	Result	Permeability
M3.4.2	973.8	Well, lateral point	Collapsed at S2 = 29.3 MPa Pressure draw-down ΔPc = 2.0 MPa	Initial 13.9 D Final 7.9 D
M9.4.2	984.1	Well, top point	Collapsed at S2 = 32.3 MPa Pressure draw-down ΔPc = 2.9 MPa	Initial 2.4 D Final 1.4 D
M1.5.1	1000.85	Well, top point	Collapsed at S2 = 27.0 MPa Pressure draw-down ΔPc = 0.5 MPa	Initial 3.5 D Final 0.8 D
M1.5.2	1000.85	Well, lateral point	Collapsed at S2 = 31.2 MPa Pressure draw-down ΔPc = 1.8 MPa	Initial 3.4 D Final 1.1 D
M10.5.1	1013.4	Well, lateral point	Collapsed at S2 = 29.9 MPa Pressure draw-down ΔPc = 1,82 MPa	Initial 5.8 D Final 3.6 D
M10.5.2	1013.4	Well, top point	Collapsed at S2 = 26.5 MPa Pressure draw-down ΔPc = 0.5 MPa	Initial 5.1 D Final 2.6 D

4 Conclusion

As a result of the studies, a significant anisotropy of strength properties of the studied rocks has been revealed. Due to anisotropy of strength properties, rock destruction along the contour of the well will be uneven—it will begin at the points corresponding to the upper and lower points of the contour of the horizontal well in the formation, and then will continue in this direction, forming a characteristic borehole collapse. Also, from the above data, we can conclude that the maximum pressure draw-down that can be safely given in the wells in the considered field should not be more than 0.5 MPa. Also, during all tests there was observed a decrease in permeability of samples, nevertheless final permeability remained big enough for effective operation of wells.

Acknowledgments The work was performed within the framework of the state assignment, registration number AAAA-A20-120011690133-1.

References

1. *Mirzajanzade, A. Kh.* Fundamentals of gas production technology. - Moscow: Nedra, 2003. - 880 p. (in Russian)
2. *Gassumov R. A., Varyagov S. A., Serebryakov E. P. et al.* Causes of destruction of unstable reservoirs and methods of removing sand plugs from wells // Problems of well workover and operation of UGSF: collection of scientific papers of SevKavNIPIgaz. - Stavropol: SevKavNIPIgaz RIC, 2001. - Vol. 34. - p. 5–13. (in Russian)
3. *Karev V.I.* Geomechanical Approach to Improving the Efficiency of the Operation of Underground Gas Storages / (eds) Physical and Mathematical Modeling of Earth and Environment Processes. Ser. Springer Proceedings in Earth and Environmental Sciences. Switzerland: Springer International Publishing. – 2018. – P. 150–158.
4. *Bashkatov A.D.* Prevention of sanding of wells. - Moscow: Nedra, 1981. - 177 p. (in Russian)
5. *Gladkov P. V., Zagornov V. F.* Optimization of methods of impact on the productive reservoir of the Orenburg OGCF during well workover // Gas Industry. - 2014. - № 6. - P. 21–23. (in Russian)
6. *Aliev Z.S., Andreev S.A., Vlasenko A. P. et al.* Technological operation mode of gas wells. - Moscow: Nedra, 1978. - 279 p. (in Russian)
7. *Kovalenko Yu.F., Ustinov K.B., Karev V.I.* Geomechanical analysis of formation of blowouts on the walls of wells // Izv. MTT. 2022. № 6. P. 148-163. (in Russian)
8. *Ustinov K.B., Karev V.I., Kovalenko Yu.F.*, Barkov S.O., Khimula V.V., Shevtsov N.I.* Experimental study of influence of anisotropy on orientation of dumps in wells // Izvestiya RAN. MTT. (accepted for publication) (in Russian)
9. *Karev V., Kovalenko Yu.* Triaxial loading system as a tool for solving problems of oil and gas production. The 12th International Congress on Rock Mechanics (Beijing, October 18 – 21, 2011).
10. *Lyav A.* Mathematical Theory of Elasticity. - Moscow - L.: ONTI NKGIP USSR. - 1935. - 676 p. (in Russian)

Assessment of Fracture and Pore Permeability in Rocks by Results of X-ray Computed Tomography

V. V. Khimulia and N. I. Shevtsov

Abstract The paper presents the results of studies of the filtration properties of rocks carried out on the ProCon X-Ray CT-MINI high-resolution scanner of the Institute for Problems in Mechanics RAS. Images of the internal structure of reservoir rocks of two different types were obtained: a highly porous coarse-grained sandstone from a gas field reservoir and a fine-grained well cemented low-permeability sandstone with a branched network of macrocracks from a gas condensate field reservoir. The approaches to the processing of tomography images in the GeoDict software are described. Numerical modeling of filtration flows through the obtained structures is performed. Filtration fields were calculated using Navier–Stokes and Stokes-only solvers of the FlowDict module of GeoDict package. Differences and difficulties in modeling depending on the type of filtration in the rock: pore permeability and fracture permeability are shown. The differences between the measured and computed permeabilities were evaluated. Conclusions are made about the possibility of using the described methods to assess the filtration properties of rocks.

Keywords Permeability · Filtration properties · Pore space structure · Navier–Stokes numerical modeling · Single phase filtration · Computed tomography (CT) · GeoDict software · Pore permeability · Fracture permeability

1 Introduction

Physical properties of reservoir rocks are essential for creating reservoir models, predicting well performance, and creating new methods to increase well productivity. To create hydrodynamic and geological reservoir models, it is necessary to conduct deep and comprehensive studies, including laboratory core analysis. One of the main characteristics of reservoir rocks requiring study and determination is permeability, because fluid flow through the pore space is one of the key processes occurring in the reservoir [1]. Until recently, rock permeability was determined through laboratory or

V. V. Khimulia · N. I. Shevtsov (✉)
Ishlinsky Institute for Problems in Mechanics RAS, Moscow, Russia
e-mail: red3991@ya.ru

V. I. Karev (ed.), *Physical and Mathematical Modeling of Earth and Environment Processes—2022*, Springer Proceedings in Earth and Environmental Sciences,
https://doi.org/10.1007/978-3-031-25962-3_52

well logging tests, or indirectly through correlations with other rock properties [2]. One of the new and promising approaches to the laboratory study of the filtration properties of rocks is the numerical simulation of filtration processes on the structures obtained by microcomputed X-ray tomography (μCT).

X-ray computed tomography is a major non-destructive method of studying the internal structure of rocks, which significantly complements the direct laboratory studies. X-ray tomography of heterogeneous media is based on reconstruction of the spatial distribution of the linear attenuation coefficient of X-rays using computer processing of the projections obtained during the radiation survey. The results of CT investigation presented as a stack of grayscale images represent a three-dimensional picture of X-ray absorption within the sample under study [3]. Linear sizes of elements, their quantity, sphericity, anisotropy are determined by means of special software [4]. For quantitative material analysis and the use of tomography results for numerical process modeling, image segmentation is necessary [5]. Segmentation is a method of analyzing images and transitioning in the sample space from X-ray absorption to the spatial distribution of the components that constitute the sample. Segmentation can be binary (two-phase) or multiphase.

The methods and devices of three-dimensional imaging developed in the last decades make it possible to penetrate inside natural and artificial materials. With the development of 64-bit technology and increased computing power, it has become possible to visualize and calculate efficient properties on large three-dimensional images. However, this requires very fast and memory-saving numerical methods [6]. Numerous researchers and engineers are interested in effective properties of homogenized materials, such as permeability, pressure drop and mean velocity, thermal and electrical conductivity, diffusion and tortuosity, stiffness, strain, stress or elastic moduli, properties that depend on saturation or compression, such as relative permeability.

Pore imaging experiments paired with modeling are valuable instruments used in industry to predict geological and physical properties [7], including porosity and permeability [8]. The calculations are performed on a segmented image, which is obtained mainly for the macropore space, that is, for pores with well resolved boundaries that can be segmented into fluid and grains on voxel basis [9]. If the rock grains are impermeable and the pore holes are very large compared to the image resolution, a fairly accurate segmentation is required to obtain a realistic estimate of flow through the rock [10, 11]. In reality, however, not all grains are non-porous, and intra-grain microporosity contributes significantly to overall rock microporosity [12].

This work presents the results of rock structure studies carried out on the ProCon X-Ray CT-MINI high-resolution scanner of the Institute for Problems in Mechanics RAS. The internal structure of reservoir rocks of two different types was scanned: a highly porous coarse-grained sandstone from the reservoir of the Cenomanian horizon gas field and a fine-grained well cemented low-permeability sandstone with a branched network of macrocracks from the reservoir of the gas-condensate field in the Caspian lowland. The approaches to the processing of tomography images in the GeoDict software are described. Numerical modeling of filtration flows through the obtained structures is performed. Filtration fields were calculated using Navier–Stokes and Stokes-only solvers of the FlowDict module of GeoDict package. Differences and difficulties in modeling depending on the type of filtration in the rock: pore permeability and fracture permeability are shown. The results of modeling are compared with the results of physical experiment on the studied rock samples. Conclusions are made about the possibility of using the described methods to assess the filtration properties of rocks.

2 Research Methodology

The ProCon X-Ray CT-MINI high-resolution micro-CT scanner [13] of the Institute for Problems in Mechanics RAS was used to perform computed tomography and obtain images of the samples (Fig. 1). The dimensions of the device are 1300 × 850 × 600 mm. The device has a rigid base (monolithic granite slab) for practical elimination of the influence of temperature drift and accurate installation of X-ray optical and other system components inside the device, which provides accuracy and stability of measurements in the whole range of working space. A design scheme with vertical placement of the core for research is provided inside: the maximum height of the loaded sample is 200 mm, the maximum diameter of the loaded sample is 200 mm. The maximum weight of the loaded sample is 5 kg. The optical scheme of the scanner allows to set a distance of 335 mm from the tube to the object. For this purpose, a precision positioning system (manipulator) is built into the design. The detector has the ability to move perpendicularly to the main axis of the system within a range of ± 25 mm. The manipulator allows for 360° rotation of the sample around its axis with an angular position reproducibility accuracy of 1.5 angular seconds. The high resolution, microfocus, closed-loop X-ray tube has an adjustable output voltage range of 20–90 kV and an adjustable current range of 10–160 μA. The maximum output power of the tube is 8 W. The smallest focal spot size is 5 μm. The high-sensitivity, low-noise X-ray detector has a pixel count of 2940 × 2304. Pixel size is 49.5 μm. The size of the active (sensitive) area is 146 × 114 mm.

Reconstruction of the scans was performed using VGSTUDIO software. 3D image processing was performed by means of Geodict Math2Market GmbH software. GeoDict is the complete solution for multiscale 3D image processing, material modeling, visualization, material property determination, material development based on modeling and process optimization. In addition to basic image processing capabilities, GeoDict offers different methods of image segmentation. GeoDict also offers different solution methods for modeling single- and two-phase fluid flow, conductivity and mechanics in porous media.

Filtration fields were calculated using the FlowDict module of the GeoDict package [14]. The FlowDict module predicts effective material properties (flow velocity, flow permeability, and flow resistivity) by simulating flow experiments and post-processing the simulation results. FlowDict predicts the physical mean flow

Fig. 1 ProCon X-Ray CT-MINI high-resolution micro-CT scanner of the institute for problems in mechanics RAS

velocity for a given pressure drop [15] and calculate permeability of a porous structure by applying Darcy's law. This module is capable of calculating incompressible, stationary Newtonian flows based on Navier–Stokes equations with various approximations. FlowDict is easy to learn and is actively used by researchers around the world to model flows in a wide variety of materials [16–19].

The LIR solver was used for numerical filtration simulations in this work [20]. LIR is the newest, very fast and memory-saving iterative finite volume method [21]. The solver calculates permeability as well as velocity and pressure fields on large three-dimensional images. The convergence speed depends on the complexity and heterogeneity of the pore space. The limitations of this approach are the simulation of very fast flows with high Reynolds numbers and the resulting turbulence and boundary layers. In addition, the simulation of slip conditions at the boundary is also very difficult. However, these limitations apply to all methods using voxel grids.

Two rock samples with different structures were selected to simulate filtration processes in FlowDict software: a highly porous coarse-grained sandstone from the reservoir of the Cenomanian horizon gas field and a fine-grained well cemented low-permeability sandstone with a branched network of macrocracks from the reservoir of the gas-condensate field in the Caspian lowland. From the core material were made cubic samples with a face of 40 mm for further permeability measurement with the TILTS unit [22, 23]. Due to limitations on tomography resolution, consideration of small pores is difficult at these sample sizes, so rock with a large number of well-defined pores was considered to calculate pore permeability. The second sample was chosen for fracture permeability modeling reasons to compare the obtained results.

3 Experimental and Simulation Results

For both samples the air permeability was measured using the TILTS facility of the Institute for Problems in Mechanics [23]. For this purpose, a cubic sample was covered with an impermeable film on four faces. Air was passed through two open faces along one of the sample axes [22]. The pressure drop did not exceed 0.1 bar. The measured permeability of the highly porous sample was 9 Darcy. Initial permeability of the second sample was less than 1 mD. After permeability measurement, the second sample was unequally loaded in the TILTS unit and a system of macrocracks formed in it, acting as new filtration channels [24]. It is not always possible to accurately measure permeability on the TILTS after the jacket is breached by cracking the sample, but the final permeability of this sample was estimated to be 150 mD.

After laboratory permeability measurements, the samples were placed in a ProCon CT-MINI high-resolution scanner. After scanning, image reconstruction in VGSTUDIO software was performed and the resulting project was processed in Geodict. For both samples we performed image alignment, cropping of air-contact areas, brightness correction by sample volume, NLM filtering for smoothing and further segmentation into two phases: air/material.

The scanning parameters of the first sample are as follows: X-ray source voltage 90 kV, current 89 μA; exposure time 0.28 s, number of averages 3, voxel size 22.326 μm.

The correct choice of segmentation method is one of the key steps for further numerical modeling, especially for porous material. In this case, we used segmentation to identify the most well-defined pores with sufficient voxel size and a relatively clear boundary. Accounting for microporosity at such scanning parameters causes an increase in the subjective impact on the image segmentation. To solve this issue, methods will be proposed further below. For comparison, segmentation was performed using two methods: Global Tresholding and AI Segmentation (trained neural network). Both segmentations showed close simulation results.

To simulate filtration flow through the resulting structure using the FlowDict module, two available filtration models were used: Stokes and Navier–Stokes [25]. To recreate the conditions of laboratory permeability measurement the following flow parameters were chosen: fluid—air, pressure drop 0.1 bar, temperature 20 °C, boundary conditions in the filtration direction—periodic with implicit region 10 voxels, in the tangential direction—No Slip. The direction of simulated flow coincided with the direction of laboratory measured permeability in the sample. LIR solver parameters: error bound 0.05–0.1, 8 $\times$ parallelization, Multigrid method, relaxation 0.3.

Attempts of numerical calculations on the cube of size 1720 $\times$ 1765 $\times$ 1670 voxels obtained during cropping and processing failed due to instability of calculations or lack of resources to start computations. It was decided to divide the cube into 4 equal parallelepipeds keeping the side length along the filtration direction. Due to homogeneity of the pore space, the calculation results for different quarters of the

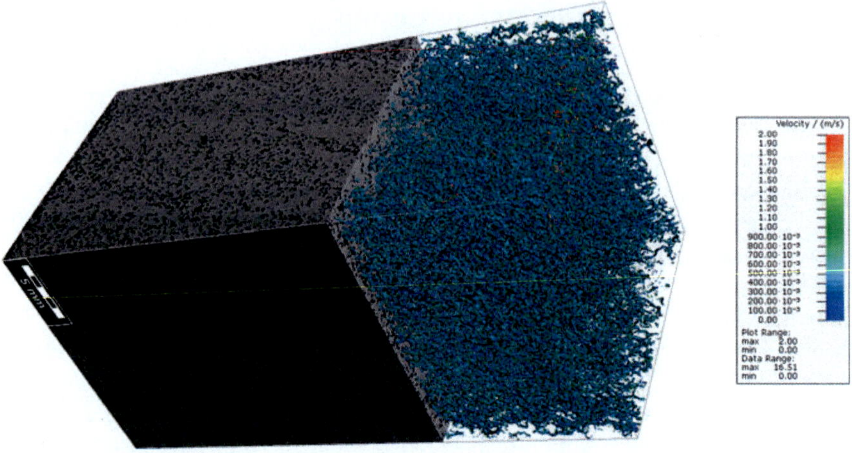

Fig. 2 The obtained structure of the quarter sample and the distribution of flow velocities

cube coincided. The calculated permeability of the quarter sample with the dimensions 860 × 1765 × 835 voxels within the Navier–Stokes model was equal to 3.5 D (with the termination condition of error bound 0.1). Visualization of the obtained structure of the quarter sample is shown in Fig. 2. The rock material is shown in gray, and the pore space is shown in black. On the right is a visualization of part of the calculated flow velocity distribution. An attempt to simulate the flow in the Stokes model showed a result of 3.8 D at an error bound of 0.1, and the calculations took three times less time.

It can be noted that the result of the calculation of both models is almost the same for this structure. At the same time, the more complex Navier–Stokes model required much more time and computer resources. However, when we tried to set the condition error bound 0.05 to increase the accuracy of the obtained result, both models did not give the final result: the calculations became unstable, the calculation accuracy at each cycle decreased to 0.07, but after that the error increased and the calculation divergence occurred. The intermediate iteration result with the highest accuracy value of 0.07 in the Stokes model showed the same permeability value as obtained earlier. However, the very fact that the cycles diverged indicates the instability of the calculations at these parameters.

A different situation was observed for a low-permeability sample with macrocracks. The Scan Parameters of the second sample are as follows: X-ray source voltage 90 kV, current 89 μA, exposure time 0.5 s, number of averages 2, voxel size 44.117 μm. Similar reconstruction and image processing methods were used, segmentation was done by Global Tresholding, because the cracks strongly contrast with the material. The conditions and parameters of the numerical simulation of the filtration process were the same as the previous ones. During processing, the cube surfaces were cropped to remove air contacts as well as to get rid of image defects.

Simulation results within the Stokes model showed a result of 1.3 Darcy at an error bound of 0.05.

Visualization of the resulting structure of the fractured sample is shown in Fig. 3. Rock material is shown in gray (with transparency for detecting fractures), and macrocracks are shown in black. On the left is a visualization of a part of the calculated filtration flow velocity distribution along the material fractures. The filtration direction is from top to bottom. Modeling within the Navier–Stokes model gave a result of 227 mD at an error bound of 0.05. Thus, the permeability values obtained for this structure differ significantly. The Navier–Stokes model gives results close to the physically measured permeability value. The Stokes model, on the other hand, is designed for small pressure drops and flow rates. Under macrocrack flow conditions, higher flow velocities develop than under pore space conditions. This may be one of the reasons for such an overestimated result. It can be noted that both types of calculations on this structure were successfully completed at error boundaries less than 0.1. For the Stokes model a result of 1.5 D at an error bound of 0.01 was obtained. Thus, the calculations appear to be more stable in estimating the fracture permeability. This may be due to less complexity and branching of the structure compared to the porous sample. At the same time, calculations within the framework of the Navier–Stokes model take much more time and resources, and also require the application of methods to optimize the calculations, such as Multigrid [6], but give a more accurate estimate of the fracture permeability.

Analyzing the results obtained for the porous sample, we can conclude that the calculated permeability values under both models for estimating porous permeability were the same within the margin of error. This suggests that both models can be used

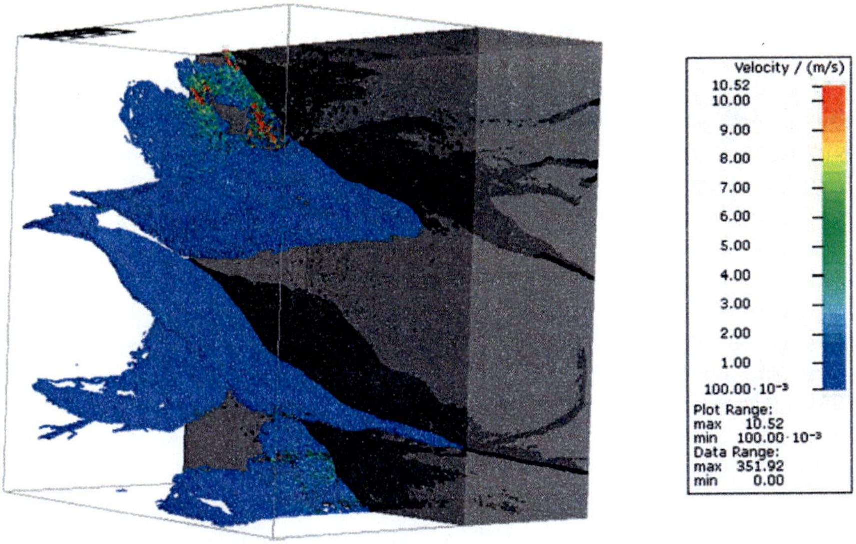

Fig. 3 The obtained structure of the sample with macrocracks, as well as the distribution of flow velocities

to estimate permeability under the described conditions. But the Stokes model has the undeniable advantage of lower consumption of computational resources and time. The preference for one or the other model should be given depending on the computational resource-intensiveness. For modeling under macrofracture conditions, the Stokes model gave highly overestimated results, in contrast to the Navier–Stokes model. However, similar comparisons for different rock types are required to make an accurate conclusion.

One can notice that the results obtained for the porous medium are very different from the physically measured value. This is predictable and caused, first of all, by the choice of segmentation model: only the largest pores were mostly selected, because the main interest in this case is the comparison of the results obtained for different mathematical models. In order to use this technique as an accurate assessment of the permeability of porous structures, it is necessary first of all to carry out a high-resolution scanning. However, this requires the use of a much smaller sample volume. This may present a problem of selecting a representative volume that correctly represents the average pore distribution in the rock. Another method is to combine micro- and nano-sampling to obtain information about porosity on different scales and its permeability. After obtaining a pore permeability value that cannot be recognized from a larger sample volume image, this permeability can be considered by adding a third segmentation phase with a customized permeability value. But the total volume of the third phase in the sample will also make a significant difference in the final permeability estimation. When conducting research on a large sample of one rock sample, it seems possible to determine the micro- and nanoporosity of the rock, selecting scanning parameters and segmentation method to accurately match the simulation results with the physically measured permeability results. After that, the created methodology can be applied to other rock samples of the same type if all scanning and image processing parameters are maintained and the device calibration is accurate and remains the same. This factor could also be affected by external conditions such as temperature. On the other hand, the results of microtomography can be complemented by other research methods, such as synchrotron radiation results. However, the use of additional expensive equipment reduces the availability of the described technique as a whole.

Consequently, the problem of the segmentation procedure and the problem of device resolution are among the main issues that need to be solved to obtain reliable results when modeling processes in rocks based on the results of computed tomography. Consequently, the problem of the segmentation procedure and the problem of device resolution are among the main issues that need to be solved to obtain reliable results when modeling processes in rocks based on the results of computed tomography. The issue of increasing the resolution of tomography is related to the computing capabilities of the data processing equipment. The problem of any segmentation procedure is that it is never strictly accurate and always contains some degree of uncertainty. Accurate segmentation will be possible only for images with almost infinite resolution. This means that each pixel/voxel usually contains a combination of different phases, but segmentation requires that it be associated with only one phase, ideally with the component that prevails in a given volume of the sample.

Objective and clear segmentation requires a very high-resolution image, a limitation of which is the characteristics of the scanning device. However, even if such an accurate image is obtained, significant computational resources are needed to process it and successfully perform simulations on the resulting structure.

4 Conclusion

The paper presents the results of studies of the filtration properties of rocks carried out on the ProCon X-Ray CT-MINI high-resolution scanner of the Institute for Problems in Mechanics RAS. Images of the internal structure of reservoir rocks of two different types were obtained: a highly porous coarse-grained sandstone from a gas field reservoir and a fine-grained well cemented low-permeability sandstone with a branched network of macrocracks from a gas condensate field reservoir. The approaches to the processing of tomography images in the GeoDict software are described. Numerical modeling of filtration flows through the obtained structures is performed. The flow through the pore space was solved using Navier–Stokes and Stokes-only solvers of the FlowDict module of GeoDict. Differences and difficulties in modeling depending on the type of filtration in the rock: pore permeability and fracture permeability are shown. The calculation data are compared with the results of physical experiment on the investigated rock samples. Conclusions are made about the possibility of using the described methods to assess the filtration properties of rocks.

Acknowledgments This work was financially supported by the Russian Federation through the Ministry of Science and Higher Education of the Russian Federation Project No. 13.1902.21.0018 (Agreement 075-15-2020-802).

References

1. Ganat T. A. A. O. Fundamentals of reservoir rock properties. Cham, Switzerland: Springer. 2020.
2. Ivanov M.K. et al. Petrophysical Methods for Study of Core Material // M.: Publishing house of Moscow State University. 2008. P. 112. (in Russian).
3. Lichun J., Mian C. 3D imaging of fractures in carbonate rocks using X-ray computed tomography technology // Carbonates and Evaporites. 2014. V. 29. I. 2, pp. 147–153.
4. Diaz M. et al. Surface roughness characterization of open and closed rock joints in deep cores using X-ray computed tomography // International Journal of Rock Mechanics and Mining Sciences. 2017. V. 98. pp. 10-19.
5. Gerke K.M. et al. Modern approaches to pore space scale digital modeling of core structure and multiphase flow // Georesources. 2021. 23(2). pp. 197–213. DOI: https://doi.org/10.18599/grs.2021.2.20.
6. Linden S., Cheng L., Wiegmann A. Specialized methods for direct numerical simulations in porous media // Technical report. Math2Market GmbH. 2018. DOI: https://doi.org/10.30423/report.m2m-2018-01.

7. Menke H.P. et al. Using Nano-XRM and High-Contrast Imaging to Inform Micro-Porosity Permeability During Stokes–Brinkman Single and Two-Phase Flow Simulations on Micro-CT Images // Frontiers in Water. 2022. V. 4. P. 935035. doi: https://doi.org/10.3389/frwa.2022.935035.
8. Blunt M. J. Multiphase Flow in Permeable Media: A Pore-scale Perspective. Cambridge: Cambridge University Press. 2017. doi: https://doi.org/10.1017/9781316145098
9. Mostaghimi P., Blunt M.J., Bijeljic B. Computations of absolute permeability on micro-CT images. Math. Geosci. 2013. V. 45. Pp. 103–125. doi: https://doi.org/10.1007/s11004-012-9431-4
10. Ar Rushood I. et al. Segmentation of x-ray images of rocks using deep learning // SPE Annual Technical Conference and Exhibition. OnePetro. 2020.
11. Alyafei N., Raeini A.Q., Paluszny A., Blunt M.J. A sensitivity study of the effect of image resolution on predicted petrophysical properties. Transp. Porous Media. 2015. V. 110. 157–169. doi: https://doi.org/10.1007/s11242-015- 0563–0
12. Cid H.E., Carrasco-Núñez G., Manea V.C. Improved method for effective rock microporosity estimation using X-ray microtomography // Micron. 2017. V. 97. P. 11-21.
13. CT-MINI by ProCon X-Ray GmbH available at https://procon-x-ray.de/en/ct-mini.
14. FlowDict: Single-Phase Fluid Flow available at https://www.math2market.com/geodict-software/geodict-base-modules/simulation/flowdict
15. Kling T. et al. Simulating stress-dependent fluid flow in a fractured core sample using real-time X-ray CT data // Solid Earth. 2016. V. 7. 1109–1124. https://doi.org/https://doi.org/10.5194/se-7-1109-2016
16. Balucan R. et al. Coal permeability stimulation by NaClO oxidation // The APPEA Journal. 2019. V. 59. №. 2. P. 846-850.
17. Steel K.M. et al. Identification of preferential pathways in the pore microstructure of metallurgical coke and links to anisotropic strength properties // Fuel. 2021. V. 296. P. 120688.
18. Herdtle T., Xue Y., Bolton J. S. Numerical modelling of the acoustics of low density fibrous media having a distribution of fiber sizes. Publications of the Ray W. Herrick Laboratories. 2017. Paper 167.
19. Soltani P. et al. Experimental and computational analysis of sound absorption behavior in needled nonwovens // Journal of Sound and Vibration. 2018. V. 426. P. 1-18.
20. Linden S. et al. The LIR space partitioning system applied to the Stokes equations // Graphical Models. 2015. V. 82. pp. 58-66.
21. Linden S. et al. The LIR space partitioning system applied to Cartesian grids // Mathematical methods for curves and surfaces. Lecture Notes in Computer Science. 2012. V. 8177. pp. 324–340.
22. Karev V., Kovalenko Y. Triaxial loading system as a tool for solving geotechnical problems of oil and gas production. True triaxial testing of rocks. 2013. Leiden: CRC Press/Balkema: 301–310.
23. Karev V., Kovalenko Y., Ustinov K. Advances in Oil and Gas Exploration and Production. Switzerland: Springer International Publishing. 2020. 166 p.
24. Barkov S.O., Shevtsov N.I. Determination of optimal parameters and modes of well operation in low-permeability reservoirs on a true triaxial loading unit // Physical and Mathematical Modeling of Earth and Environment Processes. Proceedings of 8th International Scientific Conference-School for Young Scientists. 2022.
25. Hilden J., Linden S., Planas B. FlowDict User Guide GeoDict release 2021. Published: September 18, 2020, Reviewed: June 15, 2021. Math2Market GmbH. https://doi.org/10.30423/userguide.geodict2021-flowdict.

Report "Overcoming the Limits" and New Solutions to Global Problems

K. V. Pokazeev and D. A. Solovyev

Abstract The article discusses specific aspects of the report "Overcoming the Limits", which presents the Russian view on ways to solve global problems from the standpoint of the current situation in the first quarter of the twenty-first century. The report was presented at M. V. Lomonosov Moscow State University at the end of March 2022. The features of the approaches of the authors of the report to the consideration of global problems of mankind and new ideas and approaches for their solution are considered. Some issues insufficiently reflected in the report are noted. The possibility of implementing the plans proposed by the authors to solve the global problems of mankind is discussed.

Keywords Club of Rome · Growth limits · Full world concept · Global warming · Low-carbon energy

1 Introduction

A team of researchers led by academicians (V. A. Sadovnichy, Rector of Moscow State University and A. A. Akaev) presented a report to the Club of Rome, "Overcoming the Limits" [1], timed to coincide with the 50th anniversary of the publication of the work "Limits to Growth" [2]. The report "Overcoming the Limits" summarizes the long-term scientific work of the team on the study of the dynamics of the world system [3]. The presented work develops the approaches of domestic scientists V. I. Vernadsky, P. I. Preobrazhensky, N. D. Kondratieva, P. A. Sorokina, I. R. Prigogine, N. N. Moiseev. The findings of the authors of the report "Overcoming the Limits" can be interpreted as a concretization, development of the idea of the transition of humanity to the "noospheric" stage of human evolution, based on the latest research

K. V. Pokazeev
Faculty of Physics, M.V. Lomonosov Moscow State University, Moscow, Russian Federation
e-mail: sea@phys.msu.ru

D. A. Solovyev (✉)
Shirshov Institute of Oceanology, Russian Academy of Sciences, Moscow, Russian Federation
e-mail: solovev@ocean.ru; dsolp@ya.ru

V. I. Karev (ed.), *Physical and Mathematical Modeling of Earth and Environment Processes—2022*, Springer Proceedings in Earth and Environmental Sciences,
https://doi.org/10.1007/978-3-031-25962-3_53

in several fields of science, primarily cybernetics, biology, geophysics. This model of human development allows the authors to give concrete solutions to the global problems of humanity.

The purpose of "the Overcoming the Limits" report is ambitious attempt to find solutions to the major global problems facing humanity today and the threats for all of humanity. Starting from the very title of the report, the difference between the results and conclusions of the presented work as well as the conclusions of its predecessors are emphasized. First of all, these conclusions are optimistic: the limits can be overcome, and the disasters can be avoided, however, by means of joint coordinated work of all humanity.

The study made at Moscow State University differs significantly from the work and conclusions of the Club of Rome in some ways: in the scale of the problems considered; the complexity of the approach; according to the research methodology, which is much more complex, more mathematically justified, and the analysis of global changes is based on identifying the patterns of world dynamics over a historical period of several thousand years; by the volume of testing of model calculations; according to time intervals at which the state of the Earth and humanity are considered and predicted. That is, modeling humanity dynamics is performed in more detail and at a higher level. As a consequence of all that, the work of the authors "Overcoming the Limits" differed from that of its predecessors, including the Club of Rome findings, in terms of forecasts and proposed approaches to solving global problems.

The authors of the report "Overcoming the Limits" consider global problems of humanity, not as limits but as surmountable challenges. According to the authors, there is a fundamental difference between a challenge and a limit. The challenge can be responded, but overcoming the limit, one can come force the death of the system (the biosphere, humanity) due to the loss of its stability and subsequent destruction. Therefore, ways of evolution of the system are being sought not to achieve one or another destructive limit but to cause the development of the system, which makes it possible to avoid reaching the limit.

2 The Content and Introductory Provisions of the Report "Overcoming the Limits"

The results of the authors of the report "Overcoming the Limits" can be interpreted as concretization, development of the idea of the transition of humanity to the "noospheric" stage of human evolution, based on the latest research in several fields of science, primarily cybernetics, biology, geophysics. This model of human development makes the authors possible to give concrete solutions to the global problems of humanity. The conclusions of the report, and the proposals based on them, compared with the conclusions of the works of the Club of Rome, are distinguished by optimism. Implicitly in the proposed solutions to the global problems of planet, there is a belief in the triumph of human reason.

The authors concluded that humanity is at a critical point in its history. But unlike the Western ideas of the authors of the Club of Rome, the report notes that the current global problems are no limits to growth but challenges that can and should be overcome.

First of all, in the work "Overcoming the Limits", the models of the Club of Rome of the resource type are criticized, and their limitations are noted. Unlike the authors of the Club of Rome, who focused mainly on the physical and biological limits of the planet (exhaustible natural resources bordering on the ability of the Earth to absorb industrial and agricultural pollution), the social organization of Society was considered unchanged in the forecast calculations, the possibilities of technological innovations were poorly taken into account. The conclusion of the research considers a necessity of transition to zero economic growth rates and forced population stabilization.

The study of Russian authors considers modern processes in a broader macroeconomic context, taking into account the peculiarities of long-term technological development. It is shown that a significant number of problems can be solved taking into account the development of modern technologies under the conditions of the second demographic transition (global decline in fertility).

The most challenging task facing humanity is the stabilization of the climate. The ways of solving this problem are proposed. Research conclusion: the main problem lies in the socio-political sphere, and the essential prerequisite for solving common humanitarian issues is the transition from competition between the countries of the world to joint cooperative actions. In this case, most problems become not growth limits but challenges that collective efforts and purposeful technologies can overcome. The difference between the approaches of the authors of the two reports ("Limits to Growth" and "Overcoming Limits") is shown in Fig. 1.

There is no social block in the calculations based on the World-3 model of the Club of Rome, and the failure to take social processes into account distorts dynamics of the global processes. The "Overcoming the Limits" model includes the social block and plays a significant role. However, it raises some questions, which will be discussed below. The latest reports of the Club of Rome note that capitalism has exhausted itself, and ways out of the global economic and environmental crises are proposed. True, since these proposals do not follow model calculations and are advisory, they are not as convincing as conclusions based on resource models. The proposals of the Club of Rome, outlined in the 2018 report [5] and based on the concept of "full world", should be criticized along with the forecasts of the Club of Rome based on resource models. In our opinion, the idea of a "complete world" is also not convincing. Before the world's fullness, humanity has to master the resources of some vast regions of South America and Asia, the entire World Ocean, which are not only far from being exhausted but are still far from being explored. Therefore, all forecasts based on the "full world" concept require adjustment or even denial.

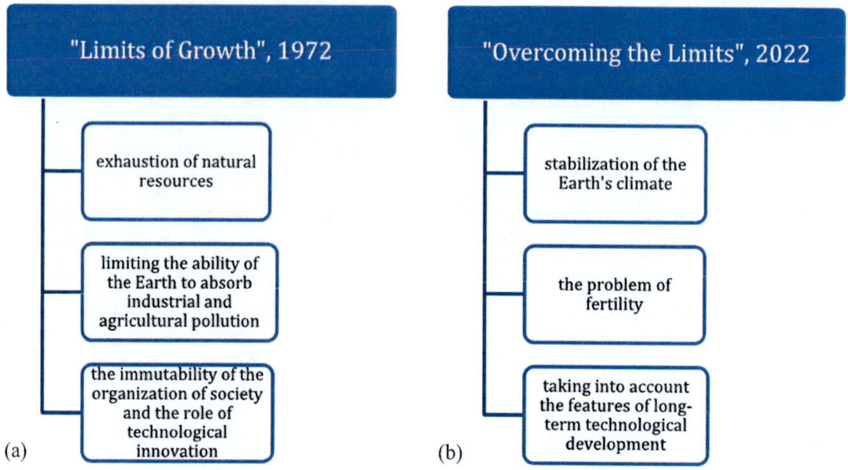

Fig. 1 The main differences in approaches to solving global problems in the reports: "Limits to Growth" (**a**) and "Overcoming the Limits" (**b**). *Source* presentation materials for the report "Overcoming the Limits", [4]

3 Discussion of the Results of the "Overcoming the Limits" Report

Using a historical approach, the authors of the "Overcoming the Limits" report believe that the most important factor in human society is technological development, which determines all spheres of human society activity, and this development takes place against the background of a certain natural background, which closely interacts with society. All the main parameters of the global system, interacting with each other, change historically.

An important distinguishing feature of the presented report is a much longer time interval of predicted calculations than previous model calculations. The models for different historical eras are based on a single cognitive scheme. The system development models are tested on extended periods, calculated in centuries and millennia. This is a striking difference from the forecasts in works of the Club of Rome based on the prolongation of the identified trends of changes in society at present for a relatively near future. However, it is not clear how far the application of the same models based on general principles to different social formations by the authors of the "Overcoming the Limits" report is, although this is given special attention to. How correct are these general principles?

In the historical dynamics modeling system, seven areas were identified for the calculations: climate (atmospheric pollution), ecology (environmental pollution), economics, demography, technology, social sphere, and politics.

The scheme of interaction between various fields of activity, which is the basis of the historical dynamics modeling system, is shown in Fig. 2. The diagram is a

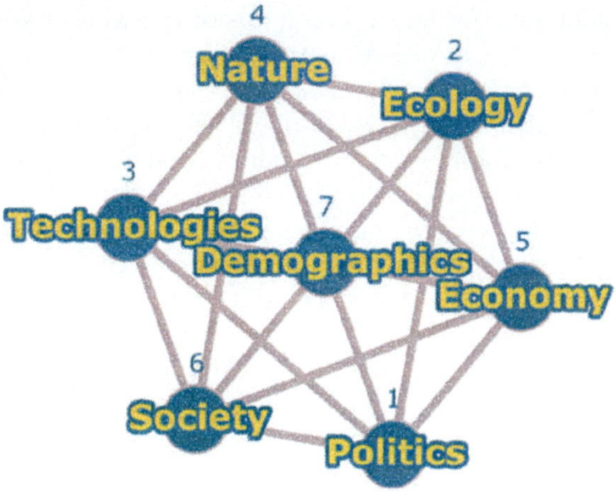

Fig. 2 Cognitive scheme for modeling historical dynamics. *Source* [6]

directed graph, where the vertices of the graph are separate spheres. The connecting lines are the influence of one sphere on another. The scheme adopted the following designations for the areas: Nature (climate, natural environment), Ecology, Technology, Demography, Economy (production, economic relations), Society (social interactions), and Politics (public administration, political interactions) [6].

The scheme of interactions in the global system of nature-humanity includes seven spheres (blocks in the terminology of the authors) of human activity: climate, natural environment; ecology; technology; demography, economy (production, economic relations); social (social interactions); politics (public administration, political interactions). For each block, basic equations were proposed to characterize these interactions. However, the specifics of the development of Society are such that at almost any stage of its development, there are various social formations in different regions of the planet. How is this diversity of social structures simultaneously taken into account in modeling? The scheme of interactions of spheres (blocks) of human activity demonstrates a significant complication compared to the model used in the "World" model. On the other hand, each area of activity (blocks) is described by common equations: for the sake of clarity and to simplify the analysis, the authors identified the main processes. It is not always clear how justified this choice is since there is no selection criterion.

For example, in the **"Demography"** block, the authors do not include migration and argue if this can be done. This statement raises doubts: when considering migration processes, it is necessary to consider the influence of all other blocks. Their methods are still far from being studied and interact non-linearly with each other. In this case, the nature of the interaction will experience substantial historical variability.

According to the authors, the fundamental equations of the "Politics" block reflect the main features of the corresponding social formations. This section cannot but remember the ideological views of the authors, so it seems to be the most artificial of all the blocks of the cognitive system. The main predictive criterion is the system stability criterion, which takes into account the interaction with the environment, as in the Club of Rome models, and the system's internal processes. Thus, it is hard to judge how correctly economic, ecological, and social interactions are taken into account if at each moment on the planet only, there are simultaneously different social systems (the method of evolution of individual systems is described in sufficient detail), interacting with each other, which complicates forecasts greatly. Also it seems wrong that influence of religion remains unaccounted for in the "Politics" block.

The **"Society"** block is probably the most challenging to describe with equations. The authors consider the conditions that ensure the stability of the functioning of the social sphere to be the main factors determining the state of the system. For some reason, religion, which plays a massive role in modern society, is not considered in this block. Although in the future, the most crucial part of religion or ideological attitudes is noted, especially during periods of reforming the economic foundations of society.

The **"Ecology"** block takes into account the man-caused CO_2. Still, it does not consider changes caused by climate variability, which can have a decisive impact on the fate of humanity, primarily during periods of transition from one stable state of the climate system to another.

An important place in the report is given to the problem of climate variability. The "nature" block includes equations describing climate change and resource endowment. The stages of the climate forecast are described in sufficient detail and include accounting for energy consumption and its structure, including during the transition to low-carbon energy and population dynamics. In nature, long-term temporal factors determine climate and temperature changes, which are not taken into account in the "nature" model. Climatic variability is described only by anthropogenic temperature change, which raises questions. How is natural variability taken into account? The role of biological factors in the development of humanity is well-known in history. Natural climatic variability contains long-term components, including hundreds of thousands of years [7, 8]. How are they counted in the block? The resource use is characterized by substantial geographic variability, with almost all other affecting blocks of the scheme. Unfortunately, this is not reflected in the model. Similar questions and remarks can be made on nearly all model blocks.

The **"Nature"** block is associated with energy problems, particularly the transition to low-carbon energy. Such a construction of a natural block seems to be too simplistic, including in historical terms. In the modern world, along with energy problems, there is an acute problem of fresh water [9], which will only get worse shortly, so the issue of fresh water should be one of the most important in the "Nature" block. Water availability determines the level and duration of life, population size, agriculture and energy, migration, etc. The limit of freshwater consumption refers to those limits that can be overcome through the collective action of several countries or the entire world community. Unfortunately, the problem of fresh water was not mentioned in the report.

The implementation of the authors' proposals for solving climate and energy problems is designed for an extended period and provides for a well-coordinated, solidary interaction of all states of the planet. The authors believe that the way out of a critical situation can be a transition to joint action by all countries of the world, and China and Russia can become leaders in the evolution of humanity to a more just society. It is not easy to imagine the problem-free possibility of implementing such a process. According to the authors' forecasts, the population of the Russian Federation, together with the CIS countries by the end of the twenty-first century, will be less than 3–5% of the world population. It is hard to imagine that a country with such a population is capable of providing not only a theoretical but also a practical example of solving global problems. Such a role was within the power of Russia only at the beginning of the previous century.

Based on the processing of statistical data for the period 1960–2021 and the trends observed in recent years in the field of energy consumption, as well as taking into account the implemented energy-efficient technologies, the authors selected and calculated three scenarios for the increase in the average global temperature of the surface atmosphere in the twenty-first century: a conservative scenario, ambitious and zero emissions scenarios. The traditional method assumes that government policies, technologies, and social preferences continue to evolve the same way as in the past. The ambitious scenario envisages introducing measures that lead to a significant reduction in carbon emissions from energy use, making it possible to limit the increase in global temperatures in the twenty-first century. The Zero Emissions scenario assumes that the measures proposed in the ambitious method are complemented and reinforced by significant societal behavior and preferences changes.

The considered scenarios of the dynamics of changes in the structure of the global fuel and energy balance in the twenty-first century indicate that a special role in the transition period to low-carbon energy will be played by gas energy as the most environmentally friendly branch of modern energy and industry provided with huge natural resources, including in the world ocean. However nuclear power, despite its high environmental friendliness, has a high cost, limiting its widest distribution, so its role in the transition period requires additional research [10].

According to the assessments, if implemented, the "zero emissions" scenario will make it possible to meet the requirements of the Paris Climate Agreement to keep global warming at 1.5–2 °C compared to the pre-industrial level. As part of the implementation of the "zero emissions" scenario, this goal is primarily expected to be achieved through the use of energy-efficient technologies, including hydrogen and the further development of renewable energy sources, as well as the simultaneous use of chemical technology for capturing and storing carbon dioxide.

Thus, it is argued that with the effective use of modern technologies and concerted actions of the world community, it is possible to avoid the climate crisis if the world acts decisively and takes the cooperative measures necessary to reduce carbon dioxide emissions. In particular, the global community needs to redirect the investment flows from the most profitable to those that will benefit society in the long term. A similar conclusion was also made concerning other global environmental problems.

Predicting the dynamics of world development, the authors of the report proceed from the reliability of the onset of global warming, which determines not only the way to solve the problem of global warming but also ways to solve the energy, environmental, and ultimately, socio-economic problems that arise due to global change of climate [11]. However, the problem of global warming is criticized by several researchers, it requires further study, and the problem of global warming would not be replaced by the problem of global cooling [12]. The prospects for "green energy" also require further research, the possibility of "green energy" to meet the basic energy needs of mankind has not been proven yet [13]. The optimism of forecasts about the prospects for "green energy" has been replaced by more balanced estimates that take into account the economic costs of developing renewable energy, its efficiency and reliability [14]. There are still many uncertainties on the way to the development of renewable energy and the search for unconventional energy sources, so we hope that there will be many promising discoveries.

The analysis carried out by the authors based on the developed cognitive scheme for modeling historical dynamics shows that modern problems are not so much in the field of "society-nature" interactions, which can be solved with the necessary coordination of the efforts of the world community, but in the field of social and political interactions.

According to the authors' ideas, the entire previous history of humanity is explained based on the assumption of the existence of social formations: X-structures and Y-structures and their periodic change from one to another. X-structures are formed under conditions of stability, and Y-structures—under growth conditions. The evolution of structures is associated with revolutionary technological changes. The role of ideology and religion significantly increases in transitional periods of the development of society [15]. The authors demonstrated the success of this approach in describing the history of humanity. It is interesting to note the importance of the authors' research on the role of inventive activity as the most crucial factor of social activity influencing the development of technologies and, as a result, the most critical indicator of the development of human society. Now the relative number of people on the planet engaged in creative work, activities associated with ingenuity, is declining everywhere.

Currently, a digital revolution is taking place, leading to the formation of a fundamentally new type of Society. The authors call it a cybernetic or W-society, which differs from the X-society and the Y-society. According to the authors, the transition to a new kind of society cannot be predicted. The change must be designed and offered to society for implementation. This is what happens in reality. Several projects of the future W-society have already been formulated—this is K. Schwab's "stakeholder capitalism" and a similar project—the "inclusive capitalism" of Pope. Both projects involve the modernization of capitalism. The "Overcoming the Limits" report is an alternative blueprint for the future of the W-society based on the importance of collaboration over competition. W-society is characterized by globalization, isolation, aging of the population and the stabilization of population, implementation of information technologies and artificial intelligence into all spheres of life, active influence of biotechnology, genetic engineering, etc. The inevitability of the transition to this type of society has been discussed repeatedly, with the transition option and the option of the emerging W-society being important. Humanity faces a bifurcation point, and the direction of further development depends on human society.

The authors model two options for the transition of humanity to the W-society and assert the existence of only these two options.

The first option is the formation of "inclusive capitalism," which will be implemented under the control of the "world government" You can name specific steps that are already being implemented in the world and direct humanity along option No. 1. This is the transition to a "basic income", gender politics, managing the mentality of the masses with the help of social media, etc. The second option is a plan for implementing the idea of the "noospheric" stage of human development, which was formulated by Vernadsky and further developed, mainly in the works of Russian scientists. However, there are no practical steps to implement this plan yet, only theoretical research. It is interesting to note that both options have many common characteristics due to the use of information—digital technologies and other scientific industries—but the created societies are strikingly different.

Thus, a fundamental conclusion is made: the type of global interactions in the nature-human system will depend on the type of social structure of society and the variant of the transition of humanity to the W-society. Also here is discussion of how the formation of the social structure will take place or how to design the choice of social structure.

4 Conclusions

The implementation of the authors' proposals to solve climate and energy problems is designed for a long period and actually provides for coordinated, solidarity interaction of all states of the planet. The authors believe that the way out of the critical situation may be the transition to joint actions of all countries of the world, and China and Russia can become the leaders of the transition of mankind to a more just society.

The current digital revolution is leading to a fundamentally new type of society. The authors call it the cybernetic or W-society, which differs from both the X-society and the Y-society that existed earlier. According to the authors, the transition to a new type of society cannot be predicted. The change must be designed and offered to the human society for implementation. This conclusion seems to be very important since there is a fundamental difference between design and forecasting. According to the authors, the W-society can be implemented in two ways. The implementation of the option depends on the type of social structure of society. The authors make a fundamental conclusion about the significance of the kind of social structure of society, conducting the transition to the society of the future. For some reason, the authors do not single out this fundamental conclusion. Nor is there any discussion of how to design the choice of social structure.

The questions and comments formulated above will contribute to the further improvement of the project "Overcoming the Limits".

Acknowledgements This work was supported by the Ministry of Science and Higher Education of the Russian Federation (State Assignment No. FMWE-2021-0018).

References

1. "Responding to challenges - overcoming limits.“ Presentation of the preprint of the report to the Club of Rome took place at Moscow State University [Electronic resource]. URL: https://scientificrussia.ru/articles/otvecaem-na-vyzovy-preodolevaa-predely-v-mgu-prosla-prezentacia-preprinta-doklada-rimskomu-klubu (accessed 20.10.2022).
2. Meadows D.H., Meadows D.L., Randers J., Behrens III WW The limits to growth - Club of Rome1972.
3. Sadovnichiy V.A., Akaev A.A., Korotaev A.V., Malkov S.Yu. Modeling and forecasting of world dynamics. Moscow: Institute for Socio-Political Research RAS, 2012
4. Sadovnichiy V.A., Askarov A.A. World development and "limits to growth" in the 21st century: modeling and forecast [Electronic resource]. 2022. URL: https://expert.msu.ru/haos22-1? (date of access: 20.10.2022).
5. Von Weizsäcker, Wijkman A., others. Come on! Springer, 2018.
6. Sadovnichiy V.A., Akaev A.A., Ilyin I.V., Korotaev A.V., Malkov S.Yu. Modeling and forecasting of world dynamics in the 21st century (preprint). Moscow: Moscow State University. M.V. Lomsonosov, 2022
7. S.P. Perov, K.V. Pokazeev. Visheratin, K.N. Kozlova G.V. In the 21st century, the Earth's climate will be colder than in the 20th // International conference "Flows and structures in liquids: physics of geospheres", Vol.2. 2009, pp. 159–162.
8. Kotlyakov V. Academician of the Russian Academy of Sciences: "The era of cooling has come, which will lead to a new ice age" [Electronic resource]. 2019. URL: https://www.business-gazeta.ru/article/451128 (accessed 20.10.2022).
9. David B.R., Spencer S., Miller J., Almahmoud S., Jouhara H. Comparative environmental life cycle assessment of conventional energy storage system and innovative thermal energy storage system // International Journal of Thermofluids. 2021. (12). C. 100116. DOI:https://doi.org/10.1016/J.IJFT.2021.100116
10. Bushuev V.V., Soloviev D.A. Climate and Energy Transition: Interaction and Interdependence // Energy Policy. 2021. No. 11(165). pp. 44–55. DOI:https://doi.org/10.46920/2409-5516_2021_11165_44.
11. Pokazeev K.V., Solovyev D.A., Nefedova L.V. New Opportunities for the Development of Renewable Sources of Hydrosphere Energy // Physical and Mathematical Modeling of Earth and Environment Processes. Springer Proceedings in Earth and Environmental Sciences. Springer, 2022, pp. 203–209.
12. Global warming and global cooling (materials of the meeting of the Presidium of the Russian Academy of Sciences) [Electronic resource]. URL: https://scientificrussia.ru/articles/globaljnoe-poholodanie (date of access: 02/04/2020).
13. Plakitkin Yu.A., Plakitkina L.S. Five basic regularities of global energy, the "green deal" as restraint factors in the development of the mining industries of the FEC // Gornaya promyshlennost. 2021. No. 4. P. 94–100.
14. Andreenko T.I., Berezkin M.Y., Bushuev V.V., Gribkov S.V., Degtyarev K.S., Zaichenko V.M., Zalikhanov A.M., Kiseleva S.V., Nefedova L.V., Nechayev A.M., Nigmatulin R.I., Nigmatulin RI, Pokazeev VK, Rafikova YY, Sinyugin OA, Solovyev A.A., Solovyev DA, Chekarev K.V., Chernova N.I., Shilova L.A. The geography of renewable energy. Moscow: Publishing House "Energia," 2021.
15. Gordienko, V.A., Starkova M.V., Pokazeev K.V. Overcoming the global ecological crisis and the possibility of forming a new type of worldview // Physical problems of ecology. No. 19. Moscow: MAKS PRESS, 2013, pp. 153–171.

Wave Dynamics of Stratified Medium with Shear Flows: Main Problems Formulation

V. V. Bulatov

Abstract The paper considers issues related to the formulation of problems for describing the dynamics of internal gravity waves in stratified media with horizontal shear flows. Physical formulations of problems in which critical levels can arise are discussed. In the first problem, the bottom oscillations are considered, which begin at some point in time, and the establishment of a critical level at large times is studied. We also consider the formulation of the problem of a stratified medium flow running on an obstacle, behind which outgoing waves can arise, with a singularity at the critical level being formed far from the obstacle.

Keywords Internal gravity waves · Shear currents · Dynamics of a stratified medium · Buoyancy frequency · Critical level

The generation of internal gravity waves by shear currents is an important example for geophysical applications of the interaction of waves with hydrodynamic flows [1–7]. When internal gravity waves propagate in shear flows, specific effects arise associated with the amplification and absorption of waves [8–14]. This occurs in the presence of resonant layers, in which the flow velocity coincides with the phase or group velocity of internal gravity waves. Critical layers play an important role in the mechanism of hydrodynamic instability of shear flows [1, 2, 15–20]. The resonant interaction of waves with critical layers is an important mechanism for the generation and absorption of various types of waves in the atmosphere, ocean, and moving plasma. A strong interaction of a propagating wave with a flow is possible in those layers where the frequency of the wave coincides with the frequencies of natural oscillations of the inhomogeneous flow. In this case, the resonance is due to the equality of the group velocity of the wave with the velocity of a certain liquid layer. As a result, the wave packet is captured in the vicinity of the resonance. The algebraic method for studying waves in shear flows is based on a piecewise linear approximation of the velocity profile. This method is unsuitable for the analysis of resonant

V. V. Bulatov (✉)
Ishlinsky Institute for Problems in Mechanics RAS, Vernadskogo Ave. 101-1, 119526 Moscow, Russia
e-mail: internalwave@mail.ru

V. I. Karev (ed.), *Physical and Mathematical Modeling of Earth and Environment Processes—2022*, Springer Proceedings in Earth and Environmental Sciences,
https://doi.org/10.1007/978-3-031-25962-3_54

features. To study waves in a flow with a continuous velocity profile, one can use direct numerical simulation or approximate asymptotic methods: the WKB method, the method of reference equations, the method of matching asymptotic expansions [21–27]. The plasma-hydrodynamic analogy may also be useful, providing a tool for the physical understanding of resonant processes. To study the dynamics of internal gravity waves, the Richardson number Ri plays a fundamental role [2, 3, 18, 19]. When the value of the number $Ri > 1/4$ in the vicinity of the critical layer, the wave packet approaches the critical point indefinitely as the wavelength decreases.

In this paper, we consider the specifics of setting the problems of wave dynamics of IGWs in a stratified medium with horizontal shear flows. Linear IGWs are described in the Boussinesq approximation by the equation [1–3, 8, 9]

$$\frac{D^2}{Dt^2}\left(\Delta+\frac{\partial^2}{\partial z^2}\right)W-\frac{D}{Dt}\left(\frac{d^2U}{dz^2}\frac{\partial W}{\partial x}+\frac{d^2V}{dz^2}\frac{\partial W}{\partial y}\right)+N^2(z)\Delta W=0$$
$$\Delta=\frac{\partial^2}{\partial x^2}+\frac{\partial^2}{\partial y^2},\ \frac{D}{Dt}=\frac{\partial}{\partial t}+U(z)\frac{\partial}{\partial x}+V(z)\frac{\partial}{\partial y} \tag{1}$$

where $(U(z), V(z))$—shear flow velocity components on the horizon z, $N^2(z)$—Brent–Väisälä frequency squared (buoyancy frequency), W—vertical component of IGWs velocity.

Уравнение (1) рассматривается в слое $-H < z < 0$. На дне $z = -H$ вертикальная компонента скорости равна нулю, на поверхности $z = 0$ используется приближение «твердой крышки», отфильтровывающее поверхностные волны. В океане ВГВ могут обмениваться энергией со средними течениями [2–7]. If the vertical velocity gradient of the currents is large, then the average currents can give energy to the waves, that is, the corresponding oscillations can be unstable. The Miles condition is known, under which there are no unstable eigenwaves, which has the form [1, 2, 16, 19]: $Ri(z) \geq 0.25$, $Ri(z) = N^2(z)/\left(\left(\frac{\partial U}{\partial z}\right)^2+\left(\frac{\partial V}{\partial z}\right)^2\right)$.

For IGWs with harmonic dependence on time and horizontal coordinates

$$W(t,x,y,z)=\exp(i(\omega t-kx))w(z) \tag{2}$$

where the direction of wave propagation is taken as the axis x, on a certain horizon $z = Z$, for which the x-component of the current velocity $U(z)$ coincides with the phase velocity $C = \omega/k$, the wave energy is absorbed, that is, part of the wave energy is transferred to the mean currents. Then the level $z = Z$ is called the critical level.

The behavior of short-wavelength IGWs packets approaching the critical level is usually considered in the WKB approximation under the assumption that $U(z)$ and $N^2(z)$ change slowly over the oscillation period of the considered wave field. However, the question of in what particular physical problems such packets can arise, as a rule, is not considered [1, 2, 19, 23].

Next, we will discuss the formulations of boundary value problems in which critical levels may arise. Typically, two-dimensional problems are considered, that is, solutions that do not depend on the horizontal variable y. Then, substituting (2) into (1) to determine the function $w(z)$, we get the Taylor–Goldstein equation [1, 2, 16, 19]

$$(\omega - kU(z))^2\frac{\partial^2 w}{\partial z^2} + \left(k^2\left(N^2(z) - (\omega - kU(z))^2\right)\right) + k\frac{d^2U(z)}{dz^2}(\omega - kU(z)) = 0 \tag{3}$$

From here, in particular, it can be seen that the horizon $z = Z$ on which $\omega = kU(z)$ is the singular point of this equation. The behavior of the function $U(z)$ on either side of the value $z = Z$ is determined by the rule for bypassing this singular point. A physically meaningful solution is a solution obtained by analytical continuation from the domain $\mathrm{Im}\,\omega > 0$.

The IGWs field $W(t, x, y, z) = \exp(i(\omega t - kx))w(z)$ near the critical level can be represented as the sum of W_+ and W_-, where W_+—corresponds to a wave that transfers energy from bottom to top, and W_-—to a wave that transfers energy from top to bottom. Each of these waves, when crossing the critical level, abruptly decreases in amplitude by a factor of $\exp(-\pi\mu)$, where $\mu = \sqrt{Ri(Z) - 0.25}$. This fact means the absorption of IGWs energy at a critical level, i.e., the transfer of energy to the average currents of part of the IGWs energy.

Next, consider the following problem statements in which critical levels may arise. It can be assumed that there are natural oscillations with critical levels, that is, solutions of the Taylor–Goldstein equation that vanish at $z = 0, -H$ and for $\omega - kU(z) = 0$ which at some $z = Z$. However, under the Miles condition and a strictly monotonic function $U(z)$, such eigenfunctions do not exist. Therefore, for a field with a given wave number k, all natural frequencies turn out to be such that there is no critical level—any of $\omega_n(k)$ satisfies one of the inequalities $\omega_n(k) > \max kU(z)$ or $\omega_n(k) < \min kU(z)$.

This result admits a simple physical interpretation: at $\mathrm{Im}\,\omega = 0$, the wave energy of any eigenoscillation $W(t, x, y, z) = \exp(i(\omega t - kx))w(z)$ must be conserved, and in the presence of a critical level, part of the IGWs energy is absorbed, transforming into the energy of medium currents. Therefore, critical levels can occur only for forced oscillations with a given frequency Ω, that is, for example, oscillations described by an inhomogeneous Eq. (1) with the right side (source distribution) of the form $\exp(i(\Omega t - kx))f(z)$ for some function $f(z)$.

Technically, it is easier to consider IGWs excited not by sources, but by boundary oscillations, that is, consider a homogeneous Eq. (1) with an inhomogeneous boundary condition in the form: $W = \exp(i(\Omega t - kx))$ at $z = -H$ and $W = 0$ at $z = 0$. Then, defining the solution in the form: $W(t, x, y, z) = \exp(i(\omega t - kx))w(z)$, one can obtain that $w(z)$ satisfies the Taylor–Goldstein equation with given values ω and k boundary conditions: $w(0) = 0$, $w(-H) = 1$. Obviously, such a solution already has a critical level.

However, such a substitution is a physical idealization, assuming that the oscillations continue indefinitely. Therefore, it seems more natural to formulate the problem, in which the bottom oscillations begin at some point in time, that is, to consider the solution of Eq. (1) that satisfies the boundary and initial conditions:

$$\begin{aligned} & W(t,x,0)=0 \\ & W(t,x,-H)=0(t<0),\ W(t,x,-H)=\exp(-i(\Omega t-kx))(t>0), \\ & W(t,x,z)\equiv 0(t<0) \end{aligned} \tag{4}$$

For $t \to \infty$, this solution should tend to a function that has a singular point at $z = Z$. For any finite value t, this solution must be a regular function z, so the question of how the solution tends to the limit $t \to \infty$, that is, how the critical level is set, is of interest. Let us further consider the mean flow $(U(z) \neq 0, V(z) = 0)$ and the variable-independent y solution of Eq. (1) $W(t, x, z)$ satisfying conditions (4). We will assume that the value Ω is within the interval $[I_-, I_+]$, where $I_\pm$ is the minimum (maximum) value of the function $kU(z)$ in the interval of the vertical coordinate.

This means that the phase velocity of the bottom oscillations coincides with the current velocity $U(z)$ at a certain horizon $z = Z$. Assuming $W(t, x, z) = \exp(ikx)u(t, z)$ for the function $u(t, z)$, you can get the boundary value problem

$$\begin{aligned} & \left(\frac{\partial}{\partial t}+ikU(z)\right)^2\left(-k^2u+\frac{\partial^2 u}{\partial z^2}\right)-ik\left(\frac{\partial}{\partial t}+ikU(z)\right)\frac{d^2U(z)}{dz^2}u-k^2N^2(z)u=0 \\ & u(t,z)\equiv 0(t<0),\ u(t,0)\equiv 0 \\ & u(t,-H)=0(t<0),\ W(t,-H)=\exp(-i\Omega t)(t>0) \end{aligned} \tag{5}$$

Then it can be expected that for $t \to \infty$, the solution $u(t, z)$ will tend to a harmonically time-dependent function $v(t, z) = \exp(-i\Omega t)V(\Omega, z)$, where the function $V(\Omega, z)$ is a solution to the Taylor–Goldstein Eq. (3).

For a particular case of a constant distribution $N^2(z)$ and a linear function $U(z)$, the solution $W(t, x, z)$ coincides with the limiting solution (2) with an arbitrarily small relative error ε for all z outside the transition zone—the vicinity of the critical level $z = Z$, whose size tends to zero at $t \to \infty$ as $A(\varepsilon)/t$. At the same time, it is obvious that $\varepsilon \to 0$ when $A(\varepsilon) \to \infty$. In this case, the most interesting question is how quickly the limiting solution is established, that is, what is the actual width of the transition zone described by the function $A(\varepsilon)$, and how the zone size depends on the parameters of the problem: the wave number, the local Richardson number.

The qualitative nature of the exit of the behavior of solutions to the limit in the presence of a critical level can be represented as follows. The function $W(t, x, z)\exp(-i(\Omega t - kx))$ outside the transition zone coincides with $U(z)$. Inside the transition zone, there should be a smooth transition from one branch of the function $U(z)$ to another. With an increase t, the width of the transition zone decreases, but the qualitative form of the transition zone, as a function of the variable $t(z - Z)$, should not change. Obviously, for arbitrary distributions of the buoyancy frequency and shear flows that satisfy the Miles conditions and the natural regularity conditions, special asymptotic methods must be used to construct solutions.

There is also another class of problems in which critical levels can arise—these are problems about lee currents. Let us consider the flow of a stratified medium running on some obstacle—a bottom ledge, a local bottom elevation. Behind this obstacle, a wave escaping from it can arise, and this wave can excite IGWs. Then the corresponding boundary value problem for Eq. (1) has the form

$$W(t, x, 0) = 0$$
$$W(t, x, -H) = 0(x < 0),\ W(t, x, -H) = \exp(-i(\Omega t - kx))(x > 0)$$

It is possible to look for a solution that depends harmonically on time, the uniqueness of which is ensured by the radiation condition, that is, assuming $\Omega = \Phi + i\Psi$, $\Psi > 0$. For a complex solution Ω, the solution is unique, since the fulfillment of the Miles condition guarantees the absence of complex eigenvalues for the Taylor–Goldstein equation with zero boundary conditions at $z = 0, -H$. Next, we look for the passage to the limit of the solution at $\Psi \to 0$. In such a formulation of the problem, the singularity of the IGWs field at the critical level is formed at $x \to \infty$, that is, far from the streamlined obstacle.

Analytical estimates and numerical results show that asymptotic constructions using model representations of the buoyancy frequency and background shear velocities can qualitatively correctly describe the amplitude-phase structure of IGWs. Taking into account the real distributions of the main hydrological parameters makes it possible to study the whole variety of generated waves, and numerical methods must be used to describe the amplitude dependences of IGWs in natural environment (ocean, atmosphere) with arbitrary background shear currents. Therefore, to study IGWs in real natural environment (ocean, atmosphere), it is necessary to combine both exact numerical methods for studying wave fields and various asymptotic approaches that allow studying the main qualitative features of excited waves, including the development of remote methods for detecting IGW by radar methods.

Acknowledgements The work is carried out with financial support from the RSF project 23-21-00194.

References

1. Lightill J (1978) Waves in fluids. Cambridge University Press, Cambridge
2. Fabrikant AL, Stepanyants YuA (1998) Propagation of waves in shear flows. World Scientific Publishing, Singapore
3. Miropol'skii YuZ, Shishkina OV (2001) Dynamics of internal gravity waves in the ocean. Kluwer Academic Publishers, Boston

4. Vlasenko V, Stashchuk N, Hutter K (2005) Baroclinic tides. New York: Cambridge University Press
5. Mei CC, Stiassnie M, Yue DK-P (2017) Theory and applications of ocean surface waves. Advanced series of ocean engineering. V. 42. World Scientific Publishing
6. Morozov TG (2018) Oceanic internal tides. Observations, analysis and modeling. Springer, Berlin
7. Velarde MG, Tarakanov RYu, Marchenko AV (Eds.) (2018) The ocean in motion. Springer Oceanography. Springer International Publishing AG, Berlin
8. Bulatov VV, Vladimirov YuV (2012) Wave dynamics of stratified mediums. Nauka, Moscow
9. Bulatov VV, Vladimirov YuV (2019) A general approach to ocean wave dynamics research: modelling, asymptotics, measurements. OntoPrint Publishers, Moscow
10. Morozov EG, Tarakanov RYu, Frey DI, Demidova TA, Makarenko NI (2018) Bottom water flows in the tropical fractures of the Northern Mid-Atlantic Ridge // J Oceanography 74(2): 147–167
11. Khimchenko EE, Frey DI, Morozov EG (2020) Tidal internal waves in the Bransfield Strait, Antarctica // Russ J Earth Science 20: ES2006
12. Slepyshev AA, Vorotnikov DI (2019) Generation of vertical fine structure by internal waves in a shear flows // Open J Fluid Mechanics 9: 140–157
13. Haney S, Young WR (2017) Radiation of internal waves from groups of surface gravity waves // J Fluid Mech 829: 280–303
14. Fraternale F, Domenicale L, Staffilan G, Tordella D (2018) Internal waves in sheared flows: lower bound of the vorticity growth and propagation discontinuities in the parameter space // Phys Rev 97(6): 063102
15. Broutman D, Rottman J (2004) A simplified Fourier method for computing the internal wave field generated by an oscillating source in a horizontally moving, depth-dependent background // Phys Fluids 16: 3682–3689
16. Gavrileva AA, Gubarev YuG, Lebedev MP (2019) The Miles theorem and the first boundary value problem for the Taylor–Goldstein equation // J Applied and Industrial Mathematics 13(3): 460–471
17. Bulatov VV, Vladimirov YuV, Vladimirov IYu (2019) Far fields of internal gravity waves from a source moving in the ocean with an arbitrary buoyancy frequency distribution // Russian J Earth Sci 19: ES5003
18. Bulatov V, Vladimirov Yu (2020) Analytical approximations of dispersion relations for internal gravity waves equation with shear flows // Symmetry 12(11): 1865
19. Bulatov VV (2022) Analytical properties of the Green's function for the equation of the internal gravitational waves in a stratified medium with shear flows // Theoretical and Mathematical Physics 211(2): 611–624
20. Lecoanet D, Le Bars M, Burns KJ, Vasil GM, Brown BP, Quataert E, Oishi JS (2015) Numerical simulations of internal wave generation by convection in water // Phys Rev E – Statistical, Nonlinear, and Soft Matter Physics 9: 1–10
21. Voelker GS, Myers P, Walter M, Sutherland BR (2019) Generation of oceanic internal gravity waves by a cyclonic surface stress disturbance // Dyn Atm Oceans 86: 16–133
22. Hirota M, Morrison PJ (2016) Stability boundaries and sufficient stability conditions for stably stratified, monotonic shear flows // Physics Letters A 380(21): 1856–1860
23. Meunier P, Dizиs S, Redekopp L, Spedding G (2018) Internal waves generated by a stratified wake: experiment and theory // J Fluid Mech 846: 752–788
24. Wang H, Chen K, You Y (2017) An investigation on internal waves generated by towed models under a strong halocline // Phys Fluids 29: 065104
25. Svirkunov PN, Kalashnik MV (2014) Phase patterns of dispersive waves from moving localized sources // Phys.-Usp. 57 (1): 80–91
26. Churilov S (2018) On the stability analysis of sharply stratified shear flows // Ocean Dynamics 68: 867–884
27. Broutman D, Brandt L, Rottman J, Taylor C (2021) A WKB derivation for internal waves generated by a horizontally moving body in a thermocline // Wave Motion 105: 102759

Changes in a Shoreline Under Influence of a Groin Field

I. O. Leont'yev and T. M. Akivis

Abstract A one-line analytical model of groin impact on a sandy beach suggested before by the authors is further developed in this work. The new version of the model considers trapping capability of groins to be dependent on shoreline displacement associated with infilling of inter-groin compartments. Thus, a feedback is implied between accumulation rate and evolution of shoreline contour which tends to its equilibrium state. It is shown that accumulation in a groin field is governed by rather fine mechanism operating in a relatively narrow range of conditions. The groin system is most efficient when the step of the groin field coincides with the length of an influence zone of a single structure. This in its turn leads to an optimal relation between the length of the structures and the distance between them. The model calculations demonstrate the ability to control the groin system operation by changing its parameters.

Keywords Groin field · Inter-groin compartment · Shoreline contour · Sandy beach · Longshore sediment flux · Accumulation volume · Down-drift erosion

1 Introduction

Groins are linear constructions connected with the shore. Usually they are combined into a system, which includes a series of elements approximately equally distanced from each other. The idea of the system is its ability to accumulate a part of sediment transported longshore by waves and currents and, as a result, to increase the beach. As shown by experience, the successful work of a groin field is possible only in the presence of a noticeable longshore sediment flux. Otherwise, the consequences for the coast may be negative [1].

The changes in nearshore morphology in the influence area of groins are related to wave diffraction at obstacles and the corresponding transformation of currents and sediment transport. This idea is a base of rather complicated numerical models

I. O. Leont'yev · T. M. Akivis (✉)
P.P. Shirshov Institute of Oceanology RAS, Moscow, Russia
e-mail: akivis@yandex.ru

V. I. Karev (ed.), *Physical and Mathematical Modeling of Earth and Environment Processes—2022*, Springer Proceedings in Earth and Environmental Sciences,
https://doi.org/10.1007/978-3-031-25962-3_55

[2–4]. Their application requires significant labor costs and is not always advisable, including for economic reasons. The relatively simple one-line analytical model [5, 6] is more convenient for operational calculations. In this model, sediment accumulation within the groins field is explained in terms of the mass conservation law. The introduced scale of the structure's area of influence allows finding an optimal ratio between the length of the groins and the distance between them.

In this work the model concept is further developed. The parameter of the groins trapping ability gets a new interpretation. It is connected directly with the coastline displacement due to infilling of inter-groin compartments. In addition, the more appropriate relationship between sediment accumulation and shoreline propagation is deduced. In the new version of the model, coastal morphology naturally evolves towards equilibrium, thus eliminating the artificial feedback function suggested before.

2 The Model Concept

First, it makes sense to recall the main backgrounds and arguments of the previous model because they are largely preserved in its new version.

We consider the groin system with the constant step λ. The construction length l_G does not exceed the width of the sediment flux. Because of this, a part of material can be transported outside the constructions (Fig. 1). It is supposed that an influence of the obstacle on the sediment flux is limited to some distance Λ (Fig. 2).

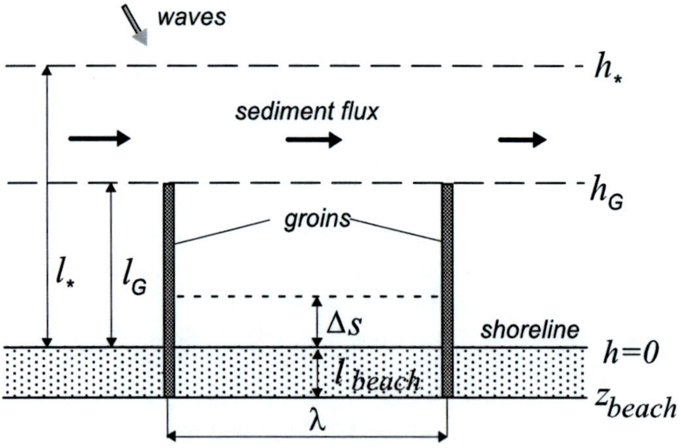

Fig. 1 A scheme of a beach and a groin system. l_*—width of sediment flux (of active part of coastal profile), h_*—closure depth, l_G—groin length, h_G—depth at head of a groin, l_{beach} and z_{beach}—width and elevation of beach, λ—distance between the groins or a system step, Δs—shoreline advancing in the inter-groin compartment

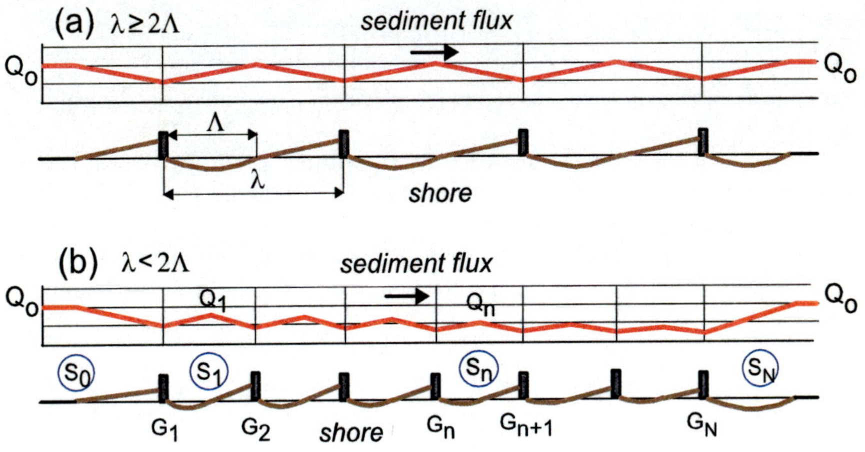

Fig. 2 Changes in sediment flux and shoreline contour for relatively long (**a**) and relatively short (**b**) inter-groin distance λ. A scale Λ is the length of the structure influence zone

If the system step λ is large enough and meets the condition $\lambda \geq 2\Lambda$ then the separate groins practically have no influence on each other. Sediment flux is unloaded before the obstacle and downstream it saturates and recovers up to initial value Q_0. Accumulation near the structure results in the shoreline propagation while sediment deficiency on the leeward side causes its retreat. However, on average, the sediment balance does not change, because of mutual compensation between accumulation and erosion (Fig. 2a).

In the case when $\lambda < 2\Lambda$, the groin system is able to accumulate material (Fig. 2b). The first structure traps a part of alongshore sediment flux $b_1 Q_0$ and initiate accumulation with a rate $Ac_0 = b_1 Q_0$. The flux increases downstream and were the distance between the groins 2Λ the growth would be $b_1 Q_0$. In other words, the flux would recover up to its initial value (Fig. 2a). But at smaller λ the increase of the flux can take only a part, namely $K b_1 Q_0$, where

$$K = \lambda/2\Lambda, \quad \lambda/2\Lambda < 1. \tag{1}$$

In the section of the flux increasing erosion occurs at a rate $K b_1 Q_0$, and on the up-drift side of the groin G_2 sediment is accumulated at a rate $b_2 Q_1$. Therefore the total accumulation rate in the first inter-groin compartment S_1 equals $Ac_1 = b_2 Q_1 - K b_1 Q_0$. Since $Q_1 = Q_0 - b_1 Q_0 + K b_1 Q_0 = [1 - (1 - K) b_1] Q_0$, we get $Ac_1 = [b_2(1 - b_1) - b_1(1 - b_2)K] Q_0$.

In a similar way accumulation rate in the next compartment (S_2) can be found: $Ac_2 = [b_3(1 - b_2) - b_2(1 - b_3)K] Q_1$, where $Q_2 = [1 - (1 - K) b_2] Q_1$. As a result we come to relations that determine the accumulation rate and the sediment flux in each inter-groin compartment S_n:

$$Ac_n = \left[b_{n+1}(1 - b_n) - b_n(1 - b_{n+1})K\right] Q_{n-1}, \quad n = 1, 2, 3, \ldots, N - 1,$$

$$Q_n = [1 - (1 - K)b_n]Q_{n-1},$$

where N is a number of structures in the groin system (so a number of inter-groin compartments is $N - 1$).

As can be seen, the sediment flux decreases downstream (Fig. 2b). Downstream from the last construction, the discharge is restored to its initial value Q_0 (Fig. 2b), which should lead to down-drift erosion.

Figures 1 and 2 reflect the situation where the sediment flux goes from left to right viewed from the shore. In the opposite direction of the flux, the groin with number N will accumulate at a rate Ac'_N and down-drift erosion will occur to the left of the first groin.

Representing the dependences found for the accumulation rates as a functions of the sediment flux at the entrance to the groin field (Q_0 and Q_N) we get the relations:

$$Ac_0 = b_1 Q_0, \quad Ac_n = a_n \prod_1^n c_n Q_0, \quad n = 1, 2, 3, \ldots, N - 1, \tag{2}$$

$$a_n = b_{n+1}(1 - b_n) - b_n(1 - b_{n+1})K, \quad c_n = 1 - (1 - K)b_n,$$

$$Ac'_N = b_N Q_N, \quad Ac'_{N-n} = a'_{N-n} \prod_1^{N-n} c'_{N-n} Q_N,$$

$$a'_{N-n} = b_{N-n}(1 - b_{N-n+1}) - b_{N-n+1}(1 - b_{N-n})K, \quad c'_{N-n} = 1 - (1 - K)b_{N-n+1},$$

where the primed values correspond to the sediment flux directed from right to left.

3 The Model Parameters

The parameter b characterizes the part of the sediment flux trapped by the groin. In the actual new version of the model it is defined by a ratio of the structure length l_G to the width of the longshore flux l_*. If $l_G/l_* \geq 1$ the groin traps all the flux and $b = 1$. But in the case $l_G/l_* \to 0$ we have $b \to 0$. In the same time, b should decrease with the shoreline advancing Δs when the groin system is being filled. It is clear that increase in Δs leads to decrease in the "working" length of the structure $(l_G - \Delta s)$. When $(l_G - \Delta s)/l_* \to 0$ then $b \to 0$. The value b also depends on permeability factor of the groin p_G which varies from 0 (fully transparent construction) to 1 (solid barrier). These considerations lead to the following relations:

$$b = \begin{cases} p_G \frac{l_G - \Delta s}{l_*}, & \frac{l_G - \Delta s}{l_*} < 1 \\ p_G, & \frac{l_G - \Delta s}{l_*} \geq 1 \end{cases}. \tag{3}$$

The parameter Λ, which characterizes the scale of the structure influence zone also depends on the ratio of l_G to l_* [7]:

$$\Lambda = \begin{cases} (l_G l_*)^{0.5}, & l_G/l_* < 1 \\ l_*, & l_G/l_* \geq 1 \end{cases}. \tag{4}$$

The value Λ essentially characterizes the distance at which the sediment flux has time to react to the presence of an obstacle. As was shown above, the inter-groin distance λ should be between Λ and 2Λ. Correspondingly, the value of the parameter K from (1) is limited by the range $0.5 \leq K < 1$ that marks the applicability limits of our model. Accumulation rate is maximal when $K = 0.5$ i.e. when $\lambda = \Lambda$. Then the optimal ratio between the groin length and the system step is

$$\lambda/l_G = (l_G/l_*)^{-0.5}. \tag{5}$$

By increasing the groin length, the distance between them λ should increase proportionally to $\sqrt{l_G}$. For short constructions, $l_G/l_* = 0.1$, the system step λ is close to $3l_G$. For relatively long groins, $l_G/l_* = 0.5$, an optimal step is $\lambda \approx 1.4l_G$. In practical use, the value λ/l_G ranges between 1 and 3 [1, 8].

The width of the sediment flux l_* can be found using the corresponding depth h_* and the average bottom slope $\beta = h_*/l_*$. For the individual storm event the value h_* refers to the depth of wave breaking h_B which marks the outer boundary of the longshore flux under the given wave action [9, 10]:

$$h_B = \left(\frac{1}{4\pi\gamma_B^2}\right)^{0.4} H_{1\%0}^{0.8}\left(gT^2\right)^{0.2}\left(\frac{\cos\Theta_0}{\cos\Theta_B}\right)^{0.4}, \tag{6}$$

where $H_{1\%0}$ is a wave height of 1% cumulative exceedance probability in deep water, T is the wave period, Θ_0 and Θ_B are the wave incidence angles in deep water and at the wave breaking point, respectively, and $\gamma_B = 0.8$ is breaking index.

For the time scale of a year or several years, the depth h_* is interpreted as a closure depth limiting the area of significant morphological changes in the coastal profile. It can be expressed in terms of wave height H_{s012}. which might be exceeded only 12 h per year (0.137% cumulative exceedance probability),

$$h_* = \overline{K} H_{s012}, \tag{7}$$

where $\overline{K}$ depends on the wave steepness and the given threshold of the bottom deformations Δh_c [11]. For typical storm waves steepness 0.03–0.04 and choosing $\Delta h_c = 0.1$ m we have $\overline{K} = (1.5 \div 1.6)$.

4 Volumes of Accumulation and Shoreline Displacement

4.1 Inter-groin Compartment

Under wave action during time t_w the volume of accumulation in the inter-groin compartment will be

$$V_n = \int_0^{t_w} Ac_n dt \quad \text{or} \quad V_n' = \int_0^{t_w} Ac_n' dt \tag{8}$$

depending on the wave angle between the wave direction and the normal to the shoreline. The accumulating material can be conditionally represented as a layer distributed in the inter-groin compartment (Fig. 3a). Formation of the layer is accompanied by appropriate propagation of the shoreline Δs. If the bottom beyond the depth h_G were horizontal the cross-sectional area of the layer would be equal to the parallelogram area with the base length Δs and the height $h_G + z_{beach}$, and the total layer volume would be $V = \Delta s(h_G + z_{beach})\lambda$.

However, since the bottom has a slope β, a certain part of sediments should be spent on the construction of the base of the layer (Fig. 3a). In the first approximation, the cross-section of the base is a triangle with bottom side $2\Delta s$, height $\beta\Delta s$ and corresponding area $\beta\Delta s^2$. So the volume of the accumulated layer is expressed as $V = \Delta s(h_G + z_{beach})\lambda + \beta\lambda\Delta s^2$. Using the approximation $(h_G + z_{beach}) \approx \beta(l_G + l_{beach})$, we get the square equation relative to Δs:

$$\Delta s^2 + (l_G + l_{beach})\Delta s - \frac{V}{\beta\lambda} = 0. \tag{9}$$

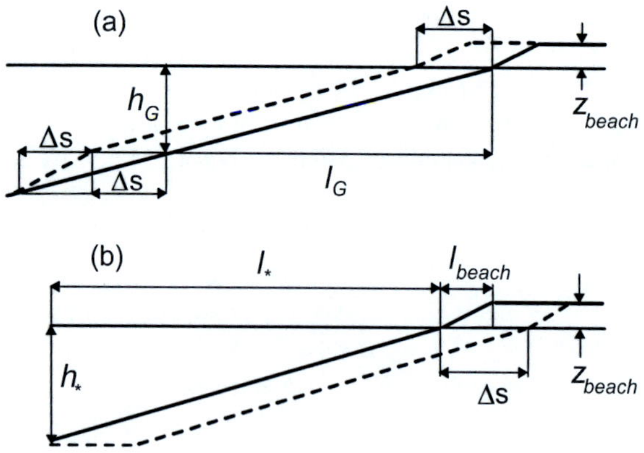

Fig. 3 Schemes explaining shoreline advancing in the inter-groin compartment (**a**) and shoreline retreat in the down-drift erosion area (**b**). Notations are given in the text

Its solution is

$$\Delta s = -\frac{1}{2}(l_G + l_{beach}) + \sqrt{\frac{1}{4}(l_G + l_{beach})^2 + \frac{V}{\beta\lambda}}. \tag{10}$$

At the initial moment when $V = 0$, we have $\Delta s = 0$. As V increases the shore advances and at certain stage the displacement of the shoreline reaches its limit $\Delta s_m = l_G$. It follows from (10) that corresponding maximal volume V_m should be

$$V_m = l_G h_G \left(2 + \frac{l_{beach}}{l_G}\right)\lambda.$$

As Δs approaches l_G the value b decreases according to (3) and the accumulation rate Ac is reduced resulting in decay of changes in volume V.

4.2 Areas Adjacent to the Groin Field

An average shoreline displacement Δs also can be determined here by (10). But instead of the distance between the groins λ in (10) the length of the accumulation zone l_A near a single obstacle should be implied. Observations show that the displacement reaches its maximum at the obstacle and decreases to zero with distance from it. Accordingly, the average displacement Δs over the length l_A is about a half of its maximal value corresponding to the groin length l_G [7, 12]. Based on this, the displacements Δs_0 and Δs_N of the shore contour before the first and the last groins are determined as follows:

$$2\Delta s \frac{l_A - x}{l_A} = \begin{cases} \Delta s_0, \; if \; -l_A \le x \le 0 \\ \Delta s_N, \; if \; 0 \le x \le l_A \end{cases}, \tag{11}$$

where x is a longshore distance.

The length Λ of an obstacle influence zone from (4) is supposed to be the scale for l_A. The accumulation area tends to expand over time proportional to $\sqrt{t}$ and for the periods of years and decades is expressed by the dependency [7, 12]:

$$l_A = c_A \Lambda \sqrt{t/t_0}, \tag{12}$$

where c_A is a proportional coefficient of order 1, $t_0 = 1$ year and t is measured by years after completion of the construction.

Changes of l_A during an individual storm event supposedly can be described by the same dependency (12). In this case, it makes sense to take the time interval t_0 as the characteristic duration of the process of forming a quasi-equilibrium nearshore morphology. This may be, for example, a 12-h interval during which the cross-shore beach profile is largely stabilized [11].

4.3 Areas of Down-Drift Erosion

Downstream from a single obstacle or the series of constructions is a down-drift erosion zone of length l_E where the longshore flux saturates and recovers to its initial value (if the shore is uniform). The erosion volume corresponds to the total sediment volume accumulated by the groin system:

$$V_{EN} = -\left(\sum_n V_n + V_0\right), \quad V_{E0} = -\left(\sum_n V_n' + V_N\right). \tag{13}$$

Imagine the eroded volume as a layer distributed within the erosion zone (Fig. 3b). The area of its cross-section is a sum of areas of two parallelograms with the base Δs and heights h_* and z_{beach}, and its volume is $V_E = \Delta s(h_* + z_{beach})l_E$. From here the shoreline retreat is

$$\Delta s = \frac{V_E}{(h_* + z_{beach})l_E}. \tag{14}$$

This value characterizes an average shoreline displacement in the erosion zone. As shown in [7, 12] the contour of the shore can be approximated by a sinusoid bounding the same area $\Delta s l_E$:

$$\Delta s_E = \frac{\pi}{2}\Delta s \sin k_E x, \quad k_E = \frac{\pi}{l_E}. \tag{15}$$

The length of the erosion zone l_E is supposed to be dependent on the scale Λ similar to the length of accumulation zone l_A and increases with time:

$$l_E = c_E \Lambda \sqrt{t_Y}, \tag{16}$$

$c_E \geq 1$ is a proportionality coefficient presumably close to c_A.

5 Comparison of Calculations with Observations

First compare the calculations with the results of the field experiment carried out on the sandy coast of the northern Yucatan, Mexico (21° N, 90° W) [13]. The aim of this experiment was an investigation of effect of a single groin. The latter was a permeable dismountable construction consisting of several dozen concrete hexapods

(60 kg each). Installation and dismantling of the 15 m long structure by a team of 20 people took about 2 h. The length of groin from its tip to the shoreline was $l_G \approx 9$ m.

The experiment lasted 24 h during which there were waves of height H_s from 0.4 till 0.8 m and period T_p of about 4 s. The wave direction varied within north-eastern sector. For this period the sand volume of $V_N = 35$ m^3 was accumulated on the east side of the groin. Figure 4a demonstrates the shoreline positions before and after the experiment. The difference between corresponding points, i.e. the shoreline displacement is represented in Fig. 4b.

Parameters for calculations were $d_g = 0.25$ mm, $H_s = 0.6$ m, $T_p = 0.4$ s, $\Theta_0 = -45°$. Corresponding height of wave breaking and sediment flux were calculated by formulae from [10] as $h_B = 1.1$ m and $Q_N = -6.6$ m^3/h. The groin permeability consistent with the mentioned above value of V_N should be $p_G = 0.6$. The sizes of the surf zone and the beach were $l_B = 20$ m, $z_{beach} = 1$ m, $l_{beach} = 15$ m.

Matching of calculated and observed changes in shoreline was achieved by choosing the value $c_A\sqrt{t/t_0}$, under the assumption of equality of coefficients $c_A = c_E$ in (12) and (16). The optimum value obtained is close to 2.5. As noted above t_0 has the meaning of a typical stabilization period of coastal morphology in an individual

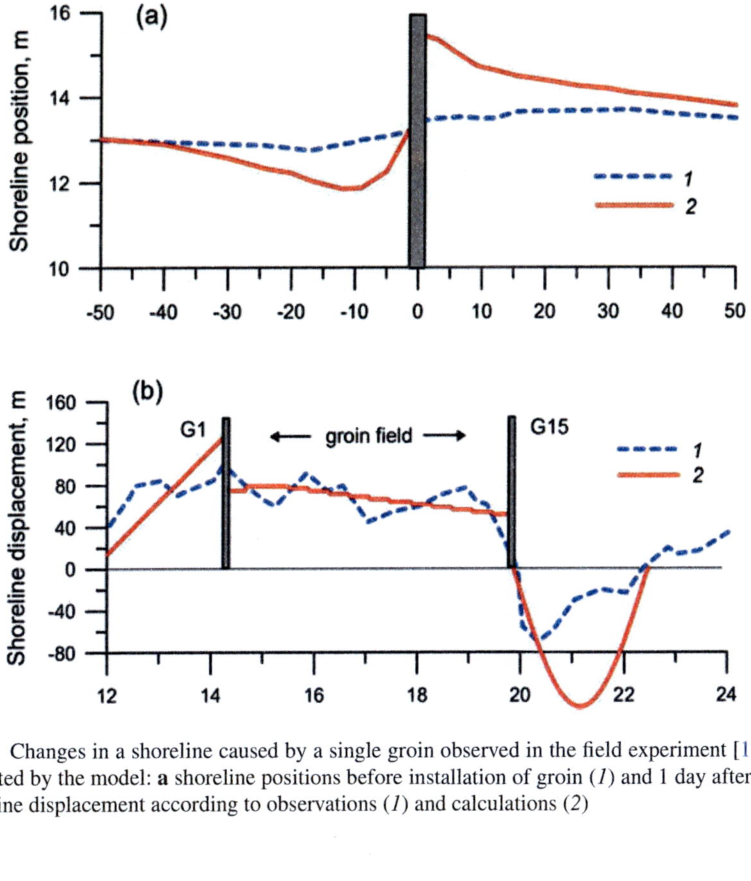

Fig. 4 Changes in a shoreline caused by a single groin observed in the field experiment [13] and predicted by the model: **a** shoreline positions before installation of groin (*1*) and 1 day after (*2*); **b** shoreline displacement according to observations (*1*) and calculations (*2*)

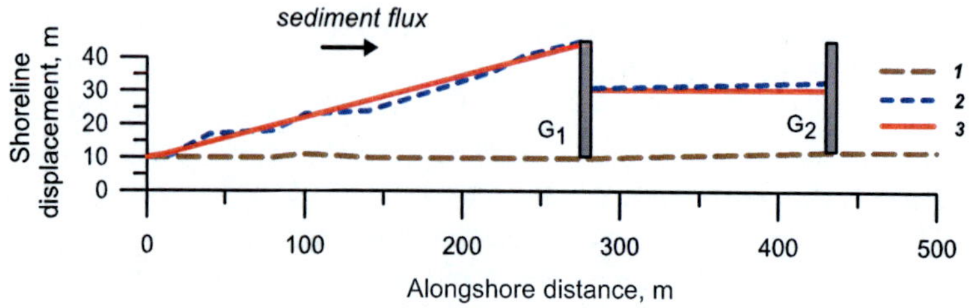

Fig. 5 Shoreline changes under influence of groins (southwest coast of India [14]). *1*—protective wall marking shoreline position in 2009, *2* and *3*—shoreline displacement 5 years after construction of groins according to observations and calculations

storm event (12 h). Since the wave action duration t in this case is 24 h we get the estimate $c_A = c_E \approx 1.8$.

The modelled changes of the shore contour are also shown in Fig. 4b. As can be seen, the model conveys the main trends of the beach changes though the actual shoreline shape is of more complicated character.

The next example, referring to an eroded site of the southwest coast of India (8° 25′ N, 76° 58′ E) [14] was used for testing the previous version of the model [5]. After the construction of the two groins, the beach quickly grew and after 5 years got advanced to the end of the first groin (Fig. 5). The protective wall in this case prevents the bank from retreating in the zone of down-drift erosion. These are the updated parameters for the calculations: $l_G = 35$ m, $\lambda = 150$ m, $Q_0 = 70$ thousands m^3/year, $h_* = 5$ m, $l_* = 200$ m, $c_A = c_E = 1.5$, $d_g = 0.3$ mm. Figure 5 shows that modelled and measured shoreline changes are nearly identical. The deviations are less than in the previous version [5].

Another modelled object is a section of Westhampton Beach, U.S.A. Atlantic coast (40° 47′ N, 72° 41′ W) [15]. A groin system including 15 elements is built here for the extension of the beach. Figure 6a shows the position of the coastline immediately before the system is constructed and after 23 years. Figure 6b shows the difference of the corresponding marks, reflecting the displacements of the coastline. Here are the results of the forecast by our model when using the following parameters: $l_G = 145$ m, $\lambda = 400$ m, $Q_0 = 150$ thousands m^3/year [4], $h_* = 8$ m, $l_* = 500$ m, $p_G = 0.8$, $c_A = c_E = 2$, $d_g = 0.25$ mm.

Inconsistencies in the volume of erosion outside the groin field were due to repeated nourishment of the beach during the period under review, which could not be factored into the calculations. However, within the bounds of the groin field, there are no significant differences in the accumulation and position of the shoreline according to observations and calculations.

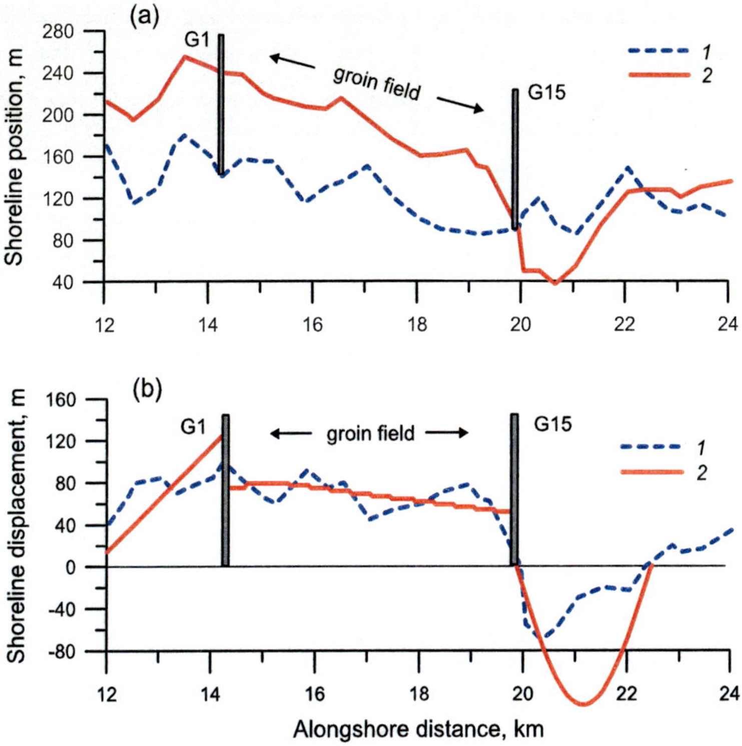

Fig. 6 Shoreline evolution at Westhampton Beach, U.S.A. Atlantic coast [15]. **a** Shorelines before and after construction of groin field: (*1*)—1966, (*2*)—1989; **b** shoreline displacement over 23-year time lapse according to observations (*1*) and calculations (*2*)

5.1 Examples of Calculations and Discussion of Results

Some examples of shoreline changes under the influence of groins demonstrate how it is possible to regulate the operation of the structures by variation of their parameters.

Consider the shore profile with the following parameters $h_* = 6$ m, $l_* = 400$ m, $z_{beach} = 1.5$ m, $l_{beach} = 30$ m. The structures are supposed impermeable ($p_G = 1$). Based on the above results we take $c_A = c_E = 2$. The conditions of the tests are shown in Table 1. There are the length and the step of the system (l_G and λ), the ratio λ/Λ, sediment fluxes (Q_0 and Q_N) and the prediction period (t_w). The results of the calculations are shown in Fig. 7.

In the tests 1 and 2 the system step is close to the optimal value $\lambda \approx \Lambda$ that causes intensive accumulation in inter-groin compartments. If the sediment flux is one-directed (test 1) then down-drift erosion is significant but for double-directed feeding (test 2) it is hardly noticeable.

In the tests 3 and 4 the distance between the groins increased and far exceeds the optimal value Λ. The volume of accumulation and the shore advancing are much less

Table 1 The conditions of the tests for calculations of shoreline changes under groins impact

Test number	l_G (m)	λ (m)	λ/Λ (–)	Q_0 (thous. m^3/y)	Q_N (thous. m^3/y)	t_w (y)	V (thous. m^3)
T1	50	150	1.06	50	–	5	51.5
T2	50	150	1.06	50	– 20	5	77.0
T3	50	220	1.56	50	– 20	5	64.3
T4	50	280	1.98	50	– 20	5	45.0
T5	100	300	1.50	50	– 20	5	135
T6	100	300	1.50	50	– 20	25	560

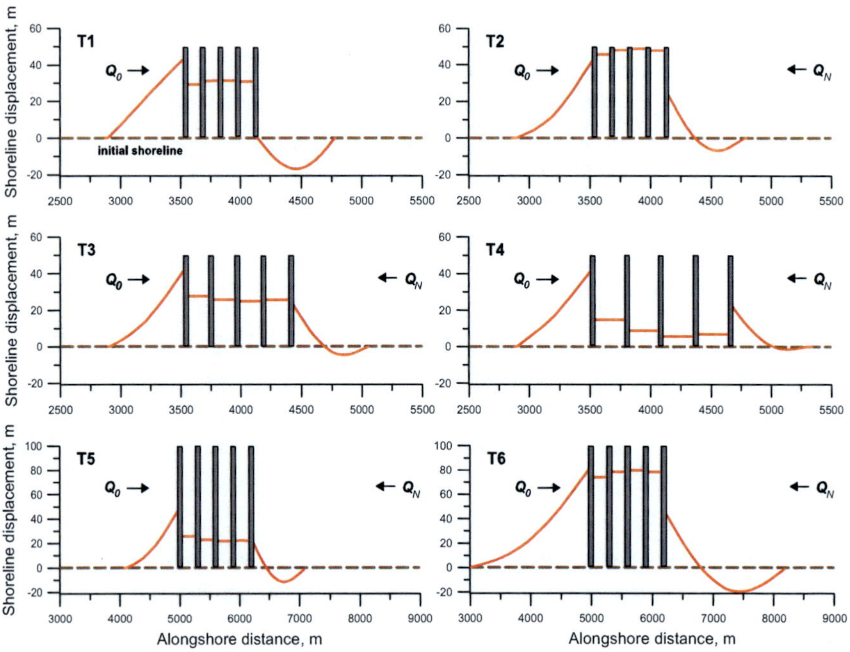

Fig. 7 Examples of computed shoreline changes under influence of a groin field for various initial conditions (given in Table 1)

than in the similar test 2. If $\lambda/\Lambda > 2$ then according to calculations accumulation in inter-groin compartments is replaced by erosion.

In the test 5 the groin length and the system step are doubled compared to test 2. The constructions trap the more significant part of longshore sediment flux.

The test 6 demonstrates the changes over the period of 25 years under the same conditions. The sediment volume in inter-groin compartments noticeably increases and accumulation and erosion areas near the outer bounds of the groin field widen significantly.

Figures 6 and 7 show that accumulation in the first (upstream) compartment is getting a little slower compared to neighboring ones (tests 1, 2 and 6). The point is that as the shoreline before the first groin advances, an increasing proportion of sand bypasses the first compartment, and the area of more intensive accumulation shifts downstream. This feature revealed in cases of relatively small distance between the groins ($\lambda/\Lambda \leq 1.5$). However, when the system step is greater, the infilling volume decreases downstream (tests 3, 4 and 5).

6 Conclusions

The represented model considers sediment accumulation within a groin field as a result of specific changes of longshore sediment flux. When entering the inter-groin compartment the flux increases causing shore erosion. However, before recovering to its original value the flux decreases before the next obstacle causing the sediment to settle. If the distance between the groins is chosen properly, the volume of accumulation exceeds that of erosion and leads to the resulting accumulation in the inter-groin compartment. This is quite a fine mechanism operating within a relatively narrow range of conditions.

Sediment accumulation in a groin system is possible if the ratio of system step λ to the scale of influence zone Λ of the structure falls within the interval $1 \leq \lambda/\Lambda < 2$ which at the same time limits the applicability of our model. The value Λ is proportional to the width of longshore flux and the structure length. The most efficient operation of the system is achieved under the condition $\lambda = \Lambda$ which determines the optimal ratio of the system step to the length of the structures λ/l_G.

For relatively short groins this ratio is close to 3 while for the long constructions it tends to 1.

In the new version of the model, the part of sediment flux trapped by the structure is characterized by the ratio of its length to the width of the active part of coastal profile. The "working" length of the groin changes with the beach advancing. This provides feedback between the rate of accumulation and the evolution of the shore contour, which tends to some equilibrium condition.

The model is applicable to different time scales including time of an individual storm event. In this case, parameters of the surf zone width l_B and the depth of wave breaking h_B are used instead of active profile width l_* and closure depth h_*.

Operation of groin system causes not only positive changes (growth of the beach), but also negative phenomena associated with the erosion of adjacent areas of the coast. Calculations on the base of the obtained dependencies allow prompt comparison of different project variants, which is shown in the given examples. The application of the model helps to choose the best option in which the negative effects can be minimized.

There is no enough examples of comparisons of modelling with observations yet to unambiguously assess calibration coefficients c_A и c_E characterizing the extent of the accumulation and erosion zones outside the groin field. The likely range of the specified coefficients from 1.5 to 2.5 can only be recommended. Wider validation of the model is still difficult due to lack of necessary data and presents the task of future research.

Acknowledgements The work was fulfilled in a scope of the Government assignment (Theme # FMWE-2021-0004) in P.P. Shirshov Institute of Oceanology RAS.

References

1. *Prushak, Z.; Ostrovsky, R.; Babakov, A,; Chubarenko, B.,* 2014. The main principles of the groins using as the coast stabilization structures. Geomorfologiya: 3, 91–104.
2. *Hanson, H.*, 1989. GENESIS: a generalized shoreline change numerical model. J. of Coastal Res.: 5(1), 1–27.
3. *Leont'yev, I.O.,* 1999. Modelling of morphological changes due to coastal structures. Coastal Engineering: 38, 143–166.
4. *Hanson, H.; Larson, M.; Kraus, N.C.*, 2010. Calculation of beach change under interacting cross-shore and longshore processes. Coastal Engineering: 57, 610–619.
5. *Leont'yev, I.O.; Akivis, T.M.*, 2020. The effect of a groin field on a sandy beach. Oceanology: 60(3), 412–420. DOI: https://doi.org/10.1134/S0001437020030042.
6. *Leont'yev, I.O.; Akivis, T.M.*, 2021. Prediction of groins impact on a sandy beach. In: Processes in GeoMedia; Olegovna C. Springer Geology. Springer, Gewerbestrasse 11, 6330 Cham, Switzerland, Volume III, 195–202. DOI: https://doi.org/10.1007/978-3-030-69040-3_19.
7. *Leont'yev, I. O.*, 2018. Changes in shoreline contour due to cross-shore structure in the sea coastal zone. Geomorfologiya: 3, 32–39. DOI: https://doi.org/10.7868/S0435428118030033.
8. *Coastal Engineering Manual* (CEM). 2002, V(3), Shore protection project. US Army Corps of Engineers. Coastal Engineering Research Center. EM 1110-2-1100.
9. *Leont'yev, I.O.*, 1988. Randomly breaking waves and surf-zone dynamics. Coastal Engineering: 12, 83–103.
10. *Leont'yev, I.O.*, 2014. Calculation of longshore sediment transport. Oceanology (Engl. Transl.): 54, 205–211.
11. *Leont'yev, I.O.*, 2022. Evaluation of depth of closure on a sandy beach. Oceanology: 62(2), 258–264. DOI: https://doi.org/10.1134/S0001437022020102.
12. *Leont'yev, I.O.*, 2007. Changes in the shoreline caused by coastal structures. Oceanology (Engl. Transl.): 47, 877–883.
13. *Torres-Freyermuth, A.; Hofman, A.; Tuz-Pech, J.C.; Medellin, G.; Roos, P.C.*, 2020. Design and performance of a permeable groins on a low-energy natural beach. J. Mar. Sci. Eng.: 8, 283–299. DOI: https://doi.org/10.3990/jmse8040283.
14. *Noujas, V.; Thomas, K.V.; Ajeesh, N.R.*, 2017. Shoreline management plan for a protected but eroding coast along the south-west coast of India. International Journal of Sediment Research: 2. DOI: https://doi.org/10.1016/j.ijsrc.2017.02.004i.
15. *Kraus, N.C.; Hanson, H.; Blomgren, S.*, 1994. Modern functional design of groin systems. In: Proceedings of 24th Conference on Coastal Engineering, Kobe, Japan, 1327–1342.

On the Issue of the Impact of Recreation on Coastal Territories

A. Yu. Sanin, O. D. Vasilyev, and V. A. Kulakovskaya

Abstract Recreational human activity, as well as other types of nature management, have a remarkable impact on natural landscapes, including in coastal territories. Its most important consequences include environmental pollution, the activation of dangerous natural processes, in particular, those related to the dynamics of the coasts, the transformation of natural landscapes, a decreasing biodiversity, the degradation of ecosystem services provided by the natural landscapes of coastal territories, visual and noise pollution, and a number of others. Similar consequences are characteristic of other types of human activity, and for some of its types, for example, for industrial human activity, they are much stronger. As a rule, an area attractive for recreation usually holds a high value for many other types of human activity, in particular, for permanent residence, which causes conflicts between them. It is necessary to know the structure of the nature management of the region and the history of its formation in order to understand the causes of conflicts—they are often caused by competition for the same territorial resources. In order to achieve the sustainable development of the territory, it is necessary to mitigate the most intense conflicts between different types of human activity. In many cases, conflict mitigation makes it possible to achieve mutual concessions to different types of economic activity to each other. Another condition for the sustainable development of territories actively used for recreation is the preservation of their tourist potential, which is impossible without control over the flows of recreants, both organized and unorganized (independent). It will allow recreationists to visit the region in the future, and will keep the attractiveness of the territories for tourists at the current level. However, other types of economic activities should also be taken into account, which also have an impact on the recreational potential of the region—its basis is largely untouched or slightly modified natural landscapes. In many cases, the only way to preserve them is to create protected natural areas within their borders and restrict economic activity on them provided by their status.

A. Yu. Sanin (✉) · O. D. Vasilyev
N.N. Zubov Russian State Oceanographic Institute, Roshydromet, Moscow, Russian Federation
e-mail: eather86@mail.ru

V. A. Kulakovskaya
Lomonosov Moscow State University, Moscow, Russian Federation

V. I. Karev (ed.), *Physical and Mathematical Modeling of Earth and Environment Processes—2022*, Springer Proceedings in Earth and Environmental Sciences,
https://doi.org/10.1007/978-3-031-25962-3_56

Keywords Seaside territories · Recreation · Tourist flows · Organized recreation · Independent recreation · Ecosystem services · Ecological framework of the territory · Environmental pollution · Conflicts of types of human activity

1 Introduction

The peculiarities of the impact of recreation, along with other types of economic use of coastal territories, are considered both as a whole [1–8] and for individual regions [9, 10].

This article focuses specifically on the effects of recreation on coastal territories. Only a small part of the Russian sea coast is popular among tourists, and we will talk about it.

Some coastal regions of Russia, which are characterized by mass recreation:

1. The Black Sea coast of the Caucasus
2. The coast of the Sea of Azov
3. The Baltic coast within the Kaliningrad and Leningrad regions
4. The Caspian coast (especially the coast of Dagestan)
5. The coast of the Sea of Japan within the southern part of Primorsky Krai in the vicinity of Vladivostok.

Anthropogenic impact on coastal recreational areas is manifested in the following:

1. Pollution of waters, coastal and bottom sediments, reduction of water quality. It is especially evident for small bays, which are characterized by a slow water exchange with the rest of the sea. An example of such water areas: Tsemesskaya Bay near the city of Novorossiysk, Krasnodar Krai, the Black Sea ([11], Sevastopol Bay, Crimea, the Black Sea, Peter the Great Bay [12], Primorsky Krai, the Sea of Japan, etc.). The main pollutants are industrial, municipal and agricultural enterprises.
2. The receipt of household garbage in the water area and the coastal zone.
3. Transformation of landscapes, which reduces their recreational potential, as well as the cost of environmental services they provide, violates the structure of the ecological frameworks of the regions.
4. "Visual pollution" of landscapes, which manifests itself in "chaotic" development of the territory, "oversaturation" of engineering structures. Noise pollution.
5. Activation of dangerous natural processes caused by man.
6. Reduction of biodiversity, invasion of new species and displacement of native flora and fauna by them.
7. Impact (dredging and shore protection works, changes in the volume of solid river flow, and a number of others) on dynamic processes peculiar to the coasts.

2 Results and Their Discussion

2.1 *Reducing the Cost of Ecosystem Services of Coastal Landscapes*

The intensification of the economic use of natural landscapes leads to a complete or partial loss of the ability of natural landscapes to provide environmental services. As a result, according to some estimates [13], over the past half century, about 60% of the world's ecosystem services have degraded.

Costanza and his colleagues have assessed global ecosystem services twice [14, 15]. As of 2014, it amounted to 125 trillion dollars per year. This sum consists of average estimates of the cost of ecosystem services provided by each natural zone of the world, and the areas of territories occupied by these natural zones. The losses of ecosystem services to the world and the economy amount to an average of about $12 trillion per year.

Apparently, the costs of preserving natural landscapes in many cases can be quite lower than the losses from a decrease in the quality of environmental services provided by natural systems. For the preservation of natural landscapes in coastal areas, which are characterized by a high intensity of anthropogenic activity, the most important role is played by the ecological framework, in particular, its cores, which, as a rule, are specially protected natural territories. The ecological framework is the whole set of undeveloped and not covered with artificial materials territories with vegetation of a different nature, which performs various ecological functions [16]. The cores of the ecological framework are characterized by the highest cost of ecosystem services that they provide.

2.2 *The Impact of Recreation on Forest Landscapes of Coastal Ecosystems*

Forest ecosystems play one of the most important roles in the conservation of natural biodiversity on our planet. However, the area of forests is constantly decreasing including in the coastal regions. The issue of forest cover conservation is on the agenda of many scientific conferences around the world. Forests are known to be important reservoirs (pools) of carbon and carbon dioxide, one of the main greenhouse gases causing global climate change. The importance of forests in reducing carbon dioxide emissions and in solving global environmental problems has been recognized at international conferences and agreements, such as the UN Framework Convention on Climate Change (UNFCCC), the Kyoto Protocol, the Paris Agreement. Preservation of forests is a necessary element for minimizing greenhouse gas emissions into the atmosphere. In addition, forests produce oxygen during photosynthesis. Thus, forests perform the most important environment-forming functions from a human point of view (services, in the context of this article, these terms can be considered synonymous)—production of oxygen and deposition of carbon dioxide [17].

At the same time, the forests of coastal areas experience a colossal anthropogenic load. Entire forests are cut down for recreation and tourist attraction. This does not create illusions about the loss of the most important ecological functions (services) of forests.

Using the reference landscapes of the Moscow Region as an example, the authors developed a methodology for cartographic assessment of the environment-forming functions of forests [17, 18]. A similar assessment technique is also valid for the coastal forests under consideration.

The authors considered forest landscapes, which are characterized by different anthropogenic influence, different forest cover and the predominance of certain tree species. So, the most valuable forest communities are coniferous, coniferous-deciduous forests. Small-leaved forest communities, as a rule, have relatively low values of oxygen production and carbon dioxide deposition [18]. It is worth noting the extreme importance of forests, which have high values of both oxygen production and carbon dioxide deposition. They need additional evaluation and protection. It is these forests that will probably be the cores of the ecological framework of both the landscape and the entire coastal territory. Within these communities, natural relationships between landscape components are preserved. Such areas have a minimum anthropogenic load. They are important for the ecological well-being of the territory and maintaining the stability of the ecosystem.

It was found that the area of just such important forests has been greatly reduced over the past quarter century. The most valuable forest communities produce oxygen by more than 6 t/ha per year and deposit carbon by more than 3 t/ha per year. The reduction of such forests in only one forest landscape leads to oxygen losses of more than 1 million tons over the past quarter of a century. For normal human life, about 0.5 million tons of oxygen is needed annually. Thus, the loss of oxygen per 1 million tons leads to its shortage for 2 million people [18]. All this negatively affects human health. Oxygen starvation leads to an increase in cardiovascular diseases.

However, it is precisely such assessments that today receive insufficient attention when planning the development of territories. The ecological value of much of the coastal forests has yet to be assessed.

2.3 *Examples of the Most Polluted Water Areas of Russia and the Post-Soviet Space and Possible Measures to Counteract Their Pollution*

Among all water areas, the greatest impact of recreation, as well as other types of human economic activity, is manifested in the open sea on small semi-enclosed bays, as well as for ponds, lakes (especially drainless) and reservoirs on adjacent land.

Table 1 Examples of the most polluted water areas of Russia and possible measures to counteract their pollution

Name of the bay	Sources of pollution	Possible measures to preserve water quality
Tsemesskaya Bay, Novorossiysk, Krasnodar Territory	Pollutants come from surface and underground waters that are polluted by industrial enterprises of Novorossiysk, not all of them are cleaned. Pollute the bay and municipal drains. An important polluter is the Commercial Seaport of Novorossiysk, which is the leader in Russia in terms of cargo turnover among all seaports of the Russia	Improvement of methods of municipal wastewater treatment, introduction of circulating water supply cycles at enterprises
Peter the Great Bay, Vladivostok, Primorsky Krai	The main sources of marine pollution are the spillway of enterprises and utilities, rivers and other watercourses, storm sewers, discharges from ships, emergencies, atmospheric precipitation. Part of the wastewater causing chemical pollution of the coastal zone is formed in the catchment area, and the other is discharged directly into Peter the Great Bay from these sources [19]	Cleaning of municipal and industrial effluents, control of all vessels, including small-size fleet

Examples of the most polluted water areas of Russia and possible measures to counteract their pollution are given in Table 1.

2.4 *Activation of Dangerous Natural Processes*

Unfavorable and dangerous natural processes appear on all coastal recreational territories, the list and intensity of which already depends on the natural conditions (and often human activity, in particular, on the percentage of development of the territory) of a particular area. For the coastal territories popular among recreationists, there is a manifestation of such dangerous processes as storms, tornadoes, mudslides, abrasive processes, landslides, intense heat, intense precipitation and floods, etc. They threaten the health and even the lives of recreational workers, leading to significant material damage to infrastructure facilities. The most dangerous natural processes are manifested in the foothill and mountainous coastal regions: the North Caucasian coast of the Black Sea, the coast of the Sea of Japan.

Recreation directly or indirectly leads to the activation of the processes of abrasion and erosion of the coasts, landslides on the slopes, floods in regions characterized by the reduction of forests and some others. The intensification of adverse and dangerous natural processes and the increase in material damage from them, which is explained, among other things, by the increased intensity of the use of the coastal zone, the involvement of new areas in the use of nature.

Since 2020, it has become an urgent need to take into account the situation that has developed since the outbreak of the coronavirus pandemic (which can also be considered as one of the dangerous natural phenomena), and especially its consequences, including economic and social.

Adverse and dangerous natural processes occur in all recreational areas, but in many cases their intensity is low, and they are rarely repeated. Because of this, it is of particular importance to identify those from them that pose the greatest threat to recreants, to counteract them, to inform recreants and local residents about them, to avoid using the sections of the coast where they manifest themselves to the greatest extent.

2.5 *Transformation of Landscapes of Coastal Territories Through Human Fault*

It is especially pronounced in the vicinity of large coastal cities: St. Petersburg, the Baltic Sea coast, Rostov near the coast of the Sea of Azov, Vladivostok on the coast of the Sea of Japan. The transformation of landscapes leads to a decrease in their aesthetic value and the cost of environmental services provided by natural systems, a decrease in biodiversity, degradation of ecological frameworks of territories.

2.6 *Environmental Pollution*

The waters of the coastal regions are polluted by industrial, domestic and agricultural runoff. Human agricultural activity leads to contamination of surface and groundwater with nitrogen, phosphorus and other pollutants. There is a decrease in water quality, especially in semi-enclosed bays and bays with relatively small volumes of water, pollution of the coastal strip with construction and household garbage, the entry of garbage, in particular plastic, into the seas.

2.7 Exceeding the Recreational Capacity of Beaches

Recreational areas are characterized by recreational capacity. Its excess jeopardizes the preservation of their natural and recreational potential, reduces the effectiveness of environmental services that they provide. In order to avoid exceeding the recreational capacity, it is necessary to control the number of all categories of recreants, including unorganized ones, and redistribute tourist flows if necessary (especially in the Krasnodar region). The most important part of the tourist capacity of recreational areas is the capacity of beaches.

According to some estimates, the recreational capacity of the beaches of the seven above-mentioned coastal territories averages several million tourists per year. However, the following important points can be noted:

1. The coastal territories are characterized by an uneven distribution of recreants over time, a significant part of them come during the "peak of the bathing season", in July–August. For some coastal territories (coast of Primorsky Krai, Kaliningrad and Leningrad regions) this is a large part of the tourist season, for others—about half or a little more. A more even distribution of tourists throughout the year can be achieved through the development of different types of recreation, in particular, sightseeing, business and others. The seaside territories have a number of natural and cultural attractions that allow the development of new types of tourism. Uneven distribution in space is one of the reasons for exceeding the recreational capacity of beaches in some cases.
2. The coastal territories are characterized by an uneven distribution of recreants in space. They are associated with different levels of infrastructure development, different number of attractions and their value for tourists, with established vacation preferences and some other reasons. Only on the North Caucasian coast of the Black Sea, the number of recreants is comparable to the recreational capacity of beaches, therefore, it will be exceeded in any case, since an absolutely uniform distribution of tourists in space and time in a market economy is unattainable. For all other coastal territories, the number of tourists is much less than their recreational capacity, but due to the uneven distribution along the coast in a number of areas, it is largely exceeded.
3. Many coastal territories are characterized by an influx of recreants from large cities located within their borders or in close proximity—first of all, these are St. Petersburg, Rostov, Krasnodar and Vladivostok—which are poorly taken into account in statistics, but also contribute to exceeding the recreational capacity of beaches, especially on weekends.

Finally, it should be noted that the recreational capacity of beaches is not identical to the maximum loads on natural landscapes, if exceeded; anthropogenic impact leads to their inevitable changes. The calculation of such loads is another issue, but it is clear that they are significantly less than the capacity of beaches.

The redistribution of tourist flows in space and time will contribute to reducing the load on natural systems where tourism is most developed, which will ensure the preservation of the recreational potential of such territories. The redistribution of tourist flows can be achieved, in particular, through the development of new types of tourism, which should be identified and planned individually for each recreational region.

2.8 Visual and Noise Pollution of Landscapes

Recreational development of coastal zones of reservoirs often occurs spontaneously, which leads to visual pollution of landscapes. As a result, the cost of aesthetic services provided by natural geosystems is noticeably reduced. It is not possible to quantify this decrease, since by now there are no proper estimates of the cost of aesthetic service of natural landscapes.

2.9 Conflicts Between Different Types of Economic Activity and Ways to Solve Them

In addition to recreation, the following types of human activity are characteristic of coastal territories (as well as for many others):

- environmental protection (functioning of national parks and other natural reserves)
- agricultural activities (crop production, animal husbandry, etc.)
- transport activities (roads and railways, sea and river ports, airports, pipelines, power lines, etc.)
- residential (use of the territory for the construction of residential buildings and accommodation of local residents in them)
- industrial
- the so-called special or military—the placement of objects necessary for the Armed Forces of the state.

A number of types of human activity can have a positive impact on recreation, in particular transport, in some cases residential (rental housing), but others can significantly limit tourist flows (industrial, military, etc.) and reduce the quality of recreation, in particular, the aesthetic value of seaside landscapes, restrict access to the reservoir, cause water pollution etc.

The potential relationship of various types of human economic activity with each other is shown in Table 2. For coastal territories, most often the relations between different types of economic activity acquire a special character.

Table 2 A potentially possible relationship between different types of human economic activity

Type	Type						
	Nature conservation	Recreational	Residential	Transport	Agricultural	Military	Industrial
Nature conservation		–	–	–	–	+–	–
Recreational	+–		+–	+–	–	–	–
Residential	–	+–		+	+	+–	+–
Transport	–	+–	+		+	+	+
Agricultural	–	–	+	+		–	0
Military	+–	–	+–	+	–		0
Industrial	–	–	+–	+	0	0	

2.10 Features of the Impact of Organized and Unorganized (Independent) Recreation on Coastal Territories

It is necessary to distinguish between organized and independent recreation. In the Russian coastal territories, both the first and the second take place, and in terms of the number of recreants for some of them—for example, for the North Caucasian coast of the Black Sea—the independent leads. Organized tourists purchase a tour, which includes accommodation on the seaside territory, and often also food and transport to and from the place of rest. An unorganized recreant organizes his rest independently, both movement and food, and accommodation.

Unorganized recreation is a source of direct income for a significant number of local residents who rent housing, open cafes and shops, organize excursions—for a representative of small and medium-sized businesses in the region visited by tourists. Organized recreants bring income mainly to a narrower circle of people—owners of travel agencies and accommodation facilities.

Both organized and unorganized recreation in the coastal zones of inland reservoirs carries certain risks for the latter. In the case of unorganized recreation, it is quite difficult to control tourist flows, the number of which is often unknown, especially if the recreation facility is located within walking distance from the places of residence of the recreants (which makes it impossible to account for their movement by car or public transport). In this case, the recreational capacity of recreation facilities is often exceeded, which leads to a decrease in their recreational potential. Visually, in such cases, such changes as the transformation of the species composition of vegetation, the formation of a "pathway" network, as well as spontaneous landfills of household garbage and simply "uniform" cluttering of recreation areas with it are recorded. Organized recreation comes to the same negative consequences, although due to better control over recreational flows, they can be somewhat smoothed out. Both unorganized and organized recreation often leads to non-compliance with the restrictions on the use of water protection zones and coastal protection strips provided for by the Water Code in Russia. Unorganized recreation is accompanied by parking of cars without a hard surface, organized often leads to the appearance of various tourist accommodation facilities that dump untreated sewage into reservoirs.

A comparative characteristic of the impact of organized and independent recreation on natural landscapes is given in Table 3.

3 Conclusions and Recommendations

The peculiarity of recreational nature management is its relatively weak impact on natural landscapes, especially in the following cases (ideally, they should take place simultaneously):

1. Recreation is represented by types that has a gentle effect on natural landscapes, for example, ecological tourism, and rural tourism.

Table 3 Comparative characteristics of the impact of organized and independent recreation on natural landscapes

Comparison criterion	Impact of organized recreation	Impact of unorganized recreation
General features of the development of the coastal zone	It can happen both according to some master plan, and spontaneously, depends on local authorities	As a rule, it occurs spontaneously, which leads to visual pollution, as well as littering. But at the same time, the impact on nature and landscape changes may be smaller in comparison with organized recreation
Impact on natural landscapes	Moderate or strong, relatively easy to control if necessary	From mild to moderate, but quite difficult to control
Impact on the water quality of the reservoir	As a rule, it is possible to organize wastewater treatment, therefore relatively weak	Often, the functioning of tourist accommodation facilities is associated with the entry of untreated wastewater into reservoirs
Impact on vegetation and wildlife	Radical transformation of natural landscapes in the places of construction of recreational infrastructure facilities, very strong local impact, in general—medium and moderate while preserving the ecological framework, outside of recreational infrastructure facilities—weak and medium	Independent tourism implies less active use of tourism infrastructure facilities; therefore, both positive and negative consequences of its presence are weaker Weak and medium impact, but widespread, often including within natural reserves, in which there are the most valuable species of plants and animals
Manageability of recreational flow	As a rule, medium or high	As a rule, low or medium
Influence on the aesthetic value of natural landscapes	Active construction of recreational infrastructure facilities, as a rule, significantly reduces it, except in cases when they manage to "fit" into the landscape	Often relatively insignificant, if unorganized recreation is not accompanied by the active construction of mini-hotels, but even in this case, it is often less impact than in the case of organized recreation. However, disorganized recreation can often be accompanied by littering of the coastal zone
Impact on the dynamic processes peculiar to the coasts	Recreational infrastructure facilities often require shore protection works, which can activate abrasion on adjacent sections of the coasts	Insignificant

(continued)

Table 3 (continued)

Comparison criterion	Impact of organized recreation	Impact of unorganized recreation
Provoking adverse and dangerous natural processes	Access roads to recreational infrastructure facilities can provoke the development of landslides	Insignificant

2. Tourists and organizations and local residents providing recreation treat natural landscapes respectfully, strive to minimize the impact on them.
3. The flows of independent recreants are predictable, controlled, and independent tourists fit under point 2.
4. The desire of recreationists to preserve the aesthetic potential of the territory when creating and ensuring the functioning of recreational facilities, to "fit" them into the landscape.

Achieving the last point can be a difficult task, solved by choosing the materials used for construction and repair, the optimal architectural design, the desired height of the object, their shape, color, etc. The problem of avoiding or at least reducing visual pollution of landscapes is quite complex; it has been considered in many publications [20, 21].

The listed types of anthropogenic impact and recreation, in particular, can ultimately significantly reduce the likelihood of achieving sustainable development of coastal recreational areas if they are not controlled and limited. In order to control the impact, it is necessary to control the flows of recreants, comply with the norms of the current environmental legislation, organize environmental monitoring in coastal areas and counteract its pollution.

In many cases, for coastal areas there is an excess of their recreational capacity which leads to a decrease in their recreational potential as well as of the level of comfort for most vacationers. However, at the same time, there are many coastal zones and their areas that are valuable for recreation which are poorly used. Thus, the solution is not to reduce the flow of recreants, but to redistribute it between different coastal territories not only in space, but also in time.

It is necessary to ensure compliance with the norms of current legislation, in particular, the restrictions on the economic use of the so-called water protection zone of the seas (according to Law of the Russian Federation, this is a strip 500 m wide from the sea's edge), this is necessary for its preservation.

To summarize, it is necessary to support both unorganized and organized recreation, provided that the flows of unorganized recreants are regulated and the negative impact of unorganized recreation on the landscapes of coastal territories is minimized.

References

1. Aybulatov N.A. Russia's activity in the coastal zone of the sea and environmental problems. IO RAS named after P.P. Shirshov. M.: Nauka, 2005, 364 p.
2. Carter R.W.G. Coastal environments. London-Toronto, 1987, 617 p.
3. Gogoberidze G.G. The economy of the coastal zone / G.G. Gogoberidze, A.S. Averkiev. St. Petersburg: Publishing House of RSMU, 2003, 156 p.
4. Dolotov Yu.S. Problems of rational use and protection of coastal areas of the World Ocean. – M.: Scientific World, 1996, 304 p.
5. Ignatov E.I. Coastal morphosystems. Moscow - Smolensk: Magenta, 2004, 352 p.
6. Peshkov V.M. Coastal zone of the sea. Krasnodar: Lacont, 2003, 350 p.
7. Robert K., Alder J. Coastal planning and management. ESF13 London and New York, 1999, 414 p.
8. Safyanov G.A. Geomorphology of sea coasts. Publishing House of the Geographical Faculty of Moscow State University, Moscow, 1996, 400 p.
9. Goryachkin Yu.N., Ivanov V.A. The level of the Black Sea: past, present and future. Sevastopol, 2006, 210 p.
10. Nadim F., Bagtzoglou A.C., Iranmahboob J. Management of coastal areas in the Caspian Sea region: Environmental issues and political challenges. Coastal Management, 2006, No. 34, P. 153–165.
11. Pashkevich M.A., Alekseenko A.V. The use of macrophytes for ecological and geochemical monitoring of the Tsemesskaya (Novorossiysk) bay // Water: chemistry and ecology. - 2016. — No. 2. — pp. 18–23.
12. Gruzinov V.M., Dyakov N.N., Mezentseva I.V., Malchenko Yu.A., Zhokhova N.V., Korshenko A.N. Sources of pollution of coastal waters of the Sevastopol district // Oceanology. - 2019. - Vol. 59. - No. 4. - C. 579–590. doi: https://doi.org/10.31857/S0030-1574594579-590.
13. Bobylev S.N., Goryacheva A.A. Identification and assessment of ecosystem services: international context // Bulletin of International Organizations: Education, Science, New Economy. — 2019. — No. 1, pp. 225–236.
14. Costanza et al. The value of the world's ecosystem services and natural capital. Nature 387, 1997, P.253–260.
15. Costanza R., de Groot R., Sutton P., van der Ploeg S., Anderson S.J., Kubiszewski I., Farber S., Turner R.K. (2014) Changes in the global value of ecosystem services // Global Environmental Change. Vol. 26.
16. Klimanova O.A., Kolbovsky E.Yu., Illarionova O.A. Ecological framework of the largest cities of the Russian Federation: modern structure, territorial planning and development problems // Bulletin of St. Petersburg University. Earth sciences. 2018. Vol. 63. Issue. 2. pp. 127–146.
17. Vasiliev O.D., Chistov S.V. Research and mapping of the environment-forming functions of the forests of New Moscow // Izvestia of higher educational institutions. Geodesy and aerial photography. 2016. Vol. 60. No. 5. pp. 128–133.
18. Vasiliev O.D. Mapping of the environment-forming functions of forests and their comparative analysis in the landscapes of the Moscow region // Bulletin of the Moscow University. Series 5: Geography. 2020. No. 6. pp. 21–31.
19. Rostov I.D., Rudykh N.I., Rostov V.I. Interannual dynamics of the level of pollution of the waters of Peter the Great Bay over the past 40 years // Bulletin of the FEB RAS, 2015, No. 6, pp. 49–63.
20. Likhacheva E.A. Relief – its essence and beauty / E.A. Likhacheva – M.: Media-PRESS. – 2015, 144 p.
21. Likhacheva E.A. City-ecosystem / E.A. Likhacheva, D.A. Timofeev, M.P. Zhidkov et al., Moscow: IGRAN, 1996, 336 p.

Serpentinization of Olivine in a Kimberlite Pipe Udachnaya

V. K. Marshintsev, V. G. Gadiyatov, and P. I. Kalugin

Abstract The article presents the results obtained in the study of the Udachnaya-Zapadnaya and Udachnaya-Vostochnaya pipes. The nature of their relationships with each other and with the host rocks, the change in the composition of kimberlite rocks, the distribution of xenogenic and deep ultrabasic and eclogitic inclusions are shown. The isochron age of serpentinites and serpentinized ultramafic rocks is given as 863 ± 45 Ma and 1203 ± 66 Ma, respectively. In the serpentinized varieties of kimberlite rocks of Yakutia, serpentine is represented by two polymorphic modifications, lizardite and chrysotile. In kimberlite breccias with a massive cement texture of the Udachnaya-Vostochnaya pipe, the serpentine of inclusions is filled with lizardite; in the autolithic kimberlite breccias of the Udachnaya-Zapadnaya pipe, it is a mixture of chrysotile and lizardite. Serpentinization of olivine occurred in the prepipe stage of kimberlite formation, allochemically, without changing the volume of olivine. Serpentinization of olivine phenocrysts occurred in the late magmatic stage and decreases with depth.

Keywords Kimberlites · Xenoliths · Serpentinization · Ultramafic rocks · Olivine · Chrysotile · Lizardite

1 Brief Description of the Kimberlite Pipe Udachnaya

The Udachnaya kimberlite pipe is located in the basin of the river. Daldyn (Western Yakutia), in the Daldyn-Alakit diamond-bearing region [1–3]. Comprises two separate bodies: Udachnaya-Zapadnaya and Udachnaya-Vostochnaya pipes, which represent a single ore body down to a depth of 200 m. The rocks of the pipe have intruded

V. K. Marshintsev
Academy of Sciences of the Republic of Sakha (Yakutia), Lenin Ave., 33, 677007 Yakutsk, Russia

Present Address:
V. G. Gadiyatov (✉) · P. I. Kalugin
Voronezh State Technical University, Voronezh, Russia
e-mail: gadiatovvg@mail.ru

V. I. Karev (ed.), *Physical and Mathematical Modeling of Earth and Environment Processes—2022*, Springer Proceedings in Earth and Environmental Sciences,
https://doi.org/10.1007/978-3-031-25962-3_57

and cut through the Lower Paleozoic sedimentary deposits. At contact with kimberlites, they are broken up by faults, folded, and brecciated. The Udachnaya-Zapadnaya pipe has an elliptical shape on the surface, in the section it is a cone-shaped tubular body. Kimberlites at the contact with the sedimentary sequence are carbonatized and ferruginated. The Udachnaya-Vostochnaya kimberlite pipe has a round-oval shape on the surface, elongated in the northeast direction. Kimberlites at contact with host sedimentary rocks are only weakly ferruginated.

Kimberlites are complex natural systems in which there are rocks consisting of minerals formed at high thermodynamic parameters (diamond, garnet-pyrope, olivine, chrome spinel, chrome diopside, ilmenite), as well as a polygenic complex of mantle inclusions (peridotites, pyroxenites, eclogites), which are formed in the stability region of diamond. In the composition of kimberlites, up to 95% or more is occupied by serpentine and calcite. Serpentinization is the main metasomatic process that is widely manifested in rocks that fill kimberlite bodies at levels accessible to observation. Serpentine is composed of pseudomorphs after olivine, as well as the bulk of kimberlites. In addition, various xenoliths of ultramafic rocks in kimberlite rocks were subjected to serpentinization. The diamond content of platform formations of various structures is associated with kimberlites and other rocks [4].

2 Isochronous Age of Xenoliths

Two groups of ultrabasic xenoliths were studied from the Udachnaya pipe by Rb–Sr isotopy: serpentinites and serpentinized ultrabasites. Calculation of isochron parameters for the first group of xenoliths gives an isochron age of 863 $\pm$ 45 Ma with a primary ratio of $^{87}Sr/^{86}Sr$ equal to 0.7083 $\pm$ 0.0003(1δ), and for the second group, respectively, 1203 $\pm$ 66 Ma and 0.7010 $\pm$ 0.0005(1δ). The rocks of the first group contain from 8.72 to 12.63% H_2O, the rocks of the second group—from 0.58 to 5.9% H_2O.

Based on this, we can conclude about the nature of the influence of the serpentinization process on the change in the Rb–Sr systems of the original rocks. It can be stated that with a strong degree of serpentinization (serpentine content from 50 to 100%), the value of the primary ratio $^{87}Rb/^{86}Sr$ increases in them.

The value of this ratio in the serpentines of the Udachnaya pipe is lower than in weakly serpentinized ultramafic rocks, which may indirectly indicate the redistribution of Rb and Sr during their serpentinization. The former are characterized by a somewhat lower content of ^{87}Rb compared to slightly altered rocks, which, respectively, is 1.2218 10^{-6} and 1.4467 10^{-6} g/g. The high linear correlation of $^{87}Sr/^{86}Sr$ and $^{87}Rb/^{86}Sr$ in both groups of rocks indicates a good homogeneity of the isotopic composition and suggests that the obtained isochron age figures correspond to the time of manifestation of specific geological events: the age of formation of ultramafic rocks is 1203 Ma and the time of their serpentinization is 863 million years The formation age of the Udachnaya pipe is 350 $\pm$ 15 Ma [5].

3 Serpentinization of Kimberlites

Most researchers consider the process of serpentinization of kimberlites as autometamorphic, due to the presence of a large amount of volatiles, primarily water. The study of hyperbasite serpentinization and the experiments performed show that serpentinization is capable of proceeding at temperatures not exceeding the upper temperature limit of serpentine stability (500°) and at any water pressure (up to 33 kbar for magnesian associations). For magnesian-ferrous compositions, the serpentinization process takes place at a temperature below 400–300 °C. Theoretically, there is no lower temperature limit for serpentinization, but in fact it occurs at a temperature not lower than 200–250 °C, and in alpine-type ultrabasites—not lower than 350 °C.

The process of serpentinization of dunites proceeds with the formation of brucite according to the following reaction:

$$2Mg_2SiO_4 + 3H_2O \rightarrow Mg_3Si_2O_5(OH)_4 + Mg(OH)_2$$

The appearance of brucite together with serpentine is confirmed by detailed studies. In experiments at high water vapor pressures under conditions of SiO_2 addition, peridotites can be serpentinized at higher temperatures (600 °C); pure olivine is not serpentinized under these conditions. Thermodynamic calculations indicate the possibility of formation in this case also of talc, which crystallizes according to the reaction:

$$Mg_2SiO_4 + 3H_2O + 3SiO_2 \rightarrow Mg_3Si_2O_5(OH)_4 + Mg_3Si_4O_{10}(OH)_2$$

However, the co-occurrence of serpentine, talc, and brucite in serpentinized rocks has not been established. This indicates the impossibility of the simultaneous presence of talc and brucite. According to thermodynamic calculations, this association is unstable, and further talc and brucite react with each other, forming serpentine. Only one of these minerals can appear in serpentines as an excess component of this reaction [6].

The serpentinization process according to the above reactions should proceed with an increase in volume. A particularly significant increase in volume occurs under the conditions of the addition of SiO_2. Detailed studies of the process of serpentinization of ultramafic Urals [7] established a change in the volume of serpentinized rocks, which is expressed in an increase in the grain size of serpentinized olivine by several times, the appearance of boudinage, and stretching of early veins composed of chrysotile. The nature of all these changes is directly dependent on the degree of serpentinization and correlates quite accurately with the grain size of serpentinized olivine. In the process of regional serpentinization, which proceeds with an obligatory increase in the volume of the rock, Mg, Fe and Si remain inert, and the ratio $RO:SiO_2$ for the initial ultramafic rocks and the serpentinites that arose after them is a constant value. For dunites it is 1.9–2.2, and for peridotites 1.4–1.8. At the same time, the preservation of the ratio indicates that brucite should appear simultaneously with serpentine.

In the case of local serpentinization with the formation of apodunitic vein serpentinites, the RO:SiO_2 ratio approaches that for pure serpentine (1.5) without an admixture of brucite, which should be accompanied by the removal of an excess part of the components under conditions of a constant volume of serpentinized olivine and rock. In this case, the elements are carried out into a mobile serpentinizing solution moving along relatively large tectonic fissures.

Thus, the process of serpentinization can proceed isochemically and allochemically. Most researchers point to the isochemical nature of the serpentinization process in kimberlites, accompanied by a significant increase in the volume of solid rock. With a content of 45% olivine in kimberlites, the increase in volume reaches 13.5% [8]. In some cases, an increase in the volume of kimberlite explains the phenomena of tearing of the layers of host rocks at the contact with kimberlites.

One of the most important indicators of the expansion of kimberlites during serpentinization can be the cracking of olivine crystals. However, when studying serpentinized kimberlites, no cracking of olivine crystals was observed [9]. This is confirmed by goniometric studies of serpentine pseudomorphs after olivine of the second group with a clear crystallographic faceting from kimberlite rocks of the Udachnaya pipe [10].

The measurements and comparison of the grain sizes of olivine and serpentine pseudomorphs on olivine show that the average grain sizes in the rocks of the western body of the Udachnaya pipe, where the degree of olivine serpentinization reaches 100%, is smaller (X = 1.28 mm, fluctuation limits 0.11–8.0 mm) than the grain sizes of the unaltered olivine of the eastern body (X = 1.47 mm, fluctuation limits 0.12–8.0 mm).

In the serpentinized varieties of kimberlite rocks of Yakutia, serpentine is represented by two polymorphic modifications, lizardite and chrysotile. The third variety of serpentine, antigorite, which is essentially a phase of a different composition, with a higher content of SiO_2 and a lower content of MgO and H_2O^+, was not found in kimberlites. Experiments have established that when olivine is transformed, antigorite and lizardite develop directly over it. An analysis of the stability of modifications of serpentine of variable iron content shows that antigorite has the largest interval of magnesium and iron isomorphism and the widest field of stability. Lizardite is a mineral of elevated pressures, developed in a limited range of conditions and at low temperatures it breaks up into an antigorite-chrysolite association. Early serpentine is usually represented by lizardite. Hence, it is assumed that, among other serpentine minerals, the replacement of kimberlite olivine by lizardite proceeded in the pre-pipe stage of its formation, under conditions of the remaining high pressure of the system. The character of the serpentine crystallization medium probably corresponded to neutral or slightly acidic. Crystallization of chrysotile is associated with further transformation of lizardite at later stages.

The serpentine of the Udachnaya pipe, which develops along large olivine porphyritic segregations in autolithic kimberlite breccias, is represented by a mixture of chrysotile (clinochrysotile) and lizardite in various proportions. In kimberlite breccias with a massive cement texture, pseudomorphases after olivine are made up of lizardite and chrysotile, which are in approximately equal proportions or only lizardite. The serpentine of the groundmass of kimberlite breccias in a massive cement texture is represented by lizardite, which is considered as an independent phase that crystallizes in the underground magmatic stage. In autolithic kimberlite breccias, the groundmass has a chrysotile–lizardite composition [11]. The composition of serpentine in serpentinized xenoliths of ultramafic rocks has a similar relationship. In kimberlite breccias with a massive cement texture of the Udachnaya-Vostochnaya pipe, the inclusion serpentine is represented by lizardite. In the autolithic kimberlite breccias of the Udachnaya-Zapadnaya pipe—a mixture of chrysotile and lizardite.

According to the data of diffraction characteristics and electron microscopy, the composition of acicular serpentine inclusions found in the olivine phenocrysts of the Udachnaya-Vostochnaya pipe turned out to be identical to the composition of serpentine of serpentinite inclusions from kimberlites, according to the data of diffraction characteristics and electron microscopy. As a result, it can be argued that the serpentinization process proceeds under similar conditions.

As noted above, the isochemical course of the serpentinization process is accompanied by the appearance of rock-forming brucite, the amount of which, obviously, should be proportional to the amount of released magnesium. When studying kimberlite rocks of the Udachnaya pipe, we noted the presence of brucite in them. In the western body, these are secant parallel veins of a transverse-fibrous structure with a thickness of 0.1 to 2.5–3.0 cm, forming at a depth of 180 m a small zone 3.5–7 cm wide. In the east (in the near-surface area, and the southeastern part)—areas of kimberlite rock, intensively impregnated with brucite. In large blocks, a series of parallel transverse-fibrous veinlets with thickness from fractions to 1–1.5 mm is observed. Microscopically, thin veinlets of brucite cut through the olivine grains, suggesting superimposed mineralization. Thus, the serpentinization of olivine in kimberlites is not accompanied by the formation of brucite, and this process proceeds allochemically.

When studying kimberlite rocks of deep horizons of the Udachnaya-Vostochnaya pipe, corresponding to the levels of olivine phenocrysts not affected by serpentinization, we noted that the bulk of the material cementing phenocryst minerals is composed of calcite. It is quite possible that it is the source for the bulk of kimberlites.

One of the possible ways of realizing the excess components of olivine serpentinization is the crystallization of pyroaurite–shegrenite–magnesian–iron hydrate–carbonate (8 $[([(Mg_{0.7}\ Fe_{0.3})\ (OH_2)]\ CO_3.\ H_2O)])$ usually present in close association with serpentine in the form of a lath-like-fibrous aggregates and constituting from 14 to 40% of the composition of the main mass of kimberlites. The mineral forms thin veinlets and small geodes. Found in the rocks of many pipes of Yakutia, it is sometimes characterized as a rock-forming mineral [12,13]. Crystallization of pyroaurite–shegrenite, judging by the nature of mineral segregations, occurs at the late stages of crystallization of minerals of the main mass of kimberlites, in the presence of CO_2.

Experimental and theoretical constructions indicate a very low water content at the levels of ultramafic melt generation, which causes small volumes of melts, in particular, small amounts of kimberlite magmas [14]. In non-serpentinized varieties of kimberlites, the water content does not exceed 2–3 wt.%, which is probably close to its true value for these rocks. Experimental studies also show that igneous melts of ultramafic composition at a pressure of 1 kbar can contain about 2% water, and at 5 kbar—already 4–5% [15]. According to Kushiro [16], at pressures of 10 and 20 kbar, the water content is 10% (at 1500 °C) and 15–23% (at 1400 °C), respectively. The tendency of magmatic masses to balance with the Earth's gravitational field contributes to the radial diffusion of water in the apical part of the magma column, where a significant concentration of water and other volatile components should be expected.

Depending on the radial extent of the magmatic column, the degree will also be different. concentration of water in the lower and upper parts of it. For a body with a length of 10 km, the difference will be 1–1.5 wt.%; at 50–100 km, the water concentration in the lower parts is 0.1% wt., and in the upper parts—10–30% wt. When certain temperatures (below 500 °C) are reached, the ascending kimberlite column, which is a kind of "fluidized porridge", probably begins serpentinization of olivine crystals, primarily in the apical part of the column, where the highest degree of olivine serpentinization is expected. It should decrease with depth. There is a "differentiation" of the kimberlite column according to the degree of serpentinization of olivine phenocrysts from complete, in the apical part (with the formation of complete pseudomorphs of serpentine after olivine) to partial (with the preservation of olivine relics) and, finally, to unaffected by serpentinization processes. The vertical extent of this process is most likely different in each specific case. The rapid rise and separation of the volatile phase with a sharp drop in pressure contributed to the fact that the process of olivine serpentinization did not have enough time to migrate down the column. The later stages of serpentinization were probably hindered by the absence of a convection influx of volatiles from the deep parts of the magma chamber.

4 Conclusion

The processes of olivine serpentinization explain the petrographic facts observed in rocks with its partial serpentinization, when there is a joint occurrence of olivine grains with varying degrees of change: from completely transformed into serpentine pseudomorphases to completely unaffected by serpentinization processes. Similar can be observed in the rocks of the pipes Udachnaya-Vostochnaya, Nagaya and a number of others. Their formation occurs during the rise of a magmatic column during the mixing of areas of kimberlite magma with varying degrees of olivine serpentinization.

Serpentinization of olivine occurs in the pre-pipe stage of kimberlite formation, allochemically, without changing the volume of olivine. Groundmass serpentine crystallizes from the residual melt enriched in excess components leached during olivine serpentinization and, at a later stage, possibly pyroaurite. Serpentinization of olivine phenocrysts proceeds to the late magmatic stage and decreases with depth.

The study of this complex natural system allows us to confidently speak about the formation of certain varieties of kimberlite rocks and the diamond content associated with them.

References

1. Kharkiv A.D., Zinchuk N.N., Kryuchkov A.I. Primary diamond deposits of the world. – M.: Nedra, 1998. – 555 p.
2. Gadiyatov V.G., Marshintsev V.K. Colored stones of Yakutia and their deposits. - Yekaterinburg, 2000. - 324 p.
3. Marshintsev V.K., Gadiyatov V.G. The wealth of the bowels of Yakutia. Minerals, mineral resource base. - Voronezh: VSU Publishing House, 2020. - 320 p.
4. Gadiyatov V.G. Geochemical methods of prospecting for gold mineralization in the greenstone strata of the Aldan shield. - Abstract. PhD dissertations. - Irkutsk, 1992. - 17 p.
5. Sarsadskikh N.N., Blagulkina V.A., Silin Yu.I. On the absolute age of kimberlites of Yakutia. - Dokl. Academy of Sciences of the USSR, 1966, vol. 168, No. 2.
6. Pugin V.A., Khitarov N.I. Experimental petrology of deep magmatism. – M.: Nauka, 1978. – 175 p.
7. Steinberg D.S., Chashchukhina I.S. Serpentinization of ultramafic rocks. – M.: Nauka, 1977. – 312 p.
8. Du Toit A. Geology of South Africa. – M.: IL, 1957. – 488 p.
9. Milashev V.A., Krutoyarsky M.A., Rabkin M.I., Erlikh E.N. Kimberlite rocks and picritic porphyrites of the northeastern part of the Siberian platform. – M.: Gosgeoltekhizdat. 1963. - 216 p.
10. Diamond deposits of Yakutia. – M.: Gosgeoltekhizdat. 1959. - 525 p.
11. Kornilova V.P., Nikishova L.V., Nikishov K.N. Minerals of the serpentine group from kimberlite rocks of Yakutia. - In the book: Parageneses of minerals of kimberlite rocks. Yakutsk: ed. YaF SO AN USSR, 1981, p. 65–81.
12. Marshintsev V.K., Nikishova L.V., Gotovtsev V.V. Serpentine filling needle channels in olivine from the Udachnaya-Vostochiya pipe. - Dokl. Academy of Sciences of the USSR, 1984, v. 277, no. 3, p. 697–700.
13. Zinchuk N.I., Melnik Yu.M., Kharkiv A.P. Pyroaurite in kimberlite rocks of Yakutia and its genesis. - Dokl. Academy of Sciences of the USSR, 1982, v. 267, no. 3, p. 722–728.
14. Lutz B.G. Chemical composition of the continental crust and upper mantle of the earth. – M.: Nauka, 1975. – 168 p.
15. Kadik A.A. The separation of water from magmas as they move towards the earth's surface. - In the book: Proceedings of the I International Geochemical Congress. T. 1. Magmatic processes. M., 1972, p. 102–115.
16. Kushiro I. Effect of water on the composition of magmas formed at high pressures. – Journ. Petrology, 1972, v. 13, No. 2.

Analytical and Laboratory Modeling of the Movement of Markers in a Vortex Flow

T. O. Chaplina

Abstract The changing the direction of floating objects rotation on the vortex surface is discovered and explained. The differences in the behavior of symmetrical and asymmetric objects on the rotating water surface are shown. The proposed model of the transfer of floating markers along the surface of a vortex trough can also be used to study the dynamics of the transfer of "islands" of debris by large-scale oceanic vortices.

Keywords Vortex trough · Marker · Solid-body rotation · Rotation angle · Spinning angle

1 Introduction

Vortex flows are widespread in natural conditions and are used in various technical applications. The problem of matter transfer in circulating flows is still very relevant and in demand for hydrophysics, ecology, and industrial technologies and has many different applications. Despite a large number of studies, this area remains insufficiently studied, since such problems, as a rule, depend on a large number of random and difficult to control factors and are very difficult for experimental study, unambiguous interpretation, and theoretical description [1–7]. In this regard, it is very important to develop a methodology for experimental laboratory modeling, designed to ensure the stationarity of flows and reproducibility of results, and the creation of an adequate theoretical model of the processes under study.

The article presents the results of experimental and theoretical studies of solid-state markers of various shapes and sizes in a vortex flow in multiphase fluids and proposes a theoretical model that explains the movement of markers placed on the surface of a vortex flow in a one-component fluid. An equation is obtained that describes the motion of the marker's center of mass and represents a logarithmic spiral on the surface of the vortex funnel, which coincides with the trajectories of

T. O. Chaplina (✉)
Ishlinsky Institute for Problems in Mechanics RAS, Moscow 119526, Russia
e-mail: tanya75.06@mail.ru

V. I. Karev (ed.), *Physical and Mathematical Modeling of Earth and Environment Processes—2022*, Springer Proceedings in Earth and Environmental Sciences,
https://doi.org/10.1007/978-3-031-25962-3_58

liquid particles near the free surface. The experimental dependence of the angle of rotation on the angle of rotation in the region of rotation of a rigid body agrees well with the theoretical dependence obtained on the basis of the proposed mathematical model.

2 Experiments and Data Representation

The methods and approaches used in the course of the work are distinguished by a high degree of originality and novelty. Laboratory experiments were carried out on three installations that allow creating a stationary vortex flow and controlling its parameters:

1. The "VTK" (Vortex Flows with Torsion) installation was created in the Laboratory of Fluid Mechanics of the IPMech RAS to study the swirling flow, and is included in the complex of unique installations of the UNU GFK IPMech RAS. The vortex flow was created by a rotating disk installed at the bottom of a transparent cylindrical container. To reduce optical distortions, the container was placed in an open rectangular pool.
2. Some of the experiments were carried out on installations for modeling flows, in which the vortex flow is created using a magnetic armature connected to a motor by a magnetic field. Heated magnetic stirrers Intllab MS-500 and ES-6120 are used as a vortex flow inductor. In addition, containers of various geometries and sizes were used. In these experiments, the free surface remained open, which makes it possible to control the introduction of impurities and record the fine structure of the flow.

A series of experiments was carried out with solid-state markers (Fig. 1) of various shapes in order to reveal the influence of controlled external parameters on the patterns of movement of a solid-state marker along the surface of a rotating liquid. Polypropylene objects of various shapes were introduced into the already established vortex flow (both in pure water and with the addition of hydrocarbons), and at the same time, the video recording of the experiment began. A technique was developed for automatic processing of the results of the behavior of the marker on the free surface of the vortex flow, based on the conversion of a raster image into a vector representation, which can significantly speed up the process of processing the results of experiments.

The displacement of the marker, which moves along the surface of the liquid involved in the complex vortex flow, can be reduced to a combination of its rotation around the vertical axis of the flow and spinning around its own center of mass. Several coordinate systems are introduced in order to describe the movement of the marker on the free surface of flow. The origin of one coordinate frame XOY (Cartesian) coincides with the geometric center of the free surface, the axes directed along the video frame sides in the recorded flow pattern, the coordinate plane with the axes coincides with the undisturbed level of the free surface (Fig. 2). Another

Fig. 1 Markers used in experiments

coordinate frame—polar (r, φ)—is located in the same plane and is most convenient to represent the marker movements around the vertical axis of the flow. In this frame, the coordinates of the center of the marker are counted depending on time.

When a marker is placed on a surface, viscous tangential stresses from the liquid begin to act on its lower submerged surface, which carry it as a whole along some trajectory lying on the surface and rotate it around its own center of mass.

Round marker after a short "involvement interval" ($\Delta t = 5$ s) is rotated around the vertical axis with almost constant angular velocity, the value of which monotonically increases with increasing of the disk rotation frequency, the estimations of which are 2.3, 6.0, 9.9 s^{-1} for the values of the inductor rotation frequencies 3.3, 9.2, 16.7 s^{-1} respectively (Fig. 3).

After being placed on a rotating free surface, the marker gradually picks up speed and moves to the center—the radial coordinate decreases, and the beat frequency increases. The processing of the experiments performed clearly demonstrates the influence of the shape of the marker on the speed of the marker and on the nature of the dependence of the change in the radius of the marker around the center. The shape of the marker is related to the rate of acceleration at the beginning of the

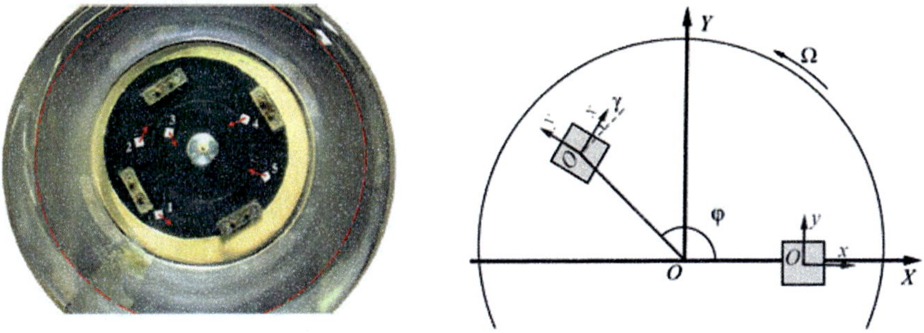

A B

Fig. 2 Marker on a vortex flow: **a** 1–5 to 1, 7, 16, 21, 35 s after start of experiment, **b** coordinate frames scheme

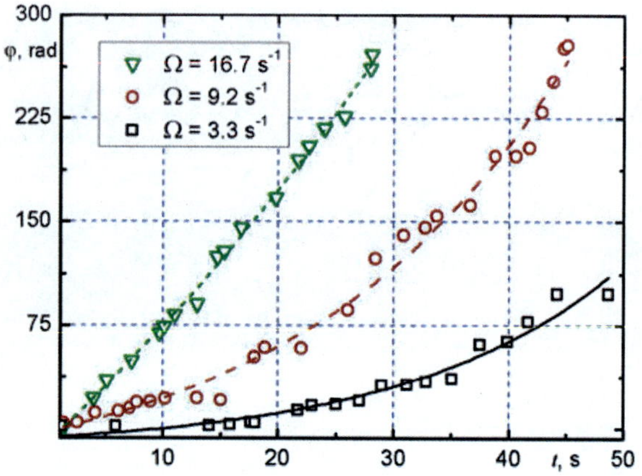

Fig. 3 The angle of rotation of the marker around its own axis from time ($H = 40$ cm, $R = 7.5$ cm, $\Omega = 3.3, 9.2, 16.7\ \mathrm{s}^{-1}$)

experiment—markers with smooth boundaries (round and oval) pick up speed more slowly and move more slowly towards the center of the current than square and cruciform markers. Predictably, the increase in speed and approach to the center of the flow are the slower, the greater the mass of the marker, even if the shape coincides. At the same time, the rate of velocity increase and approach to the center of the flow practically do not depend on the depth of the initial liquid layer involved in the vortex flow (Fig. 4).

The measurements of the radius of revolution of a large-scale solid marker from the center of the free surface show that the best agreement between the approximating function $R(t)$ and experimental data was obtained using a logarithmic dependence $R_m = A\ \ln(t) + B$ of the form free flow surface, where R_m—the marker's radial coordinate relative to the center, t—the time elapsed since the marker was placed on the free flow surface.

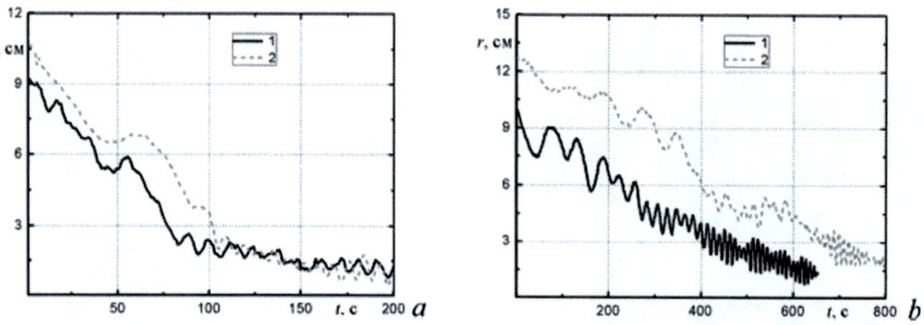

Fig. 4 Dependence of the change in the radius of circulation of markers of various shapes around the center of the rotating free surface at different depths of the liquid (R = 7.5 cm, = 200 rpm): **a** an oval marker 1.5 cm long, **b** a cruciform marker 1.5 cm, liquid depth 1—20 cm, 2—40 cm

3 Analytical Model of Marker Motion in Vortex Flow

When assessing the force impact of the water flow on the marker, the experimental fact is used that its depth of immersion in water is significantly less than the characteristic scales of the near-surface current. This makes it possible to use the approximation of the absence of flow disturbances by placing a marker on its surface. Based on this, it is assumed that the marker spreads, as it were, over the surface of the vortex, but retains its flat shape during calculations.

The marker, located on the surface of the water, experiences the action of gravity, the Archimedean force and the force of hydrodynamic origin, which is the result of the addition of several forces generated by various physical phenomena.

The hydrodynamic force is determined by the distribution of pressure over the lower and upper surfaces of the marker, capillary effects that occur at the contact boundary of water on the edge of the marker, the effects of dragging the marker with the flow and its deceleration due to its immersion in water, as well as the effects of viscous interaction with water on the entire contact surface [8].

Obtaining exact analytical expressions for all the mentioned forces in the case of a marker of an arbitrary shape is, apparently, an unsolvable problem, and therefore a number of approximations were used in the work, allowing one to obtain constructive results.

The first approximation is that the movement of the marker is always described by moving its center of mass and rotating around the instantaneous axis, which always passes through it normally to the liquid surface. Since markers were used in the experiments, the shape of which exhibits central symmetry (continuous or discrete), the component of the total capillary force acting on the marker tangent to the liquid surface is equal to zero (Fig. 5).

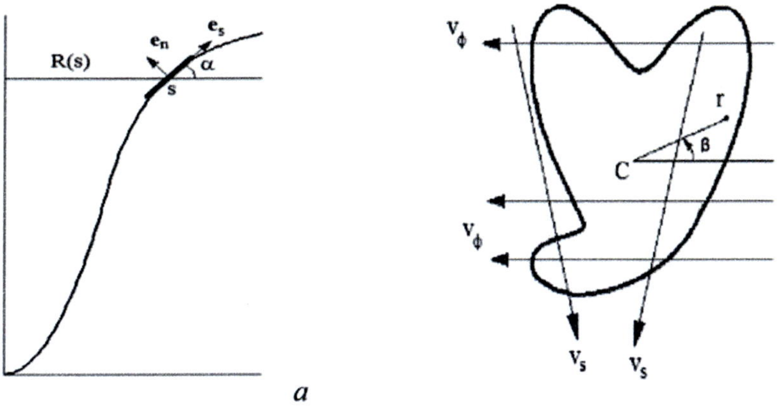

Fig. 5 Schematic representation of the marker located on the surface of the vortex (**a**) and the velocity field adjacent to its lower side (**b**)

The force acting from the side of the water flow on the immersed part of the marker edge is determined by the expression

$$\mathbf{F}_f = \oint_L c_f(L)\boldsymbol{\nu}\,(\boldsymbol{\nu}\cdot(\mathbf{v}-\dot{\mathbf{r}}_c))dL,$$

where $\mathbf{r}_c$—radius vector of the marker's center of mass; $\mathbf{v}$—the water velocity field, which, due to the small thickness of the marker in comparison with the characteristic scales, $\boldsymbol{\nu}$—is the internal normal to the contour; $c_f(L)$—local linear shape resistance coefficient (in the case of an arbitrary shape of the bounding contour L, this coefficient is a hard-to-determine value); the dots above the symbols here and below denote differentiation with respect to time.

Due to the central symmetry of the markers, the velocity head that occurs on a part of the contour and accelerates the marker is almost completely compensated by the braking effect that occurs on the remaining part of the contour. As a result, the second approximation consists in neglecting the force $\mathbf{F}_f$ generated by the action of the flow on the edge of the marker, which allows us to consider the marker to be infinitely thin, and its center of mass located on the liquid surface.

If the marker were placed on a horizontal surface, the resulting force of gravity, the Archimedean force, the component of the capillary force normal to the surface of the marker, and the force generated by the pressure difference on the upper and lower surfaces would be equal to zero. In the case of an inclined liquid surface, this resulting force is directed tangentially to the liquid surface, since the marker itself does not come off this surface, and is given by the expression

$$\mathbf{F}_{\mathrm{R}} = \lambda m g \mathbf{e}_t,$$

where m—marker weight; g—the acceleration of gravity; λ—is some dimensionless coefficient proportional to the tangent of the angle of inclination of the liquid surface to the horizon; $\mathbf{e}_t$—is the unit tangent vector to the liquid surface.

Viscous tangential stresses from the liquid side act on the lower surface of the marker, which transfer it as a whole along a certain trajectory lying on the surface and rotate around its own center of mass. Within the framework of the approximations made, the surface density of this viscous force is determined by the quantity

$$\mathbf{f}_{\mathrm{v}} = \kappa(\mathbf{v}_t - \dot{\mathbf{r}}_c - \boldsymbol{\omega}\times\mathbf{r})$$

where $\mathbf{v}_t$—tangent to the surface of the water velocity field component marker at the point with coordinate $\mathbf{r}_c + \mathbf{r}$; $\mathbf{r}$—local radius-vector of a point on the surface of the marker with origin at the center of mass, as shown in Fig. 5; κ—some coefficient characterizing the adhesion of the marker surface (depending on its material, roughness, etc.) with liquid; $\mathbf{v}_t - \dot{\mathbf{r}}_c - \boldsymbol{\omega}\times\mathbf{r}$ the speed of movement of the dot at the bottom of the marker relative to the fluid flow.

The total viscous force and the total moment of viscous forces acting on the marker relative to the axis of its rotation passing through the center of mass are determined by the expressions

$$\mathbf{F}_{\mathrm{v}} = \int_S \mathbf{f}_{\mathrm{v}}\, dS, \quad \mathbf{M}_{\mathrm{v}} = \int_S \mathbf{r} \times \mathbf{f}_{\mathrm{v}}\, dS$$

where S—area of the bottom surface of the marker.

The moment of force $\mathbf{F}_R$ applied to the center of mass is zero.

Thus, in the adopted approximate model, the equations and initial conditions describing the movement of the marker on the liquid surface have the form

$$\begin{aligned} m\ddot{\mathbf{r}}_c &= \mathbf{F}_{\mathrm{v}} + \mathbf{F}_{\mathrm{R}} = \kappa \int_S (\mathbf{v}_w - \dot{\mathbf{r}}_c - \boldsymbol{\omega} \times \mathbf{r})\, dS + \lambda m g \mathbf{e}_t \\ J\,\dot{\boldsymbol{\omega}} &= \kappa \int_S \mathbf{r} \times (\mathbf{v}_w - \dot{\mathbf{r}}_c - \boldsymbol{\omega} \times \mathbf{r})\, dS, \;\; \dot{\mathbf{r}}_c|_{t=0} = 0, \;\; \mathbf{r}_c|_{t=0} = \mathbf{r}_{c0}, \;\; \boldsymbol{\omega}|_{t=0} = 0. \end{aligned} \tag{1}$$

Here J—moment of inertia of the marker relative to the specified axis of rotation.

The shape of the surface on which the marker moves is given by the relation

$$\zeta(s) = \zeta_0 + \frac{\Omega_\varphi^2}{2g}\left[R^2(s)\theta(a - R(s)) + a^2\left(2 - a^2/R^2(s)\right)\theta(R(s) - a)\right] \tag{2}$$

and describes the emerging vortex funnel [9].

An equation was obtained that describes the movement of the center of mass of the marker in the form of a disk of radius ρ.

$$\begin{aligned} s &= \sigma(t) + \varepsilon\frac{a^2}{\sigma(t)}(1 + \lambda^2)\ln\frac{\sigma(t)}{s_0} + o(\varepsilon) \\ \varepsilon &= \Omega_s/\mu \ll 1, \quad \lambda = \Omega_\varphi/\Omega_s \\ \sigma(t) &= \sqrt{s_0^2 - 2a^2\Omega_s t}, \; \mu = \kappa S/m \end{aligned} \tag{3}$$

The rotation frequency of the marker relative to its center of mass is determined by the equation:

$$\begin{aligned} \omega &= \frac{\beta\mu}{2}\left(\varphi e^{-\mu t} - \varphi_0\right) + \frac{1}{2}\left[\frac{\dot{\varphi}_0}{1 + 4\varepsilon\dot{\varphi}_0/\Omega_\varphi} - \frac{\dot{\varphi}}{1 + 4\varepsilon\dot{\varphi}/\Omega_\varphi}\right] \\ \beta &= \frac{s_0^2\dot{\varphi}_0}{a^2\Omega_\varphi}, \quad \dot{\varphi}_0 = \varphi|_{t=0} \end{aligned} \tag{4}$$

The marker rotation angle relative to its own axis is determined by the expression

$$\gamma = \int_0^t \omega\, dt = \frac{\beta}{2}\left[\varphi_0 - \varphi e^{-\mu t} - \int_0^t \dot{\varphi} e^{-\mu t}\, dt\right] - \frac{1}{2}(\varphi - \varphi_0) \tag{5}$$

The quantity $\varphi - \varphi_0$ in Eq. (5) is the angle of rotation of the center of mass relative to the axis of rotation of the vortex funnel. Since, according to (5), we have the estimates $\int_0^t \dot{\varphi} e^{-\mu t} dt \sim \lambda\varepsilon$, $\beta \simeq 1$, then the main term in Eq. (5) has the form

$$\gamma \approx -\frac{1+\beta}{2}(\varphi - \varphi_0) + \frac{\beta}{2}\varphi\left(1 - e^{-\mu t}\right) \tag{6}$$

and this means that when the marker is in the periphery, it rotates around its own axis in the direction opposite to the direction of rotation of the center of mass of the marker relative to the axis of the vortex funnel.

An important feature of the expressions obtained is that when using the main terms in the expressions for the coordinates s and φ the center of mass of the marker, these coordinates, both in the region of solid rotation and on the periphery of the vortex funnel, are related by the relation of the general form

$$s \approx \tilde{s} \exp\left(-\frac{\Omega_s}{\Omega_\varphi}(\varphi - \tilde{\varphi})\right) \tag{7}$$

where $\tilde{s}$, $\tilde{\varphi}$—some constants.

Equation (7) describes a logarithmic spiral on the surface of a vortex funnel along which the marker's center of mass moves, and this spiral coincides with the trajectories of liquid particles near the free surface [10].

The obtained experimental data allowed to plot the dependence of the marker spinning angle ψ versus rotation angle φ (Fig. 6).

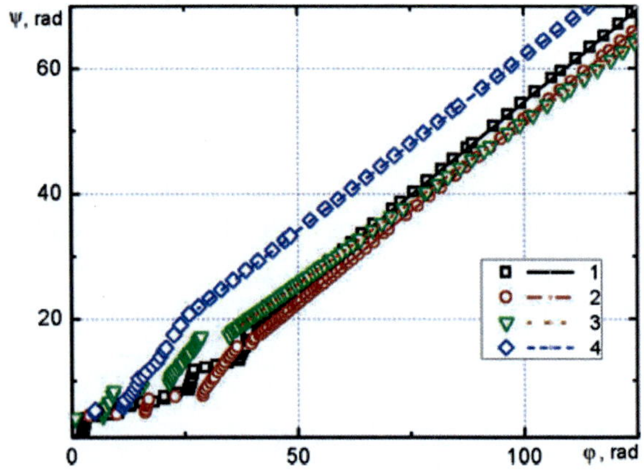

Fig. 6 Interrelation of rotation φ and spinning ψ angles for ($\Omega = 3.3\ s^{-1}$): rectangular markers *1*, *2*—0.5 cm × 1.0 cm × 0.3 cm, $H = 40$ and 20 cm, *3*, *4*—1.0 cm × 2.0 cm × 0.3 cm, $H = 20$ and 40 cm

4 Conclusions

Experimental studies of the transfer of solid-state markers of various shapes and sizes in a vortex flow in a single-component and multicomponent liquid have been carried out, a theoretical model has been proposed that explains the movement of markers, and an equation has been obtained that describes the movement of the center of mass of the marker and represents a logarithmic spiral on the surface of a vortex funnel, which coincides with the trajectories of liquid particles near the free surface. The experimental dependence of the angle of rotation on the angle of rotation in the region of rotation of a rigid body agrees well with the theoretical dependence obtained on the basis of the proposed mathematical model.

The proposed model of the transfer of floating markers along the surface of a vortex trough can also be used to study the dynamics of the transfer of "islands" of debris by large-scale oceanic vortices.

Acknowledgements The work was carried out with the financial support on the topic State assignments 123021700046-4

References

1. Helmholtz H. Uber Integrale der hydrodynamischen Gleichungen, welche den Wirbelbewegungen entsprechen // Journal fur die reine und angewandte Mathematik. – 1858. – V. 55. – P. 25–55.
2. Kelvin Lord. On vortex motion // Royal Soc. Edinburgh. – 1868. – V. 25. P. 217–260. 9. Kelvin Lord. Vortex statics // Collected works. – 1875. – V. 4. P. – 115–128.
3. Prandtl L., Tragflügeltheorie. I Mitteilung. Nachrichten von der Gesellschaft der Wissenschaften Zu Göttingen // Mathematisch-Physikalische Klasse. – 1918. – P. 151–177.
4. Poincare A. Theory of vortices. - M.: Izhevsk: RHD, 2000. - 160 p.
5. Alekseenko S.V., Shtork S.I. Traveling vortex breakdown // Technical Physics Letters 1997. V. 23(22). P. 24–28. DOI: https://doi.org/10.1134/1.1261914.
6. Zhukovsky N.E. On Associated Vortices. Full coll. soch., v. 5, M.-L.: State publishing house of technical and theoretical literature, 1937.
7. Saffman F. Vortex Dynamics - M.: Nauchny Mir, 2000. - 376 p.
8. Kistovich A.V., Chaplina T.O., Stepanova E.V. Analytical and experimental study of the substance transport in the vortex flow // Theoretical and Computational Fluid Dynamics. 2019. DOI: https://doi.org/10.1007/s00162-019-00506-x.
9. Becker R. Mesh Adaptation for Stationary Flow Control // J. Math. Fluid Mech. 2001. V. 3(4). P. 317–341. DOI: https://doi.org/10.1007/PL00000974.
10. Chaplina T.O., Stepanova E.V., Pakhnenko V.P. (2020) Features Study of the Marks Movement on the Surface and in the Depth of Vortex Flow. Processes in GeoMedia—Volume I. Springer Geology. Springer, Cham. https://doi.org/10.1007/978-3-030-38177-6_32.

The Thermohaline Structure, Hydrochemical Characteristics and Chlorophyll-A Distribution in the White Sea Based on the Expedition Data Obtained in the Summer of 2019

N. S. Frolova, E. I. Kharkhordina, and A. A. Kirilova

Abstract The article investigates the thermohaline structure and hydrochemical characteristics in the White Sea. The data were obtained during the third stage of the Transarctica-2019 expedition from July 16 to August 2, 2019. An indirect assessment of the productivity of the White Sea waters by the content of chlorophyll-a was also carried out.

Keywords Water structure · Temperature · Salinity · Biogenic elements · Chlorophyll-a

1 Introduction

The White Sea is one of the Arctic seas that belongs to the Arctic Ocean basin and borders the Barents Sea along the line joining Cape Kanin Nos and Cape Svyatoy Nos. The total area of the sea is about 90 thousand km^2, the average depth is 60 m, but the maximum depths are observed in the central part of the White Sea (Basin) and the northeast part of Kandalaksha Bay, where they are about 350 m. Due to geomorphological structure the Voronka (Funnel), the Gorlo (Throat), the Basin and four large bays are distinguished: Onega, Dvina, Mezen and Kandalaksha [1, 2].

The waters of the White Sea are formed from two main sources: fresh and salty waters. Fresh waters are transported to the sea from various sources, including the following: river runoff, subsurface drainage, distributed slope runoff, and atmospheric precipitation. A certain portion of fresh water is removed from the sea by evaporation. The river runoff is the main input component of the freshwater budget in the White Sea. The salty water source is incoming waters of the Barents Sea. The main volume of river runoff is concentrated in the eastern part of the sea with rivers: the Northern Dvina and Onega. The volume of their runoff is more than two-thirds of the total volume. In the western part of the sea, the river flow is much smaller and

N. S. Frolova (✉) · E. I. Kharkhordina · A. A. Kirilova
Russian State Hydrometeorological University, Saint Petersburg, Russia
e-mail: natagr86@mail.ru

V. I. Karev (ed.), *Physical and Mathematical Modeling of Earth and Environment Processes—2022*, Springer Proceedings in Earth and Environmental Sciences,
https://doi.org/10.1007/978-3-031-25962-3_59

distributed over many small rivers. For example, the Northern Dvina has 52.5% of the annual runoff during the high water period (May–June) [3].

The ocean waters enter the White Sea from the southeastern part of the Barents Sea. They are characterized by high salinity (about 34‰) and seasonally varying temperatures from 3 to 8 °C. Advection of these waters takes place in the western part of the open sea boundary adjacent to the Tersk coast [3]. Passing the Voronka and Gorlo, the Barents Sea waters are involved in the general cyclonic cycle within the Basin and its bays. These waters are subject to both heating and desalination. In the autumn–winter period, relatively warm water masses come from the Barents Sea, but most of the heat is lost as a result of heat exchange with the atmosphere in the ice-free northern parts of the White Sea (in the Voronka and Mezen Bay). The White Sea waters warm up more strongly than the Barents Sea in summer [1, 2, 4].

The main sources of the formation of water masses of the White Sea are the Barents Sea and mainland waters, as a result of which secondary water masses are formed. In summer, six water masses can be distinguished: water of the Barents Sea, well-mixed water of the Gorlo, sea surface water of the Basin, intermediate water, deep water and desalinated waters of bays.

The main thermohaline feature of the White Sea is the presence of two types of the temperature and salinity vertical distribution: quasi-homogeneous ("Gorlovsky") and stratified ("Basin") types [5].

The "Gorlovsky" type of stratification is peculiar to the Voronka, Gorlo, Mezen and Onega Bays. Almost the entire Voronka, from the surface to the bottom, is occupied by the homogeneous Barents Sea waters with a temperature of about 5.7 °C and salinity of 34.16‰. The freshened waters of the bays occupy the upper layer in the top of the bays of the southern part of the sea and gradually wedge out toward the exit from the bays.

The "Basin" type is peculiar to the Basin, Dvina and Kandalaksha Bays. Its features are: the vertical salinity (5–7‰) and temperature (15 °C) gradients and the presence of a thin (10–30 m) upper desalinated layer and a much thicker (100–300 m) lower layer. These layers are separated by a thermohalocline. The vertical structure of the Basin waters is approximately as follows: the surface waters of the Basin are located within 20 m, then there is a transition layer with a thickness of about 40 m. There is an intermediate water mass between the depth levels of 40 and 60 m, and, starting from a depth of 100 m, deep waters spread. The surface water mass is characterized by a temperature of 10–12 °C and salinity of 25.9–27.0‰. Intermediate water mass is common in the Basin, Dvina and Kandalaksha bays. Its temperature varies from − 0.2 °C in the area adjacent to Dvina Bay and the Gorlo to 0.7–0.9 °C in Kandalaksha Bay and the central part of the Basin, salinity ranges from 28.0 to 28.2‰. A deep water mass fills the deep-water areas of Kandalaksha Bay and the Basin with a temperature of − 1.4 °C and salinity 29.7–30‰. In the bay areas adjacent to the Basin, the vertical structure of the waters is somewhat different. The upper layer is occupied by the freshened waters of the bays, below, at depths of 10–30 m, the water mass of the upper layers of the Basin is located, and after the transition zone the intermediate waters of the Basin spread.

The scientific investigations of the White Sea phytoplankton have shown that the spatial and temporal heterogeneity of planktonic algae is determined by local hydrophysical conditions [6–8]. The most significant component of the White Sea microflora is diatoms. The development of phytoplankton varies from season to season. The beginning of vegetation occurs during melting of ice and snow, when lighting is good, and the content of biogenic elements reaches the maximum values.

The purpose of this research is to assess hydrological and hydrochemical processes in five areas of the White Sea (Kandalaksha Bay, Basin, Dvina Bay, Gorlo and Voronka) and to analyze the spatial distribution of chlorophyll-a in connection with the considered features.

2 Materials and Methods

In-situ data were obtained during the third stage of the Transarctica-2019 expedition from July 16 to August 2, 2019. The expedition was organized by the Russian Hydro-Meteorological Service together with the Arctic and Antarctic Research Institute, Northern (Arctic) Federal University named after M.V. Lomonosov, Moscow State University, Russian State Hydrometeorological University, St. Petersburg State University, Russian Federal Research Institute of Fisheries and Oceanography (VNIRO), N. Laverov Federal Center for Integrated Arctic Research of the Ural Branch of the Russian Academy of Sciences (FECIAR UrB RAS). SBE 32 equipped with 12 bathometers with a capacity of 5 l, and CTD Probe SBE 19 plus V2 for measuring temperature, pressure and salinity of the sea water, as well as dissolved oxygen level and pH. Samples for chlorophyll-a were taken at 17 biological stations in 5 areas of the White Sea (stations 29, 130, 129, 126, 119, 117, 116, 20, 21a, 60, 59, 52, 51a, 6–9) (Fig. 1). To study chlorophyll-a, an integral sample was taken. It contained water from three depth levels—0, 5 and 10 m. The water was filtered using a vacuum filtration device through a membrane filter. Then the precipitate on the filter was covered with a thin layer of magnesium carbonate. Filters with precipitation were stored for subsequent analysis in a freezer at a temperature of − 18 °C. Hydrological characteristics were determined at standard depth levels at each of the stations. Hydrochemical analysis for the presence of biogenic elements, including determination of the concentration of ammonium nitrogen, nitrates, nitrites and phosphates, was carried out at 12 stations. On the studied oceanographic sections, hydrochemical analysis data are available on the surface and at a depth of 10 m, as well as at the bottom, where the deepest sampling point is taken for the bottom. Hydrochemical analysis was carried out only at 30 and 50 m in some areas.

Data on biogenic elements were obtained in accordance with the standard methods of hydrochemical analysis for marine waters, namely according to the guidance documents [9–11]. Continuous meteorological observations were conducted along the vessel's track. The average wind speed during the research work was 7.2 m/s, and the wave height was 0.5 m. However, on July 29–31, the White Sea was under the influence of the northern cyclone periphery with strong baric gradients. Wind gusts

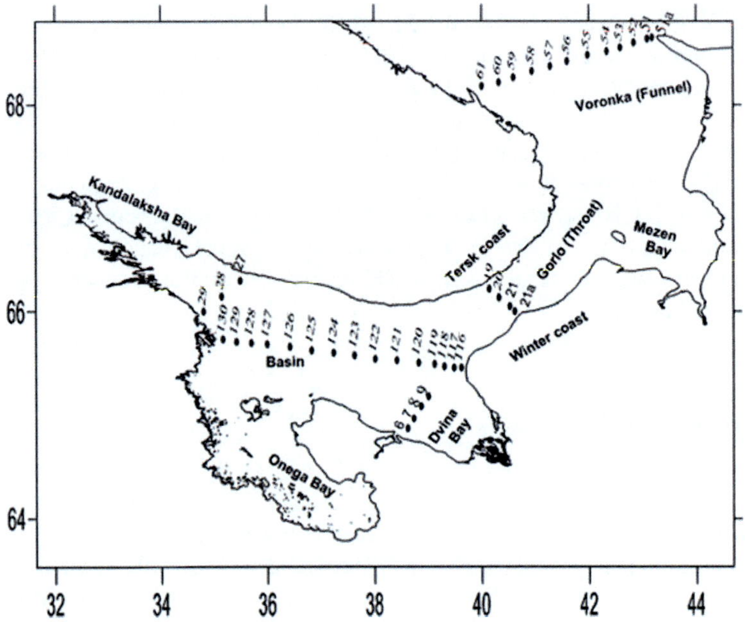

Fig. 1 Positions of the oceanographic stations in the White Sea

these days reached 22–27 m/s, and the wave height was from 2.9 to 4.5 m. Water samples in Dvina Bay (6–9 stations) were taken on the 1st of August 2019, when the storm subsided and the wind speed was less than 9 m/s.

3 Results

3.1 Hydrophysical and Hydrochemical Characteristics

The depth of visibility of a white disk varied within 4–9.2 m, with the greatest transparency of sea water recorded in the Voronka on the border with the Barents Sea (Table 1). The analysis of the materials of the deep-sea observations was carried out according to the measurements on centuries-old and standard oceanographic sections.

The following vertical distribution of hydrological characteristics can be observed in Kandalaksha Bay (Table 2):

1. The water temperature changes from 11.5 °C on the surface to − 0.78 °C at the bottom.
2. The salinity of the sea varies from 23.66 to 28.4‰.

Table 1 Data on sea water transparency, air temperature, wind speed and direction, wave height at the stations

Station	Date	Transparency, m	Air temperature, °C	Wind speed, m/s	Wind direction, °	Wave height, m
29	18.07.19	5	12.2	9	99	0.7
130	19.07.19	7	11.4	7.4	61	1
129	19.07.19	6	10.8	5.9	84	1
126	19.07.19	5	9.9	7.4	84	1
119	20.07.19	5	8.2	7.6	36	0.5
117	20.07.19	5	8.9	7.6	343	0.25
116	20.07.19	4	6.6	9.9	314	0.5
21a	20.07.19	4	6.5	5.9	295	0.4
20	21.07.19	–	5.3	3	20	0.5
60	22.07.19	9.2	6.1	6.3	106	0.3
59	22.07.19	9	6.9	5.2	120	0.2
52	23.07.19	8.5	6.5	4.8	45	0.25
51a	23.07.19	7	5.7	11.1	9	0.2
9	01.08.19	6	4.9	9.2	341	0.54
7	01.08.19	5	7.4	8	354	0.5
6	01.08.19	4	7.8	7.3	360	0.2

Analyzing the distribution of hydrochemical characteristics, it can be concluded that the waters of the White Sea in this area are well aerated. The surface oxygen content (7.02 mg/l) is slightly lower compared to the deep layers (7.98 mg/l). This indicates an increase in oxidative processes in the surface layer. The hydrogen index (pH) decreases slightly with depth. The concentration of phosphates in the area of Kandalaksha Bay increases with depth, while at a depth of 10 m there is a minimum content in this area. This situation is typical for the active development of phytoplankton in the summer. Ammonium nitrogen concentration in this area decreases with depth. The highest values are recorded up to a depth of 10 m. There are significant increase of the concentrations of nitrates and nitrites in the surface layer. This may be associated with river runoff. According to the vertical distribution of the nitrate concentration, another local maximum can be distinguished at a depth of 10 m. The results have shown that NO_3 concentrations < 5 µg/l were observed in the bottom layer.

In the Basin of the White Sea, the water temperature in the surface layer varies in a fairly wide range from 11.7 °C (station 130) to 5.7 °C (station 116) near Cape Zimnegorsky (Fig. 2). It can be also noted that there is a rapid temperature change to a depth of approximately 50 m from 11.7 to 0.5 °C, and then there is a gradual decrease in the temperature value to – 1.15 °C, which is the minimum temperature value on all sections.

Table 2 Temperature, salinity, pH, concentration of dissolved oxygen and biogenic elements

Area	St.	Depth levels, m	T, °C	S, ‰	pH	O_2, mg/l	PO_4, μg/l	NH_4, μg/l	NO_2, μg/l	NO_3, μg/l
Kandalaksha Bay	27–29	0	10.4–11.5	23.66–25.47	7.83–7.98	7.02–7.5	3.16–15.23	8.81–22.7	0.5–1.27	5–32.6
		10	9.64–10.9	24.81–26.09	7.95–7.98	7.26–7.51	2.01–3.74	3.88–8.51	0.63–1.63	5–31.7
		100	– 0.78 to – 0.61	28.3–28.4	7.8–7.9	7.53–7.98	18.11–20.98	2.24–11.2	0.5–0.8	< 5.0
Basin	116–130	0	5.7–11.7	22.35–28.19	7.45–8.08	7.2–8.12	3.45–10.06	3.44–13.44	1.54–4.16	5–66.16
		10	1.25–10.43	23.48–26.6	7.4–8.1	7.24–8.12	2.59–10.92	3.29–16.58	1.81–3.07	5–67.4
		100	– 0.95 to – 0.64	28.3–28.4	7.86–7.96	7.63–8.19	21.56 (st. 120)	5.33 (st. 120)	0.99 (st. 120)	172. 55 (st. 120)
		200	– 1.15 to – 1.13	29.06–29.11	7.78–7.85	7.1–7.72	29.89 (st. 125)	13.59 (st. 125)	1.36 (st. 125)	233.74 (st. 125)
Gorlo	19–21a	0	6.6–7.4	26.49–26.88	7.63–7.98	7.42–8.01	9.2–10.63	23.9–27.48	3.25–3.35	54.14–70.84
		10	6.44–7.26	26.29–27.5	7.69–8.01	7.32–7.9	9.48–10.35	18.07–25.69	3.44–3.62	52.54–61.42
		30	6.5–7.26	26.29–27.63	7.96–8.00	7.33–7.85	10.06	22.7–31.07	2.89–3.53	51.55–62.75
		90	6.52 (st. 20)	27.65 (st. 20)	7.98 (st. 20)	7.41 (st. 20)	–	–	–	–
Voronka	51a–61	0	5.6–9.6	28.17–31.94	8.13–8.22	7.56–8.17	–	0–13.89	–	–
		10	4.74–7.81	30.94–33.92	8.11–8.22	7.67–8.37	–	1.34–2.69	–	–
		30	3.32–5.34	32–33.04	8.14–8.22	7.58–8.06	–	15.39 (st. 51a)	–	–
		50	3.34–5.31	33.56–34	8.16–8.22	7.58–8.16	–	–	–	–

(continued)

Table 2 (continued)

Area	St.	Depth levels, m	T, °C	S, ‰	pH	O_2, mg/l	PO_4, μg/l	NH_4, μg/l	NO_2, μg/l	NO_3, μg/l
Dvina Bay	6–9	0	3.4–9.6	24.46–25.79	–	–	–	–	–	–
		10	3.2–9.5	24.8–26.8	–	–	–	–	–	–
		30	3.11–7.42	25.75–26.84						
		50	0.32–1.89	27.01–27.4						
		90	− 0.24 (st. 8)	28.11 (st. 8)						

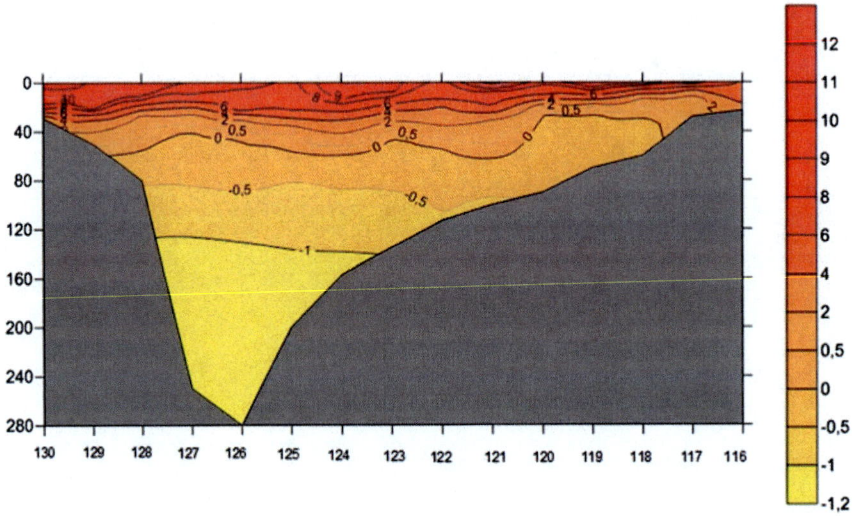

Fig. 2 Temperature distribution on the sections 130–116 (vertical axis—depth, horizontal axis—stations)

The upper layer is occupied by less salty waters and formed under the influence of currents carrying the freshened waters of bays into the Basin (Fig. 3). Salinity values grow rapidly to a depth of approximately 50–60 m, and then gradually increase to 29.3‰.

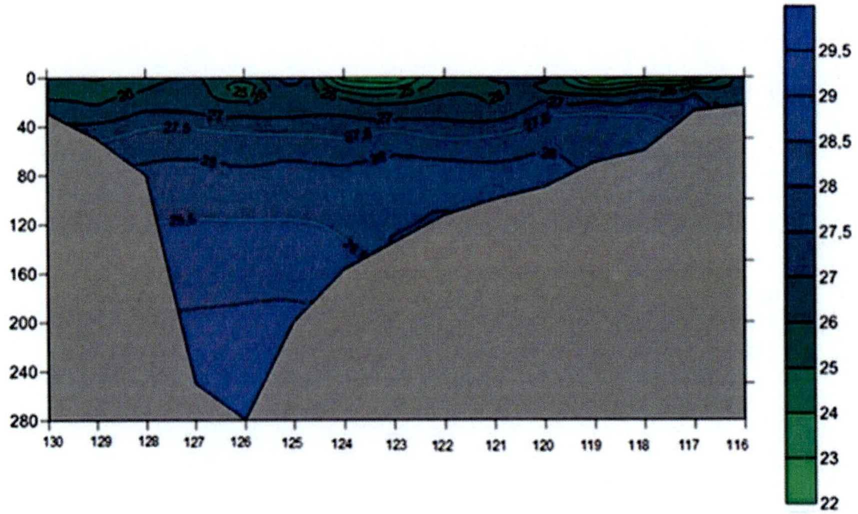

Fig. 3 Distribution of salinity in sections 130–116 (vertical axis—depth, horizontal axis—stations)

The distribution of dissolved oxygen in this section is fairly uniform and typical for the summer period. The area of the lowest concentration of dissolved oxygen falls on the deepest part of the section occupied by the deep-sea water mass, where the absolute minimum values are at the station 126 (6.43 mg/l).

The pH value depends on the salt composition of the water, the content of dissolved gases and organic compounds. The highest hydrogen index value (8.09 pH units) was recorded at the station 123 (depth level—10 m). The lowest value (7.33 pH units) was defined at the station 128 (depth of 10 m).

According to the vertical distribution of phosphates, it is possible to note a reduced content of this element in surface waters, which is associated with its consumption by aquatic organisms.

Ammonia nitrogen (NH_4^+) accumulates in water when ammonia gas (NH_3) is dissolved in water, which is formed during the biochemical decomposition of nitrogen-containing organic compounds. The main sources of dissolved ammonia (ammonium ion) are surface and underground runoff, atmospheric precipitation, as well as industrial wastewater. In the surface waters of the White Sea saturated with oxygen, under the action of nitrifying bacteria, ammonium is rapidly oxidized to an unstable nitrite (NO_2^-), and then to a stable nitrate (NO_3^-) form. Due to the fact that the waters of the White Sea are saturated with dissolved oxygen to the bottom, oxidative conditions are created. In turn, for the investigated elements that are stable under oxidizing conditions (PO_4, NO_3), an increase in their concentration to the bottom is natural.

In the Gorlo, we can trace the water exchange between the colder and saltier waters of the Barents Sea and the warmer and less salty waters of the White Sea. There is a weak stratification of waters by temperature and salinity (Fig. 4) at the Tersk coast of the Gorlo (station 19). In the deepest part of the section, the distribution of temperature, as well as salinity from a depth of 10 m, becomes close to uniform. Such an approximate homogeneous vertical distribution of temperature and salinity of the Gorlo waters from the surface to the deep level is caused by the presence of intense turbulent mixing.

The waters near the eastern coast are more desalinated in relation to the western ones, and higher water temperature values are also observed here. In contrast to the weakly stratified waters of the western part, there are rather large horizontal temperature and salinity gradients in this area, which once again confirms that the White Sea waters flow from the Basin and Dvina Bay along the Winter Coast, and the waters of the Barents Sea penetrate along the Tersk coast.

According to the vertical distribution of pH and dissolved oxygen, the interaction between the Barents Sea and White Sea waters is also well traced. At the eastern coast, elevated concentrations of dissolved oxygen are observed in the Gorlo waters and amount to 8.01 mg/l. In the middle of the section, at a depth of 20 m, the minimum dissolved oxygen content is observed, reaching 7.25 mg/l. Minimum pH values are also noted near the station 21a.

Increased concentrations of ammonium nitrogen and nitrates are observed at 30 m in this area, while reduced concentrations of nitrites are recorded at the same depth.

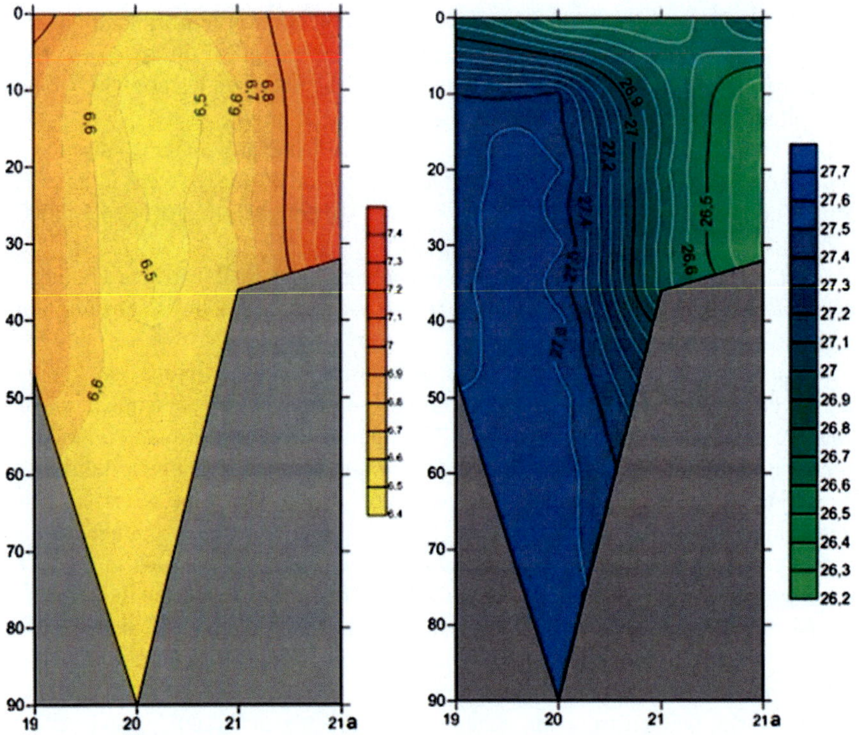

Fig. 4 Distribution of temperature (left) and salinity (right) in sections 19–21a (vertical axis—depth, horizontal axis—stations)

In the Voronka on the border with the Barents Sea, it is also possible to trace the interaction between waters of the two seas. The analysis of the temperature and salinity vertical distribution in this section has shown the following: the stratified waters are located near the Kanin coast (Fig. 5). The White Sea discharge current is clearly traced here.

Almost half of the length of the section to the Tersk coast is occupied by the Barents Sea waters. Their salinity from the surface to the bottom is almost homogeneous (S = 34‰). Due to summer warming, the incoming surface waters of the Barents Sea are temperature-stratified.

In the north-eastern part of the section, in the area of Cape Kanin Nos, in the near-surface layer, more dramatic changes in the values of the hydrogen index are observed, and there is a relatively smooth decrease in the pH value with depth, while in the south-western part the value of the hydrogen index shows a uniform depth distribution. The areas of the highest dissolved oxygen content in this section are mainly located near the surface, its concentration decreases with depth. Higher concentrations of dissolved oxygen are observed near the western coast of the Voronka, in

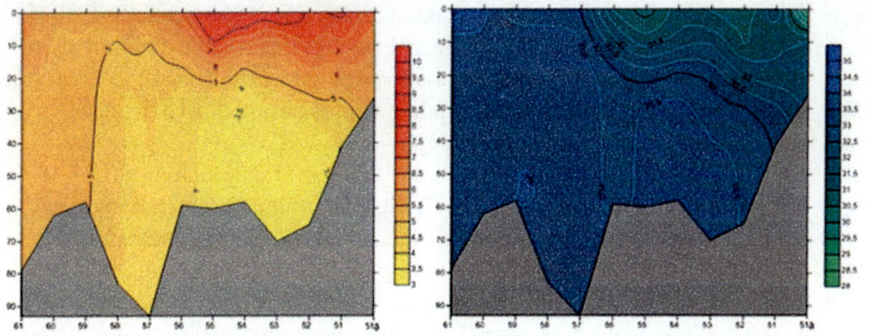

Fig. 5 Distribution of temperature (left) and salinity (right) in sections 61–51a (vertical axis—depth, horizontal axis—stations)

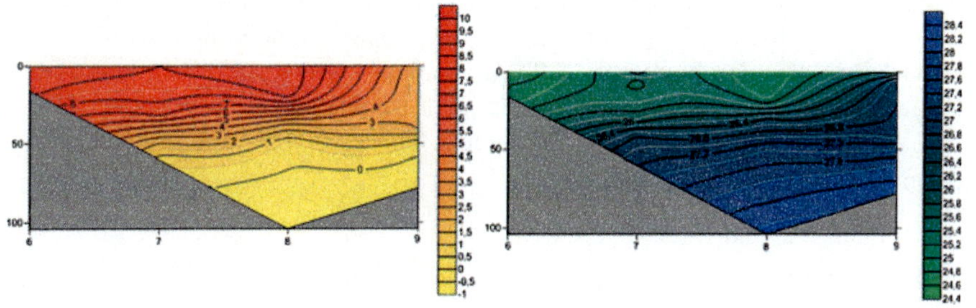

Fig. 6 Distribution of temperature (left) and salinity (right) in sections 6–9

the area of Cape Svyatoy Nos, where the maximum value of dissolved oxygen is observed at the oceanographic station 61 (a depth of 10 m), 8.37 mg/l.

The maximum concentrations of ammonium nitrogen are still observed at the depth of about 30 m in the area of Cape Kanin Nos. The concentrations of nitrates, nitrites and phosphates were not determined in this area.

The Dvina Bay distribution of temperature and salinity is the characteristic of a bay—the surface layer is desalinated and well warmed up under the influence of abundant river runoff (Fig. 6). This is especially noticeable at the stations 6, 7 and 8. The station 9 is located at a distance of 50 km from the coast, so the influence of river runoff is not so strong in comparison with other stations of this section. There are no data of other characteristics in the Dvina Bay section.

3.2 *The Concentration of Chlorophyll-A*

The thickness of the photic layer in the White Sea is 15–20 m on average [12]. According to other data, the thickness of the photic layer varies from 2 to 5 m in

Dvina Bay to 15–25 m in the deep part of the sea and averages 10–15 m [8, 13, 14]. The transparency data measured with a Secchi (white) disk (Table 1) allow us to conclude that the thickness of the photic layer varies markedly from 12 to 27 m in different parts of the sea. Therefore, the depth of the greatest photosynthetic activity of phytoplankton varies within the same limits.

According to the results of the research, the content of chlorophyll-a (Chl "a") ranges from 0.4 to 3.71 μg/l (Table 3, Fig. 7). The maximum concentrations during the observation period were found in the Basin (station 126) and Kandalaksha Bay (station 29). In Dvina Bay the measured values of chlorophyll-a concentration indicate the active development of phytoplankton at stations 6 and 7, the concentration of chlorophyll-a is 1.53 μg/l near the coast. The reduced concentrations of chlorophyll-a 0.4 μg/l were noted at the station 9. The chlorophyll-a concentration, as an indicator of trophic state, does not give very accurate estimates because the intensity of photosynthesis depends not only on the concentration of chlorophyll, but also on many other factors: light, temperature, water transparency, composition and physiological activity of cells, etc. [15]. Consequently, with one and the same trophic status concentrations of chlorophyll may differ dramatically. However, the trophic state classification of reservoirs based on a chlorophyll-a concentration makes it possible to give the White Sea preliminary assessments. This reservoir has a trophic status between the oligo- and mesotrophic seas.

Table 3 Chlorophyll "a" concentrations

Station	Chl "a", μg/l
29	2.43
130	1.06
129	1.89
126	3.71
119	2.01
117	0.73
116	0.75
21a	1.06
20	0.43
60	2.22
59	2.18
52	1.53
51a	1.27
9	0.4
7	1.53
6	1.53

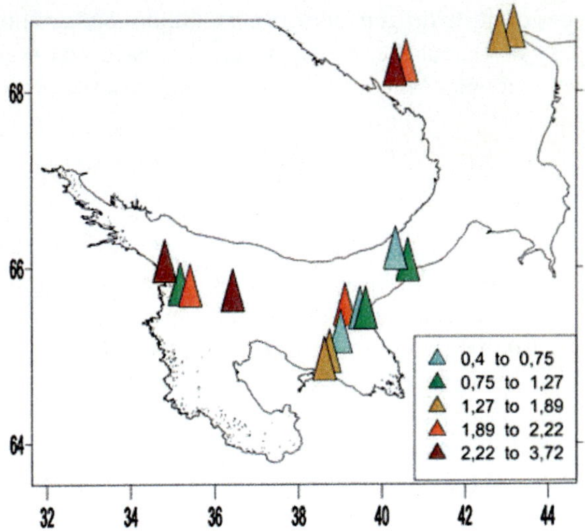

Fig. 7 Map of chlorophyll-a concentrations in the White Sea

4 Discussion of the Results

This scientific research presents the results of the White Sea vertical thermohaline structure. A water exchange zone has been identified between the waters of the White and Barents Seas, and it is also noted that the waters of the White Sea pass along the eastern coast of the Voronka, and the waters of the Barents Sea—along the western one. The interaction of the waters of these seas can be traced only on two sections (in the Gorlo and Voronka). It is impossible to see the influence of the Barents Sea on the other sections. In general, the results obtained confirm the summer feature of the White Sea hydrological structure, described in [1, 2]. The vertical structure of the Basin waters becomes apparent on the age-old section "Ivanovy Ludy–Cape Zimnegorsky": surface waters are located from the surface to a depth of 20 m, then there is a transitional layer with a thickness of about 20–30 m. After the transition zone from 50 to 100 m, there is an intermediate water mass and, starting from 100 m, a deep-sea waters spread.

Phytoplankton activity is also reflected in the vertical distribution of nutrients. The main factor limiting the development of phytoplankton in the White Sea is a sharp decrease in the nitrate nitrogen concentration [16], but in some situations phosphates can also play this role. Analysis of the obtained data has shown that, in general, the amount of biogenic elements on the surface is much less than at the bottom in summer period. It is largely due to active phytoplankton development. The complexity of the vertical distribution of biogenic elements in the White Sea is determined by a number of factors, which include, first of all, river runoff and vertical stratification in the deep-water areas of the Basin. Most of the White Sea is characterized by increased concentration of dissolved oxygen in the surface layer. In general, dissolved oxygen concentrations decrease with depth. The increase in the values of the hydrogen index occurs in those parts of the water area that are mostly filled with Barents Sea waters, where the dissolved oxygen concentration decreases. There is a significant decrease in pH values in areas of mixing of salt and fresh water.

5 Conclusions

Having studied the obtained data for five regions of the White Sea, it can be noted that the spatial distribution of biogenic elements mainly depends on the water dynamics. River runoff waters are richer in biogenic elements than surface sea waters. Therefore, the areas affected by river runoff are characterized by the increased concentrations of these elements [17].

The distribution of dissolved oxygen is closely related to the solubility of gases in water. Thus, with a decrease in temperature, an increase in the gas dissolution is observed and, consequently, an increase in the concentration of dissolved oxygen can be detected. In general, there is a slight decrease in the concentration of dissolved oxygen with a depth. However, in the summer, the waters of the White Sea, even in the bottom layer, are saturated with oxygen, which is in good agreement with the data presented in [18].

A comprehensive analysis of the distribution of biogenic elements makes it possible to find a minimum concentration of nitrates and phosphates in the surface layer and a maximum in the bottom layer. The obtained distribution of these elements is in good agreement with the concentration of chlorophyll-a. The increased concentration of chlorophyll-a indicates the intensive development of phytoplankton, which in turn reduces the content of biogenic elements. The concentration of nitrites decreases with depth, while a local maximum of concentration in the surface layer and at a depth of 50 m are observed on the section. The content of ammonium nitrogen decreases with depth, while a local maximum is recorded on some sections, most often at a depth of about 30 m.

References

1. Sistema Belogo morya. T. II. Vodnaya tolshcha i vzaimodeistvuyushchie s nei atmosfera, kriosfera, rechnoi stok i biosfera. [The White Sea System, Vol. 2: Water Column and Interacting Atmosphere, Cryosphere, the River Run-Off, and Biosphere]. Lisitzin, A.P., Ed., Moscow: Nauchnyi Mir, 2012, 784 p. (In Russian).
2. Filatov N.N., Terzhevik A.Yu. (Ed.) Beloe more i ego vodosbor pod vliyaniem klimaticheskih i antropogennyh faktorov. [The White Sea and its catchment area under the influence of climatic and anthropogenic factors]. Institute of Northern Water Problems, KarRC RAS, Petrozavodsk, 2007. 349 p. (In Russian).
3. Pantyulin A.N. Dynamics, structure and water masses. [Dynamics, structure and water masses]. Sistema Belogo morya. T. II. Vodnaya tolshcha i vzaimodeistvuyushchie s nei atmosfera, kriosfera, rechnoi stok i biosfera. [The White Sea System, Vol. 2: Water Column and Interacting Atmosphere, Cryosphere, the River Run-Off, and Biosphere]. Lisitzin, A.P., Ed., Moscow: Nauchnyi Mir, 2012, pp. 309–379. (In Russian).

4. Berger V.Ya. Produktzionnyy potentzial Belogo morya. Issledovaniya fauny morey. [The production potential of the White Sea. Studies of the fauna of the seas]. Vol. 60 (68). Saint-Petersburg, ZIN RAN, 2007. 292 p. (In Russian).
5. Zimin A.V. Subprilivnye processy i yavleniya v Belom more. [Sub-tidal processes and phenomena in the White Sea]. Moscow, GEOS, 2018. 220 p. (In Russian).
6. Ilyash L.V., Radchenko I.G., Shevchenko V.P., etc. Prostranstvennoe raspredelenie fitoplanktona Belogo morya v kontze leta v svyazi so strukturoy i dinamikoy vod. [Spatial distribution of phytoplankton of the White Sea at the end of summer in connection with the structure and dynamics of waters]. Okeanologiya. [Oceanology]. 2011, vol. 51, no. 6, pp. 1054–1063. (In Russian).
7. Ilyash L.V., Radchenko I.G., Shevchenko V.P., etc. Kontrastnye soobshchestva letnego fitoplanktona v stratifitzirovannyh i peremeshannyh vodah Belogo morya. [Contrasting communities of summer phytoplankton in stratified and mixed waters of the White Sea]. Okeanologiya. [Oceanology]. 2014, vol. 54, no. 6, pp. 781–790. (In Russian).
8. Ilyash L.V., Belevich T.A., Stupnikova A.N., etc. Vliyanie lokal'nyh gidrofizicheskih usloviy na prostranstvennuyu izmenchivost' fitoplanktona Belogo morya. [The influence of local hydrophysical conditions on the spatial variability of the phytoplankton of the White Sea]. Okeanologiya. [Oceanology]. 2015, vol. 55, no. 2, pp. 241–251. (In Russian).
9. Mass concentration of phosphates in marine waters. The method of measurement by photometric method. Moscow, 2010. 27 p.
10. RD 52.10.740-2010 Mass concentration of nitrite nitrogen in marine waters. Method of measurement by photometric method with Griss reagent. Moscow, 2010. 24 p.
11. RD 52.10.773-2013 Mass concentration of ammonium nitrogen in sea waters. Photometric Measurement Technique with Nessler's Reagent. Moscow, 2014. 21 p.
12. Ilyash L.V., Radchenko I.G., Kuznetzov L.L., etc. Prostranstvennaya variabel'nost' sostava, obiliya i produktzii fitoplanktona Belogo morya v kontze leta [Spatial variability of the composition, abundance and production of phytoplankton of the White Sea at the end of summer]. Okeanologiya. [Oceanology]. 2011, vol. 51, no. 1, pp. 24–32. (In Russian).
13. Mordasova N.V., Wentzel M.V. Osobennosti raspredeleniya fitopigmentov i biomassy fitoplanktona v Belom more v letniy sezon. [Distribution features of phytopigments and phytoplankton biomass in the White Sea in the summer season]. Kompleksnye issledovaniya ekosistemy Belogo morya. [Comprehensive studies of the ecosystem of the White Sea]. Moscow, VNIRO, 1994. pp. 83–92. (In Russian).
14. Bobrov Yu.A., Maksimova M.P., Savinov V.M. Pervichnaya produktziya fitoplanktona. [Primary production of phytoplankton]. Biologicheskie resursy i problemy ih racional'nogo ispol'zovaniya. Ch. I (The White Sea. Biological Resources and Problems of Their Rational Use. Part I), St. Petersburg, ZIN RAN, 1995, pp. 92–114. (In Russian).
15. Neverova-Dziopak E. Otzenka troficheskogo sostoyaniya poverhnostnyh vod. [Assessment of the trophic state of surface waters]. Saint-Petersburg, SPBGASU, 2020. 176 p. (In Russian).
16. Ilyash L.V., Zhitina L.S., Fedorov V.D. Fitoplankton Belogo morya. [Phytoplankton of the White Sea]. Moscow, Yanus-K, 2003. 168 p. (In Russian).
17. Maksimova M. P. Gidrohimiya Belogo morya. Diss. dokt. geogr. nauk. [Hydrochemistry of the White Sea. Dr. Sci. (Geography) thesis]. Moscow, 1990. 51 p. (In Russian).
18. Netzvetaeva O.P., Makedonskaya I.Y., Korobov V.B., Zmetnaya M.I. Zavisimost' kislorodonasyshcheniya ot soderzhaniya hlorofilla «a» v poverhnostnom sloe vod Belogo morya. [Dependence of oxygen saturation on the "a" chlorophyll content in the surface layer of the White Sea]. Arktika: ekologiya i ekonomika. [Arctic: ecology and economy]. 2018, no. 3 (31), pp. 31–41. DOI: https://doi.org/10.25283/2223-4594-2018-3-31-41. (In Russian).

Features of the Formation of the Chemical Composition of the Hydrosphere of the Buzuluk Pine Forest

I. V. Kudelina, T. V. Leonteva, and M. V. Fatyunina

Abstract Buzuluk forest is a specially protected area with the status of a National Park. Hydrocarbon deposits have been explored in the region: Gremyachovskoye, Vorontsovskoye, Mogutovskoye, Elshanskoye, Zhuravlevskoye, Neklyudovskoye, Nikiforovskoye. The first three are located on the territory of the forest. Separate wells drilled outside the forest are operated. Settlements are located along the river valleys. They are supplied with water at the expense of the Upper Tatar complex and the Lower Pliocene-Holocene horizon, fed by atmospheric precipitation. Groundwater is practically not protected from pollution. Areas are identified from hazardous to permissible in terms of the content of harmful components, taking into account the lithological composition of the rocks, the results of standard logging, test pumping with measurements of temperature and water level until its restoration. The results of the conducted research should be used for planning and conducting monitoring work on a unique territory for its further prosperity.

Keywords Buzuluk forest · National Park · Uvalistaya plain · Upper Tatar complex · Lower Pliocene-Holocene horizon · Spring flood · Hydrocarbon deposits

1 Introduction

Buzuluk forest is a specially protected area with the status of a National Park [1]. It is located in the southeast of the Russian platform in the Buzuluk depression, in its central part, within the Buzuluk and Buguruslan regions of the Orenburg and Bor, Bogatovsky, Neftegorsk regions of the Samara region. The study area has an area 2477.7 km^2 (Fig. 1).

The territory is occupied mainly by pine forests, but in addition to pine, birch, alder, aspen and oak grow here. In the southern and partly northeastern parts of the territory, a steppe zone is developed with a herbage of a forb-fescue-feather grass association with a small admixture of wormwood.

I. V. Kudelina (✉) · T. V. Leonteva · M. V. Fatyunina
Orenburg State University, Orenburg, Russia
e-mail: kudelina.inna@mail.ru

V. I. Karev (ed.), *Physical and Mathematical Modeling of Earth and Environment Processes—2022*, Springer Proceedings in Earth and Environmental Sciences,
https://doi.org/10.1007/978-3-031-25962-3_60

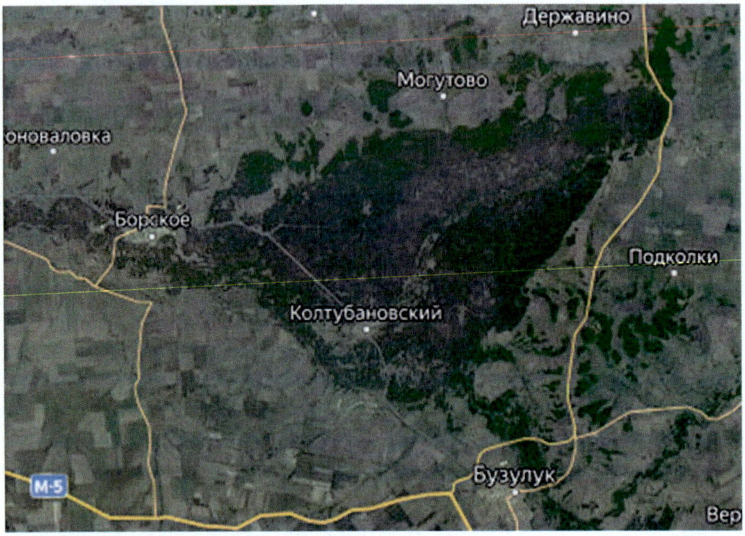

Fig. 1 The territory of the Buzuluk pine forest

Orographically, the area is confined to the ridged plain of the High Trans-Volga region [2]. It is dissected by deep river valleys, ravines and gullies and is inclined to the southwest. Relief marks decrease from the watershed of the rivers Borovka and Mal. Kinel (+ 265 m) to the northeast to + 51 m at the water's edge in the mouth of the river. Samara. Within the Buzuluk forest along the right bank of the river Borovki is a developed massif of eolian sands. They are semi-fixed, creating a cellular relief character. Its length is 16–20 km in the northeast direction. Here there are dunes with a relative height of up to 20 m. Hydrographically, the territory belongs to the basin of the river. Samara, the left tributary of the river. Volga, and only in the northeast is the river basin. Bol. Kinel.

2 Materials and Methods

In the course of the research, an analysis of materials collected in the field, in the funds of geological organizations, as well as literary sources was used. The methodological approach includes zoning and mapping of the territory, regime observations of the quality of natural waters, sources of pollution, the use of ground and remote methods of research.

3 Results and Discussion

Hydrocarbon deposits have been explored in the region: Gremyachovskoye, Vorontsovskoye, Mogutovskoye, Elshanskoye, Zhuravlevskoye, Neklyudovskoye, Nikiforovskoye. The first three are located on the territory of the forest. Separate wells drilled outside the forest are operated.

The territory is confined to the artesian East Russian basin of the 1st order, with reservoir type waters and to the basins: the Syrtovskaya of the 2nd order and the East Syrtov of the 3rd order. According to geodynamic indicators, 9 hydrogeological divisions have been identified: the aquifers of the Lower Pleistocene-Holocene and Lower Triassic, as well as the Upper Tatar complex and the impervious Upper Pliocene horizon. Domestic and drinking water supply is carried out here mainly due to the Upper Tatar aquifer and the Lower Pliocene-Holocene aquifer. They feed on atmospheric precipitation. The formation of groundwater is influenced by the physical and geographical features of the territory, the lithological composition of rocks, the degree of their drainage and depth, as well as the nature and speed of water circulation. The composition of the waters is formed as a result of leaching of rocks and their ion-salt complex, as well as the dissolution of gypsum and anhydrite. The vertical hydrogeodynamic and hydrogeochemical zonality is well traced in the section.

As a result of the research, the natural protection of the first aquifers from the surface was studied, as well as the possibility of penetration of pollutants from the underlying aquifers. An attempt was made to make an integral assessment of the degree of groundwater protection with the identification of three areas that differ in the degree of protection: weakly protected, relatively protected and protected [2, 3]. The alluvial horizon and the aeration zone are classified as weakly protected. This is evidenced by the results of studying the ecological and geochemical conditions of waters, soils and rocks in the aeration zone. The study of the concentrations of normalized components relative to the MPC made it possible to identify areas with moderately dangerous, hazardous and permissible content of harmful components, as well as areas from extremely and moderately hazardous to permissible content of normalized components. The waters of the Upper Tatar (Malokine-Kutuluk) aquifer complex are the most polluted. Of the pollutants, oil products and bromine predominate. As a result of the assessment of the geological environment, areas with a relatively favorable state, a very unfavorable and a relatively unfavorable one were identified [3, 4]. A very unfavorable condition is typical for the south-west of the territory, in the interfluve of Samara and Borovka, as well as within the watershed of the Borovka and Kutuluk rivers in the extreme north-west of the territory. A monitoring system has been proposed and substantiated, and recommendations have been issued for the protection and rational use of natural resources.

The ecological state of groundwater depends on both natural and man-made factors. According to the characteristics of natural environmental conditions, their suitability for household and drinking purposes is judged in accordance with SaNPiN 2.1.4.1074-01 [5, 6]. The suitability of water is assessed according to the following indicators: dry residue, content of sulfates, chlorides, nitrates, hardness parameters, concentration of hydrogen ions, etc.

The state of water in aquifers generally meets the requirements of SaNPiN, that is, the normalized indicators are below the Maximum Permissible Concentration (MPC), with the exception of the content of Fe and Mn. The source of iron in the waters are Perm red flowers. When the reducing environment of the oxidizing environment changes, iron ions (Fe^{2+}) begin to migrate freely [7, 8], as well as manganese ions (Mn^{2+}). They are common in the Lower Triassic and Neopleistocene-Holocene deposits due to the presence of organic matter.

The concentration of iron in the Upper Tatar aquifer is up to 3 MPC (well 50), and in the well. 53 of the lower Neo-Pleistocene-Holocene horizon reaches 1.85 mg/l, i.e. 6 MPC.

In areas where the Lower Permian deposits lie close to the surface, the concentration of iron reaches 88 MPC. Elevated concentrations of iron are found in almost all wells. Even in single wells in the southern part of the study area, its content is from 2 to 15 MPC. And at the Krasnogvardeisky water intake, the iron content in the waters of the eolian-alluvial lower-Neopleistocene-Holocene horizon, which provides water to the city of Buzuluk, the iron concentration exceeded the MPC by 29 times. The iron comes from the underlying Upper Permian rocks. With an increase in water withdrawal, the content of iron in water increases. This indicates that when drilling production wells, it is necessary to avoid opening the Permian deposits.

Elevated concentrations of manganese (0.1–0.35 mg/dm^3, i.e. up to 3 MPC, and at some points up to 8 MPC) were found in the waters of the Lower Triassic and the aeolian-alluvial horizon, which is due to the presence of organic matter in the form of fauna, buried soils and plant roots. At the Krasnogvardeisky water intake, manganese concentrations exceed the MPC by 8–10 times (wells 16 and 28). The increased oxidizability in the waters of the western part of the study area (5.4–7.2 mgO/dm^3, which is somewhat higher than the MPC (5 mgO/dm^3) is also associated with organic matter in the rocks of the Triassic and Jurassic).

In the waters of individual wells in the waters of the eolian-alluvial horizon, the MPC in terms of hardness was exceeded up to 10.8 mmol/l at the Krasnogvardeisky water intake and 10.8–20.4 at the water intake of the city of Buzuluk with a MPC of 7 mmol/l. It is located in a residential area. In the waters of the Lower Triassic horizon, the total volumetric activity of alpha radionuclides was determined in the amount of 0.2–0.56 Bq/l. It is possible that this is due to the presence of organic matter in the sediments, which contributes to the creation of a reducing environment with the formation of a gley barrier concentrating radionuclides [9].

The background and elevated concentrations of toxic substances in the aeration zone, in soils, rocks and bottom sediments, as well as their relationship with water pollution and vegetation cover, were determined. Concentrations above MPC were obtained for chromium, barium and nickel. These elements have an impact on the life of people, but do not exceed the background in the rocks of this territory. Areas with hazardous, moderately hazardous and extremely hazardous concentrations of these components, as well as areas with moderately hazardous, hazardous and permissible concentrations of normalized components were identified.

The highest concentrations of chromium, barium and nickel (hazard classes II and III) are associated with the Upper Tatar and Lower Triassic deposits, which are distinguished by their increased background. Alluvial and eolian deposits were formed as products of their destruction.

Chromium throughout the entire territory, except for the south-west of the reserve with Neogene deposits, is characterized by high and extremely dangerous and hazardous concentrations in soils and rocks. But its concentrations differ: in eolian deposits they exceed the MPC by 8–10 times, and in the Triassic—by 4–6 times; in Permian deposits it is slightly less (by 3–5), and in Neogene deposits it is only 2–3 times. In bottom sediments, MPC for chromium was taken as 100 mg/kg at Clarke in soils, according to Vinogradov, – 200 mg/kg. In bottom sediments, it is widely distributed throughout the territory at concentrations from 400 to 1500 mg/kg. This is 4–15 times higher than the MPC. Its maximum content of 1500 mg/kg was established in the silts of the river Koltubanks. Elevated concentrations of chromium in the regional plan are due to the genesis of rocks and man-made emissions in connection with the operation of thermal power plants and railways.

Barium is a product of eolian deposits and accumulates during the weathering of the Lower Triassic and Upper Permian rocks. Its concentrations reach 4–5 MPC, and even 7 MPC, that is, a moderately hazardous and dangerous level, which is much higher than within other breeds, where its concentration is moderately hazardous, not exceeding 1.5–2 MPC. And only on the eastern and western outskirts of the city of Buzuluk, at two points, its concentrations exceed the MPC by 10 and 20 times. In bottom sediments, the MPC of barium is taken as 200 mg/kg, and in all watercourses of the reserve, its content is 1.5–4.0 times higher than the MPC. Clark of barium in the earth's crust according to Vinogradov is 500 mg/kg.

Nickel is evenly distributed in moderately hazardous concentrations exceeding the MPC by 1.5–2 times. A slight increase in its concentration up to 3 MPC was found in the soils of the high floodplain along the right bank of Samara and up to 4 MPC in the area of the village of Gremyachiy on the right bank of the river. Borovki. In bottom sediments, the nickel content averages 50 mg/kg, with MPC 45 mg/kg. In all rivers, the content of nickel in bottom sediments varies from 30 to 60 mg/kg, with a minimum content in the river Koltubank (30–40 mg/kg).

The wide development of the hydrographic network ensures the removal of microelements from watersheds and slopes and accumulation in bottom sediments [9]. Therefore, as in bedrock, the content of chromium, nickel, and barium also increases in bottom sediments.

The growth of the technogenic load on the geological environment within the Buzuluk pine forest significantly transforms the geological environment. This is due to a large number of wells for various purposes and other economic activities of people. As a result, the soil layer is disturbed, the vegetation dies off, the quality of natural waters changes. The negative impact of work on oil and gas on the natural complex of boron [10, 11] and the formation of man-made landscapes with "man-made solonchaks" and the absence of vegetation have been proven.

According to the negative impact of oil wells on the forest and its ecosystem, three zones have been outlined: (1) successional overgrowth of deforestation around the mouths of wells with chemical pollution of the geological environment, water bodies and atmospheric air with technological waste with the formation of man-made landscapes contaminated with construction debris and metal structures. (2) In the second zone, treelessness is formed around the wells [12, 13]. Its area does not exceed 170–200 ha (0.3% of the forest area), but is distinguished by anthropogenic-successional plant communities. (3) The third zone is the underground environment under the exhausted pine forest. When wells are put into production, reservoir pressure changes, setting the entire ecosystem in motion, and changing not only the environment near oil fields, but also areas of the boron remote from the wells. When opening the productive layers of a number of fields, oil flowed with leakage of drilling fluid and produced oil. The surface of the earth, the aeration zone and aquifers were polluted [14].

In all wells of the tested horizons, MPC for oil products was exceeded up to 5 times and for bromine up to 9 MPC [3]. In some wells in the Lower Pliocene horizon, MPC for oil products was exceeded by 20–22 times, and for bromine by 3 times. The environment is most polluted in the central part of the paleovalley. The transport of pollutants and their accumulation in the center of the paleovalley is favored by sands and pebbles of the basal layer in the Neogene rocks and a large flow slope in the overdeep part of the valley. In surface waters, the concentrations of oil products do not exceed MPC.

Huge damage to the ecology of the Buzuluk pine forest was caused by pesticides (DDT), which were widely used until 1985 in the fight against forest pests. As a result, the animal world was on the verge of extinction. The livestock of animals and birds has not yet been restored, and in places the smell and taste of dust is felt in plants and mushrooms. Where there were warehouses of pesticides within the boron, the content of DDT in soils is 8 times higher than the MPC, and in the waters of the Upper Pleistocene-Holocene horizon near the former warehouse of pesticides, the content of nitrates is 2 times higher than the MPC, and the content of DDT is 0.002 mg/dm^3. Within the city of Buzuluk and its industrial enterprises, soils and soils contain elevated concentrations of lead, zinc and barium.

4 Conclusion

The pine forests of the Buzuluk forest have the status of a National Park. They are confined to the river basin. Samara and within the boron, groundwaters of the Upper Tatar complex and the Lower Pliocene-Holocene horizon are developed. They are practically immune to contamination. According to the content of harmful components within the boron, areas are distinguished from dangerous to permissible. Inspection of technogenic objects makes it possible to substantiate the network of monitoring observations and issue recommendations for the protection and rational use of water resources. Ecological studies help to characterize the ecological conditions of the territory by zoning it with the construction of a set of ecological maps. It became possible to evaluate and specify the role of technogenic load in the transformation of the geological environment. The most important tendencies of anthropogenic transformations in the Buzuluk pine forest with the formation of technogenic landscapes are revealed: with "technogenic solonchaks" and the death of vegetation. Three zones of the negative impact of oil exploration on the environment have been identified: (1) successional overgrowth of deforestation; (2) the formation of treelessness in an area of up to 200 ha around the wells; (3) negative changes in the underground environment under technogenically disturbed boron.

The results of the conducted research should be used for planning and conducting monitoring work on a unique territory for its further prosperity.

References

1. Klimentiev, A.I. Buzuluk pine: soils, landscapes and geographic environmental factors // Ekaterinburg: Ural Branch of the Russian Academy of Sciences, 2010. - 401 p.
2. Kovtun, S.Yu. Geoecological problems of the Buzuluk pine forest // S.Yu. Kovtun, V.P. Petrishchev // Vestnik Orenb. State University - 2007. - No. 67. - S. 207–214.
3. Ivanov, V.V. Ecological geochemistry of elements. M., Nedra, 1994, 304 p.
4. Lukinykh, E.N. Assessment of the state of the well stock in the area of the OTsLM "Buzuluksky Bor". E.N. Lukinykh, A.M. Pampushka, G.M. Vavilina and others. Report on the contract 303. Orenburg, 2000. - 100 p.
5. Chernova O.A. Information risks in the management of the region's water management complex // Upravlenets. - 2018. - T. 9. - No. 5. - P. 40–47.
6. Management of the efficiency of water supply of the regional system / Kosolapov A.E., Matveeva L.G. // Regional economy. South of Russia. - 2018. - No. 4. - P. 170–181.
7. Levchuk A.A., Barabash A.Yu., Aleksandrova A.V., Levchuk A.A. Aspects of environmental safety management in the water management complex / A.A. Levchuk [et al.] // Bulletin of Eurasian Science. - 2018. - No. 2. - Access mode: https://esj.today/PDF/44NZVN218.pdf.
8. Petrishchev, V.P. Peculiarities of anthropogenic transformation of the landscapes of the Buzuluksky Bor National Park / Petrishchev V.P., Kovtun S.Yu., Yudichev E.N. // Proceedings of the Orenburg State Agrarian University, 2011. - Vol. 1, No. 29-1. - S. 220–224. - 5 s.
9. Petrishchev V.P., Chibilev A.A. Landscape-ecological study, landscape photography and mapping sheet N-39-XXIX and Buzuluk pine forest. Final report on topic No. 6 (02/01/2000–12/31/2001). Orenburg. 2003.
10. Chibilev, A.A. National Park "Buzuluksky Bor" - a natural phenomenon of the steppe zone of the Orenburg region / A.A. Chibilev, P.V. Velmovsky, N.O. Kim et al. // Orenburg: Institute of the Steppe of the Ural Branch of the Russian Academy of Sciences, 2008. - 60 p.
11. Protected natural areas of the Orenburg region: cannot be liquidated? / Chibilev A.A., Petrishchev V.P., Udovenko I.N., Yakovlev I.G. // Actual trends of regional and local development: Sat. Art. based on materials II (X) Intern. conf., May 14–15, 2015, Samara / City administration. env. Samara, Municipal. budget. educate. institution of higher education "Samar. academic. state. and municipal. upr."; ed. V.K. Semenycheva. - Electron. Dan. - Samara: SAGMU, 2015. - S. 304–308. - 5 s.
12. Jie C, Qiwei H, Yaling L, et al. Hydrogeochemical Characteristics and Quality Assessment of Groundwater in an Irrigated Region // Northwest China [J]. Water, 2019, 11(1).

13. Deng, X. Risk assessment of sources of groundwater pollution, groundwater ulnerability, and groundwater value: A case study in the plain area of Northeast Beijing / Xipeng Deng, Yanying Li, Xiao Fan, Mingtan Zhu, Yujiao Tu, Li Qin // Research Square, September 2022. DOI: https://doi.org/10.21203/rs.3.rs-2075440/v1.
14. Xue L, Changyuan T, Yingjie C, et al. A multiple isotope (H, O, N, C and S) approach to elucidate the hydrochemical evolution of shallow groundwater in a rapidly urbanized area of the Pearl River Delta, China [J]. Science of the Total Environment, 2020, 724 (prepublish).

About the Possible Relationship of the Iceland Hot Spot and the Surrounding Ocean Bottom Relief with the System of Intense Intra-mantle Vortexes

S. Yu. Kasyanov

Abstract The paper proposes a hypothesis that the Icelandic hot spot with a mantle plume was formed as a result of the contact of an intense intra-mantle vortex with a liquid core. In the area of contact with the stationary liquid core, the rapidly rotating substance of the vortex is decelerated and heated. As a result, a jet of hot mantle matter arises above the contact area between the core and the vortex, ascending to the ocean lithosphere and forming the Icelandic hot spot. The spreading of hot matter rising from the mantle under the oceanic lithosphere leads to the formation of a boundary layer in which the Görtler instability occurs with the formation of rolls with axes oriented along the direction of the main flow. The movement of matter in the structure of swells in the boundary layer under the lithosphere forms the corresponding topography of the ocean floor surrounding the Icelandic hot spot.

Keywords Icelandic hot spot · Mantle plume · Earth's liquid core · Oceanic lithosphere · Rift zone · Intense vortex · Intramantle vortex · Convective jet · Boundary layer · Görtler instability · Taylor vortices · Görtler vortices · Gravity anomalies · Two-phase flow

1 Introduction

The Icelandic hotspot is one of the most geologically active places on Earth. The concept of the formation mechanism of the Icelandic hot spot with a mantle plume (see, for example [1]) and the complex structure of the bottom topography of the surrounding ocean area has not been finally established. In this paper, we propose a hypothesis that the Icelandic hotspot with a mantle plume was formed due to the contact of an intense intra-mantle vortex with a liquid core. In the area of contact with the stationary liquid core, the rapidly rotating substance of the vortex is decelerated and heated. As a result, a convective jet of hot mantle matter arises above the area of contact between the core and the vortex, ascending to the ocean lithosphere and

S. Yu. Kasyanov (✉)
N. N. Zubov's State Oceanographic Institute, Roshydromet, 6, Kropotkinskiy Lane, 119034 Moscow, Russian Federation
e-mail: skas53@yandex.ru

V. I. Karev (ed.), *Physical and Mathematical Modeling of Earth and Environment Processes—2022*, Springer Proceedings in Earth and Environmental Sciences,
https://doi.org/10.1007/978-3-031-25962-3_61

forming the Icelandic hot spot. The hot matter rising from the mantle spreads in the boundary layer formed under the ocean lithosphere, in which the Görtler instability occurs with the formation of rolls with axes oriented along the main flow. The movement of matter in the swells in the boundary layer under the lithosphere forms the structure of the bottom topography of the ocean area surrounding the Icelandic hot spot.

2 Hypothesis About the Formation of the Icelandic Hot Spot During the Contact of an Intense Intramantle Vortex and Liquid Core

According to our earlier hypothesis [2–5], a quasi-stationary system of intense vortices has formed in the Earth's mantle, in which the hot mantle matter moves at high speeds in a two-phase gas–liquid state. In this system of eddies, one can distinguish a system of three energy-carrying intense hot large-scale eddies (Fig. 1), which includes eddies (1) Himalaya + Tien Shan—Kolyma, (2) Kolyma—Yellowstone and (3) Yellowstone—Mediterranean Sea. The vortex (1) comes into contact with the liquid core of the Earth under Lake Baikal, which manifests itself in the form of the Baikal rift [4].

The axis of the vortex (3) currently passes under Iceland (Fig. 1), where the geologically active Icelandic hotspot is located. We assume that the vortex (3) as well as the vortex (1) is in contact with the liquid core of the Earth and the result of the contact is manifested on the day surface in the form of a relief of the lithosphere in the vicinity of the Icelandic hot spot. Upon contact with the liquid core, the substance of the vortex slows down, heats up and partially evaporates. According to [4], the density of matter in vortex (3) can be estimated as 4180 kg/m^3.

Thus, in contrast to vortex (1), the density of matter at the periphery of vortex (3) is comparable to the density of the surrounding mantle. Therefore, the heating of the vortex substance (3) leads to the fact that the density of the vortex substance (3) becomes less than the density of the mantle surrounding it, and a powerful convective jet of hot substance is formed in the mantle in the vicinity of the contact point, ascending to the oceanic lithosphere. Note that due to the high density of matter at the periphery of the vortex (1), a similar phenomenon is not observed in the mantle under Lake Baikal. The rotation of the vortex gives the ascending hot jet a horizontal velocity component directed to the southwest.

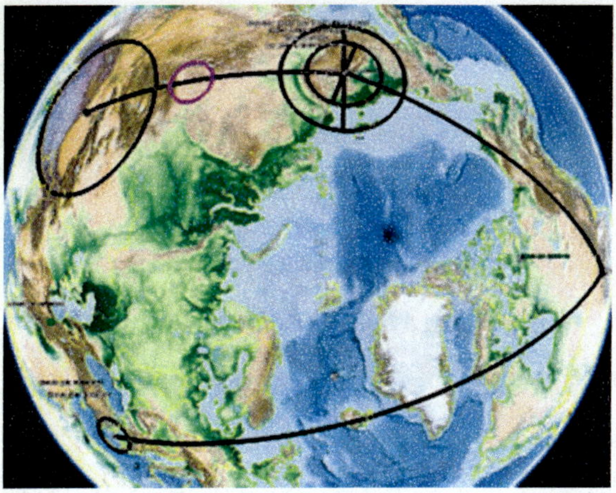

Fig. 1 A system of three energy-carrying intense large-scale vortices [4]. Eddies and estimates of velocity and density values at the lateral boundaries of eddies: (1) Himalayas + Tien Shan—Kolyma 4131 m/s, 11,800 kg/m^3, (2) Kolyma—Yellowstone – 1245 m/s, 4600 kg/m^3, (3) Yellowstone—Hudson Bay—the eastern part of the Mediterranean Sea (Greece) 2941 m/s, 4180 kg/m^3. Estimates of the values of velocity and density in eddies were obtained from the jump in geophysical parameters in 1997–1998 [4]. The drawing was made by the author using the Google Earth Pro program and its cartographic basis [6]

3 The Hypothesis of the Rotation of the Eurasian Plate

In a report made in 1995 at the Lomonosov Readings at the Faculty of Physics of Moscow State University, combined with the fiftieth anniversary of the Department of Geophysics of the Physics Department of Moscow State University, the head of the Department of Geophysics, Academician Vladimir Magnitsky reported the following results, obtained by him when processing the data of the very first edition of Ulugbek's zij of 1437: since the time of astronomical observations, the data of which are given in Ulugbek's zij, the Eurasian plate has rotated at a grate angle, significantly exceeding that which could be given by possible Ulugbek's measurement errors.

Following the fact of rotation of the Eurasian plate established by Magnitsky, we put forward a hypothesis according to which the gyroscopic moment of the vortex (3), partially located in the body of the North American continent, led to the rotation of North America when sliding along the melt layer in the asthenosphere relative to the underlying mantle. As shown earlier [2], the melt layer, bounded from below by the modern seismic boundary of 220 km, was formed at the site of the modern asthenosphere layer, which has partially solidified by now, under the influence of mantle tides caused by the revolution of a large temporary Earth satellite.

According to the hypothesis put forward, at the initial moment, Eurasia was pressed against Greenland. North America, being linked in the region of Beringia with Eurasia, brought it into rotation relative to the pole of rotation, located south of the Caspian Sea (Fig. 2). The position of the pole of rotation of Eurasia together

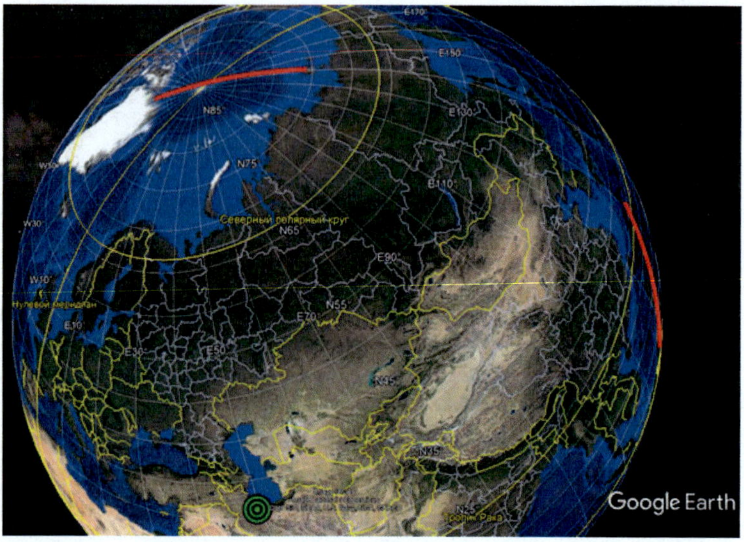

Fig. 2 Rotation of Eurasia by 25.38° around the pole of rotation with coordinates 35.88° N, 49.389° E (green circle). Pairs of points are connected in red, which were aligned with each other before the turn of Eurasia. Point on the shore Greenland was combined with the coast of the New Siberian Islands. Point on the shore Borneo was combined with the point on the coast of Minamidaito island. The drawing was made by the author using the Google Earth Pro program and its cartographic basis [6]

with Africa is determined by the alignment of the points on the shores of Borneo and Minamidaito islands and points on the coasts of the bay of Kotelny island and the Cape in Greenland (see Fig. 2). At the same time, it turns out that Eurasia has rotated clockwise by an angle of about 25.38°. Note that such a turn of Eurasia geometrically corresponds to the opening of the Arctic with the formation, in particular, of the Lomonosov Ridge between points on the Kotelny island and the Cape in Greenland, as well as the Gakkel Ridge.

Thus, two interconnected rotations of the continents took place—the rotation of North America and the rotation of Eurasia as a whole together with Africa. The rotation of North America relative to Eurasia, together with Africa, took place around a mobile pole of rotation, rigidly connected with Eurasia and located south of Beringia. At the initial moment, before the start of the turn, Eurasia was pressed against Greenland. At the same time, North America was rotated clockwise by about 34° around its mobile pole of rotation, which moved along with Eurasia. In the present position, the pole around which North America rotated lies near the point (49° N 170° E). The angle of rotation of North America is determined from the angle of opening of a characteristic trace in the lithosphere of the Pacific Ocean, coming from an area with a weakened lithosphere located near the 176° E meridian (Fig. 3).

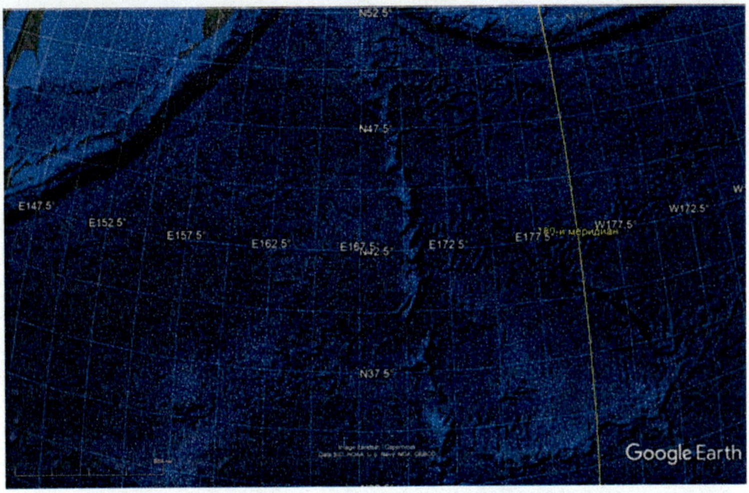

Fig. 3 A characteristic trace in the lithosphere of the Pacific Ocean with an opening angle of about 34°, coming from an area with a weakened lithosphere located near the 176° E meridian. The drawing was made by the author using the Google Earth Pro program and its cartographic basis [6]

4 Formation of the Icelandic Hot Spot and the Topography of the Surrounding Ocean Floor Under the Influence of a Hot Jet from the Area of Contact Between the Intramantle Vortex and the Liquid Core

Let us repeat that in the current position of Eurasia, the axis of vortex 3 passes under Iceland island. In the initial position, that is, before the turn of Eurasia, when Eurasia was pressed against Greenland, the axis of the vortex (3) passed along the southwestern coast of the Svalbard archipelago (Fig. 4). Perpendicular to this position of the vortex axis 3 is directed along the axial zone of the Mona Ridge, which is part of the Mid-Atlantic Ridge. Passing through Iceland, it then follows the Reykjanes Ridge.

The substance of the vortex is heated due to deceleration upon contact with the substance of the liquid core. Therefore, a jet of hot matter arises above the area of contact between the vortex and the liquid core, ascending to the day surface and leading to the formation of the Icelandic hot spot.

The hot jet rising from the area of contact between the vortex and the liquid core to the ocean floor reaches the oceanic lithosphere and is partially decelerated. As a result, an elevation of the ocean floor relief appears above the region of jet deceleration, which corresponds to the region of positive gravity anomalies in free air in Fig. 5, which is less expressed in the field of isostatic gravity anomalies in Fig. 6.

The largest uplift occurs in the region corresponding to the central part of the jet, which is formed near the center of the region of contact between the vortex and the

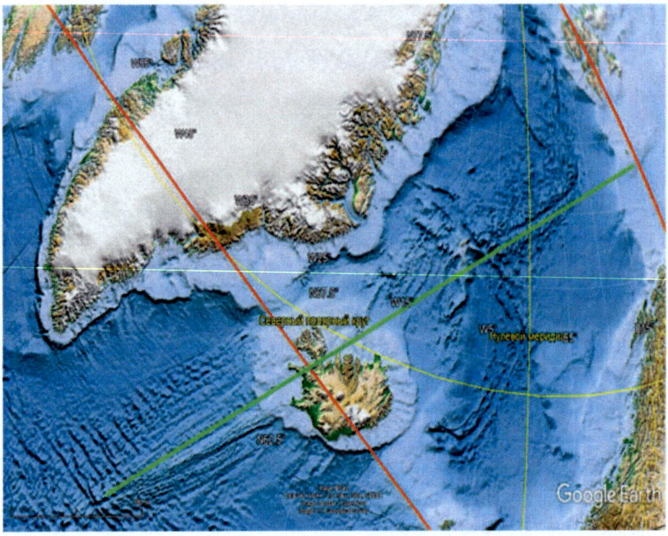

Fig. 4 The current position of the axis of vortex 3 (orange line on the left) relative to Eurasia and Greenland and the position of the axis of vortex 3 relative to Eurasia and Svalbard archipelago before the Eurasian turn (orange line on the right). The drawing was made by the author using the Google Earth Pro program and its cartographic basis [6]

liquid core. At present, this uplift corresponds to the area of Iceland, which turned out to be above the center of the ascending jet during the final turn of Eurasia to its present position, and therefore rose above the surface of the ocean.

In the field of gravitational anomalies, a wedge-shaped the Reykjanes Ridge stretching from Iceland island is clearly visible. It corresponds to the characteristic structure of a fluid flow around an obstacle in a horizontal boundary layer. This applies to both free air anomalies (Fig. 5) and isostatic anomalies (Fig. 6).

The ascending flow, resting against the oceanic lithosphere, generates a horizontal current with a maximum speed under the Iceland. This flow goes around the solidified area under the uplifted earlier lithosphere of Iceland from both sides and forms a wedge-shaped structure, which is clearly visible in the bottom topography. The axis of the wedge corresponds to the axis of the horizontal current in the boundary layer under the oceanic lithosphere, flowing around Iceland from the southwest and northeast. Both flows join where the width of the central wedge formed behind the obstruction in the form of a solid lithosphere Iceland island, decreases to zero.

A jet of hot matter ascending from the area of contact between the intramantle vortex and the liquid core to the day surface causes the divergence of matter on the ocean floor, the formation of faults, and the separation of the formed parts of the plate. The movement of Eurasia over the ascending hot jet led to the formation of extended faults in the northern part of the Mid-Atlantic Ridge. The movements of Eurasia occurred unevenly, by means of several fast turns, separated by periods of stationary positions of Eurasia, reflected in the fields of gravitational anomalies.

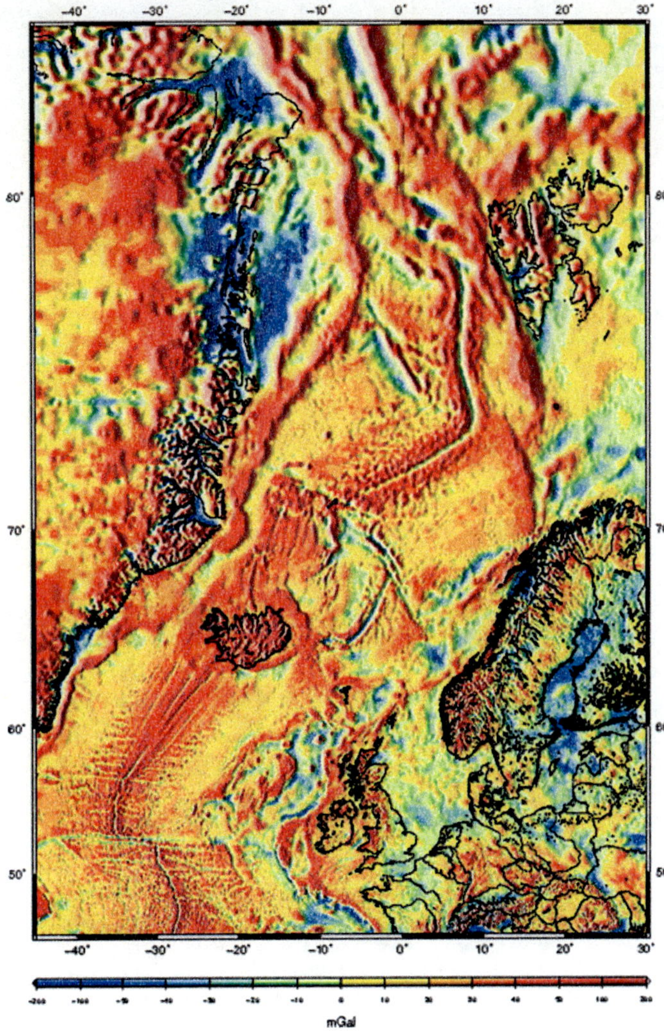

Fig. 5 WGM2012—surface free air anomaly (mGal) [7]

The consequences of the contact between the intramantle vortex and a denser liquid core are manifested in the field of gravitational anomalies in Figs. 5 and 6 in the form of traces near the Svalbard archipelago, characteristic of the intersection of the regions of a spherical liquid core and a cylindrical vortex. The substance of the counterclockwise rotating vortex 3 (when viewed from the end of the vortex) displaces the denser substance of the liquid core from the area of intersection of the spherical liquid core and the cylindrical vortex. As a consequence of this, according to isostasy, an elevation of the relief of a characteristic shape is formed on the day surface. With the subsequent rotation of Eurasia, the upper layer of the mantle,

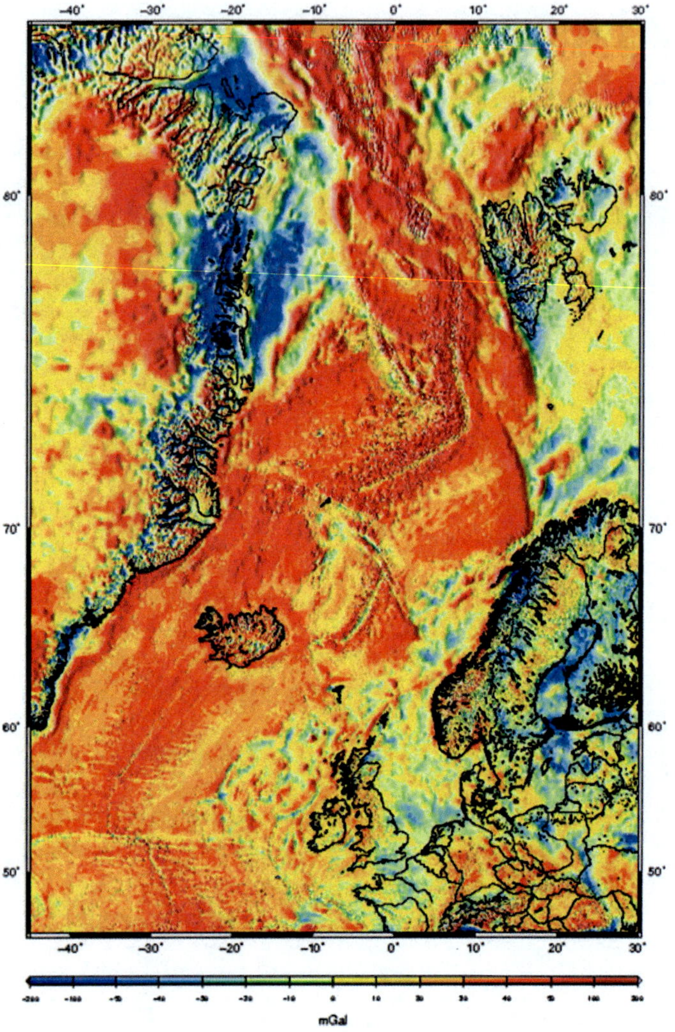

Fig. 6 WGM2012—complete spherical isostatic gravity anomaly (Airy-Heiskanen, Tc = 30 km) [7]

together with the elevation in the relief of the lithosphere, shifts, and in a new position, the resulting elevation of the relief turns out to be above the undisturbed spherical surface of the liquid core. Therefore, over this elevation of the relief, there is an excess of mass and the corresponding positive gravitational anomalies in free air (Fig. 5). **In the field of isostatic gravity anomalies, the positive perturbation of gravity near** the Svalbard archipelago, which has a shape characteristic of the intersection of a sphere and a cylinder, is even more noticeable (Fig. 6).

The hot matter, having risen in an ascending flow from the area of contact between the vortex and the liquid core, continues its movement in the horizontal boundary layer under the oceanic lithosphere. The flow boundary layer under the lithosphere is a special case of the boundary layer on a concave plate, in which Görtler vortices are formed (see Fig. 7). Görtler vortices are similar to Taylor vortices in the cylindrical Couette current (see Fig. 7) and form the topography of the ocean floor like swells with axes along the direction of the main flow.

The movement of hot matter occurs in a spiral in the structure of hydrodynamic swells under the thin oceanic lithosphere. With this movement, the hot substance gradually cools, becomes denser, and at a certain moment sinks down, closing the circulation cell. The area of subsidence of matter appears on the ocean floor in the form of a fault oriented across the direction of the rift. The extent of structures such as swells in the horizontal boundary layer under the oceanic lithosphere can significantly exceed the diameter of the intramantle vortex, which serves as the source of their movement.

The direction of the main flow in the boundary layer under the lithosphere relative to the lithosphere of Eurasia changed with the change in the position of Eurasia

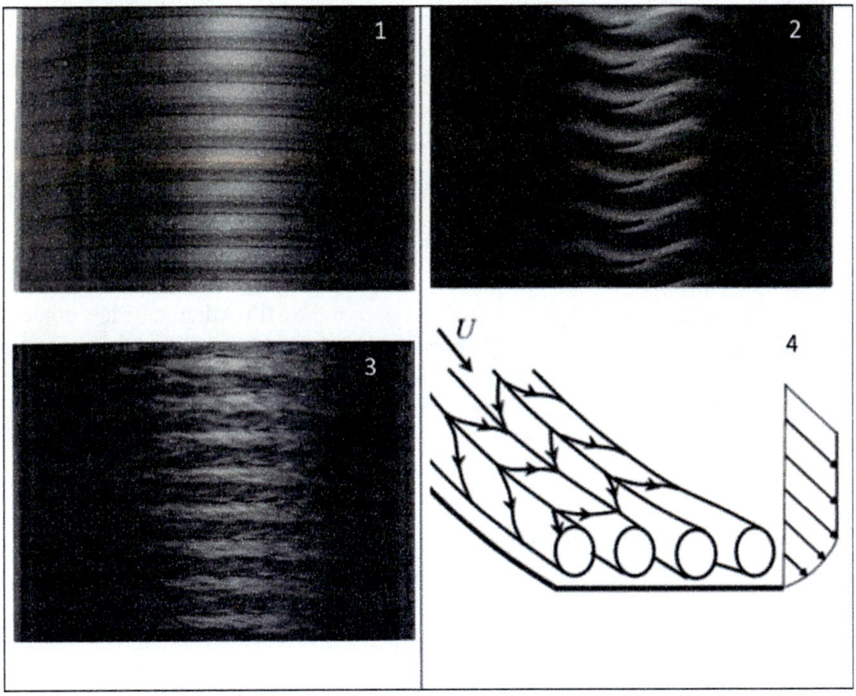

Fig. 7 Taylor vortices in the cylindrical Couette flow (1–3) [8] and Görtler vortices in the boundary layer on a concave plate (4) [9]. Hierarchy of instabilities in Taylor instability: 1—formation of laminar Taylor vortices, 2—appearance of waves on laminar Taylor vortices, 3—turbulent Taylor vortices

relative to the intramantle vortex (3) and was different in different stationary positions of Eurasia. In the stationary positions of Eurasia, the updrafts over the area of contact between the vortex and the liquid core propagated in the mantle up to the oceanic lithosphere. They formed under it hydrodynamic structures such as swells with axes oriented along the directions of the main flow in the boundary layer under the lithosphere, corresponding to the stationary positions of Eurasia, and caused the formation of characteristic ridges and depressions of the rift zones of the northern part of the Mid-Atlantic Ridge on the ocean floor. This applies, in particular, to the rift zones of the Reykjanes and Mona ridges.

Of particular interest is the difference between free-air gravity anomaly patterns in the rift zone of the Reykjanes Ridge and in the rift zone of the Mona Ridge. The smooth pattern of free air anomalies in the rift zone of the Reykjanes Ridge shows that the rift zone of the Reykjanes Ridge could be formed by Taylor laminar vortices (photo 1 in Fig. 7) or Görtler. At the same time, the irregular structure of anomalies in the rift zone of the Mona Ridge, similar to the flow pattern in turbulent Taylor eddies (photo 3 in Fig. 7), indicates turbulent motion in the boundary layer created by an intramantle eddy before or at the very beginning of the Eurasian turn. This allows us to speak of a decrease in the velocity of the large-scale intramantle vortex with time.

5 Conclusion

The formation of the Icelandic hot spot and the topography of the ocean floor surrounding it could have occurred due to the contact of one of the eddies in the system of three energy-carrying intense large-scale intramantle vortices with the liquid core of the Earth.

The Mona and Reykjanes Ridges, which are part of the Mid-Atlantic Ridge, could be formed by a jet of heated mantle matter rising from the central part of the contact area of a large-scale intra-mantle vortex with the Earth's liquid core.

Flows of heated mantle matter rising to the lithosphere from the area of contact of a large-scale intra-mantle vortex with the liquid core of the Earth form a movement in the mantle with a horizontal boundary layer of friction under the lithosphere, in which Görtler instability occurs.

The relief of the bottom of the rift zones of the ocean area adjacent to Iceland could be formed by a system of Görtler vortex oriented along the Mid-Atlantic Ridge in the boundary layer under the lithosphere.

References

1. Rickers, Florian; Fichtner, Andreas; Trampert, Jeannot (1 April 2013). The Iceland–Jan Mayen plume system and its impact on mantle dynamics in the North Atlantic region: Evidence from full-waveform inversion. Earth and Planetary Science Letters. 367: 39–51. Bibcode: 2013E&PSL.367...39R. https://doi.org/10.1016/j.epsl.2013.02.022
2. Kasyanov S, Samsonov V (2018) Effect of a tidal wave caused by large gliding satellite on formation of 220 km seismic boundary and split of the mantle into blocks. In: Karev V, Klimov D, Pokazeev K (eds) Physical and mathematical modeling of earth and environment processes. PMMEEP 2017. Springer Geology. Springer, Cham, pp 360–370. https://doi.org/10.1007/978-3-319-77788-7_37
3. Kasyanov, S.Yu. (2021). Formation of System of Intense Vortices in the Mantle When a Large Temporary Earth Satellite is Immersed. In: Chaplina, T. (eds) Processes in GeoMedia—Volume III. Springer Geology. Springer, Cham, pp 217–232. https://doi.org/10.1007/978-3-030-69040-3_21
4. Kasyanov, S.Y., Samsonov, V.A. (2021). About the Modern System of Three Energy-Carrying Intensive Vortices in the Earth's Mantle. In: Chaplina, T. (eds) Processes in GeoMedia - Volume II. Springer Geology. Springer, Cham, pp 273–291. https://doi.org/10.1007/978-3-030-53521-6_31
5. Kasyanov, S.Yu. (2022). Hypothesis of a Possible Cause of Warming in the Arctic Due to Energy Dissipation of Intensive Two-Phase Vortices in Earth's Mantle. In: Chaplina, T. (eds) Processes in GeoMedia—Volume IV. Springer Geology. Springer, Cham, pp 253–269. https://doi.org/10.1007/978-3-030-76328-2_26
6. https://www.google.com/earth/about/versions/#download-pro
7. **Bonvalot, S., Balmino, G., Briais, A., Kuhn, M., Peyrefitte, A., Vales N., Biancale, R., Gabalda,** G., Reinquin, F., Sarrailh, M., 2012. World Gravity Map. Commission for the Geological Map of the World. Eds. BGI-CGMW-CNES-IRD, Paris. http://bgi.obs-mip.fr/data-products/outils/wgm2012-maps-visualizationextraction/
8. **Van Dyke, Milton. An album of fluid motion. Stanford, Calif.: Parabolic Press, 1982, 176 p.:** LC: 81083088, ISBN: 0915760037 (hdc), ISBN: 0915760029 (pbk)
9. Egorov A.G. Lectures on hydrodynamic stability: textbook. Kazan: Kazan State University, 2009. - 170 p

Investigation of Vortex Flows Formed When Ice Melting in Multicomponent Liquids

T. O. Chaplina and O. A. Glebova

Abstract The results of experimental studies of the spontaneous rotation of ice blocks of different sizes placed both on a solid surface (aluminum, polymethyl methacrylate, glass, foam plastic, ceramics) and on the surface of a pool with water of a certain depth at a given temperature are presented. Observations have shown that if an ice block is placed on the water surface, a pronounced vertical flow is formed under the ice-water interface, which causes rotation. During the experiments, it was found that the speed of rotation of the ice block increases with the temperature of the water on the surface of which the block is placed. It has been experimentally established that with an increase in the mass of an ice block, its rotation speed increases. All observed flow patterns are stably reproduced by repeating the experiments and maintaining the conditions within the accuracy of the experiments.

Keywords Ice · Vortex flow · Convection · Arctic · Ice melting

1 Introduction

Horizontal ice-water interfaces are ubiquitous in our country and are typical for the winter period, as well as throughout the year for the polar regions. Such boundaries are located at the bottom of ice shelves, as well as under sea, river and lake ice. It should be noted that the melting of ice shelves along the lower edge in recent decades has led to a significant increase in the mass flow from glaciers to the ocean. Of interest are the currents that arise at the water-ice boundary in various natural and man-made systems.

The melting of ice shelves along the lower edge in recent decades has led to a significant increase in the flow of mass from the glaciers into the ocean. The study of the mechanisms that influence the formation of currents and processes, mixing in water bodies in different periods of the year is one of the most important areas of hydrophysical research. In recent years, more attention has been paid to studying

T. O. Chaplina (✉) · O. A. Glebova
Ishlinsky Institute for Problems in Mechanics RAS, 119526 Moscow, Russia
e-mail: tanya75.06@mail.ru

V. I. Karev (ed.), *Physical and Mathematical Modeling of Earth and Environment Processes—2022*, Springer Proceedings in Earth and Environmental Sciences,
https://doi.org/10.1007/978-3-031-25962-3_62

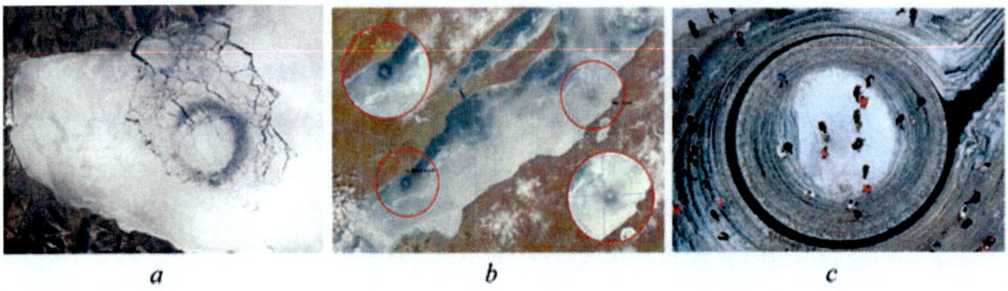

Fig. 1 Ice circles on Baikal (a, b) [4] and in the center of the Liao River (c) [5]

the real picture of ice melting dynamics in connection with the development of oil production and transportation in high latitudes—the Arctic and Antarctic.

In 2009, satellite images were the first to reveal giant rings in the ice cover on the Baikal ice [1]. And in subsequent years, such rings were regularly recorded on satellite images on the ice of Lake Baikal. A careful study of satellite images from previous years showed that the rings in the images appeared even earlier, starting from 1970s [2]. Moreover, similar rings were also fixed in different years (starting from 1975) on Lake Khubsugul in Mongolia. The rings are almost round in shape and are 5–7 km in diameter, with the width of the darker and thinner ice ring itself being about 1 km. Under-ice hydrological measurements of currents and the thermal structure of waters in the region of the rings showed [3] that in the geostrophic region of the lake, an anticyclonic eddy sits under the ring, which delivers warmer deep water to the lower surface of the ice, which leads to the melting of ice from below (see Fig. 1). But the question why the ice melts from below along the ring remains open.

Despite the large number of works, the processes of formation of convective vortex flows resulting from the melting of ice have not been fully studied, a number of formulated hypotheses have not been confirmed in experiments, which is explained by the complexity of the theoretical description and methodological difficulties in performing experiments.

The relevance of research in this field of science is justified by the need to develop analytical and physical models of the processes of formation of convective vortex flows to solve physical, hydrophysical and environmental problems.

The results of the experiments will be important, first of all, for a better understanding of the formation of convective currents that arise in the process of ice melting; they can be used to predict the thermal and dynamic regime of oceans, seas and lakes.

At the moment, there is not enough literature on the description of this movement of ice in water bodies, as well as the processes occurring under it, however, a sufficient number of works are devoted to the study of the features of convective flows under ice [6–9]. Ice disks placed on the surface of a pool of water at a given temperature are studied in [10]. It is established that the disk rotates around its own axis and the rotation speed increases with the temperature of the reservoir.

Spontaneous rotation of an ice disk during melting on a solid plate was studied in [11] and it was found that the disk rotation speed increases with an increase in the temperature of the plate and with the load applied to the ice disk. The authors proposed a model that explains the spontaneous rotation of the ice disk.

In [12], the evolution of the melting front between the solid and liquid phases of a pure incompressible material is studied, where the motion of the liquid is due to unstable temperature gradients.

The presence of pairs of counter-rotating eddies under ice is described in [13]. The numerical model in this work contains a layered ocean covered with a layer of ice. Beneath the ice layer, salt water generates an upper cyclone and a submerged anticyclone, while a fresh water source generates an upper anticyclone and a submerged cyclone.

The presence of rings on Lake Baikal is associated by the author of [14] with the rise of deep waters, which are caused by eruptions of mud volcanoes. At the same time, the temperature rises in the central part of the future ring structure (on average, by half a degree compared to other parts of the lake), and the so-called anticyclonic current (an annular counterclockwise current) is formed. The current enhances the vertical water exchange, as a result of which the ice cover is more strongly destroyed above the zones of maximum current velocities.

Another interpretation of this phenomenon is the release of the natural combustible gas methane from the sedimentary strata of the bottom of Lake Baikal. Methane, rising to the surface, cools, but manages to warm the surrounding cold water. As a result, convection is formed in the water column in the form of a toroidal figure of rotation around the release of natural gas, which brings warm water to the surface (the lower edge of the ice) away from the natural gas column. Thus, a warm vortex begins to gradually destroy the ice shell from below, and its melted structure begins to be saturated with water, which leads to the formation of those same dark rings. The author of [15] believes that such processes occur due to seismic activity and tectonic movements in the Baikal rift system.

The features of the processes of natural and forced convection under ice in a deep lake are studied in [16] using a mathematical model. Numerical experiments show that natural convection under ice intensifies the process of renewal of deep waters, triggered by random events of methane hydrates rising and decomposition.

Laboratory experiments were carried out to study free small-scale convection under a growing ice cover. The structure of convection is represented by isolated elements "solids" and jets [17]. The latter are responsible for the formation of stalactites and the formation of salt stratification of the near-ice water layer. The conditions for the implementation of congelation and intra-water (with supercooling) ice formation are registered. The elements of similarity are considered in the application of the experimental results to natural observations of thermohaline features in the surface water layer of Arctic waters.

The features of currents in the reservoir in spring are associated with the formation of an intense convective vortex, which contributes to the formation of a column of cold water with a large horizontal temperature gradient, which is a barrier to the movement of the thermal bar, as revealed in [18].

The influence of turbulence arising due to heat flows on the process of ice melting was modelled mathematically and studied experimentally in [19].

2 Study of Spontaneous Rotation of Ice on the Water Surface in Laboratory Experiments

The study of flows that occur when an ice marker is placed on the free surface of a fluid at rest requires taking into account a host of factors, such as the composition and mineralization of ice and water, temperature, salinity, the shape and size of the ice marker, the roughness of its surface, and the existing underwater and surface currents. When conducting an experiment in laboratory conditions, it is possible to eliminate or minimize the influence of most poorly controlled factors, to study a simplified problem by reducing its thermodynamic part to the problem of free convection under a smooth interface. In our case, a flat horizontal interface is considered. A hydrodynamically simpler formulation, due to the absence of most random factors, also allows us to focus most fully on the flows that arise during ice melting.

The experiments were carried out in vessels of various depths and shapes, differing in size (Table 1).

As ice markers, pieces of ice of various shapes (circle, rectangle, square, star, pentagon, complex shape) were used, differing in size and height. Circular cylinders were used most frequently, with diameters ranging from 1.5 to 11 cm and heights from 0.5 to 3.0 cm.

Ice markers with frozen-in markers were placed on the free surface of still water with both zero salinity and salinity changing water density in the range from 1.003 to 1.120 g/cm^3, the water temperature also changed from 20 to 5 °C. A number of experiments were carried out with the placement of an ice marker at the bottom of

Table 1 Containers used in laboratory studies

Container material	Form	Horizontal size, cm	Depth, cm	
			From	To
Polypropylene	Rectangle	37	1	13
	Cylinder			
	Square			
Polymethyl methacrylate	Cylinder	25	1	60
	Octagon	10,5	1	13
	Square	30	1	50
	Triangle	30	1	50
Glass	Cylinder	12	3	18
	Cylinder	30	1	25
	Square	30	1	30

a container with water (negative buoyancy of the marker is achieved by freezing a weighting weight near the center of the volume).

The flow that occurs during the melting of an ice block located on the surface of water at rest has a pronounced density structure, which can be easily traced using the shadow visualization method. Figure 2 shows photographs of an incipient vertical current formed under a hemisphere-shaped ice block placed on the surface of salt water.

For better visualization of emerging flows, the ice was selectively tinted by introducing dyes traditional for hydrodynamics (ink of various colors, aniline dyes, uranyl in small quantities) into water before making ice blocks. This relatively simple experiment makes it possible to observe the formation of a flow directly under the ice block (Fig. 3).

A vertically directed jet with a velocity vy carries a continuous mass flow directed along the gravity vector. Such a flow in the direction from the disk deep into the liquid serves as the basis for the emergence of a vortex directly under the melting ice block, as a result of which the ice block itself begins to rotate.

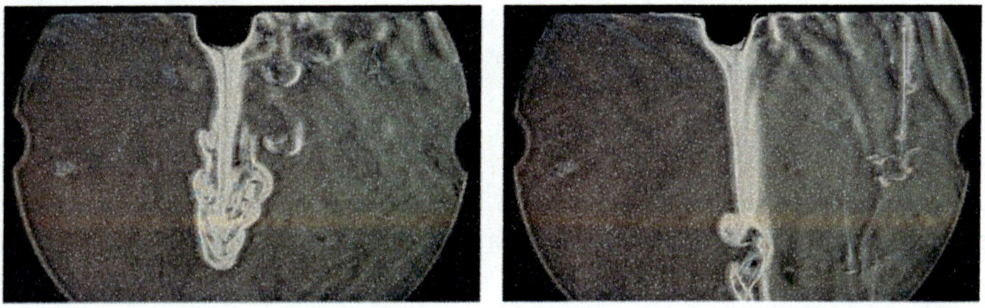

a *b*

Fig. 2 Shadow registration of the current under an ice hemispherical marker (diameter 2.5 cm) at times of 0.7 and 28 s after the marker is placed on the water surface

a b c d

Fig. 3 Flow under the ice disk for different colors at different times. Disk diameter 1.0 cm height 0.3 cm, time since the beginning of the experiment, **a–d** 5, 21; 1, 10 s, respectively

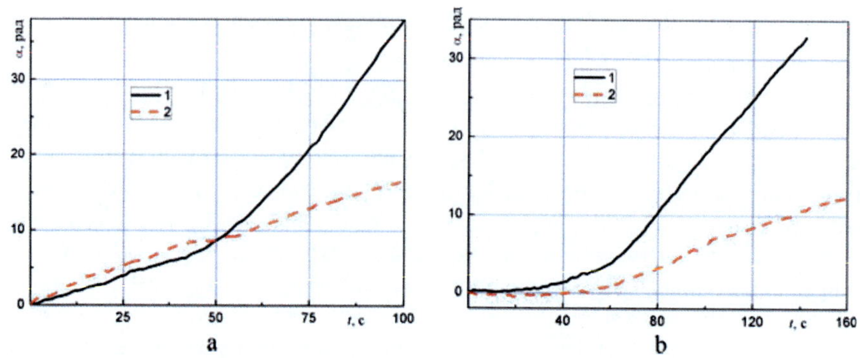

Fig. 4 Differences in the rotation speed of ice blocks 1.0 cm high on the surface of a layer of fresh (1) and salt (2) water (1.033 g/cm^3) 3 cm deep in a rectangular container (32 * 37 cm): **a** the horizontal size of the block 12 mm, **b** 25 mm

In the course of the experiments, the rotation of ice blocks of different sizes on the surface of the resting liquid was registered (Fig. 4), and the regularity of the change in the rotation frequency of the ice block with the change in the thickness of the layer of the resting water was traced. When an ice marker of any shape and size is placed on the surface of a vessel with water at rest (the area of the free surface exceeds the area of the upper face of the ice block by at least 2 times), the rotation of the block is recorded, the speed of which depends nonlinearly both on the depth of the water layer and on the size marker. In particular, in experiments with a liquid depth of 3 cm, the rotation of the ice block practically stopped.

A noticeable slowdown in the rotation of ice blocks was also observed when they were placed on the surface of resting salt water (Fig. 4b). With an increase in the depth of the liquid layer (3, 5, 10, 15 cm), on which the ice block is placed, its rotation speed increases. With a further increase in the depth of the water layer (20, 30, 40, 50 cm), the rotation speed decreases. The most characteristic example is the rotation of an oval block (Fig. 5).

3 Conclusion

Experiments were carried out during which the rotation of ice blocks of various sizes, placed both on a solid surface and on the surface of a pool with quiescent water of a certain depth and density at a given temperature, was recorded.

Observations have shown that if an ice block is placed on the water surface, a pronounced vertical flow is formed under the ice-water interface, which causes rotation. During the experiments, it was found that the speed of rotation of the ice block increases with the temperature of the water on the surface of which the block is placed.

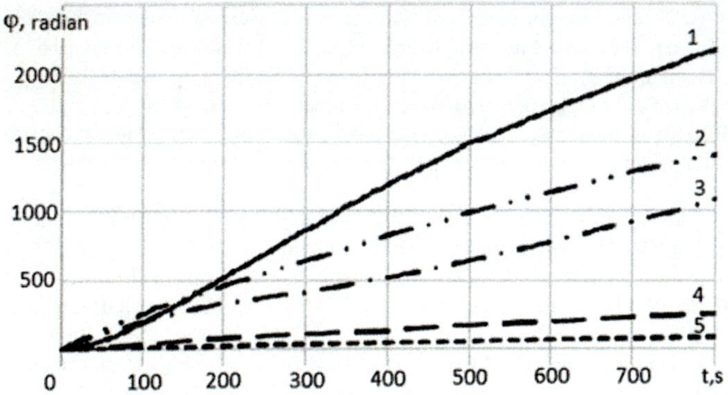

Fig. 5 Rotation speeds of an oval-shaped ice block (2.2 * 1.0 cm, height 0.5 cm) on the water surface at room temperature in a cylindrical container at large liquid depths. Lines: 1—depth 50 cm, 2—40 cm, 3—30 cm, 4—20 cm, 5—15 cm

It has been experimentally established that with an increase in the mass of an ice block, its rotation speed increases. Also, when placing an ice block on the surface of fresh water, the rotation is slower than in experiments with salt water. All observed flow patterns are stably reproduced by repeating the experiments and maintaining the conditions within the accuracy of the experiments.

Research and economic activities in the Arctic are directly dependent on ice conditions. The study carried out in this work is of great importance for monitoring the situation in the Arctic latitudes, including the ecological one. In the event of an unforeseen oil spill, studying the movement of ice floes will help improve response to accidents.

The work was supported with the financial support on the topic State assignments 123021700046-4

References

1. Granin N.G. Ringed Baikal // Science First Hand. 2009. V. P. S. 22–23.
2. Kouraev A.V., Zakharova E.A., Rémy F., Kostianoy A.G., Shimaraev M.N., Hall N.M., and Suknev A.Ya. Giant ice rings on Lakes Baikal and Hovsgol:inventory, associated water structure and potential formation mechanism // Limnology and Oceanography. 2016. Vol. 61. pp. 1001–1014. DOI: https://doi.org/10.1002/lno.10268.
3. Zyryanov V.N., Kuraev A.V., Kostyanoy A.G. Ice rings of Baikal: observations, hypotheses, theory // Water resources, 2022, V. 49, № 2, P. 123–131 DOI: https://doi.org/10.31857/S0321059622020122
4. https://www.gismeteo.ru/
5. https://meteoprog.ua/
6. Sparrow E.M., Lee L., Shamsundar N. Convective instability in a melt layer heated from below // American Society of Mechanical Engineers. 1976. (76-HT-BB)

7. Gorelikov A.V. Natural convective heat transfer in ice-water systems. Abstract diss. for the competition uch. step. cand. Phys.-Math. Sciences. 1998. Institute of Mechanics of Multiphase Systems, Tyumen. 23 p.
8. Zyryanov V.N. Seishi under the ice // Water resources. 2011. V. 38, № 3. P. 259 – 271.
9. Golovin P.N. Structural features of under-ice convection in the Arctic winter break: comparison of the results of field and laboratory experiments // Meteorology and Hydrology. 2004. No. 6. P. 72 - 87.
10. Dorbolo S., Adami N., Dubois C. et al. Rotation of melting ice disks due to melt fluid flow // Phys. Rev. E. 2016. Vol. 93. pp. 033112. DOI: https://doi.org/10.1103/PhysRevE.93.033112
11. Dorbolo S., Vandewalle N., Darbois Texier B. Spontaneous rotation of an ice disk while melting on a solid plate // Physics of Fluids. 2016. Vol. 28. № 12. P. 123601. DOI: https://doi.org/10.1063/1.4967399
12. Keitzl T., Mellado J. P., Notz D. (2016) Impact of Thermally Driven Turbulence on the Bottom Melting of Ice // Journal of Physical Oceanography. 2016. Vol. 46. P. 1171.
13. Shenn-Yu Chao, Ping-Tung Shaw Fission of Heton-like Vortices under Sea Ice // Journal of Oceanography. 1999. Vol. 55. P. 65 – 78.
14. Vyusht A. // IV Vereshchagin Baikal Conference. September 26 – October 1, 2005. P. 52–53
15. Petrov E. // World of Baikal: Popular Science Journal. 2011. № 5(29). P. 64 - 66.
16. Tsvetova E.A. Peculiarities of convection under ice in a deep lake//Journal "Interexpo Geo-Siberia". 2016, V. 4, № 1, P. 110 – 114
17. Dikarev S.N., Poyarkov S.G., Chuvilchikov S.I. Laboratory modeling of small-scale convection under the growing ice cover in the winter Arctic waters // Oceanology. 2004. V. 44. № 1. P. 70 - 79.
18. Blokhina N.S. Effect of convection and currents on ice melting from below during spring warming of a reservoir," Meteorologiya i gidrologiya. 1990. № 2, P. 100–105.
19. Keitzl T., Mellado J. P., Notz D. Reconciling estimates of the ratio of heat and salt fluxes at the ice-ocean interface // Journal of Geophysical Research: Oceans. 2016. Vol. 121. P. 8419–8433.

Response of an Extended Narrow Bay to Long-Wave Disturbances at Resonant Frequencies on the Example of the Sevastopol Bay

Yu. V. Manilyuk and D. I. Lazorenko

Abstract The response of a real bay to long-wave disturbances coming from the open sea, which have periods of potentially extreme seiche modes is studied by based on the ADCIRC hydrodynamic finite element model.

Keywords Seiches in bays · Extreme seiches modes · Sevastopol bays · ADCIRC model

1 Introduction

Seiches in partially enclosed basins (bays, bays, harbors) connected to the sea are a special kind of movement 'harbor oscillations'. They are most often generated by long-wave disturbances penetrating into them from the open sea through the entrance. The loss of wave energy in this case occurs mainly due to its radiation through the open boundary.

The values of seiche periods in the bays are determined by the coastline profile and bathymetry. They are close to the eigenperiods of the bay, which can be established by solving the eigenvalue problem. For basins of a simple shape, there are analytical solutions [1, 2] that make it possible to obtain initial estimates of seiche parameters in real water bodies. For example, such estimates, which are in good agreement with the data of field observations were made in [3] for the Petrozavodsk Bay of Lake Onega, and in [2] for the Sevastopol bays.

It was shown in [4] that the largest amplitudes in bays has resonant modes with narrow peaks in the amplification diagrams. In [5], it was proposed to call them '*extreme modes*'. The amplitudes of such modes increase significantly with an increase in the duration of the perturbation [4]. The spatial structure of extreme modes is such that it is difficult to radiate the energy of water vibrations through a bay entrance [4, 5]. For example, these can be transverse or longitudinal-transverse modes having nodal lines perpendicular to the bay entrance. An attempt to excite

Yu. V. Manilyuk (✉) · D. I. Lazorenko
Marine Hydrophysical Institute RAS, Sevastopol, Russia
e-mail: uvmsev@mhi-ras.ru

V. I. Karev (ed.), *Physical and Mathematical Modeling of Earth and Environment Processes—2022*, Springer Proceedings in Earth and Environmental Sciences,
https://doi.org/10.1007/978-3-031-25962-3_63

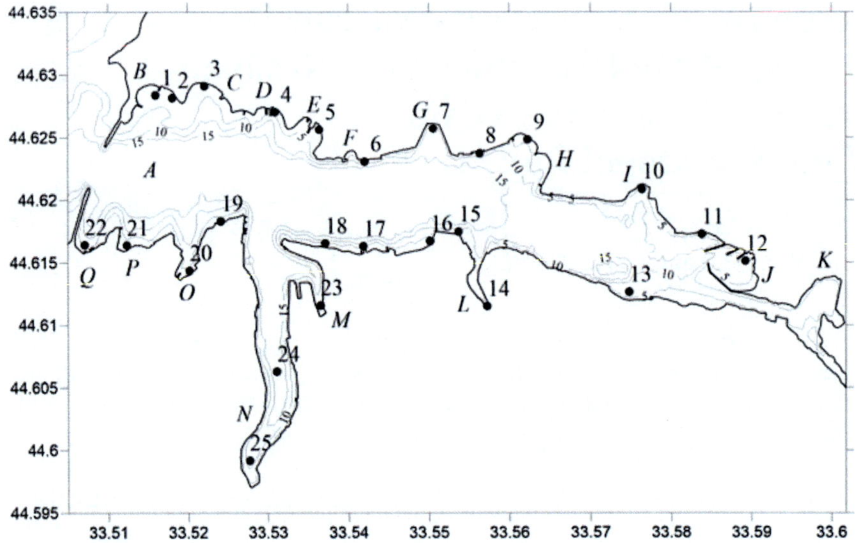

Fig. 1 Location of control points and bays included in the Sevastopol Bay (A): B-Konstantinovskaya, C-Matyushenko, D-Staroseveraya, E-Northern, F-Inzhenernaya, G-Dock (Panaiotova), H-Holland, I-Sukharnaya, J-Lighthouse, K-Grafskaya, L-Kilen-bay, M-Ship, N-South, O-Artillery, P-Alexandrovskaya, Q-Martynova

extreme modes under laboratory conditions using monochromatic waves was made in [6]. But this did not work. According to the authors, it was not possible to fine-tune the wave generator, and the excitation of such a wave requires a long time of hundreds and even thousands of wave periods.

The general regularities of extreme modes have not yet been established. The studies were carried out for a small number of model basins. In this regard, it is of considerable interest to study the possibilities of the emergence of extreme modes in real bays. Sevastopol bay is the largest in the system of Sevastopol bays and includes a number of small bays (Fig. 1).

The purpose of this work is to study the response of the Sevastopol Bay to long-wave disturbances at resonant periods belonging to the potentially extremal modes of this bay.

2 Hydrodynamic Model and Description of the Technique of Numerical Experiments

A series of numerical experiments was carried out based on the ADCIRC finite element model (Advanced Circulation Model for Shelves Coasts and Estuaries) [7] to study the response of the Sevastopol Bay to external long-wave disturbances. The linear version of the model described in [8] is used in this paper. The computational area [8] included the system of Sevastopol bays and part of the coastal zone in the form of a sector with a radius of 8 km. The computational grid was condensed in bays (the lengths of the sides of the triangles here were ~ 50 m). Time integration step $\Delta t = 0.025$ s.

Numerical experiments were carried out in two stages. The periods of the transverse and longitudinal-transverse modes of the Sevastopol Bay were refined at the first stage. According to analytical estimates [9], the values of these periods are 2.8 min. To refine the parameters of the seiche modes, the calculations were carried out for a perturbation, which is a superposition of harmonic waves generated at the liquid boundary, with periods in the range of 2.4–3.0 min. Under zero initial conditions, waves were excited by disturbances on the open boundary of the computational domain in the form [9]:

$$\eta = \frac{a_0}{N} \sum_{n=1}^{N} \sin(\omega_n t), \tag{1}$$

where $a_0 = 0{,}1$ m—oscillation amplitude; N—amount of harmonics in series; n—harmonic number; $\omega_n = 2\pi/T_n$—oscillation frequency; T_n—oscillation period. The oscillation period of each harmonic of series (1) is determined by the expression $T_n = T_{min} + (T_{max} - T_{min})\delta_n$, where T_{min}, T_{max}—minimum and maximum oscillation period, $\delta_n = (n - 1)/(N - 1)$. Number of harmonics $N = (T_{max} - T_{min})/t_d + 1$ (step was used in the calculations $t_d = 10$ s).

Then the boundary condition (1) was replaced by the condition of free passage, and free oscillations of water were modeled. Two maxima were identified at periods of 2.5 and 2.9 min, as a result of the spectral analysis of level fluctuations. The first belongs to the transverse mode, the second to the longitudinal-transverse mode. At the second stage, waves were generated at the obtained resonant periods of the Sevastopol Bay, and an analysis of forced oscillations in the Sevastopol Bay was carried out. Below is an analysis of the results obtained at the second stage of numerical experiments.

3 Results and Discussion

The response of the Sevastopol Bay to long-wave disturbances coming from the open sea with periods equal to the periods of potentially extreme seiche modes: 2.5; 2.9 min was studied. The period equal to 2.5 min belongs to a mode with a spatial structure close to transverse, and the period equal to 2.9 min belongs to a longitudinal-transverse mode.

3.1 Response of the Sevastopol Bay to a Disturbance with a Period of 2.5 min

On Fig. 2 shows the calculated mareograms for points located in different parts of the water area of the Sevastopol Bay, reflecting the process of wave generation.

As the analysis of these mariograms shows, at the initial stage, lasting about 6 h, a transient process is observed. The oscillation amplitude increases during the first 4 h (95 wave periods) of the action of the disturbance. Then their value decreases (in some cases by more than 2 times: points 2; 19; 20) (Table 1, Fig. 2) depending on the location of the control point in the bay water area, and a stationary regime sets in. The amplitude of sea level fluctuations does not change in the stationary mode. Its onset is due to the achievement of a balance between the energy coming from the open sea and its radiation through the entrance to the bay. The spatial structure of the transverse mode hinders not only the emission of energy into the open sea, but also its assimilation. Therefore, a catastrophic increase in the amplitude of level fluctuations in the case of a disturbance with a period of 2.5 min does not occur, with

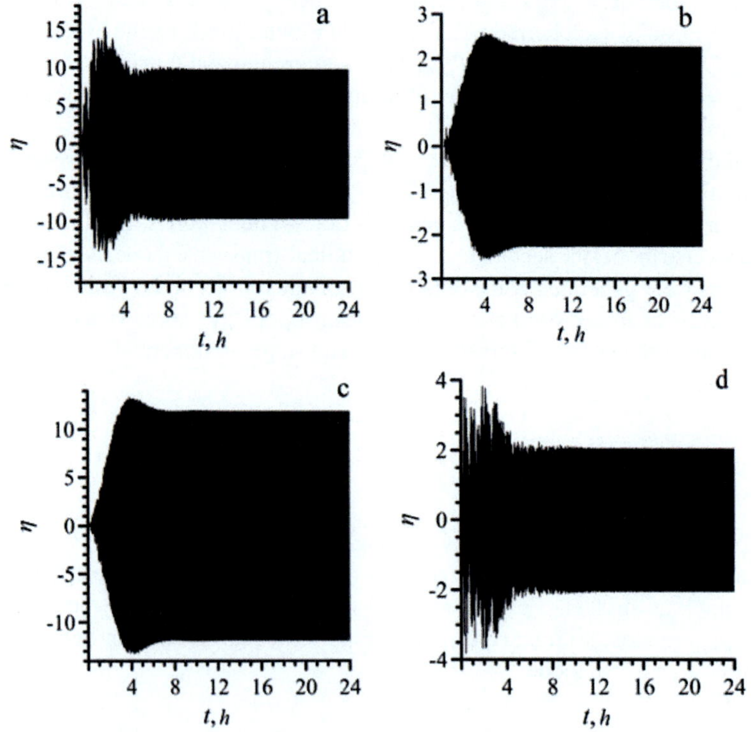

Fig. 2 Calculated mareograms normalized to the amplitude of incoming waves $a_0 = 10$ cm: Northern Bay (point 5) **(a)**, Mayachnaya Bay (point 12) **(b)**, Kilen Bay (point 14) **(c)**, Martynova Bay (point 22) **(d)**

Table 1 Amplification of sea level fluctuations in the Sevastopol Bay depending on the periods of long-wave disturbance

Control point	2.5 min		2.9 min	
	Transition process	Stationary mode	Transition process	Stationary mode
1	11.2	8.8	7.3	6.7
2	5.5	2.5	3.2	1.7
3	8.4	5.3	20.0	18.7
4	9.3	8.7	9.7	8.9
5	15.3	9.7	65.8	61.8
6	1.5	1.2	4.3	3.9
7	9.0	8.4	8.1	7.4
8	1.2	0.9	1.3	1.2
9	8.0	7.1	10.4	9.4
10	1.0	0.8	0.9	0.6
11	8.4	7.5	0.7	0.2
12	2.6	2.3	0.5	0.1
13	4.9	4.3	0.4	0.03
14	13.3	11.9	6.3	5.3
15	6.7	6.0	4.4	4.0
16	2.2	1.1	1.1	0.7
17	1.6	1.2	5.3	4.9
18	3.2	2.7	2.2	1.8
19	3.0	1.1	10.2	9.4
20	5.9	2.1	8.9	3.2
21	30.2	23.3	23.7	22.0
22	3.8	2.1	23.4	21.8
23	3.9	2.0	41.5	38.9
24	1.0	0.6	5.1	4.8
25	3.0	2.1	8.1	7.4

the exception of point 21, located in the Aleksandrovskaya Bay, where the waves are amplified 30 times (Table 1).

3.2 *Response of the Sevastopol Bay to a Disturbance with a Period of 2.9 min*

The mode with a period of 2.9 min has a longitudinal-transverse structure. The amplitude increase of this wave occurs for about 3 h (60 wave periods) from the onset of the disturbance. In this case, the incoming waves amplify more significantly than under the action of a disturbance with a period equal to 2.5 min (Table 1, Fig. 3). Thus, in Northern Bay (point 5) the waves intensify 66 times, in Korabelnaya Bay (point 23)-42 times, and in Martynova Bay (point 22)-23 times. A more significant increase in the amplitude of level fluctuations for this mode, compared to the transverse mode with a period of 2.5 min, can be explained by the fact that this mode has both a longitudinal component, which easily assimilats the energy coming from the open sea, and a transverse component that accumulates energy and prevents it from radiating into the open sea.

Thus, the longitudinal-transverse mode with a period of 2.9 min poses a serious danger to the Sevastopol Bay, because for 2 h of disturbances in some parts of its water area, level fluctuations can increase by 10 times, and with an increase in the duration of the disturbance, by 20–66 times. This also takes place in the bays used for mooring passenger boats carrying out transportation between the Northern and Southern parts of the Sevastopol, and for ship repair.

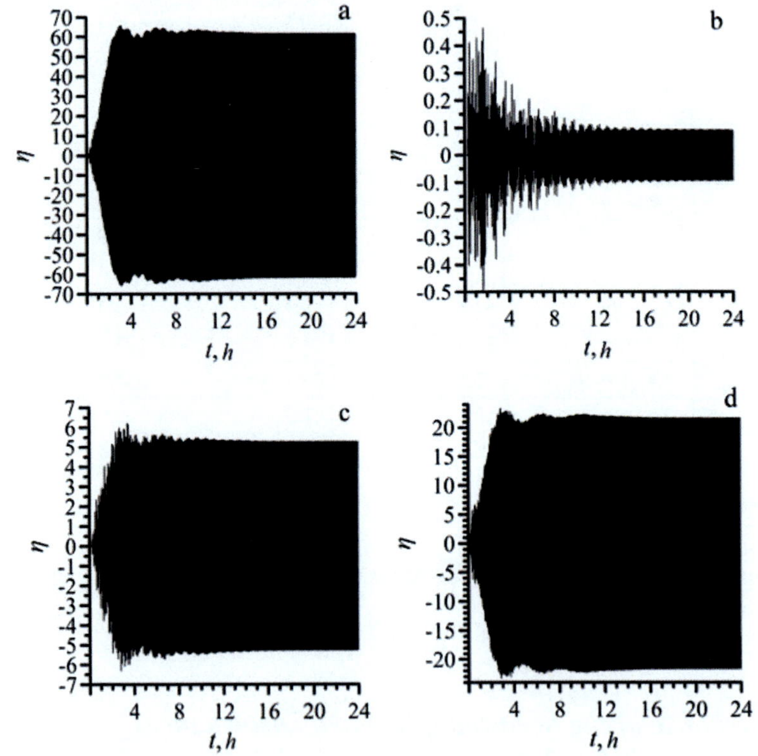

Fig. 3 Calculated mareograms normalized to the amplitude of incoming waves a_0 = 10 sm: Northern Bay (point 5) **(a)**, Mayachnaya Bay (point 12) **(b)**, Kilen Bay (point 14) **(c)**, Martynova Bay (point 22) **(d)**

4 Conclusion

The maximum oscillation amplitudes are observed at the end of the transient process caused by the action of a long-wave disturbance, the duration of which is from 3 to 5 h, depending on the period of the disturbance. In the stationary mode, the amplitude of the level oscillations is less than during the transient process; in some places—up to 2 times.

The longitudinal-transverse mode of the seiches of the Sevastopol Bay with a period of 2.9 min is a serious danger for the Sevastopol Bay, because in most of the points considered, the fluctuations increase by more than 3 times, and in Severnaya Bay-66 times, in Korabelnaya Bay-42 times.

The transverse seiche mode with a period of 2.5 min is less dangerous for the Sevastopol Bay than the longitudinal-transverse mode, despite the fact that its amplitude increases over 95 wave periods, because amplification of waves at most points occurs no more than 2–3 times. However, in the Severnaya, Kilen, Aleksandrovskaya bays, level fluctuations can increase by more than 10–30 times.

Source of Financing The work was carried out within the framework of the state assignment on topic No. FNNN-2021-0005 "Comprehensive interdisciplinary studies of oceanological processes that determine the functioning and evolution of ecosystems in the coastal zones of the Black and Azov Seas."

References

1. *Rabinovich A.B.* Seiches and Harbor Oscillations (Chapter 9) // Handbook of Coastal and Ocean Engineering / Ed. Y.C. Kim. – Singapoure: World Scientific Publ., 2009. – P. 193– 236.
2. *Ivanov V.A., Manilyuk Yu.V., Sannikov V.F.* Seiches in a basin with an open entrance // Journal of Applied Mechanics and Technical Physics. 2018. V. 59. № 4. PP. 594–600. DOI: https://doi.org/10.1134/S0021894418040041.
3. *Ivanov V.A., Palshin N.I., Manilyuk Yu. V.* Seiches in Petrozavodsk Bay, Lake Onega // Water Resources, 2019, Vol. 46, No. 5. P. 709–717. DOI: https://doi.org/10.1134/S0097807819050117.
4. *Bellotti G.* Transient response of harbours to long waves under resonansce conditions // Coastal Engenieering. 2007. V. 54. Iss. 9. P. 680–693. doi:https://doi.org/10.1016/j.coastaleng.2007.02.002.
5. *Ma X., Zheng Z., Zhang X, Dong G.* Numerical investigation on special modes with narrow amplification diagram in harbor oscillations // Ocean Dynamics. 2020. V. 70. Iss. 1. P. 1–19. https://doi.org/10.1007/s10236-019-01325-8/
6. *Dong G., Zheng Z., Gao J., Ma X., Dong Y., Wu H.* Experimental investigation on special modes with narrow amplification diagrams in harbor oscillations // Coastal Engineering. 2020. V. 159, August 2020, 103720. https://doi.org/10.1016/j.coastaleng.2020.103720/
7. *Luettich R.A., Westerink J.J.* Formulation and Numerical Implementation of the 2D/3D *ADCIRC*; 2004. http://adcirc.org/adcirc_theory_2004_12_08.pdf/

8. *Manilyuk Yu.V., Lazorenko D.I., Fomin V.V.* Investigation of seiche oscillations in the adjacent bays by the example of the Sevastopol and the Quarantine bays // Physical oceanography, [e-journal], 2020. V. 27, Iss. 3. PP. 242-256. doi:https://doi.org/10.22449/1573-160X-2020-3-242-256.
9. *Manilyuk Yu.V., Lazorenko D.I., Fomin V.V.* Seiche Oscillations in the System of Sevastopol Bays // Warter Resources. 2021. V. 48. No. 5. PP. 726–736. DOI: https://doi.org/10.1134/S0097807821050122.

Assessment of the Impact of the Construction of Underground Structures with the Use of Special Methods on Underground Pedestrian Passengers

D. L. Neguritsa, G. V. Alekseev, E. A. Medvedev, and A. A. Tereshin

Abstract During the construction of multifunctional complexes in megacities, the task of geomechanical support and ensuring the safety of construction, as well as the operation of transport infrastructure falling into the zone of influence of construction, is very acute. Multifunctional complexes, due to their uniqueness, require scientific support, which includes verification of the correctness of design solutions, control of construction work, geomechanical and geotechnical support, as well as mine surveying and geodetic monitoring of the displacement and deformation processes in the enclosing soil and rock massif. In the process of construction, buildings, structures, communications, as well as capital objects of transport infrastructure fall into the zone of influence, assessing the impact on which this article is devoted, which presents the results of modeling changes in the stress-strain state of the soil mass during the construction of a multifunctional center by the finite element method using elastic and models of hardening soil in a spatial formulation.

Keywords Scientific support of construction · Modeling · Deformation processes · Soil and rock massif · Building structures · Transport infrastructure · Underground pass

The integrated use of the territory of megacities is currently moving towards the construction of multifunctional complexes that allow efficient use of limited areas for development with active development of underground space [1, 2]. In the central part of cities, this allows the most efficient use of relatively small accessible areas with favorable transport accessibility and transport infrastructure for the construction of commercial residential and non-residential real estate that is in demand today, which also dictates the need to solve the problem of providing parking spaces, mainly due to their placement in underground space allocated territory [3, 4, 6].

A multifunctional complex (hereinafter MFC) is a complex architectural and planning and engineering facility, consisting of four volumes of different heights

D. L. Neguritsa (✉) · G. V. Alekseev · E. A. Medvedev · A. A. Tereshin
National Research Moscow State University of Civil Engineering (NRU MGSU), Moscow, Russia
e-mail: neguritsadl@mgri.ru

V. I. Karev (ed.), *Physical and Mathematical Modeling of Earth and Environment Processes—2022*, Springer Proceedings in Earth and Environmental Sciences,
https://doi.org/10.1007/978-3-031-25962-3_64

and purposes, namely, a high-rise 36-storey hotel block, an 8-storey office block, a 4-storey connecting block between the hotel and office.

Blocks of the underground part with a parking lot, technical and service rooms are located under all three blocks and the boulevard.

During the construction of the high-rise part of the complex, a pile foundation is used in a soil-rock mass to a depth of more than 50 m from the surface, and an underground enclosing structure "wall in the ground" with a depth of up to 27 m is made along the perimeter of the object.

An analysis of the hydrogeological conditions, the geomechanical state of the soil and rock mass and the adopted technological solutions for the construction of a multifunctional complex showed that the main factors determining the impact of new construction on the existing transport infrastructure facilities—underground passages and overpasses, are changes in the stress-strain state of the soil and rock mass of the foundation caused by the construction of a pit near them, additional load, as well as changes in hydrogeological conditions and technological factors, such as dynamic impacts, the influence of deep foundations and the enclosing structures of the pit, the influence of special types of work. Possible manifestation of negative processes in the enclosing soil mass associated with the performance of geotechnical works, such as suffusion processes, the formation of quicksand, thixotropic deconsolidation.

The construction area has a heterogeneous structure, and in the process of anthropogenic activity, the soil-rock mass has repeatedly undergone changes, which during the construction process can lead to an uneven change in its properties at the base of the structure, can cause activation of the development of processes of deformation of the base of the structure [5].

To assess the impact of the construction of a multifunctional center, mathematical modeling was carried out by the finite element method using geomechanical soil models, which makes it possible to assess the change in the stress-strain state of the soil massif, the structures of the structure and structures that fall into the zone of influence of transport infrastructure facilities.

For scientific support of the implementation of the project for the construction of a multifunctional center, mathematical modeling was carried out in a specialized computer complex based on the finite element method under conditions of three-dimensional setting in the PLAXIS 3D program (Fig. 1).

It should be noted that the modeling was carried out both for the multifunctional complex and for all underground and surface facilities falling within the zone of influence of the construction. But this article presents the results of the forecast for transport infrastructure facilities—two pedestrian crossings and a road overpass. Elastic model and hardening soil model—Hardening soil model was used to simulate the mechanical behavior of the structures of the excavation fence and the structures of existing buildings, and the elastic model is used to simulate the operation of structural materials such as concrete, brickwork, metal, elements of the excavation fence and the structures of existing buildings.

The use of the Hardening-Soil model makes it possible to take into account the dependence of soil stiffness on stresses and the ability to distinguish between soil

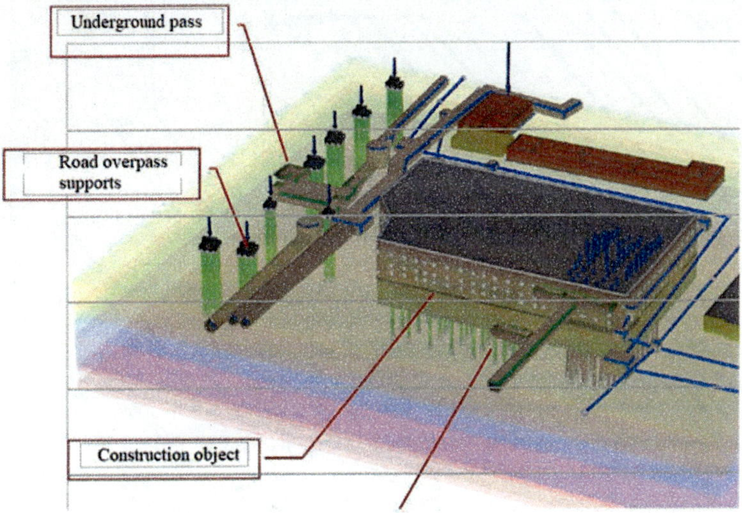

Fig. 1 Construction site multifunctional center with adjacent transport infrastructure

stiffness during loading and unloading, which most reliably describes the actual behavior of the soil-rock massif under these conditions.

An integrated geomechanical model of the soil-rock massif and the surrounding buildings provides for the forecast of the stress-strain state of the soil-rock massif and buildings and structures falling into the zone of influence for all technological stages of construction.

Underpass tunnel 1 is located at a distance of 11 m from the projected excavation. The structures of the tunnel and its staircases were made of monolithic reinforced concrete in 2005–2006. The survey performed showed that the state of the transition is assessed as satisfactory. In the course of the survey and study of the design documentation for tunnel No. 1, the maximum allowable additional subsidence was not established. Since no defects that reduce the bearing capacity of the transition were recorded, and the condition is assessed as satisfactory, to assess the impact, the maximum additional settlements can be taken as for a frameless building according to the II category of technical condition: maximum settlement-3.0 cm, relative difference of settlement-0.001 [7].

Underpass tunnel No. 2 is located at a distance of 4 m from the projected excavation. The structures of the tunnel and its staircases were made of precast concrete elements in 1968. According to the survey, the state of the transition is assessed as satisfactory. During the survey and when studying the design documents for this underground passage No. 2, the maximum allowable additional subsidence was not indicated. Since there are no defects that reduce the bearing capacity of the transition, and the condition is assessed as satisfactory, for assessing the impact, it is possible to accept the maximum additional settlements as for a frameless building

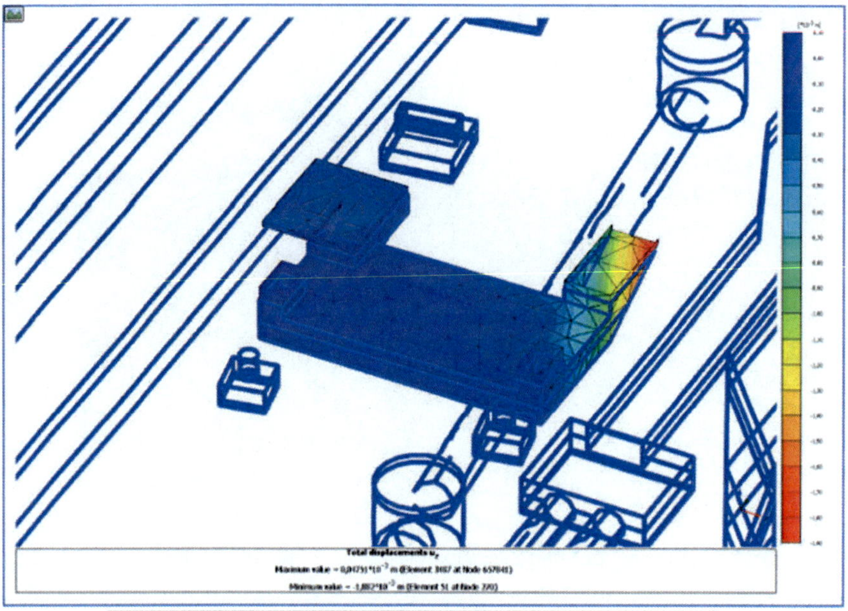

Fig. 2 Subsidence of structures of the underground passage No. 1

according to the II category of technical condition: maximum settlement-3.0 cm, relative difference of settlement-0.001 [7].

The nearest support of the road overpass is located at a distance of 19 m from the projected excavation. Intermediate supports 10, 11, 12 and 13 are made of monolithic reinforced concrete with a diameter of 1.5 m and are part of the road overpass with a total length of more than 900 m.

According to the executive documentation for the construction of the overpass, the foundations of the supports are bored piles with a diameter of 1200 mm and a length of 20–25 m. After the inspection, the condition of the supports can be assessed as good. The design and as-built documentation for the overpass does not indicate the maximum allowable additional settlement of the overpass supports. Since there are no defects that reduce the bearing capacity of the supports, and the condition is assessed as good, for assessing the impact, the maximum additional settlements can be taken as for a frameless building according to the II category of technical condition: maximum settlement-3.0 cm, relative difference of settlement-0.001 [7].

An analysis of the simulation results shows the dynamics of changes in the degree of influence of the excavation device on the surrounding buildings. The boundaries of the calculated zone of influence (42–47 m) and objects of geotechnical monitoring were determined.

Deformation processes were modeled (Figs. 2, 3 and 4) for the following structures of tunnel No. 1, tunnel No. 2 and the road overpass at the time of completion of the construction of the IFC. The simulation results are presented in Table 1.

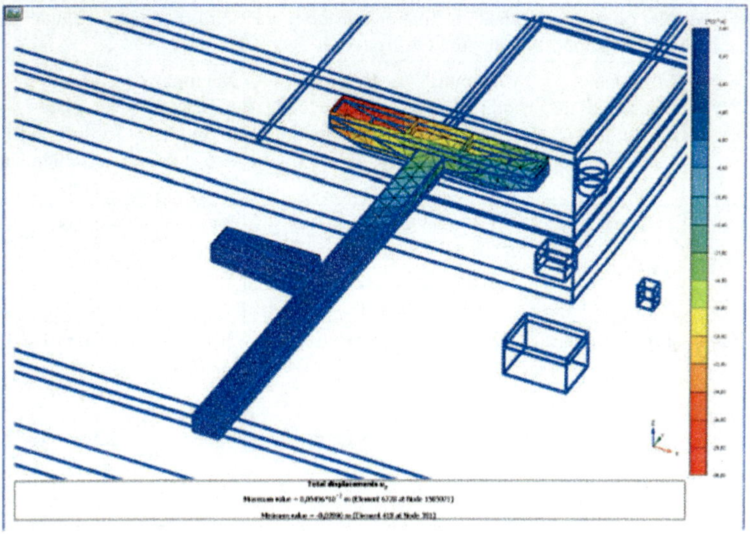

Fig. 3 Subsidence of structures of the underground passage No. 2

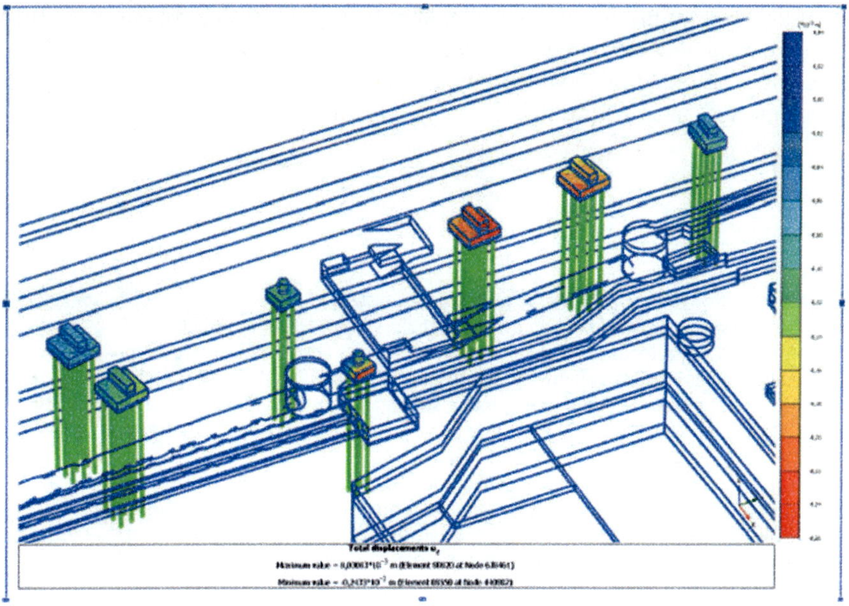

Fig. 4 Subsidence of road overpass supports

Table 1 The results of geomechanical calculations of the impact on the tunnels of underground pedestrian crossings and supports of the road overpass

Construction	Minimum distance to the object, m	Maximum additional deformations, mm	Relative difference of deformations	Maximum maximum additional deformations, mm	Limiting relative difference of settlement, mm
Tunnel of the underground pedestrian crossing №1	11	3,3	0,0003	30	0,001
Tunnel of the underground pedestrian crossing №2	4,0	28,9	0,0006	30	0,001
Road overpass supports	19,0	< 1,0	–	30	0,001

Taking into account the great responsibility of transport structures that fall into the zone of influence of the construction of transport facilities to control the processes of displacement and deformation from the influence of the construction of the IFC, it is necessary to organize geomonitoring at the observation station both in the soil and rock massif and structures [8].

References

1. Merkin V.E., Konyukhov D.S. The main problems, tasks and prospects for the development of underground space in Moscow // Metro and tunnels. 2017. № 1–2. P. 18–23.
2. Kartoziya B.A. Development of the underground space of large cities. New trends // Mining information and analytical bulletin (scientific and technical journal). 2015. № S1. P. 615–630.
3. Kulikova E.Y. Defects of urban underground structure and their prediction // IOP Conference Series: Materials Science and Engineering. 012108 (2018). DOI: https://doi.org/10.1088/1757-899X/451/1/012108.
4. Glozman O.S. Underground planning of Moscow // Housing construction. 2016. № 11. P. 14–19.
5. Neguritsa D. The problems of monitoring the deformation processes in the integrated development of the underground space of metropoli-tan cities // E3S Web of Conf. 56, 02027 (2018). DOI: https://doi.org/10.1051/e3sconf/20185602027.
6. Broere W. Urban problems – Underground solutions // Proc. 13th World Conf. of ACUUS: Advances in Underground Space Development. 2012 (2013). pp. 1528–1539.
7. SP 13-102-2003 Rules for the inspection of load-bearing building structures of buildings and structures.
8. Iofis M.A., Neguritsa D.L., Esina E.N. The displacement of rocks during the development of the bowels of the earth. Moscow, 2020. 287 p. ISBN: 978-5-209-10113-0.